ENGINEERING SATELLITE-BASED NAVIGATION AND TIMING

IEEE Press
445 Hoes Lane
Piscataway, NJ 08854

IEEE Press Editorial Board
Tariq Samad, *Editor in Chief*

George W. Arnold   Vladimir Lumelsky   Linda Shafer
Dmitry Goldgof   Pui-In Mak   Zidong Wang
Ekram Hossain   Jeffrey Nanzer   MengChu Zhou
Mary Lanzerotti   Ray Perez   George Zobrist

Kenneth Moore, *Director of IEEE Book and Information Services (BIS)*

Technical Reviewers

Jon Anderson, *Canyon Consulting*
José-Ángel Ávila-Rodríguez, *European Space Agency (ESA)*
Frank van Diggelen, *Broadcom Corporation*

Other Technical Reviewers

Michael Braasch, *Ohio University*
Alex Cerruti, *The MITRE Corporation*
Sergey Karutin, *Russian Federal Space Agency (Roscosmos)*
Phillip Ward, *Navward Consulting*
Yuanxi Yang, *China National Administration of GNSS and Applications*

ENGINEERING SATELLITE-BASED NAVIGATION AND TIMING

Global Navigation Satellite Systems, Signals, and Receivers

John W. Betz

IEEE PRESS

WILEY

The Following Material Has Been Approved by The MITRE Corporation and the U.S. Air Force Space and Missile Systems Center for Public Release; Distribution Unlimited:

Part I and Appendix A: Air Force Case Number 13-0985
Part II: Air Force Case Number 13-3073
Part III: Air Force Case Number 14-2724
Part IV: Air Force Case Number 14-4351

The author's affiliation with The MITRE Corporation is provided for identification purposes only, and is not intended to convey or imply MITRE's concurrence with, or support for, the positions, opinions or viewpoints expressed by the author.

Copyright © 2016 by The Institute of Electrical and Electronics Engineers, Inc.

Published by John Wiley & Sons, Inc., Hoboken, New Jersey. All rights reserved.
Published simultaneously in Canada.

No part of this publication may be reproduced, stored in a retrieval system, or transmitted in any form or by any means, electronic, mechanical, photocopying, recording, scanning, or otherwise, except as permitted under Section 107 or 108 of the 1976 United States Copyright Act, without either the prior written permission of the Publisher, or authorization through payment of the appropriate per-copy fee to the Copyright Clearance Center, Inc., 222 Rosewood Drive, Danvers, MA 01923, (978) 750-8400, fax (978) 750-4470, or on the web at www.copyright.com. Requests to the Publisher for permission should be addressed to the Permissions Department, John Wiley & Sons, Inc., 111 River Street, Hoboken, NJ 07030, (201) 748-6011, fax (201) 748-6008, or online at http://www.wiley.com/go/permission.

Limit of Liability/Disclaimer of Warranty: While the publisher and author have used their best efforts in preparing this book, they make no representations or warranties with respect to the accuracy or completeness of the contents of this book and specifically disclaim any implied warranties of merchantability or fitness for a particular purpose. No warranty may be created or extended by sales representatives or written sales materials. The advice and strategies contained herein may not be suitable for your situation. You should consult with a professional where appropriate. Neither the publisher nor author shall be liable for any loss of profit or any other commercial damages, including but not limited to special, incidental, consequential, or other damages.

For general information on our other products and services or for technical support, please contact our Customer Care Department within the United States at (800) 762-2974, outside the United States at (317) 572-3993 or fax (317) 572-4002.

Wiley also publishes its books in a variety of electronic formats. Some content that appears in print may not be available in electronic formats. For more information about Wiley products, visit our web site at www.wiley.com.

Library of Congress Cataloging-in-Publication Data is available.
ISBN: 978-1-118-61597-3

10 9 8 7 6 5 4 3 2 1

For Donna

CONTENTS

Preface	xv
Acknowledgments	xvii
Useful Constants	xix
List of Acronyms and Abbreviations	xxi
About the Author	xxvii

1 INTRODUCTION   1
  1.1 Satnav Revolution   2
  1.2 Basic Principles of Satnav   5
  1.3 Satnav Attributes   12
  1.4 Book Structure and How to Use This Book   12
  1.5 More to Explore   14
  Reference   15

PART I   SYSTEM AND SIGNAL ENGINEERING   17

2 SATELLITE ORBITS AND CONSTELLATIONS   19
  2.1 Kepler's Laws   20
  2.2 Orbital Deviations from Ideal   25
  2.3 Constellations   26
  2.4 Useful Geometry Calculations   30
  2.5 Summary   35
  Review Questions   35
  References   36

3 SATNAV SIGNALS   37
  3.1 Signals, Signal Processing, and Spreading Modulations   38
  3.2 Effects of Doppler and of Ionospheric Propagation   59
  3.3 Satnav Signal Characteristics   65
  3.4 Satnav Signal Structure   86
  3.5 Summary   92
  Review Questions   92
  References   99

4 LINK BUDGETS — 102
- 4.1 Free-Space Path Loss — 103
- 4.2 Calculating Maximum and Minimum Specified Received Power in Signal Specifications — 107
- 4.3 Terrestrial Link Budgets — 112
- 4.4 Building Penetration and Foliage Losses — 116
- 4.5 Summary — 119
- Review Questions — 119
- References — 120

5 CORRELATOR OUTPUT SNR, EFFECTIVE C/N_0, AND I/S — 122
- 5.1 Channel Model and Ideal Receiver Processing — 122
- 5.2 Correlator Output SNR With No Interference — 125
- 5.3 Correlator Output SNR With Interference: Spectral Separation Coefficients and Processing Gain — 127
- 5.4 Effective C/N_0 — 129
- 5.5 Interference-to-Signal Power Ratios and Effective C/N_0 — 130
- 5.6 A Deeper Look at Spectral Separation Coefficients — 130
- 5.7 Multiple Access Interference and Aggregate Gain of a Constellation — 133
- 5.8 Summary — 135
- Review Questions — 136
- References — 138

6 ERROR SOURCES AND ERROR CHARACTERIZATION — 139
- 6.1 Sources of Error in Satnav Positioning and Timing Calculation — 140
- 6.2 Dilution of Precision and Error Measures — 146
- 6.3 Positioning Errors for Standalone and Differential Satnav Receivers — 150
- 6.4 Other Error Sources — 152
- 6.5 Summary — 153
- Review Questions — 154
- References — 155

PART II SATNAV SYSTEM DESCRIPTIONS — 157

7 NAVSTAR GLOBAL POSITIONING SYSTEM — 163
- 7.1 GPS History and Plans — 165
- 7.2 GPS Description — 167
- 7.3 GPS Signals — 168

	7.4	Summary	196
		Review Questions	197
		References	198
8	**SATELLITE-BASED AUGMENTATION SYSTEMS**		**201**
	8.1	SBAS History and Plans	202
	8.2	SBAS Description	204
	8.3	SBAS Signals	205
	8.4	Summary	209
		Review Questions	210
		References	211
9	**GLONASS**		**212**
	9.1	GLONASS History and Plans	213
	9.2	GLONASS Description	214
	9.3	GLONASS Signals	215
	9.4	Summary	222
		Review Questions	224
		References	224
10	**GALILEO**		**226**
	10.1	Galileo History and Plans	227
	10.2	Galileo Description	228
	10.3	Galileo Signals	230
	10.4	Summary	248
		Review Questions	249
		References	250
11	**BEIDOU SYSTEM**		**252**
	11.1	BDS History and Plans	253
	11.2	BDS Description	254
	11.3	BDS Signals	257
	11.4	Summary	262
		Review Questions	264
		References	264
12	**QUASI-ZENITH SATELLITE SYSTEM**		**266**
	12.1	QZSS History and Plans	267
	12.2	QZSS Description	268

12.3	QZSS Signals	270
12.4	Summary	280
References		281

13 INDIAN REGIONAL SATELLITE SYSTEM — 282

13.1	IRNSS History and Plans	283
13.2	IRNSS Description	283
13.3	IRNSS Signals	284
13.4	Summary	289
References		289

PART III RECEIVER PROCESSING — 291

14 RECEIVER FRONT END — 297

14.1	Front-End Components	298
14.2	Front-End Noise Figure	321
14.3	Front-End Architectures and Frequency Plans	323
14.4	Summary	328
Review Questions		329
References		331

15 ANALOG-TO-DIGITAL CONVERSION — 333

15.1	Introduction to Analog-to-Digital Conversion and Automatic Gain Control	334
15.2	Linear Analog-to-Digital Conversion	338
15.3	Precorrelator Analog-to-Digital Conversion—The Digitizing Correlator	340
15.4	Summary	362
Review Questions		362
References		363

16 ACQUISITION — 364

16.1	Initial Conditions for Acquisition	367
16.2	Initial Synchronization Basics	370
16.3	Initial Synchronization Computation	383
16.4	Initial Synchronization Performance	392
16.5	Other Aspects of Acquisition	396
16.6	Summary	401

Review Questions 403
References 404

17 DISCRETE-UPDATE TRACKING LOOPS 406
17.1 Discrete-Update Tracking Loop Formulation 408
17.2 Discrete-Update Tracking Loop Design 412
17.3 Tracking Loop Characterization 416
17.4 Summary 426
References 427

18 CARRIER TRACKING AND DATA DEMODULATION 428
18.1 Signal Processing for Carrier Tracking 429
18.2 Frequency-Locked Loops 442
18.3 Costas Loops 447
18.4 Phase-Locked Loops 450
18.5 Data Message Demodulation 453
18.6 Summary 462
Review Questions 463
References 465

19 CODE TRACKING 467
19.1 Signal Processing for Code Tracking 468
19.2 Discriminators for Code Tracking 474
19.3 Carrier-Aided Code Tracking 480
19.4 Code Tracking Performance in White Noise 481
19.5 Code Tracking Performance in White Noise and Interference 489
19.6 Ambiguous Code Tracking 492
19.7 Summary 498
Appendix 19.A   RMS Bandwidth 499
Review Questions 502
References 502

20 POSITION, VELOCITY, AND TIME CALCULATION 504
20.1 Forming Measurements 505
20.2 Reducing Pseudorange Errors 508
20.3 Standard Point Positioning 515
20.4 Blending Solutions From Multiple Satnav Systems 520
20.5 Velocity Calculation 522

20.6	Working with Disadvantaged Receivers	524
20.7	Precise Point Positioning	527
20.8	Integrity Monitoring: Receiver Autonomous Integrity Monitoring and Fault Detection and Exclusion	529
20.9	Summary	530
	Review Questions	531
	References	534

PART IV SPECIALIZED TOPICS 537

21 INTERFERENCE 539

21.1	Interference Characteristics	540
21.2	Effects of Interference on Receiver Operation	541
21.3	Dealing with Interference	542
21.4	Summary	549
	References	550

22 MULTIPATH 551

22.1	Multipath Characteristics	552
22.2	Multipath Effects	556
22.3	Multipath Mitigation	560
22.4	Summary	567
	References	568

23 AUGMENTATIONS USING DIFFERENTIAL SATNAV 570

23.1	Overview of Differential Satnav	571
23.2	Code-Based Differential Systems	574
23.3	Carrier-Based Differential Systems	576
23.4	Summary	586
	References	586

24 ASSISTED SATNAV 588

24.1	Reducing IFU and ITU	590
24.2	Provision of Clock Corrections, Ephemeris, and Data Message Bits	591
24.3	Block Processing	592
24.4	Computing Pseudoranges and Position	592
24.5	Summary	593
	Reference	594

25 INTEGRATED RECEIVER PROCESSING — 595
    25.1 Kalman Filter Overview — 596
    25.2 Loosely and Tightly Coupled Sensor-Integrated Satnav Processing — 599
    25.3 Standalone Vector Tracking — 603
    25.4 Ultratightly Coupled Sensor-Integrated Satnav Processing — 605
    25.5 Summary — 606
    References — 607

A THEORETICAL FOUNDATIONS — 609
    A.1 Some Useful Functions and Their Properties — 610
    A.2 Fourier Transforms — 611
    A.3 Signal Theory and Linear Systems Theory — 611
    A.4 Stochastic Processes — 613
    A.5 Some Results for Keyed Waveforms — 615
    A.6 Bandwidth Measures — 619
    A.7 Matrices and Matrix Algebra — 621
    A.8 Taylor Series and Linearization — 623
    A.9 Coordinate System Overview — 624
    References — 625

Index — 627

PREFACE

The world of satellite-based navigation and timing opened for me in 1997, when Alan Moore, then the project leader of MITRE's GPS work for the Air Force, asked me a question in the corridor about how to design a new military signal that could share the same frequency band as existing GPS signals while being spectrally separated from civil signals. My off-the-cuff suggestion of a coherently modulated pair of subcarriers led to my development of Binary Offset Carrier and then involvement in other aspects of satnav. Since I had worked on spread spectrum communications, radar, sonar, and other signal-processing applications, satnav seemed to be a natural outlet for my interests and experience. There was a rich corpus of deep technical work to learn from, as well as many challenging problems still demanding innovative solutions. The GPS Joint Program Office was the place to be—full of excitement and plans for the future of GPS, with GPS legends roaming the halls.

Galileo, emerging in the early 2000s, provided an opportunity for collaboration with European colleagues to meet mutual goals of compatibility and interoperability. Japan's QZSS, Russian interest in CDMA signals, China's BeiDou, and India's IRNSS all also emerged, providing additional challenges to be addressed, as well as additional colleagues to learn from.

In 2006, Dr. Chris Hegarty put me in touch with Ms. Carolyn McDonald of NavtechGPS, and Carolyn agreed to sponsor my development and teaching of a short course emphasizing modernized satnav signals and receiver processing. Later versions of this course benefited from course blocks developed by other experts under my direction. That course, and its extensions over the years, forms the basis of this book.

As my work on GPS and other satnav systems continued, it became clear that system engineering and signal engineering interact strongly with system design and receiver design. Such thinking was innate to legends like Dr. Charlie Cahn, but not necessarily to less experienced engineers. Also, design involves continual trades between implementation complexity and performance, further complicated by the need to assess implementation complexity in the context of future technologies, when signals would be used and receivers would be developed. Yet, no textbooks existed that depicted satnav system engineering and signal engineering in an organized and comprehensive way, or that clearly portrayed complexity and performance trades. Many books summarized the history of GPS and described the original GPS signals, but no text provided a balanced description of all current and planned satnav systems and their signals, including the modernized GPS signals. Multiple texts captured decades of experience in processing the original GPS signals, but books were not available to describe explicitly the processing of new and modernized signals with their different features and technical characteristics. Further, new techniques have been developed and the satnav literature has been enriched by many excellent papers over the past decade, yet these new contributions have not been captured and integrated into a single resource.

This book is my attempt to provide a set of more comprehensive and current perspectives.

John W. Betz

ACKNOWLEDGMENTS

Long before I began working on GPS, I was benefitting from colleagues and mentors. Mr. Roger Boyell, who worked with me at RCA Government Systems, was an exemplar of how to skillfully blend technical work and technical communication. Professor John Proakis, whose clear teaching style and excellent textbooks were essential to my graduate education, was kind enough to serve as my PhD advisor. At The Analytical Sciences Corporation (TASC), working with Mr. Robert Pinto was like graduate school all over again, while Dr. Seymour Stein, through his consulting work at TASC, demonstrated how theoretical analysis could guide and affect real-world applications.

At MITRE, Mr. Alan Moore provided me with the opportunity to work on GPS, and was extremely supportive of our efforts. Dr. Kevin Kolodziejski, who originally was my graduate student, became a colleague and co-author on multiple award-winning papers. From the beginning Dr. Chris Hegarty, one of the world's premier satnav engineers, has been an extremely helpful colleague. I was fortunate to serve on two signal design teams with Dr. Charlie Cahn, whose contributions to the design of every GPS signal demonstrated his unparalleled insight, productivity, technical breadth, and technical depth, combined with admirable humility and absence of self-promotion.

Much of my work on satnav has been with or for the US Air Force, and I have benefitted from the resulting association with outstanding Air Force officers. As GPS Chief Engineer early in this century, Col. Rick Reaser (Retired) was a mentor and guide in the challenging areas of spectrum management and international interactions. Col. Jon Anderson, PhD (Retired), was the Air Force Captain in 1997 who hosted the meeting where I introduced the Offset Carrier concept; he has remained a friend and colleague over these years as we have worked in different areas of satnav together. It was a pleasure to work with Col. Mark Crews, PhD (Retired), who served as GPS Chief Engineer; Mark made fundamental decisions related to GPS Modernization while leading GPS's international outreach with Europe, Russia, and Japan during critical times. Lt. Bryan Titus was a partner during the early days of GPS–Galileo discussions, and Lt. Col. Bryan Titus remains a colleague and friend as our careers have intersected again. Col. David Goldstein, PhD, in my opinion the prime example of a technical leader in the Air Force, has been a trusted colleague.

Mr. Thomas Stansell, through his consulting work for the US Air Force and US State Department, has had tremendous effect on GPS in this century and on me. I admire his style and his influence, and appreciate what he has done for me.

The Institute of Navigation (ION) and its members have provided a welcoming, stimulating, and educational environment for me and thousands of others in the field of satnav. Thanks to Ms. Lisa Beaty and the staff at the ION National Office for all they do to make the ION a very special professional organization.

Ms. Carolyn McDonald, and her company, NavtechGPS, have been integral to GPS and to satnav for decades. NavtechGPS's early close relationship with the ION, and continuing support of instructors like me, has provided opportunities for our professional growth while literally educating a generation of satnav engineers. Thanks to Carolyn for her friendship and support over these many years, and for originally sponsoring the preparation of course notes that led to many of the chapters of this book.

More recently, I have had the distinct pleasure of working with two other giants of satnav. Dr. Pratap Misra, a gentleman in the truest sense of the word, has been as kind and thoughtful a colleague as one could ever desire. Dr. Frank van Diggelen, with his deep insights combined with entertaining and stimulating style, has been an enjoyable and thought-provoking colleague and collaborator.

My daughter, Dr. Sharon Molly (Betz) Marroquin, carefully reviewed the first 15 chapters in their original manuscript form, providing valuable corrections and suggestions before the births of Hannah Molly and, later, Joseph Daniel, rightly diverted her attention and time.

This manuscript, in its entirety, had to be reviewed by the Air Force before its public release. Thanks to the Air Force officers, especially Capt. Nate Howard and Capt. Doug Pederson, for performing these reviews in addition to all of their other duties working on GPS and serving the nation. Also, I cannot thank enough the following colleagues who reviewed the manuscript in its entirety, providing many valuable comments and corrections: Dr. Frank van Diggelen, Dr. Jon Anderson, Professor Jade Morton, and Dr. José-Ángel Ávila Rodríguez. In addition, many thanks to Mr. Phillip Ward, Dr. Sergey Karutin, Professor Yuanxi Yang, Dr. Jeffrey Hebert, Dr. Alex Cerruti, and Professor Michael Braasch for their reviews of selected chapters. The resulting book benefits considerably from the careful attention and thoughtful suggestions of these reviewers.

My father, the late Edward S. Betz, MD, who was an electrical engineer before becoming a physician, influenced me to select electrical engineering as an undergraduate major, leading me to a fascinating and rewarding professional career. Thanks to my mother, Joanna Wells Betz, who has been everything a mother should be. She has been a continual source of encouragement during this effort.

Most importantly, thanks to my wonderful and loving family, especially my wife, Donna, who endured the countless evenings and weekends required to write this manuscript and go through the challenging process of publication. Thanks also to our four children, Christopher, Sharon, Peter, and James, along with their spouses and children, for their encouragement and support.

Thanks be to God.

USEFUL CONSTANTS

Boltzmann constant: $k_B = 1.3806488 \times 10^{-23}$ Joules/K (equivalently, watts/(K-Hz)) [1]
Earth gravitational constant: $\mu_e = 3.986005 \times 10^{14}$ m³/s² [2]
Earth radius: [3]
   Equatorial radius: 6,378,137.0 m
   Arithmetic mean radius of semi-axes: 6,371,008.7714 m
   Radius of sphere of equal area: 6,371,007.1809 m
   Radius of sphere of equal volume: 6,371,000.7900 m
Earth rotation rate: $\Omega_e = 7.2921151467 \times 10^{-5}$ rad/s [2]
Pi: $\pi = 3.1415926535898$ [2]
Speed of light: $c = 2.99792458 \times 10^8$ m/s [2]
Note: Some values may vary slightly with different satnav systems and geodetic reference systems.

REFERENCES

1. The NIST Reference on Constants, Units, and Uncertainty, National Institute of Standards and Technology, http://physics.nist.gov/cgi-bin/cuu/Value?k, accessed January 17, 2015.
2. IS-GPS-200, http://www.gps.gov/technical/icwg/, accessed January 17, 2015.
3. National Imagery and Mapping Agency Technical Report, NIMA TR350.2, Third Edition, January 3, 2000, Department of Defense World Geodetic System 1984, Its Definition and Relationships with Local Geodetic Systems, available at http://earth-info.nga.mil/GandG/publications/tr8350.2/wgs84fin.pdf, accessed January 17, 2015.

LIST OF ACRONYMS AND ABBREVIATIONS

2DRMS	twice the distance root mean square
AAI	Airports Authority of India
ADC	analog to digital conversion, or analog to digital converter
AGC	automatic gain control
A-GPS	assisted GPS
ARAIM	Advanced Receiver Autonomous Integrity Monitoring
ARNS	Aeronautical Radio Navigation Service
AS	anti-spoof
AS	Authorized Service
ASCII	American Standard Code for Information Interchange
ASIC	application-specific integrated circuit
AWGN	additive white Gaussian noise
BAW	bulk acoustic wave
BCH	Bose, Chaudhuri, and Hocquenghem
BDS	BeiDou System
BDT	BeiDou Time
BGBES	BeiDou Ground Base Enhancement System
bps	bits per second
BRSD	Between Receiver Single Differencing
BSQ	bandlimiting, sampling, and quantization
BSSD	Between Satellite Single Differencing
C/A	Coarse/Acquisition
C/N_0	carrier power to noise power spectral density
CAF	cross-ambiguity function
CC	composite clock
CE50	Circular Error 50%, the radius of a circle centered at the true value containing 50% of the estimates
CE90	Circular Error 90%, the radius of a circle centered at the true value containing 90% of the estimates
CED	clock correction and ephemeris data
CEP	Circular Error Probable, the same as CE50
CFAR	constant false alarm rate
CGCS2000	China Geodetic Coordinate System 2000
CNSS	Compass Navigation Satellite System
CORS	continuously operated reference station
CRC	cyclic redundancy check
BOC	Binary Offset Carrier

CRPA	controlled reception pattern antenna
CS	commercial service
CSC	carrier-smoothed code
CSK	code shift keying
DASS	Distress Alerting Satellite System
dB	decibels
dBi	decibels referenced to an isotropic antenna
dBic	decibels referenced to an isotropic circularly polarized antenna
dBil	decibels referenced to an isotropic linearly polarized antenna
dBm	decibels referenced to one milliwatt
dBW	decibels referenced to one watt
DFT	discrete Fourier transform
DLL	delay-locked loop
DOP	dilution of precision
DRMS	distance root mean square
DSSS	direct sequence spread spectrum
ECEF	Earth-centered, earth-fixed
ECI	Earth-centered, inertial
EGNOS	European Geostationary Navigation Overlay Service
EKF	extended Kalman filter
ENU	East-North-Up coordinate system
EOP	Earth Orientation Parameters
FAA	Federal Aviation Administration (of the United States)
FDE	fault detection and exclusion
FEC	forward error control
FFT	fast Fourier transform
FIR	finite impulse response
FLL	frequency-locked loop
FPGA	field-programmable gate array
FRPA	fixed reception pattern antenna
GaAs	gallium arsenide
GAGAN	GPS And Geo-Augmented Navigation
G_{agg}	Aggregate gain of interference power
GCS	Galileo control system
GEO	geostationary
GGTO	GNSS to GPS Time Offset
GIVE	Grid Ionosphere Vertical Error
GLONASS	GLObal NAvigation Satellite System
GMS	Galileo mission system
GNSS	Global Navigation Satellite System
GoJ	Government of Japan
GPS	Global Positioning System
GST	Galileo System Time
GTRF	Galileo Terrestrial Reference Framework
HDOP	horizontal dilution of position

LIST OF ACRONYMS AND ABBREVIATIONS

HEO	highly elliptical orbit
HOW	handover word
I/S	interference to signal ratio (power ratio)
ICAO	International Civil Aviation Organization
ICD	Interface Control Document
IDFT	inverse discrete Fourier transform
IF	intermediate frequency
IGP	ionospheric grid point
IGS	International GNSS Service
IGSO	inclined geosynchronous orbit
IID	independent and identically distributed
IMES	Indoor MEssaging System
IMU	inertial measurement unit
INS	inertial navigation system
IP3	third-order intercept point
IR	image reject
IRNSS	Indian Regional Satellite System
IS	interface specification
ISRO	Indian Space Research Organization
ITRF	International Terrestrial Reference Frame
ITU	International Telecommunications Union
ITU-R	International Telecommunications Union Radio Sector
JGS	Japan satellite navigation Geodetic System
KF	Kalman filter
L2CL	long spreading code used for the GPS and QZSS L2C signals pilot component
L2CM	medium length spreading code used for the GPS and QZSS L2C signals data component
L5I	the Inphase data component of the GPS L5 signal
L5Q	the Quadraphase pilot component of the GPS L5 signal
LAMBDA	Least-squares AMBiguity Decorrelation Adjustment
LC	inductor-capacitor
LDPC	low density parity check
LEO	low Earth orbit
LEX	QZSS experimental signal
LHCP	left-hand circularly polarized
LNA	low noise amplifier
LO	local oscillator
LTI	linear time invariant
MAI	multiple access interference
MC	master clock
MDR	multipath-to-direct path ratio
MEO	medium Earth orbit
MMIC	monolithic microwave integrated circuit
MOOC	Massively Online Open Course

MS	mobile station
MSAS	MTSAT-based Satellite Augmentation System
MTSAT	Multifunctional Transport Satellite
NANU	Notice Advisory to Navstar Users
NAQU	Notice Advisory to QZSS Users
Navwar	navigation warfare
NCO	numerically controlled oscillator
NDGPS	nationwide differential GPS
NGA	National Geospatial Agency
NICT	Japan's National Institute of Information and Communications Technology
NMCT	navigation message correction table
NRC	National Research Council
OCXO	oven-controlled crystal oscillator
OLS	ordinary least squares
ONSP	Office of National Space Policy (of Japan)
OS	Open Service
P(Y)	precision(encrypted)
PAPR	peak to average power ratio
PDOP	position dilution of precision
PDP	power-delay profile
PFD	power flux density
PLL	phase locked loop
PN	pseudo-noise
PNT	positioning, navigation, and timing
ppm	parts per million
PPP	precise point positioning
PPS	precise positioning service
PRN	Pseudo-Random Number
PRS	Public Regulated Service
PSD	power spectral density
PVT	position, velocity, and time
PZ-90	Parametri Zemli (English translation, Parameters of the Earth) 1990
Q	quality factor (of a filter)
QOC	quadrature offset carrier
QPSK-R	quadrature phase shift keying with rectangular spreading symbols
QZS	Quasi-Zenith Satellite
QZSS	Quasi-Zenith Satellite System
QZSST	QZSS Time
RAAN	right ascension of the ascending node
RAIM	Receiver Autonomous Integrity Monitoring
RC	resistor-capacitor
RDSS	Radio Determination Satellite System
RF	radio frequency

RHCP	right-hand circularly polarized
RMS	root mean-squared
RNSS	radio navigation satellite service
R-S	Reed-Solomon
RS	restricted service
RSS	root sum-squared
RTK	real-time kinematic
SA	selective availability
SAIF	submeter class augmentation with integrity function
SAR	search and rescue
SAR/GPS	search and rescue GPS
SARS	search and rescue service
SAW	surface acoustic wave
SBAS	Satellite-Based Augmentation System
SC	super critical
SDCM	System for Differential Correction and Monitoring
SE50	Spherical Error 50%, the radius of a sphere centered at the true value containing 50% of the estimates
SE90	Spherical Error 90%, the radius of a sphere centered at the true value containing 90% of the estimates
SEP	Spherical Error Probable, the same as SE50
SiGe HBT	silicon-germanium heterojunction bipolar transistor
SIR	signal-to-interference power ratio
SISRE	signal in space ranging error
SNR	signal-to-noise ratio
SoL	Safety-of-Life
SPP	standard point positioning
SPS PS	SPS Performance Specification
SPS	standard positioning service
sps	symbols per second
SSC	spectral separation coefficient
SUD	Standard Under Damped
SV	space vehicle
TCXO	temperature compensated crystal oscillator
TDOP	time dilution of precision
TGP	tropospheric grid point
TLM	telemetry word
TOA	time of arrival
TOI	time of interval
TT&C	telemetry, tracking, and command (sometimes telemetry, tracking, and control)
TTIS	time to initial synchronization
UDRE	User Differential Range Error
UEE	user equipment error

UERE	user equivalent ranging error
URA	user range accuracy
USNO	United States Naval Observatory
UTC (NICT)	Coordinated Universal Time as maintained by National Institute of Information and Communications Technology
UTC	coordinated universal time
UTC	ultra-tight coupling
UTC(USNO)	Coordinated Universal Time as maintained by USNO
VDOP	vertical dilution of precision
VGA	variable gain amplifier
VLL	vector locked loop
WAAS	Wide Area Augmentation System
WGS84	World Geodetic System 1984
WLS	weighted least squares
XO	crystal oscillator

ABOUT THE AUTHOR

Dr. John W. Betz

Dr. John W. Betz is a Fellow of The MITRE Corporation, providing technical contributions and leadership to MITRE's work program, spanning research to applications. His work has involved satellite-based navigation, signal analysis and signal processing, communications, sensors, electronic warfare, and systems engineering.

With MITRE since 1989, Dr. Betz has held a variety of positions supporting the Air Force and the Department of Defense. He has led activities involving research and application of signal processing to problems in sensing, communications, navigation, and intelligence. From 2001 to 2002 he was Chief Engineer of the Intelligence, Surveillance, and Reconnaissance Integration Systems Program Office at the Air Force Electronic Systems Center, Hanscom Air Force Base.

His work on satellite-based positioning and timing (satnav) began in 1997, when he led the design of modulation and acquisition for the new GPS military M-code signal. He developed the binary offset carrier (BOC) spreading modulation selected for the GPS M-code signal and also adopted by all of the world's satellite-based navigation systems. He also has contributed to theory and practice of satellite-based navigation receiver processing, signal quality, security, and radio frequency compatibility. He also helped design the GPS L1C civil signal, and developed the multiplexed-BOC (MBOC) spreading modulation adopted for GPS L1C and for other interoperable signals on the European Galileo system and the Chinese BeiDou system, along with the time-multiplexed BOC waveform used for the GPS L1C signal. He has been a lead technical contributor to the U.S. delegation in negotiations leading to the 2004 Agreement between the U.S. and European Community concerning GPS and Galileo. Since 2004, he has contributed to U.S. activities on working groups addressing topics in compatibility and interoperability with Europe, Japan, the Russian Federation, China, and India, leading to other satnav systems' adoption of civil signals compatible and interoperable with GPS and each other.

He continues to be involved in signal and system engineering for GPS, and played a lead role in the GPS Enterprise Modernization Analysis of Alternatives that recommended substantial changes to planned military GPS, identifying more affordable and robust capabilities for warfighters. Most recently, his work has emphasized development and application of more secure and robust satnav capabilities for military and civilian applications.

He was a member of the Air Force Scientific Advisory Board (SAB) from 2004 through 2012, leading the Science and Technology Reviews of Air Force Research Laboratory, and from 2008 to 2011 was Chairman of the SAB. He has also served as a consultant to the SAB and the Defense Science Board, and since 2013 has served on the National Space-Based Positioning, Navigation and Timing Advisory Board, a Presidential advisory committee.

Before joining MITRE, he worked at The Analytic Sciences Corporation and RCA Automated Systems, and has been Adjunct Professor of Electrical and Computer Engineering and lecturer at Northeastern University.

He has authored or co-authored more than 50 research publications in journals, book chapters, and conferences, and is co-inventor on four patents and patent applications. Awards include the International Association of Institutes of Navigation's John Harrison Award (2015); Secretary of the Air Force Distinguished Public Service Award (2014), the highest public service award to private citizens by the Air Force; Institute of Navigation Satellite Division's Johannes Kepler Award (2013); Institute of Navigation Thurlow Award (2011); Fellow of the IEEE (2009); Carlton Best Paper Award, IEEE Aerospace and Electronic Systems Society (2009); MITRE Trustees' Award for international leadership in advancing global positioning, navigation, and timing (2008); named one of GPS World Magazine's "Fifty Leaders to Watch in GNSS" (2008); Fellow of the Institute of Navigation (2006); U.S. State Department Superior Honor Award (2004); MITRE President's Award for Contributions to GPS/Galileo Negotiations (2004); Burka Best Paper Award, Institute of Navigation (2001); MITRE President's Award for Contributions to GPS Modernization Design (1999); MITRE Best Paper Award (1995); Best Paper Award, IEEE Acoustics, Speech, and Signal Processing Society (1986); doctoral studies sponsored by the RCA Graduate Studies Program. He was awarded a BSEE (high honors) from University of Rochester (1976), and Masters (1979) and PhD (1984) Degrees in Electrical and Computer Engineering from Northeastern University.

1

INTRODUCTION

This book describes satellite-based navigation and timing (satnav), the engineering of systems that transmit radio frequency (RF) ranging signals from a constellation of satellites so that a passive receiver can determine time and its position. Satnav is an established field, yet it is poised to grow in terms of number of satellites and signals, number of receivers, number of innovations and applications based on satnav, and in the ways it affects people worldwide.

Like the Internet, cellular telephony, and aviation, satnav is technologically rich, benefiting from many innovations and innovators over the past decades. This book collects and describes the principles behind satnav and the theories that describe how well it works. A better understanding of how to engineer the systems, the signals, and the receivers is key for contributing to the advance of satnav technology; this material also enriches skills and backgrounds for making contributions in related fields like radar and communications.

The intent of this book is to provide a consistent and integrated depiction of the engineering behind satnav; this chapter provides an introduction and the basic background. Section 1.1 discusses the changing scene of satnav we are currently witnessing—what makes satnav so attractive, its current status, and its prospects. Section 1.2 outlines the principles behind satnav, including the basic architectures of satellites and receivers.

Engineering Satellite-Based Navigation and Timing: Global Navigation Satellite Systems, Signals, and Receivers, First Edition. John W. Betz.
© 2016 The Institute of Electrical and Electronics Engineers, Inc. Published 2016 by John Wiley & Sons, Inc.

Section 1.3 summarizes commonly employed attributes that apply to any satnav system. Section 1.4 outlines the structure of this book, and suggests how different types of readers can use it.

1.1 SATNAV REVOLUTION

Satnav is revolutionizing our concepts of positioning, navigation, and timing (PNT). Consumers expect to be able to push a button and be shown their location on a map display with accuracy of a few meters, then be guided to walk or drive to any location of interest. Professionals with more sophisticated equipment expect centimeter-level accuracies for surveying or machine control. Meanwhile, aviation worldwide is moving to satnav-guided operations for en route navigation and even airport approaches, to achieve efficiencies and to fit more flights into the increasingly congested air space. Satnav-derived timing, the less-widely recognized function of satnav, is increasingly used to synchronize communication networks and even to time stamp financial transactions. Scientists use satnav to assist with weather forecasting, environmental observations, and earthquake monitoring. Military applications include situational awareness that has substantially reduced fratricide, while highly accurate satnav-based targeting and munition guidance have enabled precision delivery of small munitions with less collateral damage and fewer civilian casualties.

The United States' Navstar Global Positioning System (GPS) has been the primary source of this revolution in what is sometimes known as space-based PNT. Since 2000, GPS modernization has been underway to provide further improved capabilities. Other systems also are contributing to the revolution. Russia's GLONASS has been revitalized in the past few years, reestablishing a full constellation and undergoing modernization. Europe is fielding Galileo to provide a worldwide system that is highly interoperable with GPS, while China has deployed a Phase 2 regional BeiDou constellation and is building out a worldwide BeiDou Phase 3 system. Meanwhile, Japan is developing a regional system called the Quasi-Zenith Satellite System (QZSS). India is fielding a regional system, and has expressed ambitions for a global system as well. Simultaneously, the international aviation community has developed standards for a Satellite-Based Augmentation System (SBAS); more and more SBASs are being fielded to augment standalone satnav systems, yielding higher integrity and higher accuracy when combined with GPS and other satnav systems.

These systems transmit signals that allow a receiver to determine time and its location. The widespread benefits of satnav come from the use of these signals, reliably present to a receiver with adequate view of the sky, by highly capable low cost receivers coupled with software for mapping, route planning, and other applications. The development of novel augmentations, combined with creative integration of satnav with existing communications infrastructure, has produced further advances in capability and user experience. The combination of GPS and these other capabilities has produced the revolution we are experiencing. The emergence of more and better satnav signals, and more satnav systems, promises to launch the next round of creative developments in receivers and application software, augmentations, and integrations with other sensors and with communications capabilities.

While the practice and applications of satnav have evolved rapidly, so have the underlying theory and principles. The original design of GPS implemented many innovations, and the designers recognized other advances that were not included then because of technology limitations in the 1970s. Some of these previously conceived advances, such as pilot components and error control coding of data messages, are now practical to implement and are widely used in new and modern signals. Theoretical advances also have been made since GPS was originally designed, spurred in some cases by the pressures of GPS modernization and the emergence of multiple satnav systems, with the resulting need for hundreds of signals to share common frequency bands. Some of these recent advances involve new techniques, while in other cases they involve a better understanding of performance and how to assess it. Furthermore, user demands for operation in challenging environments like urban canyons and indoors, where satnav operation had previously been thought impossible, and applications demanding accuracies that had not previously been considered, have led to further advances in theory and techniques, along with a better understanding of how to mitigate some of the limitations of satnav.

Satnav has been found to have important attributes that merit the interest, the investment, and the innovation. These attributes, unmatched by any other known navigation and timing technology, include:

- All-weather operation. Satnav's use of L band signals in the 1–2 GHz band makes its use for PVT calculation[1] insensitive to most weather effects, except for the heaviest rainstorms and occasional space weather events.
- Day and night utility. Satnav's use of radio frequency (RF) signals allows it to work under any conditions of light or dark.
- Worldwide utility. Global systems using satellite constellations in medium Earth orbit allow satnav to work anywhere there is an adequate view of the sky, on the surface of the Earth and even in air and space.
- Consistent performance even over featureless surfaces of the Earth.
- Absolute measurements. Measurements of position and time are made, or can be transformed into, absolute Earth-centered, Earth-fixed coordinate systems and universal coordinated time (UCT), not relative to another location or time epoch.
- Three-dimensional positioning, with accuracy of the order of meters.
- Timing accuracy to tens of nanoseconds.
- Consistent accuracy over time and location.
- Inexpensive user equipment. Receiver costs range from tens of dollars for a simple consumer device to several thousands of dollars for a professional device to tens of thousands of dollars for a certified aviation device.

---

[1] Some scientific applications of satnav, such as LEO satellite-based radio occultation, can be highly sensitive to weather.

- User equipment with small size, low mass, and modest power consumption.
- No or minimal user training, without requirement for expertise involved in using traditional means for navigation.
- Rapid updates, with time, position, and velocity measurements reported of the order of seconds or fractions of seconds.
- Passive operation. User equipment does not need to transmit, but only receive.
- No local infrastructure. Sparsely spaced monitoring stations and several ground control stations with antennas that transmit to the satellites are the only terrestrial infrastructure needed for satnav.
- No detailed local surveying or measurements, no need for local surveys of gravity, terrain, or magnetic fields, and no need to know building topography.

Two additional characteristics that have amplified the revolution are based on farsighted policies rather than technological characteristics:

- No user charges. For decades, both GPS and GLONASS have provided civil signals to the world free of user charges, funded out of the United States and Russian Federation government budgets.
- Open signal descriptions. For decades, both GPS and GLONASS have provided full technical descriptions of their civil signals, with openly available documentation on how to use them.

These policies have led to worldwide acceptance and use, and to innumerable innovations that exploit satnav systems and signals in ways never conceived by their original developers. Furthermore, more recently introduced and planned systems have adopted similar policies, with some exceptions.[2]

However, satnav has limitations. Severe space weather events can cause fading, signal distortion, and loss of tracking lock in receivers, especially those operating in equatorial and high latitude regions. Satnav signals do not propagate underground or underwater, and they do not propagate well deep indoors. In urban canyons and indoors, even when receivers are able to obtain and process the signals, measurements are often distorted by shadowing and multipath, producing errors of tens or hundreds of meters. Received satnav signals have very low power, and thus are susceptible to even relatively low power interference. Many receivers in use today do not use secure designs; like first-generation computers on the Internet, many satnav receivers have been specified, designed, and tested under the implicit assumption that no one would attack them, and consequently the receivers assume that any apparently valid signal is in fact valid.

The modernization of GPS and GLONASS and the introduction of new satnav systems will help reduce these limitations of satnav with new signal designs, higher-power signals, and new generations of receivers. These receivers will use the improved

---

[2] For example, Galileo may charge for use of civil signals that support the Commercial Service and Public Regulated Service. See Chapter 10 for a description of these services and signals.

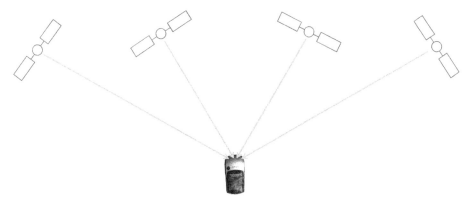

Figure 1.1.   Satnav Basic Geometry

signal characteristics to obtain better performance in stressing situations, to further extend the conditions where they can operate, and to mitigate the limitations of satnav. Readers of this book will acquire the knowledge and insights needed to contribute to this next set of advances.

1.2 BASIC PRINCIPLES OF SATNAV

Satnav goes by many different names. "GPS" is sometimes used as a generic term for any satnav system or capability. Others use "Global Navigation Satellite System," or "GNSS," despite the fact that some systems are regional. "Radio Navigation Satellite Service," or "RNSS," is also used in the communities that manage spectrum use. This book adopts "satellite-based navigation and timing" and uses the abbreviation "satnav." This is based on the widely accepted definition of navigation as going beyond merely planning and following a route, but also determining one's position and possibly velocity. Unfortunately, timing is not mentioned explicitly in "satnav," but is included implicitly.

The term "navigation" has varying definitions. Some sources define navigation as determining your own position and velocity, while others include the concept of guidance—planning and following a route—in the definition of navigation. The treatment of navigation in this book does not include guidance.[3] It does include determining position, velocity, and time (PVT), and uses the term PNT in spite of redundancy between positioning and navigation.

Figure 1.1 shows the basic geometry used in satnav. Each satellite continuously broadcasts one or more signals over the surface of the Earth in view of the satellite. A receiver processes signals from at least four satellites, measuring each signal's time of arrival relative to a clock in the receiver. The receiver also uses information encoded

---

[3] GPS commonly is blamed for inefficient or incorrect route planning in vehicle GPS systems, even though route planning is performed by terrestrial-based software not associated with GPS, which only provides the current position and velocity of the vehicle.

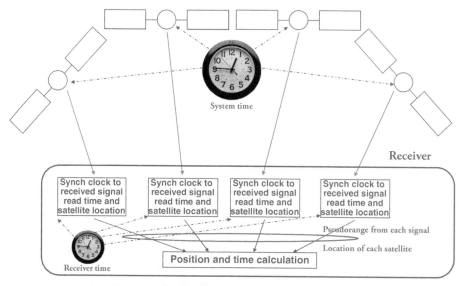

Figure 1.2. Satnav Receiver Processing Overview

in each signal to determine when that signal was transmitted by the satellite and the satellite's position at the time of transmission. To understand the underlying principle, assume initially that the receiver clock and the satellite clocks are perfectly synchronized, so that taking the difference between a signal's time of arrival and the time of transmission yields its signal's propagation delay. As long as the signals propagate at the speed of light in a straight line, the receiver can then calculate its distance from each of the satellites, and then trilaterate,[4] or calculate its location based on known distances from known locations. Clearly, in this case there is no need to calculate time, since it is assumed the receiver's clock is already synchronized to the satellite clocks.

As long as the receiver clock and satellite clocks are synchronized, trilateration can use measurements from three satellites. (The receiver is solving for three unknowns—its coordinates in three dimensions—using three independent measurements.) But attaining positioning accuracies of the order of a meter requires synchronization between the receiver clock and satellite clock of the order of several nanoseconds. Since atomic clocks that provide this type of accuracy have been large, expensive, and power-hungry, receivers would not have been compact, inexpensive, and able to operate for many hours on a small battery if this approach were used.

Instead, a variant of this approach is used, as illustrated in Figure 1.2. Here, all satellites are perfectly synchronized, as above, to system time. This synchronization could be achieved by the technique illustrated—distributing a common time to all

---

[4] Trilateration determines position by measuring distance from known locations. In contrast, triangulation determines position by measuring angles to known locations, and multilateration (also known as hyperbolic positioning) measures the differences in distances to known locations.

BASIC PRINCIPLES OF SATNAV

satellites, or with separate clocks on each satellite that are adjusted to maintain synchronization, or with separate clocks that drift, as long as the receiver can correct for the drift. This last approach is typically used in operational satnav systems, but the principle would be the same with any of these approaches. As depicted in Figure 1.2, the receiver has a number of channels, one dedicated to each satellite signal being tracked. Each channel contains signal processing circuitry that continuously measures the signal's time of arrival in terms of a time source at the receiver, or "receiver time." Although the signals are continuous, the receiver selects epochs, or reference times, on each signal to measure their time of arrival. Each signal is modulated with information receivers used to determine the system time and the satellite location when an epoch was transmitted. Comparing each signal's epoch time of arrival in terms of receiver time to the system time when the epoch was transmitted yields the propagation delay plus an offset, or bias, due to the difference between receiver time and system time. This bias is the same for each received signal, since receiver time is common to all the receiver channels, and system time is common to all the satellites. In satnav terminology, these biased delays are known as "pseudodelays," and dividing all the pseudodelays by the known speed of propagation yields a set of pseudoranges: the set of distances from the receiver to each satellite, each biased by a common amount. (Pseudodelays and pseudoranges are equivalent, and related by the speed of light.) The receiver then uses the set of pseudoranges and satellite positions to calculate its position and time in the same coordinate systems used by the system.

Figure 1.3 shows the satellite locations, receiver location, system time at a particular epoch, receiver time at the same epoch, and propagation delays. Locations may be defined in different coordinate systems, with the satellite locations described in an inertial reference system, the receiver location in a geodetic or Earth-centered, Earth-fixed (ECEF) coordinate system, and the receiver position sometimes reported in an east-north-up (ENU) or other relative coordinate system. System time is defined relative to a time standard such as UTC, and receiver time t_r is defined as system time t_s plus a receiver clock offset Δ_r. Known to the receiver are the satellite locations (x_k, y_k, z_k), system time at a particular epoch, receiver time, and the pseudodelays D_k corresponding

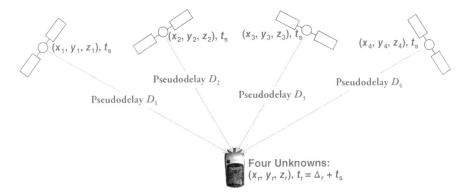

Figure 1.3. Quantities Used for Calculating Position and Time

to that epoch. Assuming the signals propagate in a straight line at the speed of light $c = 2.99792458 \times 10^8$ m/s, the pseudoranges are $D_k c$ and the true ranges are $(D_k - t_b)c$.

If there are four satellites, as shown, applying the Pythagorean theorem for each satellite yields the system of four equations in four unknowns x_r, y_r, z_r, Δ_r

$$\begin{aligned}
cD_1 &= [(x_1 - x_r)^2 + (y_1 - y_r)^2 + (z_1 - z_r)^2]^{1/2} + c\Delta_r \\
cD_2 &= [(x_2 - x_r)^2 + (y_2 - y_r)^2 + (z_2 - z_r)^2]^{1/2} + c\Delta_r \\
cD_3 &= [(x_3 - x_r)^2 + (y_3 - y_r)^2 + (z_3 - z_r)^2]^{1/2} + c\Delta_r \\
cD_4 &= [(x_4 - x_r)^2 + (y_4 - y_r)^2 + (z_4 - z_r)^2]^{1/2} + c\Delta_r
\end{aligned} \quad (1.1)$$

Numerous approaches exist for solving these equations. Some approaches yield two solutions, one of which is typically physically impossible. If there are fewer than four satellites, this approach does not yield a unique solution, although there are ways to handle that situation in some cases. If there are more than four satellites, the overdetermined system of equations can be used to advantage in different ways, including crosschecking the validity of measurements and reducing the effects of measurement errors. Chapter 20 discusses these topics in detail.

Naively applying these concepts in actual satnav would produce results with very large errors, since the preceding model is overly simplistic, with many second-order and third-order effects neglected. Satellites and receivers are not points in space, but distributed, so the positions in (1.1) actually describe the antenna phase centers. Everything is moving—satellites, receivers, and even the ECEF coordinate system, and all this motion must be taken into account in several different ways. The actual satellite locations differ from what is encoded onto the signals, the satellite clocks are not perfectly synchronized in reality, and the encoded clock corrections are not perfectly correct. The satellite speeds are high enough that relativistic effects must be taken into account. The signals do not propagate at the speed of light or in a straight line, but instead are refracted by the ionosphere and then undergo signal-specific delays in the troposphere. Reflections from nearby objects (multipath), noise, and interference contaminate the received signals, causing errors in receiver measurements. These and other details must be taken into account in order to achieve the accuracies experienced with satnav today.

Satnav engineering then consists of designing systems, signals, and receivers that provide navigation and timing capabilities, under constraints of cost and complexity. Satnav systems are generally described in terms of three segments or subsystems, portrayed in Figure 1.4. The space segment consists of the satellites that transmit the signals as electromagnetic waves. The signals are sinusoids modulated by spreading modulations and spreading codes, along with data that provides to receivers the satellite's ephemeris (location in space at different times), offset of the satellite clock relative to system time, and other system information. The user segment consists of antennas that capture the electromagnetic waves, and receivers to process the signals for measuring time of arrival, reading the data modulated onto the signal, and then calculating the receiver position, velocity, and time. While the space segment receives publicity each time a satellite is launched, and we are used to handling receivers, the relatively unknown but also essential segment is the control segment, sometimes called the ground segment. The control

BASIC PRINCIPLES OF SATNAV

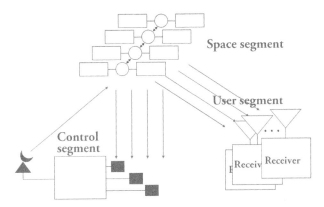

Figure 1.4. Generic Satnav System Architecture

segment makes satnav a closed-loop system by monitoring the signals from the satellites (including the same signals received by the user segment, other telemetry signals from the satellite describing health and status of its subsystems and components, and in some cases laser signals reflected from the satellites) to determine the satellites' orbits and clock alignment as well as the satellite's status. The control segment then transmits to the satellites new data message contents, with updated ephemeris and clock corrections, along with updates to other system information, to be modulated onto the signals.

Satnav signals connect the space segment to the other two segments. As explained above, these signals enable the receiver to perform two different functions:

1. Measure each signal's time of arrival (yielding pseudoranges) and frequency of arrival (yielding rate of change of the pseudorange, and from this line of sight speeds relative to the satellites),
2. Read each satellite's position and the time of transmission at a signal epoch, as well as other system information modulated onto the signal.

In this sense, satnav is much like a blend of radar, where time of arrival and frequency of arrival measurements are critical, and digital communications, involving reading digital information modulated onto the signal.

The satellite assembles the signals, including modulating the data onto the signals. An entire book could be written about satnav satellite design, describing the bus, propulsion, thermal management, power supply, communications links with ground systems, and other subsystems. The relevant focus for this book is the navigation signal generation, depicted in Figure 1.5. Key is the satellite clock, with its low phase noise and low drift rate. Satnav satellites typically transmit multiple signals at two or more different carrier frequencies; five signals are portrayed in the figure. Baseband signal generation occurs for each of these five signals, including forming data messages from information telemetered from the ground control segment and generating spreading codes to produce each baseband signal as described in Chapter 3. Multiple signals are multiplexed

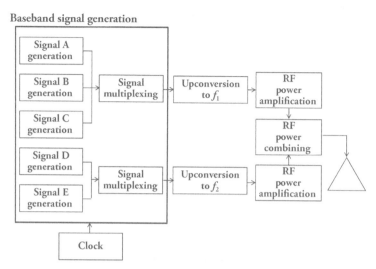

Figure 1.5. Navigation Signal Generation on a Satnav Satellite

together to be modulated onto a single carrier; in this example three signals are multiplexed being modulated onto a carrier with frequency f_1 and two signals multiplexed onto a carrier with frequency f_2, using multiplexing techniques described in Chapter 3. These low-power signals at RF are each amplified to high power (typically tens of watts of RF power for each signal) using power amplifiers tuned to each carrier frequency. The signals are then combined at high power into a single high-power signal (having effective isotropic radiated power sometimes exceeding a kilowatt) transmitted from the antenna.

Building and launching satnav satellites are both very expensive, with the cost for developing, building, and launching a satellite typically exceeding 100 million US dollars. Satellite weight is highly related to cost, and every transmitted watt of power adds to satellite weight in multiple ways, including the solar panels that supply the power to the power conversion and conditioning circuitry, power amplifiers, power combiner, antenna elements, as well as heat pipes and radiators needed to handle waste power on the satellite. The most efficient power amplifiers are highly nonlinear, operating in what is called "saturation" where the output voltage's magnitude is constant, with only the phase changing. An important aspect of satnav signals, as discussed in Chapter 3, is for the transmitted signal to have a constant magnitude, also known as "constant envelope," to enable use of efficient power amplifiers. The signals themselves must be designed to enable efficient transmission and effective receiver processing.

The receiver processes the signals to make the measurements and read the data. Receiver processing consists of four main functions:

- Perform initial synchronization: when the receiver is not already tracking signals, it detects the signals that are available to receive, then obtains an initial estimate of each signal's time of arrival and received carrier frequency to initiate tracking.

BASIC PRINCIPLES OF SATNAV

- Track the signals: the receiver estimates the time-varying time of arrival and frequency of arrival of each signal, providing periodic estimates of pseudoranges and received carrier frequencies (and possibly the carrier phases).
- Read the data messages: the receiver demodulates the encoded message bits, deinterleaves the bits, decodes the channel encoding, checks for errors, synchronizes to the message structure, and then interprets the data message to obtain satellite ephemeris and clock corrections, along with other system information.
- Calculate receiver position and velocity: the receiver uses pseudoranges, system time information, and ephemerides to calculate and smooth the receiver position and velocity.

Figure 1.6 provides a simple block diagram of receiver functions. The receive antenna transforms the incident electromagnetic field to electric signals. These are conditioned (amplified and filtered) at RF, then downconverted to an intermediate frequency (IF) or baseband. Signal conditioning involves additional amplification and filtering, as well as automatic gain control in some receivers. At some point in this process the analog signals, continuous in time and in amplitude, are processed by an analog-to-digital converter that samples them, producing outputs of discrete amplitude samples represented by digital values, at discrete points in time. Signal processing operates on these digitized signals, performing initial synchronization, tracking the signals, computing values used to demodulate the data message, and reading the data message. Signal processing for acquisition and tracking relies on correlation processing between the received signals and noise and interference with a locally generated replica. Different receiver channels perform tracking on different signals being received and processed. Navigation processing uses measurements from tracking, as well as information from the data message, to produce estimates of the receive antenna's location and velocity, and of time. The receiver processing is driven by frequencies synthesized from a reference oscillator in the receiver.

This book describes the engineering of satnav systems, signals, and receivers, with emphasis on how these receiver functions are implemented, and methods to determine and predict how well they are performed.

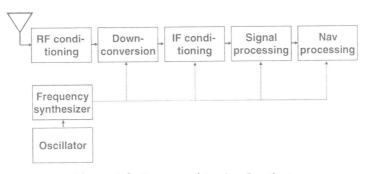

Figure 1.6. Overview of Receiver Functions

1.3 SATNAV ATTRIBUTES

The aviation community has defined four attributes that apply to any satnav system: accuracy, integrity, continuity, and availability [1]. Accuracy is defined as the degree that the navigation system's reported PVT correspond to their true values in the same reference system. Integrity indicates that the signals transmitted are valid and useful, and that their RF characteristics and data message contents are within specification and will produce valid PVT. Further, integrity indicates the system will provide timely and valid warnings when the system should not be used for the intended purpose. Any warning must be provided within a given period of time (time-to-alert) when the system should not be used. Continuity describes a navigation system's capability to perform its function without unscheduled interruptions during an intended period of operation. Availability is the fraction of time during which the service provides expected levels of accuracy, integrity, and continuity, taking into account all the reasons for the lack of service.

1.4 BOOK STRUCTURE AND HOW TO USE THIS BOOK

This book begins with a top-down view of satnav engineering. Part I describes the principles and practices of systems engineering and signal engineering for satnav, including the basic calculations that describe system operation and performance, along with the generic characteristics of satnav signals. Descriptions of satellite orbits and Kepler's laws provide insights into how ephemeris is represented and how it is used by receivers. Geometrical relationships are calculated between satellites and receivers and the resulting effects on signals propagating from satellites to receivers are derived. Concepts and terminology associated with satnav signals that link satellites to receivers are provided. The many characteristics that describe a satnav signal are discussed, with perspectives on choosing these characteristics. Mathematical representations are provided for signals and how they are processed, with definitions of terminology and concepts essential to understanding and using modern satnav signals. Link budgets are presented, showing how to calculate relationships between transmitted power and received power for space-to-Earth links used for satnav signals, how to calculate link budgets for satnav signals, and how to estimate ground-to-ground propagation loss for local interference sources, as well as how to account for additional attenuation caused by foliage and buildings. Essential descriptors of signal and interference environments and of receiver performance are introduced, along with the analytical tools to assess effects of interference on receiver performance metrics. Satnav positioning and timing error analyses are presented.

Part II then describes specific satnav systems and signals. The United States' GPS, multiple nations' SBASs, Russian Federation's GLONASS, Europe's Galileo, China's BeiDou, Japan's QZSS, and India's IRNSS are each described, with summaries of their history and characteristics. Most important from the perspective of this book, however, is the description of their signals. Using the concepts, terminology, and mathematical notation from Part I, the different chapters provide succinct yet detailed descriptions of the signal characteristics, preparing the reader to understand how to design receivers for processing these signals, as well as providing a useful introduction to each system's signal interface specifications that provide the complete set of details needed to receive the signals.

Part III describes essential aspects of receiver design and evaluating receiver processing performance, from receiver front end through position calculation. Characteristics of front-end components and various front-end architectures are described, along with associated tradeoffs. Details of analog-to-digital conversion (ADC) are described, along with models for relating ADC sampling rates and quantization to processing losses. Initial synchronization, where a receiver detects the presence of satnav signals and estimates their time of arrival and frequency of arrival, is described with focus on modern algorithms enabled by advanced digital circuitry along with associated performance. Since receiver measurements typically rely on tracking loops for estimating the time-varying carrier frequency and phase and the spreading modulation time of arrival, the theory, design, and performance of tracking loops are described using a general formulation that applies to the different types of loops in satnav receivers. The specifics of carrier tracking and code tracking are then addressed, integrated with analytical tools for assessing performance to enable design trades. Finally, position calculation techniques are introduced, along with how to account for errors introduced by the ionosphere and other phenomena. Also, variations are discussed for position calculation using signals from fewer than and more than four satellites. The material emphasizes processing of new and modern signals, using the advanced features incorporated into these new signal designs for better and more robust performance, and providing the latest results in terms of techniques and performance assessment. Numerous alternatives are described, with integrated presentation of analytical tools for predicting performance in trading off the alternatives. Along with an emphasis on modern signals, the material presented also emphasizes modern receiver processing techniques enabled by the ability to perform large and complex digital processing functions inexpensively, providing opportunities to implement techniques previously considered impractical.

Although basic receiver processing techniques are important, modern and future satnav receivers increasingly involve more advanced techniques and capabilities. Part IV introduces specialized topics that are essential for obtaining the next level of performance and enabling advanced applications. Since interference is and will be a reality, specialized techniques for dealing with interference are presented. As new and modernized multifrequency satnav systems become available, multipath and shadowing will often be the dominant source of errors, and an overview of multipath's effects and how to mitigate them is provided. Differential satnav is effective in providing submeter accuracies, and alternative architectures and approaches for differential satnav are described. Assisted satnav, another recent development that employs communications functionality to simplify and improve satnav processing, is outlined. While traditional receiver processing only uses satnav signals, and tracks each signal separately, modern satnav processing need not be limited in this way. Joint tracking of all signals is described, since its implementation is increasingly practical. In addition, other sensors are readily available at increasingly acceptable cost, size, and power consumption. Sophisticated use of information from these other sensors, combined with using satnav signals, is shown to provide important extensions and improvements in capability.

This book is structured for use as a textbook for an upper-level undergraduate course or for a graduate course in satnav engineering. For a one semester course, Part I and Part III should be emphasized, with one chapter in Part II addressed in detail—this chapter should be selected based on which satnav system is of most interest to the instructor and students. An undergraduate course would emphasize the applied problems after each chapter, while a graduate course would spend more time on the theoretical problems, helping students establish a deeper technical foundation.

For a two-semester course, much more of the book can be covered. All of Part II would be addressed so that students become familiar with the entire set of satnav systems, and there will be additional time to address details and extensions of receiver processing in Part III. Further, some or all of the specialized topics in Part IV can be addressed, and students could perform class or individual projects by accessing additional information on these topics to attain greater depth.

This book can also be a very useful supplementary text in an advanced undergraduate or graduate course, or for a short course or seminar on satnav. It is also structured to be useful for an engineer seeking to enhance his or her skills in satnav engineering, or as a reference to the practicing satnav engineer.

In addition, many of the systems engineering topics in Part I and receiver design topics in Part III apply to areas other than satnav, and the organized presentation of these topics should be of interest to students or practitioners in other RF-based technologies such as communications and sensing.

1.5 MORE TO EXPLORE

A number of books provide useful overviews of satnav, typically with emphasis on GPS. Among these are:

- *Global Positioning System: Theory and Applications*, Edited by B. W. Parkinson, J. J. Spilker, Jr., P. Axelrad, and P. Enge, American Institute of Aeronautics and Astronautics, 1996.
- P. Misra and P. Enge, *Global Positioning System: Signals, Measurements, and Performance*, Revised 2nd edition, Ganga-Jamuna Press, 2006.
- *Understanding GPS: Principles and Applications*, 2nd edition, Edited by E. D. Kaplan and C. J. Hegarty, Artech House, 2006.
- F. van Diggelen, *A-GPS: Assisted GPS, GNSS, and SBAS*, Artech House, 2009.
- *Global Navigation Satellite Systems: Signal, Theory and Applications*, Edited by S. Jin, ISBN: 978-953-307-843-4, InTech, DOI: 10.5772/29453. Available from: http://www.intechopen.com/books/global-navigation-satellite-systems-signal-theory-and-applications, accessed 25 October 2014.

A nice introduction to key aspects of satnav (although some of the specifics and numerical values are out of date), also with emphasis on GPS, is:

- R. Langley, The Mathematics of GPS, *GPS World Magazine*, July/August 1991.

For decades, a column called "Innovation" in GPS World Magazine, edited by Richard Langley, has provided excellent technical detail on focused topics. A more recent column called "GNSS Solutions" in InsideGNSS Magazine, edited by Mark Petovello, is similarly an excellent resource.

In addition, there are numerous tutorials available on the Internet that are useful but do not necessarily benefit from the same quality control as formal publications. Two examples are:

- http://en.wikipedia.org/wiki/Gps
- http://www.navipedia.net

There are also free online courses available on the Internet. For example, the Massively Online Open Course (MOOC) "GPS: An Introduction to Satellite Navigation, with an Interactive Worldwide Laboratory using Smartphones," by Professor Per Enge and Dr. Frank van Diggelen of Stanford University can be found at:

- https://www.coursera.org/course/gpslab

REFERENCE

1. Civil Aviation Authority, CAA PAPER 2003/9, April 2004, www.caa.co.uk, accessed 24 October 2014.

PART I

SYSTEM AND SIGNAL ENGINEERING

Like other engineering applications, satnav has evolved a rich set of concepts and terminology, and a powerful set of tools for design and performance assessment. The exposition employed in this book begins with this introduction of essential concepts, terminology, notation, and tools, emphasizing those for systems engineering and system performance assessment.

Starting literally at the top, Chapter 2 provides an overview of satellite orbits and constellations. This chapter also includes a section that provides useful geometrical relationships between satellite and terrestrial receivers, as well as the equations and results showing the effect of these geometrical relationships on received satnav signals.

Working down from the satellites, Chapter 3 describes satnav signals—the link between satellites and receivers. This detailed chapter is an essential foundation for the remainder of the book since it introduces concepts, terminology, notation, and characteristics that are used throughout subsequent chapters. Another important function of this chapter is to provide insight into the considerations that drive signal characteristics and consequently why signals are designed the way they are; a dozen specific signal characteristics are discussed for this purpose. Fundamental notation for direct sequence spread spectrum (DSSS) signals is introduced and applied to satnav signal designs. Also, imperfections and variations due to phenomena such as polarization, Doppler shifts, and dispersive propagation are rigorously described for use in later chapters.

Still focusing on the link between satellites and receivers, Chapter 4 describes link budget calculations. The basic free space link budget equation is developed, along with insights concerning the physical meaning of the resulting equations, and how concepts such as wavelength, frequency, transmit antennas, and receive antennas are involved. A useful application and extension of free space link budgets is presented, showing conventions used to calculate and specify minimum received power and maximum received power from a satnav system. While satnav itself primarily involves propagation of signals from space to users, where free space link budgets often apply, systems engineering often also involves consideration of terrestrial link budgets for reasons ranging from terrestrial interference sources to use of ground-based augmentations.

Engineering Satellite-Based Navigation and Timing: Global Navigation Satellite Systems, Signals, and Receivers, First Edition. John W. Betz.
© 2016 The Institute of Electrical and Electronics Engineers, Inc. Published 2016 by John Wiley & Sons, Inc.

Consequently, terrestrial link budgets are discussed as well, along with considerations of when and how to apply them. Finally, since receivers are often used under foliage or inside buildings, some basic approaches are provided for including excess losses due to foliage and building penetration.

Chapter 5 begins with the consideration of a receiver, its processing, its performance, and systems engineering analysis of interference effects on receiver performance. It introduces a simple channel model, analogous to those considered in communications engineering, as well as an ideal correlator that models the essential signal processing approach encountered in satnav receivers. It then introduces the fundamental operation of correlating received signal plus noise plus interference against a replica of the desired signal, and the associated concept of correlator output signal-to-noise ratio (SNR). The concept of effective signal power to noise spectral density, or effective C/N_0, is introduced and related to correlator output SNR. The effects of in-band interference on correlator output SNR and effective C/N_0, including multiple access interference from other DSSS signals, are taken into account using spectral separation coefficients (SSCs).

Ultimately, positioning and timing accuracies are fundamental considerations in satnav systems engineering, so Chapter 6 describes system-level accuracy measures and how they are computed. Error sources and error models are introduced, as are the assumptions and models used to obtain them. The terminology used in describing system-level accuracy, and the different accuracy measures and relationships among them, are presented.

The reader is likely to find, as the author has, that these topics are essential in discussing satnav systems, signals, and receivers, and in performing system-level assessments that shape understanding and design considerations.

2

SATELLITE ORBITS AND CONSTELLATIONS

Satnav begins with satellites. To a satellite engineer, a navigation satellite is a wonderful combination of solar cells, bus and structural elements, communications antennas and modems for tracking and telemetry and control signaling, cabling, atomic clocks, navigation signal generation hardware, navigation signal power amplifiers, microwave plumbing, antennas for navigation signals, heat pipes and radiators and other thermal management devices, thrusters and momentum wheels and other mechanisms for staying at the correct orientation in the same orbit, or maneuvering to a new one. For purposes of this chapter, however, a navigation satellite is represented as a point in space occupying a designated orbit about the Earth, typically moving at several kilometers per second. The emphasis here is the motion of navigation satellites orbiting the Earth, the configuration of a constellation of navigation satellites, and the relative geometry and dynamics between a satellite and a receiver. Description of satellite position and motion inevitably involves coordinate frames, which are introduced in the Appendix Section A.9.

Entire books are written about astrodynamics, addressing the design of orbits and constellations, and the representation, prediction, and analysis of satellite motion [1, 2]. The intent of this section is merely to provide an overview of the concepts and issues

Engineering Satellite-Based Navigation and Timing: Global Navigation Satellite Systems, Signals, and Receivers, First Edition. John W. Betz.
© 2016 The Institute of Electrical and Electronics Engineers, Inc. Published 2016 by John Wiley & Sons, Inc.

encountered, to establish familiarity with the terminology used, and to provide the perspectives needed for satnav systems engineering and receiver processing.

Section 2.1 begins with Kepler's laws, equations that describe the motion of Earth-orbiting satellites under idealized conditions using a set of Keplerian parameters. Section 2.2 discusses the deviation of actual satellite orbits from those predicted by Kepler's laws, and how these deviations from ideal are handled in satnav. Section 2.3 introduces some basics of satnav constellations, since satnav requires multiple satellites both to provide coverage over large portions of the Earth's surface and to provide the needed number of measurements. Section 2.4 presents some useful calculations of geometry between satellites and receivers, showing how the geometry maps into characteristics of received signals. Section 2.5 summarizes key elements of the chapter.

2.1 KEPLER'S LAWS

Kepler's laws apply to a two-body orbital model, where one point mass orbits a second point mass in a vacuum with no other bodies or forces present. Restated for a satellite orbiting Earth, Kepler's laws state:

1. The orbit of every satellite is an ellipse with the Earth at one of the two foci,
2. A line joining the satellite and the Earth sweeps out equal areas during equal intervals of time,
3. The square of the orbital period of a satellite is directly proportional to the cube of the semimajor axis of its orbit.

Kepler's laws generalize to elliptical orbits, the model of Copernicus, which applies to circular orbits.

The equations of Keplerian motion involve seven parameters. The first five describe the orbit, and the last two describe the location of the satellite in the orbit. In the summary below, the notation conforms to that commonly employed in orbital mechanics.

Two Keplerian parameters describe the size and shape of the orbital ellipse, portrayed in Figure 2.1:

- Semimajor axis, a,
- Eccentricity, e,

where

$$e = \sqrt{1 - \frac{b^2}{a^2}} \qquad (2.1)$$

and b is the semiminor axis of the ellipse, shown in Figure 2.1 along with other Keplerian parameters introduced below.

KEPLER'S LAWS

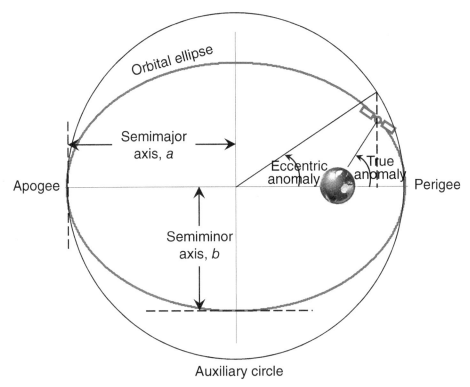

Figure 2.1. Orbital Ellipse

Three Keplerian parameters, depicted in Figure 2.2, describe the orientation of the ellipse relative to Earth. The first two of these describe the orbital plane's (the orbital plane passes through the center of the Earth and contains the orbital ellipse) orientation, and the third describes how the ellipse is oriented in the orbital plane.

- Inclination of the orbit is the angle (between 0° and 180°) of the ellipse relative to the equatorial plane at the ascending node, where the satellite passes from south to north through the equatorial plane. For example, an inclination of 90° is a polar orbit, while an inclination of 0° is an equatorial prograde orbit, with the satellite moving in the same direction as the Earth's rotation. An inclination of 180° is an equatorial retrograde orbit, with the satellite moving in the opposite direction as the Earth's rotation.
- Right ascension of the ascending node, or RAAN, describes the point on the equator where the ascending node occurs, as an angle in the equatorial plane, measured at the center of the Earth from a reference point defined as 0°. Since the Earth is spinning, an ECEF coordinate system would be spinning relative to the RAAN, so an ECI coordinate system is employed. The reference point for the RAAN is the vernal equinox, the direction from center of Earth to the sun

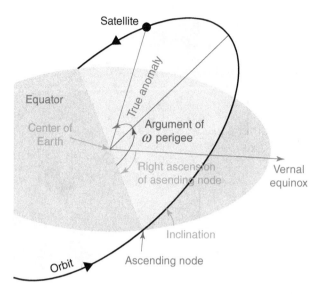

Figure 2.2. Orientation of Orbital Ellipse

at the instant the sun intersects the equator as the sun moves into the northern latitudes at springtime in the northern hemisphere.
- Argument of perigee is the angle of the ellipse's semimajor axis relative to the equatorial plane, measured at the focus of the ellipse where Earth is centered (in the terminology of orbital elements, an "argument" is simply an angle). Argument of perigee describes the ellipse's orientation in the orbital plane. For example, when the argument of perigee is 90° or 270°, the satellite is closest to the Earth as it crosses the equator.

The final two Keplerian parameters describe the location of the satellite on the orbital ellipse:

- The epoch is a reference time when the satellite's orbit and location on the orbit are specified.
- Anomaly describes satellite's position at the epoch. Three different anomaly parameters can be employed, all related to each other. The true anomaly, v, is the angle of the satellite location at a particular epoch, relative to the direction of periapsis (the direction of the line connecting center of Earth to the point on the ellipse corresponding to perigee), at the focus of the ellipse where Earth is centered. The eccentric anomaly, E, is also an angle defined as follows. The orbital ellipse is inscribed in an auxiliary circle. A line is drawn from the satellite location at an epoch to the major axis of the ellipse, perpendicular to the major axis. The eccentric anomaly is the angle between the major axis and the point where this

line intercepts the auxiliary circle. Figure 2.1 shows the true anomaly and the eccentric anomaly. The mean anomaly, M, is the fraction of the orbital period elapsed after perigee, increasing uniformly as the true anomaly increases from 0 to 2π radians. By Kepler's second law, eccentric anomaly is proportional to the area swept by the line joining Earth's center to satellite after periapsis. Whichever anomaly parameter is used, it is the only one of the Keplerian parameters that varies with time for an ideal orbit. Note $v = E = M = 0$ when the epoch is the time of perigee passage.

The anomaly parameters are related by

$$E = 2\arctan\left[\sqrt{\frac{1-e}{1+e}}\tan\left(\frac{v}{2}\right)\right] \quad (2.2)$$

$$M = E - e\sin E$$

The mean anomaly is also given by

$$M - M_0 = \sqrt{\frac{\mu}{a^3}}(t - t_0) \quad (2.3)$$

where μ is the standard gravitational parameter for Earth, whose modern[1] value is 398,600.4418 km^3/s^2, and M_0 is the mean anomaly at the epoch t_0, with $M_0 = \sqrt{\frac{\mu}{a^3}}(t_p - t_0)$ where t_p is the time of perigee passage.

The orbital period, in units of seconds, is

$$T = 2\pi\sqrt{\frac{a^3}{\mu}} \quad (2.4)$$

The apogee radius, which is the distance from the center of the Earth to the satellite at apogee, is

$$r_{\text{apogee}} = a(1+e) \quad (2.5)$$

While the perigee radius, which is the distance from the center of the Earth to the satellite at perigee, is

$$r_{\text{perigee}} = a(1-e) \quad (2.6)$$

Each satnav system provides satellite ephemerides in its signals' data messages (and, increasingly commonly now, also on the Internet), so the receiver can calculate

---

[1] Note the value used for GPS calculations is 398,600.5 km^3/s^2; an example of a system's physical constant value differing from the currently known best value.

satellite position at an epoch. While these seven Keplerian orbital parameters or their equivalents are used, the Keplerian parameters are not sufficient for representing satellite position to the high level of accuracy needed for satnav, as discussed in the Section 2.2. Consequently additional parameters are also used to account for the complexities of real-world orbital mechanics not included in Kepler's laws.

Each satnav system's orbital parameters, along with the values of physical constants used for calculations associated with that system (e.g., Earth radius, speed of light, Earth gravitational parameter), are defined in the system's signal interface specifications (ISs), also known as interface control documents, or ICDs. ICDs can be considered user's manuals for receiver processing of the signal, providing the exquisite detail needed to obtain full performance from that signal. Users should be careful to apply the orbital parameters and numerical values of physical constants for each specific system, since there may be subtle differences based on how and when the conventions were adopted for the system. (Values of physical parameters are often refined over time based on improved science, but the original values used in a satnav system are usually retained for backward interoperability.) Part II of this book reviews today's operating and evolving satnav systems and provides references to each system's ICDs (in the remainder of this book, we refer exclusively to ICDs even though some systems call them interface specifications).

More detailed derivations, discussion, and applications of Kepler's Laws can be found in standard texts such as References 1 and 2. However, the discussion and expressions provided in this section are sufficient for purposes of this book.

Different satellite orbits [3] are often succinctly discussed in terms of their orbital altitude (the satellite's height above mean Earth radius) and ellipticity. If the orbit's size is chosen so the satellite's period is one sidereal day,[2] the orbit is called a geosynchronous orbit. The ground track (the set of points on the Earth's surface where the Earth's surface intersects the line connecting the satellite during its orbit with the center of the Earth) of a satellite with inclined geosynchronous orbit (IGSO) consists of two loops. If the orbit is circular the two loops are equal sized and are shaped like a figure eight; if the orbit is elliptical one loop (the one under apogee) is smaller than the other loop (the one under perigee). Highly elliptical orbits (HEO), with proper altitude and inclination, can allow a satellite to spend a large fraction of time over a specific region of the Earth, which can be useful for communications satellites intended to provide regional service.

Circular orbits are most commonly used in satnav and other satellite applications where service is needed over large portions of the Earth. The orbital altitude of a satellite in circular geosynchronous orbit is approximately 35,786 km, as shown in Review Question QA2.1. A circular geosynchronous orbit with inclination zero degrees (in the equatorial plane and orbiting in the same direction as Earth's rotation) is called a geostationary orbit (GEO); its ground track is a point on the equator.

---

[2] A sidereal day is the time it takes the Earth to make one rotation about its axis so that the distant stars appear in the same position in the sky. Its duration is approximately 23 hours, 56 minutes, and 4 seconds—slightly less than a mean solar day.

Satellites whose orbital altitudes are in the range from several hundred kilometers to approximately one thousand kilometers are said to be in low Earth orbit (LEO). Satellites with orbital altitudes below several hundred kilometers encounter enough of the atmosphere that orbital drag tends to de-orbit the satellite relatively quickly; considerable fuel would need to be consumed to counteract this drag at low orbital altitudes. LEO satellites have orbital periods of the order of 90 minutes, so to a terrestrial observer they rise, stay in view overhead, and set in a few minutes. Only a relatively small portion of the Earth's surface is in view at any time to a LEO satellite.

Satellites whose orbital altitudes are in the range from several thousand kilometers to near GEO are said to be in medium Earth orbit (MEO). MEO orbits tend to expose satellites to especially harsh radiation environments because of the naturally occurring radiation belts at those altitudes. Satellites with semisynchronous orbits have an orbital period of half a sidereal day (the Earth revolves once in a sidereal day, so the satellite overflies the same point on Earth every second orbit.

2.2 ORBITAL DEVIATIONS FROM IDEAL

The derivation of Kepler's laws involves assumptions that are approximate but not exact in practice. Consequently, while Kepler's laws provide a good approximate description of satellite motion in Earth orbit, these simple expressions do not describe the motion of real satellites closely enough for purposes of satnav.

There are many differences between the assumptions used to derive Kepler's Laws and reality. Some of the differences that most affect satellite orbits are:

- Space is not a perfect vacuum, so satellites experience drag.
- The Earth is not spherical, nor is it even ellipsoidal, and its mass distribution is uneven and also time varying due to tidal variations. Consequently, Earth's gravitational forces on the satellite differ from those used to derive Kepler's laws.
- Gravitational forces from other bodies in space, particularly the moon and sun, perturb the motion that is calculated based on a two-body (satellite and Earth) model.
- Solar radiation pressure exerts forces on the satellite, in addition to gravitational forces.
- Satellites undergo orbital maneuvers occasionally to keep them in their desired orbits, known as stationkeeping maneuvers. In addition, satellites are sometimes moved into new orbital positions.

Consequently, the orbital elements used to describe locations of satnav satellites differ in two ways from those introduced in Section 2.1. First, there are more than seven parameters, since the equations of motion involve additional terms beyond Kepler's laws, with terms added to the standard Keplerian parameters to produce a better fit to the orbit. Second, all the parameter values change over time.

The resulting parameters define an expression that describes the satellite position over time. The expression consists of two types of terms:

- Secular terms describe long-term behavior that does not vary with time (until the parameters change),
- Periodic terms have sinusoidal characteristics over time.

Typically, these orbital descriptions are provided in an inertial coordinate frame.

The parameters used to describe satnav satellite orbits are known as osculating orbital elements. At a specific point in time, the osculating orbit is tangent to its actual orbit and has the same curvature as the orbit would have in the absence of perturbing forces; thus, the osculating orbit is said to kiss, or osculate, the actual orbit. Osculating orbital elements accurately describe a segment of the current orbit—with no error at the point of osculation, but increasing error as the satellite moves away from this point. Parameter values that describe the secular and periodic terms must be updated (typically, of the order of hours) so the satellite location can be calculated to within a few centimeters or tens of centimeters of accuracy over the entire validity interval when the orbital element set applies.

Values of these osculating orbital elements are provided in each satnav signal's ephemeris message. Each associated ICD defines the orbital parameters used and provides a detailed algorithm for computing satellite position using these orbital elements, and also for transforming the satellite location from inertial coordinates to Earth-centered, Earth-fixed (ECEF) coordinates.

The accuracy of this satellite position information in the ephemeris message improves with the number of parameters, the number of bits used to represent numerical values, and the rate that updates are provided. However, all of these improvements to ephemeris accuracy add burden to the control segment, and either require greater capacity in the data message or increase the time a receiver may need to wait before it can access the parameters from the data message. Consequently, ephemeris representation is an important topic in the design of data messages for new signals, involving tradeoffs that include accuracy, latency, and data message bit rate. The reader can examine an ICD and compare the representation of satellite location in the ephemeris message (having the precision needed to calculate position and time) with the representation of satellite location in the almanac message (having the precision to assess which satellites might be in view and what the associated Doppler shifts might be) in order to see how the same system represents the same information with different precisions for different purposes.

2.3 . CONSTELLATIONS

The previous two sections focused on describing motion of an individual satellite, but a constellation of satellites is needed for satnav, both to provide the minimum of four satellites in view needed to determine position and time, as discussed in Section 1.2, and also to provide continuous coverage over the service region, which is the portion of the Earth where the system provides specified performance. The service region is the entire Earth (and sometimes parts of space as well) for a global system, or a portion of the Earth for a regional system. The need for four or more satellites in view differs from most other satellite applications, such as remote sensing or communications, where typically only one satellite must be in view of the point on Earth to be serviced.

Satnav constellations are designed to satisfy a number of objectives. The most fundamental objective is providing enough satellites consistently in view to a receiver in the service region, based on some elevation mask, satellite geometry specification, and availability requirement. The elevation mask describes a minimum elevation angle relative to horizontal, where the quality of a satnav system and its signals is specified. Different satnav systems use different elevation masks to describe the conditions under which the system's service is specified. When a satellite is at low elevation, its signal reception is more often blocked by terrain or buildings, and degraded by propagation through long paths in the atmosphere or by multipath conditions. Since low elevation angle satellites are further from the receiver, path loss is larger and more transmit power is needed to deliver the same power to a receive antenna. Consequently, specifying performance at a higher minimum elevation angle enables lower satellite transmit power. For example, GPS specifies its signal characteristics at a minimum elevation angle of 5°, while QZSS uses 10°. The minimum elevation angle is only used to specify satnav signal characteristics; receivers can be designed to use either a larger or a smaller elevation angle.

While satnav system specifications are typically stated in terms of an elevation mask over 360° in the horizontal plane, some studies of satellite visibility to ground users employ visibility wedges, with some angular regions having higher elevation masks and others having lower elevation masks, in order to better represent visibility constraints in urban canyons or mountainous terrain.

The satellite geometry specification defines a minimally acceptable number and spatial diversity of satellites at or above the elevation mask. While typically a minimum of four satellites must be in view, some more advanced receiver processing, such as fault detection and exclusion (FDE) processing described in Section 20.8, can require six or more satellites in view. In contrast, receivers only used for timing can use signals from as few as one satellite, and under some conditions techniques exist for positioning using three or fewer satellites, as introduced in Section 20.6.1.

Spatial diversity of satellites is typically quantified in terms of metric based on the dilution of precision (DOP), as discussed in Section 6.2. The constellation of a global system might be required to deliver a specific DOP value or less 99% of the times and locations on Earth. In some cases, availability numbers are computed for a constellation with one or two satellite failures, either averaging over all possible satellite failures or, more conservatively, with the satellite outages selected to most degrade a selected DOP metric.

Since augmentation systems, described in Chapters 8, 23, and 24, are designed to improve or extend the performance of receivers also using signals from another satnav system, augmentation systems typically are not required to meet the same geometry specifications as stand-alone systems; a single satellite in view is often sufficient.

Satnav satellite constellations are typically specified in terms of the number of orbital planes occupied by the satellites, the size, shape, and orientation of these planes in terms of the first five Keplerian parameters, and the locations of satellites in these planes using the last two Keplerian parameters. Placing satellites in multiple orbital planes allows a constellation to provide simultaneous coverage over more of the Earth, and for a receiver at a given point on Earth to see more satellites simultaneously, with more diverse geometry than if all the satellites were in a given orbital plane.

Constellations are typically designed to provide graceful degradation of satellite geometry in the absence of a few satellites, as well as to accommodate additional satellites that may be launched before an on-orbit satellite has failed and needs replacement. Some constellation designs are also selected to ease stationkeeping needs, reducing on-orbit fuel consumption and increasing satellite availability. In addition, constellations with fewer planes and more satellites per plane are more amenable to launching multiple satellites on a single launcher, since launching multiple satellites into different locations on the same orbital ellipse tends to be easier and use less fuel or less time for the satellites to reach their designated orbital position, compared to placing satellites from the same launch into different orbital planes.

The goal is to satisfy these constellation objectives affordably—with the smallest possible satellites, the fewest satellites in the constellation, and the lowest number of launches possible.

From early in the design of GPS through today, many different studies have been performed to determine the best constellation design, where "best" involves multiple and often conflicting criteria. These studies have consistently demonstrated the advantages of a MEO constellation with circular orbits when global coverage is needed, since even lower orbital altitudes require many more satellites to provide adequate satellite geometry, and even higher orbital altitudes require high satellite costs and without saving much on the number of satellites, due to needs for geometry and redundancy. Different global satnav systems have selected different MEO constellation designs as summarized below. Regional systems typically use a constellation with some satellites in geostationary Earth orbit (GEO) and others in IGSO. Some regional augmentations employ strictly circular orbits, while others use satellites in elliptical orbit to maximize their time at high elevation angles over a selected service region.

Families of constellations have been defined based on specific parametric forms, with a specific constellation defined by selected parameter values. One such family encountered in satnav is a Walker constellation, defined for circular orbits, equally spaced planes, and an equal number of satellites in each plane with equal spacing in true anomaly. A Walker constellation has the very compact description of orbital radius and the four parameters $i : t/p/f$,[3] where i is the inclination, t is the total number of satellites, p is the number of orbital planes in the constellation, and f describes the relative spacing between satellites in adjacent planes, where the difference in true anomaly between the closest satellites in adjacent planes is $360 \times f/t$ degrees. In general, a constellation of satellites in circular orbits is not a Walker constellation; Walker constellations are very special cases. While planned Galileo and BeiDou satnav systems describe their constellations

---

[3] In subsequent chapters addressing signal descriptions and signal processing, the symbol i is used to denote the imaginary unit, $i = \sqrt{-1}$. Similarly, t typically denotes time elsewhere in this book, and f denotes frequency. Here, i, t, and f are used as parameters of a Walker constellation in order to conform to common use; context should avoid confusion between different uses of the same symbols.

TABLE 2.1. Summary of Nominal Constellation Characteristics

Satnav System	GPS	GLONASS	Galileo	BeiDou Global Constellation
Number of satellites	24–36	24–30	24–30	24–30
Orbital altitude (km)	20,200	19,100	23,222	21,400
Orbital altitude (mi)	12,550	11,867	14,428	13,296
Orbital radius (km)	26,578	25,478	29,600	27,778
Orbital Radius (mi)	16,513	15,829	18,390	17,258
Eccentricity	0	0	0	0
Orbital period: hour/minute/second (total s)	11/58/2 (43,082 s)	11/14/30 (40,472 s)	14/4/41 (50,681 s)	12/52/4 (46,324)
Time between ground track repeats (sidereal days)	1	8	10	7
Walker constellation description	n/a	n/a	56°: 24/3/1 plus 6 spares	55°: 24/3/1 plus 3 spares
Number of planes	6	3	3	3
Number of satellites per plane	4–6	8–10	8–10	9–10
Separation of right ascending nodes (°)	60	120	120	120
Inclination (°)	55	64.8	56	55
Satellite speed (km/s) in ECI coordinates	3.9	4.0	3.7	3.8
Maximum satellite speed (km/s) in ECEF coordinates*	3.2	3.6	3.0	3.1

*Dr. Frank van Diggelen, Private Communication, 3 February 2015.

as Walker constellations, neither of the two operational global satnav systems, GPS and GLONASS, employ Walker constellations—they have found operational benefits to nonuniform spacing of satellites in each plane, and constellation management often results in different numbers of satellites in different planes. It remains to be seen the extent to which Galileo and BeiDou will actually be operated as Walker constellations.

Table 2.1 summarizes nominal characteristics of four current or planned global satnav constellations. GPS is the only six-plane constellation. Each GPS satellite makes two complete orbits each sidereal day, so each satellite nominally passes over the same location on Earth every sidereal day. These characteristics made it easier to test and use GPS at regular times when the constellation was incomplete. GLONASS has the highest inclination, providing best coverage of polar regions. Galileo has the largest orbital radius, thus having the slowest satellite speed and the longest orbital period.[4] BeiDou also has a constellation of GEO and IGSO satellites providing regional service

---

[4] In August 2014, the fifth and sixth operational Galileo satellites were mistakenly launched into incorrect orbits. They are being maneuvered into final orbits, which will not have the nominal characteristics of Table 2.1.

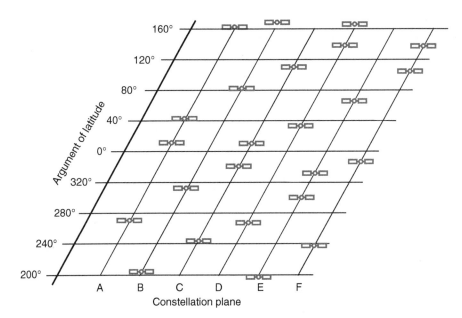

Figure 2.3. Nominal Locations of Satellites in a 24-Satellite GPS Constellation

over China and the surrounding area. The maximum satellite speed in ECEF coordinates accounts for the motion in ECI coordinates of a receiver that is stationary on the Earth's surface, and depends upon the satellite orbital radius and inclination. These results, calculated with a computer simulation, account for the fact that the Earth rotation is in the same general direction as the satellite velocity, so the maximum satellite speed relative to receiver speed, for a receiver stationary on Earth's surface, is between 10% and 20% smaller than the satellite speed in ECI coordinates.

Both GPS and GLONASS use nonuniform spacing of satellites in each plane. Figure 2.3 depicts a nominal 24 satellite GPS constellation [4]. The different planes are labeled A through F, and nominal satellite positions, or "slots," are shown in each plane. Observe the unequal spacing of satellites in each plane, with two of the four satellites in each plane spaced closely. This clearly is not a Walker constellation. In recent years, the GPS constellation has often consisted of more than 30 operating satellites, with others termed "residual"—functional but not broadcasting usable signals, in part since the current GPS control segment cannot support more than 31 satellites. Each plane then contains five or six satellites, unequally spaced around the circular orbit, rather than the four shown in Figure 2.3, and the constellation varies over time as satellites are launched, retired, brought into service, retained as spares, or maneuvered.

2.4 USEFUL GEOMETRY CALCULATIONS

This section provides some geometrical relationships between satellites and receivers that are useful in systems engineering calculations. For these calculations, orbits are modeled as circular and the Earth is modeled as spherical.

USEFUL GEOMETRY CALCULATIONS

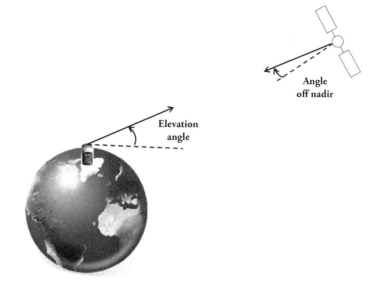

Figure 2.4. Geometry Conventions Showing Elevation Angle and Angle Off Nadir

Figure 2.4 depicts the geometric conventions used in this section. As indicated in the figure, the angle off nadir is defined as the angle, measured at the transmit antenna, between the direction the antenna is pointed (assuming the transmit antenna is pointed toward the Earth's center of mass) and the line connecting the transmit antenna and the receive antenna; angle off nadir is typically used for satellite-centric geometry calculations. Elevation angle is defined as the angle in the vertical plane, measured at the receive antenna, between the line from receive antenna to transmit antenna, and the plane tangent to Earth; elevation angle is typically used for receiver-centric geometry calculations. As defined, elevation angle is always between 0 and $\pi/2$ radians. As shown in Review Question QT2.1, the relationships between angle off nadir and elevation angle, for a receiver on the surface of the Earth, are

$$\eta = \arccos\left[\frac{r_s}{r_e}\sin(\alpha)\right]$$
$$\alpha = \arcsin\left[\frac{r_e}{r_s}\cos(\eta)\right]$$
(2.7)

where η is the elevation angle, α is the angle off nadir (both angles in radians), r_e is Earth's radius, and r_s is the orbital radius of the satellite. If the receiver is above the Earth's surface, r_e is replaced by the sum of Earth radius and altitude of the receiver above Earth's surface. However, the remainder of these calculations apply to a stationary receiver on the Earth's surface.

The two-dimensional geometry defined in Figure 2.5 applies, using an Earth-centered coordinate system with all objects located in the same plane that

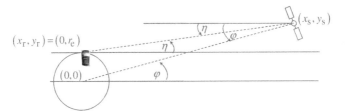

Figure 2.5. Geometry for Receiver-Satellite Calculations

intersects the center of the Earth. For calculations of relative geometry between satellite and receiver, the satellite can be located at any point in the plane (x_s, y_s), so without loss of generality the receiver can be placed at $(0, r_e)$. The angles are given by $\eta = \tan^{-1}((y_s - r_e)/x_s)$ and $\phi = \tan^{-1}(y_s/x_s)$. A satellite at a specific orbital radius and elevation angle has coordinates

$$x_s = \frac{-r_e \tan(\eta) + \left[r_s^2 \sec^2(\eta) - r_e^2\right]^{1/2}}{\sec^2(\eta)} \quad (2.8)$$

$$y_s = x_s \tan(\eta) + r_e$$

for elevation angle η.

As shown in Review Question QT2.2, the law of cosines yields the range from satellite to receiver, in terms of elevation angle and in terms of angle off nadir, as

$$r_{s2r} = -r_e \sin(\eta) + \left[r_s^2 - r_e^2 \cos^2(\eta)\right]^{1/2}$$
$$r_{s2r} = r_s \cos(\alpha) - \left[r_e^2 - r_s^2 \sin^2(\alpha)\right]^{1/2} \quad (2.9)$$

This range directly converts to signal propagation delay, assuming direct propagation along the line of sight, when the range is divided by the speed of light.

Figure 2.6 shows the signal propagation delay from satellite to receiver in the orbital plane for different elevation angles and the different constellations summarized in Table 2.1. Over the different MEO constellations and elevation angles, propagation delays range from about 64 ms to about 96 ms. For a given constellation, the difference between the shortest propagation delay and longest propagation delay is slightly less than 20 ms. As discussed in Section 1.2, trilateration in satnav involves measuring the time for the signal to propagate from transmit antenna to receive antenna; Figure 2.6 indicates the range of values to be measured, and how much the value changes as the satellite passes from overhead to on the horizon.

To estimate the line of sight velocity between satellite and receiver, let the satellite speed be s_s, and assume the receiver is in the orbital plane, where the line of sight velocity between satellite and receiver has the greatest magnitude. Assume initially that

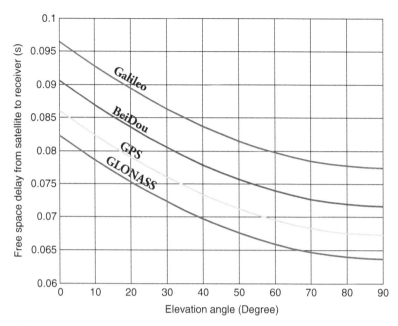

Figure 2.6. Signal Propagation Delay Dependence on Elevation Angle for Different MEO Constellations

the Earth is not rotating. In that case, the line of sight velocity between satellite and receiver is (as shown in Review Question QT2.3)

$$v_{s2r} = s_s \sin(\tan^{-1}(y_s/x_s) - \eta) \qquad (2.10)$$

where (2.8) can be substituted to obtain an expression in terms of elevation angle, Earth radius, and orbital radius.

Based on the results in Section 3.2.1, this line of sight velocity causes a transmitted signal at frequency f_0 to be received with a Doppler shift of $f_0 v_{s2r}/c$. Figure 2.7 shows the Doppler shift of a signal transmitted at 1575.42 MHz at a receiver stationary in inertial coordinates, for the different MEO constellations summarized in Table 2.1. At 10° elevation, received Doppler shifts range from slightly above 4 kHz for Galileo (the constellation having the largest orbital radius) to slightly above 5 kHz for GLONASS, the constellation with the lowest orbital altitude and consequently largest satellite speed.

Figure 2.8 shows analogous results for a signal transmitted at 1176.45 MHz. The lower frequency causes the Doppler shifts to be about 75% of those at 1575.42 MHz, so the maximum values are approximately 1 kHz lower.

While (2.10), Figures 2.7, and 2.8 are all based on the model of a receiver stationary in an inertial frame, in reality a terrestrial receiver is nearly stationary in ECEF coordinates. Table 2.1 shows the maximum relative speed between a satellite and a stationary terrestrial receiver is 10–20% smaller than the maximum computed for a stationary

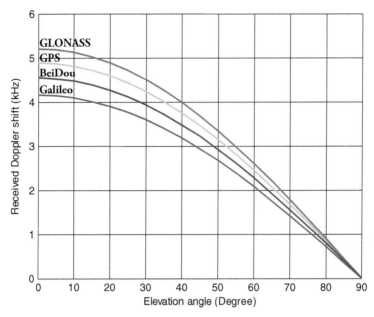

Figure 2.7. Doppler Shift versus Elevation Angle for a Signal Transmitted at 1575.42 MHz to a Terrestrial Receiver, Stationary in an Inertial Frame, in the Orbital Plane for Different MEO Constellations

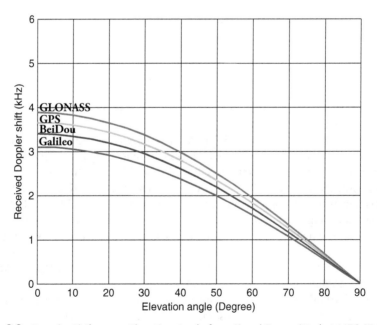

Figure 2.8. Doppler Shift versus Elevation Angle for a Signal Transmitted at 1176.45 MHz to a Terrestrial Receiver, Stationary in an Inertial Frame, in the Orbital Plane for Different MEO Constellations

receiver in an ECI coordinate frame, since Earth rotation is in the same general direction as satellite motion. Consequently, Figures 2.7 and 2.8 overstate the maximum Doppler to a receiver stationary on the Earth's surface by 10–20%, and tighter range of Doppler values can be determined using the maximum relative speed in the lowest row of Table 2.1.

2.5 SUMMARY

The classical description of Earth-orbiting satellites is based on Kepler's laws, leading to the Keplerian parameters that together describe the orbit and the satellite's location in the orbit. Since the Keplerian parameters are included in the parameter set used in satnav satellite ephemeris descriptions, precise definitions of the parameters are provided in this chapter. But satnav satellite motion is more complicated than predicted using Kepler's Laws, so descriptions of satellite ephemeris used in satnav positioning are more complicated in two ways. First, additional terms are added to the equations of motion, providing descriptions of the rates of change of Keplerian parameters and also adding periodic terms. Second, parameter values are updated over time, with a set of parameters that apply only over a validity interval typically lasting a small number of hours. Since satnav requires a constellation of satellites to provide enough satellites in view over the service region, considerations of constellation design are discussed, and numerical values provided for constellations of four existing and planned global satnav systems. Geometrical relationships between satellites and receivers on the surface of the Earth are provided, then translated into characteristics of received satnav signals, with numerical examples provided for the four global systems. These numerical results are used in subsequent chapters to characterize the distance over which satnav signals propagate, and the extent of delays and Doppler shifts that are encountered in receiver processing.

REVIEW QUESTIONS

Theoretical Questions

QT2.1. Derive (2.7).

QT2.2. Derive (2.9).

QT2.3. Derive (2.10).

Application Questions

QA2.1. Find the orbital radius of a geosynchronous satellite.

QA2.2. Find the semimajor axis of the orbit for a satellite with semisynchronous orbit.

QA2.3. Find the semimajor axis of the orbit for a satellite that has a repeating ground track every day.

QA2.4. If the moon has an orbital radius of 60 times Earth radius, what is the moon's orbital period?

QA2.5. Use the orbital radii in Table 2.1 to calculate the orbital period (in units of hours to two decimal digits) for each satnav system in the table.

REFERENCES

1. R. Bate, D. Mueller, and J. White, *Fundamentals of Astrodynamics*, Dover, New York, 1971.
2. D. Vallado, *Fundamentals of Astrodynamics and Applications*, 2nd edition, Space Technology Laboratory, 2001.
3. D. Wright, L. Grego, and L. Gronlund, *The Physics of Space Security: A Reference Manual*, American Academy of Arts and Sciences, 2005, ISBN: 0-87724-047-7.
4. Global Positioning System Standard Positioning Service Performance Standard, 4th edition, available at www.gps.gov/technical/ps/2008-SPS-performance-standard.pdf, accessed 30 May 2013.

3

SATNAV SIGNALS

Satnav signals link the satellites to receivers in the user segment and the control segment. Like many other aspects of a satnav system, signals must be exquisitely engineered and carefully implemented in order for the system to provide the expected high levels of performance. Also like other aspects of a satnav system, satnav signals have undergone a revolution in this century; while some aspects of new and modernized signals are identical to those designed for the original GPS, other aspects are strikingly different. Some of these changes are based on new theories and understanding related to satnav, and other changes reflect improvements in technology that enable satellites and receivers to implement more complicated processing. Other changes reflect the increasing need to share limited spectral bands with more signals and systems, and yet other changes respond to user interests in higher performance and more robust operation under more challenging conditions.

This long and dense chapter provides the conceptual and mathematical foundation for satnav signals used in the remainder of this book. Section 3.1 provides an overview of signals and their mathematical representations, with an emphasis on direct-sequence spread spectrum (DSSS) signals and specific spreading modulations used for satnav. Section 3.2 introduces mathematical models for the effects of Doppler shift and propagation through the dispersive ionosphere. Section 3.3 provides a primarily qualitative discussion of satnav signal characteristics, identifying tradeoffs, and providing perspectives involving signal design choices. Section 3.4 then establishes the detailed mathematics of satnav signals, identifying characteristics, parameters, and models employed in descriptions of specific signals and in definition of receiver processing. Section 3.5 provides a summary of the chapter. The results in this chapter generalize and integrate those provided in many earlier publications, including [1].

3.1 SIGNALS, SIGNAL PROCESSING, AND SPREADING MODULATIONS

Basic representations of DSSS signals are provided in Section 3.1.1. Correlation processing, which is central to receiver processing of DSSS signals, and power spectral densities (PSDs) for DSSS signals are explored in Section 3.1.2. Section 3.1.3 describes DSSS signal structures, and Section 3.1.4 addresses specific spreading modulations used for satnav.

DSSS signals are used because of their wide instantaneous bandwidths, which provide better measurement accuracies as shown in Chapter 19, better resist interference as shown in Chapter 5, and also cause less multiple access interference (MAI), where signals from other satellites interfere with reception of a desired signal, also as shown in Chapter 5.

3.1.1 Signal Models

In this book, signals are represented using complex envelope notation, with the RF signal described by

$$s(t) = \Re\{x(t)\exp[i2\pi f_0 t]\} \tag{3.1}$$

Here, $s(t)$ is a real-valued signal so its Fourier transform (if the signal is deterministic and its Fourier transform exists) or (if it is random and in the wide sense stationary) its PSD is symmetric in positive and negative frequencies. $x(t)$ is the complex envelope of the signal and f_0 is the carrier frequency. Complex envelope notation is most useful when $s(t)$ is a bandpass signal—the carrier frequency is much greater than any of the frequencies in the complex envelope, a condition that holds for satnav signals. The quantity in braces in (3.1) is called the analytic signal and is denoted $\tilde{s}(t)$.

It is often useful to describe the complex envelope in terms of its in-phase component $x_I(t)$ and quadrature component $x_Q(t)$

$$x(t) = x_I(t) + i\,x_Q(t) \tag{3.2}$$

where $x_I(t)$ and $x_Q(t)$ are both real signals. It is often convenient in analysis to deal with the complex envelope of the signal, and even to call the complex envelope the "signal," neglecting the carrier. Dealing with the complex envelope also models what happens in receiver processing when the carrier is removed, or "wiped off."

Alternative and equivalent representations of a bandpass signal found elsewhere in the literature include

$$s(t) = x_I(t)\cos(2\pi f_0 t) - x_Q(t)\sin(2\pi f_0 t)$$
$$s(t) = r(t)\cos(2\pi f_0 t + \phi(t)) \quad (3.3)$$
$$s(t) = \Re\{r(t)e^{i\phi(t)}e^{i2\pi f_0 t}\}$$

In (3.3) as in (3.2), all the quantities are real valued except the complex envelope $x(t)$ and $i = \sqrt{-1}$. $r(t) = |x(t)|$ is the amplitude envelope. When the amplitude envelope does not vary with time, $r(t) = |x(t)| = a$, the signal is known as a constant-envelope signal.

All satnav signals considered in this book are or include digitally modulated signals represented as keyed signals, whose complex envelope is represented as

$$x(t) = \sum_{k=-\infty}^{\infty} a_k g_k(t - kT_c) \quad (3.4)$$

In general, both $\{a_k\}$ and $\{g_k(t)\}$, known as symbols, are complex valued, with the subscripts indicating they can take on different values at different indexes. However, (3.4) is more general than typically needed. Many of the signals encountered in satnav are biphase keyed signals whose complex envelopes can be represented as

$$x(t) = \sum_{k=-\infty}^{\infty} a_k g(t - kT_c) \quad (3.5)$$

where $a_k \in \{-1, +1\}$, T_c is the reciprocal of the spreading code chip rate, and the symbols $g(t)$ are real valued, with the same symbol used repeatedly. Some signals are quadraphase keyed signals whose complex envelopes can be represented as

$$x(t) = \frac{1}{\sqrt{2}} \sum_{k=-\infty}^{\infty} a_k g_I(t - kT_{c_1}) + i \frac{1}{\sqrt{2}} \sum_{n=-\infty}^{\infty} b_n g_Q(t - nT_{c_2}) \quad (3.6)$$

with $a_k \in \{-1, +1\}$ and $b_k \in \{-1, +1\}$, and the in-phase symbols $g_I(t)$ and the quadrature symbols $g_Q(t)$ each real valued. Typically, the spreading codes are designed such that $\{a_k\}$ and $\{b_k\}$ are orthogonal. Specific examples of biphase keyed signals relevant to satnav are provided in Section 3.2.

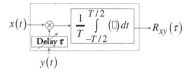

Figure 3.1. Correlation Processing

3.1.2 Correlations and Power Spectral Densities

Receiver processing of DSSS signals, and satnav signals specifically as seen in Part III, relies on correlation processing, where the correlation between two time series is defined as

$$R_{xy}(\tau) = \lim_{T \to \infty} \left[\frac{1}{T} \int_{-T/2}^{T/2} x(t) y^*(t+\tau)\, dt \right] \quad (3.7)$$

Correlation processing is generally implemented as shown in Figure 3.1, where the correlation integration time T is large (relative to the support of the correlation functions), but not infinite as in (3,7). Often, the correlation of complex envelopes is used instead of the correlation of the signal (3.1). The autocorrelation of $x(t)$ with itself can be written as $R_{xx}(\tau)$ but is commonly abbreviated $R_x(\tau)$.

Theoretically, the correlation (3.7) often depends on time t as well as delay τ, as addressed in the Appendix Section A.5. However, the correlation's dependence on time can often be neglected in analysis and receiver processing related to satnav. Appendix Section A.5 provides more detailed exploration of this topic for keyed signals.

When the correlation depends only on delay, the Fourier transform of the correlation function is the PSD

$$\Phi_x(f) = \mathcal{F}\{R_x(\tau)\} \quad (3.8)$$

PSDs are commonly used to describe the frequency content of random or unstructured signals whose Fourier transform is not defined or cannot be readily calculated, and are used extensively throughout this book.

As discussed in Appendix Section A.5, it is common to derive the second-order statistics of satnav signals under the assumption that $\{a_k\}$ and $\{b_k\}$ are ideal long spreading codes—taking on values of $+1$ and -1 with equal probability, mutually uncorrelated, and with autocorrelations given by a Kronecker delta: $E\{a_k a_j\} = E\{b_k b_j\} = \delta_{kj}$. Under such conditions, Appendix Section A.5 shows the stationary part of the complex envelope's correlation functions, biphase keyed signals and quadraphase keyed signals

SIGNALS, SIGNAL PROCESSING, AND SPREADING MODULATIONS

(when the spreading sequences in each phase are orthogonal) defined in (3.5) and (3.6), respectively, are

$$R_x(\tau) = \frac{1}{T_c} \int_{-\infty}^{\infty} g(t) g(t+\tau) \, dt \quad \text{(biphase keyed signals)}$$

$$R_x(\tau) = \frac{1}{2T_c} \int_{-\infty}^{\infty} g_I(t) g_I(t+\tau) \, dt + \frac{1}{2T_c} \int_{-\infty}^{\infty} g_Q(t) g_Q(t+\tau) \, dt \quad (3.9)$$

(quadraphase keyed signals)

and the corresponding PSDs, defined to have unit area, with correlation functions evaluating to unity at zero lag, are

$$\Phi_x(f) = \frac{1}{T_c} |G(f)|^2 \quad \text{(biphase keyed signals)}$$

$$\Phi_x(f) = \frac{1}{2T_c} |G_I(f)|^2 + \frac{1}{2T_c} |G_Q(f)|^2 \quad \text{(quadraphase keyed signals)} \quad (3.10)$$

where $G(f) = \mathcal{F}\{g(t)\}$. With the normalization used, the keyed signals have symbols with energy $\int_{-\infty}^{\infty} |g(t)|^2 \, dt = T_c$.

While cryptographically generated spreading codes are well modeled as ideal long spreading codes, many signals use spreading codes that repeat, typically every thousand to ten thousand values. In addition, these spreading codes have nonzero, but small, autocorrelation values $E\{a_k a_j\}$ at index values other than $k = j$. Both of these effects cause correlation functions and PSDs to differ from their ideal values. As shown in Reference 2, the effect of repeating codes is diminished by the biphase modulation used to encode the data message or as an overlay code, as described in Section 3.4. Of the signals addressed in this book, the GPS C/A code signal's correlation function and PSD are least appropriately modeled as resulting from an ideal long spreading code, since the C/A spreading code repeats 20 times within a data bit; the effect on MAI calculations is addressed in Section 5.6, and the effect on the C/A code signal PSD is discussed in Section 7.3.1.

3.1.3 Direct Sequence Spread Spectrum Signals

Given a sequence of symbols to be transmitted $\cdots, d_{-3}, d_{-2}, d_{-1}, d_0, d_1, d_2, d_3, \cdots$ with $d_k \in \{+1, -1\}$, define the data symbol waveform

$$\beta(t) = \sum_{\ell=-\infty}^{\infty} d_\ell \operatorname{rect}_{T_d}\left(t - \ell T_d - T_d/2\right) \quad (3.11)$$

as the complex envelope of a biphase modulated signal, defined to have unit power. The rect function is defined in Appendix Section A.1, and (3.11) is defined so there is a transition at $t = 0$. The $\{d_\ell\}$ are called data symbols since they represent the data bits of information, but they may include redundancy for channel encoding, bits for error detection, and bits for message framing or other messaging overhead. In (3.11), the data symbol rate is $1/T_d$, which, for satnav signals, is typically tens of symbols per second to hundreds of symbols per second; T_d is the data symbol duration.

Results in Appendix Section A.5 show that, when the data symbols are modeled as ideal long spreading sequences (a reasonable model especially when error control coding is used), applying (3.10) to (3.11) yields the PSD of the data symbol time series:

$$\Phi_\beta(f) = T_d \text{sinc}^2(\pi f T_d) \tag{3.12}$$

which has a null-to-null bandwidth of $2/T_d$, of the order of tens of hertz to several hundred hertz. The sinc function is defined in Appendix Section A.1.

For a spreading code bit sequence $\cdots, a_{-3}, a_{-2}, a_{-1}, a_0, a_1, a_2, a_3, \cdots$ with $a_k \in \{+1, -1\}$ and a spreading symbol $g(t)$, form a spreading waveform

$$\sigma(t) = \sum_{k=-\infty}^{\infty} a_k g(t - kT_c) \tag{3.13}$$

where the chip rate $1/T_c$ is an integer multiple N of the data symbol rate so $T_d = NT_c$; in satnav the value of N is typically in the thousands. (3.13) employs the convention that the spreading time series is defined to have unit power.

Now form the DSSS signal (actually, the complex envelope of a DSSS signal) by taking the product of the spreading time series (3.13) and the data symbol time series (3.11):

$$\begin{aligned} x(t) &= \sigma(t)\beta(t) \\ &= \sum_{\ell=-\infty}^{\infty} \sum_{k=-\infty}^{\infty} a_k d_\ell \text{rect}_{T_d}(t - \ell N T_c) g(t - kT_c) \\ &= \sum_{k=-\infty}^{\infty} b_k g(t - kT_c) \end{aligned} \tag{3.14}$$

This expression also has unit power, since the spreading time series and the data symbol time series each have unit power. The last line in (3.14) results from the observation that a single value of the data symbol d_ℓ is multiplying a segment of N adjacent values of $\{a_k\}$, forming a new sequence $\{b_k\}$ that takes on the same values as $\{a_k\}$ for segments when the data symbol takes on a value of $+1$, and the inverted values $\{-a_k\}$ for segments when the data symbol takes on a value of -1, as portrayed in Figure 3.2.

```
dₗ :  ⋯              +1                            -1
aₖ : ⋯ +1 -1 -1 -1 +1 +1 -1 +1 -1 -1  +1 +1 +1 -1 -1 -1 +1 +1 +1 -1 ⋯
bₖ : ⋯ +1 -1 -1 -1 +1 +1 -1 +1 -1 -1  -1 -1 -1 +1 +1 +1 -1 -1 -1 +1 ⋯
```

Figure 3.2. Example of Data Symbol Sequence, Spreading Sequence, and their Product for $N = 10$

If $\{a_k\}$ are an ideal long spreading code, then $\{b_k\}$ also form an ideal long spreading code, so the PSD of the DSSS complex envelope (3.14) is

$$\Phi_x(f) = \frac{1}{T_c} |G(f)|^2 \tag{3.15}$$

The preceding discussion of spreading code bits and data symbols uses what is called algebraic representation, where binary values are represented as $\{-1, +1\}$ and different binary values are combined using multiplication. An equivalent logical representation is also employed, where binary values are represented as $\{1, 0\}$ and binary values are combined using an exclusive OR operation. Table 3.1 compares these two equivalent representations, both encountered in this book. Typically, hardware employs a logical representation for implementation simplicity, while signal processing analysis uses algebraic values for ease of analysis.

The simplest and most common spreading symbol is $g(t) = rect_{T_c}(t - T_c/2)$, where the delay by half a chip duration is introduced to maintain the convention that spreading symbols begin at time values that are integer multiples of T_c. This spreading symbol has the Fourier transform $G(f) = T_c e^{-i\pi f T_c} \sin c(\pi f T_c)$, so in this case the PSD of the DSSS complex envelope is

$$\Phi_x(f) = T_c \text{sinc}^2(\pi f T_c) \tag{3.16}$$

which has a null-to-null bandwidth of $2/T_c$, N times greater than that of the data symbol sequence, and maximum value T_c, $1/N$ that of the data symbol sequence. This PSD is independent of the data symbol rate, and does not require the data symbols to be modeled as an ideal long code as long as the spreading sequence itself is an ideal long

TABLE 3.1. Equivalence of Algebraic and Logical Representations

Algebraic Representation			Logical Representation		
First Input	Second Input	Product of First and Second Inputs	First Input	Second Input	Exclusive ORing of First and Second Inputs
+1	+1	+1	0	0	0
+1	-1	-1	0	1	1
-1	+1	-1	1	0	1
-1	-1	+1	1	1	0

spreading code. The direct sequence spreading has broadened the PSD by a factor of N while reducing its peak amplitude by a factor of N when the power remains the same.

The wider bandwidth of a DSSS signal is critical for satnav. As shown in Chapter 19, the accuracy of measuring the signal's time of arrival improves with wider bandwidth. Additionally, the spectrum spreading makes receiver processing inherently more resistant to interference, a particularly important characteristic of satnav signals since their received power is low. Finally, spreading the signal power over a very wide bandwidth makes the received signal's PSD much less than the PSD of noise in a receiver. Consequently, until there is a large number of interfering signals, receiver noise dominates MAI, allowing multiple satnav signals and even multiple satnav systems to share the same frequency bands with little degradation due to MAI.

In digital communications using keyed signals, the choice of spreading symbol or spreading symbol rate does not influence performance in an additive white Gaussian noise (AWGN) channel, since the bit error rate depends on the energy in the received data symbol, not the shape of the spreading symbol or its correlation function or PSD. (Of course, many practical factors involving transmitter and receiver design, difficult channel conditions, and spectral containment are influenced by symbol shape in digital communications, so symbol design does remain an important consideration in practical digital communications.) In contrast, it will be shown in Part III that some aspects of satnav performance are significantly influenced by the choice of spreading symbol even at the theoretical level.

3.1.4 Spreading Modulations for Satnav

A satnav signal's spreading modulation is defined by its spreading symbol(s) and the chip rate. The spreading modulation establishes the shape of the spectrum, including the signal's bandwidth and the distribution of power inside and outside the spectral band to be used. By defining the frequencies where signal power is concentrated, the spreading modulation influences how much RF interference it causes to satnav receivers, and also affects its receiver's susceptibility to interference from other signals.

The spreading modulation also establishes the shape of a signal's autocorrelation function, influencing (as shown in Chapter 19) the receiver's ability to make accurate measurements in the presence of noise, interference, and multipath, as well as the ability to maintain the tracking of signals under various stresses. It also influences a number of practical factors in implementing the signal's transmitter and receiver, including the clock rate used for sampling and processing, the number of amplitude levels needed in signal generation and processing, and how sensitive performance is to distortions caused by hardware imperfections.

Figure 3.3 introduces important terminology related to spreading modulations. Segments of three different spreading time series are portrayed, all performing biphase modulation, all at the chip rate 1.023 MHz (originally selected for the GPS C/A code signal and now a standard fundamental clock rate in most satnav systems), all using the same spreading code bit sequence listed at the top of the figure, and all using spreading symbols that do not return to a value of zero in the interior of the symbol—literally non-return to zero (NRZ) symbols. The spreading symbol in the upper trace is a half

SIGNALS, SIGNAL PROCESSING, AND SPREADING MODULATIONS 45

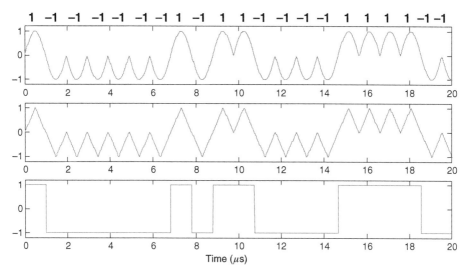

Figure 3.3. Spreading Time Series Using Different NRZ Symbols

sinusoid; the spreading symbol in the middle trace is an isosceles triangle, and the spreading symbol in the lower trace is a rectangle. Properly scaled in amplitude, each symbol could have the same energy, and then would provide the same bit error performance for digital communications in an AWGN channel. However, the PSD for each spreading time series is very different, producing very different performance in satnav, as detailed in Chapter 19. Consequently, satnav terminology must go beyond describing spreading symbols as NRZ, and more specifically describe their shape.

The lower trace, with rectangular symbols, was used for all original satnav signal designs, and is called BPSK-R, for binary phase-shift keying with rectangular symbols. Since different signals use different chip rates, a complete description of the spreading modulation is BPSK-R(n), where n is usually an integer designating the chip rate as a multiple of 1.023 MHz: $f_c = 1/T_c = n \times 1.023$ MHz. Sometimes the notation BPSK-R(f_c) is also used, with the difference being obvious by whether the quantity in the parentheses is a small integer or a large number of the order of 1 MHz or greater.

A signal with BPSK-R(n) spreading modulation has spreading time series

$$x_{\text{BPSK-R}(n)}(t) = \sum_{k=-\infty}^{\infty} a_k g_{\text{BPSK-R}(n)}(t - kT_c) \qquad (3.17)$$

with spreading symbol

$$g_{\text{BPSK-R}(n)}(t) = rect_{T_c}(t - T_c/2) \qquad (3.18)$$

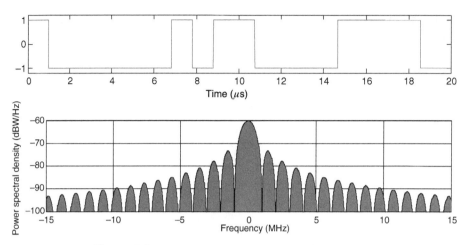

Figure 3.4. Spreading Time Series and PSD for BPSK-R(1)

having duration $T_c = 1/f_c$ where the chip rate is $f_c = n \times 1.023$ MHz. If the spreading code is an ideal long spreading code, the signal's PSC is

$$\Phi_{\text{BPSK-R}(n)} = T_c \, \text{sinc}^2(\pi f T_c) \quad (3.19)$$

and correlation function is

$$R_{\text{BPSK-R}(n)}(\tau) = \begin{cases} (T_c + \tau)/T_c, & -T_c < \tau \leq 0 \\ (T_c - \tau)/T_c, & 0 < \tau \leq T_c \\ 0, & \text{elsewhere} \end{cases} \quad (3.20)$$

Observe the spreading time series complex envelope is real valued and constant envelope, with unit power, like many of the spreading time series examined in this book. Unless specified otherwise, the discussion of spreading symbols in this section is based on infinite bandwidth; Appendix Section A.5 derives the complex envelope and spreading symbols after filtering.

Figure 3.4 shows a segment of the spreading time series and the PSD for a BPSK-R(1) spreading modulation, Figure 3.5 shows the same for a BPSK-R(10) spreading modulation, and Figure 3.6 compares their correlation functions.

A straightforward extension of BPSK-R spreading modulations is quadrature phase-shift keying with rectangular symbols, denoted QPSK-R,

$$x_{\text{QPSK-R}(n)}(t) = \frac{1}{\sqrt{2}} \sum_{k=-\infty}^{\infty} a_k g_{\text{BPSK-R}(n)}(t - kT_c) + i\frac{1}{\sqrt{2}} \sum_{n=-\infty}^{\infty} b_n g_{\text{BPSK-R}(n)}(t - nT_c)$$

$$(3.21)$$

SIGNALS, SIGNAL PROCESSING, AND SPREADING MODULATIONS

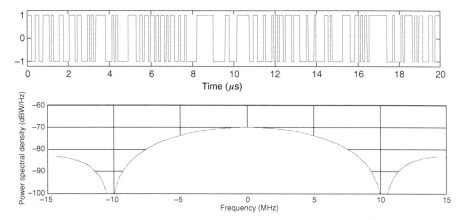

Figure 3.5. Spreading Time Series and PSD for BPSK-R(10)

When they use the same spreading symbols, BPSK-R and QPSK-R, spreading modulations have the same correlation function and PSD. They only offer one design parameter: the chip rate. The only way to move the BPSK-R spectral peak is to change the carrier frequency, so multiple signals with BPSK-R spreading modulations transmitted on the same carrier will all have spectral peaks at the same frequency. Low chip rates, attractive for some practical reasons, produce broad correlation function peaks that limit the performance of receiver processing as discussed in Chapters 5 and 19.

Binary offset carrier (BOC) spreading modulations were developed [3] to address these limitations of BPSK-R, while retaining the advantages of a real-valued (biphase), constant envelope, spreading time series. Conceptually, a BOC spreading time series is formed by multiplying a BPSK-R spreading time series by a square wave

$$x_{\text{BOC}(m,n)}(t) = x_{\text{BPSK-R}(n)}(t) \text{sgn}\left[\sin\left(2\pi f_s t + \theta_{\text{BOC}}\right)\right] \quad (3.22)$$

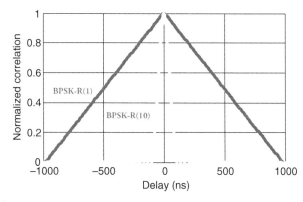

Figure 3.6. Correlation Functions for Infinite-Bandwidth BPSK-R(1) and BPSK-R(10)

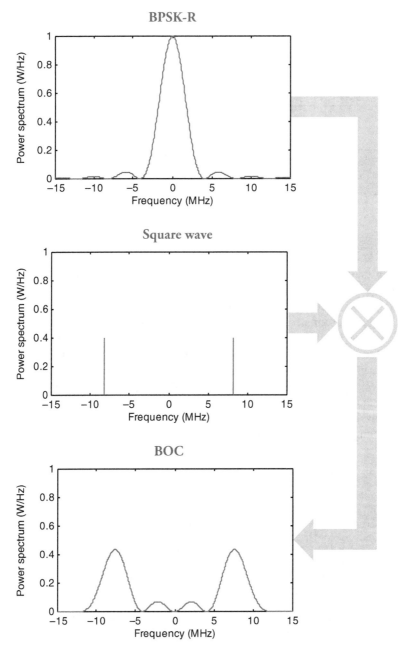

Figure 3.7. Intuitive Frequency-Domain Interpretation of BOC Generation

Here, sgn(·) is the signum function whose output is the algebraic sign of the input, and BOC(m, n) indicates a BOC spreading modulation with subcarrier frequency $f_s = m \times 1.023$ MHz spreading code chip rate $f_c = n \times 1.023$ MHz, and BOC phase θ_{BOC}, with $\theta_{BOC} = 0$ for sine-phased BOC (typically assumed if no phasing is indicated explicitly, but sometimes indicated explicitly by BOC_s), and $\theta_{BOC} = \pi/2$ for cosine-phased BOC, indicated by BOC_c. The notation BOC(f_s, f_c) is also used.

Figure 3.7 provides an intuitive interpretation of the resulting PSD, for a BOC(8,4) spreading modulation. The BPSK-R(4)'s sinc-squared PSD is on the left, while the center figure shows the PSD of the square wave, with the spectral lines corresponding to the fundamental frequency of 4×1.023 MHz. (Other spectral lines occur at odd integer multiples of the fundamental frequency.) Since (3.22) indicates a product in the time domain, which corresponds to a convolution in the frequency domain; this convolution centers a sinc-squared PSD at the frequency of each spectral line. While this interpretation is not rigorous since the frequency-domain convolution actually does not occur in the PSD domain but in the domain of the Fourier transform of the time series, the intuition is useful.

Figure 3.8 shows a block diagram for generating a BOC signal, including data message modulation and upconversion to the carrier frequency.

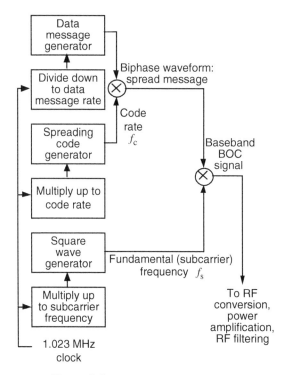

Figure 3.8. BOC Signal Generation [3]

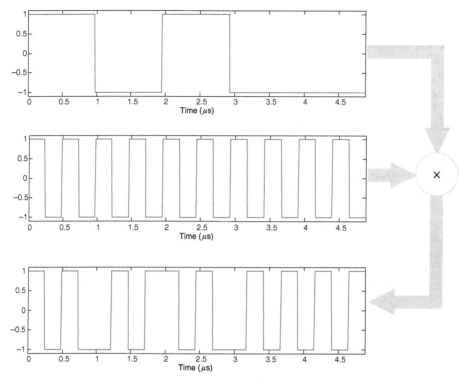

Figure 3.9. Example of BOC(2,1) Spreading Time Series Formation

BOC introduces a second design parameter: the subcarrier frequency. BOC spectral peaks do not occur at the carrier frequency, and their locations are controlled by the selection of the subcarrier frequency. The BOC ratio, defined as

$$k_{\text{BOC}} = \frac{2f_s}{f_c} = \frac{2m}{n} = \frac{T_c}{T_s} \quad (3.23)$$

where $T_c = 1/f_c$, and $T_s = 1/(2f_s)$ is essential for characterizing BOC modulations. For practical BOC modulations, k_{BOC} is an integer, indicating an integer number of square wave half periods in the reciprocal of the chip rate.

Figure 3.9 provides an example of forming a BOC(2,1) spreading time series. A BPSK-R(1) spreading time series is in the upper plot, and a 2.046 MHz square wave is in the middle plot. Their product is the BOC(2,1) spreading time series in the lower plot. Observe the sine phasing, with the rising edge of the square wave aligned with the leading edge of a BPSK-R symbol.

Figure 3.10 shows the contrasting development of a $\text{BOC}_c(2,1)$ spreading time series. Compared to Figure 3.9, the square wave is shifted by one-quarter of a period, so the rising edge of the square wave occurs $0.25/f_s$ seconds after the leading edge of a BPSK-R symbol. Staggering the square wave transitions creates shorter segments of

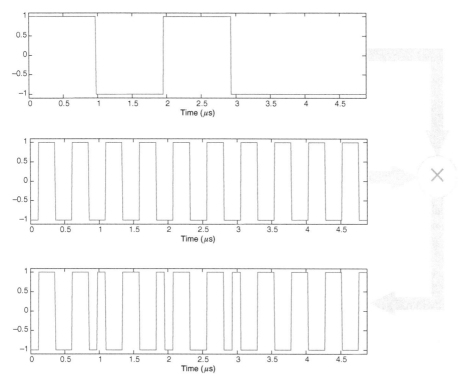

Figure 3.10. Example of BOC$_c$(2,1) Spreading Time Series Formation

waveform before sign transitions; the shorter segments contribute more high frequency content that is reflected in the PSD.

A BOC spreading time series can also be represented as a keyed waveform:

$$x_{BOC(m,n)}(t) = \sum_{k=-\infty}^{\infty} a_k g_{BOC(m,n)}(t - kT_c) \quad (3.24)$$

where the BOC spreading symbol is

$$g_{BOC(m,n)}(t) = \begin{cases} \text{sgn}\left[\sin\left(2\pi f_s t + \theta_{BOC}\right)\right], & 0 < t \leq T_c \\ 0, & \text{elsewhere} \end{cases} \quad (3.25)$$

The time support of a BOC spreading symbol is the reciprocal of the chip rate $T_c = 1/f_c$, the BOC phase θ_{BOC} is 0 for sine phasing and $\pi/2$ for cosine phasing, and the subcarrier frequency defines the rate of sign transitions within the spreading symbol.

The BOC parameter (3.23) defines whether there is a balanced spreading symbol whose average is zero ($\int_{-\infty}^{\infty} g(t)dt = 0$) or not. If k_{BOC} is even, the symbol has an integer number of square wave periods and is balanced; if k_{BOC} is odd, the symbol is

TABLE 3.2. Algorithm for Computing Correlation Function of Infinite-Bandwidth BOC Spreading Modulation

Spreading Modulation	Number of Positive and Negative Peaks in Autocorrelation Function	Delay Value	Autocorrelation Function at $\tau = jT_s/2$	
			j even	j odd
$BOC_s(m,n)$	$2k_{BOC} - 1$	$\tau = jT_s/2$, $-2k_{BOC} \leq j \leq 2k_{BOC}$	$(-1)^{j/2}(k_{BOC} - \lfloor j/2 \rfloor)/k_{BOC}$	
$BOC_c(m,n)$	$2k_{BOC} + 1$	$\tau = jT_s/2$, $-2k_{BOC} \leq j \leq 2k_{BOC}$	$(-1)^{j/2}(k_{BOC} - \lfloor j/2 \rfloor)/k_{BOC}$	$(-1)^{(\lfloor j \rfloor+1)/2}/(2k_{BOC})$

unbalanced. The PSD of the complex envelope of a BOC spreading modulation with balanced spreading symbols has a null at band center, corresponding to the RF signal having a null at the carrier frequency. Conversely, the PSD of the complex envelope of a BOC spreading modulation with unbalanced spreading symbols has a spectral sidelobe at band center.

Table 3.2 summarizes the calculation of a BOC correlation function with infinite bandwidth. The delay values listed are the values of τ where there are points of inflection, with straight lines connecting these points. The rightmost two columns provide values of the correlation function at these points of inflection.

Figure 3.11 shows examples of correlation functions computed using this algorithm. The largest BOC parameter is shown in the lowest plot; with large BOC parameters there is only a slight difference between the sine-phased and cosine-phased correlation functions as long as the bandwidth is infinite.

As an example of applying Table 3.2, use the $BOC_s(x,x)$ example in the top axes of Figure 3.11. This signal has BOC parameter $k_{BOC} = 2x/x = 2$, so it has three positive and negative peaks in the autocorrelation function according to the table; these peaks can also be seen in the figure. The delay values are $\tau = jT_s/2$ for $j = -4, -3, \ldots, 4$. Take as an example the delay $\tau = T_s = T_c/2$, which corresponds to $j = 2$. For this delay value, the autocorrelation has value $(-1)^{j/2}(k_{BOC} - \lfloor j/2 \rfloor)/k_{BOC} = (-1)^{2/2}(2 - |2/2|)/2 = -1/2$, which is the value in the top plot at the delay value $\tau/T_c = 0.5$.

For bandlimited signals, the correlation function cannot be found in closed form. The PSD can be found analytically using the following expressions, then the effects of bandlimiting or other filtering needs to be applied, and then the inverse Fourier transform must be found numerically to determine the correlation function. Review Question QT3.7 explores the effect of bandlimiting on sine-phased and cosine-phased correlation functions.

The PSD for sine-phased BOC spreading modulations is given by [2]

$$\Phi_{BOC_s(f_s,f_c)}(f) = \begin{cases} \dfrac{1}{f_c}\mathrm{sinc}^2(\pi f/f_c)\tan\left(\dfrac{\pi f}{2f_s}\right), & k_{BOC} \text{ even} \\ \dfrac{1}{f_c}\dfrac{\cos^2(\pi f/f_c)}{(\pi f/f_c)^2}\tan\left(\dfrac{\pi f}{2f_s}\right), & k_{BOC} \text{ odd} \end{cases} \quad (3.26)$$

SIGNALS, SIGNAL PROCESSING, AND SPREADING MODULATIONS

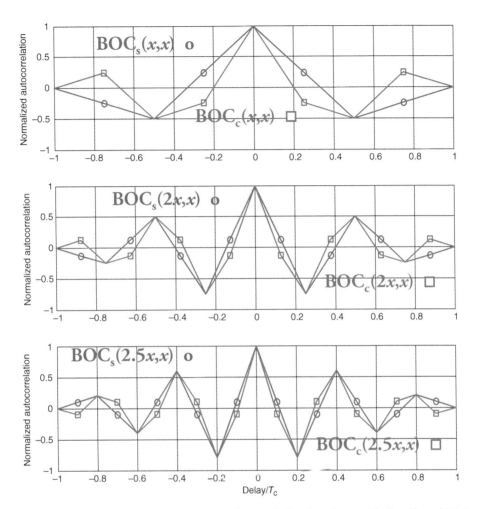

Figure 3.11. Examples of Infinite-Bandwidth Correlation Functions, with Sine-Phased BOC Colored Red and Cosine-Phase BOC Colored Blue

and the PSD for cosine-phased BOC spreading modulations is given by [4]

$$\Phi_{BOC_c(f_s,f_c)}(f) = \begin{cases} \dfrac{4}{f_c}\text{sinc}^2(\pi f/f_c) \left(\dfrac{\sin^2\left(\dfrac{\pi f}{4f_s}\right)}{\cos\left(\dfrac{\pi f}{2f_s}\right)} \right)^2, & k_{BOC} \text{ even} \\[2em] \dfrac{4}{f_c}\dfrac{\cos^2(\pi f/f_c)}{(\pi f/f_c)^2} \left(\dfrac{\sin^2\left(\dfrac{\pi f}{4f_s}\right)}{\cos\left(\dfrac{\pi f}{2f_s}\right)} \right)^2, & k_{BOC} \text{ odd} \end{cases} \quad (3.27)$$

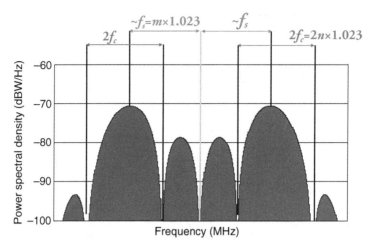

Figure 3.12. Example of BOC PSD

Figure 3.12 shows an example plot of a BOC PSD. Although the specific PSD shown is for a BOC(10,5) spreading modulation, the following characteristics apply for different phasings and BOC parameters. The null-to-null width of each of the tallest spectral peaks (each main lobe) is $2f_c$, and the center point between these nulls is f_s from the center of the spectrum. The null-to-null width of each other spectral peak (the sidelobes) is f_s. The spectral peaks of the main lobes occur near, but not exactly at f_s from the center of the spectrum. The BOC parameter is equal to two plus the number of sidelobes within the main lobes. If the BOC parameter is even, there is a null at band center; if the BOC parameter is odd there is a sidelobe at band center.

In comparison to the sine-phased BOC PSD in Figure 3.12, a cosine-phased PSD differs qualitatively in several ways. With sine-phased BOC, the PSD sidelobes between the main lobes are larger than the sidelobes outside the main lobes. Also, while the peak of the main lobes falls inside the midpoint of the main lobe nulls for sine phasing, it falls outside the midpoint of the main lobe nulls for cosine phasing.

If a signal comprises two BOC(m,n) waveforms in phase quadrature, it can be referred to a quadrature offset carrier, or QOC(m,n), analogous to QPSK-R. Such a QOC signal has the same correlation function and PSD as its constituent BOC(m,n) waveforms.

Many other families of spreading modulations have been developed recently, including binary coded symbols (BCS) [5], time-multiplexed BOC (TMBOC) [6–8], composite BOC (CBOC) [9], and alternative BOC (AltBOC) [10–15]. The remainder of this section provides an overview of these spreading modulations. This material is meant as an introduction; further exploration is available through the Review Questions at the end of the chapter and in the referenced literature. Further discussion is also provided in the descriptions of specific satnav systems in Part II.

SIGNALS, SIGNAL PROCESSING, AND SPREADING MODULATIONS

BCS [5] provide a generalized family of spreading modulations, denoted BCS($[c_0, c_1, \ldots, c_{N-1}], f_c$) and given by

$$g_{BCS}(t) = \sum_{n=0}^{N-1} c_n \text{rect}_{T_c/N}\left(t - \frac{nT_c}{N} - \frac{T_c}{2N}\right) \quad (3.28)$$

where $c_k \in \{+1, -1\}$, and the spreading symbol is defined on $t \in [0, T_c]$. While this representation appears similar to that for DSSS signals in (3.14), there are two significant differences. One is that the summation extends over infinite limits for DSSS signals, while for BCS the number of segments N is finite and typically less than 20. The second is that for BCS spreading modulations, the c_k are designed and not pseudorandom. Most importantly, (3.28) only defines a single spreading symbol of a DSSS waveform.

The symbol (3.28) generalizes BPSK-R, for which all values of c_k are unity, and also generalizes BOC, which can be represented by (3.28) using a sequence of c_k that produces a segment of a square wave. For example, a BOC(2,1) could be represented as BCS ($[1, -1, 1, -1], 1$) or BCS ($[1, 1, -1, -1, 1, 1, -1, -1], 1$), and a $BOC_c(2,1)$ could be represented as BCS ($[-1, 1, 1, -1, -1, 1, 1, -1], 1$).

The PSD of a BCS spreading time series using (3.28) for spreading symbols is found using (3.15) with the Fourier transform of the spreading symbol given by

$$G_{BCS}(f) = \frac{T_c}{N}\left[\sum_{n=0}^{N} c_n e^{-i2\pi f n T_c/N}\right] \text{sinc}(\pi f T_c/N) \quad (3.29)$$

The definition of BCS also includes time multiplexing and phase multiplexing of different spreading symbols. A simple example of time multiplexing of different BCS spreading symbols is

$$x(t) = \sum_{k=-\infty}^{\infty} [a_{2k}g_1(t - 2kT_c) + a_{2k+1}g_2(t - 2kT_c - T_c)] \quad (3.30)$$

for two different spreading symbols $g_1(t)$ and $g_2(t)$, producing a spreading time series that alternates between the two different spreading symbols. When the $\{a_k\}$ is an ideal long spreading code, the resulting PSD is

$$\Phi_x(f) = \frac{1}{2T_c}[|G_1(f)|^2 + |G_2(f)|^2] \quad (3.31)$$

Different proportions of spreading symbols can be used, and the resulting unit-power PSD is the convex sum of the individual unit-power PSDs in the same proportion.

Phase quadrature multiplexing places different spreading symbols on the in-phase and quadrature components. If there are two different unit-power spreading symbols $g_1(t)$ and $g_2(t)$ either time multiplexed with an average fraction α and $1 - \alpha$, respectively, or phase multiplexed with relative powers α and $1 - \alpha$, respectively, both with $0 \leq \alpha \leq 1$, then, assuming the spreading sequence is an ideal long spreading code, the PSD is

$$\Phi_x(f) = \frac{1}{T_c} \left[\alpha \left| G_1(f) \right|^2 + (1-\alpha) \left| G_2(f) \right|^2 \right]$$
$$= \alpha \Phi_1(f) + (1-\alpha) \Phi_2(f) \tag{3.32}$$

Note that the schedule for time multiplexing different symbols should be deterministic and documented in the ICD, so that receivers can use this schedule to process the signal most efficiently.

AltBOC extends the concept of BOC to permit different signal characteristics in the upper and lower sidelobes. Two distinct DSSS signals could be placed on symmetrically spaced subcarriers using conventional signal sideband techniques:

$$x(t) = \sqrt{\alpha} x_L(t) e^{-i 2\pi f_s t} + \sqrt{1-\alpha} x_U(t) e^{i 2\pi f_s t} \tag{3.33}$$

where $0 \leq \alpha \leq 1$, $x_L(t)$ is the lower sideband signal, $x_U(t)$ is the upper sideband signal, and f_s is the subcarrier frequency, with $x_L(t)$ and $x_U(t)$ unit-power signals. Unfortunately, (3.33) is not constant envelope, even if $x_L(t)$ and $x_U(t)$ are real and constant envelope.

Observe that (3.33) can be rewritten as $x(t) = \left[\sqrt{\alpha} x_L(t) + \sqrt{1-\alpha} x_U(t) \right] \cos(2\pi f_s t) + i \left[x_U(t) - \sqrt{\alpha} x_L(t) \right] \sin(2\pi f_s t)$ so there is both a sine-phased and a cosine-phased subcarrier. Now replace the linear subcarriers with binary subcarriers to produce an AltBOC signal

$$x(t) = \left[\sqrt{\alpha} x_L(t) + \sqrt{1-\alpha} x_U(t) \right] \text{sgn} \left[\cos(2\pi f_s t) \right]$$
$$+ i \left[\sqrt{1-\alpha} x_U(t) - \sqrt{\alpha} x_L(t) \right] \text{sgn} \left[\sin(2\pi f_s t) \right] \tag{3.34}$$

If $x_L(t)$ and $x_U(t)$ are both real valued and constant envelope, then (3.34) is constant envelope. If $\alpha = 1/2$ and $x_L(t)$ has a BPSK-R(f_c) spreading modulation with $x_L(t) = x_U(t)$, then (3.34) is BOC$_c$(f_s, f_c), while if $\alpha = 1/2$ and $x_L(t)$ has a BPSK-R(f_c) spreading modulation with $x_L(t) = -x_U(t)$, then (3.34) is BOC(f_s, f_c). But if $x_L(t)$ or $x_U(t)$, or both, are complex valued or not equal to each other or to the negative of each other, then (3.34) is not constant envelope.

Constant-envelope AltBOC uses additional terms, often called intermodulation components, added to (3.34) [10–12] to produce an AltBOC signal having BPSK-R in-phase and quadrature components using mutually orthogonal spreading codes and the same chip rate. When the upper and lower sideband signals use a BPSK-R(10) spreading modulation on in-phase and quadrature components with mutually uncorrelated spreading codes, and the subcarrier frequency is 15×1.023 MHz sampled at 10 times the subcarrier frequency, the resulting symmetric-spectrum constant-envelope AltBOC(15,10) complex envelope is given by

$$x(t) = x_A(t) + x_B(t) + x_C(t) + x_D(t)$$

where

$$\begin{aligned}x_A(t) &= [x_{L,I}(t) + ix_{L,Q}(t)]\psi_1^*(t)\\ x_B(t) &= [x_{U,I}(t) + ix_{U,Q}(t)]\psi_1(t)\\ x_C(t) &= x_{U,Q}(t)x_{U,I}(t)[x_{L,I}(t) + ix_{L,Q}(t)]\psi_2^*(t)\\ x_D(t) &= x_{L,Q}(t)x_{L,I}(t)[s_{U,I}(t) + is_{U,Q}(t)]\psi_2(t)\end{aligned} \quad (3.35)$$

Here, the lower sideband signal is $x_{L,I}(t) + ix_{L,Q}(t)$ with $x_{L,I}(t) = \sum_{k=-\infty}^{\infty} a_k g_{\text{BPSK-R}(10)}(t - kT_{10})$ and $x_{L,Q}(t) = \sum_{k=-\infty}^{\infty} b_k g_{\text{BPSK-R}(10)}(t - kT_{10})$, and the upper sideband signal is $x_{U,I}(t) + ix_{U,Q}(t)$ with $x_{U,I}(t) = \sum_{k=-\infty}^{\infty} c_k g_{\text{BPSK-R}(10)}(t - kT_{10})$ and $x_{U,Q}(t) = \sum_{k=-\infty}^{\infty} d_k g_{\text{BPSK-R}(10)}(t - kT_{10})$. Here, the $\{a_k\}$, $\{b_k\}$, $\{c_k\}$, and $\{d_k\}$ are a combination of spreading code bits, data message symbols, and overlay code bits as defined in Section 3.4, $T_{10} = 1/(10 \times 1.023 \times 10^6)$ is the chip duration and the BPSK-R(10) spreading symbol is defined in (3.18).

Two different subcarriers are involved in constant-envelope AltBOC(15,10), each with period $T_{15} = 1/(15 \times 1.023 \times 10^6)$ seconds. $\psi_1(t)$ is a digitized sinusoid taking on eight values over the period:

$$\psi_1(t) = [(1 + \sqrt{2}) - i, 1 + i, -1 + i(1 + \sqrt{2}), -(1 + \sqrt{2}) + i(1 + \sqrt{2}),$$

$$(1 + \sqrt{2}) - i, 1 + i, -1 + i(1 + \sqrt{2}), -(1 + \sqrt{2}) + i(1 + \sqrt{2})]/(4\sqrt{2})$$

and $\psi_2(t)$ is a digitized waveform taking on eight values over the period:

$$\psi_2(t) = [(1 - \sqrt{2}) + i, 1 + i(1 - \sqrt{2}), -1 + i(1 - \sqrt{2}), (\sqrt{2} - 1) + i,$$

$$(\sqrt{2} - 1) - i, -1 + i(\sqrt{2} - 1), 1 + i(\sqrt{2} - 1), (1 - \sqrt{2}) - i]/(4\sqrt{2})$$

with $\psi_1(t)$ and $\psi_2(t)$ both constant envelope. $x_A(t)$ and $x_B(t)$ are the useful signals, respectively, downconverted and upconverted by the subcarrier frequency, while $x_C(t)$ and $x_D(t)$ are intermodulation components not useful to a receiver of $x_A(t)$ or $x_B(t)$, but causing (3.35) to be constant envelope. More details are provided in the discussion of Galileo E5 signals, in Section 10.3.5.

For the more general case of an AltBOC signal having BPSK-R(f_c) signals on both carrier phases of both subcarriers, the spreading modulation is denoted AltBOC(f_s,f_c) or AltBOC(m,n) where $f_s = m \times 1.023$ MHz and $f_c = n \times 1.023$ MHz. The PSD of the resulting constant-envelope AltBOC(f_s,f_c) signal, when $k_{BOC} = 2f_s/f_c = 2m/n = T_c/T_s$ is odd valued, is [12]

$$\Phi_{AltBOC_{ce}(f_s,f_c)}(f) = \frac{4}{T_c\pi^2 f^2} \left[\frac{\cos(\pi f T_c)}{\cos(2\pi f T_c/k_{BOC})} \right]^2 \\ \times \left[\cos^2(\pi f T_s) - \cos(\pi f T_s) - 2\cos(\pi f T_s)\cos(\pi f T_s/2) + 2 \right] \quad (3.36)$$

where the parameter definitions are the same as used previously for BOC[1]: $T_c = 1/f_c$, $T_s = 1/(2f_s)$.

Additional results continue to be published concerning more general approaches for transmitting a single constant-envelope composite signal containing signals centered at upper and lower subcarriers, including References 13–15.

A different BOC variant, actually a special case of BCS with time multiplexing, involves time multiplexing two or more different BOC spreading symbols in a specific proportion to produce a TMBOC spreading time series. A unit-power multiplexed BOC (MBOC) PSD is defined as the convex sum of the PSDs of two different BOC spreading symbols, and a TMBOC spreading time series has an MBOC PSD for an ideal long spreading code. Assuming the spreading sequence is an ideal long spreading code, the MBOC PSD is given by (3.32), with $\Phi_1(f)$ the unit-power PSD corresponding to the BOC spreading symbol transmitted α of the time, and $\Phi_2(f)$ the unit-power PSD corresponding to the other BOC spreading symbol. More details are provided in the discussion of the GPS L1C signal in Section 7.3.

CBOC is a different spreading modulation that can also produce an MBOC PSD. CBOC spreading symbols are formed from the weighted sum and difference of two different BOC spreading symbols having the same spreading code rate but different subcarrier frequencies. The CBOC spreading symbols are then

$$g_{CBOC+}(t) = \sqrt{\alpha}\, g_{BOC(m_1,n)}(t) + \sqrt{1-\alpha}\, g_{BOC(m_2,n)}(t) \\ g_{CBOC-}(t) = \sqrt{\alpha}\, g_{BOC(m_1,n)}(t) - \sqrt{1-\alpha}\, g_{BOC(m_2,n)}(t) \quad (3.37)$$

where $\alpha, 0 \leq \alpha \leq 1$, is the fraction of power in BOC(m_1,n) spreading symbols, with the remaining power in BOC(m_2,n) spreading symbols. Observe

$$|G_{CBOC+}(f)|^2 = \alpha\, |G_{BOC(m_1,n)}(f)|^2 + (1-\alpha)\, |G_{BOC(m_2,n)}(f)|^2 \\ + 2\sqrt{\alpha}\sqrt{1-\alpha}\, \Re\left\{ G_{BOC(m_1,n)}(f)G^*_{BOC(m_2,n)}(f) \right\} \\ |G_{CBOC-}(f)|^2 = \alpha\, |G_{BOC(m_1,n)}(f)|^2 + (1-\alpha)\, |G_{BOC(m_2,n)}(f)|^2 \\ - 2\sqrt{\alpha}\sqrt{1-\alpha}\, \Re\left\{ G_{BOC(m_1,n)}(f)G^*_{BOC(m_2,n)}(f) \right\} \quad (3.38)$$

---

[1] These parameter definitions differ from those used in [12], and the expression is adjusted accordingly.

Assuming the spreading sequence is an ideal long spreading code, a spreading time series of only CBOC+ or CBOC− spreading symbols does not have the desired MBOC PSD because of the cross-terms in (3.38). However, if equal power CBOC+ and CBOC− spreading symbols are time multiplexed with equal occurrence, the cross-terms cancel and the time-averaged PSD is MBOC. Other ways to get the MBOC PSD without cross-terms involve forming one spreading time series using CBOC+ spreading symbols and forming a second spreading time series using CBOC− spreading symbols, with orthogonal spreading codes used for each. These two spreading time series can be added (code division multiplexing) or placed in phase quadrature (phase division multiplexing) to produce a composite signal having an MBOC spectrum, with the composite signal scaled by $1/\sqrt{2}$ to maintain unit power.

3.2 EFFECTS OF DOPPLER AND OF IONOSPHERIC PROPAGATION

Satnav signals undergo Doppler shifts caused by a combination of satellite motion and receiver motion, and Section 3.2.1 provides mathematical models and their approximations for Doppler shifts. Satnav signals also propagate through the ionosphere, and Section 3.2.2 describes how the ionosphere's dispersive characteristics affect the received signals.

3.2.1 Doppler Effects

The Doppler effect changes every frequency in the received signal, so a frequency f in the transmitted signal becomes the Doppler-shifted frequency f' at the receiver according to

$$f' = \left(\frac{c + s_r}{c - s_t}\right) f$$
$$= \left[1 + \left(\frac{s_r + s_t}{c}\right) \sum_{n=0}^{\infty} \left(\frac{s_t}{c}\right)^n\right] f \quad (3.39)$$
$$\cong (1 + \psi) f$$

where c is the speed of light, s_r is the speed of the receiver toward the transmitter along the line of sight, s_t is the speed of the transmitter toward the receiver along the line of sight, the total line of sight velocity is $s_{s2r} = s_t + s_r$, and $\psi \stackrel{\Delta}{=} s_{s2r}/c$.

The Doppler shift is the difference between the received frequency and the transmitted frequency: $f' - f = f\psi \sum_{n=0}^{\infty} \left(\frac{s_t}{c}\right)^n \cong f\psi$, and is proportional to the transmitted frequency. The speeds typically dealt with in satnav are at most of the order of kilometers per second—a small fraction of a percent of the speed of light—so $|\psi| < 10^{-4}$ and the approximate expression is suitable for all cases examined in this book and thus

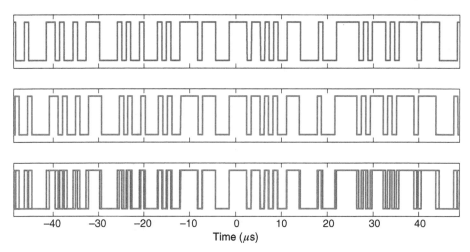

Figure 3.13. Example of Time-Companded Time Waveform

used subsequently. Under this approximation, the transmitter and receiver line of sight speeds are interchangeable.

If a waveform $y(t)$ has a Fourier transform $Y(f)$, then the Doppler-shifted signal has Fourier transform $Y'(f) = Y(f(1+\psi))$. The Doppler-shifted waveform is the inverse Fourier transform, $y'(t) = \frac{1}{1+\psi} y(t/(1+\psi)) \cong y(t(1-\psi))$. Multiplication of the time variable by $1 - \psi$ corresponds to time companding: compressing (if $\psi > 0$) or expanding (if $\psi < 0$) the time axis.

Figure 3.13 shows time companding of a BPSK-R(1) signal. The upper plot is the original signal over 100 spreading symbol periods, or approximately 97.8 μs. The middle plot shows time companding with $\psi = -0.02$ (much larger magnitude time companding than values encountered in satnav to produce a visible effect over this relatively short time), aligned with the original signal at the center of the traces. The lower plot overlays the two, showing time expansion by more than a spreading symbol duration on the edges of the plot. In practice, it is advantageous to align the two waveforms at the center of an observation time interval, interpreting any time shift needed for alignment at this point as a delay, then dealing with the companding that remains.

The Doppler-shifted bandpass signal in (3.1) is then $s'(t) = \Re\{x'(t) \exp[i2\pi f_0(1+\psi)t]\}$ where the Doppler-shifted complex envelope is a time-companded version of the transmitted one: $x'(t) \cong x(t(1-\psi))$. It is common to use the narrowband Doppler approximation $s'(t) \cong \Re\{x(t) \exp[i2\pi f_0(1+\psi)t]\}$ that neglects time companding of the complex envelope and only includes the Doppler shift of the carrier frequency. Such an approximation holds if the time-bandwidth product of the observation is small enough, but processing based on this assumption loses performance if the time-bandwidth product is too large. A conservative criterion is that the highest frequency of interest in the complex envelope, $f_{\max}$, should be phase shifted less than one half cycle over the observation time $[-T/2, T/2]$. Then if the un-companded waveform and the companded waveform are aligned at the center of the time interval, this criterion is $f_{max}T/2 - f_{max}(1 - |\psi|)T/2 < 0.5$, which yields the following criterion for neglecting time-companding of the complex envelope:

$$f_{max} |\psi| T < 1.0 \quad (3.40)$$

Apply this criterion to Figure 3.13 where $T = 97.8$ μs and $\psi = -0.02$. If the maximum frequency in the complex envelope were the chip rate of 1.023 MHz, the result would be $f_{max} |\psi| T = 0.99$, which is approximately the limit of the criterion (3.40). Given that the waveforms in Figure 3.13 contain significant power at several multiples of the chip rate (as seen by the sharp edges of the spreading symbols), the criterion is significantly exceeded and time companding of the complex envelope must be considered in this example, as is evident from the figure. If correlation processing were used on these waveforms, the cross-correlation peak between the companded and un-companded waveforms in this figure has about 0.57 the amplitude of the autocorrelation function of either waveform, indicating significant loss. When the same waveforms are generated using $\psi = -0.01$ to satisfy the criterion (3.40) with maximum frequency 1.023 MHz, the cross-correlation function peak is reduced to 0.79 of the value with no companding.

3.2.2 Ionospheric Effects

The ionosphere is an atmospheric layer of plasma, or ionized gases, extending from an altitude of approximately 80 km to an altitude as high as 1000 km above the Earth. At these altitudes, the atmosphere is so thin that free electrons can exist for short periods of time before they are captured by a nearby positive ion. Solar x-ray and ultraviolet radiation energizes the gases and produce the free electrons, so the density of free electrons is particularly affected by solar radiation as well as the Earth's magnetic field. The density of electrons varies over time (diurnally, seasonally, and with the solar cycle), with location over the Earth, and with altitude. Electron densities can vary by one to two orders of magnitude between a maximum in the early afternoon local time and a minimum at nighttime, with ionospheric fluctuations tending to be greatest near the equator and the poles. Classic references on the ionosphere and its effects include References 16 and 17]; an informative description is also provided in Chapter 2 of Reference 18, and recent new work benefitting from use of coded signals on multiple frequencies is described in Reference 19. Most of this section addresses the normal ionosphere, with only a brief mention at the end of ionospheric disturbances.

An ionized gas is a dispersive medium for radio waves, meaning the propagation delay through the medium varies with frequency and the phase shift is not linear with frequency. At a given frequency, the speed of propagation depends upon the number of free electrons along the path traveled by the wave from satellite transmit antenna to receive antenna. Signals from low elevation satellites encounter a greater number of electrons than those from satellites at zenith, because of the longer path these signals travel through the ionosphere.

The number of free electrons is known as the total electron count (TEC), defined as the number of electrons contained in a 1 m² tube from the satellite to the receiver, and measured in TEC units (TECU) where 1 TECU = 10^{16} electrons/m². Physicists map the ionosphere by measuring the TEC in the vertical direction. A typical vertical TEC value is 100 TECU, although over different times and locations the vertical TEC can range from a few TECU to hundreds of TECU, with values below 250 TECU considered "normal" ionosphere, and larger values considered a disturbed ionosphere. Geometric calculations (described in Chapter 20) adjust the vertical TEC values to account for propagation from transmitters at lower elevation angles; the length of the path traveled through the ionosphere by a signal from a low elevation satellite can be almost three times as long as the vertical path, corresponding to almost three times the dispersive effect. Maps of the ionosphere's TEC at a particular time can be obtained with accuracies typically less than 10 TECU, but with latencies up to 1 day.

The refractive index of a medium is defined as the ratio of the speed of propagation of the signal in vacuum to the speed of propagation in the medium. In the range of frequencies used for satnav, the ionosphere's refractive index for electromagnetic radiation is less than unity, meaning a sinusoidal wave propagates faster than the speed of light in vacuum, making the carrier wavelength appear shorter than it actually is. As shown in Reference 17, a sinusoid at frequency f propagating through the ionosphere arrives at an excess delay (delay different from propagation in vacuum) approximated by

$$\Delta_{\text{iono}} \cong -40.3 \times \text{TEC} / \left(cf^2\right) \quad \text{seconds} \tag{3.41}$$

with the negative sign indicating a negative delay, which is an advance. The expression in (3.41) is a first-order approximation typically accurate to a few centimeters. In (3.41), the parameter TEC is in units of electrons/m^2, not TECU. Figure 3.14 shows the computed

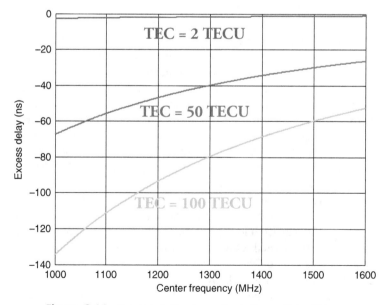

Figure 3.14. Representative Ionospheric Excess Delay Values

excess delay for different center frequencies and different TEC values. Multiplying by the speed of light yields the corresponding range error. Even at 1575.42 MHz with a TEC of 2 TECU or 2×10^{16} electrons/m², the excess delay corresponds to a pseudorange error of 32 cm—larger than desired for many satnav applications; at this frequency, TEC of 200 TECU corresponds to a pseudorange error of 32 m.

The difference in delay for signals at two different frequencies passing through the same ionospheric channel is

$$\begin{aligned} \Delta_{f_1-f_2} &= \frac{-40.3 \times \text{TEC}}{c} \left(\frac{1}{f_1^2} - \frac{1}{f_2^2} \right) \\ &= \frac{40.3 \times \text{TEC}}{c} \left(\frac{f_1^2 - f_2^2}{f_1^2 f_2^2} \right) \end{aligned} \qquad (3.42)$$

Observe from (3.42) that knowing the two frequencies and measuring the difference in delay allows calculation of the TEC. Dual-frequency satnav receivers use this observation to mitigate ionospheric effects, as discussed in Chapter 20.

More generally, propagation through a dispersive medium can be modeled by a linear time-invariant system whose output is a distorted version of the input because different frequencies of the input are delayed by different amounts. Model the ionosphere with transfer function

$$H_{\text{iono}}(f) = A \exp\left[i(-2\pi f D + \phi(f))\right] \qquad (3.43)$$

where A is the amplitude gain (which must vary slowly compared to carrier frequencies of interest), D is the delay that would have occurred had propagation been through vacuum in units of seconds, and $\phi(f)$ is the frequency-dependent phase introduced by the ionosphere, in units of radians. The expression (3.43) is a useful first-order model over the range of frequencies used for satnav, as long as the parameters are selected for a particular time and location. This transfer function is not a lowpass equivalent transfer function used to describe the response to a complex envelope, but rather describes the response to the input represented as an analytic signal.

As discussed in Appendix Section A.3, when the input is a narrowband signal, the phase delay indicates the time delay of the carrier phase, while the group delay indicates the time delay of the complex envelope, with phase delay and group delay the same for a nondispersive system. For a transfer function of the form (3.43), the phase delay and group delay, both in units of seconds, are related to the phase in units of radians:

$$\begin{aligned} \tau_{\text{phase}}(f) &= -\frac{\phi(f)}{2\pi f} \\ \tau_{\text{group}}(f) &= \frac{-1}{2\pi} \frac{d\phi(f)}{df} \end{aligned} \qquad (3.44)$$

For $H_{iono}(f)$, $\tau_{phase}(f) = -40.3 \times \text{TEC}/(cf^2)$ seconds, so $\phi(f) = -2\pi f \tau_{phase}(f) = 80.6\pi \times \text{TEC}/(cf)$ radians, with a positive sign indicating phase advance. Substituting this expression into the second equation in (3.44) yields

$$\tau_{group}(f) = \frac{-1}{2\pi}\frac{d\phi(f)}{df} = 40.3 \times \text{TEC}/\left(cf^2\right) = -\tau_{phase}(f) \qquad (3.45)$$

The result of propagation through the ionosphere, relative to propagation through vacuum, is that the carrier undergoes an apparent phase advance while the complex envelope undergoes an apparent delay. The carrier appears to be propagating faster than it would through vacuum, while the complex envelope appears to be propagating more slowly than it would through vacuum. Thus, the carrier phase appears to advance, while the complex envelope appears to be delayed.

Figure 3.15 provides an exaggerated view of ionosphere-induced carrier advance and code delay for a BPSK-R spreading modulation, with the ratio of carrier frequency to spreading code chip rate much lower than in actual signals. After propagation through the ionosphere, phase transitions no longer appear at the zero crossings of the carrier. It should be clear that this phenomenon is not uniquely related to ionospheric propagation, but rather occurs with any linear time-invariant system whose transfer function phase is proportional to the reciprocal of frequency.

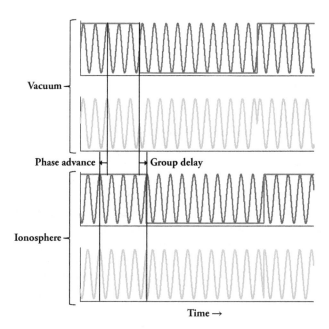

Figure 3.15. Exaggerated View of Ionospheric Effects, with First and Third Plots Showing BPSK-R Complex Envelope and Carrier Separately, and Second and Fourth Plots Showing the Modulated Carrier

Since the TEC varies with time and location, the received signal experiences time-varying TEC not only due to the time variation of TEC, but also because the signal passes through different paths in the ionosphere as the satellite and receiver move. The group delay and phase advance thus vary over time, causing the relative timing of received complex envelope and received carrier to be time varying. This phenomenon is known as code-carrier divergence, reflecting the traditional use in GPS of "code" for the complex envelope of the signal. As described in Chapters 20 and 23, unless ionospheric effects are removed from the signals being processed, the relationship between timing of code and timing of carrier in a received signal cannot be expected to remain the same over time periods approaching or exceeding 2 minutes.

Since ionospheric characteristics are spatially variable, accurate characterizations must account for this spatial variation. One common representation models the ionosphere as one or more shells of infinitesimal thickness at a specific altitude. The coordinates where the line of sight between a receiver and a satellite intersects the shell is called a pierce point. TEC is measured or calculated at different pierce points of interest.

The effects addressed in this section have focused on normal ionospheric characteristics. Severe space weather can cause the ionosphere to become very turbulent in limited angular regions, with electron densities changing rapidly. The propagating waves are diffracted, causing rapid changes in the received carrier phase (called scintillation) and in received amplitude (fading) of the received signals, stressing a receiver's ability to track the signals. Different scintillation models are described in Reference 20. Research continues into anomalous behavior of the ionosphere, and the resulting effects on signal propagation. See, for example, Reference 21.

Many of the developments in new and modernized systems, including higher signal power, coded signals on multiple frequencies, the presence of pilot components, and the availability of more satellites at diverse azimuth and elevation angles, will improve the ability of receivers to maintain track through these severe ionospheric disturbances.

3.3 SATNAV SIGNAL CHARACTERISTICS

The twelve characteristics listed below describe a satnav signal:

- Carrier frequency,
- Power,
- Polarization,
- Multiple access,
- Spreading modulation and bandwidth,
- Spreading codes,
- Data message structure,
- Data message error correction and detection,
- Data modulation,

- Pilot and data components,
- Overlay codes,
- Multiplexing.

Each of these characteristics is described in turn.

3.3.1 Signal Carrier Frequency

Numerous technical factors affect the choice of carrier frequency. Among the factors favoring lower frequencies are that nearly omnidirectional receive antennas, needed to receive signals from spatially diverse satellites distributed over the upper hemisphere, have larger effective areas and thus collect more signal power from a given received power flux density, as shown in Section 4.1. Signals at lower frequencies also tend to better penetrate structures and foliage, delivering a larger fraction of their power to a receiver operating with an impaired view of the sky. Signals at lower frequencies also refract better around obstacles such as buildings, enabling better signal reception even in difficult terrain and urban areas. In addition, lower frequency signals undergo lower excess propagation losses due to clouds and rain. Besides these factors that generally lead to a greater fraction of transmitted power being useful at the receiver, lower frequency signals also undergo proportionally smaller phase and frequency deviations due to Doppler shift, smaller phase noise in the transmitted signal for a given quality of reference oscillator, and smaller imperfections of frequency sources in receivers. These smaller deviations enable more capable receiver processing and the use of less expensive and lower power components.

Higher frequencies also have their advantages. A given amount of transmit antenna gain can be achieved at higher frequencies with a smaller transmit antenna, reducing satellite size and mass. Higher frequencies enable smaller receive antenna elements while maintaining broad angular coverage, reducing receive equipment size and weight. Beamforming and null-steering receive antennas occupy less space as well, since elements are typically spaced at some fraction of a (smaller) wavelength. Higher frequency signals are also affected less by ionospheric refraction, as discussed in Section 3.2.2. In addition, since high power terrestrial transmitters for commercial broadcast television, broadcast radio, and land mobile radio typically occupy VHF (30 MHz to 300 MHz) and low UHF (300 MHz to approximately 1000 MHz) frequency bands, placing satnav signals at higher frequencies allows receivers to operate at frequencies further from these high power transmitters, making it easier for satnav receiver circuitry to reject these high power out-of-band interfering signals as discussed in Chapter 14. Finally, since spectrum use historically began at lower frequencies and has moved to higher frequencies more recently, there is potentially more unused bandwidth available for satnav use at higher frequencies where there are fewer and less well established incumbent users.

Initially, L band (1–2 GHz) was selected as the frequency bands for GPS and GLONASS, reflecting the tradeoffs discussed above. Subsequently, L band has been recognized as "beachfront property" for mobile communications and other applications. International interest in satnav has led to designation of several parts of L band for satnav

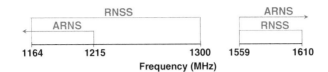

Figure 3.16. ITU Designations of L-Band Frequencies Used for Satnav

use by the International Telecommunications Union (ITU). As shown in Figure 3.16, the ITU has designated "upper L band," from 1559 MHz to 1610 MHz a Radionavigation Satellite Service (RNSS) band, as well as "lower L band," from 1164 MHz to 1300 MHz. These two portions of L band contain the frequencies primarily used for satnav signals. Also shown in Figure 3.16 is the ITU designation of frequencies from 1559 MHz to above 1610 MHz and from 1215 MHz to below 1164 MHz for Aeronautical Radio Navigation Service (ARNS)—radar, traffic collision-avoidance system, distance measuring equipment, tactical air navigation, and other safety-critical navigation uses. The combination of technical factors and spectrum allocation makes these portions of L band the frequency bands of choice for satnav, even though there is consideration of portions of S band (2–4 GHZ) and C band (4–8 GHz) for future satnav use.

New and modernized satnav systems employ signals at multiple carrier frequencies. Widely separated carrier frequencies allow receivers to compensate for ionospheric errors (see Chapter 20), and to more readily remove carrier phase ambiguities (see Chapter 23). Frequency diversity from multiple carrier frequencies also enables continued operation even if powerful interference is encountered at or near one of the carrier frequencies. In addition, distributing signals over different carrier frequencies reduces MAI, and allows some signals to be spectrally separated from the others for reasons of policy.

3.3.2 Signal Power Level

Choosing the power level for satnav signals is another fundamental consideration. Higher power received signals have numerous benefits, including improving the ability of receivers to overcome noise and interference, and allowing more economical and lower power receiver designs with less efficient antennas and greater processing losses. Higher power signals also provide margin to overcome excess propagation loss due to penetration of foliage and buildings.

Increased signal power also has disadvantages. Providing higher power requires larger or more efficient solar arrays, greater capacity in satellite power storage and conditioning, and amplifiers that can generate higher power signals, adding costs to satellites in terms of component weight and thermal management. Larger transmit antennas may be needed to provide greater gain, accompanied by the need for greater antenna pointing accuracy. Higher power signals cause greater MAI to reception of other GNSS signals, as well as greater interference to other systems (e.g., radars, radio astronomy, communications) using the same and adjacent bands.

Many current and planned signal designs produce individual signals whose received PSDs are substantially less than the noise floor, so receivers are predominantly noise limited and not interference limited. Received power levels are typically in the approximate range of -160 dBW to -150 dBW out of an antenna with essentially 0 dBi gain. Chapter 4 provides details of how these values are specified.

3.3.3 Signal Polarization

While in many aspects of this book satnav signals are treated as waveforms, polarization considerations fundamentally involve modeling satnav signals as electromagnetic waves. Far from the satellite, these electromagnetic waves can be modeled as plane waves, whose wave fronts are infinitely long and wide. In this case, the electric and magnetic fields are perpendicular to the direction of propagation and to each other, but with otherwise identical characteristics. Thus, only one of these fields needs to be described; it is conventional to deal with the electric field. This vector field is described as a simple harmonic wave comprising two components orthogonal to direction of propagation, with the two components varying sinusoidally at the same frequency. The defining characteristics of these two components are frequency, relative amplitude, and relative phase. (A satnav signal actually has a range of frequencies, but examining the polarization characteristics at the carrier frequency is adequate for narrowband signals like satnav signals.) At any instant in time, the orientation of the vector field is found from the vector sum of the two components.

Polarization describes the evolution over time of the electric field's orientation observed at a point in space. If the orientation remains the same, the field is linearly polarized in the direction of that orientation. If the orientation rotates, the field is circularly or elliptically polarized. If the orientation rotates clockwise along the direction of propagation, it is said to have right-hand polarization rotation, while if it rotates counterclockwise, it has left-hand polarization rotation.

The orientation of the field and its evolution over time are defined by the relative amplitude and phase of its two transverse components. Figure 3.17 depicts three different possible polarizations. The left column depicts circular polarization, where the two components have equal amplitudes and are in phase quadrature, with one lagging the other by 90° as they both vary sinusoidally. The upper Lissajous figure shows the components and the vector wave propagating through space with the vector sum of the components sweeping out a circle over time. The relative phasing shown produces right-hand circular polarization. The lower figure shows the view from the top, normal to the direction of propagation, making it clear that the vector field sweeps out a circle. The central column is linear polarization where the two sinusoidal components have equal amplitudes and the same phase. The upper Lissajous figure shows the components and the vector wave propagating through space. The lower figure shows the view from the top, normal to the direction of propagation, making it clear that the vector field sweeps out a line. Both the circle and the line are special cases of an ellipse; achieving a perfect circle or line requires precise matching of amplitude and phase between the two components of the electric field. When matching of amplitude or phase is imperfect,

SATNAV SIGNAL CHARACTERISTICS 69

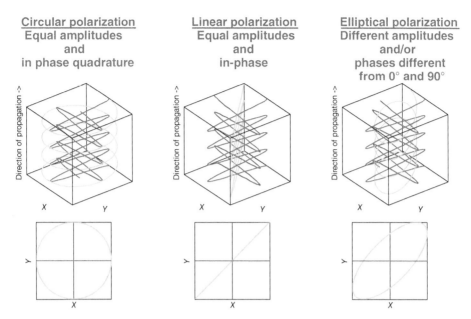

Figure 3.17. Examples of Electric Field Polarization, with the Two Vector Components in Red and Blue, and the Resultant in Green

the result is elliptical polarization, shown in the right column. Polarization orientation refers to the orientation of the semimajor axis of the polarization ellipses shown on the bottom row. Clearly orientation is irrelevant for perfectly circular polarization. Both the circular polarization and elliptical polarization examples in Figure 3.17 are right handed; right-hand circular polarization is abbreviated RHCP, while left-hand circular polarization is abbreviated LHCP.

The axial ratio is the ratio of the magnitude of the polarization ellipse's major axis to the minor axis. As a convention, the axial ratio is positive for right-hand elliptical polarization, and negative for left-hand elliptical polarization. The magnitude of the axial ratio is

$$|\alpha| = \frac{\text{Major axis length}}{\text{Minor axis length}} \tag{3.46}$$

and the axial ratio is expressed in decibels as $10\log_{10}(\alpha^2)$.

The wave's polarization depends upon the transmit antenna's design. The polarization of the receive antenna is typically defined as that of the incident field (the axial ratio, orientation, and handedness) that produces the highest power out of the antenna for a given power in the field; often the polarization of the receive antenna is determined by measuring the polarization of the field produced by the antenna when it is used to transmit. Receive antennas respond to specific polarizations; in some cases receive antennas are intended to respond only to linearly polarized fields with a specific orientation, while in other cases they are intended to respond to circularly polarized fields with a certain handedness. In practice, however, all receive antennas are elliptically polarized to a certain degree, and have some level of response to incident fields with both right-handed and left-handed ellipticity.

When the polarization of the receive antenna differs from that of the incident electromagnetic field, the antenna outputs less power; polarization loss quantifies the resulting reduction in received power due to this polarization mismatch. Polarization loss depends on the incident field's axial ratio, the receive antenna's axial ratio, and the orientation of the polarization ellipse of the incident field relative to that of the receive antenna. For given axial ratios, the polarization loss varies with relative orientation of the ellipses; maximum polarization loss occurs when major axes are orthogonal, and minimum loss occurs when major axes are aligned. Maximum polarization loss, Λ_{max}, and minimum polarization loss, Λ_{min}, are given by [22, 23]

$$\Lambda_{max} = \frac{\left(1+\alpha_r^2\right)\left(1+\alpha_i^2\right)}{\left(\alpha_r+\alpha_i\right)^2} \qquad \Lambda_{min} = \frac{\left(1+\alpha_r^2\right)\left(1+\alpha_i^2\right)}{\left(1+\alpha_r\alpha_i\right)^2} \qquad (3.47)$$

where α_i is the axial ratio of the incident field and α_r is the axial ratio of the receive antenna. Polarization loss is expressed in decibels as $10\log_{10}(\Lambda)$.

For a linearly polarized incident wave and linearly polarized receive antenna, the polarization loss is

$$\Lambda = \cos^2 \text{(Angle between major axes of the polarization ellipses)} \qquad (3.48)$$

Figure 3.18 shows examples of polarization losses over a range of incident wave axial ratios for different receive antenna axial ratios. When the incident field is perfectly circularly polarized (axial ratio of 0 dB), the polarization loss for a linearly polarized receive antenna is 3 dB, and for a perfectly circularly polarized antenna is 0 dB. Both of these values are independent of relative orientation for a perfectly circularly polarized incident wave. Polarization losses remain less than approximately 1 dB for incident field axial ratios up to 3 dB and receive antenna axial ratios as great as 6 dB. For a circularly polarized receive antenna, the minimum and maximum losses are the same, as seen from (3.47) with $\alpha_r = 1$.

It would be impractical to use linear polarization for satnav, since the linearly polarized receive antenna would need to be aligned properly with the linearly polarized fields received from multiple satellites, in order to avoid what could be very large polarization losses. Faraday rotation[2] in the ionosphere would remove any certainty

---

[2] As discussed in Section 3.2.2, the ionosphere consists of a plasma containing free electrons that, in combination with the Earth's magnetic field, cause Faraday rotation, or rotation of the field's polarization ellipse as the electromagnetic wave propagates through the ionosphere. The amount of rotation varies with location and time and is not readily predicted.

SATNAV SIGNAL CHARACTERISTICS

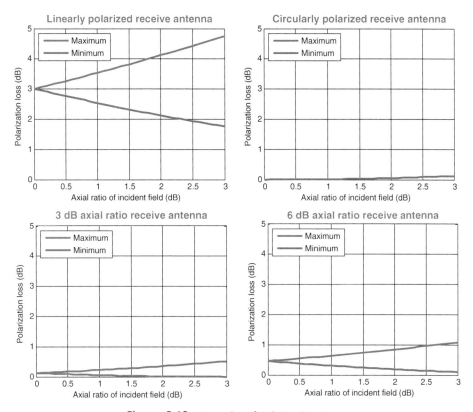

Figure 3.18. Examples of Polarization Losses

concerning the incident signal's polarization orientation at the receiver. Using linearly polarized signals and circularly polarized receive antennas, or vice versa, would impose polarization losses of 3 dB—still a considerable loss. The solution is to use circularly polarized signals, providing receivers the opportunity to use circularly polarized receive antennas and undergo no polarization loss.

All existing and currently planned satnav systems use RHCP. LHCP could have just as readily been adopted, but once GPS and GLONASS selected RHCP, later systems also selected RHCP to enhance interoperability.

In practice, satnav transmit antennas are not perfectly circularly polarized, and most ICDs indicate a worst-case axial ratio for the transmitted field. Receive antennas are often designed to be circularly polarized, although the resulting axial ratio may be less than unity, producing some degree of polarization loss. Some receiver designs intentionally use linearly polarized receive antennas because of cost or size constraints, accepting the resulting loss. In either case, (3.47) indicates how to calculate polarization losses, and this topic is further addressed with respect to link budgets in Section 4.2 and with respect to receiver design in Section 14.2.

3.3.4 Multiple Access

As discussed in Section 1.1, satnav inherently involves the reception of multiple signals, so multiple access techniques are essential. The emphasis on simple receivers, combined with limited spectrum available for satnav signals, encourages use of common frequency bands for a given type of signals from multiple satellites in a single system, for different types of signals (e.g., a civil signal and a military signal) in a single system, as well as for different signals from different satnav systems.

GPS was designed to use code division multiple access (CDMA), where all satellites transmit signals continuously at the same carrier frequency, but using DSSS with different spreading codes. CDMA permits a receiver to extract individual signals from the aggregation of signals with only modest (often insignificant) interference from the other signals, as long as all signals have comparable received power levels.

GLONASS, however, originally eschewed CDMA. Its original signals use frequency division multiple access (FDMA) where each satellite transmits signals at a different carrier frequency. With FDMA, the receiver needs to "tune" to a different center frequency to extract the desired signal, and all signals can use the same spreading code.

FDMA offers frequency diversity so that narrowband or partial band interference does not affect the reception of all signals. FDMA also significantly reduces MAI, and simplifies the receiver processing. However, if different signals in the FDMA scheme are to occupy distinct frequency bands, then each signal must have a narrow bandwidth (which is disadvantageous from a performance perspective as shown in Chapters 19 and 22), or the total occupied bandwidth must be very wide, which is inefficient in use of valuable spectrum. In addition, FDMA receivers must have wider front-end bandwidths for a given signal bandwidth, introducing a number of disadvantages discussed in Chapter 14. In contrast, CDMA provides numerous advantages including simpler receiver front ends, more flexibility in adding signals, the potential for higher accuracy since all signals experience common distortions in receiver hardware, and more efficient use of spectrum with wider bandwidth signals. Consequently, CDMA has become recognized as superior for satnav, and is being adopted for all modernized and new satnav systems, including modernized GLONASS signals.

However, as more signals fill the frequency bands, it is becoming increasingly obvious that signal designs must go beyond merely choosing between CDMA and traditional FDMA. Even for signals transmitted at the same carrier frequency, different spreading modulations are being used to concentrate different signals' powers at different frequencies. In addition, time division multiplexing is also used in some cases for simpler or more efficient transmission of multiple signals on the same carrier frequency, or for transmitting two components of the same signal, as discussed later in this chapter. In addition, different systems use carrier frequencies in different parts of L band to provide different services, as well as in different parts of L band to estimate and mitigate ionospheric errors (see Sections 3.2, 6.2, and 20.2), to assist in removing carrier phase ambiguities (Chapter 23), to reduce MAI, and for frequency diversity. Further, different systems use different spreading modulations and different carrier frequencies to limit the aggregate interference and in some cases to limit the spectral overlap between one signal type and other signals. Thus, modern satcom systems use a combination of CDMA, FDMA, and, in some cases, time division multiple access (TDMA), to fit hundreds of satnav signals into the limited spectrum.

3.3.5 Spreading Modulation and Bandwidth

All satnav signals employ DSSS, with bandwidth much greater than that needed to support the data rate as discussed in Section 3.1.3. Spreading modulations define the shape of the PSD and, through duality based on Fourier transform relationships, the shape of the autocorrelation function. Section 3.1.4 describes types of spreading modulations used in satnav.

In general, wider bandwidth spreading modulations offer advantages in enabling receivers to track signals better in noise, interference, and multipath (Chapters 19 and 22). Signals with wider bandwidths also reduce MAI and are less susceptible to narrowband interference, as demonstrated in Chapter 5. However, the choice of spreading modulation has no effect on carrier tracking or on data message demodulation performance in AWGN as long as the bandwidth is wide enough to pass most of the signal power, since the performance of both these receiver functions in AWGN depends on the received signal power, not on the shape of the signal's PSD.

Although wider bandwidth spreading modulations and wider bandwidths offer theoretical advantages, they can introduce practical disadvantages. To pass wider bandwidths without distortion, receiver antennas and analog electronics must be of a higher quality. Wider bandwidths also expose the receiver to more in-band interference, and make it more difficult to for receivers to tolerate the effects of strong interference at frequencies nearby those used for satnav. For these wider bandwidths, digital processing requires faster sampling rates and a greater receiver processing capability. Very wide bandwidth signals can suffer distortion from the ionospheric dispersion discussed in Section 3.2. Finally, spreading modulations with wider bandwidths occupy more spectrum—a precious commodity.

As seen in Part II, new and modernized signals tend to have larger bandwidths, reflecting the fundamental benefits of wider bandwidths and the benefits of technology advances that mitigate the practical disadvantages. Nonetheless, spectrum limitations and interference concerns are increasing, and implementation considerations remain. Consequently L band satnav spreading modulations are likely to continue having bandwidths that allow receivers to use from 1 MHz to a few tens of MHz of spectrum.

3.3.6 Spreading Codes

Spreading codes are also called pseudorandom noise (PRN) codes and ranging codes for GPS, and signature sequences or pseudo-noise (PN) codes for DSSS communication systems. A spreading code is a sequence of binary digits used by the transmitter to phase modulate spreading symbols in the DSSS signal. The DSSS receiver uses these codes to select and process the desired signal from the received aggregation of other signals, interference, and noise. Spreading code values can be represented either using the logical values $\{1, 0\}$ or the algebraic values $\{-1, +1\}$, as discussed in Section 3.1.3.

Open signals, provided for universal access, commonly use moderate-length (typically thousands of bits long) spreading codes that are transmitted repetitively. An essential part of the design process for an open signal using CDMA is defining a family of spreading codes, so that each signal of that type transmitted from a different satellite uses a unique spreading code from this family, with the family designed so the signals have good correlation properties.

Correlation properties are typically assessed in terms of autocorrelation and cross-correlation properties among members of the spreading code family, generally looking at both even and odd correlations. An even correlation evaluates the correlation between two repeated lengths of the spreading code and a replica of one spreading code length. An odd correlation evaluates the correlation between two repeated lengths of the spreading code, one inverted relative to the other, and a replica of one spreading code length.

Ideally, each code's even and odd autocorrelations should be a pair of Kronecker deltas separated by the code length, and all even and odd cross-correlations should be identically zero pairwise within the code family. Even in this case, there is MAI, as shown in Chapter 5. However, finite-length codes do not have these ideal correlation properties, and the distribution of sidelobe powers relative to the power in the autocorrelation peak indicates the performance of the code family. In some cases, the spreading modulation also affects the distribution of sidelobe powers, so it is wise to actually assess signal correlation performance (accounting for the spreading modulation, data bit modulation, overlay codes, and other aspects of the signal design) rather than merely the correlation performance of the spreading code family. Mathematicians tend to use the maximum relative sidelobe level of autocorrelation and cross-correlations, but these worst-case values can occur so rarely (e.g., less than one time in a billion) that they have little practical significance. It can be more useful to track higher probability correlation sidelobes, such as the 99.9999% point on their cumulative distribution.

In the terminology of digital communications, satnav uses asynchronous CDMA, where signals arrive at receivers with variable relative time alignment, rather than the nearly perfect time alignment that can occur when all the signals are transmitted from a single antenna. Similarly, satnav signals arrive at a receiver with many different Doppler shifts, due to motion of the different satellites. These factors make it more difficult to design and evaluate spreading code performance for satnav than it is for communications systems that use synchronous CDMA, where all signals are received time aligned and with the same Doppler shift. Rarely do correlation properties published for satnav account for Doppler shifts and all of the signal characteristics that influence the effect on receiver processing performance.

Spreading codes are described by different characteristics. The spreading code length describes the number of bits in the spreading code before it repeats. The spreading code type describes the mathematical structure used to define the codes. Examples of code types encountered in the satnav signals include Gold codes [24], maximal-length sequences (also known as m-sequences) [25], segments of m-sequences, Weil-based sequences [26], and memory codes, the latter being computer-generated codes that have no defined underlying mathematical structure. Spreading codes are also described by the hardware used to generate them: shift-register codes can be generated using digital shift registers, and memory codes must be read from memory in the transmitter and receiver.

Longer spreading codes tend to have better correlation performance, as do smaller families of spreading codes. If n is the length of the code and m is the number of different spreading codes in the family, the Welch bound [27] gives a lower bound on the standard deviation of the correlation sidelobes normalized by the autocorrelation peak:

$$\sigma_{\text{sidelobe}} \geq \frac{1}{\sqrt{n}} \sqrt{\frac{mn-n}{mn-1}} \cong \frac{1}{\sqrt{n}} \qquad (3.49)$$

where the approximation applies for large families of spreading codes (large m). While the Welch bounds indicates that normalized sidelobe values can decrease with the square root of the length of the spreading code, in practice normalized sidelobe values fall off more slowly than predicted by the Welch bound. A byproduct of lower spreading code sidelobes is also less susceptibility to narrowband interference. Another advantage of longer spreading codes is that they enable an assisted receiver to perform coarse-time navigation with larger initial time uncertainties, as described in Section 20.6.1.

Shorter spreading codes offer advantages for receiver processing, in particular they allow simpler or faster initial synchronization processing. The complexity of spreading code generation has been a consideration in the past, but it rarely is a significant factor in modern receivers. However, it is significantly easier to design shorter spreading codes.

Open signals' spreading codes typically use a family of codes (with signals of a given type using different spreading codes from the family) having lengths ranging from 1000 bits to 10,000 bits, although signals will be encountered in Part II with spreading code lengths both shorter and longer than this range. Depending upon the fidelity needed, three different models of spreading codes can be used to characterize the PSD and correlation function of the signal. The simplest model, often used to generate signal PSDs and correlation functions, called the ideal long spreading code, models the spreading code bits as a sequence of +1 or −1 values, each equally likely and with autocorrelation function given by the Kronecker delta function. This model is well matched to signals that actually employ very long or nonrepeating spreading codes, but does not adequately represent the fine structure of PSDs and correlation functions for signals using short or moderate-length spreading codes. A more complicated model providing higher fidelity representations of PSDs and correlation functions for signals using short or moderate-length spreading codes is called the ideal repeating spreading code. It models the spreading code bits as a sequence of +1 or −1 values, each equally likely and with autocorrelation function given by a sum of Kronecker delta functions spaced by the spreading code length. For a signal with a repeating spreading code, the most accurate representation of its PSD and correlation function is obtained by using the signal's actual spreading code bit sequence, although such an approach trades fidelity for loss of generality (numerical results must be calculated for each specific spreading code in the family) and computational complexity. These different models are introduced and employed in Reference 1.

The word "code" is overused in satnav. The original GPS signals are often called codes although this book refers to them as signals. This section describes spreading codes, while overlay codes are discussed in Section 3.3.11. Forward error control codes are used in new and modernized signals, while transmitter and receiver processing is increasingly implemented in computer code. Since five different uses of "code" is confusing at the least, throughout this book the meaning of "code" is made explicit with the appropriate descriptor.

3.3.7 Data Message Structure

The data message modulating each satnav signal contains information that receivers need to calculate position and time. Typically, for any signal using repeating spreading codes, the system time must be provided by the data message.[3] (If the spreading code never repeats, system time is uniquely related to a spreading code epoch.) Space vehicle time is typically reported relative to a specific time of origin, in terms of integer weeks since that origin, and then the integer number of messages since the most recent beginning of a week. For signals with repeating spreading codes, a receiver determines the time within a message by measuring the number of repetitions of the spreading code from the beginning of the current message.

Although precision clocks are used on satellites, their timing still drifts relative to system time. To avoid frequency adjustment of these clocks, they are allowed to drift, typically for months, before being realigned to system time. Consequently, clock corrections are provided so that the receiver can adjust satellite time to system time. Ephemeris information needed to compute the satellite location at a specific time is also provided in the data message, typically using secular and periodic orbital parameters described in Section 2.2. A constellation almanac is also included so the receiver can calculate coarse (nowhere near as accurately as with the ephemeris information) locations of other satellites in the constellation, in order to predict which other satellites might be visible or rising into visibility. The message also indicates whether the satellite is healthy and should be used; the almanac may also indicate the health status of all the satellites in the constellation.

Parameters are also provided to enable the receiver to perform various coordinate conversions, including relating system time to universal time or other time standards, and converting between ECEF and ECI coordinate systems. An approximate model of the ionosphere is provided for receivers that have no other way to remove ionospheric effects, although typically such an ionospheric model only reduces ionospheric errors by half, and even less for disturbed ionosphere conditions. Other system information, such as how to perform differential corrections between the timing of different satellites and the values of differential group delays between different signals transmitted by the satellite, are also included. The time offset between different satnav systems' times may also be provided.

The design of a signal's data message structure involves tradeoffs between accuracy of the information, latency, and robustness. Higher accuracy (of parameters describing time, ephemeris, almanac) requires more bits. While the essential information (health,

---

[3] Section 20.6.2 describes coarse-time navigation that can avoid reading time from the data message in some cases.

ephemeris, and clock corrections) is the most time sensitive since the receiver cannot calculate a fix without them, the latency of other messages can also be important. Another consideration is the ability to demodulate data messages from weak signals and in interference, since higher data rates mean less received signal energy in each bit and thus higher bit error rates. Higher data rates typically also constrain forward error control rates to provide less error correction and detection capability. Also, higher data rates often also involve shorter spreading codes having greater susceptibility to some types of interference.

Usually, the information in a data message changes slowly or predictably, so while latency is important to a receiver just beginning to track the signal, once the receiver has read the message from a given satellite, it does not need an update very often—tens of minutes to an hour or more may be adequate. The design quandary then is how often to repeat the different kinds of information.

Early GPS and GLONASS signal designs involved fixed message frame structures that locked in all contents and repetition of the data. More recent designs have added flexibility, with multiple message types defined to deliver different kinds of information, and room to define additional message types. In these more modern message designs, there typically is a commitment for the largest latency before repetition of essential information that a receiver needs to compute position and time, and more flexibility concerning other information.

Data messages can include overhead bits containing preambles or postambles (used by the receiver to find the beginning or end of a message block), other framing or synchronization bit patterns, indicators of the message type, and error checking codes appended to a message, as discussed subsequently. When data messages are discussed in this book the message size and message rate include not only the information, or payload, in the date message, but also these overhead bits.

Typically, signals use data rates of the order of 50 bps, except for signals designed to provide either low latency information or high volume information. Integrity information typically requires the lowest latency, while precise orbit information and clock corrections for a large number of satellites, and precise ionosphere information, are examples of high volume information. Such information can use data rates of many hundreds or thousands of bps. Latency of essential information typically ranges from 20 seconds to 60 seconds in most signals, while all information is typically repeated every 10–15 minutes.

3.3.8 Error Control Coding of the Data Message

New and modernized signals have employed increasingly capable techniques for correcting and detecting receiver errors in demodulating error messages, including error detection, forward error correction (FEC) encoding, and interleaving. Error detection involves redundant bits that are appended to the data message bits when the message is generated; the receiver can perform a computation on the recovered message bits and error checking bits that indicates whether the recovered message bits have random errors in them. A cyclic redundancy check (CRC) [28, 29] is used for error detection almost universally in new and modern satnav signals, due to its simplicity of operation and error detection performance. In satnav, CRC lengths are typically between 19 and 24 bits; both smaller and larger values are found in communications applications.

CRC calculations are based on finite-field polynomial division, and thus are simple to implement in digital hardware. Conceptually, the check bits are the remainder obtained by dividing the protected bits, represented as a polynomial in the binary Galois field, by a large enough prime number that the remainder has the desired length and degree of uniqueness. When the receiver has determined the message bits, it can either perform the same CRC calculation and compare the check bits to those read from the message, or it can divide the received message and the check bits, which should yield a remainder of zero. If the check bits match or the remainder is zero, the message is assumed correct; otherwise, it is assumed to be in error and the receiver must wait for the next message and try again.

In practice, many details must be defined, including how to protect any leading zeroes in the message, and the ordering of bits. These details are made explicit in each signal's ICD.

An n-bit CRC applied to a data block of arbitrary length will detect any single error burst not longer than n bits and will detect the fraction $1 - 2^{-n}$ of all longer random error bursts. For example, a 19 bit CRC provides a probability of an undetected random error burst of approximately 1.9×10^{-6}, or 1 in 524,288 messages. If each message lasts 30 seconds, an undetected random error burst occurs on average about every half year of continuous data message demodulation. The performance is even better for 24 bit CRCs. It is important to recognize that CRCs protect against random errors, but since errors that evade a CRC can be designed and introduced, CRCs provide limited protection against intentionally introduced errors.

FEC inserts much greater redundancy into the transmitter bit stream than does a CRC, allowing the receiver to correct some errors as well as detect some errors. FEC generates data message symbols from data message bits, producing an output symbol rate greater than the input bit rate. FEC employs standard techniques developed for digital communications, and the full underlying theory of FEC, design of specific techniques, and implementation of encoding and decoding are beyond the scope of this book and can be found in books dedicated to the subject such as References 30 and 31.

Block FEC, or block coding, is a batch process where a fixed number of message bits is transformed into a larger number of message symbols, represented by $(n_{\text{symbol}}, n_{\text{bit}})$ where n_{symbol} is the number of message symbols produced from a block of n_{bit} message bits. Convolutional FEC, or convolutional coding, is a process for performing FEC on a continuous stream of message symbols, producing message symbols at a code rate $k_{\text{code}} = n_{\text{symbol}}/n_{\text{bit}}$. The constraint length of the convolutional code is the number of information bits used to compute the encoder output, describing the memory length of the encoder.

Since satnav messages are not a continuous stream but instead are typically hundreds of bits long, convolutional FEC must be adjusted for the finite message length using techniques such as tail biting [32]. They are then conceptually equivalent to block codes with very large blocks, but convolutional FEC techniques are still used for encoding and decoding because of their implementation efficiency and performance.

Receivers use the redundancy introduced by FEC for both error correction and error detection, enabling the correct reading of the data message under signal and noise situations where an uncoded signal could not have been correctly read. Even when the message cannot be correctly read, the FEC can enable the receiver to recognize the existence of uncorrected errors, but not with the same capability as a CRC. Since there are more encoded symbols than message bits, the encoded symbols must be transmitted at a faster rate than the message bit rate, and thus each symbol has less energy than the message bit would have had when sent at a lower rate, energy being the integral of signal power over the duration of the symbol or bit. The redundancy in well-designed codes, however, more than compensates for the loss of energy per symbol, providing better error performance than with no FEC.

While lower rate error control codes (more data message symbols per data message bit) generally provide better error correction and detection performance, they have several disadvantages. The primary disadvantage of lower rate error control codes is they require a faster binary symbol rate to support the same message bit rate. Since an integer number of spreading code lengths (often one in modern signals) occurs in a message symbol duration, faster symbol rates yield either shorter message symbol durations and shorter spreading code lengths that limit the performance of the spreading code, or require faster spreading code rates, with the disadvantages noted in Section 3.3.6. Another disadvantage of lower error control code rates is more computationally complex decoding, an issue that becomes less important with more capable and less power-hungry processing circuitry. Finally, another disadvantage of lower rate error control codes and the resulting faster message symbol rates is the resulting need to use shorter correlation integration times in receiver processing, with corresponding reductions in performance for initial synchronization and tracking as discussed in Chapters 16, 18, and 19.

Even with the same error control code rate, the performance of block codes generally improves with longer blocks. Modern block codes, such as turbo codes [33] and low density parity check (LDPC) codes [34], can operate at very low signal-to-noise ratios, approaching the Shannon limit [35], when they are used on blocks of message bits that are thousands of bits long. However, the processing complexity of decoding such large blocks is still viewed as a concern, particularly for satnav receivers with cost and power constraints. Also, since the receiver cannot decode and use the bits until the entire block is received, using such long blocks would introduce generally unacceptable latencies. For example, a 5000 bit block transmitted at 50 bps requires 100 seconds, and if the receiver began receiving the message just after the beginning of a block, it might need to wait almost 200 seconds to receive an entire block and begin decoding it. The result would be more than 3 minutes for a user to obtain an initial position report—much longer than most users find desirable.

FEC works best when errors are uniformly distributed over the received symbols; the ability to correct or even reliably detect errors is diminished when many errors are concentrated in a small number of received symbols. Such conditions are known as burst errors, and can arise when a satnav receiver is exposed to a burst of noise or interference such as from a nearby emitter of radio frequency power, or when the signal fades or drops out as the receiver passes under a bridge or behind a tall building. Interleaving of

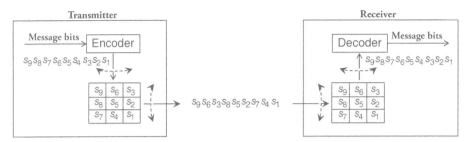

Figure 3.19. Example of Block Interleaving and Deinterleaving

the message symbols can be used to break up burst errors, distributing the errors over more symbols so that the receiver can better take advantage of the FEC in countering the effects of the burst errors. As an example of interleaving, suppose a simple FEC approach could correct any one error in three received symbols, and could detect but not correct the presence of two errors in three received symbols. If a burst error caused errors in all three adjacent symbols in a block, decoding in the receiver could neither correct nor detect the errors (although presumably a CRC would detect the errors, although it cannot correct them). Suppose instead of transmitting the symbol stream in sequence: $s_1, s_2, s_3, \ldots$ the transmitter uses a block interleaver as illustrated in Figure 3.19. After FEC, the symbols are written into columns of a matrix in memory. The symbols are then read out row by row, and transmitted in that order, as shown in the figure. The receiver assembles the symbols in a similar matrix, reading in by row and reading out by column, to form the original sequence of symbols for decoding. If a burst error causes errors in three adjacent symbols passing from transmitter to receiver, deinterleaving in the receiver causes the errors to be distributed over three different symbol triplets, with only one error in each triplet. This hypothetical FEC decoder can correct one symbol error in each three symbols, providing error-free message bits in spite of the burst error.

Not all signals use interleaving for a variety of reasons. If there is no FEC, as in the original GPS signals, interleaving provides no benefit. Some types of FEC are already optimized for burst errors, so in those cases interleaving only provides benefits if the interleaving works across separately encoded parts of the message. Also, interleaving typically adds latency. In the example shown in Figure 3.19, seven symbols must be received before the first triplet can be decoded, more than doubling the minimum latency. (Actual interleaver designs in satnav typically add much greater latency than in this simplified example.)

The original GPS and GLONASS signals use minimal FEC, making data demodulation fragile unless the received signal power is large. In addition, error checking is not robust, leading to significant probabilities of undetected errors in demodulated data. Some receivers of these early signal designs read and cross-check demodulated data from sequential messages to reduce the probability of undetected errors. As seen in Part II, convolutional encoding is commonly used in many modernized and new signals, with some newer signal designs employing even more sophisticated and capable FEC approaches.

3.3.9 Data Message Modulation

In most signals, the data message symbols are modulated using biphase shift keying, defined in (3.14). In the original GPS and GLONASS signals, message data bits biphase modulate the entire signal, introducing phase shifts unknown to the receiver that must be accommodated in receiver processing.

As discussed in the next section, many modern signals only modulate part of the signal with data, leaving the remaining "pilot" component without data modulation and thus perfectly known to a synchronized receiver. Biphase modulation is used rather than higher-order phase modulation to keep the signal's complex envelope real valued for simplicity in signal transmission and reception, to retain antipodal signaling for better bit error rate performance, and to permit more effective carrier tracking by the receiver.

The only currently known satnav signal that does not use biphase data modulation is the QZSS L6 signal, described in Section 12.3, which uses code shift keying for its very high data rate of 1000 bps. Code shift keying, described in Reference 36, uses approximately orthogonal symbols formed from different cyclically shifted versions of a finite-length spreading code. The receiver demodulates by estimating the most likely cyclic shift. Since a higher-order alphabet can be used, the data rate can be increased without increasing the symbol rate; longer symbol durations enable better signal acquisition and tracking performance as shown in Part III.

3.3.10 Pilot and Data Components

As defined in this book, a signal may consist of more than one component having the same carrier frequency, with each component providing a complementary function. This terminology is not consistent, and Galileo literature, for example, often calls different components different signals, even though pairs of components are used in complementary fashion as described below.

As introduced in Section 3.3.9, many modern signals use two different signal components, one designed for best tracking of the carrier and code, and one to provide broadcast data. As seen in Part III, receivers can obtain the best performance when they are tracking a signal with no unknown phase discontinuities caused by biphase data message modulation. If the waveform has no modulation unknown a priori to a synchronized receiver, it is called a pilot component, while if a waveform is modulated by data message symbols unknown a priori to the receiver, this waveform is called a data component.

The signal best supports tracking when all its power is applied to the pilot component, and it best supports data demodulation when all the power is applied to the data component, as long as the signal phase can still be tracked reliably to allow measurement of the data message's phase modulation.

The original GPS and GLONASS signals are entirely data components, and there are no known current or planned signal designs that are entirely pilot components (since the receiver would need to rely on other means to obtain satellite time, ephemeris, clock corrections, and other information currently provided in data messages). Most new and modernized signals are a blend of data component and pilot component, with different power ratios and different approaches for blending the two components. The increasing practicality of decoding FEC in receivers is a primary enabler of separate pilot and data components, since the FEC can recover most or all the bit error rate performance lost by diverting some signal power to the pilot component.

Another enabler for use of multiple components is the development of theory and satellite hardware for combining more signals and signal components onto the same carrier while preserving efficient amplification and transmission from the satellite. Section 3.3.12 discusses this topic in more detail.

3.3.11 Overlay Codes

For an open signal using repeating spreading codes, the pilot component, as outlined above, would simply consist of a repeated spreading time series modulating the carrier. If the code's duration were of the order of milliseconds or tens of milliseconds, there would be several disadvantages. One is time ambiguity, since when the receiver is synchronized to the pilot spreading code, time is ambiguous by multiples of the spreading code duration. Another more fundamental disadvantage would be that the PSD of the transmitted signal, and of a replica generated in the receiver and lasting more than one code period, would have strong spectral lines spaced at the reciprocal of the period [1], and these lines cause increased MAI and increased receiver susceptibility to narrowband interference.

In order to mitigate these problems, most pilot components having short- or medium-length spreading codes are biphase modulated by a sequence of overlay code bits, with the duration of each overlay code bit matched to the spreading code duration. The overlay code bits are published in the signal's ICD (and thus known to the receiver) and repeat, typically with a length of between 10 bits to many hundreds of bits.

In effect, the overlay code bits extend the length of the spreading code; the length of the overlaid spreading code is the length of the spreading code times the length of the overlay code. The resulting spreading time series only repeats at this longer duration. While the result is an effectively longer spreading code, many practical advantages of shorter spreading codes are maintained including easier design of the spreading code and faster receiver search during initial synchronization. Different terminology is used for different signals, and overlay codes are also known as secondary codes or synchronization codes.

Figure 3.20 provides a conceptual view of original signal designs that are entirely a data component, and modern signal designs with both a data component and a pilot component. Here, the logical representation of binary values and binary operations is employed for compactness. The original design only has a data component, with an integer number of spreading code durations in the data message bit duration. All the entire spreading code lengths within a data message bit are inverted when the data message bit value is 1, and not inverted when the data bit value is 0. Typically, the receiver does not know the data message bit values a priori, so the receiver does not know whether there will be a phase transition between adjacent repeats of the spreading code at transitions between data message bits.

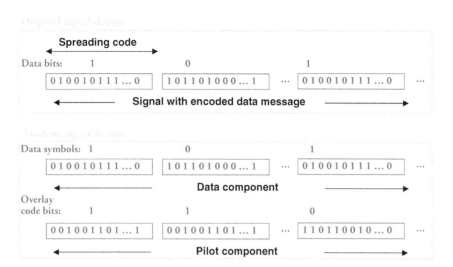

Figure 3.20. Depiction of Signal with Data Component Only, and Signal with Both Data and Pilot Components

In modern signal designs, the data component is the same as that in the original signal design, except the terminology changes to data message symbols rather than bits, reflecting the common use of error control coding. Each data message symbol is modulated onto the data component—typically with one data symbol per spreading code length (although in some cases the data symbol modulates multiple periods of a spreading code extended by a data overlay code whose length matches the data symbol).

The pilot component uses a spreading code from the pilot component spreading code family. Typically the pilot spreading codes have the same structure and length as those in the data component spreading code family, but they are designed to be mutually orthogonal. In most cases, the spreading code pilot component is biphase modulated by a pilot overlay code waveform whose bit transitions are often synchronized with the data message symbols. The pilot overlay code waveform repeats in a time period that varies from tens of milliseconds to seconds.

There is an essential but subtle difference between the data component and pilot component in the modern signal depicted in Figure 3.20: a synchronized receiver knows when phase transitions will occur in the pilot component, and does not know a priori when phase transitions will occur in the data component.[4] As described in Chapter 18, the receiver can then remove, or in satnav terminology "wipe" the overlay code from the pilot component, enabling more robust receiver processing. Further, the receiver can use the overlay code bits to resolve synchronization ambiguities associated with multiple repetitions of the underlying spreading code, achieving faster and more reliable

---

[4] In some situations such as assisted satnav described in Chapter 24, the receiver may know the values of the data message bits or symbols a priori. For example, the bit values could be provided by a separate communications link. In that case, the receiver can treat a data component like a pilot component.

synchronization to data message bits or even the data message itself, as discussed in Section 16.5.

3.3.12 Transmitter Combining of Multiple Signals or Components

All satnav systems transmit multiple signals and signal components. Some are transmitted at different carrier frequencies, but usually several are transmitted at the same carrier frequency. For example, pilot and data components of the same signal should be transmitted at the same frequency, and different signals designed to be used together, like GPS C/A code signal and GPS P(Y) code signal, are best transmitted at the same frequency to minimize radio frequency circuitry on the satellite and in receivers. Since satellite complexity, and to some degree receiver complexity, increase with the number of different carrier frequencies transmitted, it if often desirable to transmit multiple signals and components not only at the same carrier frequency, but physically modulated onto the same carrier waveform. When signals or components are transmitted on the same physical carrier from the same RF chain on the satellite, their phase relationship at the receiver can be fixed and known in advance, allowing receivers to obtain additional performance benefits by processing them coherently.

There is a strong motivation for using power amplifiers operating in saturated mode on satnav satellites, since such nonlinear power amplifiers are significantly more efficient. Constant-envelope signals must be transmitted for saturated mode operation. Most of the spreading modulations discussed in this chapter and Part II of this book are constant envelope, so transmitting any one of them would enable use of power amplifiers in saturation.

It becomes more challenging, however, to maintain a constant-envelope composite signal with increasing number of constituent biphase signals and signal components modulated onto a single carrier. Here, the term "composite signal" is used to describe the combination of more than one constituent biphase signals or signal components modulated onto the same carrier.

First-generation GPS and GLONASS satellites each transmitted two constituent signals (C/A code signal and P(Y) code signal for GPS, and what were called standard accuracy and high accuracy signals on GLONASS) having uncorrelated spreading codes on the same carrier in upper L band. Each constituent used BPSK-R spreading modulation, so was individually constant envelope. Phase quadrature combining was used, with one constituent placed in-phase on the carrier, and the other placed in phase quadrature. As confirmed by Review Question QT3.3, placing constant-envelope constituents on the in-phase and quadrature phase of a carrier yields a constant-envelope composite signal. It is not so widely recognized that the composite signal is constant envelope even if the amplitudes of the two constituents differ, but that is the case as well, demonstrated in Review Question QT3.3.

Time division multiplexing is another straightforward approach for combining two constituents. Like phase quadrature combining and unlike other approaches described subsequently, time division multiplexing incurs no combining loss, in that all the transmitted power is available for use by a receiver of one of the constituents. Time division multiplexing has been used for combining two components of the same signal, each having the same spreading code chip rate. For example, odd numbered spreading symbols can be dedicated to one component and even numbered components dedicated to the other, in order to provide both components at equal power. The multiplexing pattern is deterministic and hence accounted for in receiver processing, which distinguishes between the two components by their timing once it is synchronized to the signal. Time division multiplexing has the disadvantage of requiring a spreading code with half the length of the length that would be used if each component were transmitted continuously with the same spreading code duration. The result of the shorter spreading code length is poorer correlation sidelobes, as discussed in Section 3.3.6. The advantage, however, is that if both constituent signals or components are biphase, then the composite signal is also biphase, preserving the quadrature phase for another signal or component. Thus, two signals or components can be time division multiplexed on one carrier phase, and two additional signals or components could be time division multiplexed in phase quadrature, allowing four signals to be transmitted efficiently.

Variations on time division multiplexing have been discussed, including using segments that are longer than one spreading symbol (for example, to provide different received power levels for different constituent signals), and using time division multiplexing to add modernized signals to legacy signals. Ultimately, other combining approaches have been selected for these applications, at least in part because of concerns about effects of time division multiplexing on legacy signal spreading code performance and interoperability with existing receivers.

Two constituent signals could be linearly combined, or added algebraically. The resulting composite signal would not be constant envelope, however. Even with only two constituents, a three-level composite signal would result if the two constituents are to be received with equal power, and a four-level composite signal if the two constituents are to be received with different powers. Even more levels would result with three or more constituents and varying received power levels.

Multiplexing efficiency is the fraction of received composite signal power useful to receivers of the different constituents. For a composite signal $x_c(t)$ having total power[5] $C_c = E\{|x_c(t)|^2\}$, define the received power of the kth biphase constituent as $C_k = [E\{x_c(t)x_k(t)\}]^2$. For K constituents, the multiplexing efficiency is then $\sum_{k=1}^{K} C_k/C_c$. The efficiencies of phase multiplexing and time division multiplexing are unity, as shown in Review Question QT3.11.

Nonlinear combining is needed to produce a constant-envelope composite signal from three or more constituents when time division multiplexing is not used. Majority voting [37] is an approach developed originally for satellite communications, producing a composite signal that also takes on values of ± 1 when the inputs are three (or any odd number greater than unity) biphase constituents taking on values of ± 1. For the simplest case of three orthogonal constituents $x_1(t), x_2(t), x_3(t)$ each to be received at equal power, the composite signal $x_c(t)$ takes on the value assumed by two or more of

---

[5] To those unfamiliar with satnav notation, it may seem strange to use the variable C to denote signal power. This notation apparently originated from the concept of "carrier power"

the constituents. Majority voting for three constituents to be received at equal power can be written algebraically as

$$x_c(t) = \frac{1}{2}[x_1(t) + x_2(t) + x_3(t) - x_1(t)x_2(t)x_3(t)] \quad (3.50)$$

The intermodulation component $x_1(t)x_2(t)x_3(t)$, transmitted such that the composite signal is constant envelope, provides no useful power to any receiver of a single constituent, so the multiplexing efficiency of (3.50) is 0.75, as shown in Review Question QT3.12. Intuitively, the composite signal takes on the same value as of the constituent signals three-quarters of the time. The remaining quarter of the time, the composite signal has the opposite sign from the constituent. Correlation processing in the receiver subtracts the one-fourth portion having opposite sign from three-fourths having the correct sign, leaving half the value that would be obtained if the entire signal were dedicated to that constituent.

Review Question QT3.13 shows how to generalize majority voting so that constituents can be received at different power levels.

While majority voting produces a constant-envelope biphase signal from orthogonal biphase unit-power constituents, interplexing [38] produces a constant-envelope quadraphase unit-power composite signal $x_c(t)$ using the expression

$$x_c(t) = \left[[\alpha_1 x_1(t) + \alpha_2 x_2(t)] + i \left[\alpha_3 x_3(t) - \frac{\alpha_1 \alpha_2}{\alpha_3} x_1(t)x_2(t)x_3(t) \right] \right] / \alpha_n \quad (3.51)$$

where again the constituents are typically orthogonal, α_n is selected to normalize the power in the composite signal, and the useful power delivered to the receiver of the kth constituent is $(\alpha_k / \alpha_n)^2$.

Review Question QT3.14 shows $x_c(t)$ in (3.51) is constant envelope, $\alpha_n = \sqrt{\alpha_1^2 \alpha_3^2 + \alpha_2^2 \alpha_3^2 + \alpha_1^2 \alpha_2^2 + \alpha_3^4}/\alpha_3$, the efficiency is $(\alpha_1^2 + \alpha_2^2 + \alpha_3^2)/\alpha_n^2$, and the power in the intermodulation component (which is not useful to any receiver of any single constituent) is $(\alpha_1 \alpha_2 / \alpha_3)^2$. The most efficient implementation occurs when the constituent intended to have the largest useful power delivered to its receiver is placed in phase quadrature to the other two constituents and in-phase with the intermodulation component, since that design makes the power of the intermodulation component smallest in (3.51).

Multiplexing approaches continue to advance. New hardware capabilities for generating large numbers of different carrier phases, rather than merely four carrier phase values ($0, \pi/2, \pi, 3\pi/2$) has led to more efficient multiplexing approaches [39] that still produce constant-envelope composite signals.

3.4 SATNAV SIGNAL STRUCTURE

Building upon the earlier sections in this chapter, this section defines the terminology and mathematical models for signals used in the remainder of this book. Section 3.4.1

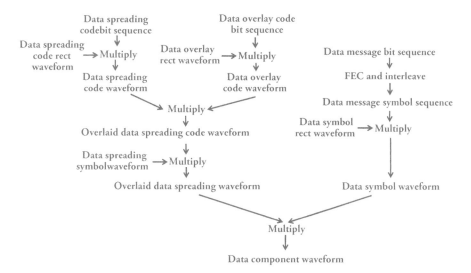

Figure 3.21. Data Component Waveform Construction

defines the data component, Section 3.4.2 defines the pilot component, and Section 3.4.3 formulates a complete satnav signal. The algebraic representation of bit values is used here.

3.4.1 Data Component Structure

Figure 3.21 shows the construction of a data component waveform at baseband. The data message bit sequence includes not only the information-bearing bits, but also other bits used for synchronization and framing, identifying message types, and error detection. The data message bit sequence (including bits for framing and error detection) is forward error encoded and interleaved, forming the data message symbol sequence $\cdots, d_{-3}, d_{-2}, d_{-1}, d_0, d_1, d_2, d_3, \cdots$ where $d_k \in \{+1, -1\}$. (If there is no forward error encoding or interleaving, the data message symbol sequence is the data message bit sequence.) Since all but one currently defined satnav signal uses biphase shift keying to modulate 2-ary symbols onto the data component, the mathematical model defined here applies for biphase shift keying. The data symbol rect sequence is $\sum_{k=-\infty}^{\infty} \text{rect}_{T_d}(t - kT_d)$ and the resulting data symbol waveform is $\beta(t) = \sum_{k=-\infty}^{\infty} d_k \text{rect}_{T_d}(t - kT_d)$, where the data symbol duration T_d is the reciprocal of the data message symbol rate $f_d = 1/T_d$.

The data spreading code bit sequence is $\cdots, b_{-3}, b_{-2}, b_{-1}, b_0, b_1, b_2, b_3, \cdots$ with $b_n \in \{+1, -1\}$. If this spreading code repeats, it has data spreading code length N_b with $b_n = b_{n+\ell N_b}$ where n and ℓ are any integers. For data spreading symbol $g_d(t)$, the data spreading symbol sequence is $\sum_{n=-\infty}^{\infty} g_d(t - nT_c)$ and the data spreading waveform (pre-overlay) is $\sum_{n=-\infty}^{\infty} b_n g_d(t - nT_c)$, with T_c the reciprocal of the data spreading code bit rate, often called the data chip rate. It is common for signals to be designed with data chip rate an integer multiple of the data message symbol rate, so $f_c/f_d = T_d/T_c$ is an integer.

For generality, let there be a data overlay code to extend the duration of the data spreading code to the data symbol duration. The data overlay code bit sequence is $\cdots, p_{-3}, p_{-2}, p_{-1}, p_0, p_1, p_2, p_3, \cdots$ where either $p_m \in \{+1, -1\}$ if there is an overlay code, or $p_m \equiv +1$ if there is no overlay code. A data overlay code bit sequence repeats with a data overlay code length N_p, where $p_m = p_{m+\ell N_p}$ and m and ℓ are any integers. The data overlay rect sequence is $\sum_{m=-\infty}^{\infty} \text{rect}_{T_p}(t - mT_p)$ with data overlay bit duration T_p, the reciprocal of the data overlay code bit rate $f_p = 1/T_p$. The data overlay code waveform is $\sum_{m=-\infty}^{\infty} p_m \text{rect}_{T_p}(t - mT_p)$.

The data spreading waveform is then the product of the data spreading waveform (pre-overlay) and the data overlay code waveform: $\sigma_d(t) = \left[\sum_{n=-\infty}^{\infty} b_n g_d(t - nT_c)\right]\left[\sum_{m=-\infty}^{\infty} p_m \text{rect}_{T_p}(t - mT_p)\right]$.

Observe that when a receiver is synchronized to the data component, it knows when values of the data symbol waveform may change, but does not know the values a priori. Conversely, the synchronized receiver does know when both the data spreading waveform and the data overlay code waveform change, and knows their values.

The data component waveform is the product of the data spreading waveform and the data symbol waveform:

$$\gamma_d(t) = \sigma_d(t)\beta(t)$$

$$= \left[\sum_{n=-\infty}^{\infty} b_n g_d(t - nT_c)\right]\left[\sum_{m=-\infty}^{\infty} p_m \text{rect}_{T_p}(t - mT_p)\right]\left[\sum_{k=-\infty}^{\infty} d_k \text{rect}_{T_d}(t - kT_d)\right]$$

$$= \sum_{n=-\infty}^{\infty} b_n \left[\sum_{m=-\infty}^{\infty} p_m \text{rect}_{T_p}(t - mT_p)\right]\left[\sum_{k=-\infty}^{\infty} d_k \text{rect}_{T_d}(t - kT_d)\right] g_d(t - nT_c)$$

$$= \sum_{n=-\infty}^{\infty} \zeta_n g_d(t - nT_c) \tag{3.52}$$

The latter expression is used when there is no need to distinguish between the spreading code bits, the spreading code bits modulated by overlay code bits, and the spreading code bits modulated by data message symbols. In that case the data spreading sequence $\cdots, \zeta_{-3}, \zeta_{-2}, \zeta_{-1}, \zeta_0, \zeta_1, \zeta_2, \zeta_3, \cdots$ is defined as the data spreading code bits modulated by the data overlay code and data message symbols.

Figure 3.22 shows the same construction of the data component waveform as Figure 3.21, but using mathematical symbols rather than descriptive terms.

SATNAV SIGNAL STRUCTURE

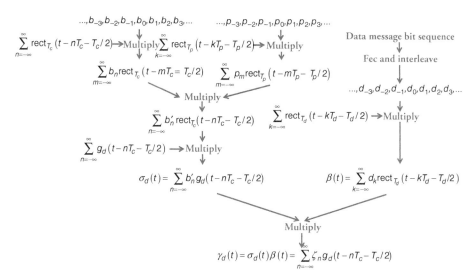

Figure 3.22. Data Component Waveform Construction Mathematical Description

When the data spreading sequence is modeled as an ideal long spreading code, the PSD of the data component waveform is

$$\Phi_{\gamma_d}(f) = |G_d(f)|^2 \qquad (3.53)$$

3.4.2 Pilot Component Structure

Figure 3.23 shows the construction of a pilot component waveform at baseband. The pilot overlay code bit sequence is $\cdots, q_{-3}, q_{-2}, q_{-1}, q_0, q_1, q_2, q_3, \cdots$ where either $q_m \in \{+1, -1\}$ if there is a pilot overlay code, or $q_m \equiv +1$ if there is no pilot overlay code.

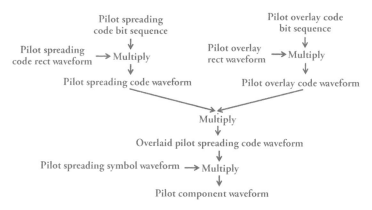

Figure 3.23. Pilot Component Waveform Construction

The pilot overlay code repeats with code length N_q with $q_m = q_{m+\ell N_q}$ where m and ℓ are any integers. The pilot overlay rect sequence is $\sum_{m=-\infty}^{\infty} \text{rect}_{T_q}(t - mT_q)$, and pilot overlay code waveform is $\alpha(t) = \sum_{m=-\infty}^{\infty} q_m \text{rect}_{T_q}(t - mT_q)$, where the overlay bit duration T_q is the reciprocal of the overlay code bit rate.

The pilot spreading code bit sequence is $\cdots, a_{-3}, a_{-2}, a_{-1}, a_0, a_1, a_2, a_3, \cdots$ with $a_k \in \{+1, -1\}$. If this spreading code repeats, it has pilot spreading code length N_a with $a_k = a_{k+\ell N_a}$ where k and ℓ are any integers. For pilot spreading symbol $g_p(t)$, the pilot spreading symbol sequence is $\sum_{n=-\infty}^{\infty} g_p(t - nT_c)$, with T_c the reciprocal of the pilot chip rate, which is typically the same as data chip rate. The pilot spreading waveform (pre-overlay) is $\sigma_p(t) = \sum_{n=-\infty}^{\infty} a_n g_p(t - nT_c)$, with T_c the reciprocal of the pilot chip rate, which is typically the same as data chip rate. It is common for signals to be designed with pilot chip rate an integer multiple of the overlay code rate, so T_o/T_c is an integer.

Observe that when a receiver is synchronized to the pilot component, it knows when values of the pilot spreading waveform change, due both to overlay code bit changes and to pilot spreading code changes.

The pilot component waveform is the product of the pilot spreading waveform and the pilot symbol waveform:

$$\gamma_p(t) = \sigma_p(t)\alpha(t) = \left[\sum_{n=-\infty}^{\infty} a_n g_p(t - nT_c)\right]\left[\sum_{m=-\infty}^{\infty} q_m \text{rect}_{T_q}(t - mT_q)\right]$$

$$= \sum_{n=-\infty}^{\infty} a_n \left[\sum_{m=-\infty}^{\infty} q_m \text{rect}_{T_q}(t - mT_q)\right] g_p(t - nT_c) = \sum_{n=-\infty}^{\infty} c_n g_p(t - nT_c)$$ (3.54)

The latter expression is used when there is no need to distinguish between the spreading code bits and the spreading code bits modulated by overlay code bits. In that case the pilot spreading sequence $\cdots, c_{-3}, c_{-2}, c_{-1}, c_0, c_1, c_2, c_3, \cdots$ is defined as the pilot spreading code bits modulated by the pilot overlay code, with length $N_q N_a$.

Figure 3.24 shows the same construction of the pilot component waveform as Figure 3.23, but using mathematical symbols rather than descriptive terms.

When the pilot spreading sequence is modeled as an ideal long spreading code, the PSD of the pilot component waveform is

$$\Phi_{\gamma_p}(f) = |G_p(f)|^2$$ (3.55)

3.4.3 Complete Satnav Signal

The analytic signal representation of the resulting satnav signal is obtained by upconverting the complex envelopes of the data component waveform and the pilot component

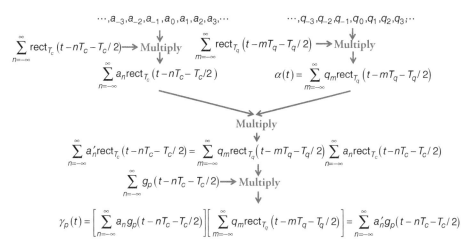

Figure 3.24. Pilot Component Waveform Construction Mathematical Description

waveform developed in the previous two sections, to the carrier frequency f_0. For phase quadrature multiplexing or code division multiplexing of the two components, the resulting satnav signal is

$$\tilde{s}(t) = \left[\gamma_p(t) e^{i\theta_p} + \gamma_d(t) e^{i\theta_d} \right] \exp\left[i 2\pi f_0 t \right] \quad (3.56)$$

The pilot component and data component phase shifts θ_p and θ_d are the same (or differ by π) for code division multiplexing, while they differ by $\pi/2$ for phase quadrature multiplexing.

In current and planned signals using time multiplexing for combining two components, spreading symbols are interleaved one after the other. A useful mathematical model for this situation is to define the pilot and data component spreading symbols so each takes on a value of zero half the time:

$$g_{p,tmux}(t) = \begin{cases} 0, & 0 \le t < T_c/2 \\ g_p(t), & T_c/2 \le t < T_c \end{cases}$$

$$g_{d,tmux}(t) = \begin{cases} g_d(t), & 0 \le t < T_c/2 \\ 0, & T_c/2 \le t < T_c \end{cases} \quad (3.57)$$

Different signals use different conventions as to whether the pilot spreading symbol or the data spreading symbol initially takes on a value of zero, so the definitions of time-multiplexed spreading symbols in (3.57) can be reversed. Here the chip rate $1/T_c$ is the chip rate for each component separately, with the overall chip rate for the signal twice that for each component separately.

3.5 SUMMARY

This chapter provides an extensive amount of information concerning satnav signal characteristics and designs. Basic signal representations, along with definitions of keyed signals, biphase data modulation, and DSSS signals establish the needed foundation. Specific spreading modulations used for satnav are defined, with expressions for their PSDs. Based on the generic signal representations, the effects of Doppler shift and ionospheric dispersion on received signals are introduced, with mathematical models provided. These effects will be revisited in Part III, since they must be accounted for in receiver processing. Twelve fundamental characteristics of satnav signals are introduced, and each characteristic is described individually to provide insight into considerations affecting the design and implementation of satnav signals. In most cases, the discussion is qualitative, but some characteristics, such as polarization, are also described quantitatively. Finally, detailed mathematical models of satnav signals are developed, accounting for the different signal characteristics and establishing a unifying set of terminology and notation that are employed throughout the remainder of the book. The mathematical models in Section 3.4, combined with the satnav spreading modulations in Section 3.1.3, provide the basis for describing specific satnav system's signals in Part II as well as the description of receiver processing and assessment of receiver processing performance in Part III.

REVIEW QUESTIONS

Theoretical Questions

QT3.1. Express the real and imaginary parts of the analytic signal in terms of the in-phase and quadrature components and the carrier frequency, and express the amplitude envelope in terms of the in-phase and quadrature components.

QT3.2. Find the amplitude envelope of a biphase keyed signal and quadraphase keyed signal in terms of the spreading symbols.

QT3.3. Show that a signal whose complex envelope is of the form $x(t) = Ay(t) + iBz(t)$, with $y(t)$ and $z(t)$ each real valued and unit power, is constant envelope even if $A \neq B$. (This contradicts the common misperception that a quadraphase signal is constant envelope only if there is the same power in the in-phase and quadraphase components.)

QT3.4. Using the result from QT3.2, show that an infinite-bandwidth signal using BPSK-R, BOC_s, BOC_c, BCS, or TMBOC spreading modulation is constant envelope. Show that a signal using CBOC is not constant envelope, but a signal that multiplexes equal power CBOC+ and CBOC− components in phase quadrature is constant envelope even though neither component is individually constant envelope. Show that the sum and difference of equal power CBOC+ and CBOC− is not constant envelope unless $\alpha = 1/2$.

QT3.5. The peak-to-average power ratio (PAPR) of a waveform $x(t)$ is defined as $\max_t \{|x(t)|^2\} / \left[\lim_{T\to\infty} \int_{-T/2}^{T/2} |x(t)|^2 \, dt \right]$ where the denominator is the RMS value of the waveform. PAPR is typically expressed in decibels as $10 \log_{10}$ the quantity above. The PAPR for a constant-envelope signal is unity, or 0 dB. Write a simple computer simulation that generates 10,000 symbols of a BPSK-R(1) waveform and use numerical techniques to find the PAPR for a signal with BPSK-R(f_c) spreading modulation bandlimited to $\pm 10\ f_c$, $\pm 5\ f_c$, $\pm 2\ f_c$, $\pm\ f_c$.

QT3.6. Find the PSD of a quadraphase keyed signal's complex envelope, when the spreading sequences in each phase are ideal long codes and orthogonal.

QT3.7. Numerically compute bandlimited correlation functions for BOC(2,1) and BOC_c(2,1), for wider bandwidth ± 10.23 MHz and narrower bandwidth ± 3.069 MHz. Observe the similarities and differences of how the correlation functions are influenced by bandwidth. Assume ideal long spreading codes.

QT3.8. Find the expression for the PSD of a BCS signal with an ideal long spreading code and spreading symbol $g_{BCS}(t) = \sum_{n=0}^{N-1} c_n \text{rect}_{T_c/N}(t - \frac{nT_c}{N} - \frac{T_c}{2N})$.

QT3.9. Consider a time-multiplexed spreading time series formed using a repeating pattern of two spreading symbols $g_1(t)$ followed by one spreading symbol $g_2(t)$, so $x(t) = \sum_{k=-\infty}^{\infty} [a_{3k} g_1(t - 3kT_c) + a_{3k+1} g_1(t - (3k+1)T_c) + a_{3k+2} g_2(t - (3k+2)T_c)]$. Write an expression for the spreading time series, its correlation function, and its PSD, assuming an ideal long spreading code.

QT3.10. Derive the approximation in (3.39).

QT3.11. Show the multiplexing efficiency is unity for phase multiplexing and for time division multiplexing of two biphase orthogonal signal components, even when they have different useful power levels.

QT3.12. Show the multiplexing efficiency of majority voting three components as in (3.50) is 0.75.

QT3.13. Generalize majority voting of three constituent biphase constant-envelope signals so that the constituents can be received at different power levels. Specifically, design a biphase constant-envelope composite signal that provides the first constituent signal at twice the useful power of the other two constituent signals. (Hint: use a combination of time multiplexing and majority voting.)

QT3.14. Show the interplexed composite signal $x_c(t) = \left[[\alpha_1 x_1(t) + \alpha_2 x_2(t)] + i \left[\alpha_3 x_3(t) - \frac{\alpha_1 \alpha_2}{\alpha_3} x_1(t) x_2(t) x_3(t) \right] \right]$ is constant envelope when the constituent signals $x_1(t), x_2(t), x_3(t)$ are constant envelope and unit power. When the constituent signals are mutually orthogonal, find the ratio between the useful powers in $x_1(t)$ and $x_2(t)$, and the ratio between the useful powers in $x_2(t)$ and $x_3(t)$.

QT3.15. Compare two different ways of achieving the objective in QT3.14: designing a composite signal that delivers one constituent signal at twice the useful power of the other two constituent signals, where all the constituent signals are biphase and constant envelope. Compare the average multiplexing efficiency of the approach for designed in QT3.14 with the most efficient interplex approach.

QT3.16. Consider two BPSK-R($1/T_c$) waveforms $x(t) = \sum_{k=-\infty}^{\infty} a_k g_{\text{BPSK-R}(1/T_c)}(t - kT_c)$ and $y(t) = \sum_{k=-\infty}^{\infty} b_k g_{\text{BPSK-R}(1/T_c)}(t - kT_c)$ with ideal long spreading codes that are statistically independent, and spreading symbol $g_{\text{BPSK-R}(1/T_c)}(t) = \text{rect}_{T_c}(t - T_c/2)$. (3.20) states its correlation function is

$$R_{\text{BPSK-R}(1/T_c)}(\tau) = \begin{cases} (T_c + \tau)/T_c, & -T_c < \tau \leq 0 \\ (T_c - \tau)/T_c, & 0 < \tau \leq T_c \\ 0, & \text{elsewhere} \end{cases}$$

There is a contradiction between the following two statements:
1. The product $z(t) = x(t)y(t)$ is a BPSK-R($1/T_c$) waveform having the same correlation function as above.
2. As stated in Appendix Section A.4, the correlation function of $z(t) = x(t)y(t)$ is $R_z(\tau) = R_x(\tau)R_y(\tau)$, so

$$R_z(\tau) = R^2_{\text{BPSK-R}(1/T_c)}(\tau) = \begin{cases} (T_c + \tau)^2/T_c^2, & -T_c < \tau \leq 0 \\ (T_c - \tau)^2/T_c^2, & 0 < \tau \leq T_c \\ 0, & \text{elsewhere} \end{cases}$$

Which statement is true, and why?

Application Questions

QA3.1. Plot a spreading symbol for the following spreading modulations, and express each as a BCS symbol: BOC(5,2), BOC$_c$(5,2), BPSK-R(2), BOC(4,1), BOC$_c$(4,1), BPSK-R(1).

QA3.2. The PSDs in Figures 3.25 through 3.28 all correspond to different versions of BOC spreading modulations using subcarrier frequencies and spreading code chip rates that are multiples of 1.023 MHz. Label each plot indicating whether it is sine phased or cosine phased, and indicating its subcarrier frequency and spreading code chip rate.

REVIEW QUESTIONS

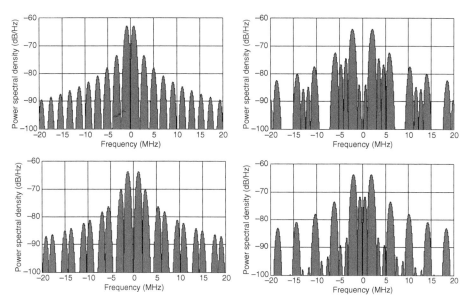

Figure 3.25. First Set of PSDs for QA3.2

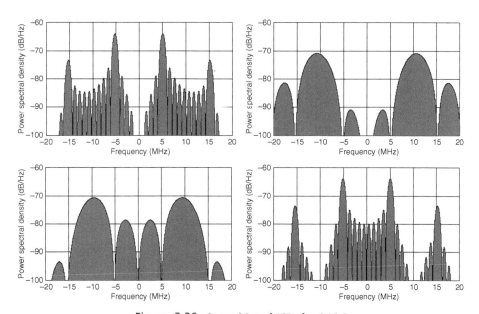

Figure 3.26. Second Set of PSDs for QA3.2

96 SATNAV SIGNALS

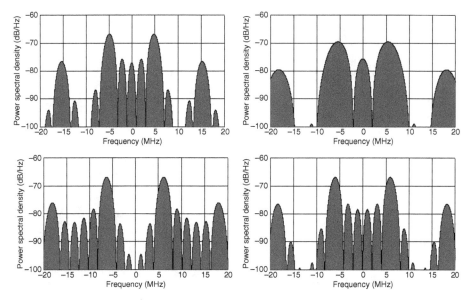

Figure 3.27. Third Set of PSDs for QA3.2

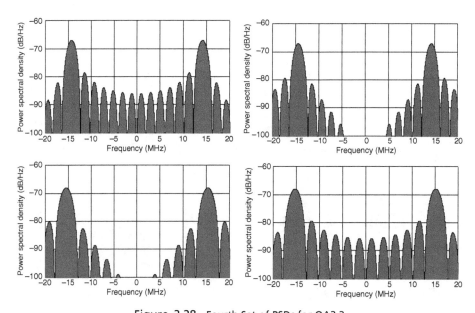

Figure 3.28. Fourth Set of PSDs for QA3.2

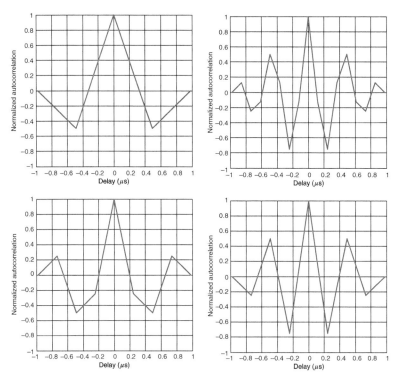

Figure 3.29. First Set of Correlation Functions for QA3.3

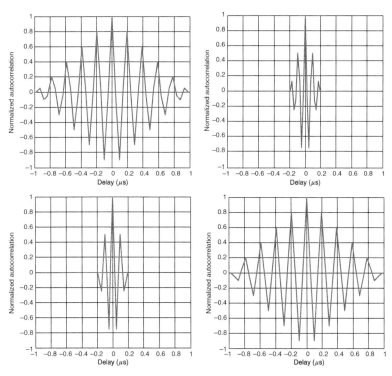

Figure 3.30. Second Set of Correlation Functions for QA3.3

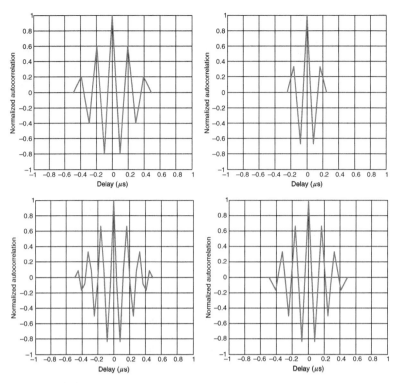

Figure 3.31. Third Set of Correlation Functions for QA3.4

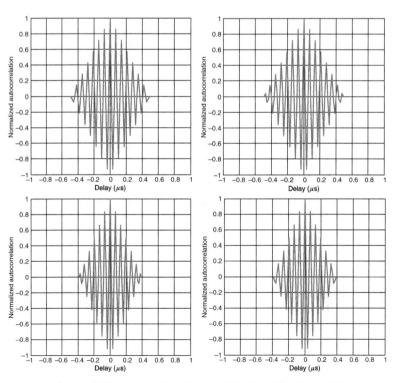

Figure 3.32. Fourth Set of Correlation Functions for QA3.5

QA3.3. The correlation functions in Figures 3.29 through 3.32 all correspond to different versions of BOC spreading modulations using subcarrier frequencies and spreading code chip rates that are multiples of 1.023 MHz. Label each plot indicating whether it is sine phased or cosine phased, and indicating its subcarrier frequency and spreading code chip rate.

QA3.4. For a 1 W infinite-bandwidth signal using the following spreading modulations, find the maximum PSD: BPSK-R(1), BPSK-R(10), BOC(1,1), BOC(10,5), $BOC_c(10,5)$, MBOC(6,1,1/11), BCS([1,1,−1,1],1), constant-envelope AltBOC(15,10).

QA3.5. Using Appendix Section A.6, find the null-to-null bandwidth, −3 dB bandwidth, −10 dB bandwidth, 90% power bandwidth, equivalent rectangular bandwidth, and RMS bandwidth of signals using the following spreading modulations: BPSK-R(1), BPSK-R(10), BOC(1,1), BOC(10,5), $BOC_c(10,5)$, MBOC(6,1,1/11), BCS([1,1,−1,1],1), constant-envelope AltBOC(15,10). Use spectra normalized over infinite bandwidths and compute the bandwidth measures over infinite bandwidth, except for the RMS bandwidth, which should be normalized and computed over ±15.345 MHz for all signals except AltBOC(15,10), for which ±25.575 MHz should be used.

REFERENCES

1. J. Á. Ávila Rodríguez, "On Generalized Signal Waveforms for Satellite Navigation," Doctoral Thesis, University FAF Munich, 2008, available at https://athene-forschung.unibw.de/doc/86167/86167.pdf, accessed 15 September 2015.
2. J. W. Betz, "On the Power Spectral Density of GNSS Signals, with Applications," Proceedings of the Institute of Navigation International Technical Meeting 2010, ION-ITM-2010, Institute of Navigation, January 2010.
3. J. W. Betz, "Binary Offset Carrier Modulations for Radio Navigation," *NAVIGATION: The Journal of the Institute of Navigation*, Vol. 48, No. 4, Winter 2001/2002, pp. 227–246.
4. P. Ward, J. W. Betz, and C. J. Hegarty, "GPS Satellite Signal Characteristics," *Understanding GPS*, 2nd edition, Edited by E. D. Kaplan and C. Hegarty, Artech House, 2006.
5. C. J. Hegarty, J. W. Betz, and A. Saidi, "Binary Coded Symbol Modulations for GNSS," Proceedings of the Institute of Navigation Annual Meeting 2004, ION-AM-2004, Institute of Navigation, June 2004.
6. G. W. Hein, J.-A. Avila-Rodriguez, S. Wallner, J. W. Betz, C. Hegarty, J. J. Rushana, A. J. Kraady, A. R. Pratt, S. Lenahan, J. Owen, J.-L. Issler, and T. Stansell, "MBOC: The New Optimized Spreading Modulation Recommended for GALILEO L1 OS and GPS L1C," Proceedings of the IEEE/ION Conference on Positioning, Location, and Navigation Systems 2006, PLANS-2006, IEEE Press, April 2006.
7. J. W. Betz, C. J. Hegarty, and J. J. Rushanan, "Time-Multiplexed Binary Offset Carrier Signaling and Processing—Group I," Application No. 11/785,571. Expressly abandoned by MITRE to allow universal free access to the invention, 26 June 2010.

8. J. W. Betz, C. J. Hegarty, and J. J. Rushanan, "Time-Multiplexed Binary Offset Carrier Signaling and Processing—Group II," Application No. 12/755,933. Expressly abandoned by MITRE to allow universal free access to the invention, 11 November 2010.
9. "European GNSS (Galileo) Open Service Signal in Space Interface Control Document," European Commission webpages, available at: http://ec.europa.eu/enterprise/policies/satnav/galileo/open-service/index_en.htm, accessed 12 January 2013, Issue 1.1, September 2010.
10. L. Lestarquit, G. Artaud, and J. L. Issler, "AltBOC for Dummies or Everything You Always Wanted To Know About AltBOC," Proceedings of the 21st International Technical Meeting of the Satellite Division of the Institute of Navigation (ION GNSS 2008), September 2008.
11. N. C. Shivaramaiah and A. G. Dempster, "The Galileo E5 AltBOC: Understanding the Signal Structure," Proceedings of International Global Navigation Satellite Systems Society IGNSS Symposium 2009, December 2009.
12. E. Rebeyrol and C. Macabiau, "Boc Power Spectrum Densities," Proceedings of the ION NTM, San Diego, CA, 24–26 January 2005.
13. L. Zhu, H. Yin, Z. Yao, M. Lu, "Non-symmetrical ALTBOC Multiplexing for Compass B1 Signal Design," Fifth European Workshop on GNSS Signals and Signal Processing, December 2011.
14. Z. Yao and M. Lu, "Constant Envelope Multiplexing Technique for Components on Different Carrier Frequencies with Unequal Power Allocation," Proceedings of the Institute of Navigation International Technical Meeting ITM-2013, January 2013.
15. Z. Yao and M. Lu, "Design, Implementation, and Performance Analysis of ACE-BOC Modulation," Proceedings of the 26th International Technical Meeting of the Satellite Division of the Institute of Navigation (ION GNSS+ 2013), Nashville, TN, September 2013, pp. 361–368.
16. J. Klobuchar, "Ionospheric Time-Delay Algorithms for Single-Frequency GPS Users," *IEEE Transactions on Aerospace and Electronic Systems*, Vol. ASE-23, No. 3, 1987, pp. 325–331.
17. J. Klobuchar, "Ionospheric Effects on GPS," Chapter 12 in *The Global Positioning System: Theory and Application*, Volume 1, Edited by B. W. Parkinson and J. J. Spilker, AIAA Press, 1996.
18. A. P. Cerruti, "Solar Radio Burst Effects on Global Positioning System Receivers," PhD Dissertation, Cornell University, January 2008.
19. C. S. Carrano, K. M. Groves, W. J. McNeil, and P. Doherty, "Direct Measurement of the Residual in the Ionosphere-Free Linear Combination During Scintillation," Proceedings of the Institute of Navigation International Technical Meeting ITM-2013, January 2013.
20. L. Deambrogio and C. Macabiau, "Vector Tracking Aiding for Carrier Phase Estimation in the Presence of Ionospheric Scintillation," International Technical Meeting of the Institute of Navigation, ION ITM 2013, San Diego, January 2013, pp. 333–342.
21. B. Carter, "Predicting Daily Space Weather Will Help Keep Your GPS On Target," *The Conversation*, 3 December 2014, available at http://theconversation.com/predicting-daily-space-weather-will-help-keep-your-gps-on-target-34750, accessed 14 December 2014.
22. T. A. Milligan, *Modern Antenna Design*, 2nd edition, IEEE Press/Wiley-Interscience, 2005.
23. C. Hegarty, "Polarization Mismatch Effects on GPS/GALILEO Minimum/Maximum Received Power Calculations," Unpublished Note, 14 November 2005.
24. R. Gold, "Optimal Binary Sequences for Spread Spectrum Multiplexing (Corresp.)". *IEEE Transactions on Information Theory,* Vol. 13, No. 4, 1967, pp. 619–621.

25. S. W. Golomb and G. Gong, *Signal Design for Good Correlation: For Wireless Communication, Cryptography, and Radar,* Cambridge University Press, 2005, ISBN 978-0-521-82104-9.
26. J. J. Rushanan, "Weil Sequences: A Family of Binary Sequences with Good Correlation Properties," Proceedings of the 2006 IEEE International Symposium on Information Theory, 2006, pp. 1648–1652.
27. L. R. Welch, "Lower Bounds on the Maximum Correlation of Signals," *IEEE Transactions on Information Theory*, Vol. IT-20, No. 3, May 1974, pp. 397–399.
28. W. W. Peterson and E. J. Weldon, *Error Correcting Codes*, MIT Press, Cambridge, MA, 1972.
29. J. McNamara, *Technical Aspects of Data Communication*, 2nd edition, Digital Equipment Corporation, Digital Press, 1982.
30. B. Sklar, *Digital Communications: Fundamentals and Applications*, 2nd edition, Prentice-Hall, 2001, ISBN: 978-0130847881.
31. S. Lin and D. J. Costello, *Error Control Coding: Fundamentals & Applications*, 2nd edition, Prentice-Hall, 2004.
32. H. Ma and J. Wolf, "On Tail Biting Convolutional Codes," *IEEE Transactions on Communications*, Vol. 34, No. 2, 1986, pp. 104–111.
33. C. Berrou, A. Glavieux, and P. Thitimajshima, "Near Shannon Limit Error – Correcting," Proceedings of IEEE International Communications Conference, Geneva, Switzerland, May 1993, pp. 1064–1070.
34. T. J. Richardson and R. L. Urbanke, "Efficient Encoding of Low-Density Parity-Check Codes," *IEEE Transactions in Information Theory*, Vol. 47, No. 2, February 2001, pp. 638–656.
35. C. E. Shannon, *A Mathematical Theory of Communication*, Urbana, IL, University of Illinois Press, 1949 (reprinted 1998).
36. G. M. Dillard, M. Reuter, J. Zeidler, and B. Zeidler, *IEEE Transactions on Aerospace and Electronic Systems*, Vol. 39, No. 3, 2003, pp. 786–798.
37. J. J. Spilker, *Digital Communications by Satellite*, Prentice-Hall, 1977.
38. S. Butman and U. Timor, "Interplex – An Efficient Multichannel PSK/PM Telemetry System," *IEEE Transactions on Communications*, Vol. 20, No. 3, 1972, pp. 415–419.
39. P. A. Dafesh and C. R. Cahn, "Phase-Optimized Constant-Envelope Transmission (POCET) Modulation Method for GNSS Signals," Proceedings of the Institute of Navigation Conference on Global Navigation Satellite Systems 2009, ION-GNSS-2009, Institute of Navigation, September 2009.

4

LINK BUDGETS

An essential aspect of satnav systems engineering is quantifying the relationship between received power of a desired signal or an interfering waveform, given the transmit power. In other cases, one wants to know the needed transmit power to assure a specific received power. Given the satellite orbits described in Chapter 2, and the signal characteristics discussed in Chapter 3, Section 4.1 provides the equations for calculating the relationship between transmit power and received power for the basic case of space-to-Earth propagation. Section 4.2 goes into more detail for satellite-to-receiver links, discussing considerations in assessing minimum received power and maximum received power and different conventions used by different systems for this purpose, taking into account transmit antenna gain and geometry. While satnav generally involves propagation from satellites to receivers, terrestrial propagation is also encountered, particularly when assessing effects of interferers on satnav receiver performance. Section 4.3 describes various models and their applicability for terrestrial propagation. Since satnav receivers often operate under foliage and in buildings, Section 4.4 provides models for additional attenuation caused by propagation through these materials. Section 4.5 provides a summary of the chapter.

4.1 FREE-SPACE PATH LOSS

Free-space propagation conditions occur in vacuum and when the receiver is in the far field of the transmitted electromagnetic wave. A receiver is in the far field when its separation from the transmitter is much greater than the wavelength, much greater than largest dimension of the transmit antenna D, and is greater than the Fraunhofer distance $2D^2/\lambda$ where λ is the wavelength of the carrier frequency [1].

In practice, free-space propagation models are employed when there is optical line of sight between transmitter and receiver. Once Earth's curvature or objects significantly block optical line of sight, L band propagation losses typically increase relative to what is predicted by free-space propagation models.

A first step is to check whether line of sight conditions exist. Assuming a smooth spherical Earth, when the transmitter has height h_t above the Earth's surface and the receiver has height h_r above the Earth's surface, the maximum range where there is optical line of sight between transmitter and receiver r_{t2r} is given by,

$$\begin{aligned} r_{t2r} &= \sqrt{h_t^2 + 2r_e h_t} + \sqrt{h_r^2 + 2r_e h_r} \\ &\cong \sqrt{2r_e h_t} + \sqrt{2r_e h_r} \\ &= 3572 \left(\sqrt{h_t} + \sqrt{h_r} \right) \text{ m} \end{aligned} \quad (4.1)$$

where the approximation holds when the heights of receiver and transmitter are much less than Earth's radius, and the numerical value in the last line uses $r_e = 6378.137$ km. For a given distance between transmitter and receiver r_{t2r}, a receiver above height h_r is said to be above the transmitter's radio horizon. Figure 4.1 shows the maximum range between transmitter and receiver for different terminal heights, implementing (4.1).

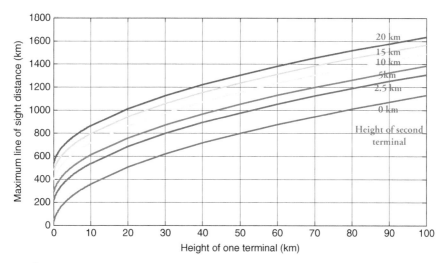

Figure 4.1. Maximum Line of Sight Range for Terminals above Earth's Surface

Under some conditions, radio waves refract through the Earth's atmosphere, causing the waves to bend downward so that propagation can exist at ranges longer than predicted by optical line of sight. A common approximation for this condition [2] is to use four-thirds Earth radius in (4.1), leading to four-thirds Earth radius model for maximum range between transmitter and receiver,

$$r_{t2r} \cong 4124(\sqrt{h_t} + \sqrt{h_r}) \, \text{m} \qquad (4.2)$$

where the approximation holds as long as transmitter and receiver heights above the Earth surface are small relative to the Earth's radius.

Propagation from space to Earth, as well as in other free-space propagation situations, involves the aspects shown in Figure 4.2. The RF transmitter power, P_T watts, drives the transmit antenna that has gain G_T in the direction of the receiver, producing a given effective isotropic radiated power (EIRP) $P_T G_T$ watts. (EIRP is the amount of power that a theoretical isotropic, or omnidirectional, antenna would emit in all directions to produce the same power flux density (PFD) at the receive antenna.) As the transmitted wave propagates through free space, its power is diminished by the path loss $4\pi r_{t2r}^2 L_{ex}$, where L_{ex} is the excess loss beyond free-space loss to account for attenuation due to oxygen, water vapor, and rain in the atmosphere, and ionized structures in the atmosphere. The resulting PFD at the face of the receive antenna is the EIRP reduced by the free-space loss and the excess loss, with units of W/m². The PFD is defined as the amount of power incident on a 1 m² surface normal to the direction of propagation

$$\text{PFD} = \frac{P_T G_T}{4\pi r_{t2r}^2 L_{ex}} \qquad (4.3)$$

Typically, the excess propagation loss is less than 1 dB at L band. Observe that the PFD at the receiver is independent of wavelength.

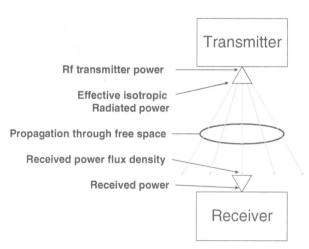

Figure 4.2.   Link Budget Components

FREE-SPACE PATH LOSS

The power output by a receive antenna boresighted in the direction of signal arrival is the product of the PFD and the effective area of the receive antenna A_R, where[1]

$$A_R = \frac{G_R \lambda^2}{4\pi L_{ant}} \quad (4.4)$$

with G_R the receive antenna gain at boresight (i.e., in the direction of maximum gain), λ the wavelength, and L_{ant} the loss associated with the receive antenna relative to ideal, the reciprocal of the antenna's efficiency. A typical value for an antenna's radiation efficiency is 0.55, corresponding to L_{ant} of 2.6 dB. Antenna gain is commonly expressed in decibels relative to an ideal isotropic antenna having unity gain, or dBi.

The expression (4.4) shows for a fixed antenna gain, the effective area of the antenna increases with the square of the wavelength. Since wavelength is related to speed of propagation and frequency by

$$\lambda = c/f \quad (4.5)$$

the effective area of the antenna decreases with the square of the frequency. Conversely, if the effective area of an ideal antenna remains fixed, its gain increases with the square of the frequency.

The gain of an antenna is related to its beamwidth. While the details vary with antenna type and design, a useful rule of thumb is

$$G \cong \frac{29000}{\beta_\theta \beta_\phi} \quad (4.6)$$

where G is the gain at boresight, and β_θ and β_ϕ are, respectively, the azimuth and elevation 3 dB two-sided beamwidths (the number of angular degrees between half power points in the antenna gain pattern) in units of degrees. The value 29,000 is a typical value, and can be as low as 20,000 for small beamwidths and antenna efficiency 0.5, and as large as 50,000 for wide beamwidths and higher antenna efficiency. This expression and the numerical values are derived in the first review problem of Chapter 14.

Combining (4.4) and (4.6) yields a rule of thumb relationship between antenna area and beamwidth:

$$\beta_\theta \beta_\phi \cong \frac{3537 \lambda^2}{A_R} \quad (4.7)$$

Consequently, increasing a receive antenna's boresight gain decreases its beamwidth, reducing its gain off boresight and limiting its ability to receive signals from satellites anywhere in the sky. A multi-element antenna with multiple beams does

---

[1] These expressions all apply to transmit antennas as well as receive antennas, but the discussion here is focused on receive antennas.

enable higher gain in multiple directions, but with increased complexity, size, and power consumption for the antenna array and supporting electronics.

The power output by the receive antenna is then

$$P_R = \text{PFD} \times A_R = \frac{P_T G_T G_R \lambda^2}{(4\pi)^2 r_{t2r}^2 L_{ex} L_{ant}} \tag{4.8}$$

The above discussion demonstrates a subtle but important result. The PFD at the receive antenna is independent of wavelength; it is the receive antenna that (because the fixed receive antenna gain makes its effective area vary with wavelength) makes the received output power depend on wavelength.

Numerical evaluation of (4.4) indicates that the effective area of a unit-gain antenna only varies by a few tenths of a decibel over the bandwidths of L band satnav signals, well within the accuracy of typical link budget calculations. Consequently, link budget calculations can typically be performed at a satnav signal's center frequency with adequate accuracy.

When the EIRP is expressed in dBW (i.e., decibels relative to 1 watt) and losses are expressed in dB, (4.8) provides the received power in dBW as

$$P_R = EIRP - L_{ex} - L_{ant} + 27.56 - 20\log_{10}(f) - 20\log_{10}(r_{t2r}) \tag{4.9}$$

with f the center frequency in MHz, and r_{t2t} the range between transmitter and receiver in meters. The corresponding free-space propagation loss in decibels is

$$L = L_{ex} - 27.56 + 20\log_{10}(f) + 20\log_{10}(r_{t2r}) \tag{4.10}$$

with f the center frequency in MHz, and r_{t2r} the range between transmitter and receiver in meters.

The columns of Table 4.1 show example link budget calculations. These numbers are representative but are not specified values for any system. Case 1 shows a typical calculation at 1575.42 MHz. The EIRP is 1 kW, even though the actual RF transmitter power is 50 W, since the directive transmit antenna concentrates the power in the direction of the receiver. For a satellite overhead the receiver at a range of 20,000 km, the received PFD is −127.5 dBW, meaning if the receive antenna had an effective area of 1 m², its output power would be −127.5 dBW. Since typical satnav receive antennas have approximately hemispherical gain patterns with maximum gain of the order of 3 dB, (4.4) shows these antennas are physically small. The receive antenna effective area for an ideal 0 dBi receive antenna is less than 0.3% of a square meter, reducing the output signal power correspondingly. When the receive antenna loss is 2 dB, the received signal power is −154.9 dBW, a representative value. The link budget calculation in Case 2 shows results for the same parameters, except using a larger range from the transmitter to receiver corresponding to a satellite at a lower elevation angle with a resulting 2.3 dB greater free-space path loss. Cases 3 and 4 use a much lower center frequency, and correspondingly 1 dB lower transmit antenna gain. The lower received PFD is only

TABLE 4.1. Example Link Budget Calculations

Case	1	2	3	4
Transmitter power (dBW) ✓	17	17	17	17
Transmit antenna gain (dBi) ✓	13	13	12	12
Effective isotropic radiated power (dBW)	30	30	29	29
Range from transmitter to receiver (km) ✓	20,000	26,000	20,000	26,000
Excess propagation loss (dB) ✓	0.5	0.5	0.5	0.5
Propagation loss (dB)	157.5	159.8	157.5	159.8
Received PFD (dBW/m^2)	−127.5	−129.8	−128.5	−130.8
Frequency (MHz) ✓	1575.42	1575.42	1176.45	1176.45
Wavelength (m)	0.19	0.19	0.26	0.26
Receive antenna area (dBW/m^2)	−25.4	−25.4	−22.9	−22.9
Receive antenna loss (dB) ✓	2.0	2.0	2.0	2.0
Received signal power (dBW)	−154.9	−157.2	−153.4	−155.7

related to the lower EIRP, and is offset by the larger effective area of an omnidirectional receive antenna at the lower frequency, resulting in a net 1.5 dB increase in received power at the lower frequency.

Inputs are indicated using a checkmark (✓) and other quantities are calculated based on the inputs.

4.2 CALCULATING MAXIMUM AND MINIMUM SPECIFIED RECEIVED POWER IN SIGNAL SPECIFICATIONS

Satnav system specifications typically indicate the minimum received power for each signal. As spectrum is increasingly shared among multiple systems and multiple signals making MAI more of an increasing concern, satnav system engineers are more aware of maximum received power, and are taking steps to calculate and control this value for each signal and indicate it in their ICDs as well. This section discusses conventions used in making these calculations, recognizing that different systems may use different conventions, and even a single system's conventions may change over time. The general expressions are developed, building upon results in Sections 2.4 and 4.1, and then illustrated using numerical examples roughly corresponding to GPS.

While users find it most relevant for signal specifications to identify the received power (i.e., at the output of the receive antenna, hence the input to the receiver), satnav system operators do not specify or control the receive antenna characteristics. Consequently, it is necessary for signal specifications to define a reference receive antenna model that is used in specification of received signal power. Typically, the reference receive antenna's gain is specified to include antenna efficiency. The gain is normal to the direction of propagation, and the reference receive antenna's gain and polarization characteristic are typically selected so that the antenna behaves like a lossless unit-gain antenna in converting a perfectly circularly polarized incident field to output power.

Two different reference receive antennas are widely used in specifying received signal power: a linearly polarized antenna with 3 dB gain (3 dBil reference receive antenna) and a circularly polarized antenna with 0 dB gain (0 dBic reference receive antenna). The 3 dBil reference antenna is used in specifications of minimum received power by GPS and GLONASS. The intent is to control both the polarization purity and the PFD of the signal at the face of the antenna, making the specified received signal power conservative. The polarization response of some receive antennas becomes nearly linear at low elevation angles, resulting from shorting of the incident electric field by a horizontal conductive surface like a ground plane. Since the received PFD is often lowest when the satellite is near the horizon (due to the longer propagation range and decreased gain of the satellite's transmit antenna at larger angles off boresight), this condition most stresses the minimum received power. If the receive antenna is linearly polarized, its orientation could be aligned with the semiminor axis of the incident signal's polarization ellipse due to relative orientation of the transmit and receive antennas, combined with Faraday rotation of the propagating field.

Consequently, a conservative specification of minimum received power uses a 3 dBil reference receive antenna, normal to the incident field so that the 3 dB gain applies, but at the worst polarization orientation where the antenna is aligned with the semiminor axis of the incident field. If the incident field is circularly polarized, Section 3.3.3 shows that there is 3 dB polarization loss, which is offset by the 3 dBi gain to produce a net gain of 0 dB. However, if the incident field is elliptically polarized, the satellite must provide additional power to compensate for the greater maximum polarization loss, as shown in the upper left plot in Figure 3.18. A 0 dBic reference receive antenna also has a net gain of 0 dB for a circularly polarized incident field, but suffers very little loss if the incident field has moderate polarization ellipticity, as shown in the upper right plot in Figure 3.18. If the incident field has 1 dB axial ratio, using the 3 dBil reference receive antenna requires approximately 0.7 dB additional received PFD to compensate for the maximum polarization loss, while using the 0 dBic reference receive antenna requires negligible additional power. Consequently, if System A specifies minimum received power into a 3 dBil reference receive antenna and System B specifies the same minimum received power into a 3 dBic reference receive antenna, when both systems' signals have 1 dB axial ratio, System A actually delivers 0.7 dB higher PFD to the receive antenna.

While many systems do not specify maximum received power, there is an increasing trend toward at least being aware of this parameter. Maximum received power is typically specified into a 0 dBic reference receive antenna for several reasons. First, maximum received power typically occurs at higher elevation angles where many receive antennas have close to 0 dB axial ratio. Second, maximum received power is often used to assess aggregate interference from a satnav system, as described in Section 5.7. Typically, the receive antenna is exposed to incident interfering signals from multiple satellites, and each incident signal's polarization ellipse is typically oriented differently. Using a 0 dBic reference antenna for maximum receive power essentially averages over the different polarization orientations.

Since the calculations of minimum received power and maximum received power are typically performed as part of characterizing the space segment of the satnav system,

CALCULATING MAXIMUM AND MINIMUM SPECIFIED RECEIVED POWER

the calculations shown here are expressed in terms of angle off boresight (see Figure 2.2)—the satellite's coordinate system. All of the equations can also be expressed in terms of elevation angle using the equations provided in Section 2.4.

From (4.8) the received power, P_R as a function of angle off boresight, α, is given by

$$P_R(\alpha) = \frac{P_T G_T(\alpha) G_R \lambda^2}{(4\pi)^2 r_{s2r}^2(\alpha) L_{ex} L_{ant}} \tag{4.11}$$

where the range between satellite and receiver is $r_{s2r}(\alpha) = r_s \cos(\alpha) + [r_e^2 - r_s^2 \sin^2(\alpha)]^{1/2}$ from (2.5) and the transmit antenna gain is also a function of angle off boresight. While generally the receive antenna gain also depends on elevation angle (which can be related to angle off boresight) received signal power specifications are based on the assumption that the reference receive antenna used to measure received signal power is bore sighted in the direction of the incident signal.

Different systems' satellites, and different satellite designs for a specific system, have different transmit antenna gain patterns. The transmit antenna gain patterns also differ for different signals at different frequencies transmitted from the same satellite. Table 4.2 shows some representative GPS satellite gain patterns used as an example. These gain patterns, plotted in Figure 4.3, show higher L1 gain at large angles off boresight in order to counteract, in part, the increased path loss, although the gain patterns for L2 and L5 do not display the same characteristic. (The edge of Earth at the altitude of GPS satellites is approximately 13.9° angle off boresight, so providing the information for angles off boresight less than 14° is adequate for terrestrial users.)

Different conventions are used for minimum received power and maximum received power. Minimum received power conventions are conservative in the sense that the actual power delivered to a user antenna should typically be greater than calculated. The excess atmospheric loss in (4.11) is set to a pessimistically high level. Originally, both GPS and GLONASS used L_{ex} of 2 dB in early signal specifications; in 2003 GPS adjusted its value to 0.5 dB as more representative. The receive antenna loss in (4.11) includes

TABLE 4.2. Representative, Not Specified, GPS Satellite Transmit Antenna Gain Patterns

	Mean GPS Transmit Antenna Gain (dBic) Examples		
Angle Off Boresight (°)	C/A, L1C	L2C	L5
0	13.6	14.3	13.6
2	13.4	14.1	13.5
4	13.2	13.6	13.1
6	13.4	13.1	12.6
8	14.3	13.0	12.5
10	15.0	13.4	12.8
12	15.1	13.8	13.3
14	14.4	13.9	13.5

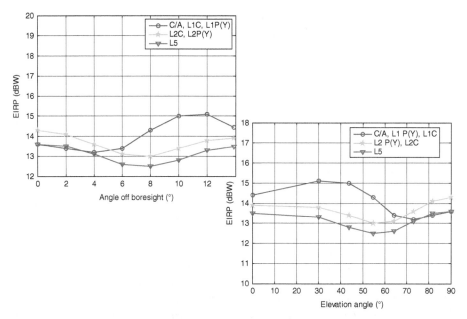

Figure 4.3. Representative, Not Specified, GPS Satellite Transmit Antenna Gain Patterns

the polarization loss described earlier. From the largest allowable ellipticity in the GPS interface specifications (see Reference 3) and the results in Section 3.3.3, the maximum polarization losses for GPS signals with a 3 dBil reference receive antenna at worst normal orientation are 4.0 dB for C/A and L1C, 4.2 dB for L2C, and 4.4 dB for L5. Accounting for the antenna gain, the net losses are 1.0 dB for C/A and L1C, 1.2 dB for L2C, and 1.4 dB for L5.

Minimum GPS received signal powers are specified at and above 5° elevation. Table 4.3 shows the minimum specified received signal power levels from GPS interface specifications as of early 2013, along with the RF power needed to provide it using the GPS parameter values in this paragraph and (4.11). Figure 4.4 shows the received power levels at different elevation angles using the same GPS minimum received power conventions, when the RF transmit powers are set to provide minimum specified power at or above elevation angle 5° as listed in Table 4.3.

TABLE 4.3. Power Levels for GPS Signals

GPS Signal	C/A	L1C	L2C	L5
Minimum specified received power above 5° elevation (dBW)	−158.5	−157.0	−160.0	−154.0
Representative RF transmit power for minimum specified received power (dBW)	12.9	14.4	10.6	17.3

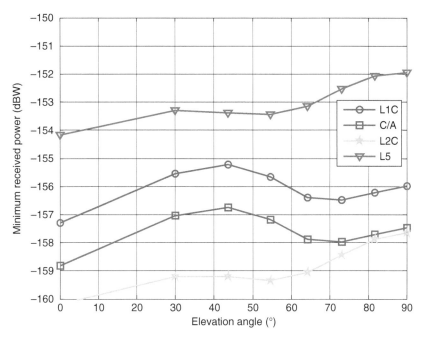

Figure 4.4. Representative Minimum GPS Received Power Levels at Different Elevation Angles

Maximum received power conventions, while less mature at this point, should be conservative in the sense that the actual power delivered to a user antenna should not be greater than calculated under the conditions assumed. The excess atmospheric loss in (4.11) is set to its minimum of 0 dB. The receive antenna polarization loss is also 0 dB, reflecting the circularly polarized reference receive antenna. Figure 4.5 shows the received power levels at different elevation angles using the GPS maximum received power conventions, when the RF transmit powers are set to meet the GPS minimum received power specifications. The higher power levels in Figure 4.5 reflect the different conventions (no excess atmospheric loss, no polarization loss) used for maximum received power as compared to minimum received power—both sets of conventions might apply simultaneously to different receivers at different locations on the Earth's surface.

However, additional margin is still needed for several purposes. Actual transmit antenna gain is not axially symmetric, and transmit antenna gain patterns also vary among different production units. The RF transmit power must be increased somewhat to ensure minimum power is delivered even with the gain pattern variations. Also, power control is not perfect—measurements of received power on the ground have errors, and there may be quantization steps in setting power on the satellites. It is prudent to set the RF transmit power 1 dB or higher than shown in Table 4.3 in order to provide margin so the minimum received power specification is met. Adding margin to the values in Figure 4.6, and comparing the resulting maximum received power at any elevation angle

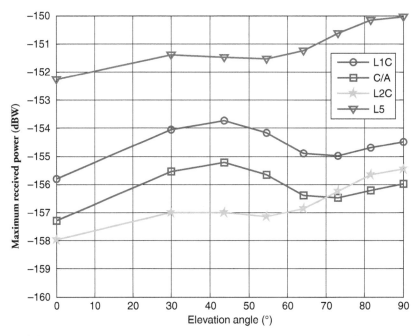

Figure 4.5. Representative Maximum GPS Received Power Levels at Different Elevation Angles with RF Power Set to Deliver Minimum Specified GPS Signal Power

to the minimum received power at or above 5° in Figure 4.5 leads to the conclusion that the maximum received power level at some elevation angles may exceed the minimum specified received power level by 5 dB or more.

4.3 TERRESTRIAL LINK BUDGETS

While satellite-to-receiver link budgets can be predicted with reasonably good accuracy using the model in Section 4.2, terrestrial link budgets (between a transmitter on or near the ground and a receiver on or near the ground) are much more variable and thus very difficult to predict accurately. Entire books are dedicated to this problem, and Reference 2 is useful.

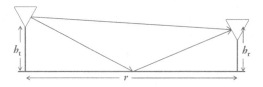

Figure 4.6. Geometry for Two-Ray Model

The choice of terrestrial link budget model depends on many considerations. A model that predicts higher loss is typically used by a service provider desiring to ensure delivery of a minimum power level, while lower loss models may be more appropriate in conservative calculations of received interference levels. Design for a specific site is often based on models that account for specific terrain and cultural features, while more generic designs can only treat such features statistically.

Often, free-space propagation loss is used as an approximate lower bound on propagation loss (i.e., an upper bound on received power), and terrestrial models are used to compute how much greater propagation loss may be encountered. There are, however, situations where propagation loss is less than predicted using free-space loss models, as seen in the two-ray model introduced below.

Considering propagation models other than free space is warranted when

- The heights above ground of transmit and receive antennas are each less than the range separating these two antennas,
- Terrain, buildings, other obstacles, or the Earth itself are in or near the optical line of sight between transmit antenna and receive antenna.

Table 4.4 compares different types of models for terrestrial propagation loss. Service providers may use a deterministic, site-specific model in determining where to site a cellular telephone base station or an antenna for wireless data services, or for guidance

TABLE 4.4. Comparison of Terrestrial Model Types

Model Type	Generic Physics Based	Deterministic, Site Specific	Stochastic	Empirical
Basis	First-principles physics based	First-principles physics based	Environment modeled as set of random variables	Field measurements
Inputs required	Few	Extensive description of site geometry and materials	Modest, general description of environment	Selection of environment class
Computation	Low	Extensive	Modest	Low
Applicability	Limited to situations where physics assumptions are met	Specific site and transmit/receive locations	Broad	Broad
Accuracy, typical standard deviations	Good when physics assumptions are met, poor otherwise	Good, <3 dB	Poor, >15 dB	Modest, <10 dB

in setting initial transmit power levels. Systems engineering calculations typically use a combination of stochastic models (based on Monte Carlo simulations over different environmental parameters) or empirical models (that fit parametric models to measurements collected in the field). This section describes a generic physics-based model, and then focuses on empirical models.

The two-ray model for propagation loss is a simple physics-based model that applies when there is no blockage, when the Earth surface is smooth relative to the wavelength, and for moderate distance between transmitter and receiver. Under these conditions, the Earth is represented as a flat reflecting plane, as shown in Figure 4.6. The resulting two-ray model for propagation loss in decibels is [2]

$$L = 20\log_{10}\left(\frac{2\pi r_{t2r}}{\lambda}\right) - 20\log_{10}\left(\sin\left(\frac{2\pi h_t h_r}{\lambda r_{t2r}}\right)\right) \quad (4.12)$$

where wavelength λ, range between transmitter and receiver r_{t2r}, transmitter height above Earth's surface h_t, and receiver height above Earth's surface h_r all have the same units. The last term in (4.12) represents the effect of the ray reflected from the Earth's surface. Depending upon the received phase of this ray relative to the received phase of the direct path, the reflected ray may either constructively add to or destructively subtract from the direct path. This expression applies for modest ranges between transmitter and receiver—tens of meters to a few tens of kilometers, where Earth's curvature can be neglected. If there is blockage in the direct path the propagation loss may be higher; if there is blockage in the reflected path the propagation loss is often lower since the reflected ray is attenuated and does not cancel the direct path as completely.

When $h_t h_r \ll \lambda r_{t2r}$, (4.12) simplifies to the two-ray model approximation:

$$L \cong 40\log_{10}(r_{t2r}) - 20\log_{10}(h_t) - 20\log_{10}(h_r) \quad (4.13)$$

This surprising result states that, under these conditions, the propagation loss model is independent of wavelength and falls off with the fourth power of range between transmitter and receiver. This condition is sometimes called a "range exponent of four," in contrast to the "range exponent of two" associated with free-space propagation.

One of the most widely used empirical propagation models is known as the COST-231 Hata model, a refinement by the COST-231 group of a model originally developed by Hata and then updated [4]. The COST-231 Hata model for propagation loss in decibels is

$$\begin{aligned}L = 46.3 + 33.9\log_{10}(f_{MHz}) - 13.82\log_{10}(h_t) - M \\ + [44.9 - 6.55\log_{10}(h_r)]\log_{10}(r_{t2r}) + C\end{aligned} \quad (4.14)$$

where

- f_{MHz} is the frequency in MHz, $500 < f < 2000$,
- h_t is the transmitter height above the Earth in meters, $30 < h_t < 200$,

TERRESTRIAL LINK BUDGETS

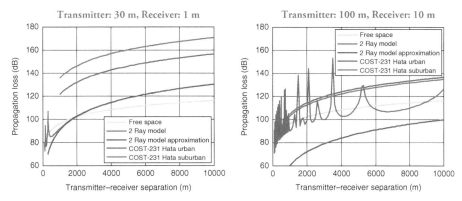

Figure 4.7. Propagation Losses Predicted by Different Models at 1575.42 MHz

- h_r is the receiver height above the Earth in meters, $1 < h_r < 10$,
- r_{t2r} is the range between transmitter and receiver in kilometers, $1 < r_{t2r} < 10$,
- M is the receiver term, $M = [1.1 \log_{10}(f_{MHz}) - 0.7]h_r - [1.56 \log_{10}(f_{MHz}) - 0.8]$,
- C is a constant that depends on the type of environment: for urban areas $C = 3$ while for flat rural or suburban areas $C = 0$.

Figure 4.7 shows predicted mean path loss using the free space, exact and approximate two-ray models, and the COST-231 Hata model for urban and for suburban or flat rural conditions. For the lower transmitter and receiver, the two-ray model predicts much lower losses than does the COST-231 Hata model. The shape of the COST-231 Hata and two-ray curves are similar, indicating a Hata loss range exponent near four, but the two-ray curves are consistently lower, since they do not account for direct path blockage included in the COST-231 Hata model. The exact two-ray model shows propagation losses less than free space for short separation distances; these are situations where the reflection from the ground constructively adds to the direct path. For the lower height transmitter and receiver, the two-ray approximation matches the exact result for ranges greater than approximately 1500 m. For the higher transmitter and receiver, the two-ray model approximation does not apply at the transmitter–receiver separations shown, but the result is still provided. For the higher transmitter and receiver, the losses predicted by COST-231 Hata for urban and suburban conditions are similar, since there is less blockage with the higher antennas. For the same reason, in this case the losses predicted by COST-231 Hata are closer to free-space losses.

Figure 4.8 shows results equivalent to those in Figure 4.7, but at the lower frequency 1176.45 MHz. The results do not change significantly, indicating that the uncertainty in these models is greater than their frequency dependence over the range of L band satnav frequencies.

Experience during open air testing shows that actual propagation loss varies within the envelope of the highest-loss curve and the lowest-loss curves in these figures,

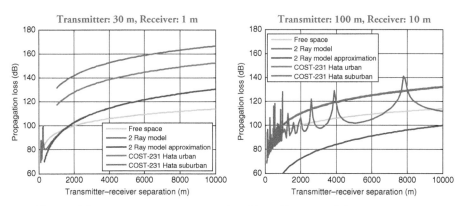

Figure 4.8. Propagation Losses Predicted by Different Models at 1176.45 MHz

depending on specific locations of the transmitter and receiver and on the resulting specific characteristics of the channel. Consequently, a conservative approach for providing service (when trying to ensure received power is greater than a minimum needed) is to use the largest of the propagation loss predictions at a given range, while the conservative approach for assessing interference (when trying to ensure received power is less than a maximum tolerable) is to use the smallest of the propagation loss predictions at a given range, except for the two-ray model approximation when it does not apply. Typically, the largest propagation loss prediction is provided by COST-231 Hata, and a 10 dB standard deviation can be added for extra conservatism; the smallest propagation loss prediction is provided by the free-space model or the two-ray model.

4.4 BUILDING PENETRATION AND FOLIAGE LOSSES

The propagation loss models discussed in previous sections all address the case where signal does not need to propagate through building materials or foliage. When the signal must propagate into or out of a building or through foliage, excess losses must be taken into account by adding them (in decibels) to the loss predicted by propagation models described in previous sections. These excess losses generally increase with frequency, favoring satnav signals at the lower L band frequencies.

Building penetration losses vary considerably with building construction, materials, structure, and the location of the receiver within building. While the discussion here only addresses propagation loss, often the received signals also undergo significant reflections, introducing errors in their time of arrival, and resulting errors in pseudorange. Further discussion of building penetration losses is found in Reference 5.

The excess loss in decibels due to building penetration is typically modeled, for signals arriving from above the building, as

$$L = L_{\text{roof}} + n_{\text{floor}} L_{\text{floor}} \tag{4.15}$$

where

- L_{roof} is the roof penetration loss, which can range from 1 dB to 30 dB at L band,
- n_{floor} is the number of floors penetrated,
- L_{floor} is the loss per floor, which can range from 1 dB to 10 dB at L band.

For building penetration through walls, the excess loss in decibels is similarly typically modeled as

$$L = L_{\text{ext}} + n_{\text{int}} L_{\text{int}} \qquad (4.16)$$

where

- L_{ext} is the exterior wall penetration loss, which can range from 1 dB to 30 dB at L band,
- n_{int} is the number of interior walls penetrated,
- L_{int} is the loss per interior wall, which can range from 1 dB to 10 dB at L band.

Table 4.5 lists representative losses for different building materials, drawing from an extensive set of measurements reported in Reference 6.

A standard model for foliage penetration losses is found in Reference 7. The loss in decibels due to foliage is given by

$$L = 3 \log_{10}(f_{\text{MHz}}) + 6 \log_{10}(d_{\text{foliage}}) - 7 \qquad (4.17)$$

where

- f_{MHz}, f is the frequency in MHz,
- d_{foliage} is the depth of the foliage traversed in meters, $d_{\text{foliage}} < 400$.

TABLE 4.5. Measured Losses in Decibels for Different Building Materials

	Frequency (MHz)	
Material	1176.45	1575.42
Brick	3–7	5–9
Composite brick/concrete	14–25	17–33
Brick/masonry block	11	10
Poured concrete	12–45	14–44
Reinforced concrete	27–30	30–35
Masonry block	11–27	11–30
Drywall	0.2–0.5	0.4–0.7
Glass	0.8–3	1.2–4
Dry lumber	3–6	3.5–8
Wet lumber	3.5–7.5	6–10

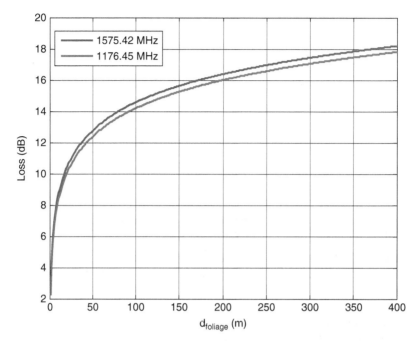

Figure 4.9. Computation of Foliage Loss Model [7]

This model does not account for the type of foliage, the foliage density or wetness, or the density of branches and trunks. Trunks and branches can actually contribute more than half (in dB) of this loss, applying even when deciduous trees have no leaves.

Figure 4.9 computes (4.17) at two different frequencies, showing very little difference in losses across the spread of L band frequencies used for satnav. As is the case for building penetration, trees can also produce significant multipath reflections, introducing pseudorange errors as well as attenuation.

An alternative model for excess propagation loss in decibels due to foliage is given by [8]

$$L = 26.6 \times f^{-0.2} \times d_{\text{foliage}}^{0.5} \qquad (4.18)$$

when vegetation is out of leaf, and

$$L = 26.6 \times f^{-0.009} \times d_{\text{foliage}}^{0.26} \qquad (4.19)$$

when vegetation is in leaf, using the same parameter definitions as in (4.17). Reference 9 provides test results to conclude that the model (4.17) is more accurate, but still has an error of 6.5 m RMS. Consequently, while using a foliage model is beneficial, simple and universal models do not accurately account for the excess loss due to foliage.

Further experimental results showing the variability associated with foliage attenuation are found in References 10 and 11.

4.5 SUMMARY

Relating transmitted power to received power is important in satnav systems engineering, for desired satnav signals and also for interfering waveforms. When free-space propagation models apply, path loss can be predicted with good accuracy, and the relationship between transmitted and received power can be assessed with good confidence. Free-space propagation models are at the heart of calculations used to compute minimum received power and maximum received power from a satellite. There are subtleties in specifying the reference receive antenna characteristics in some computations, with different conventions used for gain, excess path loss, and polarization axial ratio. Section 4.2 shows these computations, for example, GPS characteristics, including different conventions used for conservatism in assessing minimum received power and maximum received power.

Path loss in terrestrial propagation depends on many details and is much more difficult to model accurately than is free-space propagation. Different modeling approaches are available, primarily developed for communications applications. Either simple physics-based models or empirical models are best suited to satnav systems engineering applications. A two-ray reflective Earth model can be a good predictor of path loss over modest ranges (of the order of 10 kms or less) when terrain is smooth and there is little blockage from buildings or terrain. In more complicated situations, the COST-213 Hata model can be useful. Real-world measurements show wide variation in path loss, and typically large errors from predictions using physics-based or empirical models. Site-specific measurements or predictions are the only way to get good accuracy in most cases. For systems engineering purposes, physics-based and empirical models can be used to develop bounds, and then engineering judgment can be employed to choose where to operate within the bounds.

Propagation through building materials and foliage adds not only additional loss, but additional path loss variability. Simple models are provided for estimating excess losses from building and foliage penetration, but here again significant variability, not predicted by these models, is encountered in real-world conditions. The propagation models provide coarse estimates but are not useful for high-accuracy predictions of actual losses.

REVIEW QUESTIONS

Theoretical Question

QT4.1. From the exact 2-ray flat reflective Earth expression (4.11), derive the approximation (4.12).

Application Questions

QA4.1. Compute the EIRP needed to provide −155 dBW at 1575.42 MHz out of a lossless omnidirectional receive antenna, for GLONASS, GPS, SBAS, Galileo, BeiDou, and QZSS satellites directly overhead, all with 0.5 dB excess atmospheric loss. For QZSS and SBAS, use orbital radius of 42,164 km.

QA4.2. Compute the EIRP needed for a GPS satellite 26,000 km from the receiver to provide −155 dBW out of a lossless omnidirectional receive antenna. First, perform the calculation at 1575.42 MHz with 0.5 dB excess atmospheric loss. Then calculate the EIRP again at 5000 MHz with 6 dB excess atmospheric loss (rain causes greater excess losses at higher frequencies).

QA4.3. The power flux density gives the antenna output power if the receive antenna had 1 m^2 effective area. What would be the gain and beamwidth of a lossless receive antenna with a 1 m^2 effective area at 1575.42 MHz?

QA4.4. Find the EIRP at each integer value of angle off boresight between 0° and edge of Earth at 1575.42 MHz that would produce constant signal power of −155 dBW on the Earth's surface out of a lossless omnidirectional receive antenna with no excess atmospheric loss, for a satellite with orbital altitude 20,000 km. An antenna that delivers constant power over the Earth's surface is called an isoflux antenna.

QA4.5. Find the maximum power flux density in a ±0.5 MHz bandwidth centered at 1176.45 MHz for signals with the following spreading modulations: BPSK-R(10), BPSK-R(1), BOC(5,2), BPSK-R(2), BOC(5,1) each at power −150 dBW over infinite bandwidth.

REFERENCES

1. T. S. Rappaport, *Wireless Communications: Principles and Practice*, 2nd edition, Prentice-Hall, 2010.
2. J. D. Parsons, *The Mobile Radio Channel*, 2nd edition, John Wiley & Sons, 2000.
3. http://www.gps.gov/technical/icwg/, accessed 2 January 2015.
4. V. Abhayawardhana, I. Wassell, D. Crosby, M. Sellars, and M. Brown "Comparison of Empirical Propagation Path Loss Models for Fixed Wireless Access Systems," Proceedings of 61st IEEE Vehicular Technology Conference (VTC), 2005.
5. J. Doble, *Introduction to Radio Propagation for Fixed and Mobile Communications*, Artech House, 1996.
6. W. C. Stone, "Electromagnetic Signal Attenuation in Construction Materials," NISTIR 6055, NIST Construction Automation Program, Report No. 3, October 1997.
7. Federal Communications Commission, Office of Engineering and Technology, New Technology Development Division, "Millimeter Wave Propagation: Spectrum Management Implications," Bulletin Number 70, July 1997.

8. COST-235, "Radiowave Propagation Effects on Next Generation Terrestrial Telecommunications Services," Final Report, 1996.
9. E. K. Tameh and A. R. Nix, "The Use of Measurement Data to Analyze the Performance of Rooftop Diffraction and Foliage Loss Algorithms in a 3-D Integrated Urban/Rural Propagation Model," IEEE Vehicular Technology Conference, May 1998.
10. D. L. Ndzil, L. M. Kamarudin, E. A. A. Mohammad, A. Zakaria, R. B. Ahmad, M. M. A. Fareq, A. Y. M. Shaka, and M. N. Jafaar, "Vegetation Attenuation Measurements and Modeling in Plantations for Wireless Sensor Network Planning," *Progress in Electromagnetics Research B*, Vol. 36, 2012, pp. 283–301.
11. J. Goldhirsh and W. J. Vogel, "Handbook of Propagation Effects for Vehicular and Personal Mobile Satellite Systems: Overview of Experimental and Modeling Results," Report Jointly Authored by The Johns Hopkins University Applied Physics Laboratory and The University of Texas at Austin Electrical Engineering Research Laboratory, A2A-98-U-0-021 (APL) EERL-98-12A (EERL), December 1998.

5

CORRELATOR OUTPUT SNR, EFFECTIVE C/N$_0$, AND I/S

The combination of signals, noise, and interference at the receiver input affects correlator output signal-to-noise ratio (SNR)—an important measure of receiver performance. This chapter describes the underlying concept of correlator output SNR and defines it rigorously, leading to the important concepts of effective C/N$_0$ and processing gain. Since it is common to describe the severity of an interference environment in terms of the interference power-to-signal power ratio, or I/S, the conversion between effective C/N$_0$ and I/S is also described. The resulting tools can then be extended and applied to assess the effects of MAI from a given constellation, and even from a number of constellations.

5.1 CHANNEL MODEL AND IDEAL RECEIVER PROCESSING

The dominant signal processing function used in a satnav receiver is the correlator—it is used for acquisition, signal tracking, and data message demodulation. Figure 5.1 shows the channel model employed in this chapter, followed by ideal receiver processing. The channel shows the baseband biphase signal, $s(t)$, having PSD $\Phi_s(f)$ upconverted to carrier frequency f_1 and arriving at the receiver with delay D and carrier phase θ. To this

---

Engineering Satellite-Based Navigation and Timing: Global Navigation Satellite Systems, Signals, and Receivers,
First Edition. John W. Betz.
© 2016 The Institute of Electrical and Electronics Engineers, Inc. Published 2016 by John Wiley & Sons, Inc.

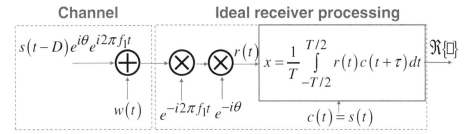

Figure 5.1. Channel Models and Ideal Receiver Processing Used for Output SNR Analysis

signal a combination of white noise and interference is added, whose sum is denoted by $w(t)$. It is assumed that $s(t)$ and $w(t)$ are uncorrelated and zero mean, so are orthogonal. The interference is modeled as circularly symmetric with PSD $\Phi_I(f)$.

This model implicitly assumes that complex envelopes of signals, noise, and interference after conversion to baseband are undistorted versions of the received signals at RF, frequency translated but otherwise only different in amplitude and delay. Such a model is valid when these waveforms are in the receiver passband, and at low enough power so distortion does not occur. When the waveforms are at carrier frequencies outside the receiver passband, or are received at high enough power to cause receiver components to respond nonlinearly, the models and results in this chapter may not apply. As a specific example, these models would not be suited to predict the effect on satnav receivers from the effects of nearby very high power transmissions at frequencies near those occupied by GPS signals, such as transmissions from the ancillary terrestrial component proposed by Light Squared [1].

When the desired signal and interference are DSSS signals, as in satnav, the assumption that $s(t)$ and $w(t)$ are uncorrelated results from the assumption the spreading codes are ideal long codes: independent identically distributed binary sequences that are statistically uncorrelated and never repeat. No spreading codes could produce less interference, but multiple access interference (MAI) still occurs.

In the ideal receiver processing, multiplying by $e^{-i2\pi f_1 t}$ removes the carrier (called "carrier wiping" or "wiping off the carrier" in satnav receiver vernacular), and multiplying by $e^{-i\theta}$ removes the carrier phase rotation, producing $r(t)$, whose real part is the sum of the delayed biphase signal, $s(t-D)$, and the real part of $w_1(t) = w(t)e^{-i2\pi f_1 t}e^{-i\theta}$, while the imaginary part of $r(t)$ is the imaginary part of $w_1(t)$. All of the signal contributions to the time-limited correlation between $r(t)$ and a replica consisting of the baseband biphase signal are in the real part since the replica used in the correlator is real valued. Consequently, the imaginary part of the correlator output contains no signal-related information and can be neglected, reducing the noise and interference power by half in the correlator output.

In practice, the input to the receiver processing in Figure 5.1 does not have infinite bandwidth, but instead is bandlimited. It is often useful to approximate this bandlimiting by a bandpass filter with rectangular magnitude transfer function whose width is called the precorrelation bandwidth. In practice such bandlimiting may be the concatenation of multiple filters and other processing in the receiver.

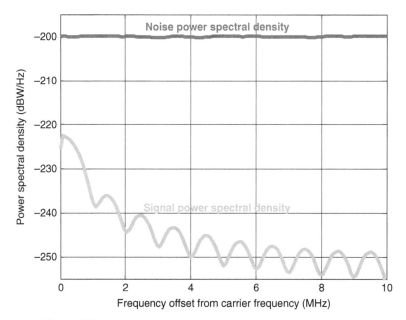

Figure 5.2. Estimated Power Spectral Densities at Receiver Input

A computer simulation illustrates this correlation processing. Figure 5.2 shows PSDs estimated from realizations of the input signal, along with input signal plus noise, using a BPSK-R signal with 1 MHz spreading code chip rate, signal power $C = -165$ dBW, and complex white noise having PSD $N_0 = -200$ dBW/Hz. The input signal power-to-noise density ratio is then $C/N_0 = 35$ dB-Hz. This quantity, whose units are equivalent to dB/s, describes the ratio of the power in the entire received signal to the power in a 1 Hz bandwidth of the complex-valued white noise. Unlike the infinite bandwidth of the ideal correlator in Figure 5.1, the simulation uses a precorrelation bandwidth of ± 10 MHz, modeled as an ideal rectangular bandpass with infinite attenuation in the stopbands.

Since white noise has infinite bandwidth, the input noise power is $\beta_r N_0$ for receiver precorrelation bandwidth $\pm \beta_r/2$. For the assumed white noise, the input SNR (ratio of input signal power to input noise power) can be made arbitrarily close to zero by choosing an arbitrarily large precorrelation bandwidth. Nonetheless, it will be shown that maximum output SNR in white noise is achieved with the largest possible precorrelator bandwidth, even though such a bandwidth makes the input SNR vanishingly small.

Passing multiple realizations of signal, interference, and noise through the correlator produces multiple time-limited correlation functions. Figure 5.3 shows ten of these, each computed with a correlation integration time $T = 5$ ms, and with $D = 0$. Time-limited correlations between the real part of the input and the real-valued replica are labeled "real" in the figure. Also shown are time-limited correlations computed between the imaginary part of the input and the replica, labeled "imaginary". Correlator output SNR is defined as the ratio of the mean squared of the real correlator output to its variance, where the mean and variance are computed at $\tau = D$. The values labeled "real" have a nonzero average approaching a triangle with peak at $\tau = D$, while the values labeled

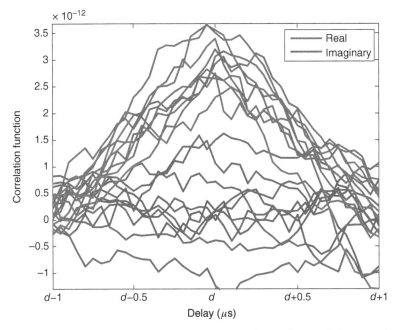

Figure 5.3. Correlation Function Estimates between the Real Part of the Input (Labeled "Real") and the Replica, and between the Imaginary Part of the Input (Labeled "Imaginary") and the Replica

"imaginary" tend to average to a value of zero everywhere. Observe from Figure 5.3 that, due to variations in each realization of signal and noise, individual time-limited correlations do not always have a maximum value at the true delay, although their average tends to do so. Correlator output SNR measures the variation of correlation peak amplitude about the mean. It applies regardless of the contamination (noise, interference, and rapidly changing multipath) that causes this fluctuation, as long as it induces random fluctuations that independently affect sequential correlations.

If the contamination is correlated with the replica over infinite time or does not cause independent effects over different correlation integration times, it may not be adequate to model its effect on correlator output SNR. If the contamination introduces slowly changing variations rather than independent fluctuations, other models should be explored. Two practical situations where contamination introduces slowly changing fluctuations are addressed in subsequent chapters: nonzero correlations between strong interfering signals and weak desired signals during weak-signal acquisition (Section 16.4), and static multipath (Chapter 22).

5.2 CORRELATOR OUTPUT SNR WITH NO INTERFERENCE

This section considers the effects of only noise; interference effects are considered in subsequent sections. Figure 5.4 shows the correlation function sample outputs corresponding to the true delay value for 1000 realizations of signal and noise, and the

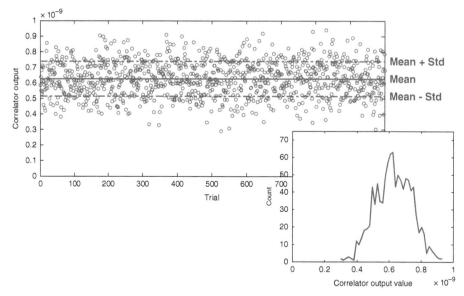

Figure 5.4. Correlation Function Values with Replica Matched in Delay and their Histogram, Using 1000 Realizations of Signal and Noise

resulting histogram, using precorrelation bandwidth of ± 10 MHz. The scaled sample mean is 6.27×10^{-10} and the corresponding sample standard deviation is 1.13×10^{-10}, resulting in an estimated output SNR of $20 \log_{10}\left(\frac{6.27 \times 10^{-10}}{1.13 \times 10^{-10}}\right) = 14.9$ dB. As shown in Reference 2, this ideal correlator with finite precorrelation bandwidth has output SNR in white noise

$$\rho = 2T\frac{C}{N_0}\eta \qquad (5.1)$$

where T is the correlation integration time and the fraction of desired signal power passed by the precorrelation filter is

$$\eta \stackrel{\Delta}{=} \int_{-\beta_r/2}^{\beta_r/2} \bar{\Phi}_s(f) df \qquad (5.2)$$

and $\bar{\Phi}_s(f)$ is the PSD of the unit-power signal with $\int_{-\infty}^{\infty} \bar{\Phi}_s(f) df = 1$. Observe when the precorrelation filter bandwidth is greater than the 90% power bandwidth (see Appendix Section A.6), the output SNR in white noise is within 0.5 dB of the result for infinite bandwidth.

The factor of 2 in (5.1) results from coherent correlation, since half the noise power is in phase quadrature to the signal component and is ignored in the correlation processing.

For the example shown in the preceding figures, the precorrelation bandwidth is large enough to pass essentially all the signal power, so $\eta \cong 1$. Using (5.1) and the numerical values from this example, the output SNR ρ expressed in decibels is $\rho = 10\log_{10}(2) + 10\log_{10}(0.005) + 35 = 15$ dB, very close to the 14.9 dB estimated from the 1000 realizations in Figure 5.4.

For this case of signal in white noise with no interference, the output SNR is proportional to the product of correlation integration time and C/N_0. Since a PSD is non-negative, η is nondecreasing as a function of precorrelation bandwidth β_r, meaning that the output SNR is maximized when the precorrelation bandwidth is infinite, even though as precorrelation bandwidths approach zero, the input SNR $C/(\beta_r N_0)$ approaches zero.

Observe the signal's PSD affects correlator output SNR only through the fraction of signal power passed by a given precorrelation bandwidth. Parseval's Theorem (see Appendix A.2) shows that the fraction of signal power passed by a given precorrelation bandwidth is equal to the height of the correlation peak. Hence, the output SNR is proportional to the height of the correlation function peak, which is only a function of signal power and not the shape of the correlation function; it does not depend on whether the correlation function has one or multiple peaks, or whether the peaks are narrow or wide.

5.3 CORRELATOR OUTPUT SNR WITH INTERFERENCE: SPECTRAL SEPARATION COEFFICIENTS AND PROCESSING GAIN

When $w(t)$ in Figure 5.1 is the sum of white noise and wide sense stationary interference that is uncorrelated to the signal having PSD $\Phi_I(f)$, Reference 2 shows that the correlator output SNR is

$$\rho = 2T \frac{C\eta^2}{\eta N_0 + I\kappa_{ls}} \quad (5.3)$$

where

$$\kappa_{ls} \stackrel{\Delta}{=} \int_{-\beta_r/2}^{\beta_r/2} \bar{\Phi}_I(f)\bar{\Phi}_s(f)df \quad (5.4)$$

and η is defined in (5.2).

Here, $\bar{\Phi}_I(f)$ is the interference PSD normalized to have unit power over infinite bandwidth, and I is the received power of the interference, so the interference PSD is $I\bar{\Phi}_I(f)$. κ_{ls} is called the spectral separation coefficient (SSC), and has units of seconds, or Hz^{-1}. The SSC typically takes on very small values and is often expressed in dB/Hz.

SSCs quantify how much the interference "leaks into" the correlator output, increasing the variance of the correlator peak amplitude relative to the square of its average value. Additional characteristics of SSCs are explored later in this chapter, and in the Review Questions at the end of this chapter.

Since the desired signal and interference are orthogonal, $E\{s(t)w(t+\tau)\} \equiv 0, \forall \tau$, the interference does not affect the mean correlator output. But when $I\kappa_{ls}$ is nonzero, the interference increases the variance of the correlator output. Only when the PSDs are orthogonal so the SSC is zero does interference not affect the variance of the correlator output. This subtlety is often misunderstood. Even with ideal long spreading codes and no code crosscorrelations within the spreading code family, there is MAI as long as the signals have overlapping frequencies.

When the interference is strong so that its effects dominate that of the white noise, $N_0\eta + I\kappa_{ls} \cong I\kappa_{ls}$. When the precorrelation bandwidth is wide enough so that $\eta \cong 1$, (5.3) becomes

$$\rho \cong 2T\frac{C}{I\kappa_{ls}} = 2\left(\frac{T}{\kappa_{ls}}\right)\left(\frac{C}{I}\right) \tag{5.5}$$

The term C/I is the input signal-to-interference ratio (SIR), and the output SNR is twice the input SIR times the processing gain T/κ_{ls}. This processing gain term generalizes the conventional definition of processing gain associated with DSSS communications systems, which is said to be the ratio of the chip rate to the data rate [3]. The processing gain T/κ_{ls} explicitly accounts for bandwidth and applies to any signal and interference PSDs. It applies even if there is no data modulated onto the signal (e.g., in the case of a pilot component), and shows how to characterize performance of receivers using correlation integration times different from the duration of a data message bit or pilot code bit.

A signal with BPSK-R spreading modulation and spreading code chip rate f_c has unit-area PSD $\bar{\Phi}_s(f) = \frac{1}{f_c}\text{sinc}^2(\pi f/f_c)$. Suppose each data bit or symbol is modulated onto the carrier at a rate $1/T_b$ using biphase modulation so the PSD of this data signal is $T_b \text{sinc}^2(\pi f T_b)$. The null-to-null bandwidth of the spread signal is $2f_c$ while the null-to-null bandwidth of the data is $2/T_b$, so the ratio of the spread bandwidth to the data bandwidth is $T_b f_c$.

Now suppose the correlation integration time in (5.5) is $T = T_b$, and evaluate T/κ_{ls} for this same situation. If the interference is a very narrowband waveform whose normalized PSD can be modeled as a Dirac delta function centered at the peak of the signal's PSD, $\bar{\Phi}_I(f) = \delta(f)$, then the SSC is readily found using the sifting property of the delta function: $\kappa_{ls} = \int_{-\beta_r/2}^{\beta_r/2} \bar{\Phi}_I(f)\bar{\Phi}_s(f)df = \frac{1}{f_c}\int_{-\beta_r/2}^{\beta_r/2} \delta(f)\text{sinc}^2(\pi f/f_c)df = 1/f_c$. The processing gain is then $T/\kappa_{ls} = T_b f_c$, consistent with the results above for the conventional definition.

However, as explored further in the Review Questions at the end of the chapter, if the signal had a BOC spreading modulation or some other spreading modulation producing a different PSD shape, or if the narrowband interference were not centered at the peak of the sinc-squared PSD, or if the interference were other than narrowband, then the conventional definition could either overstate or understate processing gain, while the ratio T/κ_{Is} explicitly accounts for signal and interference PSDs and carrier frequencies, and any correlation integration time.

Continuing to examine correlator output SNR in the presence of noise and interference, consider when the interference is expressed as the aggregation of multiple types of interference, with each type having a different spectral shape. This occurs, for example, with interference from a satnav system transmitting different signals with different spectral shapes at the same carrier frequency. In that case $I\bar{\Phi}_I(f) = \sum_{k=1}^{K} I_k \bar{\Phi}_{I_k}(f)$, and

$$\rho = 2T \frac{C\eta^2}{\eta N_0 + \sum_{k=1}^{K} I_k \kappa_{I_k s}},$$

$$\kappa_{I_k s} = \int_{-\beta_r/2}^{\beta_r/2} \bar{\Phi}_{I_k}(f) \bar{\Phi}_s(f) df, \quad \int_{-\infty}^{\infty} \bar{\Phi}_{I_k}(f) df = 1$$

(5.6)

This decomposition is not unique; any or no decomposition can also be selected based on convenience. In analyzing satnav interference, it is often convenient to represent each signal type from each system (e.g., all the E1 OS signals from Galileo) as a separate term in the sum, so the effect of each signal type can be bookkept separately.

5.4 EFFECTIVE C/N$_0$

It is clear from (5.6) that many different combinations of signal power, noise density, SSCs, and interference power can produce the same correlator output SNR. When the precorrelation bandwidth is infinite and there is no interference, however, there is a unique value of C/N$_0$ that produces a given output SNR for a given correlator integration time. For a given output SNR, this is called the effective C/N$_0$, given by $(C/N_0)_{\text{eff}} = \rho/(2T)$. Applying this definition to (5.6) yields the expression for effective C/N$_0$:

$$\left(\frac{C}{N_0}\right)_{\text{eff}} = \frac{C\eta^2}{N_0 \eta + \sum_{k=1}^{K} I_k \kappa_{I_k s}}$$

(5.7)

Effective C/N$_0$ enables an analyst to represent all the complexities of interference, bandlimiting, and even processing losses in the receiver (as discussed in Chapter 15) by an equivalent ratio of signal power to white noise PSD and an ideal correlator. Given the same correlator integration time, the effective C/N$_0$ and an ideal correlator will produce the same correlator output SNR as the actual correlator with inputs of signal, noise, and interference, hence equivalent performance for acquisition, carrier tracking, and data message demodulation.

$(C/N_0)_{\text{eff}}$ should not be used for code tracking performance assessment, however. As shown in Chapter 19, code tracking performance is affected by characteristics of signal and interference beyond those that affect correlator output SNR.

5.5 INTERFERENCE-TO-SIGNAL POWER RATIOS AND EFFECTIVE C/N₀

It is common in specifications for military receivers, and in other system-level discussions of receiver performance, to quantify the interference environment in terms of the ratio of interference-to-signal power at the input to the receiver, colloquially referred to as I/S, or, if the interference is called jamming, J/S (it is consistently called interference in this book). The results provided in this section show how to convert between $(C/N_0)_{\text{eff}}$ and I/S.

The first step in relating these two quantities is recognizing the use of two different symbols for the same quantity: C and S each represent the received power of the desired signal, so $C/I = S/I$. In (5.7), let there be only one type of interference having power I having SSC κ_{ls}. Assuming the interference has much greater effect than white noise so $N_0\eta + I\kappa_{ls} \cong I\kappa_{ls}$. (A more complicated expression can be derived if white noise is retained as shown in Review Question QT5.3, but it usually is not needed for systems engineering calculations.) Substituting C for S in (5.7) and rearranging terms yields

$$\frac{I}{S} \cong \frac{\eta^2}{\kappa_{ls}\left(\frac{C}{N_0}\right)_{\text{eff}}} \tag{5.8}$$

When I/S is expressed in decibels, κ_{ls} in dB/Hz, and $(C/N_0)_{\text{eff}}$ in dB-Hz, and β_r is large enough that η is approximately unity

$$\frac{I}{S} \cong -\kappa_{ls} - \left(\frac{C}{N_0}\right)_{\text{eff}} \tag{5.9}$$

5.6 A DEEPER LOOK AT SPECTRAL SEPARATION COEFFICIENTS

It is useful to develop approximate SSC values for representative types of spreading modulations and specific types of interference, and to calculate SSC values associated with common spreading modulations. This section provides these results, along with a discussion of when the SSC expressions presented in Section 5.3 do—and do not—apply.

Table 5.1 provides approximate expressions for SSCs in the case of two different common spreading modulation types and three different interference types. Interference having the same spectrum shape as that of the desired signal is called matched spectrum interference. For reception of BOC spreading modulations in the presence of narrowband interference, the SSC is the same when the interference is located at either spectrum

A DEEPER LOOK AT SPECTRAL SEPARATION COEFFICIENTS

TABLE 5.1. Approximate SSCs for Some Common Spreading Modulation Types and Interference Types

Interference Type	Approximate SSC for Desired Spreading Modulation Type	
	BPSK-R with Chip Rate f_c	BOC with Chip Rate f_c
Narrowband centered at peak of desired signal's spectrum	$\dfrac{1}{f_c}$	$\dfrac{0.5}{f_c}$
Matched spectrum	$\dfrac{2}{3f_c}$	$\dfrac{2\alpha}{3f_c}, \quad \dfrac{1}{3} \leq \alpha \leq 0.4$
Bandpass white over bandwidth $\pm\beta_r/2$ wide enough to pass most of desired signal power	$\dfrac{1}{\beta_r}$	$\dfrac{1}{\beta_r}$

peak or when the interference is placed at both spectrum peaks, with power divided in any ratio between them, as shown in Review Question QT5.2. For BOC spreading modulations, the SSC associated with matched spectrum interference (called the self-SSC) takes on a range of values—for larger BOC ratios where the subcarrier rate is much larger than the spreading code chip rate, the value of α is as small as $1/3$; in other cases, the value can approach 0.4. Opportunities for exploring these and other SSC values are provided in the Review Questions at the end of this chapter.

Table 5.2 provides numerical values for common spreading modulations and two different types of interference. The values are computed for signal and interference PSDs bandlimited to ±15.345 MHz, and receiver precorrelation bandwidth of the same value.

SSCs calculated using (5.4) apply in a wide range of practical cases, but not in every case. Since SSCs were introduced in Reference 4 for use in satnav, they have been studied in detail and extensively used to predict the effects of intrasystem and intersystem satnav interference, as well as other kinds of in-band interference. The derivations leading to the expressions for SSCs given in this chapter require the spectrum of either the desired signal or the interfering waveform, or both, to be smooth in the sense that it is well modeled

TABLE 5.2. SSC Values for Some Common Spreading Modulation Types and Interference Types

Interference Type	SSC Value (dB/Hz) for Desired Spreading Modulation Type			
	BPSK-R(1)	BPSK-R(10)	BOC(10,5)	BOC(1,1)
Narrowband centered at peak of desired signal's spectrum	−60.1	−69.9	−69.9	−62.8
Matched spectrum	−61.8	−71.3	−71.6	−64.7

as linear over frequency differences corresponding to the reciprocal of the correlation integration time. This condition occurs in a pilot component when its spreading code, or its overlaid spreading code, is roughly 100 ms or longer, or in a data component when the overlaid spreading code is at least as long as a data bit duration. The 100 ms applies for typical correlation integration times and dynamics associated with MEO satellites. In these cases, SSCs calculated using the PSD dot product in (5.4) provide good models for interference effects on correlator output SNR.

The conditions above apply for most currently defined satnav signals, with a few exceptions noted here. As described in Section 7.3, the spreading code on the GPS L5 signal's pilot component, sometimes termed the I5 signal, is only extended by its pilot code to 20 ms length. Consequently, the self-SSC that describes interference from other GPS L5 pilot components to reception of a GPS L5 pilot component may sometimes be a decibel or more larger than what is calculated using (5.4). However, when the interference calculation also includes equal power interference from the L5 data component as well as from other satnav systems, the aggregate effect of interference to reception of the L5 pilot component is reasonably approximated by the use of (5.4).

The same conclusion does not apply for self-interference from C/A code signals. As described in Reference 5, the repetition of 20 one ms long C/A spreading code within each data bit produces a C/A code spectrum with lines spaced by 1 kHz that can be approximately 20 dB higher than the PSD associated with an ideal long spreading code. When the correlation integration time is greater than 1 ms, if spectral lines from one or more interfering C/A signals align with the spectral lines in the correlator replica, the degradation of correlator output SNR is much greater than predicted using (5.4) and the PSD for a signal with ideal long spreading code. Since interfering signals are distributed in delay and Doppler, the much greater interference from the interfering signals whose spectral lines align with those of the replica is partly offset by those that do not align, but often the resulting SSC can be larger than predicted using (5.4).

If the spreading codes were ideal long spreading codes, a 20 ms correlation integration time would compute the correlation over 20,460 bits of spreading code, suppressing the interference's effect on correlator output SNR consistent with an SSC value of -61.8 dB/Hz. But if data bit transitions in the interfering signal and desired signal are aligned with no relative Doppler shifts, the C/A code signal's structure of 20 repeating 1023-bit long Gold codes within a data bit means that the average over 20 ms is similar to that over 1 ms, yielding an SSC that can be as large as approximately 13 dB larger than with the ideal long spreading code.

In practice, the effects of C/A on C/A interference typically involve an aggregation of interfering signals, each received with different delays and Doppler values relative to those of a desired C/A code signal. The interference effects vary significantly over time and location, as well as with the correlation integration time. Various approaches have been explored to quantify the resulting effects, including References 6–11.

The consensus appears to be that, for 20 ms correlation integration time, the SSC for interference from a single C/A code signal to reception of a C/A code signal can exceed -50 dB/Hz, consistent with the assessment above indicating the potential for an SSC 13 dB larger than the ideal long spreading code value of -61.8 dB/Hz. However, with aggregate C/A code signal interference from a full GPS constellation of GPS satellites, not all the interfering signals are received simultaneously at the delay and Doppler that produce the greatest interference effects. Thus, for a large GPS constellation with correlation integration time of 20 ms, a representative severe SSC for aggregate C/A code signal interference to C/A code signal reception is approximately −55 dB/Hz. The corresponding SSC value may be 1–2 dB larger for smaller constellations, but the aggregate interference power is lower. For a large GPS constellation with correlation integration time of 10 ms, a representative severe SSC for aggregate C/A code signal interference to C/A code signal reception is approximately −56 dB/Hz. At other times and locations, the SSC values can be −75 dB/Hz or smaller. These interference levels occur intermittently and not to all desired signals simultaneously.

For 1 ms correlation integration times, the SSC value of −61.8 dB/Hz, corresponding to an ideal long spreading code, is appropriate for C/A on C/A interference.

5.7 MULTIPLE ACCESS INTERFERENCE AND AGGREGATE GAIN OF A CONSTELLATION

The increasing number of satellites and signals sharing common parts of L band, combined with trends for increased signal power, lead to greater awareness of MAI. Intrasystem MAI is caused by interference to reception of a specific satellite's signal from other signals transmitted by the same satellite and other satellites in the same system's constellation. Intersystem MAI is caused by interference to reception of a specific satellite's signal from signals transmitted by other systems' constellations. Since the interference from a single signal at typical power levels is typically negligible, it is the aggregation of interference from many signals that makes the MAI potentially significant.

A systems engineering procedure has been developed to assess aggregate interference from a satnav constellation. For a signal of a given type (i.e., GPS C/A code signals, or GLONASS L1OC signals) transmitted by a constellation of satellites, the total interfering signal power at the input to a receiver depends upon the receiver's location, the receive antenna's gain pattern, the location of all interfering satellites, the transmit antenna gain pattern, and the transmit power of the interfering signal.

A reference receive antenna gain pattern is specified for purposes of assessing aggregate MAI. This is usually a different reference receive antenna than that discussed in Section 4.2 for purposes of defining minimum and maximum received power from a single satellite. Typically, MAI is assessed using a gain pattern that is axially symmetric and bore sighted at vertical, for a receiver located on the surface of a spherical Earth. An example reference receive antenna gain pattern for assessing aggregate MAI is provided in Table 5.3. The gain is quite low at low elevation angles, meaning little received interference from low elevation satellites. Conversely, the gain exceeds 0 dBic at higher elevation angles, meaning larger interference from satellites overhead.

For a receiver at a particular location and time, the aggregate interference power, I (location, t), from all satellites transmitting a given type of signal is

$$I(\text{location}, t) = \sum_{k=1}^{K} I_k G(\theta_k) \quad (5.10)$$

TABLE 5.3. Example Reference Receive Antenna Gain Pattern for Assessing Aggregate MAI

Elevation Angle (°)	Gain (dBic)
0	−7
5	−6
10	−1
15	0
20	1
25	1
30	1
35	1
40	1
45	1
50	1
55	2
60	2
65	2
70	2
75	2
80	2
85	2
90	2

where I_k is the power that would be output by a 0 dBi receive antenna, $G(\theta)$ is the receive antenna gain at elevation angle θ (assuming symmetry in azimuth) at a particular time and receiver location, and θ_k is the elevation angle to the kth satellite at a particular time and receiver location.

In order to be conservative, the I_k are typically calculated using the approach described in Section 4.2 for calculating maximum received power, with the RF power into the transmit antenna set to meet the maximum received signal power specification at some point on Earth. The maximum received signal power specification for this signal type is denoted $I_{\max}$.

The maximum aggregate gain of interference, denoted G_{agg} and pronounced as two syllables: "G-agg," for a particular type of signal is then defined, using (5.10), as

$$G_{\text{agg}} = \max_{\text{location},t} \{I(\text{location}, t)\}/I_{\max} \qquad (5.11)$$

G_{agg} is a dimensionless quantity typically expressed in decibels.

In (5.11), the maximum over time is typically computed using simulations over the time duration a constellation takes to repeat its ground track, and the maximum over receiver location is computed on a grid over the entire Earth's surface.

Even signals from the same satellite may have different G_{agg} values, if different signals are transmitted from different antennas, and since antenna gain patterns differ at different center frequencies. The MAI from the mth given type of signal has effective noise density at the receiver:

$$I_{0,m} = G_{agg,m} I_{max,m} \kappa_{m,s} \qquad (5.12)$$

where $G_{agg,m}$ is the G_{agg} for the mth signal type, $I_{max,m}$ is the specified maximum received power of this signal type, and $\kappa_{m,s}$ is the SSC between that interfering signal type and the desired signal. When a system transmits M different signal types whose spectra overlap that of the desired signal, that system's MAI produces effective noise density

$$I_{0,system} = \sum_{m=1}^{M} I_{0,m} \qquad (5.13)$$

This value may somewhat overstate the largest possible total effective noise, since if different signals have different G_{agg} values, each of them may occur at a different receiver location and time. Also, MAI from the same type signal as the desired signal should, strictly speaking, not include the desired signal itself in the sum (5.13), but including the desired signal simplifies the calculation considerably, and often minimally increases the calculated total effective noise.

An upper bound on the aggregate effective noise to reception of a desired signal type is obtained by summing the results of (5.13) for each system. A tighter bound is obtained if regional systems are only included at locations where its satellites are above the horizon.

While this approach overstates the possible interference, it has a number of attractive characteristics:

- It involves quantities separately computed for each signal type and each system, being independent of relative satellite phasing or other interactions among different systems,
- Once G_{agg} values are computed, interference calculations are simply computed,
- The result is conservative, providing some margin for protection of receivers from interference.

In a number of different situations, it has been observed that this approach is not excessively conservative, with aggregate effective noise assessed using this approach only 1 or so dB greater than much more complicated approaches that consider all signal types from all constellations simultaneously. This approach has been adopted by the ITU in Reference 12.

5.8 SUMMARY

This chapter presents important systems engineering techniques for assessing in-band interference and its effects on receiver performance. These results apply only when the received power of the interference is low enough that receiver hardware components pass it without unintended distortion. Correlator output SNR is the primary metric used to assess receiver performance in interference, and expressions for correlator output SNR are provided for desired signals in noise and interference, including the effect of precorrelation bandlimiting. The effects of interference depend upon both the received interference power and the SSC, which accounts for the spectrum shape of the interference and the desired signal. For a real-world situation where the correlator has inputs of signal, noise, and interference, and whose implementation introduces processing losses that affect the apparent power of signal and interference, there is an equivalent ratio of received signal power to white noise PSD, called the effective C/N_0, that produces the same output SNR from an ideal correlator, and can be used to represent the real-world situation in terms of correlator output SNR that affects performance of signal acquisition, carrier tracking, and data message demodulation. The commonly employed concepts of processing gain and interference-to-signal ratios can be rigorously defined and employed using SSCs.

SSCs are a powerful tool for assessing the effect of interference on correlator output SNR. For most of the signal designs addressed in this book, the SSC can be computed as the spectral cross product between the PSD of the desired signal and the PSD of the interfering signal type. This model is not accurate for predicting the effects of interference from C/A code signals to C/A code signal reception, and specialized, complicated calculations must be performed to obtain exact results for this case.

The effects of MAI from all signals of a given type in a system can be assessed using the interfering signal type's aggregate gain, or G_{agg}, and an upper bound on interference from all interfering signal types can readily be found using each signal type's maximum specified received power, G_{agg}, and SSC to the desired signal.

REVIEW QUESTIONS

Theoretical Questions

QT5.1. Suppose the desired signal has a rectangular PSD with width $\pm B_d/2$ and the interference has the same center frequency and a rectangular PSD with width, and the precorrelation bandwidth is greater than or equal to $\pm B_d/2$. Find the SSC between the interference and the desired signal as a function of B_d and B_I, including for the case of matched spectrum interference $B_d = B_I$. What conclusions do the results lead to concerning choosing desired signal bandwidth to reduce the effects of interference?

QT5.2. Suppose the desired signal has a PSD that is symmetric about the center frequency. Consider two different interference PSDs having the same power:

$$\Phi_{I,1}(f) = \begin{cases} \Phi_+(f), & f \geq 0 \\ 0, & f < 0 \end{cases} \quad \text{and} \quad \Phi_{I,2}(f) = \begin{cases} \Phi_+(f)/2, & f \geq 0 \\ \Phi_+(-f)/2, & f < 0 \end{cases}$$

Show that both interference PSDs have the same SSC with the desired signal.

QT5.3. Derive the relationship between I/S and effective C/N_0 when white noise is included rather than assumed insignificant, assuming the precorrelation bandwidth is wide enough to pass all signal power.

Application Questions

QA5.1. Use numerical techniques to calculate the SSCs, including self-SSCs, between signals with the following spreading modulations, all having the same center frequency assuming the signal and interference powers normalized over infinite bandwidth and the precorrelation bandwidth is infinite: BPSK-R(1), BPSK-R(5), BPSK-R(10), BOC(10,5), BOC(1,1), MBOC(6,1,1/11), BOC(5,1), BOC(5,2), BOC(14,2), $BOC_c(15,2.5)$.

QA5.2. Suppose a receiver has white noise density of -201.5 dBW/Hz. For the spreading modulations, BPSK-R(1), BPSK-R(10), BOC(10,5), BOC(1,1), what narrowband interference power (when the narrowband interference is at the frequency that causes the most interference) degrades the correlator output SNR by an amount equal to the white noise, so the effective C/N_0 is 3 dB lower than the C/N_0? Use SSCs computed in Table 5.2.

QA5.3. Suppose interference dominates white noise, and the ratio of interference to signal is 40 dB, where the interference has the same spectral shape as that of the signal (i.e., matched spectrum interference). Find the resulting effective C/N_0 for the following spreading modulations: BPSK-R(1), BPSK-R(10), BOC(10,5), BOC(1,1). Use SSCs computed in Table 5.2.

QA5.4. For desired signal power -164.0 dBW, find the effective C/N_0 to a receiver of signals having the following spreading modulations: BPSK-R(1), BPSK-R(10), BOC(10,5), BOC(1,1), when the maximum received power and G_{agg} from each type of interference is given in the following table, and the white noise density is -200.0 dBW/Hz. Use SSCs computed using signal and interference powers normalized over infinite bandwidth with infinite precorrelation bandwidth.

Interference Type	Maximum Received Power from a Single Signal (dBW)	G_{agg} (dB)
BPSK-R(1)	-154.0	12.0
BPSK-R(10)	-150.0	13.0
BOC(10,5)	-145.0	14.0
BOC(1,1)	-152.0	13.0

QA5.5. If a desired signal is received at C/N_0 of 35 dB-Hz and eight interfering signals are each received at 45 dB-Hz, and the interfering signals have SSC with the desired signal of -65 dB/Hz, what is the effective C/N_0 of the desired signal? Assume the precorrelation bandwidth is wide enough to pass all the signal power.

QA5.6. Suppose that interference has a flat spectrum over 24 MHz, a bandwidth that contains essentially all the desired signal power. What I/S level produces effective C/N_0 of 30 dB-Hz?

REFERENCES

1. National Space-Based Positioning, Navigation, and Timing Systems Engineering Forum (NPEF), "Assessment of Light Squared Terrestrial Broadband System Effects on GPS Receivers and GPS-Dependent Applications," available at: http://www.govexec.com/pdfs/061711bb1.pdf, accessed 19 January 2013.
2. J. W. Betz and K. R. Kolodziejski, "Generalized Theory of Code Tracking with an Early-Late Discriminator, Part 1: Lower Bound and Coherent Processing," *IEEE Transactions on Aerospace and Electronic Systems*, Vol. 45, No. 4, 2009, pp. 1538–1550.
3. J. D. Gibson (Ed.), *The Communications Handbook*, 2nd edition, CRC Press, 2002.
4. J. W. Betz and D. B. Goldstein, "Candidate Designs for an Additional Civil Signal in GPS Spectral Bands," Proceedings of the Institute of Navigation National Technical Meeting 2002, ION-NTM-2002, Institute of Navigation, January 2002.
5. J. W. Betz, "On the Power Spectral Density of GNSS Signals, with Applications," Proceedings of the Institute of Navigation International Technical Meeting 2010, ION-ITM-2010, Institute of Navigation, January 2010.
6. C. O'Driscoll and J. Fortuny-Guasch, "On the Determination of C/A Code Self-Interference with Application to RFC Analysis and Pseudolite Systems," ION-GNSS-2012.
7. S. Raghavan, J. Hsu, and T. Powell, "Upper Bound on C/A Code Spectral Separation Coefficient," IEEE Aerospace Conference 2010.
8. A. Cerruti, J. J. Rushanan, and D. Winters, "Modeling C/A on C/A Interference," Proceedings of the Institute of Navigation International Technical Meeting 2009, ION-ITM-2009, January 2009.
9. J. J. Rushanan, "Methods to Determine the Spectral Separation Coefficient between C/A Code Signals," MITRE Product MP120583, October 2012.
10. A. P. Cerruti, J. W. Betz, and J. Rushanan, "Further Investigations into C/A-to-C/A Interference," Proceedings of the 2014 International Technical Meeting of the Institute of Navigation, San Diego, California, January 2014, pp. 349–361.
11. C. J. Hegarty, "A Simple Model for C/A-Code Self-Interference," Proceedings of the 27th International Technical Meeting of the Satellite Division of the Institute of Navigation (ION GNSS+ 2014), Tampa, Florida, September 2014, pp. 3484–3494.
12. International Telecommunications Union, "A Coordination Methodology for RNSS Intersystem Interference Estimation," Recommendation ITU-R M.1831, 2007.

6

ERROR SOURCES AND ERROR CHARACTERIZATION

While satnav systems are exquisitely accurate relative to other sources of position and time available over large portions of the Earth, their measurements still have errors. An important aspect of satnav systems engineering involves understanding the sources of the errors, quantifying the error contributors, and estimating the resulting effect on PVT accuracy. Ultimately, the errors are unique to specific components of a particular system, at a particular time and location, and are best determined by measurement. However, it is still beneficial to develop an understanding at the system level, while recognizing that the approximations, simplifications, and generalizations involved make the results broadly applicable, but not precise in any given situation.

Section 6.1 describes sources of error, introducing different categories of errors and describing the dominant contributors to ranging error. Section 6.2 describes a simple and common model for converting ranging error to errors in estimating position and time, showing how the geometry of the satellites affects these errors and summarizing many different error measures with their relationships. Section 6.3 describes the resulting position and time errors, introducing differential satnav as a way to reduce errors significantly. Section 6.4 describes some other error sources, while Section 6.5 summarizes this chapter, wrapping up the discussion of satnav systems engineering and signals in Part I.

Engineering Satellite-Based Navigation and Timing: Global Navigation Satellite Systems, Signals, and Receivers, First Edition. John W. Betz.
© 2016 The Institute of Electrical and Electronics Engineers, Inc. Published 2016 by John Wiley & Sons, Inc.

140 ERROR SOURCES AND ERROR CHARACTERIZATION

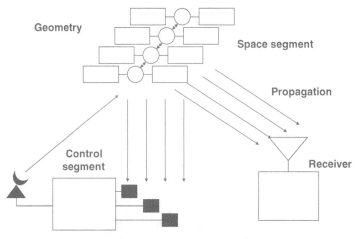

Figure 6.1. Categories of Satnav Errors

6.1 SOURCES OF ERROR IN SATNAV POSITIONING AND TIMING CALCULATION

Different satnav sources of error are portrayed in Figure 6.1. The space segment contributes errors by transmitting imperfect signals containing information that is not perfectly accurate. Propagation through the atmosphere, as well as local propagation conditions, further distort the signals arriving at the receive antenna. The received signals are corrupted by noise and interference, causing the receiver to make errors in estimating their time of arrival (TOA). While the control segment closes the loop so that each satellite transmits more accurate ephemeris and clock corrections, its corrections are also imperfect. All of the above contribute to errors in pseudorange, and the satellite geometry transforms these pseudorange errors into errors in estimating position and time.

The decomposition of error sources starts by addressing errors in estimating the pseudorange between receiver and satellite. (More precisely, the pseudorange is computed between the phase center of the receive antenna and the phase center of the satellite transmit antenna.) Overall pseudorange error, called user equivalent range error (UERE), is the ranging error along the vector between the receiver and a particular satellite. UERE is decomposed into two types of errors; the signal in space ranging error (SISRE)[1] and the user equipment error (UEE), each modeled as statistically independent. In typical use, UERE, SISRE, and UEE refer to error standard deviations, so

$$\text{UERE} = [(\text{SISRE})^2 + (\text{UEE})^2]^{1/2} \tag{6.1}$$

---

[1] SISRE is sometimes also called user range error (URE) in official GPS documents, but SISRE is employed in this book since it is a more descriptive term and more distinct from UERE.

SISRE accounts for pseudorange error due to the space segment and control segment, while UEE accounts for effects of propagation and receiver processing. (There is sometimes a philosophical debate about whether propagation effects should be included in SISRE or UEE, but since ultimately all the errors are root sum squared (RSSd) together, the outcome of this debate has no effect on UERE.) Dilution of precision (DOP), described in Section 6.2, converts UERE to positioning or timing error using the expression

$$\text{Standard deviation of positioning or timing error} = \text{DOP} \times \text{UERE} \quad (6.2)$$

As discussed in Section 6.2, different DOP measures apply for different positioning or timing measurements.

While the simplicity of the two fundamental error expressions, (6.1) and (6.2), is seductive, it also relies upon many models and assumptions that do not apply perfectly in practice. Fundamental to these equations is the assumption that all errors are zero mean and uncorrelated, with all UERE having the same variance. Since these assumptions rarely apply in practice, the resulting error models are not highly accurate, and typically do not apply to any particular situation. Nonetheless, these models do provide an overall view of system accuracy, and can be useful for general comparisons of a system's accuracy trends over time, or for comparing two systems. Moreover, the sources of error described here do apply to specific situations as well as overall system performance, so efforts that reduce these errors are generally beneficial.

Four distinct error sources—clock error, ephemeris error, satellite group delay error, and selective availability (SA)—dominate SISRE, and each is discussed in sequence. Clock error involves four error sources in describing how a particular satellite's clock deviates from system time:

- Clock estimation error characterizes the control segment's ability to estimate the offset from system time of a particular satellite's clock at a particular instant in time. These errors are caused by noise, interference, and multipath in the control segment's monitoring receivers. The resulting measurement errors are reduced by postprocessing that combines measurements from multiple receivers over time, but they never are eliminated.
- Clock prediction error reflects the fact that the control segment must forecast clock offsets, so the predictions can be uploaded to satellites and then downloaded by receivers for use at a time in the future. Even if there were no clock estimation error, the actual clock offsets in the future would depart from those that are predicted.
- Clock curve fit error is associated with the fact that the data message represents future clock errors using a parametric model (typically a quadratic polynomial), and finds the best parametric representation over a validity interval that could last hours. Even if there were no clock prediction error, there would still be a clock curve fit error since the clock variation is not perfectly described by this parametric model with finite precision parameters.

- Clock stability error reflects the fact that clock predictions account for long-term variations in a clock due to factors such as aging and temperature sensitivity. Short-term instabilities, or jitter, cause additional differences between the actual clock deviation at a particular instant in time and the projected deviation.

Ephemeris error involves three error dominant sources that affect how the broadcast ephemeris of a particular satellite differs from that satellite's position and velocity:

- Ephemeris estimation error, analogous to clock estimation error, characterizes the control segment's ability to estimate the satellite's location at a particular time. Such errors are reduced by a worldwide network of monitoring stations, a denser network of monitoring stations allowing redundant measurements by multiple stations, and the use of laser range finding rather than merely ranging using navigation signals.
- Ephemeris prediction error, analogous to clock prediction error, characterizes imperfections in the forecast of future ephemerides.
- Ephemeris curve fit error, analogous to clock prediction error, quantifies the extent the actual ephemeris over time is not represented by the combination of secular and periodic terms in the parametric model used to represent ephemeris with finite precision parameters.

There is no ephemeris stability error analogous to clock stability error, since the tremendous momentum of a satellite weighing a thousand kilograms or more and hurtling through space at several kilometers per second prohibits any such unintended abrupt maneuvers. When orbital maneuvers are intentionally caused by firing thrusters on the satellite for station-keeping or for changing orbital position, the satellite's signals are set unhealthy before the maneuver occurs and the satellite settles into a new orbit, with new ephemeris estimates by the control segment, before the signals are set healthy again and valid ephemeris data is provided.

Satellite group delay error is another contributor to SISRE. There is group delay in the signal paths within the satellite, contributing to a lag between the time at the satellite clock and the signal leaving the satellite antenna. For example, there are propagation delays as the signal passes through cabling connecting different components. These delays are carefully calibrated, and a group delay correction is broadcast as part of the data message. However, group delay can change over time and with temperature (so-called group delay stability), and also can differ between different signals, particularly at different frequencies (known as inter-signal group delay). These imperfections produce satellite group delay errors.

Before May 2000, these other contributors to GPS SISRE were dominated in standalone GPS civil signals by SA—the intentional distortion of the civil signal's clock correction and ephemeris parameters. Now that SA is turned off and not built into future GPS satellites, this fourth contributor to SISRE only merits attention for historical purposes, and much greater attention has been paid to reducing the other contributors to SISRE in GPS.

TABLE 6.1. Example Contributors to SISRE

Time Frame / Error	Legacy	Modern	Future
RMS clock error (m)	2.7	1.0	0.3
RMS ephemeris error (m)	2.7	1.5	0.3
RMS group delay error (m)	0.3	0.2	0.1
SISRE (m)	3.8	1.8	0.4

Table 6.1 provides some representative numerical values for these error contributors, showing their reduction over time from legacy (pre-2000), modern (circa 2010), and future (circa 2016+) operational systems. None of these values are specification values or associated with any particular system; they are generalizations based on a reasonably well-designed and well-operated system. The SISRE in the bottom row is merely the RSS of the terms in the preceding rows. It is sometimes surprising what factor limits accuracy as systems move into the modern and future eras. For example, there may be cases where the number of bits used to represent coefficients in the ephemeris representation can be the limiting factor in ephemeris accuracy.

The values in Table 6.1 for future are not particularly aggressive predictions, and values close to those listed for future may be observed in some satellites already launched. However, it will be some time before system operators (or satellite manufacturers) are willing to commit to such small SISRE values.

Five different sources of error dominate the propagation and receiver errors contributing to UEE.

- Ionospheric errors are caused by dispersive propagation through the ionosphere as discussed in Section 3.2. The ionosphere's frequency-dependent refractive index is less than unity, causing an advance of the carrier and a delay of the spreading modulation by as much as hundreds of nanoseconds. If the receiver were not to compensate for ionospheric errors, they could readily exceed 10 m even for a relatively quiet ionosphere. A single-frequency receiver using the broadcast ionospheric correction reduces the error by approximately a factor of two, although the actual results can vary considerably. Finally, a dual-frequency receiver that estimates TEC as described in Section 3.2 and produces ionosphere-free range measurements as described in Chapter 20 removes ionospheric errors to a factor of the ranging accuracy, which can often be a fraction of a meter. In contrast to the simple error model being employed in this chapter, ionospheric errors are rarely uncorrelated between different satellite signals, since multiple signals typically pass through regions of the ionosphere having similar TEC values.
- Tropospheric errors occur as the signal propagates through the troposphere, an atmospheric layer below the ionosphere extending to an altitude as high as approximately 50 km. Dry gases and water vapor in the troposphere refract L band satnav signals with a refractive index greater than unity, causing a delay in signal propagation typically between 5 and 10 ns for a satellite directly overhead. The

delay is independent of frequency over the range of frequencies used in satnav, so there is no dispersive effect. The refractive index depends on temperature, pressure, and relative humidity in the troposphere. Larger delays occur for lower elevation satellites, since the signal propagates through the troposphere for greater distances. The delay is typically partitioned into hydrostatic delay, caused by dry gases, which typically contributes about 90% of the delay, and the non-hydrostatic delay, caused by water vapor. Since most of the water vapor is in the lower portion of the troposphere, receivers on aircraft experience smaller delays than those on the ground. Various models of tropospheric delay exist [1 and 2], and are used by receivers to reduce the tropospheric error as described in Chapter 20.

- Multipath errors, discussed in Chapter 22, are anticipated to often be the largest source of error in modern satnav with dual-frequency receivers or single-frequency receivers that are provided accurate ionospheric models through assisted satnav (see Chapter 24) or other means. Multipath occurs when the broadcast signal's electromagnetic wave reflects from an object into the receive antenna, contributing a delayed signal component that either distorts measurements made on the direct path, or replaces a shadowed or obscured direct path signal in the receiver processing. The result can be tens or even hundreds of meters of error in code tracking, and much smaller errors in carrier phase tracking. Receiver antennas and processing can reduce the error introduced by multipaths contaminating the direct path, and such processing techniques can be aided by modern signal designs.
- Group delay errors also occur in the receiver. Receivers of CDMA signals experience very small (fractional nanosecond) group delay errors when all measurements are made using a single signal type, since all the signals of the same type have the same frequency content and consequently most group delay effects are common to all the signals. These common group delays are treated as an additional time bias in the position calculation described in Section 1.2, and do not contribute to positioning error although they do contribute to time error. Antenna variations are the dominant source of group delay errors that affect positioning for receivers of CDMA signals. FDMA signals from different satellites, however, have different center frequencies and consequently experience different group delays that can be significant. Modern high-accuracy receivers that use FDMA signals include the capability to calibrate out much of the frequency-dependent group delay and consequently reduce this error term.
- Noise and interference errors produce jitter in the estimates of a signal's TOA and carrier phase. Such errors are characterized in Part III, and depend on the power of the noise and interference, the spectral shape of the interference, the desired signal, and the receiver processing.

These contributors to UEE can vary dramatically with design of the receive antenna and the receiver processing, with the characteristics of the signal being received, and with the environmental conditions. Table 6.2 provides some representative values for errors contributing to UEE, for a single-frequency receiver using the broadcast ionospheric

TABLE 6.2. Example Contributors to UEE for a Single-Frequency Receiver

Time Frame / Error	Legacy	Modern	Future
RMS ionosphere error (m)	7.0	7.0	5.0
RMS troposphere error (m)	0.2	0.2	0.2
RMS multipath error[a] (m)	0.4	0.2	0.1
RMS group delay error (m)	0.1	0.1	0.1
RMS TOA estimation error (m)	0.2	0.1	0.1
UEE (m)	7.0	7.0	5.0

[a]Multipath error with direct path present. If the direct signal is blocked and the receiver only tracks a reflected signal, the error is the extra path length of the reflected signal, which could be tens or hundreds of meters.

model for ionospheric corrections. The same three time frames employed in Table 6.1 are repeated here. The reduced ionospheric error for the future time frame represents the prospect of better ionospheric models being employed, with an example [3]. The tropospheric error assumes the receiver compensates using a good model of the troposphere. Multipath errors could be much larger or smaller than shown. The decreasing values for modern and future time frames represent improvements in receiver processing and in modern signal characteristics, as does the decreasing TOA estimation error. Clearly the ionospheric error dominates the single-frequency receiver UEE, although in some situations multipath error could be even larger.

Table 6.3 provides corresponding representative values for errors contributing to UEE, for a dual-frequency receiver that develops ionosphere-free measurements. The only differences from Table 6.2 are in the top and bottom rows. The top row is based on errors in TEC estimation caused by multipath errors and TOA errors, as discussed in Chapter 20. As these errors diminish over time frame, the ionospheric error decreases and is a less dominant contributor to UEE, with UEE falling to less than 1 m. Here again, much larger or somewhat smaller multipath errors can occur, depending upon the receiver's environment, so the results could vary considerably, from a fraction of a meter to hundreds of meters.

Table 6.4 shows how the SISRE and UEE errors in Tables 6.1–6.3 RSS to produce UERE. The single-frequency receiver's UERE continues to be dominated by ionospheric

TABLE 6.3. Example Contributors to UEE for a Dual-Frequency Receiver

Time Frame / Error	Legacy	Modern	Future
RMS ionosphere error (m)	1.3	0.7	0.4
RMS troposphere error (m)	0.2	0.2	0.2
RMS multipath error[a] (m)	0.4	0.2	0.1
RMS group delay error (m)	0.1	0.1	0.1
RMS TOA estimation error (m)	0.2	0.1	0.1
UEE (m)	1.4	0.8	0.5

[a]Multipath error with direct path present. If the direct signal is blocked and the receiver only tracks a reflected signal, the error is the extra path length of the reflected signal, which could be tens or hundreds of meters.

TABLE 6.4. Example Calculations of UERE

Time Frame / Error Receiver Type	Legacy		Modern		Future	
	Single Freq.	Dual Freq.	Single Freq.	Dual Freq.	Single Freq.	Dual Freq.
SISRE (m)	3.8	3.8	1.8	1.8	0.4	0.4
UEE (m)	7.0	1.4	7.0	0.8	5.0	0.5
UERE (m)	8.0	4.0	7.2	2.0	5.0	0.7

errors for the situation shown, although the multipath error can be much larger and dominate. The results show the great benefit of dual-frequency open signals on all new and modern satnav systems, since receivers can substantially reduce the UERE by using ionosphere-free measurements, except when multipath errors dominate. UERE values less than 1 m, with open sky, are expected to be increasingly common in the coming years.

6.2 DILUTION OF PRECISION AND ERROR MEASURES

While they are important to understand, SISRE, UEE, and UERE are only intermediate quantities that do not themselves inform about positioning and timing errors. UERE only describes ranging errors projected onto the line of sight between receiver and satellite; geometry must be taken into account in assessing the resulting positioning and timing errors.

Figures 6.2 and 6.3 illustrate conceptually why geometry matters. Both cases show the same ranging error projected onto the lines of sight between a receiver and two satellites, along with the resulting uncertainty region. The horizontal extent of the

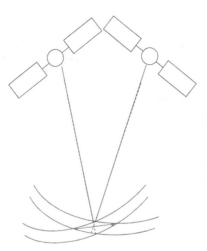

Figure 6.2. Poorer Satellite Geometry

DILUTION OF PRECISION AND ERROR MEASURES

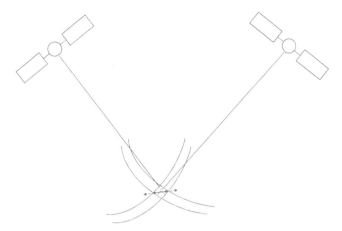

Figure 6.3. Better Satellite Geometry

uncertainty region is roughly proportional to the cosine of the angle to the two satellites, with vertex at the receiver. As the angle between the satellites increases, the horizontal uncertainty region shrinks, to the point where if the angle were 90° in this simple example, the horizontal uncertainty region would be the same size as the ranging error.

The DOP term is expressed as follows. Define a unit vector from the receiver to the kth satellite $\mathbf{u}_k = [(x_k - x_r)/r_{s2r,k}, (y_k - y_r)/r_{s2r,k}, (z_k - z_r)/r_{s2r,k}] = [u_{x,k}, u_{y,k}, u_{z,k}]$, where (x_k, y_k, z_k) is the location of the kth satellite, (x_r, y_r, z_r) is the location of the receiver, and $r_{s2r,k}$ is the range between the kth satellite and receiver. Form the $K \times 4$ geometry matrix, where K is the number of satellites whose signals are being used by the receiver

$$\mathbf{H} = \begin{bmatrix} u_{x,1} & u_{y,1} & u_{z,1} & 1 \\ u_{x,2} & u_{y,2} & u_{z,2} & 1 \\ \vdots & \vdots & \vdots & \vdots \\ u_{x,K} & u_{y,K} & u_{z,K} & 1 \end{bmatrix} \quad (6.3)$$

Denote the estimation errors for the receiver location by $(\tilde{x}_r, \tilde{y}_r, \tilde{z}_r)$ and estimation error for the receiver time bias by $\tilde{\Delta}_r$. Review Question QT6.1 shows that when the UERE to every satellite is zero mean, uncorrelated between satellites, and has the same variance σ^2_{UERE}, the covariance matrix for the estimation errors (with time bias error scaled by the speed of light so that all the quantities have the same units) is

$$E\left\{ \begin{bmatrix} \tilde{x}_r \\ \tilde{y}_r \\ \tilde{z}_r \\ c\tilde{t}_b \end{bmatrix} \begin{bmatrix} \tilde{x}_r & \tilde{y}_r & \tilde{z}_r & c\tilde{t}_b \end{bmatrix} \right\} = (\mathbf{H}^T \mathbf{H})^{-1} \sigma^2_{\text{UERE}} \quad (6.4)$$

assuming the inverse exists.

The DOP matrix is defined as

$$\mathbf{D} = (\mathbf{H}^T \mathbf{H})^{-1} = \begin{bmatrix} D_{11} & D_{12} & D_{13} & D_{14} \\ D_{12} & D_{22} & D_{23} & D_{24} \\ D_{13} & D_{23} & D_{33} & D_{34} \\ D_{14} & D_{24} & D_{34} & D_{44} \end{bmatrix} \qquad (6.5)$$

Under the assumptions used to derive the DOP matrix, having another independent measurement always reduces the diagonal elements of the DOP matrix and hence the DOP. This statement is true even if the measurement comes from a satellite having the same unit vector as another measurement.

Different DOP measures can be developed from the DOP matrix. Following the convention that the unit vectors in (6.3) are defined in a local-horizontal coordinate system, geometric DOP (GDOP), position DOP (PDOP), horizontal DOP (HDOP), vertical DOP (VDOP), and time DOP (TDOP), are each given by

$$\begin{aligned} \text{GDOP} &= \sqrt{D_{11} + D_{22} + D_{33} + D_{44}} \\ \text{PDOP} &= \sqrt{D_{11} + D_{22} + D_{33}} \\ \text{HDOP} &= \sqrt{D_{11} + D_{22}} \\ \text{VDOP} &= \sqrt{D_{33}} \\ \text{TDOP} &= \sqrt{D_{44}}/c \end{aligned} \qquad (6.6)$$

Substituting any of the first four into (6.2) yields an error having units of distance, while TDOP is divided by the speed of light in order to yield an error having units of seconds. Using PDOP provides an error in three dimensions, HDOP describes the horizontal error (useful for positioning on the ground), VDOP describes the vertical error, and TDOP describes the time error.

Typically, values of PDOP less than six are considered acceptable, and less than two are considered very good. PDOP is typically a factor of 2.1 greater than HDOP. Good values of HDOP are less than unity, and VDOP is typically a factor of 1.9 times HDOP. An intuitive explanation for VDOP being larger than HDOP is that low elevation satellites with diverse azimuths provide excellent geometrical diversity for horizontal positioning, but ground-based receivers cannot use satellites below the receiver (since the Earth blocks their signals), limiting the diversity for vertical positioning.

Inherent to the derivation of DOP is the assumption that errors are zero mean, uncorrelated, and have the same variance. When errors have positive correlation (as ionospheric errors often do), their effect on positioning error is smaller, since the positive correlation behaves in part like a common time bias that is removed in the positioning calculation. Consequently, actual positioning errors can be smaller than calculated using DOP.

Different error measures are used across different parts of the satnav community, as summarized in Table 6.5. Here, RMS errors and standard deviations are used interchangeably due to the assumption of zero-mean errors. The probabilities are not exact,

TABLE 6.5. Summary of Different Error Measures

Error	Expression	Interpretation
RMS positioning error, or RMS 3D	$\sqrt{\sigma_x^2 + \sigma_y^2 + \sigma_z^2} =$ PDOP × UERE	Radius of sphere centered at true position, containing 3D position estimate with probability ~61%
RMS horizontal position error or distance root mean square (DRMS)	$\sqrt{\sigma_x^2 + \sigma_y^2} =$ HDOP × UERE	Radius of horizontal circle, centered at true position, containing horizontal position estimate with probability ~63%
Twice DRMS (2DRMS)	$2\sqrt{\sigma_x^2 + \sigma_y^2} =$ 2 × HDOP × UERE	Radius of horizontal circle, centered at true position, containing horizontal position estimate with probability ~98%
RMS vertical positioning error	$\sigma_z =$ VDOP × UERE	Vertical extent, centered at true vertical position, containing vertical estimates with probability ~68%
RMS time error	$\sigma_t =$ TDOP × UERE	Range of time values, centered at true time, containing time estimates with probability ~68%
Circular error probable (CEP) or CE50	CE50 ≈ 0.83 × HDOP × UERE	Radius of horizontal circle, centered at true position, containing horizontal position estimate with probability ~50%
CE90	CE90 ≈ 1.5 × HDOP × UERE	Radius of horizontal circle, centered at true position, containing horizontal position estimate with probability ~90%
Spherical error probable (SEP), SE50	SE50 ≈ 0.89 × PDOP × UERE	Radius of sphere centered at true position, containing 3D position estimate with probability ~50%
SE90	SE90 ≈ 1.4 × PDOP × UERE	Radius of sphere centered at true position, containing 3D position estimate with probability ~90%

since they depend on the degree of asymmetry of the errors in different directions; for example, when the variance in one dimension is twice that in the other two dimensions, the values in this table differ from those shown here by less than 2%. Further discussion of these error measures is found in References 4–6.

Approximate conversion factors between different error measures are summarized in Table 6.6. Multiply the value in the "From" column by the table entry to obtain an approximate value in the "To" column. The values in the table are based on identical error variances in all three dimensions; the values vary by small amounts for modest asymmetries. For example, when the variance in one dimension is twice that in the other two dimensions, the values in this table vary by less than 5%.

TABLE 6.6. Approximate Conversion Factors between Different Error Measures [4 and 5]

From \ To	CE50	RMS (horizontal)	CE90	2DRMS	RMS (3D)	SE50	SE90
CE50	1.0	0.83	0.55	0.42	0.68	0.77	0.47
RMS (horizontal)	1.2	1.0	0.66	0.50	0.82	0.92	0.57
CE90	1.8	1.5	1.0	0.76	1.2	1.4	0.86
2DRMS	2.4	2.0	1.3	1.0	1.6	1.8	1.1
RMS (3D)	1.5	1.2	0.81	0.61	1.0	1.1	0.69
SE50	1.3	1.1	0.72	0.54	0.89	1.0	0.62
SE90	2.1	1.8	1.2	0.88	1.4	1.6	1.0

6.3 POSITIONING ERRORS FOR STANDALONE AND DIFFERENTIAL SATNAV RECEIVERS

Table 6.7 provides representative positioning errors, combining results from Table 6.4 and DOP. With many satellites in the sky, particularly when multiple satnav constellations are available for use, the DOP values can be even smaller than shown, producing even smaller positioning errors. Obtaining DRMS values less than 1 m through a combination of smaller UERE and smaller DOP is very impressive for a standalone receiver.

While the preceding error analysis applies to standalone receivers, most high-accuracy receivers are not standalone, but instead make use of differential satnav. Chapter 23 discusses differential satnav techniques, and the underlying concept and associated error analysis are provided here.

The differential satnav concept is portrayed in Figure 6.4. A receiver at a reference station processes the satnav signals, as does the user receiver. Since the reference station is at a surveyed location, it can compare its known position with the position calculated using satnav and estimate some of the satnav error sources that contribute to the discrepancy between its known position and calculated position. Depending upon the differential satnav technique used, the reference station either solves explicitly for the contributors to SISRE, ionospheric errors, and tropospheric errors, or finds the combined effect of these error sources on pseudorange error or its position solution. The

TABLE 6.7. Representative Positioning Errors for Standalone Receivers

Time Frame / Error Receiver Type	Legacy		Modern		Future	
	Single Freq.	Dual Freq.	Single Freq.	Dual Freq.	Single Freq.	Dual Freq.
UERE (m)	8.0	4.0	7.2	2.0	5.0	0.7
RMS positioning error (m) (PDOP=2)	16.0	8.0	14.4	4.0	10.0	1.4
DRMS (m) (HDOP=1)	8.0	4.0	7.2	2.0	5.0	0.7

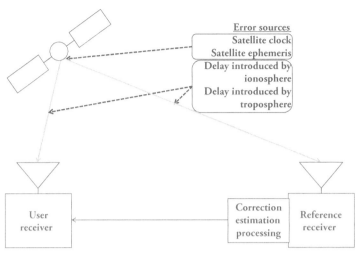

Figure 6.4. Differential Satnav

reference station then communicates corrections to the user receiver, which removes these errors from its measurements. Differential satnav relies on error sources being positively correlated between the reference receiver and the user receiver, and different differential satnav architectures provide different levels of error reduction.

Table 6.8 shows typical SISRE values for satnav in modern and future time frames. Differential techniques typically remove virtually all of the errors due to clock and group delay, and significantly reduce the errors due to ephemeris. The remaining ephemeris error, δ_1, is typically much less than a meter, and depends on the differential technique used as well as the baseline between user receiver and reference station(s).

Representative UEE budgets for differential satnav are provided in Table 6.9. The residual ionospheric and tropospheric errors depend on the differential technique used as well as the baseline between user receiver and reference station(s). Multipath error, receiver group delay error, and TOA estimation error are all specific to the user receiver and cannot be removed by differential satnav. However, it is critical that these errors be as small as possible for the receiver in the reference station, so they do not contaminate the corrections. As discussed previously, multipath errors can be much larger than shown, depending upon conditions. As described in Section 23.2, code-based differential

TABLE 6.8. Representative SISRE Values for Differential Satnav

Time Frame \ Error	Modern	Future
RMS clock error (m)	0.0	0.0
RMS ephemeris error (m)	δ_1	δ_1
RMS group delay error (m)	0.0	0.0
SISRE (m)	δ_1	δ_1

TABLE 6.9. Representative UEE Values for Differential Satnav (Code Based/Carrier Based)

Time Frame / Error	Modern	Future
RMS ionosphere error (m)	δ_2	δ_2
RMS troposphere error (m)	δ_3	δ_3
RMS multipath error[a] (m)	0.2/0.01	0.1/0.01
RMS group delay error (m)	0.0	0.0
TOA estimation error (m)	0.1/0.01	0.1/0.01
UEE (m)	$[0.05 + \delta_2^2 + \delta_3^2]^{1/2}$ / $[0.0004 + \delta_2^2 + \delta_3^2]^{1/2}$	$[0.02 + \delta_2^2 + \delta_3^2]^{1/2}$ / $[0.0004 + \delta_2^2 + \delta_3^2]^{1/2}$

[a]Multipath error with direct path present. If the direct signal is blocked and the receiver only tracks a reflected signal, the error is the extra path length of the reflected signal, which could be tens or hundreds of meters.

techniques use pseudoranges computed from the spreading code, while carrier-based differential techniques use carrier phase measurements described in Section 23.3.

The resulting differential UERE values are shown in Table 6.10. When the ephemeris, ionosphere, and troposphere errors are small, the UERE in the modern time frame is approximately 0.2 m, and in the future time frame is approximately 0.1 m. Under favorable conditions, specialized processing in the receivers can bring the UEE down to centimeter levels, yielding UERE at centimeter levels as well for short-baseline differential systems.

The result is positioning errors at the decimeter level or better for code-based differential systems, and at the centimeter level for carrier-based systems. Similar accuracies are being obtained with precise point positioning (PPP), discussed in Section 20.7, that uses independently provided high-accuracy information to reduce SISRE, and advanced receiver processing techniques that reduce UEE.

6.4 OTHER ERROR SOURCES

There are several relativistic phenomena that must be taken into account in order to avoid significant PVT errors. Reference 7 and Chapter 7 of Reference 8 describe these phenomena and compensation for them. From the General Theory of Relativity, the

TABLE 6.10. Representative UERE Values for Differential Satnav (Code Based/Carrier Based)

Time Frame / Error	Modern	Future
SISRE (m)	δ_1	δ_1
UEE (m)	$[0.05 + \delta_2^2 + \delta_3^2]^{1/2}$ / $[0.0004 + \delta_2^2 + \delta_3^2]^{1/2}$	$[0.02 + \delta_2^2 + \delta_3^2]^{1/2}$ / $[0.0004 + \delta_2^2 + \delta_3^2]^{1/2}$
UERE (m)	$[0.05 + \delta_1^2 + \delta_2^2 + \delta_3^2]^{1/2}$ / $[0.0004 + \delta_1^2 + \delta_2^2 + \delta_3^2]^{1/2}$	$[0.02 + \delta_1^2 + \delta_2^2 + \delta_3^2]^{1/2}$ / $[0.0004 + \delta_1^2 + \delta_2^2 + \delta_3^2]^{1/2}$

satellite clock frequencies must be slowed down slightly to compensate for relative time dilation caused by the combination of satellite speed and the gravitational frequency shift that is induced by the different gravitational potential at orbital altitudes relative to terrestrial altitudes. To account for these effects, GPS satellite clocks are slowed by approximately 4.5 mHz below 10.23 MHz, and clocks for other satnav systems are slowed according to their speed and orbital radius. Receivers need not account for this phenomenon since it is addressed by the satellite clocks. Even when satellite orbits are intended to be circular, the slight amount of eccentricity that occurs in practice can introduce errors of the order of meters, due to special relativity. This error must be compensated by the receiver, as described in Section 20.2.1. Moving receivers and receivers far from the surface of the Earth can benefit from compensation for additional relativistic effects, as described in Reference 6.

The Sagnac effect is also significant. As shown in Section 2.4, signals take tens of milliseconds to travel from a satellite in MEO to a receiver on the ground. During this propagation time, the Earth rotates, rotating an ECEF coordinate system. A point on Earth's equator moves 1670 km/h or 464 m/s, with points at other latitudes moving at these speeds times the cosine of the latitude angle. If the signal's propagation delay is 80 ms, a receiver at 45° latitude moves 26 m. Accounting for the Sagnac effect is straightforward but important, and described in Section 20.3.1.

6.5 SUMMARY

This book's description of satnav systems and signal engineering culminates with this chapter on errors. Like other systems engineering calculations, error analysis involves assumptions and approximations that make the results imprecise but still useful for general characterization. Sources of ranging errors are divided into SISRE, due to space and control segment, and UEE, due to propagation and receivers. Under the assumption of error sources that are uncorrelated, the RSS of SISRE and UEE forms UERE, which describes the ranging error projected onto the line of sight between receiver and a satellite. When UERE to different satellites is zero mean, uncorrelated, and has the same variance, positioning errors can be factored into the product of a dimensionless term related to geometry, called DOP, and UERE. The DOP matrix was presented, and different DOP measures presented in terms of elements in the DOP matrix. Example numerical calculations illustrate contributors to SISRE, UEE, and UERE at different time frames. Combining the resulting UERE with DOP measures yields positioning errors. Many different positioning error measures are used in satnav; the most common are summarized and their relationships are provided. The example numerical calculations continue to provide positioning errors for single-frequency and dual-frequency standalone receivers in different time frames. Differential satnav provides a way to significantly reduce most contributors to SISRE and some contributors to UEE, significantly reducing UERE. Error budgets for differential satnav show that it provides significantly higher accuracy relative to standalone satnav.

REVIEW QUESTIONS

Theoretical Questions

QT6.1. Derive the expression for DOP (6.4), using the following steps.
   a. Write (1.1) for the kth satellite, and define $c\tau_k = f(\mathbf{v})$ where the vector of unknown receiver position and time bias is $\mathbf{v} \stackrel{\Delta}{=} [x_r, y_r, z_r, t_b]$.
   b. Using the expression for multidimensional Taylor series in Appendix Section A.8, expand $f(\mathbf{v})$ around its correct value in a first-order Taylor series expansion.
   c. Use the first-order Taylor series expansion to express pseudorange error in terms of receiver-to-satellite geometry and errors in receiver position and time bias.
   d. Now suppose there are pseudorange measurements from K satellites, and write the system of equations for $K = 4$ satellites. Find the expression relating the covariance of pseudorange error to the covariance of errors in receiver position and time bias. Finally, assume pseudorange errors are uncorrelated and have the same variance, and find (6.4) for this case.
   e. Repeat part d for $K > 4$ satellites when the receiver position and time bias are found using the (unweighted) least squares solution of the overdetermined system of equations, as described in Appendix Section A.7.

QT6.2. Show that DOP never increases when more measurements are added, as long as the additional measurements' errors are uncorrelated with the others and have the same variance. (Hint: use the Sherman–Morrison formula in Appendix Section A.7.)

QT6.3. Suppose the number of satellites being used is doubled, with all measurements uncorrelated and having the same variance. The unit vectors of the additional satellites are identical to those originally used. What is the effect on DOP from adding these redundant measurements with the same geometry?

Application Questions

QA6.1. Suppose a satnav system is designed to provide perfect clocks or clock corrections, yielding an RMS clock error of 0.0 m. How much smaller would the SISRE be for the example values in Table 6.1?

QA6.2. Compute the PDOP for four satellites, with one satellite directly overhead and the remaining three in the same plane as the receiver, with the plane normal to the direction of the overhead satellite, and these receivers at the apexes of an isosceles triangle. Plot the PDOP as a function of the isosceles triangle's shape, showing the minimum PDOP occurs when the triangle is equilateral.

QA6.3. Repeat QA6.2 for HDOP and VDOP rather than PDOP.

QA6.4. Consider four satellites, with one satellite directly overhead a receiver and the remaining three satellites all in a plane normal to the direction of the overhead satellite, at the apexes of an equilateral triangle, with the line joining the overhead satellite and receiver at the center of the triangle. Compute the PDOP as a function of the elevation angle of the three satellites in the plane, finding the elevation angle that minimizes PDOP.

REFERENCES

1. J. J. Spilker, Jr., "Tropospheric Effects on GPS," Chapter 13 in *Global Positioning System: Theory and Applications*, Edited by B. W. Parkinson, J. J. Spilker, Jr., P. Axelrad, and P. Enge, American Institute of Aeronautics and Astronautics, 1996.
2. C. Satirapod and P. Chalermwattanachai, "Impact of Different Tropospheric Models on GPS Baseline Accuracy: Case Study in Thailand," *Journal of Global Positioning Systems*, Vol. 4, No. 1–2, 2005, pp. 36–40.
3. Ionospheric Correction Algorithm for Galileo Single Frequency Users, European GNSS (Galileo) Open Service, Document Issue 1.1, February 2015.
4. F. van Diggelen, "GPS Accuracy: Lies, Damn Lies, and Statistics", *GPS World*, January 1998.
5. F. van Diggelen, "GNSS Accuracy: Lies, Damn Lies, and Statistics", *GPS World*, January 2007.
6. GPS Position Accuracy Measures, Novatel APN-029 Rev 1, 2003, available at http://www.novatel.com/assets/Documents/Bulletins/apn029.pdf, accessed 3 April 2014.
7. N. Ashby and J. J. Spilker, Jr., "Introduction to Relativistic Effects on the Global Positioning System," Chapter 18 in *Global Positioning System: Theory and Applications*, Edited by B. W. Parkinson, J. J. Spilker, Jr., P. Axelrad, and P. Enge, American Institute of Aeronautics and Astronautics, 1996.
8. P. A. Kline, Atomic Clock Augmentation for Receivers Using the Global Positioning System, PhD Dissertation, University of Virginia, available at http://scholar.lib.vt.edu/theses/available/etd-112516142975720/unrestricted/etd.pdf, February 1997.

PART II

SATNAV SYSTEM DESCRIPTIONS

This part of the book builds upon the concepts and terminology provided in Part I in order to describe current and planned satnav systems. Each chapter in this part is structured similarly, providing context, insights, and technical aspects of the systems and their signals, but without the full technical detail available in formal specifications.

Part II forms the link between the technical characteristics of satnav signals provided in Part I, and the receiver processing described in Part III. Since almost every signal from every satnav system is different, Part II outlines specific characteristics for a given signal and system, so the reader is cognizant of actual signal characteristics when learning and evaluating receiver processing approaches in Part III.

Since the focus of this book is on systems engineering and receiver processing, only modest attention is given to the history and description of each system. These aspects are addressed briefly to provide context, but other resources provide much more information for tracing the detailed history of each system, and describing the specific architectures of each system.

Moreover, all these systems and their signals are evolving, so their descriptions only capture their status and plans at a particular moment in time. Once signals are described and defined by formal ICDs, however, the signal characteristics are more stable and more accessible, so the signals with published ICDs are addressed in greater detail than those not described in ICDs. There is less information in the public domain concerning some systems, especially modernized GLObal NAvigation Satellite System (GLONASS), BeiDou, and Indian Regional Satellite System (IRNSS); the level of detail in different chapters reflects this difference in publicly available information.

Each satnav system's representations of location and time are based on different geodetic datums and time references, summarized in each chapter. The geodetic datum defines the size and shape of the Earth model and the origin and orientation of the coordinate systems used to map the Earth. It includes a terrestrial reference frame that provides the coordinates of reference points located on the Earth's surface, so that the location of other points can be determined relative to these reference points. It also describes the rotation of the Earth, measured with respect to a frame tied to stellar objects, called a celestial reference frame. Earth Orientation Parameters (EOPs) connect these two frames, providing the transformation between a point defined in the terrestrial frame and its coordinates in the celestial reference frame.

Although different satnav systems use different geodetic datums, all are being modified to converge to the International Terrestrial Reference Frame (ITRF) [1], and generally are already within a few centimeters of each other. Typically, only differential satnav is concerned with that level of accuracy, and since differential satnav involves relative measurements to presurveyed points (see Chapter 23), such small errors in the datums have little effect.

The Earth's shape varies over time, changing diurnally due to tides in the oceans and the Earth crust, and evolving more gradually due to major events like earthquakes and volcanic eruptions. Thus, ITRF and the other geodetic datums will continue to change as well. The ITRF also models the time evolution of Earth's shape, allowing comparison of observations over time.

All satnav systems' time references are associated to Universal Coordinated Time (UTC), but in different ways. UTC is managed to maintain alignment of time with a mean solar day. Mean solar time, based on the spin of the Earth on its axis, is not uniform but gradually changes due to tidal friction and other factors such as motions of the Earth's fluid core. Consequently, mean solar days are slightly longer than the $24 \times 60 \times 60 = 86400$ seconds nominally contained in a day, and, leap seconds are occasionally added to UTC in order to keep it close to mean solar time. Otherwise, over many years, noon in Greenwich, England would gradually diverge from the average time the sun is directly overhead Greenwich.

Each satnav system defines its own time reference, typically with respect to a set of atomic clocks maintained by the nation or nations responsible for developing and operating the system. Relationships between system time references matter when different receivers using time from different systems need to synchronize with each other (e.g., to synchronize a communications network), or when a receiver is blending measurements from signals transmitted by satellites from different systems. If the receiver does not know the time offset between system times of the two satnav systems, each with its own system time, the receiver needs to make measurements from at least two satellites from each system, and at least five satellites in total, to determine three-dimensional position, time of one system, and the time offset to the other system. Adding a measurement from only a single satellite in a second system does not provide additional information if the time offset between two system times is not known, since the measurement also introduces an additional unknown—the systems' time offset.

Most systems will broadcast time offsets relative to other systems so that signals from a single satellite in a system are useful. In that case, even a measurement from a single satellite in a different system can be used. However, when enough satellites are in view from each constellation, receivers typically will compute the time offset rather than use the broadcast time offset, which may have significant latency associated with it.

Signals provide, in their data messages, the parameters needed for a receiver to compute UTC from satnav system time. Besides having different time references, satnav systems also differ in deciding whether to include leap seconds for alignment to UTC, or not to include leap seconds, thereby avoiding time discontinuities.

As multiple systems become operational and share parts of the frequency spectrum, interoperability and compatibility become increasingly important. Interoperability means that receivers and applications can use multiple signals and multiple systems to obtain better capability than would be obtained using only one signal or system. Compatibility means that multiple signals and systems can be used separately or together without unacceptably degrading or otherwise affecting use of an individual service or signal.

In general, compatibility is considered to be higher priority, with interoperability sought once compatibility has been assured. Compatibility implies absence of harmful radio frequency interference as well as, in some cases, spectral separation between a system's military or authorized signal and other signals. Spectral separation does not apply to all such signals. For example, the United States has indicated the need for spectral separation between other signals and the Global Positioning System (GPS) M-code signal, but other GPS signals and signals from other systems spectrally overlay the GPS P(Y) code signal, and multiple GLONASS civil signals spectrally overlay the original and planned modernized GLONASS military signals.

Interoperability is a multidimensional and subjective characteristic. Often, signals are considered more interoperable when their characteristics are similar. Table II.1 lists different technical characteristics that affect interoperability, with more important characteristics listed higher in the table. These characteristics affect performance, ease of use, and receiver complexity when using multiple systems and signals.

When signal similarity is employed to achieve interoperability, the principles of interoperability and compatibility can be in opposition—signals with similar designs at like frequencies introduce more radio frequency interference than those at distinct frequencies. At some point, adding more signals of similar design can introduce unacceptable levels of interference to the reception of existing signals. It may be necessary, therefore, in the future multi-GNSS environment, to accept less interoperability in some

TABLE II.1. Characteristics Affecting Interoperability

Characteristic	Interoperability Benefit
Common time and geodetic frames, or broadcast offsets	Navigation solutions can benefit from fewer measurements
Common center frequencies	Common antenna and radio receiver—lower power and cost; common frequency-dependent errors for higher accuracy
Similar spreading modulation spectra	Common frequency-dependent errors for higher accuracy
Common spreading code lengths and common spreading code families	Lower crosscorrelation sidelobes for better weak-signal reception; common receiver processing for acquisition and tracking
Common data message modulation, encoding, structure	Common receiver processing for demodulation, decoding, and interpretation

cases in order to maintain compatibility, or to seek use of diverse or complementary frequencies.

Chapter 7 describes the United States' GPS, the original and most widely used current satnav system. Since the turn of the century, GPS has been undergoing extensive modernization of its control segment, satellites, and signals, while receiver developers have also been developing new generations of user equipment, driving extensive innovation in applications that rely on precise PNT from GPS. This chapter describes the original system and signals as well as GPS modernization, with details of original, new, and planned signal designs. The original signal designs have influenced all other satnav signals, while the new and planned signal characteristics include a variety of legacy and novel characteristics.

Chapter 8 provides information about Satellite-Based Augmentation Systems (SBASs), a family of wide area augmentation systems originally developed for aviation safety of life applications but now used extensively for consumer applications as well. Each SBAS is a regional system, but because they conform to the same standards, they are completely interoperable. Together, they cover much of the northern hemisphere today. Coverage is slated to become almost complete by 2020 or so, as well as to extend significantly into the southern hemisphere. Currently, SBASs are very closely tied to GPS, since GPS is the only satnav system used for aviation safety of life applications today. Signal characteristics, message contents, and even the geodetic datum and time reference are based on those used for GPS. However, it is likely that future SBASs will also be tied to other global satnav systems as they mature. SBAS signal designs are described, setting the stage for development and assessment of SBAS receiver processing techniques.

Chapter 9 addresses the Russian Federation's GLONASS, the other original satnav system. While some aspects of GLONASS are similar to GPS, others are very different, from technical characteristics of signal and satellite designs to a less reliable history of providing satnav service. Like GPS, GLONASS is undergoing modernization of every aspect, moving to much greater capability and performance even while use of the original GLONASS signals is increasing. While less information is publicly available about modernized GLONASS signals than about modernized GPS signals, the chapter summarizes the available information about modernized signals while also providing an in-depth description of the original GLONASS signals.

Chapter 10 describes the European Galileo system, racing against China's BeiDou to be the third operational global satnav system. Galileo's design reflects an organized development, starting with planned services, then flowing to requirements, and then leading to detailed system and signal characteristics. Galileo's signal set includes some highly innovative characteristics that provide opportunities and perhaps challenges for receiver designers. While Galileo has provided draft ICDs for its open signals, there is still some uncertainty about their eventual characteristics due to recent changes in Galileo services that might be reflected in some signal design changes. Only very limited information is currently available for Galileo signals not designed for open service.

Chapter 11 depicts China's BeiDou navigation system (BDS). This system has undergone a rapid and significant transformation from a Phase 1 system of three geostationary satellites providing regional Radio Determination Satellite Service (RDSS) in

2002, to an operational Phase 2 regional system in 2012, consisting of approximately 14 satellites in geostationary and inclined geosynchronous orbit that provide service based on passive trilateration as well as RDSS. Before 2020 or so, BeiDou will evolve to a Phase 3 system that will provide a global system using a constellation of satellites in medium Earth orbit, along with the higher altitude constellation overlay that will continue to provide regional service. Although the BDS Phase 2 system has been declared operational, there is relatively little public information about it. Details of the control segment have not been made available to the same extent that details are available for GPS and Galileo. ICDs for the open signals are becoming available. Further, the Phase 3 BeiDou signal set involves not only the introduction of new signals, but potentially the phasing out of some Phase 2 signals. Available information about Phase 2 signals and planned Phase 3 signals is included in this chapter.

Japan's Quasi-Zenith Satellite System (QZSS) is described in Chapter 12. QZSS was originally intended to provide regional augmentation of GPS and perhaps other global systems, but is being transformed into a system that also provides regional standalone service. Most QZSS signal designs are virtual duplicates of GPS and SBAS signal designs, making it very easy for receiver designers to add QZSS functionality to a GPS and SBAS receiver. The combination of availability enhancement and accuracy enhancement provided by QZSS offers receivers new levels of performance without use of terrestrial augmentations. While QZSS has published an ICD for all of its current signals, the addition of standalone capability has motivated reconsideration of the QZSS signal set and signal designs, leaving some uncertainty concerning whether new or

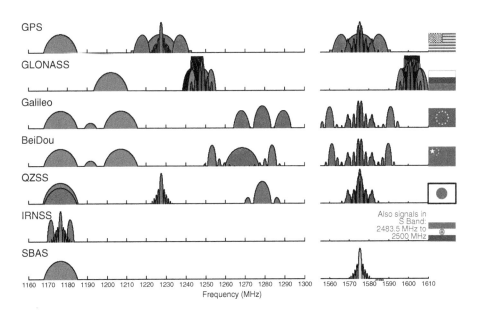

Figure II.1. Projected Satnav Signal Spectra Circa 2020

revised signal designs will result. However, the currently planned signals are well-defined and summarized in Chapter 12.

The most recent satnav system, India's IRNSS, is summarized in Chapter 13. IRNSS has been discussed for many years, and over the past few years satellites have been launched and ICDs have been provided for the open signals. Clearly, IRNSS is now moving very quickly.

Part II thus describes four global satnav systems and four regional satnav systems (counting SBAS as a regional system), all projected to be fully operational circa 2020. If the projections for each system become reality, more than 150 satnav satellites, each transmitting two or more signals, will be available for receiver designers to use, providing unprecedented performance, redundancy, and diversity, yielding the signal spectra portrayed in Figure II.1. Despite the diversity of systems, satellites, and signals, all are instances of the characteristics described in Part I of this book, and the receiver processing techniques described in Part III are the basis for making measurements from these signals and for using the measurements to determine PVT.

REFERENCE

1. Welcome to the ITRF web site, http://itrf.ensg.ign.fr/, accessed 19 May 2013.

7

NAVSTAR GLOBAL POSITIONING SYSTEM

GPS was the first operational global satnav system and, as of 2015, is the best-performing and most widely used system. GPS is intended for dual use—designed and operated to provide service to civilian and to military users. The United States offers civil GPS signals for use by anyone, anywhere its signals can be received, with no user fees. In addition, GPS Interface Specifications (ISs) [1] provide the full unrestricted technical information for anyone to construct and manufacture receivers of GPS civil signals. Historically, GPS has called the service it provides to civil users the Standard Positioning Service (SPS) and the service it provides to military users the Precise Positioning Service (PPS).[1] While the ISs provide specifications on individual signals, GPS

---

[1] Presumably, this terminology arose from the higher chip rate of the military signal, yielding potentially greater pseudorange accuracy. However, overall accuracy depends on many more factors than chip rate, and ultimately depends upon receiver processing and augmentations. Civilian receivers and augmentations have often emphasized accuracy more heavily than have military receivers. The result can be civil receivers that use C/A code signals delivering higher accuracy than military receivers that use P(Y) code signals.

Engineering Satellite-Based Navigation and Timing: Global Navigation Satellite Systems, Signals, and Receivers, First Edition. John W. Betz.
© 2016 The Institute of Electrical and Electronics Engineers, Inc. Published 2016 by John Wiley & Sons, Inc.

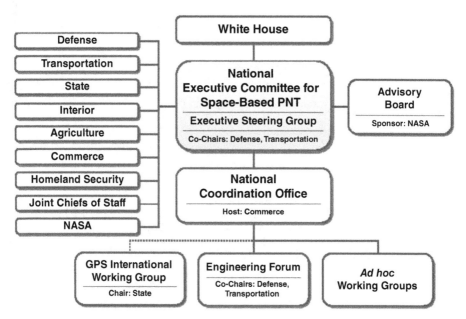

Figure 7.1. US Government Organizational Structure for GPS Governance [3]

also provides a system performance specification, the SPS Performance Standard (SPS PS) [2], specifying the system-level characteristics that GPS commits to provide to civil users.

GPS comprises a space segment, ground segment, and user segment. While the specified intended minimum constellation size is 24 satellites, a minimum of 27 satellites is likely for the foreseeable future, and 30 or more satellites have been operational for the past few years.

Given GPS's role as a dual-use system, an interagency US governance structure has been defined for space-based PNT, which includes not only GPS, but also US-developed, -owned, and/or -operated systems used to augment or otherwise improve GPS. Figure 7.1 shows the organizational structure employed for GPS.

GPS is primarily funded out of the Defense Department's budget, with a small fraction of its funding from civil agencies. The US Air Force has been made responsible for developing and operating GPS, with participation from other military services. Air Force Space Command's Space and Missile Systems Center at Los Angeles Air Force Base, California, acts as the Department of Defense's executive agent for developing and acquiring GPS satellites, the control segment, and military user equipment. The 50th Space Wing's Second Space Operational Squadron at Schriever AFB, Colorado, is responsible for operating and maintaining the GPS constellation.

This chapter provides a brief history of GPS and summarizes current plans in the first section, followed by an overview of GPS. Section 7.3 provides detailed characteristics of GPS signals, followed by a summary of the chapter.

7.1 GPS HISTORY AND PLANS

The development of GPS in the 1970s built upon previously developed and explored technical concepts such as radio navigation from terrestrial transmitters, developments of space-based navigation systems using fundamentally different approaches from GPS, maturation of space-qualified atomic clocks, and application of concepts from digital communications. Numerous books and articles [4–6] document the techno-political efforts needed to achieve the synthesis and integration of these and other technical concepts into the original GPS system whose performance, success, and endurance likely exceed even the wildest dreams of its designers. Consequently, the history discussed here emphasizes more recent events involving GPS Modernization.

The GPS program itself was established in the early 1970s, building on previous work in radio navigation including the Air Force's Project 621B, the Navy's Transit and Timation projects, and the Army's Sequential Collation of Range program. The first operational GPS satellite was launched in 1978. The next 20 years involved continuous growth and maturation of the system. In 1983, the United States formally announced provision of GPS for civil use, followed by a 1991 announcement that civil GPS would be provided free of charge to the international community. The then-specified constellation size of 24 operational satellites was achieved in 1994, transmitting the Coarse/Acquisition (C/A) code signal at the carrier frequency designated L1 (1575.42 MHz) and the Precision (Encrypted) (P(Y)) code signal at both L1 and the carrier frequency designated L2 (1227.60 MHz).

By that time, GPS had reached key levels of acceptance: professional users, such as surveyors, recognized and exploited its benefits; the aviation community was planning to adopt GPS use widely; and military organizations around the world had observed the way it changed warfighting in numerous ways, from precision maneuver, to highly accurate time distribution, to delivery of precision munitions.

Even with this degree of success, however, limitations became apparent. Military and civilian receivers were large, expensive, and power-hungry. Selective availability (SA), an intentional distortion of the civil signal's accuracy intended to ensure that the United States and its allied military users had exclusive access to the full accuracy of GPS, also limited the peaceful utility of GPS. With a civil signal at only one carrier frequency, there was no frequency diversity or redundancy, and thus the performance advantages of dual-frequency receiver processing were not available using only civilian signals. From a military perspective, some future limitations of the P(Y) code signal were evident, while emerging civilian technologies for semicodeless processing of P(Y) code signal [7] and differential systems threatened the efficacy of SA as a means for retaining military exclusivity of high accuracy GPS.

In the mid-1990s, the concept of navigation warfare (Navwar), was defined as "[d]eliberate military operations to gain and maintain a positioning, navigation, and timing information advantage," [8]. A study by the National Research Council (NRC) [9] evaluated prospects for a new military signal. Studies were initiated at the GPS Joint Program Office to consider changes to or replacement of the P(Y) code signal.

Almost simultaneously, a set of events was influencing civil GPS signals. The NRC study [9] also included a recommendation for a second civil signal at a carrier frequency other than L1, and similar recommendations followed from other groups. Ultimately, the impetus for a second civil signal evolved to a commitment for a third signal as well, as the originally planned C/A code-like civil signal at L2 was deemed unacceptable for civil aviation use, and an additional civil signal at a new carrier frequency, termed L5 (1176.45 MHz), was defined [10].

Most aspects of the new civil signal designs at L2 and L5 combined characteristics of the original C/A and P(Y) designs with characteristics recommended by the Project 621B study but not implemented in the original GPS signals, presumably due to technology limitations [11]. The same BPSK-R spreading modulations were employed, but with longer spreading codes, stronger forward error control, and distinct pilot and data components as recommended decades previously. The data message designs are the exception: the new signals' messages employ an entirely different structure that provides considerably more flexibility than did the original signals' message.

The same evolutionary design steps were not possible for the new military signal. A key element of Navwar is higher power military signals for more robust operation, yet there was significant motivation to retain the same carrier frequencies for a variety of reasons [12]. This confluence of objectives and constraints raised, for the first time in satnav, the issue of radio frequency compatibility in spectrum sharing, leading to the development of the BOC spreading modulation [13, 14] as a way to more effectively share limited satnav spectrum. The resulting M code signal was the first to employ this novel spreading modulation [15]. As BOC was further evaluated, the initial assessment of its other advantages, including improved code tracking accuracy in noise and multipath, and better resistance to interference for a given spreading code chip rate, were confirmed [16, 17].

The first tangible step in GPS modernization is often considered to be turning off SA in May 2000. In parallel, the GPS control segment was being updated through the addition of 10 more monitoring stations from the National Geospatial Agency (NGA) [18], providing more accurate tracking of GPS satellites with resulting improvements in clock corrections and ephemeris.

An independent event that indirectly accelerated the widespread use of GPS was the Federal Communications Commission's requirements (gradually phased in over the first decade of this century) [19] that mobile phones be "location capable" when making emergency calls. As service providers determined that GPS was the most cost-effective way to meet this mandate, new system technologies such as assisted-GPS (A-GPS) [20] were developed. Perhaps more importantly, the result was significant investment in small, inexpensive, and low power GPS chipsets that drastically increased the utility and utilization first of GPS and more recently other satnav systems.

The GPS constellation, originally designed as an augmented Walker 55°: 18/6 constellation with additional satellites placed in each orbital plane, has been optimized over the years to make better use of the additional satellites being operated, providing better DOP and greater resilience to the loss of any one satellite. In recent years, it has been common for more than 30 satellites to be operating simultaneously, with additional satellites on orbit in "residual" status, ready to be activated if an operating satellite is taken out of service.

With three civil signals at different frequencies and pairs of original and modernized military signals at two frequencies, some expected the development of new GPS signals to be complete. However, negotiations between the United States and Europe concerning GPS and Europe's planned Galileo system began in 2002, yielding an agreement in 2004 [21] that included a US commitment to implement an additional civil signal at L1 designed to be highly interoperable with a Galileo civil signal at the same carrier frequency. A BOC(1,1) spreading modulation was selected for both signals at the time. However, European experts subsequently expressed displeasure with the limited performance provided by the BOC(1,1) spreading modulation [22], proposing various alternatives that the United States found unsatisfactory. Finally, in 2007, enhanced spreading modulations were developed for L1C signal [23, 24], with similar ones then adopted for the Galileo E1 OS signal based on the multiplexed BOC (MBOC) spectrum [25], introduced in Section 3.1.4.

GPS designed this signal in close cooperation with international users and developers of other satnav systems, to meet a range of user needs and increase the likelihood that its key design features would be adopted by other satnav systems for best interoperability [26].

As GPS Modernization has proceeded over the second 20 years of GPS's existence, its effects have become apparent in new generations of signals, new satellites, control segment updates, and modernized military and civilian receivers. Yet modernization has been more extensive and taken longer than originally planned. Acquisition delays have been experienced on both control segment modernization and the development of new blocks of satellites. Operational satellites have lasted far longer than their planned mission lives, and for a time new satellites with new capabilities were launched only when existing satellites no longer functioned (so-called "launch on replace"). Consequently, GPS Modernization will continue to roll out after 2020.

7.2 GPS DESCRIPTION

GPS offers two services, the SPS and PPS. Eventually, SPS will include two civil signals at L1, C/A code signal and L1C, one civil signal at L2, L2C, and the L5 signal at L5. PPS will include both P(Y) code and M code signals on L1 and L2.

There is also a plan that future GPS satellites will carry the Distress Alerting Satellite System (DASS) [27], a Canadian-developed secondary payload that performs a search and rescue function. This payload, known as Search and Rescue GPS (SAR/GPS) will provide near-instantaneous detection and location of 406 MHz emergency beacons, and will be the US contribution to Cospas-Sarsat, an international consortium [28] of 39 countries and organizations. SAR/GPS is compatible with 406 MHz emergency beacons as defined in the Cospas-Sarsat beacon specification, and more than 900,000 maritime, aviation, and land-based distress beacons already use the system worldwide.

The GPS constellation, described in Section 2.3, is defined to have a minimum of 24 satellites in six Earth-centered orbital planes, with 60° separation. The orbital planes are inclined 55° relative to the equatorial plane. Orbits are circular with nominal radius 26,600 km, yielding an orbital period of 11 hours, 58 minutes, or twice each sidereal day. Satellite speed is approximately 3.9 km/s.

The size and weight of GPS satellites have increased with each satellite block, as increased capability has been provided. While Block IIA satellites, first launched in

1978, weighed 1800 kg [29], the early Block III satellites, to be launched starting in 2016, are slated to weigh more than 3600 kg at launch [30]. The weight increase is due to the need to transmit more signals at higher power, combined with increased fuel and redundancy to achieve longer satellite life. Most GPS satellites use rubidium clocks, and the Block IIF satellites use both rubidium and cesium atomic clocks, typically four in all, including backups. Currently, only one satellite can be launched at a time; studies have been performed to examine the feasibility and benefits of dual launching future satellites.

The GPS ground segment uses a global network of transmitting and receiving stations, along with a master control station at Schriever Air Force Base, CO, and an alternate master control station at Vandenberg Air Force Base, CA. Six GPS monitor stations have been augmented by 10 NGA monitor stations, so each satellite is always in view of at least two monitor stations. Five ground antennas provide telemetry, tracking, and command to the satellites, typically uploading updated messages containing updated ephemeris and clock corrections every 12 hours.

The GPS geodetic datum is based on the World Geodetic System 1984, or WGS-84 [31], the first worldwide geodetic coordinate system which was constructed to enable GPS use. WGS-84 defines a geoid, claimed to exhibit an absolute accuracy of 1.0 m or better. Improved measurement techniques, combined with continuous changes in the shape of the Earth, have led to several revisions of the WGS-84 geodetic datum since 1984.

GPS time is aligned with UTC, except that GPS time is continuous and does not include leap seconds, meaning that GPS time differs from UTC by an integer number of seconds, 16 as of early 2013. GPS system time is given by its Composite Clock (CC). The CC or "paper" clock consists of all operational GPS monitor station and satellite frequency standards. GPS system time, in turn, is referenced to the Master Clock (MC) at the United States Naval Observatory (USNO) and steered to UTC as maintained by USNO, referred to as UTC(USNO), from which GPS time will not deviate by more than 1 μs. The most recent GPS time origin, when GPS time was initialized to zero, was time at 23:59:47 UTC on August 21, 1999 (00:00:19 TAI on August 22, 1999). It resets every 1,024 weeks, or approximately 19.6 years.

GPS uses advisory messages, called Notice Advisories to Navstar Users (NANUs), to inform users of a change in the GPS constellation. These messages are released 72 hours in advance for planned maintenance, and whenever available to notify users of unscheduled outages. General NANUs can be used to disseminate general GPS information.

7.3 GPS SIGNALS

As described in Section 7.1, GPS originally transmitted three signals on two carrier frequencies. Satellites launched since 2005 have gradually added signals and then another carrier frequency. GPS III satellites scheduled to be launched starting in 2016 will transmit eight signals (three original signals and five modernized signals; four civil signals and four military signals) on three carrier frequencies. Figure 7.2 summarizes the

GPS SIGNALS

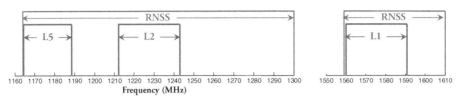

Figure 7.2. GPS Frequency Bands

frequency plan using GPS's notation. The same terminology is used for frequency bands as for carrier frequencies. L1 band's carrier frequency is 1575.42 MHz; L2 band's carrier frequency is 1227.60 MHz; L5 band's carrier frequency is 1176.45 MHz. As discussed in Section 3.3.1, frequency bands designated "RNSS" are designated for radio navigation satellite service, or satnav.

All signals use RHCP. Minimum received signal power is specified out of a 3 dBil linearly polarized antenna located near the ground, for satellites at elevation angles of 5 degrees or more, assuming 0.5 dB excess atmospheric attenuation. The receive antenna is normal to the incident wave and rotated to be aligned with the minor axis of the incident field's polarization ellipse, making this power specification conservative as discussed in Section 4.2.

The original GPS signals have historically been referred to as "codes," since the only difference between the C/A code signal and the L1 P(Y) code signal is the spreading code. However, modern signals are much more than spreading codes, as discussed in Section 3.3, and all GPS signals are called "signals" in this book.

Three different GPS ISs, all found in Reference 1, define the GPS civil signals. IS-GPS-200, the original specification on GPS signals in space, describes the C/A code signal and the L2C signal, along with publicly released aspects of the P(Y) code signal. IS GPS-705 defines the L5 signal. IS-GPS-800 defines the L1C signal. While the ICDs of some satnav systems are explicit about the signal combining used to modulate multiple signals onto the same RF carrier (see Section 3.3.12), signal combining approaches are not specified in GPS ISs, although phase relationships between signals and components may be specified. Not specifying the signal combining approaches provides flexibility for different satellite designs to perform different signal combining approaches, and limits the ability for receiver designers to employ special techniques that determine and then rely on a particular combining approach used in the design of a specific satellite block. Backward compatibility requirements ensure that the signal combining approaches used remain transparent to receivers.

The ISs for civil signals define the spreading codes used for each signal. Although the acronym "PRN" typically means "pseudo-random number," the GPS ISs also use the term "PRN" as an index into the family of spreading codes defined for each GPS signal type. Each operating satellite is assigned a PRN index that is provided in the GPS almanac. A receiver wishing to track a signal from that satellite looks up that satellite's PRN index and generates the associated spreading code.

Another somewhat arcane aspect of GPS terminology is a nonstandard code. As described in Reference 32, nonstandard codes are incorrect spreading codes—spreading

TABLE 7.1. Summary of C/A Code Signal Characteristics

Carrier frequency	L1: 1575.42 MHz = 154 × 10.23 MHz
Minimum received power	−158.5 dBW
Polarization	RHCP
Multiple access	CDMA
Spreading modulation and transmit bandwidth	BPSK-R(1), ±15.345 MHz
Spreading codes	Length 1023-bit Gold codes, duration 1 ms
Data message structure	NAV
Data rate	50 bps
Data message error correction and detection	Error detection using extended (32,26) Hamming code
Data modulation	50 sps biphase modulation
Pilot and data components	100% power data
Overlay code	None
Multiplexing with other signals	Varies with GPS satellite block, but always in phase quadrature to L1 P(Y) code signal

codes not defined in the interface specifications that are transmitted when signals are anomalous or under test. While the monitoring stations can track signals with nonstandard codes, typical receivers do not, protecting receivers from using signals not suitable for use.

The following subsections summarize each of the GPS signals in order of decreasing carrier frequency.

7.3.1 GPS C/A Code Signal

Table 7.1 summarizes key characteristics of the C/A code signal, using the same characteristics introduced in Section 3.3. The data rate shown here includes overhead bits used for framing, synchronization, and parity; the actual information data rate is somewhat lower.

Figure 7.3 shows the autocorrelation function and PSD, assuming an ideal long spreading code, yielding a triangular autocorrelation function and sinc-squared PSD. Of all the satnav signals described in this book, actual C/A code signals most differ from their approximations obtained for an ideal long spreading code, because of their spreading code structure.

As shown in Reference 33, when the C/A code signal data modulation is modeled as perfectly random, the C/A code signal PSD is

$$\Phi_{C/A}(f) = \left[\frac{1}{f_c}\mathrm{sinc}^2(\pi f/f_c)\right] \left[\frac{1}{K}\left[\frac{\sin(\pi f K N/f_c)}{\sin(\pi f N/f_c)}\right]^2\right] \left[\frac{1}{N}\left|\sum_{n=0}^{N-1} a_n e^{-i2\pi f n T_a}\right|^2\right] \quad (7.1)$$

GPS SIGNALS

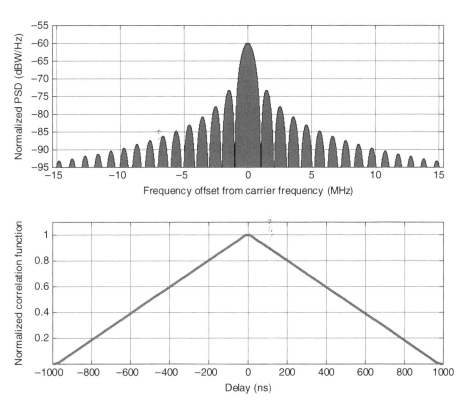

Figure 7.3. Autocorrelation Function and PSD for C/A Code Signal, Using Ideal Long Spreading Code Model

where $f_c = 1.023$ MHz, $K = 20$, $N = 1023$, $T_a = 1/(1.023 \times 10^6)$ and the $\{a_n\}$ are the actual 1023 bits of a C/A spreading code, generated as described later in this section. An alternative approximation that yields second-order statistics having greater fidelity than the ideal long spreading code model but less fidelity than using actual spreading code bits in (7.1) is called the ideal repeating spreading code model. Here, the spreading code is modeled as having a correlation function that is identically zero, except at lags corresponding to integer multiples of the code length, where the value is unity. In this case, the last term in (7.1) is identically unity. Figure 7.4 shows the PSD of the C/A code signal using the spreading code defined for PRN 1. The green curve is the PSD using an ideal long spreading code model, given by only the first term in (7.1). The red curve employs the ideal repeating spreading code model based on the first two terms in (7.1). It exhibits spectral lines 100 Hz wide (defined by the null-to-null bandwidth of the second term in (7.1), since $f_c/(KN) = 50$ Hz) at 1 kHz spacing. The blue curve uses the actual Gold code, repeated 20 times within each data bit. The specific spreading code makes some of the spectral lines larger and other smaller than shown for the ideal repeating spreading code model. While all curves have the same unit area over infinite frequency, the repeating spreading code introduces spectral peaks and nulls. The ideal repeating

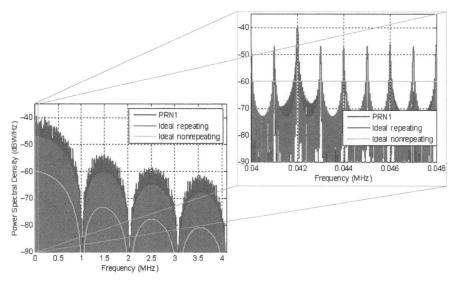

Figure 7.4. PSD for C/A Code Signal Having PRN 1 [33]

code model captures much of the essence of the PSD for a signal having repeating spreading code without the complexity of calculations using the actual spreading code.

Figure 7.5 shows a block diagram for computing the Gold codes used for C/A code spreading codes. Gold codes [34] are formed by exclusive ORing two specific maximum length sequences (or m-sequences) having specific relative phases, or delays, relative to each other.

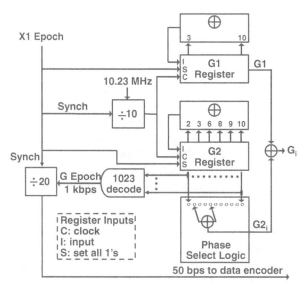

Figure 7.5. Gold Code Generator for C/A Code Signal Spreading Codes [32]

GPS SIGNALS 173

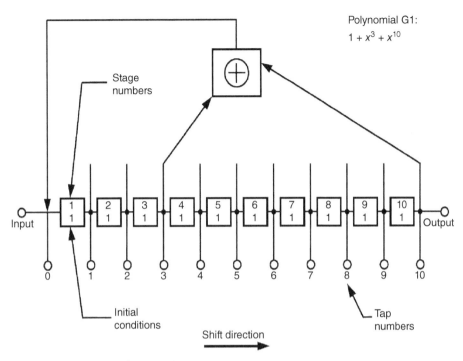

Figure 7.6. G1 m-Sequence Generator [32]

The G1 and G2 registers each produce m-sequences, and the phase selection logic takes the exclusive OR of two G2 outputs (which is mathematically equivalent to a delayed output of the G2 register) for exclusive ORing with the output of the G1 register. Gold codes [34] are formed by exclusive ORing of two m-sequences of the same length that have crosscorrelation less than $2^{(n+2)/2}$, where n is the number of delays in the linear shift register used to generate both m-sequences, each of length $2^n - 1$ bits. Each m-sequence of degree n is completely described by a specific polynomial of order n, and can be generated by a linear shift register having n delay elements whose taps correspond to the specific polynomial. There is no sequence of digits with a longer period that can be produced by a linear shift register with n delay elements.

Different Gold codes from the same family are formed by circularly shifting one of the two m-sequences relative to the other before exclusive ORing. The output of the phase selection logic is a circularly shifted version of the m-sequence.

Figure 7.6 shows the G1 m-sequence generator that implements the indicated polynomial, while Figure 7.7 shows the G2 m-sequence generator.

Figure 7.8 shows how IS-GPS-200 specifies a different spreading code for each of the first 32 PRN indices defined for the C/A code signal. The code phase selection produces the same effect as the delay ($X2_i$). The first 10 values for each C/A code are provided so developers can crosscheck results from their code generators. As noted in a footnote of the table, the two-tap combining of the output from the G2 generator in

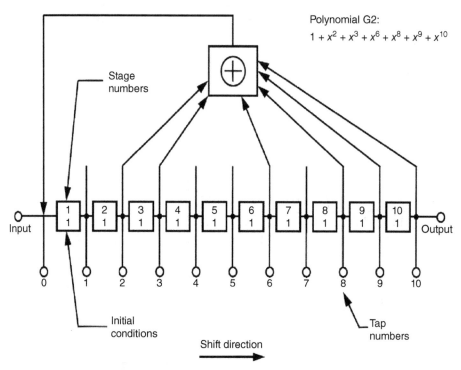

Figure 7.7. G2 m-Sequence Generator [32]

SV ID No.	GPS PRN Signal No.	Code Phase Selection C/A(G2i)****	Code Phase Selection (X2i)	Code Delay Chips C/A	Code Delay Chips P	First 10 Chips Octal* C/A	First 12 Chips Octal P
1	1	2⊕6	1	5	1	1440	4444
2	2	3⊕7	2	6	2	1620	4000
3	3	4⊕8	3	7	3	1710	4222
4	4	5⊕9	4	8	4	1744	4333
5	5	1⊕9	5	17	5	1133	4377
6	6	2⊕10	6	18	6	1455	4355
7	7	1⊕8	7	139	7	1131	4344
8	8	2⊕9	8	140	8	1454	4340
9	9	3⊕10	9	141	9	1626	4342
10	10	2⊕3	10	251	10	1504	4343
11	11	3⊕4	11	252	11	1642	
12	12	5⊕6	12	254	12	1750	
13	13	6⊕7	13	255	13	1764	
14	14	7⊕8	14	256	14	1772	
15	15	8⊕9	15	257	15	1775	
16	16	9⊕10	16	258	16	1776	
17	17	1⊕4	17	469	17	1156	
18	18	2⊕5	18	470	18	1467	
19	19	3⊕6	19	471	19	1633	4343
20	20	4⊕7	20	472	20	1715	4343
21	21	5⊕8	21	473	21	1746	
22	22	6⊕9	22	474	22	1763	
23	23	1⊕3	23	509	23	1063	
24	24	4⊕6	24	512	24	1706	
25	25	5⊕7	25	513	25	1743	
26	26	6⊕8	26	514	26	1761	
27	27	7⊕9	27	515	27	1770	
28	28	8⊕10	28	516	28	1774	
29	29	1⊕6	29	859	29	1127	
30	30	2⊕7	30	860	30	1453	
31	31	3⊕8	31	861	31	1625	
32	32	4⊕9	32	862	32	1712	
***	33	5⊕10	33	863	33	1745	
***	34**	4⊕10	34	950	34	1713	
***	35	1⊕7	35	947	35	1134	
***	36	2⊕8	36	948	36	1456	
***	37**	4⊕10	37	950	37	1713	4343

* In the octal notation for the first 10 chips of the C/A code as shown in this column, the first digit (1) represents a "1" for the first chip and the last three digits are the conventional octal representation of the remaining 9 chips. (For example, the first 10 chips of the C/A code for PRN Signal Assembly No. 1 are: 1100100000).

** C/A codes 34 and 37 are common.

*** PRN sequences 33 through 37 are reserved for other uses (e.g. ground transmitters).

**** The two-tap coder utilized here is only an example implementation that generates a limited set of valid C/A codes.

⊕ = "exclusive or"

NOTE: The code phase assignments constitute inseparable pairs, each consisting of a specific C/A and a specific P code phase, as shown above.

Figure 7.8. C/A Code Signal Spreading Code Definitions [32]

GPS SIGNALS

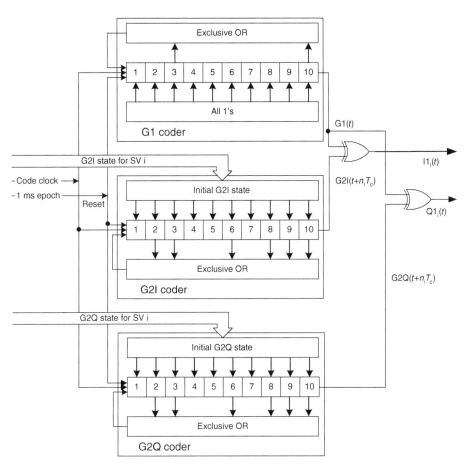

Figure 7.9. More General Alternative Gold Code Generator for C/A Code Signal Spreading Codes [32]

Figure 7.5 can only be used to generate a subset of the full number of C/A codes. A more general implementation is obtained by delaying the G2 register by a specified number of chips, leading to the C/A code generator portrayed in Figure 7.9.

An additional 173 Gold codes are defined in IS-GPS-200, yielding a total of 210 spreading codes in the code family for C/A code signals.

The NAV message is fully described in IS-GPS-200, with additional description in Reference 35. Only a brief overview is provided here. Fundamental to the use of the data message is GPS timing, often referred to as the Z count, which indicates epochs, called X1 epochs, each lasting 1.5 seconds. The 10 most significant bits of the Z count provide the modulo 1024 representation of the GPS week. The 19 least significant bits of the Z count provide the number of X1 epochs since the transition from previous week at midnight Sunday night, UTC.

Each NAV message frame is 1500 bits long, lasting 30 seconds. Each message frame is divided into five subframes, each 300 bits long and lasting 6 seconds and containing 10 words each 30 bits long and lasting 0.6 seconds. The first word in each subframe is a telemetry (TLM) word that begins with an eight-bit preamble for synchronization. The rest of the TLM word contains information for authorized P(Y) code users.

The second word in each message frame is the handover word (HOW), whose first 17 bits indicate the number of subframes since the start of the week. The HOW also contains the subframe number and information that describes that status of the P(Y) code. Each subframe also contains eight data words that provide clock correction for the satellite, satellite ephemeris, almanac, ionospheric correction information, and other system information.

A complete message requires 25 frames, lasting 12.5 minutes. Every frame contains the information needed to navigate using the signal, but all 25 frames are needed to access the complete almanac and other system information.

7.3.2 GPS L1C Signal

GPS's commitment to develop and operate a new civil signal on the L1 carrier arose from the US negotiations with the European Community in Reference 21, since Europe sought a GPS signal that would be more interoperable than C/A code signal with the Galileo open signal at this carrier frequency. The L1C design team surveyed users internationally, developed design options [26], then settled on an almost-final design [36], with a subsequent enhancement of the pilot component's spreading modulation from BOC(1,1) to TMBOC.

The resulting L1C characteristics are described in Reference 37 and summarized in Table 7.2. In many ways, L1C is the most technically sophisticated of all the signals described in this book. Providing three times the power in the pilot relative to the data improves tracking performance and robustness. The use of powerful LDPC error control coding, combined with block interleaving and a data message design that enables receivers to trade latency for lower demodulation thresholds, as described below, also improves data demodulation robustness. Introduction of the TMBOC spreading modulation enables high-accuracy receivers to track a pilot component with larger RMS bandwidth, while low cost receivers can process the pilot component as if it had a BOC(1,1) spreading modulation, allowing narrower precorrelation bandwidth and lower sampling rate than TMBOC. The combination of TMBOC and the Weil-based codes also reduce signal crosscorrelation and autocorrelation sidelobes [39].

While the general concept of TMBOC was introduced in Section 3.1, L1C's specific TMBOC design is described here. Figure 7.10 shows the spreading waveforms used for L1C. The L1C data component comprises BOC(1,1) spreading symbols, indicated by the open rectangles, each of which is one cycle of a square wave having 1.023 MHz fundamental frequency. Its power is 25% of the total signal power. In contrast, the L1C pilot component contains a mixture of time-multiplexed BOC(1,1) and BOC(6,1) spreading symbols; BOC(6,1) spreading symbols are indicated by solid rectangles in the figure. The pilot's power is 75% of the total signal power, providing more robust tracking. The 10230 spreading symbols within each pilot overlay code bit consist of

TABLE 7.2. Summary of L1C Signal Characteristics

Carrier frequency	L1: 1575.42 MHz = 154 × 10.23 MHz
Minimum received power	−157.0 dBW
Polarization	RHCP
Multiple access	CDMA
Spreading modulation and transmit bandwidth	TMBOC(6,1,4/33) Pilot, BOC(1,1) Data, ±15.345 MHz
Spreading codes	Length 10230 bit Weil-based codes, duration 10 ms
Data message structure	CNAV2
Data rate	50 bps
Data message error Correction and detection	1/2-rate LDPC coding, block interleaving, and 24-bit CRC
Data modulation	100 sps biphase modulation
Pilot and data components	75% power pilot, 25% power data, in same carrier phase
Overlay code	1800 bit pilot overlay codes, 100 bps, different for each satellite; segments of m-sequences and Gold codes
Multiplexing	Varies with GPS satellite block; pilot and data components share same carrier phase

310 repetitions of the 33 spreading symbols shown in the lower trace of Figure 7.10. In each set of 33 spreading symbols, four (the 1st, 5th, 7th, and 30th) are BOC(6,1), while the others are BOC(1,1). Review Question QT7.1 shows that the resulting L1C PSD, assuming ideal long spreading code, is given by the MBOC PSD

$$\Phi_{L1C}(f) = \frac{10}{11}\Phi_{BOC(1,1)}(f) + \frac{1}{11}\Phi_{BOC(6,1)}(f) \qquad (7.2)$$

Assuming ideal long spreading codes, the autocorrelation and PSD for the data component are displayed in Figure 7.11, while the autocorrelation and PSD for the pilot

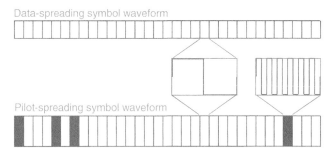

Figure 7.10.   Data and Pilot Spreading Waveforms, Scaled in Amplitude to Reflect 25%/75% Power Allocation

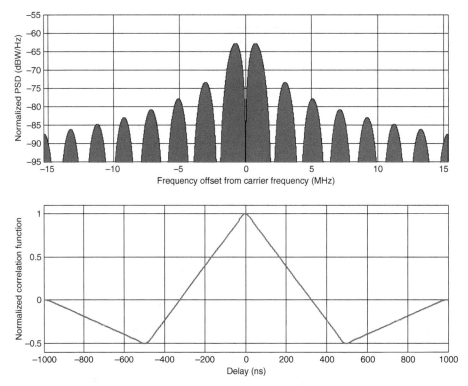

Figure 7.11. L1C Data Component PSD and Autocorrelation

component are displayed in Figure 7.12. The BOC(6,1) in the pilot component power produces additional power near frequencies offset by approximately 6 MHz from the carrier frequency.

L1C is the only signal described in this book whose spreading codes are neither memory codes nor generated by linear shift registers. Instead, each of L1C's Weil-based codes is generated as shown in Figure 7.13. First the 10223-bit Legendre sequence is generated. (An n-length Legendre sequence [38] is defined for any odd prime number n. Using logical representation of the values, the first value is always 1; the kth value for integers $1 < k \leq n - 1$ is 0 if k is a square modulo n,[2] and 1 otherwise. There is only one Legendre sequence of a given length.) This Legendre sequence can either be generated upon initialization of a satellite or receiver, or prestored in nonvolatile memory.

Next, a 10223-length Weil sequence is formed by exclusively ORing the Legendre sequence with a circularly shifted version of itself, where the number of bits in the circular shift is given by the Weil index, which ranges from 1 to the length of the Legendre

---

[2] An integer k is a square modulo n if there exists an integer m such that $m^2 = k \pmod{n}$, or equivalently so that $m^2 - n \lfloor m^2/n \rfloor = k$ with $\lfloor \rfloor$ indicating the floor function.

GPS SIGNALS

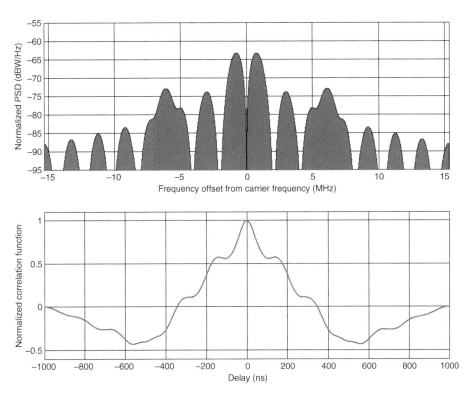

Figure 7.12. L1C Pilot Component Autocorrelation and PSD

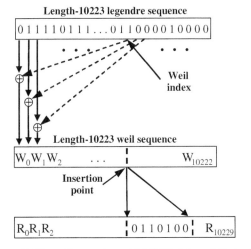

Figure 7.13. Generation of L1C's Weil-based Codes [37]

GPS PRN Signal No.	L1C_P				L1C_D			
	Weil Index (w)	Insertion Index (p)	Initial 24 Chips (Octal)	Final 24 Chips (Octal)	Weil Index (w)	Insertion Index (p)	Initial 24 Chips (Octal)	Final 24 Chips (Octal)
1	5111	412	05752067	20173742	5097	181	77001425	52231646
2	5109	161	70146401	35437154	5110	359	23342754	46703351
3	5108	1	32066222	00161056	5079	72	30523404	00145161
4	5106	303	72125121	71435437	4403	1110	03777635	11261273
5	5103	207	42323273	15035661	4121	1480	10505640	71364603
6	5101	4971	01650642	32606570	5043	5034	42134174	55012662
7	5100	4496	21303446	03475644	5042	4622	00471711	30373701
8	5098	5	35504263	11316575	5104	1	32237045	07706523
9	5095	4557	66434311	23047575	4940	4547	16004766	71741157
10	5094	485	52631623	07355246	5035	826	66234727	42347523
11	5093	253	04733076	15210113	4372	6284	03755314	12746122
12	5091	4676	50352603	72643606	5064	4195	20604227	34634113
13	5090	1	32026612	63457333	5084	368	25477233	47555063
14	5081	66	07476042	46623624	5048	1	32025443	01221116
15	5080	4485	22210746	35467322	4950	4796	35503400	37125437
16	5069	282	30706376	70116567	5019	523	70504407	32203664
17	5068	193	75764610	62731643	5076	151	26163421	62162634
18	5054	5211	73202225	14040613	3736	713	52176727	35012616
19	5044	729	47227426	07750525	4993	9850	72557314	00437232
20	5027	4848	16064126	37171211	5060	5734	62043206	32130365
21	5026	982	66415734	01302134	5061	34	07151343	51515733

Figure 7.14. L1C Spreading Code Definitions for the First 21 PRN Indices [37]

sequence. There are as many different Weil sequences as there are bits in the underlying Legendre sequence; one for each Weil index.

Finally, in order to extend the length of a 10223-bit Weil sequence to 10230 bits, seven bits using the bit pattern 0 1 1 0 1 0 0 are inserted into the Weil sequence. Thus, there are two design parameters for a Weil-based spreading code, the Weil index and the insertion point for the additional seven bits. As described in Reference 39, an extensive search over these two design parameters yielded a family of 210 spreading code pairs that exhibit very good autocorrelation and crosscorrelation properties. Even though the Weil-based codes cannot be generated with shift registers, they can be generated using simple logic.

Figure 7.14 defines the Weil index and insertion point for the pilot and data spreading codes associated with the first 21 PRN indices. Provision of the first 24 and last 24-bit values allows for ample crosschecking of a spreading code generator. Additional definitions for spreading codes associated with PRN indices 22 through 210 are available in Reference 37.

10230-length spreading codes are found in L2CM, both L5 spreading codes, both L1C spreading codes, and both spreading codes for Galileo E5a (see Section 10.3.5). Table 7.3 summarizes key performance characteristics of signals with these spreading codes [36], when there is no Doppler offset between the replica and received signal. The values are calculated using signals sampled at the rate of one sample per spreading

TABLE 7.3. Correlation Properties of Signals with Different 10230-Length Spreading Codes [36]

Code Family	Number of Codes	Max. Auto Even Sidelobe (dB)	Max. Cross Even Sidelobe (dB)	99.9999% Auto Even/Odd Sidelobe (dB)	99.9999% Cross Even/Odd Sidelobe (dB)
L1C TMBOC	420	−31.1	−27.7	−29.4	−28.7
Galileo E5a	200	−28.6	−25.5	−28.6	−26.4
L5 (I5 and Q5)	420	−28.6	−26.0	−26.9	−27.0
L2C CM	37	−27.0	−25.4	−27.0	−25.4

symbol, except that 24 samples per spreading symbol are used for TMBOC to adequately sample the BOC(6,1) parts of the waveform. Mathematicians typically look at the maximum side lobes in the crosscorrelation between a replica and a received signal that does not undergo a (data symbol or overlay code) bit transition (called "even") and an incoming signal stream that does undergo a bit transition (called "odd"). Of more practical interest may be the sidelobe levels at some higher probability point, like the 99.999% point of the cumulative distribution functions shown here. The TMBOC spreading modulation also provides benefits in signal correlation performance, and its effects are included here. As evident from the Welch bound discussed in Section 3.3.6, it is more challenging to obtain good correlation performance from code families with larger numbers of codes. Even though the L1C spreading code family is as large as or larger than the other spreading code families shown, it provides lower sidelobe levels in the absence of Doppler shift, reducing crosscorrelation effects that can degrade performance in situations where a receiver is searching for a weak signal while exposed to interference from stronger signals, as discussed in Chapter 16.

Each L1C pilot component employs a different overlay code 1800 bits long and generated at 100 bps, so the duration is 18 seconds. Some overlay codes are segments of m-sequences and others are segments of Gold codes, both generated using one or both of the 11-register linear shift register structures shown in Figure 7.15 [39]. Figure 7.16 shows the code phase assignments for the first 21 PRN indices. Reference 37 provides corresponding code phase assignments for the remaining 210 overlay codes. The top part of the figure shows one shift register used to generate a sequence called the S1 sequence while the bottom part of the figure shows a shift register used to generate a second sequence, called the S2 sequence. The first 63 are segments of m-sequences only using the S1 sequence, while the remaining overlay codes are Gold codes generated using both the S1 and S2 sequences. As described in Reference 39, overlay codes were designed to have low autocorrelation and crosscorrelation sidelobes not only over their entire length, but also over one-second (100 bits) and two-second (200 bits) segments, since some receivers align to the overlay code by correlating shorter segments of the overlay code.

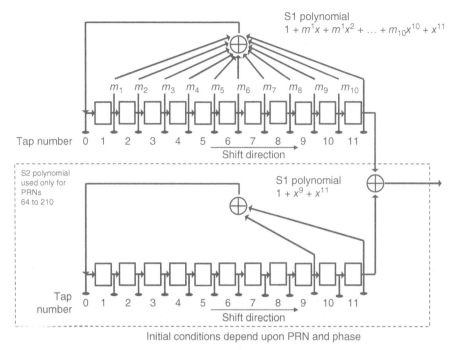

Figure 7.15. Linear Shift Registers Used to Generate L1C Overlay Codes [37]

Figure 7.17 portrays the L1C data message structure, showing three distinct subframes that total 900 bits of information, with an 18 second duration. The first subframe contains only the time of interval (TOI), providing a count of how many message beginnings have passed since the start of the GPS week. (The GPS week count is provided as part of subframe 3 in some messages.) The TOI is encoded using a BCH(52,9) code [40]. Subframe 2 provides a 576-bit representation of clock corrections and ephemeris data (CED) along with a separate 24-bit CRC, using a higher precision representation than in the legacy NAV message. It is encoded using LDPC(1200,600), 1/2-rate block encoding that transforms 600 bits into 1200 symbols. Subframe 3 contains 250 bits of various system messages (almanacs, time and coordinate conversions, group delay and time offsets, etc.) along with a separate 24-bit CRC. It is encoded using LDPC(548,274), 1/2-rate block encoding [41] that transforms 274 bits into 548 symbols. Block interleaving of encoded symbols from subframes 2 and 3, using a 38 × 46 block matrix, enhances robustness to burst errors, as discussed in Section 3.3.8.

The strong BCH encoding enables the TOI to be read at low C/N_0 levels. Moreover, since the TOI is the same in data messages broadcast from all satellites, a receiver can read the TOI from any satellite (one whose received signal is stronger than the others) and apply it to the others.

The CED must be known for the receiver to use the signal. Each subframe is separately encoded, and subframes 2 and 3 have separate CRCs, so changes in the TOI

GPS SIGNALS 183

GPS PRN Signal No.	S1 Polynomial Coefficient (Octal) * $(m_{i,j})$	Initial 11 Bits (Octal †) **	Final 11 Bits (Octal †)
1	5111	3266	0410
2	5421	2040	3153
3	5501	1527	1767
4	5403	3307	2134
5	6417	3756	3510
6	6141	3026	2260
7	6351	0562	2433
8	6501	0420	3520
9	6205	3415	2652
10	6235	0337	2050
11	7751	0265	0070
12	6623	1230	1605
13	6733	2204	1247
14	7627	1440	0773
15	5667	2412	2377
16	5051	3516	1525
17	7665	2761	1531
18	6325	3750	3540
19	4365	2701	0524
20	4745	1206	1035
21	7633	1544	3337

NOTES:
* The polynomial coefficient is given as $m_{11}, \ldots, m_0$. Thus octal 5111 corresponds to the generator polynomial $P_1(x) = 1 + x^3 + x^6 + x^9 + x^{11}$.
** The initial 11 bits also represent the initial condition, $n_{11}, \ldots, n_1$ for each PRN signal number.
† The initial and the final bit values are obtained after dropping the initial bit value 0. Thus octal 3266 corresponds to 11 values 1 1 0 1 0 1 1 0 1 1 0.

Figure 7.16. L1C Overlay Code Shift Register Phase Assignments [37]

and in subframe 3 do not introduce any changes to the encoded symbols from subframe 2. Since the CED symbols can be located and deinterleaved from the received message, they can be decoded and error checked (using the Subframe 2 CRC) independent of the subframe 1 and subframe 3 information. The L1C CED will remain the same for at least 15 minutes and up to 120 minutes, typically changing on a schedule of every 2 hours,

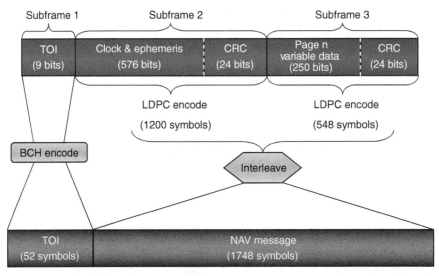

Figure 7.17. CNAV2 Data Message Structure [36]

and applies over a curve fit interval of 3 hours in an ECEF coordinate system. Except for times when the CED changes, receivers can perform data symbol combining across multiple subframe 2 data message symbols, reducing the data demodulation threshold as discussed in Section 18.5.

Subframe 3 contents are variable and do not follow a defined sequence, with different information, contained in pages, provided in sequential messages. Table 7.4 provides currently defined pages. Coordinated Universal Time parameters describe how to convert from GPS time to UTC. Ionospheric correction parameters are provided for single frequency users. GPS/GNSS time offsets provide the time differences between GPS time and the system time of other satnav systems. Earth orientation parameters provide the parameters needed for a receiver to construct the transformation between ECEF and ECI coordinate systems. The reduced almanac is a more compact but lower precision representation of the almanac. Differential correction parameters apply to the clock and

TABLE 7.4. Examples of CNAV2 Subframe 3 Pages

Page Number	Message Contents
1	UTC and ionospheric corrections
2	GPS/GNSS time offset and Earth orientation parameters
3	Reduced almanac parameters
4	Almanac parameters
5	Differential correction parameters
6	Text message
7	Signal phase for each SV

ephemeris data transmitted by other GPS satellites. Text messages contain free-form ASCII text.

7.3.3 GPS P(Y) Code Signal

The P(Y) code signal originally used the long but published P (precision) code, but an encrypted (Y) mode was added and is typically employed. The P(Y) code signal was intended for use by authorized (typically military, but some other government as well) users. However, ways of tracking the signal without access to the spreading code, known as codeless processing or semicodeless processing, have been developed [7], and consequently these techniques are widely used in receivers employed by professional users as well. The US government has announced that it will commit to supporting such semicodeless use of P(Y) only through the end of 2020 [42]. After that time, GPS users are expected to make use of the new L2C signal and the government may change P(Y) in ways that affects its tracking by semicodeless users.

Table 7.5 summarizes the characteristics of the P(Y) code signal. Two identical versions, sometimes denoted L1 P(Y) and L2 P(Y), are transmitted on two carrier frequencies. The minimum received power values in Table 7.5 are for GPS III satellites; some earlier satellite versions provide lower specified minimum received power.

Figure 7.18 shows the P(Y) code signal autocorrelation function and PSD, assuming an ideal long spreading code (which is a good model for either P code or P(Y) code).

While the P(Y) code signal's spreading code design and structure are not publicly available, the P code is described in Reference 32. P code generation is based on four 12-bit m-sequence generators denoted X1A, X1B, X2A, and X2B. The X1A and X1B registers start at the same value for all PRN indices, while the X2A and X2B registers

TABLE 7.5. Summary of P(Y) Code Signal Characteristics

Carrier frequency	L1: 1575.42 MHz = 154 × 10.23 MHz
	L2: 1227.60 MHz = 120 × 10.23 MHz
Minimum received power	−161.5 dBW
Polarization	RHCP
Multiple access	CDMA
Spreading modulation and transmit bandwidth	BPSK-R(10), ±15.345 MHz
Spreading codes	Precision (P) code or encrypted (Y) code
Data message structure	NAV
Data rate	50 bps
Data message error correction and detection	Error detection using extended (32,26) Hamming code
Data modulation	50 sps biphase modulation
Pilot and data components	100% power data
Overlay code	None
Multiplexing with other signals	Varies with GPS satellite block, but L1 P(Y) is in phase quadrature to the C/A code signal

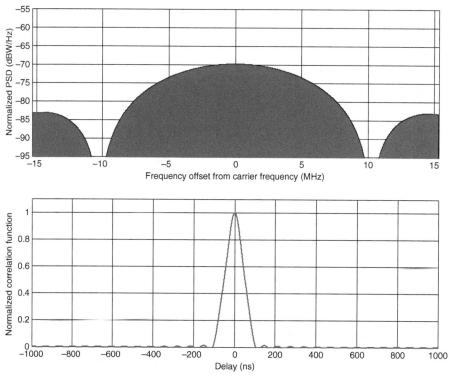

Figure 7.18. PSD and Autocorrelation for P(Y)

are delayed by different numbers of chips for each PRN index, as shown in Figure 7.8. X1A and X2A are short cycled (reset to their starting value) after 4092 bits, while X1B and X2B are short cycled after 4093 bits.

Using the logical representation of binary values summarized in Table 3.1, two sequences are formed from the four m-sequences:

$$X1 = X1A \oplus X1B$$
$$X2_k = X2A_k \oplus X2B_k$$
(7.3)

X1 is the same for all satellites, while $X2_k$ is delayed by a different number of chips for a different PRN index, so indexed by k. Every 15345000 bits, the X1A and X1B registers are reset to their starting values, so X1 repeats every 15345000 bits, or 1.5 seconds at the 10.23 MHz chip rate (leading to the terminology "X1 epoch"). $X2_k$ is reset every 15345037 bits, or approximately 3.6 μs longer than 1.5 seconds. The P-code spreading code for the kth PRN index is then $P_k = X1 \oplus X2_k$. While its potential length is $15345000 \times 15345037 = 2.3547 \times 10^{14}$ bits, or approximately 38 weeks at the 10.23 MHz chip rate, each P code is reset at the beginning of each GPS week.

TABLE 7.6. Summary of M Code Signal Characteristics

Carrier frequency	L1: 1575.42 MHz = 154 × 10.23 MHz
	L2: 1227.60 MHz = 120 × 10.23 MHz
Minimum received power	Restricted
Polarization	RHCP
Multiple access	CDMA
Spreading modulation and transmit bandwidth	BOC(10,5), ±15.345 MHz
Spreading codes	Restricted
Data message structure	MNAV
Data rate	Choice of 100 bps, 25 bps, 0 bps
Data message error correction and detection	Restricted
Data modulation	Choice of 200 bps, 50 bps, 0 bps biphase modulation
Pilot and data components	50% power pilot, 50% power data, time multiplexed spreading symbol by spreading symbol
Overlay code	Restricted
Multiplexing with other signals	Multiplexing with other signals is not defined; varies with satellite block

This algorithm for generating P-code spreading codes yields 37 different spreading codes as indicated in Figure 7.8. An additional 173 P-code spreading codes, for a total of 210, are obtained from circular shifts of these sequences by numbers of chips equal to 1, 2, 3, 4, or 5 days, as defined in Reference 32.

The NAV message for the P(Y)-code signal is identical to that for the C/A code signal, as described in Section 7.3.1.

7.3.4 GPS M Code Signal

Table 7.6 summarizes key characteristics of the M code signal, the GPS modernized military signal, which uses the same carrier frequencies as the P(Y) code signal. Many details of the M code signal are not publicly available. Some additional information concerning the M code signal's design is available in References 15 and 43. Time multiplexing of the pilot and data components is on a spreading symbol by spreading symbol basis, just as described for the L2C signal in Section 7.3.5. The GPS control segment selects the data rate for each carrier frequency on each satellite, allowing for a blend of lower latency and more robust data messages. Figure 7.19 shows the M code signal's autocorrelation function and PSD, assuming an ideal long spreading code.

7.3.5 GPS L2C Signal

When the decision was made to place a civil signal at the L2 carrier frequency, the initial plan was to merely replicate the C/A code signal, so it would be transmitted at

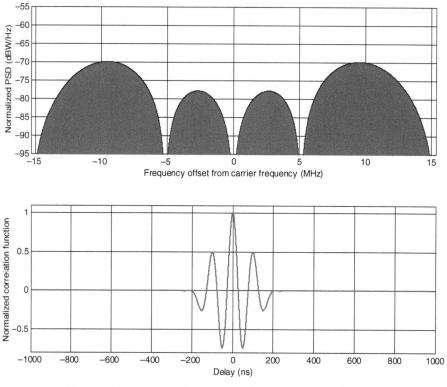

Figure 7.19. PSD and Autocorrelation for the M Code Signal

two carrier frequencies. However, there was also the desire to employ a better signal design, including drawing from the recommendations from the Project 621B report [44] that had not been implemented in C/A code signal. Due to contractual constraints and time constraints, it was decided to make L2C differ from the C/A signal in only four primary ways: pilot and data components, error control, spreading codes, and data message design. Each of these changes is discussed subsequently.

Distinct pilot and data components, recommended in Reference 44 and described in Section 3.3.10, were incorporated into the L2C design. In order to reduce changes to the satellite's signal generator, L2C needed to remain a biphase signal. Consequently, equal-power pilot and data components were time-multiplexed by alternating spreading symbols, as described later in this subsection.

Data message modulation and error control needed to be redesigned to provide more robustness, in part to match the increased tracking robustness provided by the pilot component, and also to compensate for the 50% loss of power in the data component, compared to the C/A code signal. Consequently, the data message bit rate was reduced from 50 bps to 25 bps, increasing the energy in each bit by 3 dB. Further, 1/2-rate convolutional coding was used, further reducing the data demodulation threshold and providing a 50 sps data message symbol rate—the same as that of C/A code signal. In addition, a CRC was added to virtually assure detection of uncorrectable random errors.

Longer spreading codes were developed to mitigate the disadvantages of C/A's short and repeating codes as discussed in Sections 5.6 and 7.3.1. While in retrospect overlay codes could have been used to extend the 1023-bit Gold codes, instead it was decided to use longer spreading codes. A medium length code, whose duration is matched to the duration of a data message symbol, was designed for the data component. A long code, whose duration is matched to the X1 epoch of 1.5 seconds, was designed for the pilot component.

The structure of the C/A code signal's NAV message is rigidly defined in IS-GPS-200, with few spare bits and little flexibility for enhancement or changes. A new message design, called CNAV, was developed for the L2C signal. CNAV features a flexible message structure, where different message types are defined to provide different information, and the opportunity exists to add new message types as the need arises. While the repetition structure for some fundamental messages, such as those containing clock corrections and ephemeris, is defined in Reference 32, the sequence of other message types can be adjusted, allowing for growth and adaptation of the L2C data message as needs arise.

Table 7.7 summarizes key characteristics of the L2C signal. Its lower rate data message trades increased robustness for longer latency, matching the needs of professional users who do not have stringent requirements for fast acquisition or high data rates; this

TABLE 7.7. Summary of L2C Signal Characteristics

Carrier frequency	L2: 1227.60 MHz = 120 × 10.23 MHz
Minimum received power	−158.5 dBW
Polarization	RHCP
Multiple access	CDMA
Spreading modulation and transmit bandwidth	BPSK-R(1), ±15.345 MHz
Spreading codes	Pilot component: Length 767250-bit L2CL code, duration 1.5 secondsData component: Length 10230-bit L2CM code, duration 20 ms
Data message structure	CNAV
Data rate	25 bps
Data message error correction and detection	$\frac{1}{2}$-rate convolutional coding with constraint length 7 with 24-bit CRC
Data modulation	50 sps biphase modulation
Pilot and data components	50% power pilot, 50% power data, time multiplexed spreading symbol by spreading symbol
Overlay code	None
Multiplexing with other signals	Multiplexing with other signals is not defined in Reference 32; L2C may be either in phase quadrature to L2 P(Y) or in the same phase

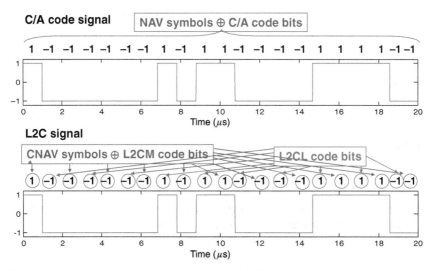

Figure 7.20. Example Comparing Data Modulation on C/A Code Signal and L2C Signal

tradeoff was discussed in Section 3.3.7. The minimum received power shown is for GPS III satellites; lower specified minimum received power is available on GPS Block IIR-M and IIF satellites. As indicated in Reference 32, the L2C data message can alternatively be the NAV message, at 25 bps and convolutionally encoded just like the CNAV message. However, this alternative mode was defined in case it was needed during initial transmission of L2C, is not expected to be used, and thus is not captured in the table below. When the spreading codes are modeled as ideal and long, the autocorrelation function and PSD of the L2C signal are the same as those shown in Figure 7.3.

Figure 7.20 provides a more detailed examination of the time multiplexing used to combine L2C's pilot and data components. The figure shows a hypothetical case where the bit sequences modulating the spreading symbol waveforms for C/A code signal and L2C are the same, even though the process for forming these bit sequences is very different. The upper trace shows a segment of C/A code signal, where the data spreading symbol waveform is modulated by the combination of data message symbols and spreading code bits. The lower trace shows a segment of the L2C signal. The L2C long (L2CL) and L2C medium (L2CM) spreading codes are each generated at a rate of 0.5115 Mbps (each spreading code is generated at the rate 1.023/2 MHz, yielding a multiplexed rate of 1.023 MHz), and data message symbols are combined with the L2CM spreading code bits, but not the L2CL spreading code bits. In the lower trace of this figure, the data spreading symbol waveform comprises rectangular symbols with duration $1/1.023$ μs interspersed with zero values with duration $1/1.023$ μs. The pilot spreading symbol waveform comprises zero values interspersed with rectangular symbols, also each with duration $1/1.023$ μs. The combination of data message symbols and L2CM spreading code bits modulates the data spreading symbol waveform, while the L2CL spreading code bits modulate the pilot spreading symbol waveform. The resulting data component waveform and pilot component waveform are summed, and

GPS SIGNALS

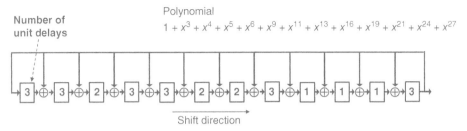

Initial conditions depend upon PRN and spreading code type (moderate, long)

Figure 7.21. m-Sequence Generator for L2CM and L2CL Spreading Codes [32]

since each is nonzero when the other is not, the resulting composite signal is the time division multiplexing of the data spreading symbols modulated by the combination of L2CM bits and data message symbols, with pilot spreading symbols modulated by L2CL bits.

All L2CL and L2CM spreading codes are segments of the order 27 m-sequence generated by the shift register and polynomial in Figure 7.21. This m-sequence has more than 134 million bits, and each L2CM and L2CL spreading code is a different segment, each generated by starting the shift register with a unique initial state, then short-cycling it after the needed number of bits has been generated. Different shift registers having identical structures but different starting states are used for L2CL and L2CM, each clocked at 511.5 kHz.

Figure 7.22 defines the spreading codes for different PRN indices assigned to different space vehicles (SVs), showing initial and end states for the shift registers. This table shows the 37 pairs (L2CM and L2CL) spreading codes; a total of 210 pairs of

SV ID No.	GPS PRN Signal No.	Initial Shift Register State (Octal) L2 CM	Initial Shift Register State (Octal) L2 CL	End Shift Register State (Octal) L2 CM *	End Shift Register State (Octal) L2 CL **
1	1	742417664	624145772	552566002	267724236
2	2	756014035	506610362	034445034	167516066
3	3	002747144	220360016	723443711	771756405
4	4	066265724	710406104	511222013	047202624
5	5	601403471	001143345	463055213	052270433
6	6	703232733	053023326	667044524	761743665
7	7	124510070	652521276	652322653	133015726
8	8	617316361	206124777	505703344	610611511
9	9	047541621	015563374	520302775	352150323
10	10	733031046	561522076	244205506	051266046
11	11	713512145	023163525	236174002	305611373
12	12	024437606	117776450	654305531	504676773
13	13	021264003	606516355	435070571	272572634
14	14	230655351	003037343	630431251	731320771
15	15	001314400	046515565	234043417	631326563
16	16	222021506	671511621	535540745	231516360
17	17	540264026	605402220	043056734	030367366
18	18	205521705	002576207	731304103	713543613
19	19	064022144	525163451	412120105	232674654
20	20	120161274	266527765	365636111	641733155
21	21	044023533	006760703	143324657	730125345
22	22	724744327	501474556	110766462	000316074
23	23	045743577	743747443	602405203	171313614
24	24	741201660	615534726	177735650	001523662
25	25	700274134	763621420	630177560	023457250
26	26	010247261	720727474	653467107	330733254
27	27	713433445	700521043	406576630	625055726
28	28	737324162	222567263	221777100	476524061
29	29	311627434	132765304	773266673	602066031
30	30	710452007	746332245	100010710	012412526
31	31	722462133	102300466	431037132	705144501
32	32	050172213	255231716	624127475	615373171
***	33	500653703	437661701	154624012	041637664
***	34	755077436	717047302	275636742	100107264
***	35	136717361	222614207	644341556	634251723
***	36	756675453	561123307	514260662	257012032
***	37	435506112	240713073	133501670	703702423

* Short cycled period = 10230
** Short cycled period = 767250
*** PRN sequences 33 through 37 are reserved for other uses (e.g. ground transmitters).

NOTE: There are many other available initial register states which can be used for other signal transmitters including any additional SVs in future.

Figure 7.22. Definition of L2C Spreading Codes [32]

spreading codes have been defined in Reference 32, enabling other uses or systems to employ spreading codes from the same family.

L2C's CNAV message structure differs significantly from the legacy NAV message used on C/A code and P(Y) code signals. Each message has fewer bits, in part to compensate for the slower message bit rate. The message format was originally designed to enable flexible sequences of messages, although some rigidity has been introduced subsequently to assure users of a maximum latency for essential information. Because the message type is not implied by its location in a rigid sequence of messages, message type IDs are provided. The CRC reduces the probability of undetected error to near zero.

Each message comprises 300 bits. The first 38 bits of the message are as follows: an eight-bit preamble of 10001011, followed by the 6-bit PRN index number of the transmitting SV, a 6-bit message type ID, a 17-bit message time of week (TOW) count, and a single bit "alert" flag that indicates to the user that the signal accuracy may be worse than indicated in the respective message types. Multiplying the TOW count by six yields satellite time in seconds at the start of the next 12-second message. These 38 bits are followed by 238 bits of message payload, followed by a 24-bit CRC covering the message contents.

Table 7.8 summarizes some of the CNAV message types defined in IS-GPS-200. The default message consists of alternating ones and zeros. The ephemeris representation is a higher precision representation than in the NAV message, and applies over a curve fit interval of three hours in an ECEF coordinate system. Ephemeris requires two messages, which also provide indicators of "health"—which signals are valid to use, and their SISRE. The reduced almanac is a more compact but lower precision representation of the almanac. Differential correction parameters apply to the clock and ephemeris data transmitted by other GPS satellites. Clock correction parameters are distributed

TABLE 7.8. Examples of CNAV Message Types

Message Type Number	Message Contents
0	Default
10	Ephemeris 1 and health
11	Ephemeris 2 and health
12	Reduced almanac
13	Differential correction parameters
14	Differential correction parameters
15	Text message
30	Clock correction, ionosphere correction, group delay
31	Clock correction, reduced almanac
32	Clock correction, Earth orientation parameters
33	Coordinated universal time parameters
34	Differential correction parameters
35	GPS-to-GNSS Time Offsets (GGTO)
36	Clock corrections and text message
37	Almanac

across two messages. Text messages contain free-form ASCII text. Ionospheric correction parameters are provided for single frequency users, while differential group delay parameters describe time offsets among L1, L2, and L5 signals for multifrequency users. Earth orientation parameters provide the parameters needed for a receiver to construct the transformation between ECEF and ECI coordinate systems. Coordinated Universal Time parameters describe how to covert from GPS time to UTC. GPS/GNSS time offsets provide the time difference between GPS time and the system time of other satnav systems.

The sequence of message types is arbitrary but they follow certain restrictions: under the command of the control segment, messages providing clock corrections and ephemeris will be broadcast at least once every 48 seconds, defining the maximum latency before a receiver can begin to calculate position and time from the satellite. When all satellites in the GPS constellation are transmitting L2C, the reduced almanac will be transmitted at least every 20 minutes, and the full almanac every 120 minutes. Time offsets to other satnav systems will be broadcast at least every 288 seconds.

7.3.6 GPS L5 Signal

The L5 signal was designed primarily to meet the aviation safety-of-life community's needs for a second signal at a different carrier frequency [10]. It introduced a new carrier frequency to GPS, and 1176.45 MHz has become the second carrier frequency (along with L1) adopted by most satnav systems. Table 7.9 summarizes the L5 signal's characteristics, which are described in Reference 45. Its two components, phase multiplexed, are sometime referred to as two separate signals, L5I (the in-phase data component) and L5Q (the quadraphase pilot component). The minimum received power shown is for GPS III satellites; lower specified minimum received power is available on GPS Block IIF satellites. L5 documentation calls the overlay codes "synchronization sequences." The autocorrelation function and PSD of the L5 signal, assuming an ideal long spreading code, are the same as for the P(Y) code signal in Figure 7.18, since both use BPSK-R(10) spreading modulations.

Figure 7.23 shows L5 signal generation. The upper trace shows the data component, with formation of the data message, generation of data message symbols, and modulation by the overlay code and the spreading code. The lower leg shows the pilot component, with modulation by the overlay code and the spreading code.

The code spreading codes for the L5 signal, denoted XI_k for the data spreading code and XQ_k for the pilot spreading code, are formed by

$$\begin{aligned} XI_k &= XA \oplus XBI_k \\ XQ_k &= XA \oplus XBQ_k \end{aligned} \quad (7.4)$$

where XA, XBI, and XBQ are m-sequences, each of order 13. XA is the same for all satellites, while XBI and XBQ are different for each satellite. All XBI and XBQ sequences are generated using the same linear shift register design, but with different phasings. Figure 7.24 shows the linear shift registers that are used.

TABLE 7.9. Summary of L5 Signal Characteristics

Carrier frequency	L5: 1176.45 MHz = 115 × 10.23 MHz
Minimum received power	−154.0 dBW[a]
Polarization	RHCP
Multiple access	CDMA
Spreading modulation and transmit bandwidth	QPSK-R(10), ±12 MHz
Spreading codes	Pilot component: Length 767250-bit L2CL code, duration 1.5 seconds
	Data component: Length 10230-bit L2CM code, duration 20 ms
Data message structure	CNAV
Data rate	50 bps
Data message error correction and detection	$\frac{1}{2}$-rate convolutional coding with constraint length 7 with 24-bit CRC
Data modulation	100 sps biphase modulation
Pilot and data components	50% power pilot, 50% power data, phase quadrature multiplexed with pilot component on the in-phase component
Overlay codes	10 bit, 1 kbps Neuman–Hoffman code (1111001010) on I5, 20 bit, 1 kbps Neuman–Hoffman code (00000100110101001110) on Q5. Same on all satellites
Multiplexing with other signals	Not needed

[a]The value is −154.9 dBW for Block IIF satellites, but is planned to be −154.0 for GPS III and all subsequent satellites.

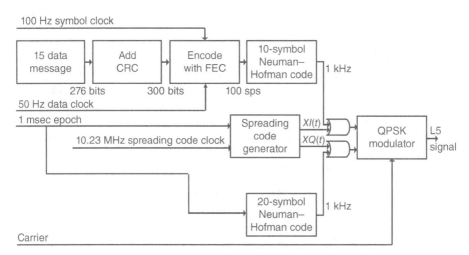

Figure 7.23.   Formation of L5 Signal [45]

GPS SIGNALS

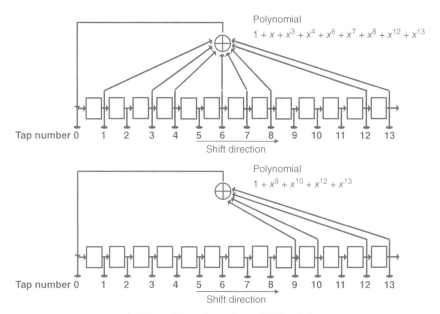

Figure 7.24. m-Sequence Generation for L5 Spreading Codes

Since the m-sequence from a length 13 shift register is only 8191 bits long, the XBI and XBQ registers are restarted after 8191 bits and used to generate an additional 2039 bits (which are the same as the first 2039 bits of the sequence) to produce a total of 10230 bits, then reset to their starting value. The XA register is short cycled after

GPS PRN Signal No.*	XB Code Advance – Chips** I5	XB Code Advance – Chips** Q5	Initial XB Code State*** I5	Initial XB Code State*** Q5	GPS PRN Signal No.*	XB Code Advance – Chips** I5	XB Code Advance – Chips** Q5	Initial XB Code State*** I5	Initial XB Code State*** Q5
1	266	1701	0101011100100	1001011001100	20	5443	1523	0110101101101	1100001110001
2	365	323	1100000110101	0100011110110	21	5641	4548	0010000001000	0110110010000
3	804	5292	0100000001000	1111000100011	22	5816	4484	1110111101111	0010110001110
4	1138	2020	1011000100110	0011101101010	23	5898	1893	1000011111110	1000101111101
5	1509	5429	1110111010111	0011110110010	24	5918	3961	1100010110100	0110111110011
6	1559	7136	0110011111010	0101010101001	25	5955	7106	1101001101101	0100010011011
7	1756	1041	1010010011111	1111110000001	26	6243	5299	1010110010110	0101010111100
8	2084	5947	1011110100100	0110101101000	27	6345	4660	0101011011110	1000011111010
9	2170	4315	1111100101011	1011110100011	28	6477	276	0111101010111	1111101000010
10	2303	148	0111111011110	0010010000110	29	6518	4389	0101111100001	0101000100100
11	2527	535	0000100111010	0001000000101	30	6875	3783	1000010110111	1000001111001
12	2687	1939	1110011111001	0101011000101	31	7168	1591	0001010011110	0101111100101
13	2930	5206	0001110011100	0100110100101	32	7187	1601	0000011111001	1001000101010
14	3471	5910	0100000100111	1010000111111	33	7329	749	1101010000001	1011001000100
15	3940	3595	0110101011010	1011110001111	34	7577	1387	1101111111001	1111001000100
16	4132	5135	0001111001001	1101001011111	35	7720	1661	1111011101100	0110010110011
17	4332	6082	0100110001111	1110011001000	36	4924	7777	1111100001110	1011011100100
18	4924	6990	1111000011110	1011011100100	37	8057	708	0011010010000	0010011010001
19	5343	3546	1100100011111	0011001011011					

* PRN sequences 33 through 37 are reserved for other uses (e.g. ground transmitters).
** XB Code Advance is the number of XB clock cycles beyond an initial state of all 1s.
*** In the binary notation for the first 13 chips of the I5 and Q5 XB codes as shown in these columns. The rightmost bit is the first bit out. Since the initial state of the XA Code is all 1s, these first 13 chips are also the complement of the initial states of the I5 or Q5-codes.

NOTE: The code phase assignments constitute inseparable pairs, each consisting of a specific I5 and a specific Q5-code phase, as shown above.

Figure 7.25. Definition of First 37 Spreading Code Pairs for L5 Signals

8190 bits, and used to generate an additional 2040 bits (which are the same as the first 2040 XA bits) for a total of 10230 bits.

A family of 210 L5 spreading code pairs is defined in IS-GPS-705 [45]. Figure 7.25 shows the definition of the first 37 spreading code pairs, with the XB registers defined both in terms of the number of chips advanced from a starting input of all ones, and as initial code states. The XA register always starts with register values of all ones.

The L5 CNAV message is the same flexible message design as used for the L2C signal, as described in Section 7.3.3, with the same 300-bit length and underlying structure. Some of the message types have the same numbers as in Table 7.8, while other message numbers are different.

7.4 SUMMARY

GPS provides a global system with civil signals on L1, L2, and L5. As GPS is modernized, its signal set is growing from three original signals to eight signals—four civil signals on three carrier frequencies, and four military signals on two carrier frequencies. The spectra of these evolving signals are shown in Figure 7.26, for spreading codes modeled as ideal and long, and key signal characteristics are summarized in Table 7.10.

The C/A code signal has been joined by additional civil signals at L2 and L5, paving the way for civilian users to access coded signals on up to three carrier frequencies. The L2C signal is intended primarily for professional users, while the L5 signal's characteristics are tuned to aviation safety of life applications. The L1C signal will provide significantly more capability than the C/A code signal for a variety of user types, while requiring greater receiver complexity due to its longer spreading codes, wider bandwidth, and modern forward error control techniques.

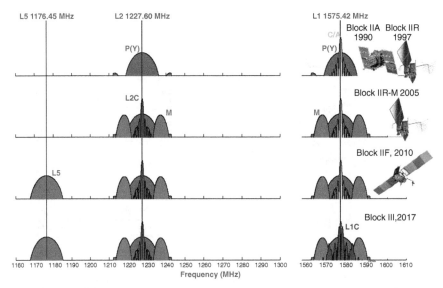

Figure 7.26. PSDs of GPS Signals at Different Stages of GPS Modernization

TABLE 7.10. Summary of GPS Signals

Signal Name	Carrier Frequency (MHz)	Spreading Modulation	Service
C/A code	1575.42	BPSK-R(1)	Standard positioning
L1C	1575.42	MBOC(6,1,1/11)	Standard positioning
L1 P(Y) code	1575.42	BPSK-R(10)	Precise positioning
L1 M code	1575.42	BOC(10,5)	Precise positioning
L2C	1227.60	BPSK-R(1)	Standard positioning
L2 P(Y) code	1227.60	BPSK-R(10)	Precise positioning
L2 M code	1227.60	BOC(10,5)	Precise positioning
L5	1176.45	QPSK-R(10)	Standard positioning

REVIEW QUESTIONS

Theoretical Questions

QT7.1. Show that the L1C PSD is given by the MBOC PSD $\Phi_{L1C}(f) = \frac{10}{11}\Phi_{BOC(1,1)}(f) + \frac{1}{11}\Phi_{BOC(6,1)}(f)$, independent of whether the pilot and data components are in the same carrier phase or in phase quadrature. Assume ideal long spreading codes and signal power computed over the bandwidths defined for each signal in Section 7.3.

QT7.2. List characteristics of C/A code and L2C signals that are similar, and characteristics that are different.

QT7.3. List characteristics of C/A code and L1C signals that are similar, and characteristics that are different.

QT7.4. List characteristics of L2C and L5 signals that are similar, and characteristics that are different.

QT7.5. In 2007 GPS announced the decision to adopt a TMBOC(6,1,4/33) spreading modulation for the L1C pilot component, instead of the BOC(1,1) spreading modulation that had been previously planned. Compute and compare the RMS bandwidths (See Appendix Section A.6) of TMBOC(6,1,4/33) and BOC(1,1) for precorrelation bandwidths of ± 2.5 MHz, ± 5.0 MHz, ± 15.0 MHz, ± 10.0 MHz, and ± 15.0 MHz. Assume ideal long spreading codes and signal power computed over the bandwidths defined for each signal in Section 7.3.

Application Questions

QA7.1. If the received power of a GPS C/A code signal is -153 dBW, what is the maximum power spectral density of the received signal? Assume ideal long spreading codes and signal power computed over the bandwidths defined for each signal in Section 7.3.

QA7.2. If the received power from a constellation of interfering GPS C/A code signals is -141 dBW, what is the maximum power spectral density of the aggregate C/A code interference? If a receiver's thermal noise density is −201.5 dBW/Hz, what is the maximum power spectral density of the combined aggregate C/A code interference and thermal noise? Assume ideal long spreading codes and signal power computed over the bandwidths defined for each signal in Section 7.3.

QA7.3. What power of aggregate M code interference would produce the same maximum power spectral density as interfering GPS C/A code signals at −141 dBW? Assume ideal long spreading codes and signal power computed over the bandwidths defined for each signal in Section 7.3.

QA7.4. For the following calculations, define C to be the minimum specified received power for all components of the relevant GPS signal, and let N_0 be −201.5 dBW/Hz. Define E_b as the energy per information bit, with $E_b = C_d/R_b$, with C_d the minimum specified power in the data component, and R_b the information bit rate in the data message, before FEC. For the following signals and E_b/N_0 thresholds, find the minimum C/N_0.

Signal	E_b/N_0 Threshold (dB)	Minimum C/N_0 (dB-Hz)
C/A code	10.0	
L1C	2.4	
P(Y) code	10.0	
M code, high data rate	4.5	
M code, low data rate	4.5	
L2C	4.5	
L5	4.5	

REFERENCES

1. GPS.gov: Official U.S. Government information about the Global Positioning System (GPS) and related topics, http://www.gps.gov/technical/, accessed 3 August 2013.
2. *Global Positioning System Standard Positioning Service Performance Standard*, 4th edition, September 2008, available at www.gps.gov/technical/ps/2008-SPS-performance-standard.pdf, accessed 3 March 2013.
3. GPS.gov: Official U.S. Government information about the Global Positioning System (GPS) and related topics, http://www.gps.gov/governance/excom/, accessed 3 March 2013.
4. W. H. Guier and G. C. Weiffenbach, "Genesis of Satellite Navigation," *Johns Hopkins Applied Physics Laboratory Technical Digest*, Vol. 19, No. 1, 1998, pp. 1–17.
5. S. Pace, G. Frost, I. Lachow, D. Frelinger, D. Fossum, D. Wassem, and M. Pinto, "Appendix B: GPS History, Chronology, Budgets," The Global Positioning System—Assessing National Policies, RAND Corporation Report, 1995.
6. K. D. McDonald, "Global Positioning System: Origins, Early Concepts, Development, and Design Success," Chapter in *Success Stories in Satellite Systems*, Edited by D. K. Sachdev, American Institute of Aeronautics and Astronautics, 2009.
7. K. T. Woo, 'Optimum Semi-Codeless Carrier Phase Tracking of L2," Proceedings of the 1999 International Technical Meeting of the Satellite Division of the Institute of Navigation, September, 1999.

REFERENCES

8. Department of Defense Instruction Number 4650.06, 24 November 2009, available at http://www.dtic.mil/whs/directives/corres/pdf/465006p.pdf, accessed 9 March 2013.
9. Committee on the Future of the Global Positioning System, *The Global Positioning System: A Shared National Asset—Recommendations for Technical Improvements and Enhancements*, National Academy Press, Washington, DC, 1995.
10. C. J. Hegarty, "A Brief History of GPS L5," Stanford's 2010 PNT Symposium November, 2010, available at http://scpnt.stanford.edu/pnt/PNT10/presentation_slides/18-PNT_Symposium_Hegarty.pdf, accessed 9 March 2013.
11. J. W. Betz, "Something Old, Something New—Signal Structures for Satellite-Based Navigation: Past, Present, and Future," *InsideGNSS Magazine*, July/August, 2013, pp. 34–42.
12. J. Anderson, J. W. Betz, and J. Clark, "Advanced GPS Signal Development for a Future Information Warfare Environment," Joint Electronic Warfare Conference, Colorado Springs, April 1998.
13. J. W. Betz, "An Offset-Carrier Military Signal for Spectrum Sharing," PowerPoint presentation to the New Military Signal Design Team, 21 August 1997.
14. J. W. Betz, "The Offset Carrier Modulation for GPS Modernization," Proceedings of the 1999 National Technical Meeting of the Institute of Navigation, San Diego, CA, January 1999, pp. 639–648.
15. B. C. Barker, J. Betz, J. Clark, J. Correia, J. Gillis, S. Lazar, K. Rehborn, and J. Straton, III, "Overview of the GPS M Code Signal," Proceedings of the Institute of Navigation National Technical Meeting 2000, ION-NTM-2000, Institute of Navigation, January 2000.
16. J. W. Betz, "Effect of Jamming on GPS M Code Signal SNIR and Code Tracking Accuracy," Proceedings of the Institute of Navigation National Technical Meeting 2000, ION-NTM-2000, Institute of Navigation, January 2000.
17. J. W. Betz, "Design and Performance of Code Tracking for the GPS M Code Signal," Proceedings of ION-GPS-2000, Institute of Navigation, September 2000.
18. B. Wiley, B. Wiley, D. Craig, D. Manning, J. Novak, R. Taylor, and L. Weingarth, "NGA's Role in GPS," Proceedings of the 2006 Institute of Navigation Conference on Global Navigation Satellite Services, ION GNSS 2006, September 2006.
19. Federal Communications Commission, CC Docket Number 94-102, Report and Orders, adopted 12 June 1996.
20. F. van Diggelen, *A-GPS: Assisted GPS, GNSS, and SBAS*, Artech House, 2009.
21. White House Fact Sheet, "U.S.-EU Summit: Agreement on GPS-Galileo Cooperation," June 2004.
22. G. W. Hein, J.-A. Avila-Rodriguez, L. Ries, L. Lestarquit, J.-L. Issler, J. Godet, and T. Pratt, "A Candidate for the Galileo E1 OS Optimized Signal", Proceedings of ION GNSS 2005, Long Beach, CA, 13–16 September 2005.
23. J. W. Betz, C. J. Hegarty and J. J. Rushanan, "Time-Multiplexed Binary Offset Carrier Signaling and Processing—Group I," Application No. 11/785,571 . Expressly abandoned by MITRE to allow universal free access to the invention, 26 June 2010.
24. J. W. Betz, C. J. Hegarty, and J. J. Rushanan, "Time-Multiplexed Binary Offset Carrier Signaling and Processing—Group II," Application No. 12/755,933. Expressly abandoned by MITRE to allow universal free access to the invention, 11 November 2010.
25. G. W. Hein, J.-A. Avila-Rodriguez, S. Wallner, J. W. Betz, C. J. Hegarty, J. J. Rushanan, A. Kray, A. Pratt, S. Lenahan, J. Owen, J.-L. Issler, and T. Stansell, "MBOC: The New Optimized Spreading Modulation Recommended for Galileo L1 OS and GPS L1C," *InsideGNSS Magazine*, May/June, 2006, pp. 57–65.

26. J. W. Betz, C. R. Cahn, P. A. Dafesh, C. J. Hegarty, K. W. Hudnut, A. J. Jones, R. Keegan, K. Kovach, L. S. Lenahan, H. H. Ma, J. J. Rushanan, T. A. Stansell, C. C. Wang, and S. K. Yi, "L1C Signal Design Options," Proceedings of the Institute of Navigation National Technical Meeting on Global Navigation Satellite Systems 2006, ION-NTM-2006, Institute of Navigation, January 2006.
27. NASA Goddard Space Flight Center Search and Rescue Mission Office, http://searchandrescue.gsfc.nasa.gov/dass/index.html, accessed 9 March 2013.
28. COSPAS-SARSAT.INT, International Satellite System for Search and Rescue, www.cospas-sarsat.org, accessed 9 March 2013.
29. Space Technology, GPS III, http://www.spaceandtech.com/spacedata/constellations/navstar-gps-block2a_conspecs.shtml, accessed 9 March 2013.
30. Global Positioning System III (GPS III) Military Satellite, United States of America, http://www.airforce-technology.com/projects/gps-iii-military-satellite/, accessed 9 March 2013.
31. National Geospatial Intelligence Agency, NGA: DoD World Geodetic System 1984, http://earth-info.nga.mil/GandG/publications/tr8350.2/tr8350_2.html, accessed 19 May 2013.
32. IS-GPS-200, http://www.gps.gov/technical/icwg/IS-GPS-200G.pdf, accessed 9 March 2013.
33. J. W. Betz, "On the Power Spectral Density of GNSS Signals, with Applications," Proceedings of the Institute of Navigation International Technical Meeting 2010, ION-ITM-2010, Institute of Navigation, January 2010.
34. R. Gold, "Optimal Binary Sequences for Spread Spectrum Multiplexing (Corresp.)". *IEEE Transactions on Information Theory*, Vol. 13, No. 4, 1967, pp. 619–621.
35. J. J. Spilker, Jr., "GPS Navigation Data," Chapter 4 in *Global Positioning System: Theory and Applications*, Edited by B. W. Parkinson, J. J. Spilker, Jr., P. Axelrad, and P. Enge, American Institute of Aeronautics and Astronautics, 1996.
36. J. W. Betz, M. A. Blanco, C. R. Cahn, P. A. Dafesh, C. J. Hegarty, K. W. Hudnut, V. Kasemsri, R. Keegan, K. Kovach, L. S. Lenahan, H. H. Ma, J. J. Rushanan, D. Sklar, T. A. Stansell, C. C. Wang, and S. K. Yi, "Description of the L1C Signal," Proceedings of the Institute of Navigation Conference on Global Navigation Satellite Systems 2006, ION-GNSS-2006, Institute of Navigation, September 2006.
37. IS-GPS-800, http://www.gps.gov/technical/icwg/#is-gps-800, accessed 9 March 2013.
38. M. R. Schroeder, *Number Theory in Science and Communication*, Springer-Verlag, 1993.
39. J. J. Rushanan, "The Spreading and Overlay Codes for the L1C Signal," *Navigation: The Journal of the Institute of Navigation*, Vol. 54, No. 1, 2007, pp. 43–51.
40. S. Lin and D. J. Costello, *Error Control Coding: Fundamentals and Applications*, 2nd edition, Prentice-Hall, 2004.
41. T. J. Richardson and R. L. Urbanke, "Efficient Encoding of Low-Density Parity-Check Codes," *IEEE Transactions in Information Theory*, Vol. 47, No. 2, 2001, pp. 638–656.
42. http://www.space.commerce.gov/tag/semi-codeless/, accessed 10 March 2013.
43. P. Dafesh, J. K. Holmes, C. R. Cahn, and T. Stansell, "Description and Analysis of Time-Multiplexed M-Code Data" Proceedings of the Institute of Navigation 2002 Annual Meeting, April 2002.
44. C. R, Cahn, M. M. Goutmann, and G. P. Haefner, "System 621B Signal Definition Study," Technical Report Space and Missile Systems Office TR-72-248, Vol. 1, October 1972.
45. IS-GPS-705, http://www.gps.gov/technical/icwg/IS-GPS-705C.pdf, accessed 3 August 2013.

8

SATELLITE-BASED AUGMENTATION SYSTEMS

A satellite-based augmentation system (SBAS) is a type of wide area differential sathav system [1], described in Chapter 23. While SBAS was developed for aviation safety of life applications, it provides benefits in many other applications as well. SBAS is a generic term; different regional SBASs have been and are being developed to cover much of the northern hemisphere and parts of the southern hemisphere as well.

SBASs augment, but do not replace, standalone satnav systems. SBASs provide three main benefits to users of the augmented satnav system: integrity, accuracy, and availability, all defined in Section 1.3. Integrity provides the trust needed for safety of life operations, either assuring that the signals are valid and useful and within specification, or providing an alarm that they are not. SBAS provides a six second "time to alarm," meaning an SBAS receiver will receive an integrity alert—identifying an invalid signal transmitted by the augmented satnav system—within six seconds of any uncorrectable out-of-specification characteristics in a transmitted signal. Observe that SBAS integrity applies to the transmitted signals, not to the received signals. Degradations of the received signal that occur after transmission, such as propagation distortions like multipath and shadowing, or reception of invalid signals transmitted by others, are not addressed by SBAS integrity indications.

Engineering Satellite-Based Navigation and Timing: Global Navigation Satellite Systems, Signals, and Receivers, First Edition. John W. Betz.
© 2016 The Institute of Electrical and Electronics Engineers, Inc. Published 2016 by John Wiley & Sons, Inc.

Along with providing integrity information, SBASs provide the accuracy benefits of differential GNSS, enhancing the accuracy of the augmented system. SBASs provide corrections to the clock and ephemeris broadcast from GPS (and other satnav systems in the future), and also provide improved ionospheric information. The clock and ephemeris corrections from SBAS can be used by a receiver in any location. The ionospheric information describes ionospheric characteristics at specific pierce points in the ionosphere; receivers in the service area of the SBAS can select and use applicable ionospheric information for each satellite signal being used.

An SBAS's integrity and accuracy functions both involve communications, and thus could be provided to a receiver by any communications link. In contrast, an SBAS's availability enhancement is a ranging function. If the SBAS signal is designated for use as a ranging signal, a receiver can also make pseudorange and carrier phase measurements on the signal, resulting in an increased number of satellite signals available for the receiver to use to improve DOP and help overcome limited signal visibility. Not every SBAS signal is designed or certified for ranging.

Each SBAS comprises a space segment and ground segment. Typically, the space segment consists of transponders on two to four geostationary satellites, while the ground segment involves a network of ground reference stations, master stations that process measurements from the ground reference stations, and uplink stations that transmit the SBAS messages to the transponders. While the coverage of most satnav systems is dictated by signal availability, coverage of an SBAS is also limited to the region near ground reference stations—unless a ground reference station is nearby, the region is not considered in the coverage area since ionospheric corrections do not apply to all signals received in that region.

Each SBAS is regional, and its integrity and accuracy information only applies to satellites from satnav systems that it is designed to augment. Even then it only applies to satellites in that system within view of the ground reference stations. All currently operational SBASs augment GPS. Some SBASs are being designed to augment other standalone satnav systems as well, and SBAS standards may be developed so that all SBASs work with some other satnav systems in the future.

This chapter provides a brief history of SBAS, followed by an overview of SBAS and of some specific SBASs. Section 8.3 provides detailed characteristics of SBAS signals, and Section 8.4 a summary of the chapter.

8.1 SBAS HISTORY AND PLANS

The United States' Federal Aviation Administration (FAA) created the Wide Area Augmentation System (WAAS) program in 1992 primarily to provide the necessary integrity to utilize GPS signals for precision approach. Later in the 1990s, the International Civil Aviation Organization (ICAO), an agency of the United Nations, adopted standards for SBAS based on WAAS. The original and current specifications on SBAS define a signal having carrier frequency at 1575.42 MHz, and similar to the GPS C/A code signal in some other respects as well, as described in Section 8.3.

Current ICAO standards address augmentation of GPS and GLONASS, although currently operational SBASs only work with GPS. Before SBASs can be employed for aviation safety of life applications, they need to go through extensive certification and approval by aviation authorities, particularly to validate their performance in handling the safety-critical functions of promptly and reliably reporting system faults.

The first satellite with a WAAS transponder was launched in 1996, and WAAS became the world's first SBAS system approved for aviation applications in July 2003 [2]. Since that time, additional WAAS ground reference stations have been installed, and satellite transponders have been replaced and added to provide additional coverage and other enhancements.

Other similar space-based augmentation systems, either in operation or under development, include Japan's Multifunctional Transport Satellite (MTSAT)-based Satellite Augmentation System (MSAS), the European Geostationary Navigation Overlay Service (EGNOS), India's GPS And Geo-Augmented Navigation (GAGAN) system, Russia's System for Differential Correction and Monitoring (SDCM), and China's BeiDou.

MSAS transponders are currently hosted on MTSATs, geostationary satellites developed by Japan to provide imagery for meteorological purposes as well as to host SBAS payloads. The first MTSAT was lost in a launch failure in 1999. A replacement satellite was launched in 2005 and a second MTSAT was launched in 2006. MSAS began broadcasting test signals in 2005 [3]. By the end of 2007, MSAS was certified to provide aircraft with navigation service from en route through non-precision approach all over Japan, although it was not yet certified for precision approach and landings. With the planned expansion of QZSS, described in Chapter 12, future QZSS geostationary satellites, rather than the MTSATs, will host Japan's SBAS transponders [4].

The first satellite with an EGNOS transponder was launched in 1996, and the second in 1998 [5]. For some time there was uncertainty concerning the long-term role of EGNOS, since Galileo's Safety of Life service (see Section 10.2) would have been redundant with SBAS, although it would have required use of Galileo. However, in 2003 Europe announced EGNOS would remain an integral part of the European satnav capability, and the first EGNOS master control center was completed. In 2009, the European Commission declared that EGNOS's basic navigation signal was operationally ready as an open and free service. EGNOS was certified for safety of life applications in 2011. Ground reference stations are being added to extend coverage across Europe, as well as southward into Africa.

GAGAN is being developed jointly by the Airports Authority of India (AAI) and Indian Space Research Organization (ISRO) [6]. Preliminary acceptance testing of the ground infrastructure, including 15 initial ground reference stations, was completed in 2010. The first GAGAN navigation payload was flown on GSAT-8, launched in 2011, and the second on GSAT-10, launched in 2012. GAGAN was certified for en route navigation and precision approach without vertical guidance early in 2014.

The initial network of ground reference stations for SDCM was put into operational testing in 2007, with the first SDCM transponder launched in 2011, hosted on one of the Luch series of communication satellites. SDCM is unique among the SBASs, in that from the start it will provide SBAS functionality to GLONASS as well as to

GPS. A second SDCM transponder was launched on another Luch satellite in 2012, and additional SDCM-equipped Luch satellite launches apparently was launched in 2014. SDCM is still in development, and has not yet been certified for aviation safety of life use.

China is planning to add an SBAS-type service to BeiDou, described in Chapter 11. The transponders would presumably be hosted on some of the geostationary satellites that are part of the regional BeiDou constellation. As of early 2015, however, it is unclear whether this service will use an SBAS-compliant signal with 1575.42 MHz carrier frequency, or instead it will transmit SBAS signals at 1561.098 MHz—the same carrier frequency used by the BeiDou Phase 2 B1I signal described in Section 11.3.1.1. Presumably, from the start this system will provide corrections to BeiDou signals as well as to GPS signals.

SBAS coverage continues to expand, as existing systems add more satellites and ground reference stations, and additional systems are implemented. For more than a decade, there has been discussion of adding an SBAS signal at the GPS L5 carrier frequency of 1176.45 MHz, providing dual-frequency safety of life service. Newer WAAS satellites have begun to transmit a test version of such a signal, in parallel with development of the ICAO standards for this signal. Section 8.3.2 describes prospective characteristics of this signal.

8.2 SBAS DESCRIPTION

As a wide area differential GNSS system, described in more detail in Chapter 23, an SBAS uses a network of ground reference stations, precisely surveyed, to collect measurements from the satellite signals being augmented [7]. Ground reference stations currently receive GPS L1 and L2 signals: the C/A code signal as well as P(Y) code signal on L1 and L2. The P(Y) code signal is received without use of the spreading code bits using codeless or semicodeless techniques [8]. As portrayed in Figure 8.1, the ground reference stations' measurements are communicated, via a terrestrial communications network or satcom network, to one or more master stations that use the dual-frequency ionospheric measurements to compute a grid of ionospheric corrections. In addition, the master stations evaluate the discrepancies between measurements and the known locations of the ground reference station antennas to compute corrections to the satellites' ephemeris broadcasts and clocks. The master stations also determine each signal's integrity (whether the signal's RF and data message characteristics are within specification, data message contents are valid, and pseudorange errors are correctable), and the measurement quality available from the signal. The master stations then construct augmentation messages containing ionospheric corrections and (for each individual signal from each satellite) ephemeris and clock corrections, integrity, and measurement quality indicators, and pass them to uplink stations for transmission to a transponder on a geostationary satellite. Uplink messages are transmitted at higher carrier frequencies than L band, with C band, K band, and Ku band being commonly used. The GEO transponder translates the signal at the uplink carrier frequency to the desired L band carrier frequency, amplifies the signal, and emits it from an Earth-coverage antenna.

SBAS SIGNALS

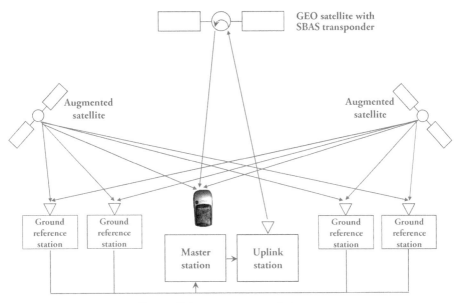

Figure 8.1. SBAS Functional Diagram

This simple transponder operation, consisting of a frequency shift and amplification, is often called "bent pipe" operation. It is conceivable that some future SBAS satellites might use signal-generating payloads rather than bent-pipe operation.

In future, SBAS ground reference stations will also monitor the GPS L5 signal, and also use the L5 signal to obtain more robust ionospheric measurements than available with semicodeless processing of P(Y) code signals. SBASs will also begin to monitor and to augment other satnav systems; initial expansions of each SBAS will presumably involve the corresponding satnav system: Galileo for EGNOS, GLONASS for SDCM, etc.

SBAS time is aligned to GPS time, and the SBAS coordinate frame is WGS-84—the same as used for GPS.

8.3 SBAS SIGNALS

The SBAS frequency plan, shown in Figure 8.2, is a subset of GPSs. The L1 SBAS signal shares the same carrier frequency, spreading modulation, and spreading code family with GPS C/A code signal, in order to obtain the greatest possible interoperability between the signal types. As the GPS L5 signal is transmitted from more satellites, SBAS will also adopt an L5 SBAS signal, sharing the same carrier frequency, spreading modulation, and spreading code family as the GPS L5 signal. Both of these SBAS signals are placed in aeronautical radio navigation service (ARNS) bands (see Section 3.3.1) that provides additional regulatory protection for safety-critical uses of the spectrum such as

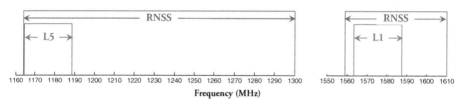

Figure 8.2. SBAS Frequency Bands

radar, traffic collision-avoidance system, distance measuring equipment, and tactical air navigation, as well as safety-critical satnav.

The SBAS signals all use RHCP. Minimum received signal power is specified out of a 3 dBil linearly polarized antenna located near the ground, for satellites at elevation angles of 5° or more, with 0.5 dB excess atmospheric attenuation. The receive antenna is normal to the incident wave and rotated to be aligned with the minor axis of the incident field's polarization ellipse, making this power specification conservative as discussed in Section 4.2. Since these minimum received power conventions are the same as for GPS, SBAS minimum received power of a given level corresponds to the same value for GPS. Section 8.3.1 describes the SBAS L1 signal, while Section 8.3.2 provides insights about the SBAS L5 signal.

8.3.1 SBAS L1 Signal

Table 8.1 summarizes key characteristics of the SBAS L1 signal, using the same characteristics introduced in Section 3.3. This signal uses the same carrier frequency as GPS C/A code signal and many other planned systems' civil signals in upper L band,

TABLE 8.1. Summary of SBAS L1 Characteristics

Carrier frequency	L1: 1575.42 MHz = 154 × 10.23 MHz
Minimum received power	−158.5 dBW
Polarization	RHCP
Multiple access	CDMA
Spreading modulation and transmit bandwidth	BPSK-R(1), ±10 MHz
Spreading codes	Length 1023-bit Gold codes, duration 1 ms, from the GPS C/A code family
Data message structure	L1 SBAS
Data rate	250 bps
Data message error correction and detection	$1/2$-rate convolutional coding with constraint length 7, no interleaving
Data modulation	500 sps biphase modulation
Pilot and data components	100% power data
Overlay code	None
Multiplexing with other signals	None required

PRN	G2 Delay (Chips)	Initial G2 Setting (Octal)	Fiist 10 SBAS Chips (Octal)
120	145	1106	0671
121	175	1241	0536
122	52	0267	1510
123	21	0232	1545
124	237	1617	0160
125	235	1076	0701
126	886	1764	0013
127	657	0717	1060
128	634	1532	0245
129	762	1250	0527
130	355	0341	1436
131	1012	0551	1226
132	176	0520	1257
133	603	1731	0046
134	130	0706	1071
135	359	1216	0561
136	595	0740	1037
137	68	1007	0770
138	386	0450	1327

Figure 8.3. Spreading Code Definitions for L1 SBAS Signals (Note: first digit in right 2 columns is a bit, and not an octal character)

enhancing interoperability. The transmit bandwidth shown in this table is the widest used in current SBAS transponders; some use narrower bandwidths, and in some cases tracking of these narrower bandwidth signals is not authorized for aviation use, and only the data message is employed. The data rate shown here includes extra bits used for framing, synchronization, and decoder flushing; the actual information data rate is somewhat lower. Full signal interface specifications are found in Reference 9.

Figure 8.3 shows the definitions of spreading codes for L1 SBAS signals. These spreading codes are from the C/A code family, and the spreading codes are generated using the approach in Figure 7.9.

The L1 SBAS message structure had duration of 1 second, containing 250 bits encoded as 500 symbols [9]. Each message begins with an eight bit preamble; there are three different preambles, each used in sequential messages. The following six bits indicate the message type, so 64 message types can be supported. The next 212 bits contain the message payload; they are followed by a 24-bit CRC.

The L1 SBAS message uses a flexible message design like CNAV and CNAV2. Table 8.2 summarizes current L1 SBAS message types. While some of the message contents are similar to that for GPS signals described in Chapter 7, additional concepts and terminology are introduced for SBAS [10]. PRN mask assignments identify the PRNs for different GPS satellites being augmented, so the receiver knows the sequence of PRNs for which corrections are provided. Fast corrections describe rapid clock variations, and are broadcast every 6 seconds. Long-term corrections describe slower drift errors in

TABLE 8.2. L1 SBAS Message Types

Message Type Number	Message Contents
0	Do not use for safety applications (system testing)
1	PRN mask assignments
2–5	Fast corrections
6	Integrity information
7	Fast correction degradation factor
8	Reserved
9	GEO navigation message
10	Degradation parameters
11	Reserved
12	WAAS network time/UTC offset parameters
13–16	Reserved for future messages
17	GEO satellite almanacs
18	Ionospheric grid point masks
19–23	Reserved
24	Mixed fast corrections/long-term satellite error corrections
25	Long-term satellite error corrections
26	Ionospheric delay corrections
27	SBAS service message
28	Clock–ephemeris covariance matrix message
29–61	Reserved
62	Internal test message
63	Null message

almanac and clock, and are broadcast every 2 minutes. The covariance error message describes the variance on the user differential range error (UDRE), characterizing the remaining error after corrections are applied.

8.3.2 SBAS L5 Signal

Table 8.3 summarizes key characteristics of the SBAS L5 signal, using the same characteristics introduced in Section 3.3. There is not yet a final definition of the signal, and since some aspects are not yet fully defined, different alternatives are included in the table. Some aspects: the carrier frequency, multiple access, spreading modulation, and spreading codes will almost certainly remain as described in Table 8.3. It appears that, in phase quadrature to the data component, there optionally may be no second component, or a pilot component, or a data component. While the data rate on the baseline data component is likely to be 250 bps, the optional component will be allowed to have different data rates. Different overlay codes will be used for the optional component, depending upon data rate.

The correlation function and PSD for the SBAS L5 signal are the same as for the GPS P(Y) code signal shown in Figure 7.18.

TABLE 8.3. Summary of Tentative L5 SBAS Signal Characteristics

Carrier frequency	L5: 1176.42 MHz = 115 × 10.23 MHz
Minimum received power	−157.0 dBW
Polarization	RHCP
Multiple access	CDMA
Spreading modulation and transmit bandwidth	BPSK-R(10), ±10 MHz
Spreading codes	Length 10230-bit codes, duration 1 ms, from the GPS L5 code family
Data message structure	L5 SBAS
Data rate	Baseline component: 250 bps
	Optional component: 250, 125, 100, 50, or 0 bps
Data message error correction and detection	$1/2$-rate convolutional coding with constraint length 7, no interleaving
Data modulation	Baseline component: 500 sps biphase modulation
Pilot and data components	Baseline: 100% power data
	Optional: 50% power baseline data and 50% power optional component
Overlay code	1000 bps with structure that depends on selected data rate:
	Manchester code (0 1) for 250 bps
	4-bit Willard code (1 1 0 0) for 125 bps
	5-bit Barker code (1 1 1 0 1) for 100 bps
	I5 Neuman–Hoffman code (1 1 1 1 0 0 1 0 1 0) for 50 bps
Multiplexing with other signals	Baseline component inphase; optional component in phase quadrature

8.4 SUMMARY

While each SBAS is a regional system, the increasing number of SBASs is getting closer to global coverage, with most of the northern hemisphere likely to be within the service area of at least one SBAS by 2020, and increasing amounts of the southern hemisphere are covered as well.

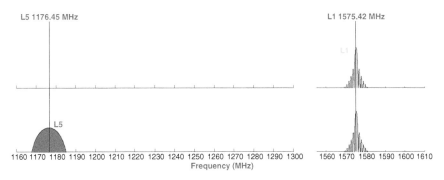

Figure 8.4. PSDs of SBAS Signals: Current (Upper) and Prospective (Lower)

TABLE 8.4. Summary of SBAS Signals

Signal Name	Carrier Frequency (MHz)	Spreading Modulation	Service
SBAS L1	1575.42	BPSK-R(1)	SBAS
SBAS L5	1176.45	BPSK-R(10)	SBAS

SBAS signals are designed primarily to provide data that communicates the integrity of augmented signals, as well as corrections that improve the accuracy of augmented signals; only some SBAS signals are currently approved for ranging. The high degree of interoperability between SBAS signals and corresponding GPS signals enables SBAS functionality to be added to a GPS receiver with minimal additional complexity. Thus, while SBASs were primarily intended for safety of life use, many consumer receivers use the SBAS-provided corrections to provide higher accuracy. Figure 8.4 shows the PSDs of the current SBAS L1 signal and the planned SBAS L5 signal, and key signal characteristics are summarized in Table 8.4.

SBAS signals employ BPSK-R spreading modulations, no pilot components, and other design characteristics very similar to the GPS C/A code signal and L5 signal.

REVIEW QUESTIONS

Theoretical Questions

QT8.1. List characteristics of GPS C/A code and L1 SBAS signals that are similar, and characteristics that are different.

QT8.2. List characteristics of GPS L5 and L5 SBAS signals that are similar and characteristics that are different.

Application Questions

QA8.1. For the following calculations, define C to be the minimum specified received power for all components of the relevant SBAS signal, and let N_0 be -201.5 dBW/Hz. For the following signals and E_b/N_0 thresholds, find the minimum C/N_0.

Signal	E_b/N_0 Threshold (dB)	Minimum C/N_0 (dB-Hz)
SBAS L1	4.5	
SBAS L5, 250 bps	4.5	
SBAS L5, 125 bps	4.5	
SBAS L5, 100 bps	4.5	
SBAS L5, 50 bps	4.5	

REFERENCES

1. C. Kee, B. W. Parkinson, and P. Axelrad, "Wide Area Differential GPS," *Navigation, Journal of the Institute of Navigation*, Vol. 38, No. 2, Summer 1991.
2. Federal Aviation Administration, Navigation Programs – History – Satellite Navigation, https://www.faa.gov/about/office_org/headquarters_offices/ato/service_units/techops/navservices/history/satnav/index.cfm.
3. "Overview of MSAS," Presentation for ICG, 2007, available at http://www.oosa.unvienna.org/pdf/icg/2007/icg2/presentations/04_01.pdf, accessed 23 March 2013.
4. H. Tashiro, "MSAS Status," Civil Aviation Bureau of Japan Presentation given to IWG#28, April 2015.
5. EGNOS Program Information, http://www.egnos-portal.eu/discover-egnos/programme-information/, accessed 23 March 2013.
6. Indian Space Research Organization, Satellite Navigation, http://www.isro.org/spacecraft/satellite-navigation, accessed 23 March 2013.
7. GPS.gov, Official U.S. Government information about the Global Positioning System (GPS) and related topics, Augmentation Systems, http://www.gps.gov/systems/augmentations/
8. K. T. Woo, "Optimum Semi-Codeless Carrier Phase Tracking of L2," Proceedings of the 1999 International Technical Meeting of the Satellite Division of the Institute of Navigation, September, 1999.
9. International Standards and Recommended Practices, Aeronautical Telecommunications, Annex 10 to Convention of International Civil Aviation, Volume 1 (Radio Navigation Aids).
10. "Global Positioning System Wide Area Augmentation System (WAAS) Performance Standard," 1st edition, available at http://www.gps.gov/technical/ps/2008-WAAS-performance-standard.pdf, accessed 9 March 2013.

9

GLONASS

GLObal NAvigation Satellite System (GLONASS) was the second functioning global satnav system, after GPS. Originally developed by the Soviet Union, its design involved several choices fundamentally different from those made for GPS, including different constellation design, the use of FDMA rather than CDMA, pressurized satellite payloads, and the deployment of ground monitoring stations only regionally on the territory of the former Soviet Union. The Russian Federation now develops and operates GLONASS, and is modernizing GLONASS, departing from many of these original design choices.

Like GPS, GLONASS is a dual-use system with some signals and services intended for military use, and others for civilian use by anyone, anywhere its signals can be received, with no user fees. In addition, the GLONASS Interface Control Document (ICD) [1] provides the full unrestricted technical information for anyone to construct and manufacture receivers of the original GLONASS civil signals; ICDs have not yet been published for more recently announced planned signals. While a complete GLONASS constellation has 24 satellites, up to 28 operational satellites may be employed in future.

GLONASS is developed and operated by Roscosmos, the Russian Federal Space Agency. It comprises a space segment, a ground segment, and a user segment.

This chapter provides a brief history of GLONASS and summarizes current plans in Section 9.1, followed by an overview of the GLONASS system in Section 9.2. Section 9.3

Engineering Satellite-Based Navigation and Timing: Global Navigation Satellite Systems, Signals, and Receivers, First Edition. John W. Betz.
© 2016 The Institute of Electrical and Electronics Engineers, Inc. Published 2016 by John Wiley & Sons, Inc.

provides detailed characteristics of GLONASS signals, followed by a summary of the chapter in Section 9.4.

9.1 GLONASS HISTORY AND PLANS

GLONASS's history parallels GPS's history, but with a few years' lag. Its first operational satellites were launched in 1983, and in 1988 the Soviet Union offered civil GLONASS free of charge to the international community. By 1995 there was a full constellation of 24 satellites in space. With a full constellation, international use of GLONASS began to increase, in part because GLONASS's accuracy approached or exceeded the standalone accuracy of GPS C/A code signals, due to the errors introduced by SA.

However, the need to continue launching GLONASS satellites at a high rate due to their relatively short lifetimes, combined with economic difficulties in Russia, led to a deterioration of the number of satellites and the quality of GLONASS service. The number of fully operating satellites in space fell to single digits near the turn of the century.

Starting in 2003, modernized satellites, denoted GLONASS-M, were launched instead of the original GLONASS satellites (81 original GLONASS satellites were launched). GLONASS-M satellites promised better clocks and longer satellite life. Also, they transmit civil signals on two carrier frequencies, enabling receivers to perform dual-frequency compensation for ionospheric errors. In January 2006, the Russian government issued a GLONASS modernization directive, making significant investments in GLONASS with a plan to return to full functionality by 2011.

Like the original GLONASS satellites, GLONASS-M satellites operate key components within pressurized vessels, adding loss of pressure to other causes for satellite failure. As of September 2012 no operating satellite had been operational for as long as 7 years [2], in contrast to GPS, some of whose satellites have operated for more than 21 years. Yet GLONASS satellite life has been increasing and the quality of satellite clocks has improved. GLONASS also has improved its monitoring stations, and added laser reflectors to GLONASS-M satellites for more accurate ephemeris estimation.

The number of operational GLONASS satellites has been at or near 24 since 2011. GLONASS-M satellites will be replaced by GLONASS-K satellites, the first GLONASS satellites with an unpressurized payload. GLONASS-K satellites also provide more stable clocks, COSPAS-SARSAT search and rescue functionality, and a design life of 10 years. A prototype of this next generation GLONASS satellite block, termed GLONASS-K1, was launched in 2011. This satellite transmits the first GLONASS CDMA signal, denoted L3OC. The next generation of GLONASS-K satellites, termed GLONASS-K2, will add more CDMA signals. Modernized GLONASS-K satellites, designated GLONASS-KM, are also being planned, but the distinction between GLONASS-K2 and GLONASS-KM capabilities is not clear.

Recently launched GLONASS-M satellites have been augmented with the ability to transmit the L3OC signal, in order to continue fielding this modernized signal while design problems are being resolved with GLONASS K satellites. In addition, more

GLONASS-K1 satellites may be launched than had been originally planned, with transition to GLONASS-K2 satellites deferred until late in this decade.

Recognizing that GLONASS accuracy will be inherently limited as long as satellite monitoring stations are only located in a small segment of the Earth, Russia is moving to add worldwide monitoring stations. The Russian government's financial commitments, funding development of an improved ground segment, new satellite designs, and new signals, are leading to a new level of consistent performance from GLONASS.

Ultimately, the performance of a satnav system depends not only on its hardware and software, but also upon its operation. The April 2014 problems with GLONASS broadcast ephemeris [3] raised international concerns about the reliability and quality of GLONASS service.

9.2 GLONASS DESCRIPTION

GLONASS offers two navigation services, each supported by multiple signals. GLONASS has developed new designations for its signals in recent years, indicating that the service intended for universal free use is called the Open Service, while the government/military service is called the Secure Service.

The baseline GLONASS constellation uses 24 satellites in three orbital planes. Each orbit contains eight satellites, equally spaced in each plane with 45° displacement in argument of latitude. The three Earth-centered orbital planes have 120° separation, with nominal inclination 64.8° relative to the equatorial plane. The orbits are nominally circular, with radius 25,500 km, producing an orbital period of approximately 11 hours, 15 minutes, and leading to an approximately repeating ground track every 8 days. Satellite speed is approximately 4.0 km/s. With eight satellites in each plane and uniformly spaced, one of the satellites will be at the same location in the sky at the same sidereal time each day.

Descriptions of the GLONASS satellites are provided in Reference 4. GLONASS-M satellites, with their pressurized structure, have mass approximately 1400 kg, while GLONASS-K1 satellites, being unpressurized, have mass approximately 970 kg. GLONASS-K2 satellites, with their additional functionality, are projected to have mass greater than 1600 kg. GLONASS-M satellites employ three Cesium clocks, while GLONASS-K satellites employ two Cesium and two Rubidium clocks. Up to three satellites can be launched at a time.

GLONASS position information is referenced to the Parametri Zemli (English translation, Parameters of the Earth) 1990, or PZ-90, geodetic datum. In 2012, the Russian Federation Government Regulation about National Reference Frames was passed declaring the following reference frames as national:

- The Geodetic Reference Frame 2011 (GSK-2011) will be used for geodetic and map-making purposes;
- The Geocentric Reference Frame PZ-90.11 will be used for orbital flights and navigation tasks geodetic provision.

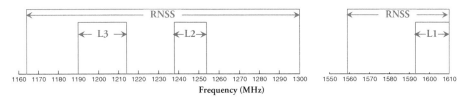

Figure 9.1. GLONASS Frequency Bands

The Geodetic Reference Frame 1995 (SK-95) and National Geodetic Reference Frame 1942 (SK-42) will be used until 1 January 2017 (S. Karutin, Private Communication, 10 March 2015). GLONASS receivers typically provide position coordinate outputs transformed into WGS-84.

GLONASS time is based on an atomic time scale maintained by Russia, denoted UTC(SU). The GLONASS time scale is not continuous, since it is adjusted occasionally for leap seconds, unlike GPS. It is also 3 hours ahead of UTC.

Additional improvements are planned to the GLONASS reference framework, including geodesy, time, and coordinate conversion parameters. As ground monitoring stations are installed around the world, the GLONASS System Control Center, located in Moscow, will be able to compute clock corrections and ephemeris with greater accuracy. It remains unclear whether ground antennas for uplinking new messages to the satellites will be installed beyond the borders of the Russian Federation, or if satellites will need to be within view of Russian territory in order to receive updates.

9.3 GLONASS SIGNALS

The GLONASS signal frequency plan continues to evolve. Figure 9.1 presents the currently defined frequency plan using GLONASS's notation, which is the same as GPS notation even though the meaning is different. The upper L band segment is called L1, spanning frequencies above the GPS L1 band and having a nominal carrier frequency of 1602.0 MHz. The GLONASS L2 band also occupies frequencies primarily above GPS L2, with a nominal carrier frequency of 1246.0 MHz, although there is a slight overlap with GPS L2 band. GLONASS L3 is between GPS L2 and GPS L5, with a carrier frequency of 1202.025 MHz.

All signals use RHCP. Minimum received signal power is specified out of a 3 dBil linearly polarized antenna located near the ground, for satellites at elevation angles of 5° or more, with 2.0 dB excess atmospheric attenuation. The receive antenna is normal to the incident wave and rotated to be aligned with the minor axis of the incident field's polarization ellipse, making this power specification conservative as discussed in Section 4.2. Since the excess atmospheric loss is 1.5 dB greater than that used for minimum received power conventions in GPS, GLONASS minimum received power of a given level corresponds to 1.5 dB higher than the same value for GPS.

GLONASS satellite ephemeris is described in terms of position, velocity, and acceleration vectors in the PZ-90 ECEF coordinate system at a reference epoch, rather than

TABLE 9.1. Summary of L1OF and L2OF Characteristics

Carrier frequency	L1OF: 1598.0625–1605.375, spaced by 0.5625 MHz
	L2OF: 1242.9375–1248.625, spaced by 0.4375 MHz
Minimum received power	L1OF: −161.0 dBW
	L2OF: −167.0 dBW
Polarization	RHCP
Multiple access	FDMA
Spreading modulation and transmit bandwidth	BPSK-R(0.511 MHz), ±0.511 MHz
Spreading codes	The same 511 bit m-sequence is used for all signals
Data message structure	GLONASS
Data rate	50 bps
Data message error correction and detection	Error detection using extended (85,81) hamming code
Data modulation	50 sps biphase modulation
Pilot and data components	100% power data
Overlay code	"Meander sequence" (1 1 1 1 1 0 0 0 0 0) at 1 kbps
Multiplexing with other signals	Quadrature phase multiplexed with L1SF signal and L2SF signal respectively

the Keplerian-based ephemeris parameters and their perturbations used for GPS. The GLONASS almanac, however, uses Keplerian-based parameters.

The names of GLONASS signals have evolved over time, and the notation used here reflects what GLONASS has adopted in recent years. In general, each signal is designated by two characters indicating the GLONASS frequency band occupied, an "O" or "S" indicating whether the signal is part of the open service or secure service, and a fourth character of either "F" or "C" indicating whether the signal uses FDMA or CDMA. GLONASS has published an ICD for its FDMA signals [1], but no official description is yet available for the CDMA signals. However, some information has been provided unofficially, allowing preliminary descriptions of these signals.

9.3.1 GLONASS L1OF and L2OF Signals

Table 9.1 summarizes key characteristics of the L1OF and L2OF signals, using the same characteristics introduced in Section 3.3. L1OF is one of the original GLONASS signals, and L2OF was added on GLONASS-M satellites—both signals use the same design on different carrier frequencies. The spans of carrier frequencies divided by the spacing, shown in the first row of Table 9.1, only yields 14 different carriers in each frequency band.[1] Twice this many satellites can be accommodated by assigning the same carrier frequency to antipodal satellites in each orbital plane. The GLONASS FDMA signals

---

[1] The original GLONASS ICD listed additional carrier frequencies up to 1609.3125 MHz in upper L band and 1251.6875 MHz in lower L band, totaling 21 carrier frequencies in each band. However, signals at these higher carrier frequencies in upper L band caused interference to radio astronomy receivers in the band

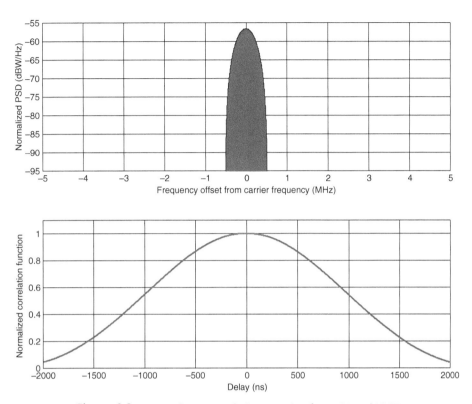

Figure 9.2. PSD and Autocorrelation Function for L1OF and L2OF

are the only ones described in this book whose carrier frequency and spreading code chip rate are not multiples of 1.023 MHz. The data rate shown here includes extra bits used for framing and synchronization; the actual information data rate is somewhat lower. While GLONASS-K2 satellites will transmit FDMA signals and also CDMA signals in the same L1 and L2 bands but at different carrier frequencies, how the FDMA and CDMA signals in each band will be combined for transmission on the satellite remains to be seen.

Figure 9.2 shows the autocorrelation function and power spectral density of the L1OF and L2OF signals, computed using the approximation of the spreading code as ideal and long. Compared to C/A code, the autocorrelation function is approximately twice as wide (providing less accuracy) and the power spectral density is approximately half as wide (making it more susceptible to in-band interference). These are both averages, not reflecting detailed structure introduced by the repeated spreading code. While the actual spreading code repeats 20 times in a data message symbol duration, like GPS C/A code, the GLONASS signal designers imposed a "meander sequence" on the

---

1610.6–1613.8 MHz, so GLONASS has agreed not to use these highest seven carrier frequencies except over Russian territory.

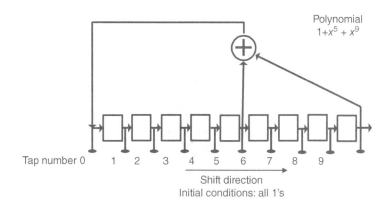

Figure 9.3. Spreading Sequence Generation for L1OF and L2OF Signals [1]

repeating spreading code to invert the phase of the second 10 repetitions of the spreading code. This simple overlay code, while not as effective as other overlay codes would be, reduces the spectral lines compared to the GPS C/A code signal, which uses no overlay code.

The m-sequence generator used to produce the spreading code for all L1OF and L2OF signals is shown in Figure 9.3. Figure 9.4 shows baseband signal generation, driven by 100 Hz clock pulses. Each message string has 2 second duration, the first 1.7 seconds (85 bits) are data bits with framing and parity, and the last 0.3 seconds (30 half-bits) are what the GLONASS ICD calls the time mark (1 1 1 1 1 0 0 0 1 1 0

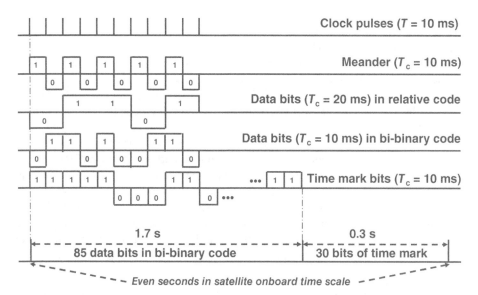

Figure 9.4. Baseband L1OF and L2OF Signal Generation [1]

1 1 1 0 1 0 1 0 0 0 0 1 0 0 1 0 1 1 0) used by the receiver to maintain synchronization to the data message. Each data bit is multiplied by one period of the square wave (the "meander sequence"; the product is called a bi-binary code). Observe that 15% of the data message capacity (a high percentage) is used by the time mark bits.

The superframe structure of a GLONASS message consists of five frames, each containing 15 message strings. The first four strings of each frame contain "immediate data" for the satellite, including time, health, ephemeris, and clock corrections. The final 11 strings in a frame contain almanac data for five of the satellites, with the last two strings of the fifth string designated as reserve bits, since they are not needed for a 24 satellite constellation's almanac. Different frames contain almanac data for different satellites. Each string's duration is two seconds, and each frame's duration is 30 seconds, so the immediate data repeats every 30 seconds. The entire superframe repeats every 150 seconds.

According to Karutin (Private Communication, 10 March 2015), there are ongoing studies evaluating alternative message structures for the L1OF and L2OF signals. A modernized structure might accommodate 30 satellites, as well as a flexible message structure like that developed for the modernized GPS signals. Such an approach would allow for different types of message strings or frames, and less rigidity in the sequence of strings or frames that is transmitted.

9.3.2 GLONASS L1SF and L2SF Signals

Table 9.2 summarizes key characteristics of the other two GLONASS FDMA signals, designated L1SF and L2SF signals, using the same characteristics introduced in

TABLE 9.2. Summary of L1SF and L2SF Characteristics

Carrier frequency	L1OF: 1598.0625–1605.375, spaced by 0.5625 MHz
	L2OF: 1242.9375–1248.625, spaced by 0.4375 MHz
Minimum received power	L1OF: −161.0 dBW
	L2OF: −161.0 dBW
Polarization	RHCP
Multiple access	FDMA
Spreading modulation and transmit bandwidth	BPSK-R(5.11 MHz), bandwidth unspecified
Spreading codes	Restricted
Data message structure	GLONASS
Data rate	50 bps
Data message error correction and detection	Error detection using extended (85,81) hamming code
Data modulation	50 sps biphase modulation
Pilot and data components	100% power data
Overlay code	None
Multiplexing with other signals	Phase multiplexed with L1OF signal and L2OF signal respectively

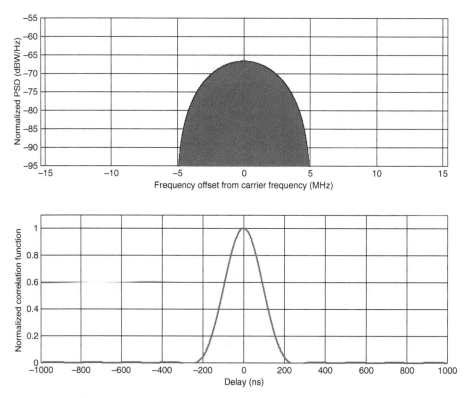

Figure 9.5. PSD and Autocorrelation Function for L1SF and L2SF

Section 3.3. Each L1SF and L2SF signal has the same carrier frequency as the L1OF and L2OF signals transmitted from the same satellite. Analogous to GPS P(Y) code and C/A code signals, the L1SF and L2SF signals have 10 times the chip rate of the L1OF and L2OF signals, and all signals share the same data message.

Figure 9.5 shows the autocorrelation function and power spectral density of the L1SF and L2SF signals, computed using the approximation of the spreading code as long and nonrepeating. Compared to P(Y) code, the autocorrelation function is approximately twice as wide (providing less accuracy) and the power spectral density is approximately half as wide (making it more susceptible to in-band interference). These are both averages, not reflecting detailed structure introduced by the repeated spreading code.

9.3.3 GLONASS L3OC Signal

The L3OC signal is the first CDMA GLONASS signal, transmitted first by GLONASS-K1 satellites. While no official ICD has been published, Reference 5, written by the signal designers, provides an unofficial description, summarized in Table 9.3. The autocorrelation function and PSD of the L3 signal, assuming an ideal long spreading

TABLE 9.3. Summary of Preliminary L3OC Characteristics

Carrier frequency	L3: 1202.025 MHz = 1175 × 1.023 MHz
Minimum received power	−158.0 dBW
Polarization	RHCP
Multiple access	CDMA
Spreading modulation and transmit bandwidth	QPSK-R(10), bandwidth unspecified
Spreading codes	Length 10230 truncated Kasami sequences, duration 1 ms
Data message structure	Modernized GLONASS message
Data rate	100 bps
Data message error correction and detection	$\frac{1}{2}$-rate convolutional coding
Data modulation	200 sps biphase modulation
Pilot and data components	50% power data, 50% power pilot
Overlay code	5 bit, 1 kbps Barker code (0 0 0 1 0) on data component, 10 bit, 1 kbps Neuman–Hoffman code (0 0 0 0 1 1 0 1 0 1) on pilot component
Multiplexing with other signals	Not needed

code, are the same as for the P(Y) code signal in Figure 7.10, since both use BPSK-R(10) spreading modulations. A family of 31 spreading codes is defined.

The Kasami sequences used for spreading codes are shortened to 10,230 bits from their full length of 16,383 bits. A Kasami sequence, sometimes known as a Kasami code, is a binary sequence of length $2^n - 1$, where n is an even integer for the small-set Kasami sequences considered here [6,7]. Kasami sequences are generated from a long m-sequence of length $2^n - 1$, along with a sequence of the same length formed by $2^{n/2} - 1$ repeats of a decimated version of this m-sequence, called the short m-sequence. The decimation is by the factor $2^{n/2} + 1$ (observe $2^n - 1 = (2^{n/2} + 1)(2^{n/2} - 1)$). For example, a 16,383 length Kasami sequence is generated using $n = 14$, and the decimation is by the factor $2^{14/2}+1 = 129$, with $16,383/129 = 127$, so 127 repeats of the short sequence forms another 16,383 length sequence. The original m-sequence and cyclic shifts of the repeated decimated sequence are exclusive ORed, with each member of the Kasami sequence defined by a different cyclic shift. (The different cyclic shifts of the shorter sequence are equivalent to starting the cyclic decimation of the longer sequence at a different point, so the decimation can be started at any point of the longer sequence.) Consequently, there are $2^{n/2} + 1$ members of the Kasami sequence code family for each m-sequence. While Kasami sequences have well-defined and excellent correlation sidelobe properties, these properties may be degraded when they are shortened or with Doppler differences.

As an overview of the data message structure, the navigation message uses one of the two fixed formats, depending upon whether GLONASS has only 24 operational satellites or between 24 and 30 operational satellites in its constellation. Each superframe consists of either 8 or 10 navigation frames, with the smaller number of frames used for the smaller constellation. Each navigation frame has 15 second duration, containing five strings each with 3 seconds duration. Each navigation frame contains clock corrections and ephemeris for the satellite, along with almanac for three other satellites. The full constellation almanac is broadcast in each superframe. Time is represented by a time marker at the beginning of each string.

9.3.4 Future GLONASS Signals: L1OC, L1SC, L2OC, L2SC

Signal designs in the upper portion of upper L band are constrained by stringent PFD limits in the radio astronomy band 1610.6–1613.8 MHz. These limits affect not only the signals themselves but also any intermodulation products generated in signal combining. Reflecting this constraint and the desire to use identical signals in L1 band and L2 band, GLONASS appears to be leaning toward adopting the CDMA signals described here for transmission by GLONASS-K2 satellites.

The L1OC and L1SC signals will use a carrier frequency of 1565×1.023 MHz = 1600.995 MHz. The L1OC signal will have a BPSK-R(1) data component and a BOC(1,1) pilot component, time multiplexed to form a composite biphase signal. The L1SC signal will use a BOC(5,2.5) spreading modulation for both pilot and data components, time multiplexed to form a composite biphase signal and modulated onto the carrier in phase quadrature to the composite L1OC signal. This use of time multiplexing with carrier phase quadrature multiplexing generates constant envelope signals without introducing intermodulation components that could produce undesirable power in the 1610.6–1613.8 MHz radio astronomy band.

Similarly, the L2OC and L2SC signals will use a carrier frequency of 1220×1.023 MHz = 1248.06 MHz. Their signal structures are identical to those for L1OC and L1SC. It is not known whether the spreading codes and data messages will be identical for L1OC and L1SC.

9.4 SUMMARY

GLONASS's history makes evident the challenges of developing and sustaining high-quality satnav service. Its rejuvenation in the first decade of this century has restored its status as the second global satnav system. Figure 9.6 shows the modernization of GLONASS signals; all operational satellites are already transmitting civil signals on two carrier frequencies.

Modernized GLONASS signals will use CDMA and a clock based on 10.23 MHz, with carrier frequencies and chip rates that are multiples of 1.023 MHz, unlike the original GLONASS using FDMA with a clock based on 5 MHz and chip rates that are multiples of 0.511 MHz. Modernized GLONASS satellites are displaying better clock stability, and all aspects of the GLONASS system are also undergoing modernization. Like GPS, GLONASS's current plans are to continue transmitting its original signals while adding modernized signals. Key characteristics of original and modernized GLONASS signals are summarized in Table 9.4.

SUMMARY

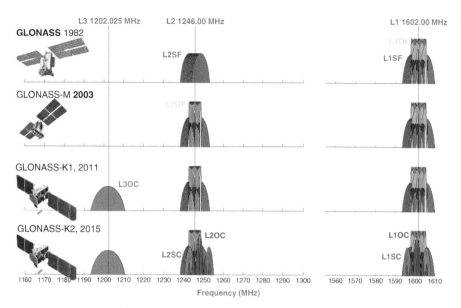

Figure 9.6. Modernization of GLONASS Signals

Receiver processing of the original GLONASS signals is similar to that for the original GPS signals, except the receiver tunes to different carrier frequencies for signals from different satellites, and uses the same spreading code for all original signals. Since the original GLONASS signals use BPSK-R spreading modulations and have no pilot component, the greatest conceptual departure from GPS is in demodulating and

TABLE 9.4. Summary of GLONASS Signals

Signal Name	Carrier Frequency (MHz)	Spreading Modulation	Service
L1OF	1598.0625–1605.375, spaced by 0.5625 MHz	BPSK-R(0.511 MHz)	Open
L1SF	1598.0625–1605.375, spaced by 0.5625 MHz	BPSK-R(5.11 MHz)	Secure
L1OC	1600.995 MHz	BOC(1,1)/BPSK-R(1)	Open
L1SC	1600.995 MHz	BOC(5,2.5)	Secure
L2OF	1242.9375–1248.625, spaced by 0.4375 MHz	BPSK-R(0.511 MHz)	Open
L2SF	1242.9375–1248.625, spaced by 0.4375 MHz	BPSK-R(5.11 MHz)	Secure
L2OC	1248.06 MHz	BOC(1,1)/BPSK-R(1)	Open
L2SC	1248.06 MHz	BOC(5,2.5)	Secure
L3OC	1202.025 MHz	QPSK-R(10)	Open

interpreting the GLONASS data message. Techniques described in Chapters 13 through 20 can readily be applied to the original GLONASS signals. Details of processing modernized GLONASS signals await publication of these signals' ICDs.

REVIEW QUESTIONS

Theoretical Questions

QT9.1. List characteristics of GPS C/A code and GLONASS L1OF signals that are similar, and characteristics that are different.

QT9.2. List characteristics of GPS P(Y) code and GLONASS L1SF signals that are similar, and characteristics that are different.

QT9.3. Section 3.3.1.4 of the GLONASS ICD indicates the L1OF and L2OF signals are tightly bandlimited to ± 0.511 MHz about the carrier frequency. Suppose a receiver uses a precorrelation filter with ± 10 MHz bandwidth. Compute and compare the RMS bandwidths (See Appendix Section A.6) for the L1OF or L2OF signals if they are bandlimited to ± 0.511 MHz, ± 5.0 MHz, and ± 10.0 MHz.

Application Questions

QA9.1. Compare the maximum PSD of a GPS C/A code signal and a GLONASS L1OF signal. Assume ideal long spreading codes and signal power computed over the bandwidths defined for each signal in Sections 7.3 and 9.3.

QA9.2. Compare the maximum PSD of a GPS P(Y) code signal and a GLONASS L1SF signal, assuming its transmit bandwidth is ± 5.11 MHz. Assume ideal long spreading codes and signal power computed over the bandwidths defined for each signal in Sections 7.3 and 9.3.

REFERENCES

1. Global Navigation Satellite System (GLONASS) Interface Control Document, Navigational radiosignals in Bands L1, L2, Edition 5.1, 2008, available at http://www.glonass-ianc.rsa.ru/, accessed 3 August 2013.
2. S. Revnivykh, "GLONASS Status and Modernization," ION GNSS 2012, Nashville, Tennessee, 19 September 2012.
3. A. Cameron, "GLONASS Gone...Then Back," *GPS World*, April 2014, available at http://gpsworld.com/glonass-gone-then-back/, accessed 15 December 2014.
4. R. Fatkulin, V. Kossenko, S. Storozhev, V. Zvonar, and V. Chebotarev, "GLONASS Space Segment: Satellite Constellation, GLONASS-M and GLONASS-K Spacecraft Main Features," ION GNSS 2012, 21 September 2012.

REFERENCES

5. Y. Urlichich, V. Subbotin, G. Stupak, V. Dvorkin, A. Povalyaev, S. Karutin, "GLONASS: Developing Strategies for the Future," *GPS World*, April 2011, pp. 42–49.
6. T. Kasami, "Weight Distribution Formula for Some Class of Cyclic Codes," Technical report R285, University of Illinois, United States Defense Technical Information Center, 1966, DTIC AD0632574.
7. A. Chandra and S. Chattopadhyay, "Small Set Orthogonal Kasami Codes for CDMA System," Proceedings of the 2009 IEEE International Conference on Computers and Devices for Communication.

10

GALILEO

Galileo is one of the four global satnav systems currently deployed or being developed. It is the only system being developed by a consortium of nations. Europe's stated motivations for developing Galileo include independence from GPS as well as economic benefits. Civilian organizations lead Galileo's development and control, but military considerations are integral to aspects of Galileo's development, and Galileo is also slated to be used for military applications. Galileo is intended to be highly interoperable with GPS and GLONASS, and presumably with BeiDou as well.

Galileo comprises a space segment, a ground segment, and a user segment. The fully deployed system will consist of a minimum of 24 satellites and extensive worldwide ground infrastructure.

The establishment of organizations responsible for Galileo and allocation of responsibilities among these organizations has evolved, but appears to have stabilized [1]. The European Commission (EC) is responsible for program supervision of Galileo, including funding, political interactions, and high-level requirements across all three segments, along with efforts addressing legal, standardization, certification, and regulatory issues. The European Space Agency (ESA) is responsible for deployment of Galileo program, including program management of developmental activities, establishing and managing contracts for design and procurement of the space segment and the ground segment,

---

Engineering Satellite-Based Navigation and Timing: Global Navigation Satellite Systems, Signals, and Receivers, First Edition. John W. Betz.
© 2016 The Institute of Electrical and Electronics Engineers, Inc. Published 2016 by John Wiley & Sons, Inc.

as well as defining and developing these segments through on-orbit validation. The European GNSS Agency (GSA) is responsible for Galileo exploitation, which includes operation of the space and control segments. Other GSA responsibilities include providing and marketing services, improvement and evolution of the space and ground infrastructure, research and development of receiver technology, developing follow-on capabilities, and cooperation with other satnav system providers.

This chapter provides a brief history of Galileo and summarizes current plans in Section 10.1, followed by an overview of the Galileo system in Section 10.2. Section 10.3 provides detailed characteristics of Galileo signals, followed by a summary of the chapter in Section 10.4.

10.1 GALILEO HISTORY AND PLANS

Development of initial design concepts for a European satnav system began in the 1990s, with selection of an initial system design in 2002. Galileo In-Orbit Validation Element (GIOVE) satellites were launched in 2005 and 2008. The first of these, called GIOVE-A, transmitted Galileo signals in time to meet the requirements for using the spectrum that had been reserved by filings at the ITU. GIOVE-A also provided experience with space-based atomic clocks, operation in the MEO space environment, and transmitting and receiving the signal designs selected for Galileo. The second satellite, called GIOVE-B, was capable of broadcasting the final signal set established for Galileo. Both GIOVE satellites were retired in 2012.

The Galileo constellation design was originally planned based on a three-plane Walker constellation with a minimum of nine active satellites in each plane. Relative to GPS, Galileo satellites are at higher orbit, where orbital perturbations are said to be lower, resulting in less need for station-keeping and thus lower fuel consumption. Also, Galileo's orbital planes are at slightly greater inclination to the equatorial plane than GPS, yielding somewhat better coverage at high latitudes.

Galileo's services, described in Section 10.2, were defined early, and the system architecture and signal characteristics were then developed to meet the needs of the services. Recent changes to the definitions of Galileo services could affect some of the motivations for established signal designs; it is not clear whether signal designs might be changed in response to the evolved requirements.

Since 2002, Europe and the United States have worked together to establish a special relationship concerning Galileo and GPS, influencing both systems' signal designs. Initial European efforts on signal design in the 1990s focused on approaches drawn from digital communications—good for efficient use of spectrum but lacking the large RMS bandwidths that yield good code tracking performance as described in Chapter 19. When technical experts from the European Community and United States began working together in 2002, the Galileo frequency plan had been established and an initial set of spreading modulations had been selected at different carrier frequencies to meet the needs of Galileo services. The United States opposed Galileo's planned signal designs in upper L band, since they unacceptably overlaid the frequencies used by the GPS M code signal. Eventually European experts selected a different design for the Public Regulated Service

(PRS) signal in upper L band, moving its spectrum outside the frequencies used by M code. A BOC(1,1) spreading modulation was initially selected for the Open Service signal, and the United States committed to transmit a new civil signal (L1C) at the same frequency, using the same BOC(1,1) spectrum. Based on this conclusion, the United States and the European Community signed an agreement on GPS and Galileo in 2004 [2]. Subsequently, European experts expressed displeasure with the limited performance provided by the BOC(1,1) spreading modulation [3], proposing various alternatives that the United States found unsatisfactory. Finally, in 2007, enhanced spreading modulations were developed for L1C signal, with similar ones then adopted for the Galileo E1 OS signal based on the multiplexed BOC (MBOC) spectrum [4].

Throughout the development of Galileo, European nations have debated the need and cost of its various services and characteristics, as well as the sources of this funding. During the first decade of this century, the Galileo business plan transitioned from a public–private partnership to be funded substantially by the private sector and others obtaining benefits from Galileo, to the current plan for government funding. Budgets are currently in place for establishment of the needed ground infrastructure along with approximately 26 satellites. Four of these satellites are In-Orbit Validation (IOV) satellites, two of which were launched in 2011, with the remaining two launched in 2012. An ambitious launch schedule has been published, leading to a projected full operational capability with 30 satellites on orbit by 2020.

Besides changes in the planned Galileo services described in Section 10.2, other changes to the Galileo space segment became evident in 2014. As described in Reference 5, the specified constellation now has a minimum of 24 satellites with up to six active spares, rather than the 27 satellites committed to earlier, with up to three spares. Also, in August 2014, the first two Full Operation Capability satellites (intended to be the fifth and sixth operational Galileo satellites) were mistakenly launched into incorrect orbits. They are being maneuvered into final orbits, but will not have the same orbits as was planned. It is likely that these satellites will be used as part of the operational Galileo constellation.

10.2 GALILEO DESCRIPTION

Galileo was designed to offer four navigation services as well as a search and rescue service. The Open Service (OS) provides signals for timing and positioning to anyone, free of user charges. OS signals are intended for interoperability with GPS, providing enhanced geometry and availability from the two constellations. OS is defined for mass market applications, with no integrity information and no service guarantee or liability provided by the Galileo operator. Users are responsible for determining the quality of OS signals.

The PRS is the government-authorized robust and access-controlled service using encrypted signals only available to authorized users. PRS provides protection against jamming and spoofing, and is intended to support critical transport and emergency services, law enforcement, border control, peace missions, and military operations.

The Safety-of-Life service (SoL) had originally been developed to provide assured-integrity signals for timing and positioning. It had been intended to provide a worldwide high-integrity capability for safety-critical applications such as transport applications, certified and guaranteed by the Galileo operator, with performance obtained by certified receivers. In 2012 the SoL service was "reprofiled"—replaced by an Integrity Monitoring Service (IMS) that is intended to support the needs of safety of life users in conjunction with other satnav systems. As of this writing, Galileo has not revealed the specific functionality added by the IMS beyond what is already provided by the other Galileo services and EGNOS. The I/NAV data message used by two of Galileo's signals was specifically designed to meet SoL service requirements; it remains to be seen whether and how the designs for these signals will change. One possibility would be for a lower data rate even if the symbol rate remains the same; another possibility would be to use the data rate for other navigation purposes such as navigation message authentication or the provision of high accuracy corrections to support precise point positioning, as described in Section 20.7.

Galileo's Commercial Service (CS) is defined for applications that need higher performance than provided by OS, offering fee-based value-added services for professional high-precision users. Access to the CS signal is controlled by encryption. Service providers will buy from the Galileo operator the right to use the CS signals for applications that include data broadcasting and resolving ambiguities in differential applications. It is expected that CS users would also use Galileo OS signals, and perhaps external communication systems. Possible applications include service guaranteed, precise positioning and timing services through the provision of an additional signal at a different carrier frequency, high-resolution ionospheric delay models, and local differential correction signals. Considerations and plans for the CS are discussed in Reference 6.

Galileo's fifth service is not a navigation service, but the Search And Rescue Service (SARS). It represents Europe's contribution to Cospas-Sarsat, an international satellite-based search and rescue distress alert detection system. SARS allows a Galileo satellite to receive signals from emergency beacons carried on ships, planes, or persons, and then report the emergency to rescue authorities, providing the location of the transmitter. A Galileo satellite can transmit a response to the beacon, providing feedback that its emergency transmission has been received.

The baseline Galileo constellation uses 24 satellites in a Walker 24/3/1 design, along with up to six operating spares—two in each plane. The eight satellites in each plane are equally spaced; locations of the spares will be determined. The three Earth-centered orbital planes have 120° separation, with inclination 56° relative to the equatorial plane. The orbits are nominally circular orbits, with radius 29,600 km, producing a nominal orbital period of more than 14 hours, leading to 17 orbits in 10 days. Satellite speed is approximately 3.7 km/s.

Galileo satellites nominally have a mass of 700 kg, measuring $2.7 \text{ m} \times 1.1 \text{ m} \times 1.2 \text{ m}$, with deployed solar arrays spanning 13 m. The primary clock uses passive hydrogen maser technology, with a rubidium secondary clock and duplicate backups of each clock. Depending upon the launcher, either two or four satellites can be launched at a time.

The Galileo ground segment uses a global network of transmitting and receiving stations, along with two control centers located in Fucino, Italy and Oberpfaffenhofen, Germany. Within each control center are parts of the Galileo Control System (GCS) whose functions involve control and management of the satellites as well as parts of the Galileo Mission System (GMS) that controls mission functions associated with the timing and positioning of payloads. The GCS includes five globally distributed telemetry, tracking, and control (TT&C) stations for communication with the satellites and a communications network that interconnects the TT&C stations and the two control centers. The GMS uses a global network of Galileo Sensor Stations (GSS) to monitor Galileo downlink signals for two purposes: measurement of orbit and clock offsets from each satellite, and monitoring the integrity of each signal. Measurements of clock and orbit offset are processed in the control centers to determine updated data messages that are uploaded to each satellite nominally every 100 minutes from a global network of five uplink stations.

Galileo's terrestrial reference frame is known as the Galileo Terrestrial Reference Framework (GTRF), an independently developed reference framework. The three-dimensional differences of positions represented in GTRF are specified to be within 3 cm (2-sigma) of those represented in the most recent ITRF. The Galileo Geodetic Service Provider (GGSP), a consortium of seven institutions, is developing the Galileo Terrestrial Reference Frame (GTRF) and establishing products and information for Galileo users.

Galileo System Time (GST) is established at the Galileo Control Centre in Fucino, Italy, based on the average of multiple atomic clocks. The Galileo starting epoch is 00:00:00 UTC (midnight) on Sunday, August 22, 1999 (midnight between 21 and 22 August). Like GPS, Galileo uses continuous time and thus differs from UTC by an integer number of leap seconds.

10.3 GALILEO SIGNALS

The Galileo frequency plan, established early in this century, involves multiple signals at different carrier frequencies. Figure 10.1 summarizes the frequency plan using what has become Galileo's notation. The upper L band segment had been denoted E2-L1-E1, but is now commonly called E1, a designation that is also used for the Galileo carrier frequency in upper L band of 1575.42 MHz, the same carrier frequency GPS denotes L1. The next lower band, designated E6, is in the upper part of lower L band, with a carrier frequency of 1278.75 MHz. E5 designates the lower part of lower L band, subdivided into E5b and E5a. The E5 carrier frequency is 1191.795 MHz; E5b and E5a carrier frequencies are, respectively, 1207.14 MHz and 1176.45 MHz, the latter the same frequency as GPS L5.

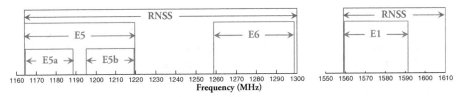

Figure 10.1. Galileo Frequency Bands

TABLE 10.1. Relationship between Galileo Services and Galileo Signals

Navigation Service	Signals						
	E1 OS	E1 PRS	E6 CS	E6 PRS	E5	E5b IMS	E5a OS
Open service (OS)	X				X	X	X
Integrity monitoring service (IMS)	X				X	X	X
Commercial service (CS)	X		X		X	X	X
Public regulated service (PRS)		X		X			

All signals use RHCP. Minimum received signal power is specified out of a 0 dBic circularly polarized antenna, located near the ground, for satellites at elevation angles of 10° or more, with an estimated minimum power for satellites at elevation angles of 5°. Currently there is no excess atmospheric attenuation accounted for in the specification of minimum received power. There is some indication that Galileo is considering changes that would specify the same minimum received power levels at 5° elevation and including some excess atmospheric loss. Such a change would require more power from space than required by the current conventions, if the minimum received power values were to remain the same. Since neither excess atmospheric loss nor polarization loss is included in the Galileo link budget conventions for minimum received power, and the path loss at 10° is less than at 5° used by GPS conventions, Galileo minimum received power values would be reduced by as much as 2 dB if the GPS conventions were employed.

The names of Galileo signals have evolved over time. Much of the current Galileo signal terminology deals with individual signal components, while the focus here is on signals that in many cases consist of multiple components. Consequently, in this book Galileo signals are named by their carrier frequency and by the primary service they support, recognizing that some signals serve the needs of multiple Galileo services.

Table 10.1 shows the Galileo signals that support different services. The E1 OS signal had been designed to support SoL (now IMS) and CS in addition to OS, requiring a high data rate as a result. While the E5b signal had been designed to support SoL service, it now appears that civil aviation users will use the E5a signal for IMS, since E5a has the same carrier frequency as signals used for aviation safety of life by GPS and SBAS. It is not clear whether the data rate for the E5a signal will be increased as a result, or what applications will employ the E5b signal rather than the E5a signal. Sections 10.3.1 through 10.3.5 summarize characteristics of different Galileo signals, as described in the Galileo ICD [5].

Single-frequency use of Galileo signals is enhanced with the use of a new ionospheric model [7] that provides more accurate corrections than does the Klobuchar model employed by GPS.

10.3.1 Galileo E1 OS Signal

Table 10.2 summarizes key characteristics of the E1 OS signal, using the same characteristics introduced in Section 3.3. This signal uses the same carrier frequency as the GPS

TABLE 10.2. Summary of E1 OS Characteristics

Carrier frequency	E1: 1575.42 MHz = 154 × 10.23 MHz
Minimum received power	−157.0 dBW
Polarization	RHCP
Multiple access	CDMA
Components	50% power pilot, 50% power data, both components in the same carrier phase
Spreading modulation and transmit bandwidth	Multiplexed BOC spectrum, MBOC(6,1,1/11); implemented as CBOC− on pilot component and CBOC+ on data component, ±12.276 MHz
Spreading codes	Length 4092 bit "memory" codes, different for pilot and data components
Overlay code	Length 25 bit, 250 bps on pilot, same for all signals (0011100000001010110110010)
Data message structure	I/NAV
Data rate	125 bps
Data modulation	250 sps biphase modulation
Data message channel encoding	1/2-rate convolutional coding with constraint length 7, block interleaving
Multiplexing with other signals	Interplex-type multiplexing with both components on in-phase part of carrier, and E1 PRS in phase quadrature

C/A code signal and the L1C signal, enhancing interoperability. The data rate shown here includes extra bits used for framing, synchronization, and decoder flushing; the actual information data rate is somewhat lower.

As discussed in Section 3.1, Composite BOC (CBOC) is a multilevel spreading modulation designed to implement the MBOC spectrum. For E1 OS, the CBOC spreading symbols are given by (3.37), with $\alpha = 10/11$, $m_1 = 1$, $m_2 = 6$, and $n = 1$. Figure 10.2 shows segments of the resulting waveforms. Since CBOC− has slightly higher RMS bandwidth than CBOC+, giving it better potential tracking performance, the pilot component uses CBOC−. As described in Section 3.1.4, transmitting both CBOC− and CBOC+ simultaneously produces the desired MBOC spectrum. Figure 10.2 shows formation of the CBOC waveforms used for Galileo E1 OS, using terminology from Section 3.4.

As described in Chapter 18, receiver processing must account for the four-level signal components. Review Question QT.10.4 shows how these two multilevel components are combined with the PRS E1 signal to produce a constant-envelope transmitted E1 signal.

Figure 10.3 shows the autocorrelation function and PSD of CBOC+, and Figure 10.4 shows the same for CBOC−, assuming ideal long spreading codes. The CBOC− correlation function peak is slightly narrower, and its PSD has wider sidelobes near ±6 MHz, both corresponding to larger RMS bandwidth.

Figure 10.5 shows the signal generation flow adapted from the Galileo ICD [5]. The upper leg can be explained using the terminology and notation defined in Figures 3.22 through Figure 3.25. $C_{E1-B}(t)$ is the data spreading waveform, denoted in Section 3.4.1

GALILEO SIGNALS

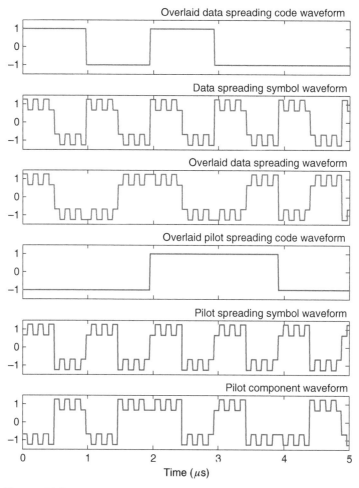

Figure 10.2. Example Waveforms for E1 OS Data and Pilot Components

by $\sigma_d(t)$, with spreading code chip duration $T_c = 1/1.023$ μs so the spreading code chip rate is 1.023 MHz. The data spreading code bit sequence, denoted in Section 3.4.1 by $\{b_n\}$, has a length of 4092 bits, or a 4 ms duration, and there is no data overlay code, since the data spreading code's duration is the same as the duration of a data message symbol. $D_{E1-B}(t)$ is the data symbol waveform, denoted in Section 3.4.1 by $\beta(t)$, with $T_d = 4092 \times T_c = 4$ ms, so the data symbol rate is 250 Hz. $sc_{E1-C,a}(t)$ is a BOC(1,1) spreading symbol waveform, while $sc_{E1-C,b}(t)$ is a BOC(6,1) spreading symbol waveform, and their weighted sum forms the data spreading symbol waveform $\sum_{n=-\infty}^{\infty} g_d(t - nT_c - T_c/2)$, where $g_d(t)$ is a CBOC+ spreading symbol. The output at the right-hand side of the upper leg is the data component waveform.

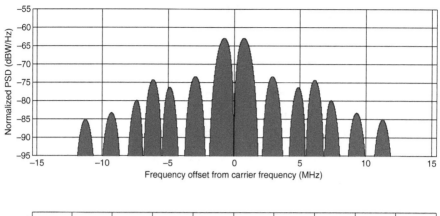

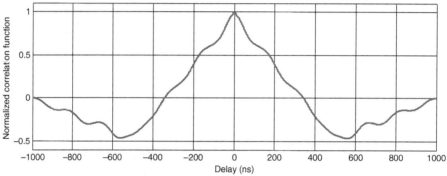

Figure 10.3. PSD and Autocorrelation Function for CBOC+

The lower leg, which forms the pilot component, begins with the pilot spreading code waveform, $C_{E1-C}(t)$, which is then multiplied by the pilot overlay code waveform $O_{E1-C}(t)$. The overlay code has a length of 25 bits, or a 100 ms duration, so the overlaid pilot spreading waveform, has a 100 ms duration. The resulting product $e_{E1-C}(t)$, is then multiplied by the pilot spreading symbol waveform $\alpha\, sc_{E1-C,a}(t) - \beta\, sc_{E1-C,b}(t) = \sum_{n=-\infty}^{\infty} g_p(t - nT_c)$, where $g_p(t)$ is a CBOC− spreading symbol. The resulting pilot component waveform is subtracted from the data component waveform, forming a unit power biphase multilevel signal.

Galileo documentation refers to tiered codes, formed by taking the product of the entire primary code with each bit of a secondary code whose rate is the reciprocal of the primary code duration. The primary code is equivalent to what is called a spreading code in this book, the secondary code corresponds to what is called an overlay code in this book, and the tiered code is equivalent to what is called an overlaid spreading code in this book.

E1 OS spreading codes are termed memory codes, found by computer search and then painstakingly listed bit by bit in the Galileo ICD. There are no identified

GALILEO SIGNALS

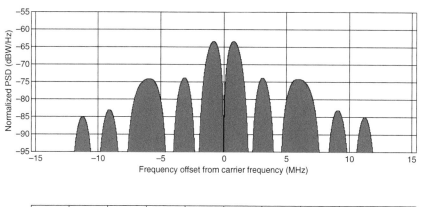

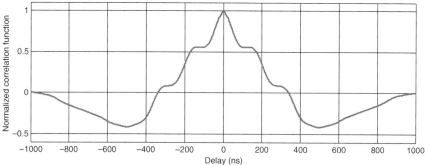

Figure 10.4. Autocorrelation Function and PSD for CBOC–

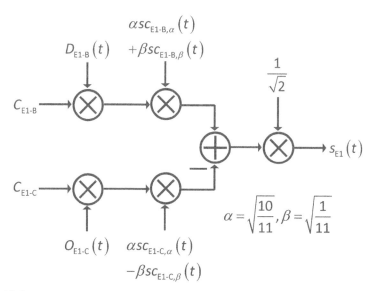

Figure 10.5. Description of E1 OS Generation Based on Galileo ICD Description and Notation [5]

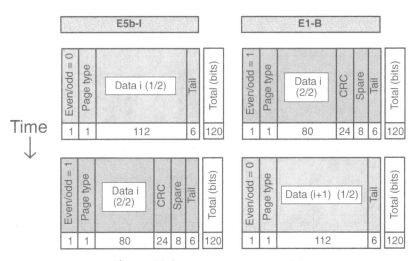

Figure 10.6. I/NAV Page Structure [5]

algorithms for generating memory codes; instead, they must be stored in memory and then accessed as needed. Like all the Galileo spreading codes, these spreading codes have been explicitly selected to provide good correlation properties in Doppler, as well as to address other criteria, as described in Reference 8.

The E1 OS signal was designed to provide the upper L band signal used for SoL service. Consequently, it uses the I/NAV data message structure designed to support the integrity services planned for SoL. Figure 10.6 summarizes the I/NAV page structure. The message comes in two half-pages that are transmitted simultaneously on E1 OS and E5b SoL signals. A dual-frequency E1/E5b receiver can then read the entire pages simultaneously from the two signals; single-frequency receivers read one half-page, and then wait to read the second half in sequence from the signal on a single frequency.

Within the message structure shown, the number of bits allocated is indicated in the lower row. The Even/odd field indicates the first or second half of the page (0 = even/ 1 = odd) being broadcast, and the page type is set to 0 to indicate the nominal page type. The data field contains a word of 192 bits (split between the two half-pages; the even half-page has 112 of the data bits while the odd half-page has 80 of the data bits); in addition to the 192 bits of data, the message structure includes 8 spare bits for reserved fields not yet defined. Tail bits are zero value bits used to flush the decoder. The CRC is computed over both the Even/odd fields, Page type fields, and Data fields. The CRC is computed for the Even and odd parts of a page of the same frequency and is always broadcast on the second half of the page on each frequency. Spare field bits and tail bits are not protected by the CRC.

Each page comprises 120 bits before encoding, so the half-rate encoding produces 240 symbols. Each encoded page is preceded by the 10-bit synchronization pattern 0101100000 for a total of 250 symbols transmitted in 1 second at 250 sps. Figure 10.7 shows the frame, subframe, and page structure of the I/NAV message.

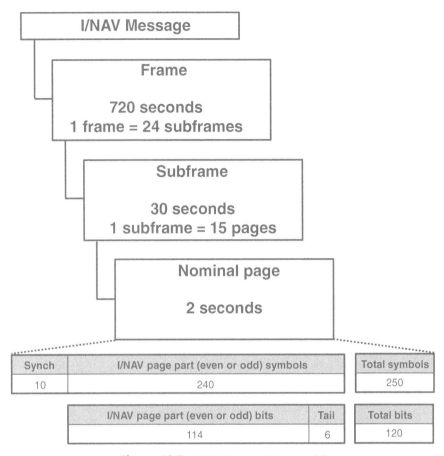

Figure 10.7. I/NAV Message Structure [5]

When needed, an SARS return link message having separate format described in Reference 5 is transmitted only on the E1 signal, providing the return link to an SARS terminal broadcasting an emergency message.

10.3.2 Galileo E1 PRS Signal

Since aspects of PRS are not made public, less information is available concerning this signal, but the available characteristics are summarized in Table 10.3. The transmit bandwidth value is an estimate based on the null-to-null bandwidth of the spreading modulation; the actual value may be larger than this.

Figure 10.8 shows the E1 PRS autocorrelation function and PSD, assuming an ideal long spreading code. The high BOC ratio of 6 produces high-amplitude correlation function subpeaks, and the high subcarrier frequency produces subpeaks with close spacing from the main peak in delay. The cosine phasing produces very low spectral

TABLE 10.3. Summary of E1 PRS Characteristics

Carrier frequency	E1: 1575.42 MHz = 154 × 10.23 MHz
Minimum received power	−157 dBW
Polarization	RHCP
Multiple access	CDMA
Components	Single biphase signal, may have time-multiplexed pilot and data components
Spreading modulation and transmit bandwidth	$BOC_c(15,2.5)$, ±17.9025 MHz
Spreading codes	Restricted
Overlay code	Restricted
Data message structure	Restricted
Data rate	Restricted
Data modulation	Restricted
Data message channel encoding	Restricted
Multiplexing with other signals	Interplex-type multiplexing placing E1 PRS on quadrature phase of the carrier with E1 OS Components on in-phase

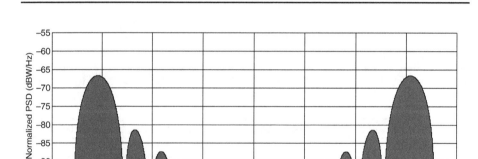

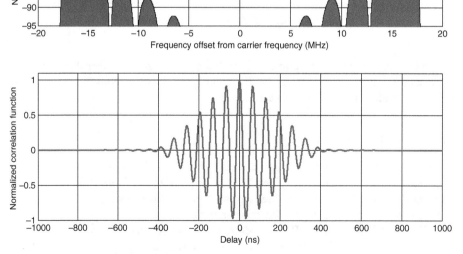

Figure 10.8. PSD and Autocorrelation for E1 PRS

sidelobes inside the main lobe, with very good spectral separation (small SSCs) from signals with spectrum concentrated near the carrier frequency.

The mathematical model of the baseband composite E1 signal is

$$x_{E1}(t) = \left[\sum_{k=-\infty}^{\infty} a_k g_{CBOC+}(t - kT_1 - T_1/2) - \sum_{k=-\infty}^{\infty} b_k g_{CBOC-}(t - kT_1 - T_1/2) \right]$$
$$- i \left[\sum_{k=-\infty}^{\infty} c_k g_{BOC_c(15,2.5)}(t - kT_{2.5} - T_{2.5}/2) + m(t) \right] \quad (10.1)$$

where the $\{a_k\}$ are E1 OS data spreading code bits multiplied by data message symbols, taking on values ± 1, $g_{CBOC+}(t)$ and $g_{CBOC-}(t)$ are, respectively, the CBOC+ and CBOC− spreading symbols given by (3.37), with $\alpha = \sqrt{10/11}$, $m_1 = 1$, $m_2 = 6$, and $n = 1$; the $\{c_k\}$ are E1 PRS spreading code bits multiplied by any data message symbols and overlay codes used for E1 PRS; $g_{BOC_c(15,2.5)}(t) = \text{sgn}[\cos(30\pi t/T_1)], 0 \le t \le T_{2.5}$ and zero elsewhere; $T_{2.5} = T_1/2.5$; and $m(t)$ is an intermodulation component designed so that $x_{E1}(t)$ has constant envelope. Review Question QT.10.5 shows that that $x_{E1}(t)$ has constant envelope.

10.3.3 Galileo E6 CS Signal

Table 10.4 summarizes key characteristics of the E6 CS signal. This signal shares the same carrier frequency with the E6 PRS signal. The high rate data message can provide high volumes of data with short latency to support the CS functions, but this high data sacrifices E_b/N_0 and thus raises the CS data demodulation threshold.

TABLE 10.4. Summary of E6 CS Characteristics

Carrier frequency	E6: 1278.75 MHz = 125 × 10.23 MHz
Minimum received power	−155 dBW
Polarization	RHCP
Multiple access	CDMA
Components	50% power pilot, 50% power data, both components in the same carrier phase
Spreading modulation and transmit bandwidth	BPSK-R(5), ±20.46 MHz
Spreading codes	Restricted
Overlay code	100 bit, 1000 bps on pilot, same for all signals
Data message structure	C/NAV
Data rate	500 bps
Data modulation	1000 sps biphase modulation
Data message channel encoding	1/2-rate convolutional coding with constraint length 7, block interleaving
Multiplexing with other signals	Interplex-type multiplexing with both components on in-phase part of carrier, and E6 PRS in phase quadrature

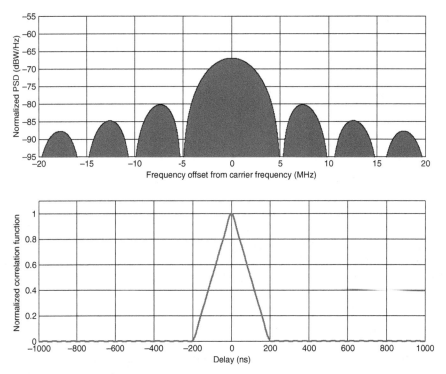

Figure 10.9. Autocorrelation and PSD for E6 CS

Figure 10.9 shows the E6 CS autocorrelation function and PSD, assuming an ideal long spreading code. These have conventional shapes based on the BPSK-R(5) spreading modulation.

The signal generation flow for E6 CS is shown in Figure 10.10; it is similar to that for E1 OS shown in Figure 10.5, but simpler because no subcarrier is involved. Here,

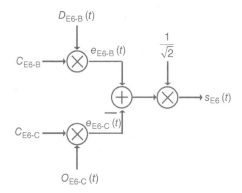

Figure 10.10. Description of E6 CS Generation [5]

$C_{E6-B}(t)$ is the overlaid data spreading code waveform using a BPSK-R(5) spreading modulation, and $C_{E6-C}(t)$ is the pilot spreading code waveform also using a BPSK-R(5) spreading modulation. $D_{E6-B}(t)$ is the data symbol waveform at 1000 sps, while $O_{E6-B}(t)$ is the pilot overlay code waveform at 1000 bps. As for all the Galileo pilot signals, the overlaid spreading code duration is 100 ms. The lower leg, which forms the pilot component, is subtracted from the data component in the upper leg, and the result is scaled to produce a unit power composite signal.

10.3.4 Galileo E6 PRS Signal

Table 10.5 summarizes key characteristics of the E6 PRS signal, which uses the same carrier frequency as the E6 CS signal. Like the E1 PRS signal, little is published about this signal except for its spreading modulation. While the spreading modulation is similar to that for the GPS M-code signal, the PRS uses cosine phasing, which increases the sidelobe levels outside the main spectral peaks relative to the sidelobes within the main peaks, causing greater opportunity for ambiguity in code tracking of the bandlimited signal, but providing greater spectral separation from BPSK-R signals at the same center frequency. Figure 10.11 shows the autocorrelation function and PSD, assuming an ideal long spreading code.

The mathematical model of the baseband composite E6 signal is

$$x_{E6}(t) = \left[\sum_{k=-\infty}^{\infty} a_k g_{\text{BPSK-R}(5)}\left(t - kT_5 - T_5/2\right) - \sum_{k=-\infty}^{\infty} b_k g_{\text{BPSK-R}(5)}\left(t - kT_5 - T_5/2\right) \right]$$
$$- i \left[\sum_{k=-\infty}^{\infty} c_k g_{\text{BOC}_c(10,5)}\left(t - kT_5 - T_5/2\right) + m(t) \right] \quad (10.2)$$

TABLE 10.5. Summary of E6 PRS Characteristics

Carrier frequency	E6: 1278.75 MHz = 125 × 10.23 MHz
Minimum received power	−155 dBW
Polarization	RHCP
Multiple access	CDMA
Components	Single biphase signal, may have time-multiplexed components
Spreading modulation and transmit bandwidth	BOC$_c$(10,5), ±20.46 MHz
Spreading codes	Restricted
Overlay code	Restricted
Data message structure	Restricted
Data rate	Restricted
Data modulation	Restricted
Data message channel encoding	Restricted
Multiplexing with other signals	Interplex-type multiplexing placing E6 PRS on quadrature phase of the carrier with CS Components on in-phase

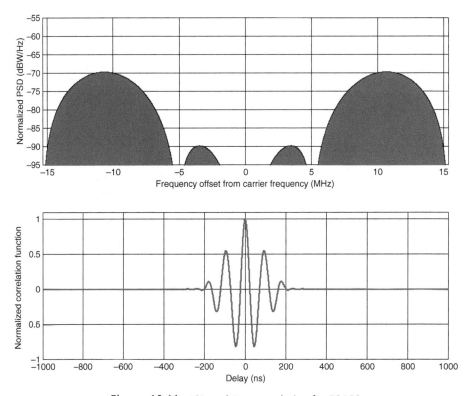

Figure 10.11. PSD and Autocorrelation for E6 PRS

where the $\{a_k\}$ are E6-CS data spreading code bits multiplied by data message symbols, taking on values ± 1; $g_{\text{BPSK-R}(5)}(t)$ is the BPSK-R(5) spreading symbol given by (3.18) with $T_c = T_5 = 1/(5 \times 1.023 \times 10^6)$ seconds; the $\{b_k\}$ are E6-CS pilot spreading code bits multiplied by overlay code bits, taking on values ± 1; the $\{c_k\}$ are E6-PRS spreading code bits multiplied by any data message symbols and overlay code bits used for E6-PRS, taking on values ± 1; $g_{\text{BOC}_c(10,5)}(t)$ is given by (3.25); and $m(t)$ is an intermodulation component designed so $x_{E6}(t)$ has constant envelope.

10.3.5 Galileo E5 Signals

By virtue of its constant-envelope AltBOC(15,10) spreading modulation, defined in (3.37), the E5 signal is at once a single signal and a triplet of signals. All three signals, denoted E5, E5a, and E5b, are described in this section and summarized in Table 10.6, and the detailed relationship of these three signals is delineated later in this subsection. Observe that overlay codes are used on the data components as well as the pilot components to extend the duration of the overlaid spreading code to that of the data symbols, as on the GPS L5 signal.

TABLE 10.6. Summary of E5, E5a, and E5b Signal Characteristics

	E5 Signal	E5b Signal	E5a Signal
Carrier frequency	E5: 1191.795 MHz = 116.5 × 10.23 MHz	E5b: 1207.14 MHz = 118 × 10.23 MHz	E5a: 1176.45 MHz = 115 × 10.23 MHz
Minimum received power	−152 dBW	−155 dBW	−155 dBW
Polarization	RHCP	RHCP	RHCP
Multiple access	CDMA	CDMA	CDMA
Components	Two pilot components, each 25% power pilot; two data components, each 25% power	50% power pilot, 50% power data	50% power pilot, 50% power data
Spreading modulation and transmit bandwidth	AltBOC(15,10), ±25.575 MHz	QPSK-R(10)), ±10.23 MHz	QPSK-R(10)), ±10.23 MHz
Spreading codes	See E5b, E5a	10,230-bit codes, different for pilot and data components	10,230-bit codes, different for pilot and data components
Overlay code	See E5b, E5a	4 bit, 1000 bps (data); 100 bit, 1000 bps (pilot); different for each satellite	20 bit, 1000 bps on data; 100 bit, 1000 bps on pilot; different for each satellite
Data message structure	See E5b, E5a	I/NAV	F/NAV
Data rate	See E5b, E5a	125 bps	25 bps
Data modulation	See E5b, E5a	250 sps biphase modulation	50 sps biphase modulation
Data message channel encoding	See E5b, E5a	1/2-rate convolutional coding with constraint length 7, block interleaving	1/2-rate convolutional coding with constraint length 7, block interleaving
Multiplexing with other signals	Constant-envelope AltBOC	Constant-envelope AltBOC	Constant-envelope AltBOC

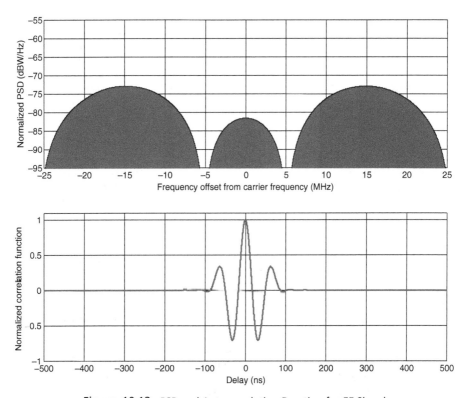

Figure 10.12. PSD and Autocorrelation Function for E5 Signal

Figure 10.12 shows the autocorrelation function and PSD of the E5 signal, assuming an ideal long spreading code. This is the widest bandwidth of any satnav signal currently defined, having the narrowest correlation function peak.

The baseband E5 signal is defined in terms of its four components, respectively, the E5a data and pilot components, and the E5b data and pilot components:

$$s_{\text{E5a-d}}(t) = \sum_{k=-\infty}^{\infty} a_k g_{\text{BPSK-R(10)}}(t - kT_{10} - T_{10}/2)$$

$$s_{\text{E5a-p}}(t) = \sum_{k=-\infty}^{\infty} b_k g_{\text{BPSK-R(10)}}(t - kT_{10} - T_{10}/2)$$

$$s_{\text{E5b-d}}(t) = \sum_{k=-\infty}^{\infty} c_k g_{\text{BPSK-R(10)}}(t - kT_{10} - T_{10}/2)$$

$$s_{\text{E5b-p}}(t) = \sum_{k=-\infty}^{\infty} d_k g_{\text{BPSK-R(10)}}(t - kT_{10} - T_{10}/2)$$

(10.3)

TABLE 10.7. E5 Subcarrier Values

Time	$\psi_1(t)$	$\psi_2(t)$
$0 \leq t < T_{15}/8$	$(1 + \sqrt{2} - i)/(4\sqrt{2})$	$(1 - \sqrt{2} + i)/(4\sqrt{2})$
$T_{15}/8 \leq t < 2T_{15}/8$	$[1 - i(1 + \sqrt{2})]/(4\sqrt{2})$	$[1 + i(1 - \sqrt{2})]/(4\sqrt{2})$
$2T_{15}/8 \leq t < 3T_{15}/8$	$[-1 - i(1 + \sqrt{2})]/(4\sqrt{2})$	$[-1 + i(1 - \sqrt{2})]/(4\sqrt{2})$
$3T_{15}/8 \leq t < 4T_{15}/8$	$(-1 - \sqrt{2} - i)/(4\sqrt{2})$	$(-1 + \sqrt{2} + i)/(4\sqrt{2})$
$4T_{15}/8 \leq t < 5T_{15}/8$	$(-1 - \sqrt{2} + i)/(4\sqrt{2})$	$(-1 + \sqrt{2} - i)/(4\sqrt{2})$
$5T_{15}/8 \leq t < 6T_{15}/8$	$[-1 + i(1 + \sqrt{2})]/(4\sqrt{2})$	$[-1 + i(-1 + \sqrt{2})]/(4\sqrt{2})$
$6T_{15}/8 \leq t < 7T_{15}/8$	$[1 + i(1 + \sqrt{2})]/(4\sqrt{2})$	$[1 + i(-1 + \sqrt{2})]/(4\sqrt{2})$
$7T_{15}/8 \leq t < T_{15}$	$(1 + \sqrt{2} + i)/(4\sqrt{2})$	$(1 - \sqrt{2} - i)/(4\sqrt{2})$

where the $\{a_k\}$ are E5a data spreading code bits multiplied by overlay code bits and data message symbols, taking on values ± 1; $g_{\text{BPSK-R(10)}}(t)$ is the BPSK-R(10) spreading symbol given by (3.18); $T_{10} = 1/(10 \times 1.023 \times 10^6)$ seconds; the $\{b_k\}$ are E5a pilot spreading code bits multiplied by overlay code bits, taking on values ± 1; the $\{c_k\}$ are E5b data spreading code bits multiplied by overlay code bits and data message symbols, taking on values ± 1; and the $\{d_k\}$ are E5b pilot spreading code bits multiplied by overlay code bits, taking on values ± 1.

AltBOC construction involves two subcarriers and four waveforms formed from the components in (10.3), as described next. The subcarriers, denoted $\psi_1(t)$ and $\psi_2(t)$, each has period $T_{15} = 1/(15 \times 1.023 \times 10^6)$ seconds, taking on the eight discrete amplitude values in each period given in Table 10.7. Two periods of these subcarriers are plotted in Figure 10.13, showing $\psi_1(t)$ is a quantized-amplitude complex sinusoid, while $\psi_2(t)$ has a more complicated structure. Inspection of the values in Table 10.7, confirmed by the plots in Figure 10.13, shows that both of the subcarriers are constant envelope, with magnitudes $\sqrt{2 + \sqrt{2}}/4 \approx 0.4619$ and $\sqrt{2 - \sqrt{2}}/4 \approx 0.1913$, respectively.

To see how the subcarriers are used to form the E5 signal, define the four waveforms

$$\begin{aligned} x_A(t) &= \left[s_{\text{E5a-d}}(t) + is_{\text{E5a-p}}(t)\right] \psi_1^*(t) \\ x_B(t) &= \left[s_{\text{E5b-d}}(t) + is_{\text{E5b-p}}(t)\right] \psi_1(t) \\ x_C(t) &= s_{\text{E5b-d}}(t) s_{\text{E5b-p}}(t) \left[s_{\text{E5a-p}}(t) + is_{\text{E5a-d}}(t)\right] \psi_2^*(t) \\ x_D(t) &= s_{\text{E5a-d}}(t) s_{\text{E5a-p}}(t) \left[s_{\text{E5b-p}}(t) + is_{\text{E5b-d}}(t)\right] \psi_2(t) \end{aligned} \quad (10.4)$$

In $x_A(t)$, the multiplication by the complex exponential produces a single-sideband signal shift to a higher carrier frequency by 15×1.023 MHz, while analogously $x_B(t)$ is a single-sideband signal shifted to a lower carrier frequency by 15×1.023 MHz. The sum $x_A(t) + x_B(t)$ is not constant envelope, but the composite E5 signal

$$x_{E5}(t) = x_A(t) + x_B(t) + x_C(t) + x_D(t) \quad (10.5)$$

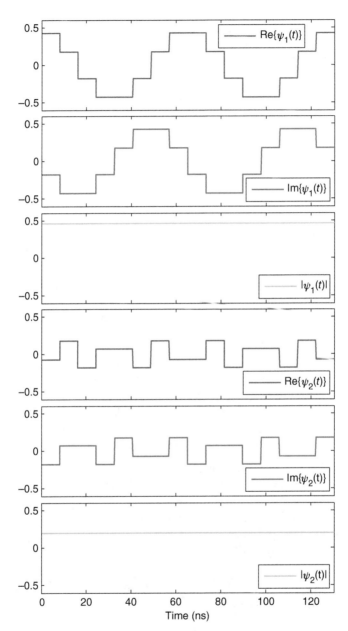

Figure 10.13. Galileo E5 Signal Subcarriers

GALILEO SIGNALS

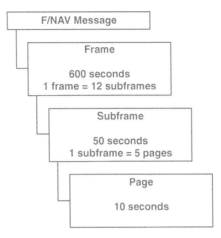

Figure 10.14. F/NAV Message Structure [5]

is constant envelope and unit power, as demonstrated by simulation in Figure 10.13. In (10.4), $x_A(t)$ and $x_B(t)$ are the useful components corresponding to E5a and E5b, respectively, while $x_C(t)$ and $x_D(t)$ are intermodulation terms that produce the constant-envelope composite signal.

Spreading codes for the E5a and E5b signals are generated using two linear shift registers of length 14, short-cycled to provide only the needed 10,230 bits. All shift registers for the E5a signals use a given set of feedback taps, and all shift registers for the E5b signals use a different set of feedback taps. Each spreading code for a different signal is produced by initializing one of the shift registers by a different amount, then XORing the outputs of the two shift registers. The Galileo ICD [5] provides the specific polynomials and start values used for each of the 50 pairs of spreading codes.

The data message for E5b is the I/NAV message also used for E1 OS and described in Section 10.3.1. The data message on E5a, denoted F/NAV for freely accessible navigation, has the structure shown in Figure 10.14. Each subframe contains clock corrections and ephemeris for the satellite, so the 50 second subframe duration determines the latency for obtaining this information. The F/NAV page layout, shown in Figure 10.15, is similar to that for I/NAV. Up to 64 page types are supported, each with a payload of 208 bits, CRC, and the same tail of all zeroes used to flush the decoder. The 12 bit synch pattern is 101101110000.

Synch	Symbols	Total symbols
12	488	500

	Word		Tail	Total bits
Page type	Navigation data	CRC		
6	208	24	6	244

Figure 10.15. F/NAV Page Layout [5]

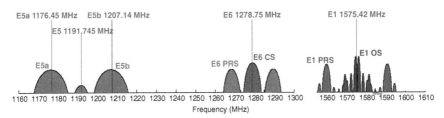

Figure 10.16. PSDs of Galileo Signals

10.4 SUMMARY

Galileo plans a global system, including open signals that are highly interoperable with GPS civil signals on L1 and L5. Figure 10.16 shows the PSDs of the Galileo signals, assuming ideal long spreading codes. The PRS signals are spectrally separated from the other Galileo signals, and the E5 signal occupies a very wide bandwidth in the lower part of lower L band.

Key signal characteristics are summarized in Table 10.8.

The E1 OS signal uses the multilevel CBOC spreading modulation to produce the MBOC spectrum. Chapter 18 shows how to extract pilot and data components from the composite signal, how one-bit correlators can be used to process the multilevel signal, and other variants of CBOC processing. E1 OS's spreading codes must be read from memory, since there is no algorithm for generating the spreading codes using digital hardware. E1 OS's data message has the high data rate of 125 bps, reducing the E_b/N_0 and causing a high data demodulation threshold, particularly since the convolutional encoding is not as capable as more modern encoding techniques. The high symbol rate restricts the correlation integration time used on the data component, increasing squaring loss in carrier and code tracking as explained in Chapters 18 and 19. The high symbol rate also restricts the length of the spreading code, causing higher correlation sidelobes. However, the high data rate can be used to achieve lower latency for urgent information like integrity assurances, clock correction, and ephemeris.

Little detailed information is available on the PRS signals and the E6 CS signal, which will only be available for authorized users.

The E5 signal uses the innovative AltBOC spreading modulation that permits processing of the composite signal or separate components designated E5a and E5b. The

TABLE 10.8. Summary of Galileo Signals

Signal Name	Carrier Frequency (MHz)	Spreading Modulation	Service
E1 OS	1575.42	MBOC(6,1,1/11)	OS, IMS, CS
E1 PRS	1575.42	$BOC_c(15,2.5)$	PRS
E6 CS	1278.75	BPSK-R(5)	CS
E6 PRS	1278.75	$BOC_c(10,5)$	PRS
E5	1191.795	Constant-envelope AltBOC(15,10)	OS, IMS, CS
E5b	1207.14	QPSK-R(10)	OS, IMS, CS
E5a	1176.45	QPSK-R(10)	OS, IMS, CS

E5a signal is highly interoperable with the GPS L5 signal and other BPSK-R(10) and QPSK-R(10) signals centered at 1176.45 MHz. With higher minimum received power than the E1 OS signal and a 25 bps data message, the E5a signal should enable robust performance under challenging conditions such as indoor or urban operation, and higher interference environments. Chapter 18 describes how the E5 signal is processed to use the entire signal, or extract E5a and/or E5b components.

REVIEW QUESTIONS

Theoretical Questions

QT10.1. List characteristics of GPS C/A code and Galileo E1 OS signals that are similar, and characteristics that are different.

QT10.2. List characteristics of GPS L1C and Galileo E1 OS signals that are similar, and characteristics that are different.

QT10.3. Compute the RMS bandwidths (see Appendix Section A.6) of the E1 OS pilot and data components using the bandwidth in Table 10.2 and, based on the results, provide a conjecture as to why CBOC − was selected for the pilot component. Compute the RMS bandwidth numerically using

$$\bar{\beta}_{\text{rms}} = \left[\frac{\int_{-\beta_r/2}^{\beta_r/2} f^2 \Phi_x(f)\, df}{\int_{-\beta_r/2}^{\beta_r/2} \Phi_x(f)\, df} \right]^{1/2} = \left[\int_{-\beta_r/2}^{\beta_r/2} f^2 \bar{\Phi}_x(f)\, df \right]^{1/2},$$

$$\int_{-\beta_r/2}^{\beta_r/2} \bar{\Phi}_x(f)\, df = 1$$

Assume ideal long spreading codes and signal power computed over the bandwidths defined for each signal in Section 10.3.

QT10.4. Show that the composite E1 signal given by (10.1) is constant envelope when

$$m(t) = \left[\sum_{k=-\infty}^{\infty} a_k g_{CBOC+}\left(t - kT_1 - T_1/2\right) \right]$$

$$\times \left[\sum_{k=-\infty}^{\infty} b_k g_{CBOC-}\left(t - kT_1 - T_1/2\right) \right]$$

$$\times \left[\sum_{k=-\infty}^{\infty} c_k g_{BOC_c(15,2.5)}\left(t - kT_{2.5} - T_{2.5}/2\right) \right]$$

Assume ideal long spreading codes and signal power computed over the bandwidths defined for each signal in Sections 10.3.

Application Questions

QA10.1. If the received power of a Galileo E1 OS code signal is −152 dBW, what is the maximum power spectral density of the received signal? Assume ideal long spreading codes and signal power computed over the bandwidths defined for each signal in Section 10.3.

QA10.2. If the maximum received power from a constellation of Galileo E1 OS code signals is −140 dBW, what is the maximum power spectral density of the aggregate E1 OS signals? If a receiver's thermal noise density is −201.5 dBW/Hz, what is the maximum power spectral density of the combined aggregate E1 OS signals and thermal noise?

QA10.3. What power of aggregate E1 PRS code signals would produce the same maximum power spectral density as aggregate Galileo E1 OS signals at −140 dBW? Assume ideal long spreading codes and signal power computed over the bandwidths defined for each signal in Section 10.3.

QA10.4. For the following calculations, define C to be the minimum specified received power for all components of the relevant Galileo signal, and let N_0 be −201.5 dBW/Hz. For the following signals and E_b/N_0 thresholds, find the minimum C/N_0.

Signal	E_b/N_0 Threshold (dB)	Minimum C/N_0 (dB-Hz)
E1 OS	4.5	
E5b	4.5	
E5a	4.5	

REFERENCES

1. GSA and EC Sign Agreement on Galileo Service Provision, *InsideGNSS*, 9 October 2014.
2. White House Fact Sheet, "U.S.-EU Summit: Agreement on GPS-Galileo Cooperation," June 2004.
3. G. W. Hein, J.-A. Avila-Rodriguez, L. Ries, L. Lestarquit, J.-L. Issler, J. Godet, and T. Pratt, "A Candidate for the Galileo L1 OS Optimized Signal", Proceedings of ION GNSS 2005, Long Beach, CA, 13–16 September 2005.
4. G. W. Hein, J.-A. Avila-Rodriguez, S. Wallner, J. W. Betz, C. Hegarty, J. J. Rushana, A. J. Kraady, A. R. Pratt, S. Lenahan, J. Owen, J.-L. Issler, and T. Stansell, "MBOC: The New Optimized Spreading Modulation Recommended for Galileo L1 OS and GPS L1C," *InsideGNSS*, May/June 2006, pp. 57–65.

5. European GNSS (Galileo) Open Service Signal in Space Interface Control Document, Issue 1.2, available at http://ec.europa.eu/enterprise/policies/satnav/pubconsult-2/files/galileo_os_sis_icd_v1-2-3_short_version270614_en.pdf, accessed 4 January 2015.
6. I. Fernández-Hernández, I. Rodríguez, G. Tobías, J.D. Calle, E. Carbonell, G. Seco-Granados, J. Simon, R. Blasi, "Galileo's Commercial Service; Testing GNSS High Accuracy and Authentication," *InsideGNSS*, January/February 2015, pp. 38–48.
7. Ionospheric Correction Algorithm for Galileo Single Frequency Users, European GNSS (Galileo) Open Service, Document Issue 1.1, February 2015.
8. F. Soualle, M. Soellner, S. Wallner, J.-A. Avila-Rodriguez, G. W. Hein, B. Barnes, T. Pratt, L. Ries, J. Winkel, C. Lemenager, P. Erhard, "Spreading Code Selection Criteria for the Future GNSS Galileo," Proceedings of the 2005 GNSS Conference, 2005, available at http://www.researchgate.net/publication/266185051_Spreading_Code_Selection_Criteria_for_the_future_GNSS_Galileo, accessed 15 March 2015.

11

BEIDOU SYSTEM

BeiDou is the first operational regional satnav system, and the first satnav system with plans to operate as both a regional system and a global system, as explained subsequently. Like GPS and GLONASS, the BeiDou System (BDS) is intended for dual use—designed and operated to provide service both to civilian users and to authorized users. China has indicated it will offer open BeiDou signals for use by anyone, anywhere its signals can be received, with no user fees. China has also committed to providing ICDs containing the full unrestricted technical information needed to construct and manufacture receivers of BeiDou open signals. Currently, official information concerning BeiDou is limited; one useful source of information is the unofficial web site [1]. Additional information is also provided in the official web site [2], as well as in References 3–5.

The name "BeiDou" refers to the Big Dipper constellation, used to locate the North Star. Because of the Big Dipper's use in navigation, "BeiDou" is used by China as the name of its satnav system, and is also translated into English as "Compass." While China has previously referred to this system as Compass, or the Compass Navigation Satellite System (CNSS), China now calls it the BeiDou Navigation Satellite System, abbreviated BeiDou, or BDS. Consequently the terms BeiDou and BDS are used exclusively in this book.

Engineering Satellite-Based Navigation and Timing: Global Navigation Satellite Systems, Signals, and Receivers, First Edition. John W. Betz.
© 2016 The Institute of Electrical and Electronics Engineers, Inc. Published 2016 by John Wiley & Sons, Inc.

Like other satnav systems, BDS comprises a space section, a ground section, and a user section. However, its design differs from those of other systems in several ways:

- In addition to enabling user terminals to perform positioning through passive trilateration, regional BeiDou also provides a radio determination satellite system (RDSS) positioning capability, described in Section 11.2;
- Associated with the RDSS capability is a regional two-way short messaging capability;
- BeiDou includes a constellation of GEO and IGSO satellites for regional service; it has launched some MEO satellites already and plans to provide a full constellation of MEO satellites for global service;
- BeiDou plans to provide SBAS-type capability integrated into BeiDou, rather than as a separate system.

This additional functionality makes all three segments of BDS differ from those of other satnav systems, with satellites in three different types of orbit transmitting signals for passive trilateration; the GEO satellites also provide RDSS, short messaging, and SBAS functionality. The ground segment supports regional and global constellations as well as regional RDSS, short messaging, and SBAS functionality. User terminals support any or all of this functionality.

BDS is being developed and operated by the China National Space Administration. BDS's development and deployment span three phases. Each phase uses a different constellation, a different set of signals, and a different ground segment. The number of satellites increases with each phase; the intended BeiDou Phase 3 constellation size will be 35 satellites with 27 in MEO orbit, 5 in GEO orbit, and 3 in IGSO. It appears this will be the number of operational satellites; there is no current indication whether on-orbit spares will also be present, either as active (transmitting) spares or inactive spares.

This chapter provides a brief history of BeiDou and summarizes current plans in Section 11.1, followed by an overview of BDS in Section 11.2. Section 11.3 describes what is known about BeiDou signals, followed by a summary of the chapter in Section 11.4.

11.1 BDS HISTORY AND PLANS

What is now known as BeiDou Phase 1 was developed in the 1990s, and was also called the BeiDou Experimental Satellite Navigation System. It was designed to implement RDSS, a fundamentally different positioning approach than the passive trilateration employed by the other satnav systems described in this book. RDSS was proposed in China in 1983 [1] and demonstrated in 1989. The RDSS concept appears to have originated with Dr. Gerard O'Neill [6] who in 1980 filed a patent in the United States that was later awarded [7], then extended in additional patents [8,9]. Geostar Corporation, founded by Dr. O'Neill to implement RDSS, originally worked closely with China to develop an RDSS system in China, then went bankrupt in late 1991 [6]. China then continued to develop the RDSS capability on its own.

The BeiDou Phase 1 ground segment was constructed in the 1990s, and the two requisite GEO satellites were launched in 2000. The system was declared operational in 2001. A third GEO satellite, serving as a backup, was launched in 2003.

Meanwhile, in 2003, China completed an agreement with Europe to participate in the Galileo system. Chinese engineers began participating in cooperation projects involving Galileo development, and it was widely assumed that China would use Galileo to meet its satnav needs. Within several years, however, China began to withdraw from involvement in Galileo support, apparently being dissatisfied with restrictions on its role.

China subsequently announced that it intended to develop and operate its own satnav system, and it rapidly became apparent that China was committed and moving rapidly. BeiDou Phase 2 satellites were launched in GEO, IGSO, and MEO orbits starting in 2007.

By the end of 2012, the BeiDou Phase 2 satellites launched included six GEO satellites, five IGSO satellites, and five satellites in MEO. China then announced its regional BeiDou Phase 2 system to have achieved full operational capability.

American and French researchers observed the satellite signals and reverse-engineered some aspects, including spreading codes, for some signals. However, complete and official descriptions of the signals have been slow in coming. In December 2012, an ICD was released for a Phase 2 signal known as B1I, and in January 2014 an updated ICD was published to describe both Phase 2 open signals [10]. However, less information has been provided publicly about other Phase 2 signals, or about planned Phase 3 signals.

China has indicated plans to transition to BeiDou Phase 3 by 2020. The plans for BeiDou Phase 3 indicate a global system consisting of 27 MEO satellites, along with a regional system of 3 IGSO satellites and 5 GEO satellites. BeiDou Phase 3 will provide four navigation-related services: a worldwide open service, a worldwide authorized service, a regional wide area differential service, and a regional position reporting service apparently based on RDSS. In addition, it appears BeiDou Phase 3 will continue to provide the regional text messaging service associated with RDSS.

The signals used for passive trilateration in BeiDou Phase 3 are slated to be somewhat different from those in BeiDou Phase 2, with some new signals added in Phase 3, and some Phase 2 signals slated to be phased out. How and when these signal changes occur, and the specific designs of Phase 3 signals, remains to be seen.

In 2013, the China Satellite Navigation Office published an open service performance standard [11], making BDS the second system, along with GPS, to provide such a performance commitment.

11.2 BDS DESCRIPTION

Since all three BeiDou phases are distinctly different, each is described separately in the three subsections below. Even though BeiDou Phase 1 is now history, it is described since its RDSS functionality is maintained in BeiDou Phase 2 and planned to be expanded in BeiDou Phase 3.

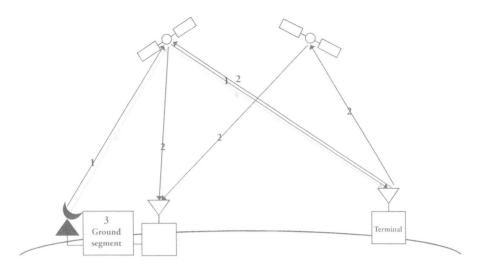

Figure 11.1. Basic RDSS Architecture

11.2.1 BeiDou Phase 1 Description

BeiDou Phase 1 provided regional positioning and short messaging service using RDSS. As described in References 1 and 8, an RDSS system comprises a ground station, at least two geostationary satellites, and a user terminal. As illustrated in Figure 11.1, determining the terminal location requires four steps, numbered in the figure.

1. The ground station transmits an interrogation signal to the user terminal through a transponder on one of the geostationary satellites. In BeiDou, the downlink from satellite to user terminal is a direct sequence spread spectrum signal using a BPSK-R spreading modulation, a chip rate of approximately 4 MHz, and S band carrier frequency of 2491.75 MHz.
2. The user terminal transmits a response that is relayed by both geostationary satellites to the ground station. In BeiDou, this uplink signal is a direct sequence spread spectrum signal with 1615.68 MHz carrier frequency. This response includes data describing the terminal's altitude if it is not on the Earth's surface.
3. The ground station measures the round-trip interrogation–response time through the two different satellites. Knowing these times, the altitude of the user terminal (inferred from a digital terrain map for terminals on the surface of the Earth), and the satellite locations enables the ground station to compute the terminal's location in three dimensions.
4. The ground station then provides a message to the user terminal, via one of the satellites, informing it of the terminal's location.

Both the interrogation message and the response message can be augmented with additional characters used for messaging, up to 160 bytes. Thus, RDSS provides an integrated positioning and communications capability.

The Phase 1 GEO satellites were positioned at 80°E and 140°E, with the backup satellite at 110.5°E. The service area was between Longitude 70°E and 140°E, and Latitude 5°N to 55°N. Three ground tracking stations for determining satellite orbits were located at Jiamushi, Khashi, and Zhanjiang in China [12].

RDSS positioning accuracy depends on a number of factors, including the accuracy of measuring the round-trip interrogation and response, knowledge of the satellite locations, knowledge of the terminal's altitude, and the geometry. Reportedly, BeiDou Phase 1 provided positioning accuracy of 100 m. Ground correction stations at surveyed locations can be used to calibrate some of the delays in the system, such as delays in satellite transponders, improving positioning accuracy to the order of 20 m.

RDSS has the advantage of using fewer satellites than passive trilateration systems like GPS, and of using satellites only in geostationary orbits. However, it requires the user terminal to be active—transmitting the response. Also, the user terminal only knows its location if the ground station provides it. In addition, the resulting positioning accuracy is nowhere near as good as with a well-operated passive trilateration system. Finally, unlike passive trilateration, whose broadcasts can service an unlimited number of user terminals, RDSS capacity is limited since each terminal must be serviced separately. Consequently, a terminal typically only obtains its position intermittently, not on the rapid once-per-second rate available from GPS.

11.2.2 BeiDou Phase 2 Description

BeiDou Phase 2 was declared operational at the end of 2012. As of 2013, the operational satellites in the Phase 2 system consisted of five GEO satellites, five IGSO satellites, and four MEO. All of these satellites transmit L band signals for passive trilateration, as described in Section 11.3. In addition, all of the GEO satellites provide the RDSS capability.

GEO satellites are denoted BeiDou-G, IGSO satellites are designated BeiDou-IGSO, and MEO satellites designated BeiDou-M. The BeiDou-G and BeiDou-IGSO satellites are similar, with a design life of 8 years. Their mass exceeds 4000 kg, and they are launched one at a time. As of 2013, five GEO satellites are operational, positioned at 58.75°E, 84.7°E, 144.0°E, and 160°E. The IGSO satellites are in a 55° inclination geosynchronous orbit, with an orbital period of 1436.2 minutes, or approximately 23 hours and 56 minutes. BeiDou-M satellites have a gross mass of 800 kg, with a design life of 8 years. Two or more can be launched on a single launch vehicle. All satellites are said to be equipped with laser reflectors, enabling highly accurate orbit determination from ground stations. Early reports of positioning performance are found in Reference 13.

The ground system is known as the BeiDou Ground Base Enhancement System (BGBES). It is being developed to include a network of 30 ground base stations, an operating system, and a precision satellite positioning system [14].

11.2.3 BeiDou Phase 3 Description

BeiDou Phase 3 is projected to be operational around the year 2020. Phase 3 represents a significant transition from Phase 2 with some signals added and some Phase 2 signals

BDS SIGNALS

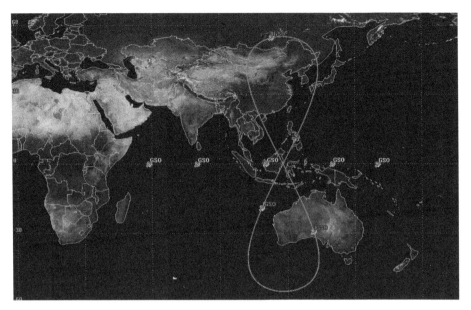

Figure 11.2. BeiDou Phase 3 Regional Constellation Ground Tracks

eliminated, a different regional constellation consisting of five GEO satellites and three IGSO satellites, and a constellation of 27 MEO satellites for worldwide coverage. Ground tracks for the Phase 3 regional constellation are shown in Figure 11.2.

The BeiDou MEO constellation, described in Section 2.3, has 27 satellites in 3 Earth-centered orbital planes, with 120° separation. The orbital planes are inclined 55° relative to the equatorial plane. Orbits are circular with nominal radius 27,900 km yielding an orbital period of 12 hours, 52 minutes. Satellite speed is approximately 3.8 km/s.

The BDS datum is based on the China Geodetic Coordinate System 2000 (CGCS2000). BeiDou time (BDT) is aligned with UTC (within 100 ns), except BDT is continuous and does not include leap seconds, meaning that BDT differs from UTC by an integer number of seconds. The BeiDou starting epoch is 00:00:00 UTC on January 1, 2006 (midnight between December 31, 2005 and January 1, 2006).

11.3 BDS SIGNALS

Since BeiDou signals have different characteristics in each phase of BeiDou, and there is no ICD publicly available, at least in English, describing the RDSS signals, no further description of RDSS of BeiDou Phase 1 signals is provided here. Section 11.3.1 provides the available information about BeiDou Phase 2 signals, and Section 11.3.2 summarizes what is known about BeiDou Phase 3 signals, not including the signals used for RDSS.

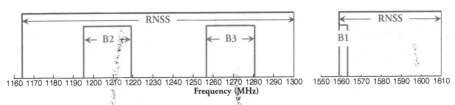

Figure 11.3. BeiDou Phase 2 Frequency Bands

11.3.1 BeiDou Phase 2 Signals

BeiDou Phase 2 transmits six signals on three carrier frequencies; two of which have a signal intended for open use is in-phase, with a signal for authorized use on the corresponding quadrature component. Figure 11.3 summarizes the Phase 2 frequency plan using BeiDou notation. The same terminology is used for frequency bands as for carrier frequencies. B1 band's carrier frequency, 1561.098 MHz, is called B1. B2 band's carrier frequency, 1207.14 MHz, is called B2. B3 band is between the other two bands; the B3 frequency, 1268.52 MHz, is called B3.

All signals use RHCP, with ellipticity no greater than 2.9 dB. Minimum received signal power is specified out of a 0 dBi circularly polarized antenna located near the ground, for satellites at elevation angles of 5° or more, with no mention that excess atmospheric attenuation is taken into account. Since neither excess atmospheric loss nor polarization loss is included in the BDS link budget conventions for minimum received power, BeiDou minimum received power values would be reduced by as much as 1.5 dB if GPS conventions were employed.

Characteristics of the various Phase 2 signals are summarized in Table 11.1.

A summary of the two open signals are given below. Specifics of the authorized signals, beyond the information in Table 11.1, are not publically available.

Table 11.2 summarizes the B1I and B2I signal characteristics, drawn from Reference 10. Both signals have identical characteristics except for carrier frequency. The data rates shown here include overhead bits used for framing, synchronization, and parity; the actual information data rate is somewhat lower. Signals from the GEO satellites have different data message structure and contents than from the IGSO and MEO satellites, producing some differences in other signal characteristics as well. The symbol rates are

TABLE 11.1. Summary of BeiDou Phase 2 Signals

Signal Name	Carrier Frequency (MHz)	Spreading Modulation	Service
B1I	1561.098	BPSK-R(2)	Open
B1Q	1561.098	BPSK-R(2)	Authorized
B3	1268.52	QPSK-R(10)	Authorized
B2I	1207.14	BPSK-R(2)	Open
B2Q	1207.14	BPSK-R(10)	Authorized

TABLE 11.2. Summary of B1I and B2I Signal Characteristics

B1I carrier frequency	B1: 1561.098 MHz = 1526 × 1.023 MHz
B2I carrier frequency	B2: 1561.098 MHz = 1180 × 1.023 MHz
Minimum received power	−163 dBW
Polarization	RHCP
Multiple access	CDMA
Spreading modulation and transmit bandwidth	BPSK-R(2), ±8 MHz (B1I), ±18 MHz (B2I)
Spreading codes	Length 2046 bit segments of 2047 length Gold codes, duration 1 ms
Data message structure	MEO and IGSO satellites: BeiDou D1 NAV
	GEO satellites: BeiDou D2 NAV
Data rate	MEO and IGSO satellites: approximately 37 bps
	GEO satellites: approximately 367 bps
Data message error correction and detection	BCH(15,11) code with block interleaving over two blocks
Data modulation	MEO and IGSO satellites: 50 sps biphase modulation
	GEO satellites: 500 sps biphase modulation
Pilot and data components	100% power data
Overlay code	The same 1 kbps overlay code for MEO and IGSO satellites:
	Neuman–Hoffman code:
	0 0 0 0 0 1 0 0 1 1 0 1 0 1 0 0 1 1 1 0
	No overlay code for GEO satellites
Multiplexing with other signals	Phase quadrature to B1Q signal

clearly defined, with data rates given by 11/15 of the symbol rate, as dictated by the BCH(15,11) code.

Figure 11.4 shows the autocorrelation function and PSD, assuming an ideal long spreading code, yielding a triangular autocorrelation function and sinc-squared PSD.

The structure for generating the B1I and B2I Gold codes is described in Reference 10, but permission was not granted to allow reproduction of this graphic here. As described in Section 7.3.1, two different 11-stage m-sequence generators are employed, each implementing a different polynomial and having the same starting phase of 0 1 0 1 0 1 0 1 0 1 0. Different spreading codes for different satellites are produced by using different phase selection logic on the output taps of the one register, producing an output equivalent to different initial phases of this register. Phase assignments for 37 different spreading codes in the family are given in Reference 10, with the first five assigned to GEO satellites and the remaining spreading codes assigned to IGSO or MEO satellites. These spreading codes are actually segments of Gold codes with the final bit removed, since the length of the corresponding Gold code is actually 2047 bits, but the spreading code length is only 2046 bits. The relationship between satellites, ranging code numbers, and phase selection logic is documented in Reference 10, but permission was not obtained to reproduce this information here.

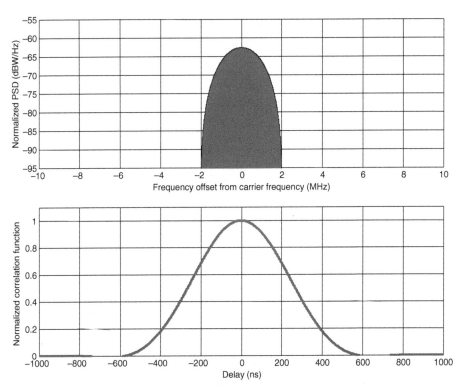

Figure 11.4. PSD and Autocorrelation Function for B1I and B2I Signals, Using Ideal Long Spreading Code Model

The IGSO and MEO satellites transmit B1I and B2I signals modulated by the D1 NAV message, while the GEO satellites transmit B1I and B2I signals modulated by the D2 NAV message. Both messages are described here. The D1 NAV message contains basic information that the receiver needs for navigation. The D2 NAV message contains the same information as the D1 NAV message, and also includes augmentation service information including integrity, differential corrections, and more detailed ionospheric corrections. Both message structures are discussed in sequence. While Reference 10 refers to the message contents as bits, it actually is describing encoded bits, or symbols. This book is consistent in distinguishing between information bits and encoded symbols, even when the encoded bits consist of information bits combined with parity symbols.

The D1 NAV message superframe consists of 36,000 symbols with 12 minutes duration. Each superframe is composed of 24 frames, each frame consisting of 1500 symbols with duration 30 seconds. There are five subframes in each frame; each subframe has 300 symbols and 6 second duration. There are 10 words in each subframe, each word with 30 symbols and lasting 0.6 second. Each word contains data and parity bits. The first word in each subframe includes a preamble, a subframe ID, and the number of seconds from the beginning of the week to the preamble first bit. This structure is portrayed in Reference 10.

The first three subframes of the D1 NAV message contain the basic information needed to navigate: time, user range accuracy index, satellite health flag, ionospheric delay model parameters, satellite ephemeris and its age, satellite clock correction and its age, and equipment group delay differential. In the first word of each subframe, some information, such as the preamble 1 1 1 0 0 0 1 0 0 1 0, is not encoded. The last 11 bits in the first word are encoded into 15 symbols. Combined, the last five words in the first three subframes contain all the information except time.

In the D1 NAV message, all words except the first word in the first subframe are BCH(15,11) encoded. All subframes except the first contain two 15-symbol blocks per word with block interleaving of alternate symbols from the first and second blocks in each word. The first three subframes provide the same information every 30 seconds, except when it is updated once an hour.

The fourth and fifth subframes in each frame of the D1 NAV message contain almanac data and time offsets from other satnav systems. Ephemeris and almanac information is represented using Keplerian parameters, with sinusoidal correction terms also used in the ephemeris. Clock corrections are expressed by a constant and a linear term. The information in the fourth and fifth subframes is repeated every 24 frames, except when it is updated; this information is updated at least every 7 days. The contents of subframe 5 in frames 11 through 24 are reserved.

The D2 NAV message superframe consists of 180,000 symbols with 6 minutes duration. Each superframe is composed of 120 frames, each frame consisting of 1500 symbols with duration 3 seconds. There are five subframes in each frame; each subframe has 300 symbols and 0.6 second duration. There are 10 words in each subframe, each word with 30 symbols and lasting 0.06 second. Like the D1 NAV message words, each D2 NAV message word contains data and parity bits. The first word in each subframe includes a preamble, a subframe ID, and the number of seconds from the beginning of the week to the preamble first bit. This structure is portrayed in Reference 10.

The first subframe of the D2 NAV message contains the basic information needed to navigate: time, user range accuracy index, satellite health flag, ionospheric delay model parameters, satellite ephemeris and its age, satellite clock correction and its age, and equipment group delay differential. In the first word of each subframe, some information, such as the preamble 1 1 1 0 0 0 1 0 0 1 0, is not encoded. The last 11 bits in the first word are encoded into 15 symbols. In the D2 NAV message, navigation information is provided with greater precision, so 10 words are required for the navigation information rather than the 5 words used in the D1 NAV message. Consequently, 10 sequential repetitions of subframe 1 in 10 sequential frames provides the navigation information, with the same 30 second repetition rate as in the D1 NAV message.

In the D2 NAV message, all words except the first word in the first subframe are BCH(15,11) encoded. All subframes except the first contain two 15-symbol blocks per word with block interleaving of alternate symbols from the first and second blocks in each word. Subframes 2, 3, and 4 contain integrity and differential correction information for BeiDou satellites. Six sets of subframes are required to provide all of this information, so it repeats every 18 seconds. Subframe 5 provides almanac, ionospheric grid points, and time offsets from other systems. 120 different versions of subframe 5 are required for the transmission of this information, which then repeats every 6 minutes.

TABLE 11.3. Summary of BeiDou Phase 3 Signals

Signal Name	Carrier Frequency (MHz)	Spreading Modulation	Service
B1-C	1575.42	MBOC	Open
B1-A	1575.42	QOC(14,2)	Authorized
B3	1268.52	QPSK-R(10)	Authorized
B3-A	1268.52	BOC(15,2.5)	Authorized
B2	1191.795	Constant modulus AltBOC(15,10)	Open
B2b	1207.14	QPSK-R(10)	Open
B2a	1176.45	QPSK-R(10)	Open

11.3.2 BeiDou Phase 3 Signals

There is currently no ICD available for the BeiDou Phase 3 signals. Some descriptions of carrier frequencies and spreading modulations are available from material provided in international forums, and Table 11.3 summarizes that information. The B1-C signal is a migration of the Phase 2 B1I signal, changing the carrier frequency to one that is common with GPS L1 and Galileo E1, and adopting a yet-unspecified waveform producing the MBOC PSD. The B1-A signal generalizes the Phase 2 B1Q signal, since its QOC(14,2) spreading modulation centered at 1575.42 includes a BPSK-R(2) component at 1561.098 MHz that could be interoperable with the B1Q signal. The B3 signal is the only one that apparently remains the same from Phase 2 to Phase 3. Interestingly, Phase 3 adds a second authorized signal, denoted B3-A, at the same carrier frequency as the B3 signal. The Phase 2 authorized signal B2Q vanishes in Phase 3, and the Phase 2 open signal B2I is replaced by the open AltBOC B2 signal that has the same design as the Galileo E5 signal, including signals on the same carrier frequency as B2I. Like the Galileo E5 signal described in Section 10.3, the BeiDou B2 signal can be treated as a single signal, or can be decomposed into two separate signals, B2b and B2a, summarized in the Table 11.3.

As of the end of 2015, China has launched several Phase 3 satellites and is reportedly experimenting with some different signal variants at B2 and B3. The final designs may have somewhat different PSDs or different implementations from those described here. See Reference 15 for further discussion of some of the options.

How BeiDou satellites and receivers will transition from Phase 2 to Phase 3 signals is not yet clear, although there are many Chinese publications on generalized AltBOC signal concepts that might allow satellites to transmit Phase 2 and Phase 3 signals simultaneously during a transition period.

11.4 SUMMARY

BeiDou is an operational regional system growing into a global system. Its growth from Phase 1 based on RDSS to operational Phase 2 based on passive trilateration has been rapid, and its planned transition to Phase 3 is aggressive. The Phase 2 and Phase 3

SUMMARY 263

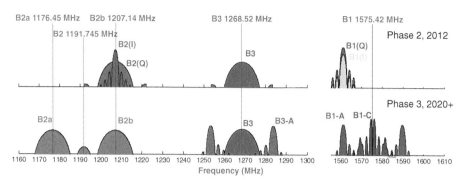

Figure 11.5. PSDs of Phase 2 and Phase 3 BeiDou Signals

signals are portrayed in Figure 11.5, showing the evolution planned for this decade, and Table 11.4 summarizes BeiDou Phase 2 and Phase 3 signal characteristics.

The Phase 2 signals are based on traditional designs, with BPSK-R spreading modulations and no pilot components. Phase 3 signals are slated to include a blend of BOC, MBOC, and AltBOC spreading modulations, many with pilot components. Information concerning Phase 2 and Phase 3 signals is incomplete at this point, with an ICD available only for the open civil Phase 2 signals, B1I and B2I. Their design is very similar to that of GPS C/A code signal, but with twice the spreading code chip rate, and an overlay code that mitigates issues associated with repeating spreading codes within a data bit.

TABLE 11.4. Summary of BeiDou Phase 2 and Phase 3 Signals

BeiDou Phase	Signal Name	Carrier Frequency (MHz)	Spreading Modulation	Service
2	B1I	1561.098	BPSK-R(2)	Open
2	B1Q	1561.098	BPSK-R(2)	Authorized
2	B3	1268.52	QPSK-R(10)	Authorized
2	B2I	1207.14	BPSK-R(2)	Open
2	B2Q	1207.14	BPSK-R(10)	Authorized
3	B1-C	1575.42	MBOC	Open
3	B1-A	1575.42	QOC(14,2)	Authorized
3	B3	1268.52	QPSK-R(10)	Authorized
3	B3-A	1268.52	BOC(15,2.5)	Authorized
3	B2	1191.795	Constant modulus AltBOC(15,10)	Open
3	B2b	1207.14	QPSK-R(10)	Open
3	B2a	1176.45	QPSK-R(10)	Open

Phase 3 BeiDou, with its blend of regional and global coverage, its provision of regional RDSS capability and text messaging, and its modern signal designs, promises to be a valuable contribution to navigation and timing around the world.

REVIEW QUESTIONS

Theoretical Questions

QT11.1. List characteristics of GPS C/A code and BeiDou B1I signals that are similar, and characteristics that are different.

QT11.2. List characteristics of Galileo E1 OS and BeiDou B1I that are similar, and characteristics that are different.

QT11.3. List characteristics of SBAS L1 and BeiDou B1I that are similar, and characteristics that are different.

Application Questions

QA11.1. For the following calculations, define C to be the minimum specified received power for all components of the relevant Galileo signal, and let N_0 be -201.5 dBW/Hz. For the following signals and E_b/N_0 thresholds, find the minimum C/N_0.

Signal	E_b/N_0 Threshold (dB)	Minimum C/N_0 (dB-Hz)
B1I, D1 NAV message	7.5	
B1I, D1 NAV message	7.5	

REFERENCES

1. "Dragon in Space: COMPASS (BeiDouj-II)," http://www.dragoninspace.com/navigation/compass-beidou2.aspx, accessed 1 May 2013, archived at http://archive.is/iB6fh.
2. "BeiDou Satellite Navigation System," http://en.beidou.gov.cn/index.html, accessed 4 May 2013.
3. China Satellite Navigation Office, "Report on the Development of BeiDou Navigation Satellite System (Version 2.1)," December 2012.
4. C. Han, Y. Yang, and Z. Cai, "BeiDou Navigation Satellite System and Its Time Scales," *Metrologia*, Vol. 48, 2011, pp. 1–6.
5. Y. YuanXi, L. JinLong, X. JunYi, T. Jing, G. HaiRong, and H. HaiBo, "Contribution of the Compass Satellite Navigation System to Global PNT Users," *Chinese Science Bulletin*, Vol. 56, No. 26, 2011, pp. 2813–2819.

REFERENCES

6. "RDSS LLC," www.rdss.com, accessed 1 May 2013.
7. "Satellite-Based Vehicle Position Determining System," United States Patent 4,359,733, Published November 1982.
8. "Position Determination and Message Transfer System Employing Satellites and Stored Terrain Map," European Patent EP0171751 A2, Published February 1986.
9. "Position Determination and Message Transfer System Employing Satellites and Stored Terrain Map," United States Patent 4,839,656, Published June 1989.
10. China Satellite Navigation Office, "BeiDou Navigation Satellite System Satellite Signal in Space Interface Control Document Open Service Signal (Version 2.0)," December 2013, available at http://en.beidou.gov.cn/, accessed 15 January 2014.
11. China Satellite Navigation Office, BeiDou Navigation Satellite System Open Service Performance Standard (Version 1.0), December 2013, available at http://en.beidou.gov.cn/, accessed 15 January 2014.
12. "eoPortal Directory: CNSS (Compass/BeiDou Navigation Satellite System)/BDS (BeiDou Navigation System)," https://directory.eoportal.org/web/eoportal/satellite-missions/c-missions/cnss, accessed 14 February 2015.
13. H. He, J. Li, Y. Yang, J. Xu, H. Guo, and A. Wang, "Performance Assessment of Single- and Dual-frequency BeiDou/GPS Single-Epoch Kinematic Positioning," *GPS Solutions: The Journal of Global Navigation Satellite Systems*, September 2013, available at http://rd.springer.com/article/10.1007/s10291-013-0339-3#page-1.
14. "BeiDouj Ground System Approved," http://gpsworld.com/bds/, accessed 20 September 2015.
15. Z. Zhou, J. Wei, Z. Tang, T. Yan, and T. Li, "Extended Time-Division AltBOC Modulation Techniques for GNSS," Proceedings of ION-GNSS+2015, Institute of Navigation, September 2015.

12

QUASI-ZENITH SATELLITE SYSTEM

The Quasi-Zenith Satellite System (QZSS) is the first announced regional satnav system that is not an SBAS. It had been intended as an augmentation to GPS and other global satnav systems, and now is planned to be a standalone regional system as well. QZSS takes its name from the quasi-zenith orbit planned for its original satellites, designed to provide an additional satellite near zenith for users in Japan as well as at high elevation angles over a large part of the Asia-Pacific region. QZSS is strictly a civilian system, with a signal set that includes close adaptations of all GPS civil signals, as well as additional signals.

QZSS is being developed by the Government of Japan (GoJ). Its early development was led by the Japan Aerospace Exploration Agency (JAXA), which sponsors a web site [1] providing comprehensive information about QZSS, including links to the QZSS IS [2].

QZSS comprises a space segment and a ground segment, with the user segment not considered to be officially part of the system. Japan currently plans [3] a constellation of satellites in geostationary and inclined non-geostationary orbits covering the Asia-Pacific region. A seven-satellite constellation will be provided, with the potential for up to two additional active backup satellites, four in geostationary orbit and five in inclined non-geostationary orbits all having the same ground track.

Engineering Satellite-Based Navigation and Timing: Global Navigation Satellite Systems, Signals, and Receivers, First Edition. John W. Betz.
© 2016 The Institute of Electrical and Electronics Engineers, Inc. Published 2016 by John Wiley & Sons, Inc.

Along with this expanded plan for QZSS, a new governance has come for QZSS development and operation. Japan has established a cabinet-level Office of National Space Policy (ONSP). QZSS is said to be the highest priority program for the ONSP. QZSS is funded by the GoJ, although the initial contract issued for development of the expanded QZSS reportedly includes private financing [4].

This chapter provides a brief history of QZSS and summarizes current plans in Section 12.1, followed by an overview of QZSS in Section 12.2. Section 12.3 provides detailed characteristics of QZSS signals, followed by a summary of the chapter in Section 12.4.

12.1 QZSS HISTORY AND PLANS

QZSS was originally announced in 2002 as a regional system that would provide enhanced communications and PNT capability to Japan and the surrounding Asia-Pacific region. The plan for QZSS development and operation included joint financing by the GoJ and private industry, with GoJ financing for the PNT aspect and industry funding for the communications aspects. In 2004, the GoJ issued a new national space policy calling for efforts leading to a long-term autonomous satellite-based navigation system. It was determined that the GoJ would take responsibility for funding the PNT functions, and the communication functionality was eliminated. JAXA took the lead for development of the QZSS satellites and ground segment, while management of QZSS was based on a complicated set of committees involving multiple cabinet offices and ministries, none of which had clear responsibility for funding.

By 2004, JAXA had already begun an intense development process, working in parallel on many aspects of system design. While different system concepts were explored, an early decision was for QZSS, at least initially, to serve as an augmentation and enhancement to other satnav systems. Analyses showed the benefits of providing another satellite at high elevations to improve performance in Japan's extreme urban canyons and mountainous terrain, compared to only a standalone global satnav system.

The QZSS design needed to focus initially on augmenting and enhancing one global satnav system. At that time, GLONASS's difficulties with constellation sustainment made it a less desirable prospect. While Galileo was considered strongly, QZSS eventually decided to place initial emphasis on augmenting the more proven and mature GPS. QZSS experts worked closely with GPS experts on the design of L1C in order to support both system's needs, and QZSS experts also developed designs of QZSS signals that would be as interoperable as possible with each of GPS's civil signals, and with GPS as a system. GPS allocated spreading codes from the GPS families of spreading codes for QZSS use, in order to enhance interoperability and compatibility with the corresponding QZSS signals. Two classes of QZSS signals were defined: availability enhancement signals improve the availability (visibility and geometry) of GPS signals, while performance enhancement signals improve the accuracy and reliability of GPS signals. The plan is to eventually improve the availability and performance of other satnav systems' signals as well.

In 2008 the GoJ reorganized Japanese government space activities, more clearly defining the funding responsibility for programs like QZSS. The ONSP was created at the cabinet level and given primary decision-making authority for Japanese space activities. In 2011, the ONSP identified QZSS as the highest of its three primary priorities.

In the meantime, the first QZSS satellite was being designed and constructed, and ground stations for monitoring and controlling the satellite were being constructed and integrated. The first Quasi-Zenith Satellite (QZS), denoted QZS-1 and named Michibiki, was successfully launched in 2010. Its availability enhancement signals were successfully checked out and declared operational in 2011, while numerous experiments have been conducted to help refine the design and application of the performance enhancement signals.

The first QZSS IS was issued in 2007, and several updates have been published. All can be found in Reference 1, with the most recent as of this writing named IS-QZSS Version 1.6. The QZSS ISs provide summary descriptions of QZSS and its segments, along with definitions of the QZSS signal set.

In 2013, the Japanese Cabinet Office announced a contract award to build one geostationary satellite and two additional QZSs. These three satellites are scheduled to be launched near the end of 2017. In addition, another contract was also signed to fund the design, construction, and operation of the ground control system for the nine-satellite constellation. QZSS will continue providing availability enhancement and performance enhancement to GPS, as well as possibly to other global satnav systems in future. It is also on the path toward being a standalone regional system by sometime in the 2020s.

12.2 QZSS DESCRIPTION

QZSS is a regional satnav system developed to service the Asia-Pacific region. Initially, its space segment will consist of satellites in a highly inclined elliptical orbit with modest eccentricity and an orbital period of one sidereal day, with orbital planes selected so that all the satellites share the same ground track. Eventually, satellites in geostationary orbit will be added. All of its signals are intended for open civilian use. Currently, it augments GPS with two classes of signals: those that provide availability enhancement and those that provide performance enhancement to GPS signals and to QZSS availability enhancement signals. As the geostationary satellites are added to its constellation, QZSS will also provide standalone service with its availability enhancement signals, as well as possibly SBAS functionality. At some point, the performance enhancement signals will work with other global satnav systems as well.

The QZSS space segment will grow from one satellite in highly inclined orbit to a constellation of as many as nine satellites, including two active backups. The highly inclined orbits have nominal values as follows: semi-major axis: 42,164 km, eccentricity: 0.075, and inclination: 43°. Figure 12.1 shows the ground track of these satellites, along with the planned locations of the GEO satellites.

Although the first QZS was developed by JAXA, subsequent satellites will be developed under contract; it appears that these later satellites will be similar to QZS-1. Geostationary satellites will have more functionality than those in quasi-zenith orbit, with an ability to broadcast emergency messages at S band to users in Japan.

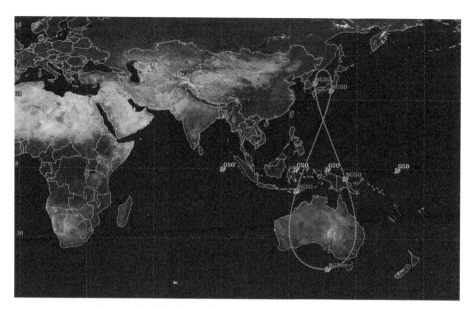

Figure 12.1. Planned Ground Tracks of QZSS Satellites

The current experimental QZSS ground segment uses a regional network of nine monitoring stations and a tracking control station, all coordinated by a master control station. Figure 12.2 depicts their locations. The monitoring stations monitor signals from QZSS satellites as well as GPS satellites, and will be expanded to monitor

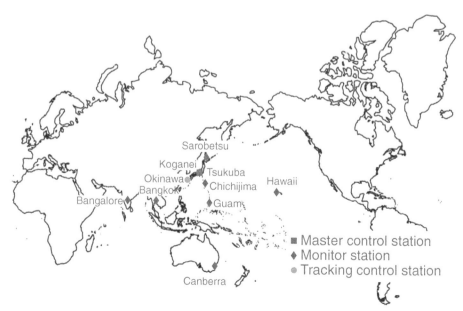

Figure 12.2. Current QZSS Ground Segment Laydown [2]

satellites from other systems in future. The original experimental QZSS ground segment will be replaced for the operational system, with more monitoring stations extending over a wider geographical region. Since QZSS is a regional system with a dense configuration of ground stations, all signals from all satellites can be monitored and contacted continuously from the ground, enabling rapid detection of any anomalies and frequent updates of clock and ephemeris parameters in the signals' data messages.

The QZSS geodetic datum is known as the Japan satellite navigation Geodetic System (JGS). Like all geodetic coordinate systems used in satnav, it is defined to approach the ITRS very closely.

QZSS time, known as QZSST, is very similar to GPS time, except it conforms to the Japan timing standard UTC (NICT) provided by Japan's National Institute of Information and Communications Technology.

The offset between QZSS time and GPS time is carefully controlled to within approximately 6.7 ns of GPS time with a probability of 95% or better. Like GPS time, QZSST is continuous and does not include leap seconds. The starting point of the week number for QZSST is the same as GPS: 23:59:47 UTC on August 21, 1999 (00:00:19 TAI on August 22, 1999). To maintain time on the satellite, QZSS satellites will use rubidium atomic clocks.

QZSS is also taking steps that could bring communications/navigation integration to mass-market users. One such step, summarized in Section 12.2, involves the pioneering provision of information on the Internet. Its web site http://qz-vision.jaxa.jp/, in both Japanese and English, provides operational information regarding interruption or degradation of service via Notification Advisories to QZSS Users (NAQUs), analogous to NANUs for GPS. While Michibiki is being used for experiments, its experimental schedule (primarily affecting the L1S and L6 signals) is also posted. Recent system performance information is also available. Also significant are tools for user support that predict QZSS and GPS orbits and provide the latest almanac and ephemeris data. Also, precise orbit and clock information can be downloaded from the web site. It is possible that applications will be developed to run on a receiver, automatically downloading orbit and clock information from the QZSS web site and generating long-lasting ephemeris for the receiver to use. Such an application in a vehicle navigation system would provide faster initial position (avoiding latency in reading a data message) and more robust operation (enabling position fix under conditions when the data message cannot be read from the satellite signal).

12.3 QZSS SIGNALS

QZSS will broadcast eight signals on four carrier frequencies, using the frequency plan portrayed in Figure 12.3. QZSS uses the same terminology as GPS for most of its carrier frequencies and frequency bands: L1 band's carrier frequency is 1575.42 MHz; L2 band's carrier frequency is 1227.60 MHz; and L5 band's carrier frequency is 1176.45 MHz. QZSS uses an additional frequency band that is the same as Galileo's E6; QZSS calls it the L6 band and its carrier frequency is 1278.75 MHz.

QZSS SIGNALS

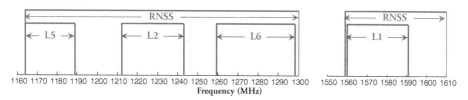

Figure 12.3. QZSS Frequency Bands

All signals use RHCP. Minimum received signal power is specified out of a 0 dBi circularly polarized antenna located near the ground, for satellites at elevation angles of 10° or more, with no excess atmospheric attenuation loss included. Since neither excess atmospheric loss nor polarization loss is included in the QZSS link budget conventions for minimum received power, and the path loss at 10° is less than at 5° used by GPS conventions, QZSS minimum received power values would be reduced by as much as 2 dB if the GPS conventions were employed.

All QZSS signals are defined in a single IS [2]. IS-QZSS also provides a thorough overview of the system and its coverage, availability, and performance. Like the GPS ICDs and ISs, IS-QZSS does not commit to specific signal combining approaches used to modulate multiple signals onto the same RF carrier (see Section 3.3.12), although phase relationships between signals and components are specified. Not specifying the signal combining approaches provides flexibility for different satellite designs to perform different signal combining approaches, and limits the ability for receiver designers to employ special techniques that rely on a particular combining approach.

The availability enhancement signals are L1-C/A, L1C, L2C, and L5. As discussed below, the designs of these signals are very similar to the designs of corresponding GPS signals. The performance enhancement signals are L1Sa, L1Sb, L6, and L5s. The designs of L1Sa and L1Sb are very similar to the design of L1 SBAS signals, however L1Sa is not intended for safety-of-life applications but rather to provide high-accuracy corrections to QZSS availability enhancement signals and GPS signals. L1Sb, transmitted only from the geostationary satellites, will be an ICAO-compliant safety-of-life signal, so that the QZSS geostationary satellites eventually replace the MTSAT satellites for transmitting Japan's SBAS signals [5]. The L6 signal is an original QZSS design, using the same carrier frequency and spreading modulation as the Galileo E6 signal, but with distinctly different spreading code data message structure, and data message modulation. The L5s signal's design is expected to be similar to the design of the L5 SBAS signal; and presumably will also be an ICAO-compliant safety-of-life signal for SBAS.

QZSS has also introduced the Indoor MEssaging System (IMES) as an approach for deep indoor navigation where satellite signals have limited utility. IMES employs very low power beacons, deployed inside buildings, each transmitting its location. Since the signal is very low power, a receiver that reads the data message must be within several (typically 10) meters of the transmitter, and thus knows its location is approximately the same as that of the beacon. IMES uses signals similar to the C/A code signal (i.e., BPSK-R(1) spreading modulation with carrier frequency near 1575.42 MHz), but with a distinct spreading code drawn from the same Gold code family used for C/A code and different data message structure. Consequently IMES functionality can be added to a satnav receiver merely through additional message processing, since the same receiver front end, acquisition, tracking, and data demodulation processing hardware can be employed. Since IMES is not a satnav signal, it is not described in detail here, but it does offer a potential extension for modified satnav receivers to operate in deep indoor applications, if interference concerns can be addressed.

Since the designs of the availability enhancement signals are so similar to those of corresponding GPS signals, readers can refer to Chapter 7 for many of the details. However, QZSS signals do have some common characteristics summarized here that are different from GPS. Since QZSS does not transmit a P(Y) code signal, some bits in the C/A code signal data messages associated with P(Y) code signal status are permanently set to default values. Also, group delay values describe timing relationships among the QZSS signals, and do not include the relationship to P(Y) code signals since QZSS has none. QZS ephemeris values are relative to a different orbital radius, and have different reference values and parameter ranges. The almanac reference eccentricity, rate of change in right ascension of ascending node, inclination, and argument of perigee, are different. Eccentricity values can be larger than those for GPS. Ephemeris and clock parameters are referenced to the QZS antenna phase center, as is the almanac data. Signal health is represented differently and provided for QZSS satellites and also for GPS satellites. The QZSS curve fit interval is also different than GPS's for clock corrections and ephemeris. The navigation message correction table (NMCT) also includes information for QZSs, and the NMCT user algorithm is different. UTC parameters also describe the relationship between QZSST and GPS time. Ionospheric parameters for QZSS satellites are optimized for Japan and its surrounding area, rather than the world. SV health is judged at QZSS monitoring stations, which monitor all QZSS signals continuously. GNSS-to-GPS time offset (GGTO) parameter IDs differ. Also, the QZSS L1-C/A code signal message length may extend beyond 25 pages (12.5 minutes) in order to provide parameters for multiple satnav systems. The following subsections summarize each of the QZSS signals.

12.3.1 QZSS L1-C/A Signal

Table 12.1 summarizes key characteristics of the L1-C/A code signal, using the list of characteristics introduced in Section 3.3. This signal is virtually identical to the GPS C/A code signal, enhancing interoperability. The data rate shown here includes overhead bits used for framing, synchronization, and parity; the actual information data rate is somewhat lower.

Other characteristics are found in Section 7.3.1, except for the detailed message characteristics found in IS-QZSS [2].

12.3.2 QZSS L1S Signals

The L1S signal was previously known as the L1-SAIF signal, where SAIF stood for submeter class augmentation with integrity function. The two L1S signals, L1Sa and

TABLE 12.1. Summary of QZSS L1-C/A Signal Characteristics

Carrier frequency	L1: 1575.42 MHz = 154 × 10.23 MHz
Minimum received power	−158.5 dBW
Polarization	RHCP
Multiple access	CDMA
Spreading modulation and transmit bandwidth	BPSK-R(1), ±15.345 MHz
Spreading codes	Length 1023-bit Gold codes, duration 1 ms
Data message structure	NAV
Data rate	50 bps
Data message error correction and detection	Error detection using extended (32,26) Hamming code
Data modulation	50 sps biphase modulation
Pilot and data components	100% power data
Overlay code	None
Multiplexing with other signals	Varies with QZSS satellite block, phase relationship to other QZSS L1 signals not defined

L1Sb, have identical designs, except L1Sb is designed and operated to satisfy ICAO standards for safety-of-life integrity requirements. Consequently, L1Sb signals will only be transmitted by GEO satellites, while L1Sa signals will be transmitted by all satellites. Except for the messages, the signal designs are based on the L1 SBAS signal described in Section 8.3.1. Table 12.2 summarizes the characteristics of the L1S signals.

The L1S signals employ three different preambles designated Pattern A, Pattern B, and Pattern C, all of which are forward error encoded and protected by the CRC.

TABLE 12.2. Summary of L1S Code Signal Characteristics

Carrier frequency	L1: 1575.42 MHz = 154 × 10.23 MHz
Minimum received power	−158.5 dBW
Polarization	RHCP
Multiple access	CDMA
Spreading modulation and transmit bandwidth	BPSK-R(1), ±15.345 MHz
Spreading codes	Length 1023-bit Gold codes, duration 1 ms, from the GPS C/A code family
Data message structure	L1 SBAS
Data rate	250 bps
Data message error correction and detection	1/2-rate convolutional coding with constraint length 7, no interleaving, with 24-bit CRC
Data modulation	500 sps biphase modulation
Pilot and data components	100% power data
Overlay code	None
Multiplexing with other signals	Varies with QZS satellite block, phase relationship to other QZSS L1 signals not defined

TABLE 12.3. L1S Message Types

Message Type Number	Message Contents
0	Test mode
1	PRN mask assignments
2–5	Fast corrections and user differential range error (UDRE)
6	Integrity data
7	Fast correction degradation factor
10	Degradation parameters
11	Reserved
12	Timing information
18	Ionospheric grid point (IGP) masks
24	Mixed fast /long-term corrections
25	Long-term corrections
26	Ionospheric delay and grid ionosphere vertical error (GIVE)
28	Clock–ephemeris covariance matrix
40–51	Reserved for application demonstrations using L1S unique messages
52	Tropospheric grid point (TGP) mask
53	Tropospheric delay corrections
54, 55	Atmospheric delay corrections
56	Inter signal bias corrections
57	Reserved for orbital corrections
58	QZS ephemeris
59	QZSS almanac data
60	Regional information/maintenance schedule
62	Internal test message
63	Null message

Messages rotate through the different patterns in sequence. Different information is provided in messages having different patterns, allowing more satellites and ionospheric grid points to be described.

Table 12.3 summarizes the L1S message types. There is a significant overlap with L1 SBAS message types, but the L1S messages provide additional corrections in accord with the GPS performance improvement function, and also support more satellites.

12.3.3 QZSS L1C Signal

QZSS is the first system to transmit the L1C signal; its Michibiki satellite transmits it with a BOC(1,1) spreading modulation while subsequent satellites will use the same combination of TMBOC and BOC(1,1) employed by GPS L1C. The QZSS L1C signal is very similar to the GPS L1C signal, except that fields in the data message have some different meanings as needed to represent QZSS-unique characteristics like orbits. Table 12.4 summarizes the QZSS L1C characteristics.

Other characteristics are found in Section 7.3.6, except for detailed message characteristics found in IS-QZSS [2].

TABLE 12.4. Summary of QZSS L1C Signal Characteristics

Carrier frequency	L1: 1575.42 MHz = 154 × 10.23 MHz
Minimum received power	−157.0 dBW
Polarization	RHCP
Multiple access	CDMA
Spreading modulation and transmit bandwidth	TMBOC(6,1,4/33) pilot, BOC(1,1) data, ±15.345 MHz
Spreading codes	Length 10,230-bit Weil-based codes, duration 10 ms
Data message structure	CNAV2
Data rate	50 bps
Data message error correction and detection	1/2-rate LDPC coding, block interleaving, and 24-bit CRC
Data modulation	100 sps biphase modulation
Pilot and data components	75% power pilot, 25% power data, in same carrier phase
Overlay code	1800-bit pilot overlay codes, 100 bps, different for each satellite; segments of m-sequences and Gold codes
Multiplexing with other signals	Varies with QZSS satellite block, phase relationship to other QZSS L1 signals not defined

12.3.4 QZSS L6 Signal

The QZSS L6 signal, formerly known as the LEX or experimental signal, has a unique design. It has a very high data rate, presumably to support precision users, and employs 8-ary code shift keying (CSK) data modulation [6] rather than the BPSK data modulation employed by every other signal described in this book.

Table 12.5 summarizes key characteristics of the QZSS L6 signal. The overlay code on the pilot component is a simple square wave that doubles the period of the pilot component in order to suppress spectral lines.

Figure 12.4 portrays generation of the L6 signal. Construction of the 250 eight-bit sps message is as follows. Each message begins with a 32-bit preamble (00011010110011111111110000011101). Next comes the PRN of the satellite transmitting the message, represented in eight bits. Next is an eight-bit message type, followed by a one bit alert flag. Since definition of message types and use of the alert flag are still experimental at this time, these details are not provided here. These 49 bits precede the 1695-bit payload, totaling 1744 bits, or 218 eight-bit words.

The R-S code is implemented as a systematic code, so the encoded symbols consist of the input 1744 bits followed by 256 parity bits computed as follows. The input to the encoder is 72 zeros followed by the eight-bit satellite PRN, the eight-bit message type, the one-bit alert flag, and the 1695-bit payload, totaling 1784 bits or 223 eight-bit symbols. These symbols are input to the R-S encoder, and the resulting 32 eight-bit parity symbols are added to the 218 eight-bit words (1744 bits) defined in the preceding paragraph, producing the 250 eight-bit symbol (2000-bit) data message that is transmitted in 1 second.

TABLE 12.5. Summary of QZSS L6 Signal Characteristics

Carrier frequency	L6: 1278.75 MHz = 125 × 10.23 MHz
Minimum received power	−155.7 dBW
Polarization	RHCP
Multiple access	CDMA
Spreading modulation and transmit bandwidth	BPSK-R(5), ±19.5 MHz
Spreading codes	Pilot component: Long code, a 1,048,575-bit segment of a small Kasami sequence, duration 0.410 s
	Data component: Short code, a 10,230-bit segment of a small Kasami sequence, duration 4 ms
Data message structure	L6
Data rate	1744 bps
Data message error correction and detection	Reed–Solomon (255,223)
Data modulation	8-ary CSK modulation, 250 sps or 2000 binary sps
Pilot and data components	50% power pilot, 50% power data, time multiplexed
Overlay code	2-bit pilot overlay code on the pilot component, 1/0.41 bps, (0,1) for all satellites
Multiplexing with other signals	None required

The L6 data message structure is designed to be flexible, with multiple message types and no predetermined commutation of the message types. The message types and use of alert flag are not yet finalized; experiments are being conducted with Michibiki to determine their final design. Consequently, the preliminary details are not described here but can be found in IS-QZSS.

The spreading code generators, described below, produce the long code and the short code, each at a rate of 2.5575 Mchip/s. Each eight-bit symbol output from the

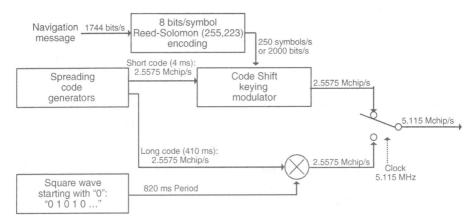

Figure 12.4. L6 Signal Generation [2]

QZSS SIGNALS

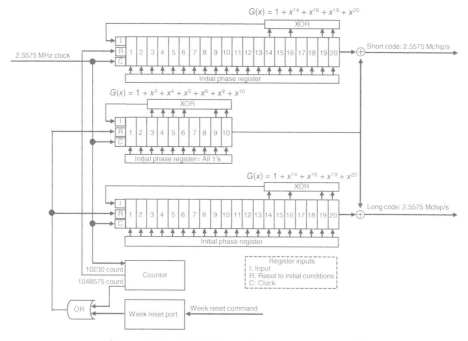

Figure 12.5. QZSS L6 Spreading Code Generation [2]

R-S encoder is used to circularly shift the short code by the corresponding number of bits in the CSK modulator. The result is encoding of an 8-ary symbol every 4 ms, for a data rate of (eight binary symbols)/(0.004 seconds) = 2000 binary sps. The long code is overlaid with the square wave, then the CSK-modulated short code is spreading symbol-by-spreading symbol time multiplexed with the overlaid long code to produce the baseband signal. Since the spreading symbols are rectangular, they do not need to be shown explicitly in the figure.

Both spreading codes are segments of Kasami sequences, defined in Section 9.3.3. Spreading code generation is illustrated in Figure 12.5. Two identical length-20 m-sequence generators are used, one for the long m-sequence and one for the short m-sequence used in generating a Kasami sequence. Different PRNs employ different initial phase values for these m-sequence generators. The same length-10 m-sequence generator, which produces a decimated version of the longer sequences, is used for both the long code and the short code, with all m-sequence generators clocked at 2.5575 MHz. Both code generators are reset at the beginning of each week. In addition, the short code generator is short cycled every 10,230 bits (4 ms) while the long code generator is short cycled every 1,048,575 bits (0.410 seconds), so that every other reset of the long code generator occurs at the same time as a reset of the short code generator, with 2 resets of the long code generator every 205 resets of the short code generator.

It appears the primary role of the QZSS L6 signal will be to serve high precision users in applications like surveying, machine control, and precision farming. The

TABLE 12.6. Summary of QZSS L2C Signal Characteristics

Carrier frequency	L2: 1227.60 MHz = 120 × 10.23 MHz
Minimum received power	−158.5 dBW
Polarization	RHCP
Multiple access	CDMA
Spreading modulation and transmit bandwidth	BPSK-R(1), ±15.345 MHz
Spreading codes	Pilot component: Length 7,67250-bit L2CL code, duration 1.5 seconds
	Data component: Length 10,230-bit L2CM code, duration 20 ms
Data message structure	CNAV
Data rate	25 bps
Data message error correction and detection	1/2-rate convolutional coding with constraint length 7 with 24-bit CRC
Data modulation	50 sps biphase modulation
Pilot and data components	50% power pilot, 50% power data, time multiplexed
Overlay code	None
Multiplexing with other signals	None required

data message will provide wide area differential corrections with greater accuracy than SBAS, enabling real-time precise point positioning (PPP) described in Section 20.7. Experimental messages are also defined that provide accuracy enhancement for Galileo and GLONASS signals as well as GPS and QZSS. Since the L6 signal is biphase and the only signal on this carrier, there is an opportunity to add another signal in phase quadrature. However, no public announcement has been made concerning such a signal.

12.3.5 QZSS L2C Signal

The QZSS L2C signal's characteristics are very similar to those of the GPS L2C signal described in Section 7.3.5. Table 12.6 summarizes key characteristics of the QZSS L2C signal. Table 12.7 lists QZSS L2C message types. These are very similar to those for GPS L2C, but message types 28, 46, 47, 49, 51, and 53 are added to retransmit the following GPS messages, respectively: reduced GPS almanac, ionospheric parameters, reduced GPS almanac, UTC parameters, GGTO, and midi almanac.

12.3.6 QZSS L5 Signal

The QZSS L5 signal's characteristics are very similar to those of the GPS L5 signal described in Section 7.3.6. Table 12.8 summarizes key characteristics of the QZSS L5 signal. Table 12.9 lists QZSS L5 message types. These are very similar to those for QZSS L2C and to GPS L5, but message types 28, 46, 47, 49, 51, and 53 are added to retransmit the following GPS messages, respectively: reduced GPS almanac, ionospheric parameters, reduced GPS almanac, UTC parameters, GGTO, and midi almanac.

TABLE 12.7. QZSS L2C Message Types

Message Type Number	Message Contents
0	Default
10	Ephemeris 1 and health
11	Ephemeris 2 and health
12, 28	Reduced almanac
13	SV clock performance enhancement
14	Ephemeris performance enhancement
15	Text message
30, 46	Clock correction, ionosphere correction, group delay
31, 47	Clock correction, reduced almanac
32	Clock correction, Earth orientation parameters
33, 49	Coordinated universal time parameters
34	SV clock performance enhancement
35, 51	GPS/GNSS time offsets
37, 53	Midi almanac

TABLE 12.8. Summary of QZSS L5 Signal Characteristics

Carrier frequency	L5: 1176.45 MHz = 115 × 10.23 MHz
Minimum received power	−154.0 dBW
Multiple access	CDMA
Polarization	RHCP
Spreading modulation and transmit bandwidth	QPSK-R(10), 24 MHz
Spreading codes	Pilot component: Length 767,250-bit L2CL code, duration 1.5 s
	Data component: Length 10,230-bit L2CM code, duration 20 ms
Data message structure	CNAV
Data rate	50 bps
Data message error correction and detection	1/2-rate convolutional coding with constraint length 7 with 24-bit CRC
Data modulation	100 sps biphase modulation
Pilot and data components	50% power pilot, 50% power data, phase quadrature multiplexed with pilot component on the in-phase component
Overlay codes	10 bit, 1 kbps Neuman–Hoffman code (1111001010) on I5, 20 bit, 1 kbps Neuman–Hoffman code (00000100110101001110) on Q5. Same on all satellites
Multiplexing with other signals	Not needed

TABLE 12.9. QZSS L5 Message Types [2]

Message Type Number	Message Contents
0	Default
10	Health, user range accuracy (URA), ephemeris 1
11	Ephemeris 2
12, 28	Reduced almanac
13	SV clock performance enhancement
14	Ephemeris performance enhancement
15	Text message
30, 46	Clock correction, ionosphere correction, group delay
31, 47	Clock correction, reduced almanac
32	Clock correction, Earth orientation parameters
33, 49	Clock correction, coordinated universal time parameters
34	Clock correction, performance enhancement
35, 51	Clock correction, GPS/GNSS time offsets
37, 53	Clock correction, midi almanac

12.3.7 QZSS L5S Signal

No information is currently publicly available concerning the QZSS L5S signal. However, its design is probably similar to that of the SBAS L5 signal described in Section 8.3.2. If that is the case, it remains to be seen whether this signal will be used as a performance enhancement signal or an ICAO-complaint safety-of-life signal.

12.4 SUMMARY

QZSS provides an augmentation and enhancement to GPS and potentially other global systems as it grows into a standalone regional system serving the Asia-Pacific region. It bridges original and modernized GPS by providing all the legacy GPS civil signals and all the modernized GPS civil signals. It also connects to Galileo by providing an L6 signal similar to the Galileo E6 signal, and connects to SBAS with SBAS-like signals on L1 and L6. The resulting eight signals on four carrier frequencies are shown in Figure 12.6, for spreading codes modeled as ideal and long.

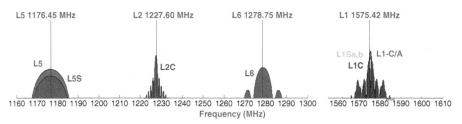

Figure 12.6. PSDs of QZSS Signals

TABLE 12.10. Summary of QZSS Signals

Signal Name	Carrier Frequency (MHz)	Spreading Modulation	Function
L1-C/A Code	1575.42	BPSK-R(1)	Availability enhancement
L1C	1575.42	MBOC(6,1,1/11)	Availability enhancement
L1Sa	1575.42	BPSK-R(1)	Performance enhancement
L1Sb	1575.42	BPSK-R(1)	Performance enhancement
L6	1278.75	BPSK-R(5)	Performance enhancement
L2C	1227.60	BPSK-R(1)	Availability enhancement
L5	1176.45	QPSK-R(10)	Availability enhancement
L5S	1176.45	BPSK-R(10)	Performance enhancement

Key signal characteristics are summarized in Table 12.10.

REFERENCES

1. "QZ-Vision," http://qz-vision.jaxa.jp/, accessed 3 July 2013.
2. Japan Aerospace Exploration Agency, "Quasi-Zenith Satellite System Navigation Service: Interface Specification for QZSS, IS-QZSS V1.6, November 2014," available at http://qz-vision.jaxa.jp/USE/is-qzss/index_e.html, accessed 15 March 2015.
3. "Recommendation ITU-R M.1787-1; Questions ITU-R 217-2/4 and ITU-R 288/4," Document 4C/TEMP/62, Preliminary draft revision of recommendation ITU-R M.1787-1, Description of systems and networks in the radionavigation-satellite service (space-to-Earth and space-to-space) and technical characteristics of transmitting space stations operating in the bands 1 164-1 215 MHz, 1 215-1 300 MHz and 1 559-1 610 MHz, International Telecommunications Union, Radiocommunications Study Groups.
4. "InsideGNSS: Japan Awards Contracts for QZSS Space, Ground Segments," http://www.insidegnss.com/node/3510, accessed 3 July 2013.
5. H. Tashiro, "MSAS Status," Civil Aviation Bureau of Japan Presentation given to IWG#28, April 2015.
6. G. M. Dillard, M. Reuter, J. Zeidler, and B. Zeidler, "Cyclic Code Shift Keying: A Low Probability of Intercept Communication Technique", *IEEE Transactions on Aerospace and Electronic Systems,* Vol. 39, No. 3, 2003, pp. 786–798.

13

INDIAN REGIONAL SATELLITE SYSTEM

The Indian Regional Satellite System (IRNSS) is the newest of all currently deployed or announced satnav systems. It is intended to provide Indian-controlled regional satnav service to the Indian subcontinent and surrounding regions up to 1500 km from India's border.

IRNSS is being developed by the Indian Space Research Organization (ISRO), the part of the Indian government responsible for development of space technology and its application to various national tasks [1]. After being discussed for a number of years, IRNSS promises to become operational by 2016.

IRNSS comprises a space segment, a ground segment, and a user segment. When complete, the planned constellation will have a minimum of seven satellites, with three in geostationary orbit and four in inclined non-geostationary orbits. IRNSS will offer a Standard Positioning Service (SPS) for open, unrestricted use as well as a Restricted Service (RS) for use by authorized users.

This chapter provides a brief history of IRNSS and summarizes current plans in Section 13.1, followed by a description of IRNSS. Section 13.3 provides an overview of IRNSS signals, followed by a summary of the chapter in Section 13.4.

13.1 IRNSS HISTORY AND PLANS

The first public disclosure of IRNSS plans occurred in 2008 [2], marking the initiation of IRNSS satellite design and the deployment of the IRNSS ground segment. From the start, it had been determined that IRNSS' signals would use S band as well as L band, presumably in an attempt to reduce ionospheric effects, which can be severe near India.

By 2013, ISRO had deployed the IRNSS ground segment and completed the design of the first IRNSS satellite, which was launched into an inclined geosynchronous orbit in 2013 and designated IRNSS-1A. Successful checkout of this satellite made IRNSS the first system to use S band signals for passive trilateration. IRNSS-1A is in a geosynchronous orbit with 29° inclination, crossing the equator at 55°E. More recently [3] a second IRNSS satellite, designated IRNSS-1B, was launched in 2014 into an orbit with the same ground track as IRNSS-1A, but antipodal phasing. The first IRNSS GEO satellite, IRNSS-1C, was also launched in 2014. ISRO has announced plans to launch four more IRNSS satellites in 2015—two more IRNSS GEO satellites and two more satellites in inclined orbit but with equator crossing at 111.75°E. If these plans are implemented, a complete seven-satellite constellation will be operational by 2016.

An ICD for the IRNSS OS signals was issued in 2014 [4] describing the open service signals and user algorithms for employing these signals.

13.2 IRNSS DESCRIPTION

IRNSS is a regional satnav system developed to service the Indian subcontinent region. The IRNSS space segment will consist of a minimum of seven satellites, with three geostationary satellites and four satellites in inclined geostationary orbits. Ground tracks for a seven-satellite constellation are shown in Figure 13.1, with the non-geostationary satellites in orbits having 29° inclination. All satellites have an orbital radius of 42,164 km, and orbital period of one sidereal day. It is possible that, in future, geostationary IRNSS satellites might also host SBAS transponders in order to avoid the expense of leasing SBAS transponders on commercial satellites for GAGAN.

The IRNSS satellites have a navigation payload and two ranging payloads. The navigation payload transmits S band and L band signals as described in Section 13.3. One ranging payload uses a C band transponder for accurate determination of the range of the satellite from the ground. The other ranging payload consists of corner cube retroreflectors for laser ranging from the ground. Rubidium clocks are used for timekeeping on the satellites. The satellites' dry mass is 614 kg and liftoff mass is 1425 kg [5]. The bus, not including solar cells, is approximately a cube having 1.5 m sides. The solar cells are reported to generate 1660 W, and satellite design life is 10 years.

The IRNSS ground segment, shown in Figure 13.2, is built around the ISRO Navigation Center at Byalalu, where satellite tracking data is processed and navigation messages are generated. Fourteen IRNSS range and integrity monitoring stations perform continuous one-way ranging on the IRNSS navigation signals transmitted at S band and L band, and also determine integrity of the IRNSS navigation signals. In addition, four IRNSS CDMA ranging stations make range measurements on the IRNSS satellites

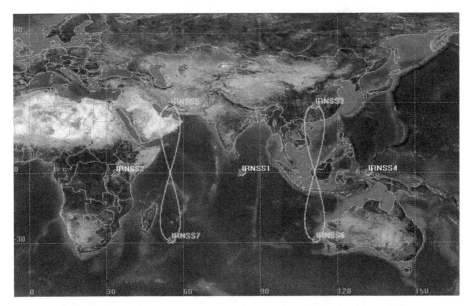

Figure 13.1. Planned Ground Tracks of Seven IRNSS Satellites [2]

using the C band transponders. The IRNSS network timing center in Byalalu generates, maintains, and distributes IRNSS time. The spacecraft control facility commands the IRNSS satellites through TT&C networks, and also uplinks the navigation message parameters computed at the ISRO Navigation Center. In future, laser ranging stations will be added to the ground segment as well.

The IRNSS geodetic datum is WGS-84. The start epoch of IRNSS System Time, when IRNSS time was initialized to zero, was the time at 23:59:47 UTC on August 21, 1999 (00:00:19 TAI on August 22, 1999), the same as the most recent rollover of GPS time. IRNSS time resets every 1024 weeks, or approximately 19.6 years. Like GPS time, IRNSS System Time is continuous and does not include leap seconds. However, it was initialized to 13 leap seconds at its start epoch.

13.3 IRNSS SIGNALS

IRNSS will broadcast signals at two or three carrier frequencies, using the frequency plan portrayed in Figure 13.3. IRNSS uses the same terminology as GPS for its carrier frequencies and frequency bands in L band: L1 band's carrier frequency is 1575.42 MHz, and L5 band's carrier frequency is 1176.45 MHz. IRNSS uses an additional frequency band between 2483.5 MHz and 2500 MHz, with carrier frequency 2492.028 MHz, that is simply called S band. The first IRNSS satellites only transmit in S band and L5 band, but some information provided by IRNSS suggests an interest in future use of L1 band, so that band is also shown in the figure.

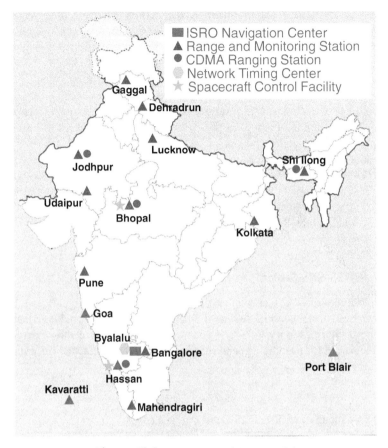

Figure 13.2. IRNSS Ground Segment [5]

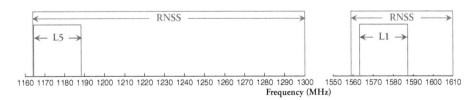

Figure 13.3. IRNSS Frequency Bands

All signals use RHCP. Minimum received signal power is specified out of a 0 dBi circularly polarized antenna located near the ground, for satellites at elevation angles of 5° or more, with no excess atmospheric attenuation loss included. Since neither excess atmospheric loss nor polarization loss is included in the IRNSS link budget conventions for minimum received power, IRNSS minimum received power values would be reduced by as much as 1.5 dB if the GPS conventions were employed.

IRNSS' use of S band is an innovation, and should provide better performance than L band signals under a range of ionospheric conditions. As discussed in Section 3.2.2, ionospheric refraction and dispersion decrease with the square of carrier frequency, meaning that ionospheric distortions of IRNSS S band signals could be a factor of $(1575.42/2492.028)^2 \approx 0.4$ relative to signals at L1.

All IRNSS OS signals are defined in a single ICD [4]. Identical signal structures are employed at S band and L5. The OS signals are described in Section 13.3.1 while what is known about the RS signals is described in Section 13.3.2.

13.3.1 IRNSS SPS Signals

Table 13.1 summarizes key characteristics of the IRNSS SPS signals, using the list of characteristics introduced in Section 3.3. The signals, denoted L5-SPS and S-SPS, have strong similarities to the GPS C/A code signal. The data rate shown here includes overhead bits used for framing, synchronization, and parity; the actual information data rate is somewhat lower.

TABLE 13.1. Summary of IRNSS SPS Signal Characteristics

Carrier frequency	L5-SPS L5: 1176.45 MHz = 115 × 10.23 MHz
	S-SPS S: 2492.028 MHz = 2436 × 1.023 MHz
Minimum received power	L5-SPS −159.0 dBW
	S-SPS −162.3 dBW
Polarization	RHCP
Multiple access	CDMA
Spreading modulation and transmit bandwidth	BPSK-R(1), ±12.0 MHz (L5), ±8.25 MHz (S)
Spreading codes	Length 1023-bit Gold Codes, duration 1 ms
Data message structure	SPS
Data rate	25 bps
Data message error correction and detection	1/2-rate convolutional coding with constraint length 7, block interleaving, with 24-bit CRC
Data modulation	50 sps biphase modulation
Pilot and data components	100% power data
Overlay code	None
Multiplexing with other signals	Interplexed on in-phase channel, along with pilot component of RS signal, in-phase quadrature to data component of RS signal

TABLE 13.2. SPS Signal Spreading Code Definitions [4]

PRN ID	SV Location	L5-SPS		S-SPS	
		Initial Condition for G2 Register	First 10 Chips in Octal2	Initial Condition for G2 Register	First 10 Chips in Octal2
1	55°E	1110100111	130	0011101111	1420
2	55°E	0000100110	1731	0101111101	1202
3	83°E	1000110100	0713	1000110001	0716
4	111.75°E	0101110010	1215	0010101011	1524
5	111.75°E	1110110000	0117	1010010001	0556
6	32.5°E	0001101011	1624	0100101100	1323
7	131.5°E	0000010100	1753	0010001110	1561

The 1023-bit Gold codes use the same shift registers as described for GPS C/A code signal in Section 7.3.1, with the more general alternative Gold code generator portrayed in Figure 7.9. Table 13.2 shows the PRN IDs, their assignments based on space vehicle (SV) location, and the initial conditions for the G2 registers. Interestingly, different spreading codes are used at the two different carrier frequencies.

The SPS navigation message is 2400 symbols long, having 48 seconds duration. It is divided into four subframes, each 600 symbols long with duration 12 seconds. Subframes 1 and 2 always provide the same navigation data (week number, clock correction and ephemeris, satellite and signal health status, user range accuracy, and total group delay) while subframes 3 and 4 provide variable system data such as almanac, ionosphere data, time offset with respect to UTC and other satnav systems, text messages, differential corrections, and Earth orientation parameters).

All subframes contain 8 bits of TLM (telemetry, reserved for future), 17 bits of time of week count (TOWC) that describes the number of counts from the beginning of the week to the start of the next subframe, 1 bit each for the Alert flag (indicating that use of the navigation data is at the user's risk) and AutoNav flag (set to indicate that the satellite is operating during the 7 day AutoNav period during which it gets no uplink from the ground and operates with stored ephemeris and clock data), 2 bits of subframe ID, and 1 spare bit. 232 bits are available for the data payload, followed by a 24-bit CRC and a 6-bit tail to flush the decoder. All these bits are FEC encoded and interleaved to produce 584 encoded symbols, and then preceded by an unencoded 16-bit synchronization word, 1111110010010000.

Subframes 3 and 4 can contain the different message types listed in Table 13.3, each with a 220-bit payload. The payload is preceded by a 6-bit message type, and 6 bits are appended to provide the PRN ID of the satellite, for a total of 232 bits as in subframes 1 and 2.

IRNSS provides high-fidelity ionospheric data for the region of the Earth near the Indian subcontinent. A total of 90 ionospheric grid points (IGPs) are defined at 350 km altitude over the Indian subcontinent region. The ionospheric correction for L5 signals is provided in each message type 5 for 15 of the IGPs, requiring six messages to contain

TABLE 13.3. IRNSS Message Types [4]

Parameter	MSG ID
Ionospheric grid parameters for 15 grid points (1 of 6 regions)	5
Almanac parameters	7
UTC and time sync parameters with respect to GPS	9
EOP and ionosphere coefficients	11
Differential corrections for one satellite	14
Text message	18
UTC and time sync parameters with respect to GNSS	26
Null message	0
Reserved for future	1–4, 6, 8, 10, 12, 13, 15–17, 19–25, 27–63

all of the information. At each point, the vertical delay estimate is provided, allowing a receiver to calculate the delay for satellites at other elevation angles through the same pierce point. Also, the delay values can be scaled by the square ratio of center frequencies to obtain the smaller delays at S band.

13.3.2 IRNSS RS Signals

Less information is known about the IRNSS RS signals, but some can be inferred from Reference 4 and is summarized in Table 13.4.

The total RS power is almost 5 dB larger than the power of the SPS signal.

TABLE 13.4. Summary of IRNSS RS Signal Characteristics

Carrier frequency	L5-SPS	L5: 1176.45 MHz = 115 × 10.23 MHz
	S-SPS	S: 2492.028 MHz = 2436 × 1.023 MHz
Minimum received power	L5-SPS	−159.0 dBW
	S-SPS	−162.3 dBW
Polarization	RHCP	
Multiple access	CDMA	
Spreading modulation and transmit bandwidth	BOC(5,2) pilot and data, ±12.0 MHz (L5), ±8.25 MHz (S)	
Spreading codes	Restricted	
Data message structure	Restricted	
Data rate	Restricted	
Data message error correction and detection	Restricted	
Data modulation	50 sps biphase modulation	
Pilot and data components	67% power data, 33% power pilot	
Overlay code	Restricted	
Multiplexing with other signals	Pilot component interplexed on in-phase channel, along with SPS signal and data component in phase quadrature	

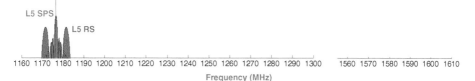

Figure 13.4. PSDs of IRNSS Signals in L Band

TABLE 13.5. Summary of IRNSS Signals

Signal Name	Carrier Frequency (MHz)	Spreading Modulation	Service
S SPS	2492.028	BPSK-R(1)	Standard positioning
S RS	2492.028	BOC(5,2)	Restricted
L5 SPS	1176.45	BPSK-R(1)	Standard positioning
L5 RS	1176.45	BOC(5,2)	Restricted

13.4 SUMMARY

IRNSS is an emerging standalone regional system that could contribute significantly to satnav service over and near the Indian subcontinent. Its innovative use of S band, coupled with the broadcast of detailed ionospheric models, should provide better performance under a range of ionospheric conditions. Currently planned IRNSS signals in L band are shown in Figure 13.4, for spreading codes modeled as ideal and long. Signals having the same PSDs and designs are also transmitted at S band. Table 13.5 summarizes the key IRNSS signal characteristics.

REFERENCES

1. "Indian Space Research Organization: Satellite Navigation," http://www.isro.gov.in/spacecraft/satellite-navigation, accessed 20 September 2015.
2. P. K. Jain, Indian Satellite Navigation Programme: An Update, 45th Session of S&T Subcommittee of UN-COPUOS, Vienna, 11–22 February 2008.
3. W. Graham, "Indian PSLV Lofts IRNSS-1C Satellite," nasaspaceflight.com, 2014, http://www.nasaspaceflight.com/2014/10/indian-pslv-irnss-1c-satellite/, accessed 4 January 2015.
4. "Indian Regional Navigation Satellite System Signal in Space ICD for Standard Positioning Service," ISRO-IRNSS-ICD-SPS-1.0, June 2014.
5. "PSLV-C22/IRNSS/1A," http://www.isro.gov.in/sites/default/files/pdf/pslv-brochures/PSLVC22.pdf, accessed 20 September 2015.

PART III

RECEIVER PROCESSING

Receiver processing extracts from the received satnav signals the information needed to estimate position, velocity, and time (PVT). Such processing techniques are mature for the original GPS and GLONASS signals, where decades of innovation have resulted in impressive receiver capabilities. Further advances are limited in many cases by basic characteristics of these signals and systems, and in other cases by the evolving but still limited ability of semiconductor technology to implement known techniques within size, cost, and power constraints. Augmenting and assisting satnav has helped address some of these limitations. A vast number of resources, including those identified in Chapter 1 of this book, provide excellent references for understanding and designing processing for the original GPS signals.

Many applications, services, and users of satnav have apparently insatiable demands for availability (including accessing satnav in difficult environments such as urban canyons and indoors) and accuracy. It is sometimes said that availability and accuracy are addictive—as observed performance improves, expectations and demands increase. New signals and systems described in Part II can help meet these demands, but only to the extent that receiver processing exploits these signals and systems.

In spite of the advances already achieved, many challenges remain in satnav processing, including those listed in Table III.1. Multipath, or multiple paths, involves reflected signals, typically from terrestrial objects, that enter the receive antenna (see Chapter 22). Since these reflected signals travel additional distance, they are received with excess delays relative to the direct path signal from the satellite. If the direct path signal is also received, the multipath signals can distort measurements made in receiver processing. If the direct path signal experiences shadowing, or blockage, a multipath signal may be mistaken for the (absent) direct path signal, introducing a significant measurement error. Signals with narrower bandwidths experience greater measurement errors from multipath, and fewer satellites in view make it more difficult for a receiver to discern errors arising from a combination of multipath and shadowing. Narrowband interference (see Chapter 21) at specific frequencies can degrade measurements and data message demodulation performance. Signals with narrower bandwidths and repeating spreading codes within a data bit are more susceptible to narrowband interference. Ionospheric errors, discussed in Chapter 3, can be a dominant source of measurement errors.

Engineering Satellite-Based Navigation and Timing: Global Navigation Satellite Systems, Signals, and Receivers, First Edition. John W. Betz.
© 2016 The Institute of Electrical and Electronics Engineers, Inc. Published 2016 by John Wiley & Sons, Inc.

TABLE III.1. Challenges in Satnav Receiver Processing

Multipath and shadowing
Narrowband interference
Ionospheric errors
Limited satellite visibility
Fragile carrier tracking
Fragile code tracking
Detecting and correcting data message errors
Disambiguating carrier phase
Long time to initial fix
Achieving bit synchronization and frame synchronization
Avoiding false correlations
Accommodating self-interference

Using of worldwide ionospheric models broadcast in the messages from global satnav systems removes only approximately 50% of the ionospheric errors; instead, ionospheric errors can be substantially reduced by measuring the delay using signals on two carrier frequencies (Chapter 20), and by using more accurate ionospheric models provided by augmentations (Chapter 23). Limited satellite visibility, especially when only a portion of the sky is in view due to obstructions, can prevent a receiver from having enough measurements to calculate position and time, or yield poor DOP (see Chapter 6) that causes large errors in position and time measurements. Poor satellite visibility also can limit the number of redundant measurements that can be used to identify and remove flawed measurements (see Chapter 20). Fragile carrier tracking—caused by a combination of oscillator phase noise, uncompensated dynamics, lower power signals, and signals having only a data component (see Chapter 3), requiring use of more fragile Costas loops (Chapter 18), particularly with high data rates—can cause loss of lock in carrier tracking loops (see Chapter 18). Fragile code tracking—caused by uncompensated dynamics, lower power signals, the need to track data components with high data rates—can cause loss of lock in code tracking loops (see Chapter 19). Detecting and correcting data message errors is more difficult with weak error control coding, absence of CRC, and lower power signals. Correcting carrier phase ambiguities is more of a challenge when using signals only on one carrier frequency, and with fragile carrier tracking. Long time to first fix is dominated, in many cases, by the latency associated with reading clock and ephemeris data from the data message. Achieving bit synchronization and frame synchronization is more complicated when spreading codes repeat within a data bit, and when spreading codes or overlay codes repeat within a message frame. False correlations occur when the desired signal is weak and interfering signals have high power; the problem is exacerbated by short spreading codes or other spreading code designs producing higher crosscorrelation sidelobes. Self-interference by signals of the same or similar designs can be significant for some original signals with narrow bandwidths and repeating spreading codes, as discussed in Chapter 5, but is not as significant for other signals.

The emergence of new signal designs and new satnav systems enables receiver processing to move to a higher level of capability. As described in Chapter 3 and made

TABLE III.2. New Signal Features, with Challenges They Potentially Address and Exacerbate

Features Found in New Signal Designs	Challenges Potentially Addressed	Challenges Potentially Exacerbated
Wider bandwidth spreading modulations	Multipath and shadowing	
	Narrowband interference	
	Fragile code tracking	
	False correlations	
	Self-interference	
Longer spreading codes with better designs	Achieving bit synchronization	Long time to initial fix
	False correlations	
Separate pilot and data components	Fragile carrier tracking	Detecting and correcting data message errors
	Fragile code tracking	
	Carrier phase ambiguities	
Dual-frequency and three-frequency signals	Narrowband interference	
	Ionospheric errors	
	Carrier phase ambiguities	
More power error control coding and CRC	Detecting and correcting data message errors	
Lower rate data messages and code combining	Detecting and correcting data message errors	
Shorter clock and ephemeris latency	Long time to initial fix	
Overlay codes on pilot channel	Achieving bit synchronization and frame synchronization	
	False correlations	
Higher power	Narrowband interference	Self-interference, false correlations
	Fragile carrier tracking	
	Fragile code tracking	
	Detecting and correcting data message errors	
	Long time to initial fix	
More satellites	Limited satellite visibility	Interference
	Multipath and shadowing	

evident throughout Part II, new and modernized signals provide many features that receivers can exploit to address the challenges listed in Table III.1. Table III.2 summarizes features found in many new signal and system designs enabled by more capable and complicated satellites, and exploited by more sophisticated receiver processing, and lists the challenges they help receivers to address as well as challenges the features could exacerbate.

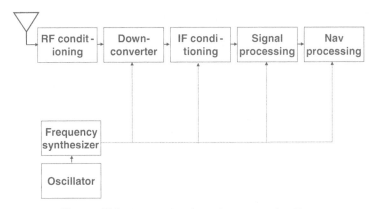

Figure III.1. Conventional Receiver Processing Flow

Exploiting these capabilities requires moving beyond the receiver processing techniques used for the original GPS and GLONASS signals. Design trades for these characteristics require tools that apply to new signal designs. This part of the book describes these receiver processing techniques and associated performance predictions.

Figure III.1 shows a canonical block diagram of receiver processing used to structure this part of the book. This block diagram outlines a conventional receiver, and also represents the sequence of chapters in this part of the book. An oscillator drives a frequency synthesizer that distributes frequency and timing references to different parts of the receiver. The signal enters the receiver via the antenna, which may include filtering to suppress out-of-band signals. The preamplifier may be integrated with the antenna (called an active antenna) or distinct from the antenna, in which case the antenna is called a passive antenna. This preamplifier increases the signal voltage to overcome subsequent insertion losses and cable losses, and also includes circuitry that protects the electronics from overvoltage such as from a lightning strike. It is typically followed by additional amplification as well as additional filtering. Downconversion translates the signal from its original carrier frequency to an intermediate frequency or to baseband. Intermediate frequency (IF) conditioning includes additional filtering and amplification, automatic gain control (AGC), and analog to digital conversion (ADC). Signal processing, which is described over several chapters, includes generating replicas of each desired signal and using those replicas for initial synchronization, and then signal tracking. Following initial synchronization, separate receiver channels, each dedicated to a different desired signal, perform carrier tracking and code tracking to make measurements from the received signal, along with demodulating data message symbols from each signal. The measurements and data message information from each channel are employed in navigation processing, which calculates smoothed PVT.

Chapter 14 describes the receiver front end, including antenna, preamplifier, and analog downconversion, along with oscillator and frequency synthesizer. This chapter is written to help readers understand the selection of appropriate components. It is not at the level needed to design any of these components, since the background for design of each component could involve an entire separate book. It also introduces a number of different receiver front-end architectures that are employed, and discusses the demands that the different architectures place on components and on other aspects of receiver processing.

Chapter 15 addresses ADC. The concepts of ADC are introduced, along with associated bandlimiting, sampling, and quantization (BSQ). The concept of a digitizing correlator is introduced, and compared to conventional considerations for ADC followed by linear processing. The effects of sampling and quantization on signal, noise, and interference are described, as is an extended parametric model for effective C/N_0 that includes BSQ effects. An analytical approach for computing BSQ losses is introduced and used to develop quantitative results for many different spreading modulations, precorrelation filter bandwidths, sampling rates, and numbers of bits used in sampling.

Chapter 16 describes the processing associated with signal acquisition, focusing on initial synchronization processing, where the receiver must detect the presence of a signal and obtain an initial estimate of the received signal's time and carrier frequency relative to time and frequency references in the receiver. Once an initial synchronization is achieved, the receiver tracks the signal and begins to make measurements, while reading ephemeris and clock corrections from the signal's data message, and synchronizing with other signals, including those on other carrier frequencies. Initial synchronization processing is the greatest signal processing challenge in acquisition.

Chapter 17 describes the design and performance of tracking loops, setting the stage for carrier tracking loops and code tracking loops in subsequent chapters. The theory of what are often called digital tracking loops, more accurately known as discrete-update tracking loops, is presented. Fundamental characteristics such as loop order, loop bandwidth, and loop damping are introduced, an approach for loop parameter selection is described, and loop performance is illustrated. Key to this chapter is the design and characterization using the discrete-update approach, rather than older techniques based on mapping analog loop (continuous-update) designs to discrete-update designs. The tracking loop theory and implementation are defined in terms of generic approaches that apply to tracking loops for different parameters.

Chapter 18 then addresses carrier tracking loops, drawing upon the tracking loop design and characterization in Chapter 17. It focuses specifically on tracking different carrier parameters, frequency and phase, with the associated signal processing and discriminator designs. Carrier phase tracking using a Costas loop, required when the signal is biphase data modulated, is described along with coherent phase-locked loop tracking that can be used when there is no unknown biphase modulation. Linearized models are provided for tracking loop accuracy in noise and interference, and loss of lock criteria are provided. This chapter ends with a brief discussion of data demodulation techniques and performance, including the unique capability to demodulate the L1C data message at reduced thresholds through data symbol combining.

Chapter 19 described code tracking loops, also known as delay-locked loops. The associated signal processing is described, leading to coherent and noncoherent discriminators and linearized models for code tracking accuracy in noise and interference, showing the relevance of a signal's RMS bandwidth to its code tracking performance. Loss of lock criteria are also introduced. The effects of different spreading modulations are evaluated, along with performance trades and considerations involved in selection of other code tracking loop parameters. Also introduced is carrier-aided code tracking, which uses measurements of carrier Doppler to remove dynamics from a code tracking loop to allow smaller loop bandwidths for code tracking accuracy. Some modern spreading modulations introduce the possibility of ambiguous code tracking points; this phenomenon is discussed along with various mitigation approaches. Reference 1 provides additional information concerning signal tracking described in Chapters 17, 18, and 19.

Chapter 20 completes Part III with the description of algorithms for position and time calculation using measurements from the code and carrier tracking loops along with information from signals' data messages. Approaches for correcting and smoothing measurements are described, leading to an algorithm for standard point positioning. Next, advanced position calculation techniques are introduced for situations when there is no sufficient number of satellites in view, or when system time cannot be read from the data message. More recently developed approaches for precise point positioning are also introduced. Finally, the concept of integrity checking using an excess of measurements is introduced.

Together, these chapters equip the reader with receiver design knowledge applying to present and future signals, since the material covers algorithms, techniques, and performance predictions that apply to original GPS and GLONASS signals as well as to the characteristics and capabilities of new and modernized signals.

REFERENCE

1. T. Pany, *Navigation Signal Processing for GNSS Software Receivers*, Artech House, 2010.

14

RECEIVER FRONT END

The receiver front end comprises the antenna, amplification, filtering, mixing, and conversion to quantized samples needed to convert received electromagnetic waves at one or more carrier frequencies into digitized waveforms at baseband or an intermediate frequency (IF) for signal processing. In addition, the receiver's oscillator and frequency synthesizer are discussed in this chapter. In some references, the receiver front end is defined to end at the mixer output, but here the definition is extended through digitization and downconversion, in order to address in a unified way all the receiver functions before digital signal processing.

> This chapter is based in part on instructional materials prepared by Mr. Phillip W. Ward, President of Navward GPS Consulting. These materials were prepared as part of the NavtechGPS course "Using Advanced GPS/GNSS Signals and Systems." The author thanks Mr. Ward for permission to use these materials in the preparation of this chapter, and also thanks Ms. Carolyn McDonald, President of NavtechGPS, for sponsoring the original development of these materials. Any errors or omissions, however, are the author's.

Engineering Satellite-Based Navigation and Timing: Global Navigation Satellite Systems, Signals, and Receivers, First Edition. John W. Betz.
© 2016 The Institute of Electrical and Electronics Engineers, Inc. Published 2016 by John Wiley & Sons, Inc.

A variety of technologies is available for implementing each component, and entire books are written about the design of each type of component. Consequently, this chapter does not provide the details needed to design components, but rather addresses considerations in their selection and use. Section 14.1 provides subsections on each class of component, describing their key characteristics and parameters, and the uses of different classes of devices. Section 14.2 introduces the important characteristic of a receiver's noise figure, and shows how it is calculated from component characteristics. Section 14.3 introduces some common alternative architectures for receiver front ends, describing various considerations in their selection. Section 14.4 summarizes the chapter.

14.1 FRONT-END COMPONENTS

Front-end components set the stage for satnav receiver processing. Properly selected and integrated, these components can enhance the desired signal and mitigate interference and noise, helping to achieve the desired receiver performance. The components employed in a receiver front end do not directly benefit from well-known technology advances in digital semiconductors, but their performance does improve over time with advances in materials and semiconductor technologies and manufacturing practices.

Figure 14.1 provides an expanded view of a conventional receiver front end. There is tremendous variation in different receiver designs, and only some of these are described in this chapter. Not only do different architectures exist, but some receiver designs do not include all of the components described here, and others use additional components or multiple stages of components. The receive antenna transforms power in incident electromagnetic waves to electrical signals, outputting the resulting electrical signals at the carrier frequency (often termed "RF" for radio frequency). The RF conditioning can include a limiter, prefilter, and RF amplifier, commonly called a low noise amplifier (LNA). The limiter protects subsequent circuitry from overvoltage conditions caused by lightning. It typically uses a diode that shorts out the input leads when a high voltage occurs that exceeds the levels associated with normal inputs. The prefilter suppresses

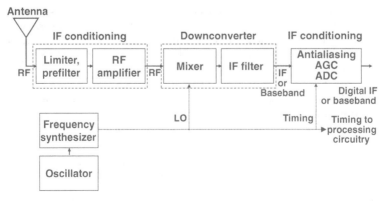

Figure 14.1. Expanded View of Receiver Front End Processing Flow

strong interfering waveforms outside the frequency band(s) of interest to protect subsequent electronic components, and the LNA boosts the voltage to overcome losses and electronic noise in subsequent components. The preamp output is still at RF, and has a wider bandwidth than that of the desired signals. Downconversion consists of one or more stages of mixing, amplifying, and filtering. Each mixer uses a local oscillator (LO) output to translate the frequency of its input to new center frequencies, then an intermediate frequency (IF) filter to select the signal (along with noise and interference) at one of these center frequencies, which could be at or near 0 Hz (in which case it is referred to as baseband). Interspersed amplifiers, not shown here, provide gain to offset losses from mixing and filtering. The output is at IF or baseband, and still an analog waveform. IF conditioning provides antialiasing filtering, automatic gain control (AGC), and analog-to-digital conversion (ADC), producing a digital output (bandlimited, sampled, and quantized) at IF or baseband for signal processing in the receiver. While the above description applies for a conventional receiver front end, different designs and receiver architectures can deviate from this description, with some components absent, added, or configured differently. Some of these different architectures are discussed in Section 14.3. The oscillator in Figure 14.1 ideally provides a stable sinusoid at a standard frequency; the frequency synthesizer uses this sinusoid to produce various frequency-related and timing signals used in the receiver, including one or more local oscillator (LO) signals, as well as clock signals used by the ADC and the remaining processing.

Section 14.1.1 discusses antennas, while Section 14.1.2 addresses amplifiers, especially LNAs. Section 14.1.3 discusses filters, and Section 14.1.4 addresses mixers. Oscillators and frequency synthesis are described in Section 14.1.5. A recent survey of front-end components and their characteristics is found in Reference 1. An excellent tutorial on front-end design issues and solutions is provided in Reference 2. Additional details are also found in Reference 3.

14.1.1 Antennas

Antennas convert electrical power into electromagnetic waves and vice versa. This chapter focuses on receive antennas, which convert incident electromagnetic waves into electrical current and voltage at terminals that are connected to the receiver inputs. Among the electromagnetic waves incident on the antenna are desired signals and interference, so the receive antenna outputs the sum of desired signals, interfering signals, and thermal noise produced by electronics in the antenna. The reciprocity principle states that electrical characteristics of a transmit antenna, such as gain, efficiency, directivity, and frequency response, are the same when an antenna is used for transmitting and for receiving. Reciprocity only holds when the antenna behaves linearly and without hysteresis, and is commonly assumed and approximated in practice.

An antenna acts as a filter in three dimensions: direction of arrival, frequency, and polarization of the incident electromagnetic waves. The design and characterization of antennas for satnav receivers is art as well as science, and until recently has not been addressed in comprehensive references. Two recent publications, References 4 and 5, provide more information than had previously been available in integrated references. A very helpful application note is also available [6].

An antenna consists of an arrangement of conductors (antenna elements) configured so that an incident electromagnetic wave induces corresponding currents in the elements. The antenna consists of these resonant conductors, structures that hold the elements in place, circuits and networks that connect the antenna elements to output terminals, and reflective or directive elements (such as ground planes or cavities) that interact with the incident electromagnetic wave, along with their packaging. In addition, other materials, such as dielectrics and metamaterials, may be included to obtain desired performance [5]. An amplifier, sometimes preceded by a filter to suppress interference at unwanted frequencies, may also be included in the receive antenna package. In that case, the resulting component is termed an active antenna, even though strictly speaking the antenna is still passive, it is just that an amplifier has been included within the antenna package. In contrast, a passive antenna does not include an integral amplifier. The passive portion of an active antenna consists of all antenna components before the amplifier.

For satnav applications, incident electromagnetic waves are typically modeled as plane waves [7], so the antenna response to them can be expressed in terms of their direction of arrival relative to the antenna. The direction of arrival is described in polar coordinates, relative to a specified orientation of the antenna. Antenna characteristics are often parameterized by elevation angle, using the approximation that they are invariant with azimuth. More precise characterization is obtained by reporting results by elevation angle for different azimuth values, sometimes called "azimuth cuts."

Antenna properties are typically defined for transmit antennas, and applied to receive antennas using reciprocity. Here, however, the properties are defined directly for receive antennas. For an electromagnetic wave incident at a particular elevation angle θ and azimuth angle ϕ, the antenna gain pattern is defined at a specific frequency for a passive antenna or the passive portion of an active antenna to be

$$g(\theta, \varphi) = \frac{P_{out}(\theta, \phi)}{P_{in}} \tag{14.1}$$

where P_{in} is the power in the incident electromagnetic wave at the specific frequency,[1] defined to be the same at all azimuths and orientations, and $P_{out}(\theta, \phi)$ is the output electrical power at the passive antenna terminals resulting from the incidence of this electromagnetic wave at a specific elevation angle and azimuth. The gain pattern also depends upon the polarization of the incident wave, as discussed in Section 3.3.3. Often, the gain pattern is expressed in decibels (taking $10 \log_{10}$ of the quantity in (14.1)) and denoted "dBi" for decibels relative to an isotropic antenna.

Antenna gain is a single value that is the maximum of the antenna gain pattern: $G = \max_{\theta, \phi} g(\theta, \phi) = g(\theta^+, \phi^+)$, where (θ^+, ϕ^+), the direction of maximum antenna gain pattern, is sometimes called the boresight angle. Antenna gain is also expressed in terms of the effective area of the antenna in (4.4).

---

[1] Antennas are usually characterized using incident waves that are sinusoidal in time, making it straightforward to characterize how performance depends on frequency.

The antenna efficiency is defined as the ratio of the average output power to the input power over all angles of incidence, for constant input power:

$$\eta \stackrel{\Delta}{=} \frac{P_{out,average}}{P_{in}}$$

$$P_{out,average} = \frac{1}{4\pi} \int_{\phi=0}^{\phi=2\pi} \int_{\theta=0}^{\theta=\pi} P_{out}(\theta,\phi) \, d\theta \, d\phi \tag{14.2}$$

so (14.1) becomes

$$g(\theta,\phi) = \eta \frac{P_{out}(\theta,\phi)}{P_{out,average}} \tag{14.3}$$

If the antenna gain pattern is approximated as constant within the beam and zero outside the beam, then the gain is

$$G = \eta \frac{\text{Area of sphere}}{\text{Area of beam}} = \eta \frac{4\pi}{\alpha} \tag{14.4}$$

for a beam of solid angle α (in radians), so the gain is the antenna efficiency divided by the fraction of the sphere containing the beam. A common approximation for this case is provided in (4.6), $G \cong \frac{29000}{\beta_\theta \beta_\phi}$ where β_θ and β_ϕ are, respectively, the azimuth and elevation beamwidths in degrees. This expression is derived in Review Question QT14.1.

Different types and designs of antennas have very different efficiencies ranging from as large as 0.8 0.1 or smaller. A typical value of efficiency for a well-designed antenna is $\eta = 0.55$ or a loss of 2.6 dB. A number of different factors contribute to the efficiency, including:

- Losses in the antenna elements due to dielectric and surface wave losses, as well as resistive effects,
- Polarization mismatch between the incident wave and the antenna,
- Losses between the antenna and the feed network, and in the feed network that extracts RHCP.

Since (14.2) includes all of these factors, η is sometimes called the antenna total efficiency.

Antenna directivity pattern at a given elevation angle θ and azimuth angle ϕ is defined at a specific frequency for a passive antenna or the passive portion of an active antenna to be the power at the antenna output from an incident field having a given power, divided by the output power averaged over all angles of incidence when the incident field of constant power is applied at every elevation and azimuth angle:

$$d(\theta, \phi) = \frac{P_{out}(\theta, \phi)}{P_{out,average}} \tag{14.5}$$

The difference between the directivity pattern, $d(\theta, \phi)$, and the antenna gain pattern, $g(\theta, \phi)$, is that directivity pattern compares output power at a specific elevation angle and azimuth angle to average output power, whereas antenna gain compares the output power to input power, thus accounting for the antenna efficiency. From (14.2) it is clear that the antenna efficiency does not depend on elevation and azimuth, so

$$g(\theta, \phi) = \eta \, d(\theta, \phi) \tag{14.6}$$

Antenna directivity is defined as the maximum of the antenna directivity pattern, occurring at boresight:

$$D \stackrel{\Delta}{=} d(\theta^+, \phi^+) = \max_{\theta, \varphi} d(\theta, \phi) \tag{14.7}$$
$$D = \eta \, G$$

The gain and directivity parameters are usually expressed in decibels.

Common usage does not carefully distinguish among the terms antenna gain pattern, antenna gain, antenna directivity pattern, and antenna directivity, but they are actually different, though related as seen above. For a perfectly efficient antenna, however, the antenna gain pattern and antenna directivity pattern are identical, as are the gain and directivity.

Often, the polarization of the incident wave is indicated explicitly in describing gain and directivity. If the incident wave is perfectly linearly polarized, the gain in decibels is expressed as dBil. If the incident wave is perfectly circularly polarized, the gain in decibels is expressed as dBic. (When the polarization sense of the receive antenna is best aligned with that of the incident wave the resulting gain pattern is called "copolarization gain.") Crosspolarization gain describes the gain if the receive antenna's polarization orientation is least aligned with that of the incident wave: for a linearly polarized wave, the antenna is oriented such that the minor axis of its polarization ellipse is aligned with the polarization of the incident wave; for a circularly polarized wave, the antenna has the opposite polarization sense from that of the incident wave.

Figure 14.2 shows a conceptual gain pattern for a satnav receive antenna, for a single azimuth cut. The (copolarization) gain is greater than 0 dB at high elevation angles, and approximately hemispherical until it drops off near the horizon at 0° and 180°. Ideally, this gain pattern is the same at every azimuth angle. Beamwidth is not a precisely defined quantity, but often is considered to be the total angle where the antenna gain pattern is always within 3 dB of the gain at boresight.

A gain pattern for typical satnav applications has several desirable characteristics. The gain should be as small as possible in the backlobes—elevation angles below the

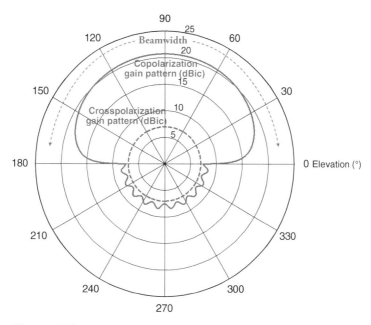

Figure 14.2. Conceptual Gain Patterns for a Satnav Receive Antenna

horizon. Also, the crosspolarization gain should be as small as possible at every elevation angle to reduce antenna's response to undesired signals having different polarizations. Different objectives exist for low elevation angles. Since lower DOP values can be obtained from signals from low elevation satellites, it can be helpful for gain to be as large as possible at elevation angles above 5° in order to use signals from such low elevation satellites. However, it is challenging to design and install antennas so that they have high gain at low elevation angles. In practice, signals from such low elevations are generally less desirable since they are received at lower power, and tend to suffer from more distortion (ionospheric, tropospheric, multipath) than do signals from higher elevation satellites. Also, in many applications interference is more common and more powerful at low elevation angles; consequently increased gain at low elevation angles can increase the received interference. Furthermore, some antennas, particularly when mounted on a metal ground plane, become increasingly elliptically polarized at low elevation angles, approaching vertical polarization, since the horizontally polarized component is shorted out. The result is the same response to incident RHCP and LHCP signals, and an inability to discriminate against multipath reflections that often are LHCP.

Different idealized antenna gain patterns are encountered in satnav. An isotropic antenna is an idealized antenna that has unit antenna gain pattern in all directions. Such an antenna cannot be built, but it has useful conceptual properties. Another such antenna has a hemispherical gain pattern with gain of 3 dB in one hemisphere (typically pointed up in satnav applications) and no response in the backlobes.

The cardioid gain pattern shown in Figure 14.2 corresponds to what is sometimes termed a fixed reception pattern antenna (FRPA), corresponding to an approximately hemispherical gain pattern with maximum gain at 90° elevation. As shown, the copolarization and crosspolarization gain patterns are approximated as symmetric around 90° elevation (the same at every azimuth), although real antennas in real environments typically do not exhibit perfect azimuthal symmetry.

In contrast to a FRPA, a controlled reception pattern antenna (CRPA) uses an array of antenna elements, with associated antenna electronics that combines the antenna outputs with different amplitudes, delays, or phases in order to adapt the resulting antenna gain pattern in a way that reduces the power of directional interference and noise while increasing the power of desired signals, performing a combination of spatial filtering and frequency-domain filtering. Chapter 4 of Reference 5 addresses CRPAs and adaptive antenna processing in more detail.

While antennas are commonly considered, specified, and tested as standalone components, an antenna's surroundings can greatly affect its gain pattern. Mounting an antenna on a ground plane, either on an aircraft, ground vehicle, or other metal surface or for testing in a laboratory, can change the gain pattern significantly. As an example, the horizontal component of incident waves at angles that graze the ground plane is approximately shorted out by the conductive surface, leaving an approximately vertically polarized wave incident on the antenna. The radiation patterns of antennas in handheld devices like mobile phones are affected by other components in the device, and also by the hand and body of a nearby person, as well as other surroundings. As a rough rule of thumb, conductive objects within the Fraunhofer distance affect the antenna's gain pattern, while those further than the Fraunhofer distance introduce multipath.

The feed network and other cables or wires that extract current and voltage from the antenna elements and provide it at the antenna terminals also can have many subtle effects on antenna performance. They can act as receive antennas themselves, disturbing the desired properties of the antenna elements. They may radiate very low power electromagnetic waves that induce additional currents in the antenna elements. They may not be symmetrically connected to the antenna elements (the so-called "unbalanced feed") producing asymmetry of the antenna gain pattern and other electrical characteristics. A balun, which stands for "balanced to unbalanced transformer" can be used to help achieve balance, but at the cost and complexity of an additional component in the antenna assembly.

Antennas are resonant, meaning they have greater gain at some frequencies than others. Wideband antennas provide consistent gain and low signal distortion over a wide frequency range (with a bandwidth to center frequency ratio of 2:1 or larger). However, this characteristic also means that wideband antennas are less frequency selective, and can pass interfering waveforms even at frequencies other than those used by desired signals. When this interference has high power, either it must be suppressed by filters after the antenna, or else the amplifiers, mixers, and filters must have the dynamic range needed to pass this interference without distortion, as discussed in Section 14.1.3.

Furthermore, physics requires an antenna have a minimum antenna electrical size (the maximum dimension of an antenna element, measured in wavelengths) to provide a given gain over a given beamwidth and frequency range, as defined by the

Chu–Harrington limit [8]. Consequently, maintaining the same gain performance over a wider frequency range necessarily requires a larger antenna. In applications like handheld devices where size is at a premium, or aviation where existing form factors must be retained, increased size may not be an option, and gain must be traded against frequency range.

The introduction of new and modernized systems described in Part II, with their wider bandwidth signals and center frequencies spread over more than 400 MHz (greater than $\pm 15\%$ bandwidth), can significantly change the demands on satnav receive antennas. Advances in electronics capabilities allow receivers to cost effectively process more signals, but antenna performance may still be a constraint. Guided by the Chu–Harrington limit, one way to relax the demand on antennas is to use a narrower beamwidth since more satellites will be available at higher elevation angles. It has been suggested [4] that while multiple carrier frequencies will require antennas with wider frequency coverage, the larger number of satellites in the GPS constellation and the increased availability of signals from other satnav systems, will enable future receiver antennas having narrower beamwidths with less gain at lower elevation angles.

Other considerations affect some classes of receive antennas. Precision receivers that provide centimeter-level accuracy through carrier phase processing are sensitive to antenna phase center offset and phase center variation. Phase center offset occurs when the antenna phase center (the apparent location where an incident wave is converted to electrical current) is different from the physical center of an antenna. Phase center variation occurs when the phase center occurs at a different point in space for waves at different angles of incidence and different frequencies. Antennas for precision applications are designed to have small antenna phase center variation, and then calibrated to determine the phase center offset and to mitigate remaining phase center variations, since these effects can introduce errors that are significant relative to the needed accuracy. Since these high-precision systems employ differential corrections, where a user station makes measurements relative to a reference station (see Chapter 23), phase center variations from the user station antennas and the reference station antenna must be taken into account. One way to do this is to use matched antennas, all with the same orientation.

An extensive discussion of different antenna types and their characteristics is found in Reference 5, and GPS World Magazine provides an extensive annual survey of satnav antenna designs and capabilities. A recent such survey is found in Reference 9, with a brief summary provided here. Where space allows, planar crossed bow tie antennas and drooping crossed bow tie antennas can provide good performance over the frequency range associated with new and modernized satnav signals. However, they are three-dimensional structures and require ground planes, consuming significant volume. Helix antennas can also provide a good combination of gain and frequency coverage, and low-profile versions can approximate the desired wide beamwidth shown in Figure 14.2. Filling the volume inside the helix of antenna elements with a high dielectric constant material allows a smaller antenna to provide equivalent gain, but tends to reduce the bandwidth. Ground planes are not essential for helix antennas, but improve performance. Patch antennas are commonly used, since they are inexpensive and well suited for flat surfaces. They require a ground plane that is typically 70 mm on a side at satnav frequencies. Multiple patch antennas having different resonant frequencies can

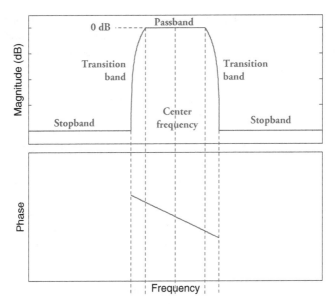

Figure 14.3. Ideal Filter Characteristics

be stacked to receive signals at different frequency bands. Their beamwidth is not as large as that of a helix antenna.

In handheld devices, there are significant incentives to using a single antenna to cover not only satnav frequencies, but also those used for Bluetooth, wireless local area networks, and various frequency bands used for cellular telephone. More compact antennas are obtained using microstrip antennas and chip antennas, although both technologies are linearly polarized, and it is difficult to get good gain over a broad frequency range. Some microstrip antennas using fractal structures are claimed to approach or go beyond the Chu–Harrington limit. Other compact and wide bandwidth antenna structures include inverted F-antennas, and antennas using ceramic and low temperature cofired ceramic substrates. However, some of these are linearly polarized, and their emphasis on small volume, low cost, and wide bandwidth penalize gain, and gain values approaching −10 dBic are found in some cellular telephones.[2]

14.1.2 Filters

Filters are placed at multiple locations in a typical receiver front end, with characteristics carefully selected for each location. Before the ADC, filters are physical devices whose behavior is dictated by their materials, size, and manufacturing. An ideal filter has transfer function shown in Figure 14.3. The magnitude should be perfectly flat within

---

[2] It has been said that a paper clip would provide better gain than a typical antenna used in cellular telephones... but it is too expensive. Clearly this is an exaggeration, but the constraints on cost and size for antennas in consumer devices should not be underestimated.

the filter passband where frequencies associated with desired signals are located, taking on a value of unity (0 dB) so that the desired signals are not attenuated. Outside the passband, the magnitude should diminish rapidly in narrow transition bands to the stopbands, which should provide significant attenuation. The phase response should be linear over the passband and transition bands. As described in Appendix Section A.3, group delay is proportional to the derivative of phase with respect to frequency, and thus the ideal group delay is constant over the passband and transition bands.

Unfortunately, such ideal filters cannot be built, and filters with the smallest size and lowest cost tend to deviate most from ideal. The magnitude has ripples in the passband. Transition bands are wider than desired, forcing compromises between the width of the passband and the ability to attenuate adjacent channel interference. Neither magnitude nor phase responses are symmetric around the center frequency. In addition, the phase is not perfectly linear in the passband, and becomes nonlinear even at frequencies inside of the transition band. The effects of these distortions on receiver processing can be evaluated using the approach described in Reference 10.

The ratio between the passband squared magnitude and the stopband squared magnitude is called the filter's selectivity. The selectivity describes how much power of a waveform in the stopband is attenuated relative to the power in the passband. A filter's insertion loss describes how much the filter reduces the power of a desired signal in the passband. Often, minimum gain in the passband is used as a conservative approximation for the insertion loss.

A filter's quality factor, or Q, characterizes the relationship between the passband and center frequency. If the lower edge of the passband occurs at frequency $f_0 - \beta/2$ and the upper edge of the passband occurs at frequency $f_0 + \beta/2$, then the quality factor is the ratio between the center frequency and the two-sided bandwidth of the passband,

$$Q \stackrel{\Delta}{=} \frac{f_0}{\beta} \tag{14.8}$$

Higher Q filters are typically physically larger and more expensive, and may exhibit greater insertion loss and phase nonlinearities (group delay variation) in the passband. It is challenging to find a filter at L band with Q as large as 100. It is easier to find filters with a given narrow passband bandwidth when the center frequency is lower. Further, some analog filter technologies have characteristics that vary significantly from one component to another component from the same production lot, and also vary with temperature variation and aging. Consequently, even if a very narrowband filter could be constructed using these technologies, filter variations, especially in center frequency, could make it difficult to use such a filter.

Different compromises must be made for different classes of receivers and for different filters in the receiver front end. Section 14.2 shows that for a prefilter before the first LNA, low insertion loss is critical. Phase linearity over the signal frequencies that will eventually be used is also important, especially for wider bandwidth signals and for high accuracy receivers. For prefilters, since the signal is still at RF, the compromise typically made is to use wider passband and wider transition bands than desired. After the first LNA, some receivers use interspersed additional filtering and gain stages before downconversion, sometimes termed distributed gain filtering. As shown in Section 14.2, the larger insertion loss can be more readily accommodated after the first LNA; accepting larger insertion loss can allow sharper transition bands and deeper stopbands.

Demands on IF filters after mixing are often relaxed for several reasons. If the IF is lower than the RF, the Q factor can be lower for the same bandwidth. In addition, since an IF filter can be preceded by and followed by amplifiers, greater insertion loss can be accommodated.

Antialiasing filters provide additional stopband suppression, so the inevitable aliasing of noise and interference and signal power at frequencies beyond the Nyquist rate is acceptably low. It is important to ensure that the sampling rate is large enough relative to the passband, but also relative to the combination of the passband and transition bands. This higher sampling rate avoids the aliasing of power at frequencies in the transition bands. As a general statement, the passband of the antialiasing filters determines the useful frequency range in the receiver, while the stopband of the antialiasing filter determines the minimum sampling rate.

In addition to analog filtering, digital filtering may also be performed in a receiver, including for antialiasing when decimation (sampling rate reduction) takes place at some stage in the digital processing. Since finite impulse response (FIR) filters allow careful control of magnitude as well as perfectly linear phase, such filtering can typically be implemented in a way that has no deleterious effects on signal quality, although the associated power consumption can be an issue in some receivers.

The combination of all the above filters can be represented as the product of their lowpass equivalent transfer functions, yielding a composite transfer function for what is called the precorrelation filter. This precorrelation filter is a useful way to model all of the filtering that happens in the receiver. The equivalent rectangular bandwidth of this precorrelation filter is the precorrelation filter bandwidth used in numerous equations found in this book.

A relatively recent overview of various filter technologies and their attributes is found in Reference 11, and summarized here. One important consideration is variability of the filter characteristics both over temperature and aging for a specific component, and over different components from a given lot.

While cavity filters provide the best performance as prefilters, they tend to be larger (with dimensions approaching or exceeding 10 cm), heavier, and more expensive than most other filter technologies. They can provide insertion loss less than 1 dB, and transition bands whose slopes exceed 2 dB per MHz in L band. In some cases when there is a separate antenna module, it may be feasible to incorporate a cavity filter in this module as a prefilter.

Dielectric resonator filters are commonly used at RF. Higher Q L band dielectric resonator filters have insertion losses that can exceed 2 dB, and transition bands whose slopes approach 1 dB per MHz. Their largest dimension is often less than 2 cm. Lower Q L band dielectric resonator filters provide a wider passband with insertion losses that may be less than 1 dB, but transition band slopes of less than 1 dB per MHz.

Surface acoustic wave (SAW) filters tend to be the smallest and least expensive filters, with different versions that are available for use at RF and at IF. RF SAW filters can have insertion losses less than 1 dB, but wide passbands—many tens of MHz wide before much attenuation occurs. Their passband upper and lower limits can change by multiple MHz with temperature, and group delay distortion can also vary significantly with temperature. For these reasons, SAW filters at RF are only used for low-end receiver applications where accuracy is not a driver.

IF SAW filters provide good selectivity (of the order of 50 dB or more) but insertion loss exceeding 10 dB. However, since amplification can be used before and after the IF filters, the insertion loss can be tolerated. Their transition bandwidths are very narrow, with slopes greater than 5 dB per MHz, but group delay distortion can still be a significant issue for IF SAW filters.

Bulk acoustic wave (BAW) filters can be an attractive alternative to SAW filters for RF filtering. While their package size is slightly larger than SAW packages, the largest dimension is only a few millimeters, facilitating their placement on a circuit board. They offer Q values as large as 100, and insertion losses of the order of 2–3 dB. They can provide narrow transition bandwidths with slopes greater than 5 dB per MHz. Group delay tends to be controlled better for BAW filters than SAW filters.

Other IF filter technologies offer generally poorer performance and thus more limited utility. They include passive inductor–capacitor (LC) filters, constructed using inductors and capacitors mounted externally to the receiver integrated circuit. Their Q tends to be of the order of 10, and their group delay can be well controlled. Active resistor–capacitor (RC) filters employ an operational amplifier. Since components used for active RC filters are small, they can be included as part of the receiver integrated circuit.

14.1.3 Amplifiers

Amplifiers are essential for increasing the received signals' voltage relative to thermal noise in the electronics, and for counteracting losses in cables, filters, and other components, as discussed in Section 14.2. There are typically multiple amplifiers in the receiver front end, interspersed among lossy components to maintain adequate voltage levels. However, the first amplifier after the antenna, called the RF amplifier or LNA, is the most critical. Although the following discussion applies to amplifiers, the discussion of dynamic range also applies to mixers.

The amplifiers in a satnav receiver front end are intended to be linear, with the relationship between input voltage and output voltage[3] given by

$$v_{\text{out}}(t) = k_1 v_{\text{in}}(t) \tag{14.9}$$

where k_1 is the voltage gain of the amplifier and the output voltage is generally measured across some standard impedance—often 50 Ω. However, (14.9) is actually a small-signal

---

[3] Section 14.1 is the only section in this book where real-valued signal representations, rather than analytic signal representations, are used for describing and analyzing signal processing—in this case the signal processing effects of amplifiers and mixers. The bookkeeping involved in assessing nonlinear effects on analytic signals is excessively complicated for these topics, and the results are easier to apply for real-valued signals.

approximation that applies only when $|v_{in}(t)|$ is small enough. The tools provided in this subsection apply for amplifiers that are intended to operate linearly, but become increasingly nonlinear as $|v_{in}(t)|$ becomes larger. Such behavior is sometimes termed weakly nonlinear. The characterizations provided here do not apply to severe nonlinearities associated with limiters or with amplifiers operating in saturated mode.

Amplifiers in satnav receivers are typically modeled as weakly nonlinear and memoryless, with an odd-symmetric input–output voltage relationship. In this case, the input–output voltage relationship can be expressed as a Taylor series with all coefficients of even-order terms taking on values of zero. A fifth-order polynomial is often used, with the input–output voltage relationship approximated by

$$v_{out}(t) \approx k_1 v_{in}(t) + k_3 v_{in}^3(t) + k_5 v_{in}^5(t) \tag{14.10}$$

Here, k_1 is the small-signal gain, k_3 establishes the degree of nonlinearity, and k_5 describes the saturation point: the magnitude of the input where the slope of the input–output relationship is zero. The parameter k_1 is dimensionless, k_3 has dimension volts^{-2}, and k_5 is has dimension volts^{-4}. The small-signal gain represents a voltage gain, and is often characterized in dB as $20\log_{10} k_1$.

Review Question QT4.3 shows that an amplifier modeled by (14.10) saturates (the magnitude of the output does not increase even if the magnitude of the input increases by an infinitesimal amount) when $k_3^2 \geq 20 k_1 k_5 / 9$, and in that case the input voltage at the saturation point is given by

$$v_{in,sat} = \pm \left[-0.3 \left(\frac{k_3}{k_5} \right) - \sqrt{0.09 \left(\frac{k_3}{k_5} \right)^2 - 0.2 \left(\frac{k_1}{k_5} \right)} \right]^{1/2} \tag{14.11}$$

Observe from (14.11) that $v_{in,sat}$ is real valued (the amplifier has a saturation point) only when $k_3 < 0$.

Figure 14.4 shows the input–output voltage relationship for a specific example, along with the ideal linear input–output voltage relationship having the same small-signal gain. Evaluating (14.11) yields $v_{in,sat} = 0.0085$ and $v_{out,sat} = 0.0483$.

The 1 dB compression point is defined as the value of $v_{in}^2(t)$ where $v_{out}^2(t)$ is 1 dB lower than it would be from an ideal linear amplifier with the same small-signal gain. The fact that k_3 is negative causes this compression—a smaller output value at larger values of $v_{in}^2(t)$ than would occur with the linear model.

Review Question QT14.4 shows that the 1 dB compression point for an input–output voltage relationship of the form (14.10) occurs only if $k_3^2 \geq 0.4348 k_1 k_5$.[4] This review

---

[4] When this inequality is not satisfied, or the sign of k_3 is positive, the gain of the nonlinear amplifier never falls 1 dB below that of the ideal linear amplifier with the same small-signal gain. Such an amplifier is not physically realistic.

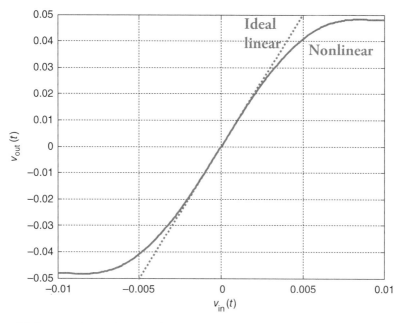

Figure 14.4. Example Nonlinear and Ideal Linear Input–Output Voltage Relationships with $k_1 = 10$ (Corresponding to a 20 dB Small-Signal Gain), $k_3 = -8 \times 10^4$ V^{-2}, and $k_5 = 2.8 \times 10^8$ V^{-4}

question also shows that, for this fifth-order model, the input voltages corresponding to the 1 dB compression point are

$$v_{\text{in},1\,\text{dB},\text{5th order}}(t) = \pm \left[-0.5 \frac{k_3}{k_5} - \sqrt{0.25 \left(\frac{k_3}{k_5}\right)^2 - 0.1087 \left(\frac{k_1}{k_5}\right)} \right]^{1/2} \quad (14.12)$$

Figure 14.5 shows the input–output voltage-squared relationships for the same case as in Figure 14.4. At lower input voltages, the 20 dB small-signal gain of the amplifier is evident. Evaluating (14.12) indicates the 1 dB compression point occurs at an input voltage of 0.0038 V. The magnitude of the 1 dB compression point is often termed the input dynamic range of the amplifier.

It is important to understand the conditions that can cause $|v_{\text{in}}(t)|$ to be large enough for nonlinear behavior to occur. The input voltage is the sum of desired signal and interference received through the antenna, summed with thermal noise generated by electronics in the receiver front end. The thermal noise power spectral density can never be less than $k_B T$, where $k_B = 1.381 \times 10^{-23}$ J/°K (equivalently, watts/(°K-Hz)) is the Boltzmann constant and T is the temperature in degrees Kelvin. For room temperature of 290°K,[5] the thermal noise power spectral density is then 4.00×10^{-21} W/Hz or -204.0

---

[5] By convention, this is the standard room temperature used for computing thermal noise, equal to 16.8°C or 62.3°F.

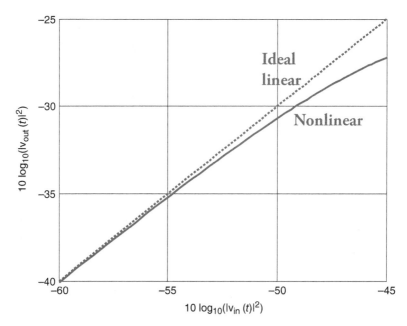

Figure 14.5. Nonlinear and Ideal Linear Input–Output Power Relationships in Figure 14.4

dBW/Hz. For a precorrelation filter with bandwidth $\pm\beta_r/2$, the total noise power is then $k_B T \beta_r$ watts, and even for a narrow precorrelation filter bandwidth with β_r of 2 MHz, the total noise power at room temperature is -141.0 dBW.

Many receivers have larger precorrelation bandwidths than used to obtain this value, and as seen in Section 14.2, real receivers have higher thermal noise density than calculated here. Consequently, the thermal noise power is typically greater than -140 dBW. Even this value, however, is significantly larger than the received signal powers for satnav systems described in Part II of this book, even when they are received with several dBi of antenna gain. Consequently, it is clear that thermal noise power is much greater than the power of any desired received signal.[6] Thus, if there were no interference, the maximum value of $|v_{in}(t)|$ would be set by the thermal noise, and would be well known and easily accounted for in the amplifier design.

Received interference power can vary widely, including interference at frequencies outside the range of those used by the desired signal. In multifunction devices such as mobile phones, device interference is produced by other circuitry. Interference can also be transmitted by terrestrial systems near the receive antenna. Signals transmitted for broadcast radio or television, or for high capacity mobile communications such as cellular data networks, can have EIRP values exceeding 40 dBW at the transmitter. These

---

[6] This condition holds even for military signals that potentially might have higher power than discussed in Part II, since these signals typically require at least 10 times the precorrelation bandwidth used in the above calculation, increasing the total thermal noise by 10 dB to greater than -130 dBW.

interfering signals, even though they are not in the same frequency band as the desired satnav signals, can cause nonlinear behavior of amplifiers in a satnav receiver's front end. Designing for adequate dynamic range combined with adequate filtering before each amplifier is essential for retaining the desired linear operation of amplifiers.

When interference, either in-band or out-of-band but inadequately suppressed, causes $|v_{in}(t)|$ to be large enough that significant nonlinear effects occur, the interference is sometimes called a blocker, and the interference is said to desensitize the amplifier since the gain is reduced. Several effects occur at this point. When the magnitude of the input voltage reaches the 1 dB compression point, the gain is 1 dB less than the small-signal gain, producing a lower desired signal power for subsequent processing. Alternatively, as will be seen in Section 14.2, this reduction in gain can be seen as increasing the noise figure of the receiver. Another effect is that the interference and signal are distorted, so the amplifier's output waveform is not a scaled replica of the input, and the output's spectrum differs from that of the input, with interference at new frequencies. This distortion can make the effects of interference worse, and degrade the performance for receiver processing of the desired signal. These distortions may also produce additional losses in downconversion, ADC, linear filtering, and correlation processing. Since the interference PSD produced differs from the input interference PSD, subsequent filtering designed based on the input interference PSD may not be effective in reducing the interference.

To better understand how nonlinearities generate waveforms at additional frequencies, consider a sinusoidal input $v_{in}(t) = a\cos(2\pi f_0 t)$ applied to an amplifier characterized by a cubic odd-symmetric nonlinearity

$$v_{out}(t) = k_1 v_{in}(t) + k_3 v_{in}^3(t) \qquad (14.13)$$

Following the same approach as used to obtain (14.12), but for this third-order model, the input voltages corresponding to the 1 dB compression point are

$$v_{in,1\,dB,3rd\,order}(t) = \pm\sqrt{-0.1087\left(\frac{k_1}{k_3}\right)} \qquad (14.14)$$

Using the trigonometric identity $\cos^3\alpha = \frac{3}{4}\cos\alpha + \frac{1}{4}\cos 3\alpha$ yields the output from (14.13)

$$\begin{aligned}v_{out}(t) &= k_1 a \cos(2\pi f_0 t) + \frac{3}{4}k_3 a^3 \cos(2\pi f_0 t) + \frac{1}{4}k_3 a^3 \cos(2\pi 3 f_0 t) \\ &= \left[k_1 a + \frac{3}{4}k_3 a^3\right]\cos(2\pi f_0 t) + k_3 a^3 \frac{1}{4}\cos(2\pi 3 f_0 t)\end{aligned} \qquad (14.15)$$

This result implicitly assumes there is no delay in the amplifier, and that there is no delay or that the delay is constant over frequency (linear group delay) so that the output sinusoids combine coherently. This is a conservative assumption. Typically, there is some delay variation with frequency; a less conservative approach is to apply different phases at each frequency, so the different components combine noncoherently (in power) rather than coherently (in voltage).

Two effects are apparent from (14.15):

- The output amplitude of the cosine at the desired frequency f_0 is diminished (recall k_3 is negative for a realistic amplifier) relative to what would occur with an ideal linear amplifier,
- Even though the input is a pure sinusoid at frequency f_0, the output also includes a waveform at an additional frequency $3f_0$.

The sinusoidal 1 dB compression point for (14.13) is found by determining the value of a where the coefficient of the $\cos(2\pi f_0 t)$ term in (14.15) is 1 dB lower than what would be produced by the small-signal gain $k_1 a$. This condition occurs when (see Review Question QT14.5)

$$a_{1\text{ dB,3rd order,sine}} = \pm\sqrt{-0.1449\frac{k_1}{k_3}} \qquad (14.16)$$

The third-order intercept point, often denoted IP3, is defined here as the input voltage where the coefficient of the $\cos(2\pi f_0 t)$ term in (14.15) becomes zero. Algebra shows the IP3 is

$$a_{\text{IP3}} = \pm\sqrt{-\frac{4k_1}{3k_3}} \qquad (14.17)$$

Comparing (14.16) and (14.17) leads to the conclusion that $a_{1\text{dB}} = 0.3298 a_{\text{IP3}}$ or the 1 dB compression point is 9.6 dB less than the IP3. Since the simple cubic model (14.13) used to derive this result is typically not a good model of the amplifier nonlinearity at input values as large as the IP3, the IP3 is merely an indicator of the amplifier's nonlinearity, and input magnitudes should be kept approximately 10 dB below the IP3.

Comparing (14.12), (14.14), and (14.16) shows different 1 dB compression points for different models of the amplifier nonlinearity and different inputs. The resulting numerical values are similar, and reflect the fact that these are rules of thumb used to provide general characterizations of amplifier behavior, not precise characterizations based on exact models.

A definition of third-order intercept point that is different from (14.16) is sometimes used. In this case, the IP3 is defined as the value of a where the magnitude of the output third harmonic is equal to the output of the first harmonic for a perfectly linear amplifier, or

$$a'_{\text{IP3}} = \pm\sqrt{-\frac{4k_1}{k_3}} \qquad (14.18)$$

Since a_{IP3} and a'_{IP3} differ by 4.8 dB, it is important to understand which definition is being used.

While the analysis in this subsection has focused on odd-symmetric input–output voltage relationships, some amplifiers lack this symmetry; in this case, even-order polynomial terms must be included as well. Then, a second-order intercept point, or IP2, can be defined as in Review Question QT14.6.

The development and definitions provided here describe the 1 dB compression point and third-order intercept points in terms of input voltage or power. Sometimes these characteristics are instead described in terms of output voltage or power. Knowing the input–output voltage or power relationship allows conversion from one to the other, but it is important to clarify any ambiguity in definition to avoid misinterpretation.

While it can be seductive to use the analytical approaches outlined in this section to precisely characterize an amplifier's nonlinear performance, in most cases it is not practical to do so. Taylor series models are only approximations; accurately modeling the output may require more precise models that include delays and other effect of memory in the amplifier. Also, the polynomial coefficients are generally not known exactly, and even if they were known they would apply only to a specific device at a given temperature. The most important practical message from this subsection is to pay attention to the input voltage magnitudes that an amplifier will be exposed to, and then ensure the amplifier operates linearly (i.e., the input magnitudes are below the 1 dB compression point and 10 dB below the IP3) for the combination of anticipated interference conditions, filtering, and amplifier specifications. Then the linear small-signal model applies and behavior will be approximately as predicted.

As discussed in Reference 11, an LNA may be implemented in discrete monolithic microwave integrated circuit (MMIC) or included on a larger integrated circuit along with other functions. Gallium Arsenide (GaAs) or silicon-germanium heterojunction bipolar transistor (SiGe-HBT) devices are commonly used. Sometimes two devices are placed in cascade to obtain the desired gain; in that case the characteristics of each device need to be selected carefully as discussed in Review Questions QT14.9 and QT14.10.

There are several practical considerations that cause receiver designers to select amplifiers with limited dynamic range. Amplifiers with a higher dynamic range tend to have lower gain, higher noise figure, higher cost, and higher power consumption; they also tend to be more difficult to integrate onto a monolithic integrated circuit with other functions.

Clearly, the amplifier must provide the dynamic range to accommodate unmitigated in-band interference. But in many cases the dominant blocking occurs from out-of-band interference, so the designer has the opportunity to trade how much filtering is used to protect the amplifier from this interference against how much dynamic range is needed in the amplifier.

14.1.4 Mixers

A mixer, typically portrayed as in Figure 14.6, changes the carrier frequency of a bandpass waveform. A mixer has two inputs. One is the bandpass waveform, and the other is a sinusoid often called the LO, or local oscillator. The output is a bandpass waveform having the same complex envelope as the input signal but a different carrier frequency.

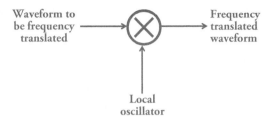

Figure 14.6. Mixer Block Diagram

Typically, a downconversion mixer is used in receivers, with ideal representation using complex envelope notation:

$$x(t)e^{-i2\pi f_{IF}t} = k_1 x(t)e^{-i2\pi f_{RF}t} e^{-i2\pi(f_{IF}-f_{RF})t} \quad (14.19)$$

Here, $x(t)e^{-i2\pi f_{RF}t}$ is the RF input, $e^{-i2\pi(f_{IF}-f_{RF})t}$ is the LO input with LO frequency $f_{LO} = f_{IF} - f_{RF}$, and $x(t)e^{-i2\pi f_{IF}t}$ is the IF output, using the convention that $f_{IF} \geq 0$. An upconversion mixer is used in transmitters to produce the output at RF. Its two inputs are the IF port and the LO port, and its output is the RF port.

Employing a real-valued signal model for a mixer makes it easier to analyze practical considerations:

$$\begin{aligned} v_{out}(t) &= k_1 v_{in}(t) \cos[2\pi f_{RF}t + \theta(t)] \cos[2\pi(f_{RF}-f_{IF})t] \\ &= \frac{k_1 v_{in}(t)}{2}[\cos[2\pi f_{IF}t + \theta(t)] + \cos[2\pi(2f_{RF}-f_{IF})t + \theta(t)]] \end{aligned} \quad (14.20)$$

and as above the LO frequency is $f_{LO} = f_{IF} - f_{RF}$. When $f_{IF} = 0$, the signal is downconverted to baseband; this is called direct conversion.

The mixer is followed by a filter (bandpass when f_{IF} is much larger than the bandwidth of $v_{in}(t)$ and lowpass for direct conversion or when f_{IF} is of the order of the bandwidth of $v_{in}(t)$) that selects the IF term $\frac{k_1 v_{in}(t)}{2} \cos[2\pi f_{IF}t + \theta(t)]$ while suppressing the much higher frequency term $\frac{k_1 v_{in}(t)}{2} \cos[2\pi(2f_{RF}-f_{IF})t + \theta(t)]$. The result of mixing and filtering is then approximated by $v_{out,filtered}(t) = \frac{k_1 v_{in}(t)}{2} \cos[2\pi f_{IF}t + \theta(t)]$.

The model (14.20) is called low-side mixing or low-side injection, since the LO is less than the RF. High-side mixing, also known as high-side injection, uses an LO greater than the RF: $f_{LO} = f_{IF} + f_{RF}$.

In (14.19) and (14.20), $k_1 > 0$ is the small-signal gain. Passive mixers do not provide amplification, and have $k_1 < 1$. Active mixers include an amplifier and typically provide $k_1 > 1$. Mixers cannot be implemented by linear time-invariant circuits; mixer circuits must be either time varying or nonlinear, or both.

An important problem even with ideal mixers involves image frequencies. The real mixer (14,20) takes an input waveform at frequency f and produces an output waveform involving two cosines, one at carrier frequency $f_{IF} = f - f_{LO}$ and one at carrier frequency $f + f_{LO}$.

First consider a low-side mixer that uses $f_{LO} = f_{RF} - f_{IF}$ so for input waveform at carrier frequency f, one output cosine is at carrier frequency $f - f_{RF} + f_{IF}$ and the second is at carrier frequency $f + f_{RF} - f_{IF}$. When the input is at carrier frequency $f = f_{RF}$, the output is at carrier frequency f_{IF} and at carrier frequency $2f_{RF} - f_{IF}$, just as desired—one cosine is at IF, and the other cosine is at a much higher frequency that can be suppressed by filtering the mixer output. (Typically, f_{IF} is selected to be low enough to permit use of a filter with modest Q; any insertion loss can be compensated by amplification before or after the filter.)

If there is an additional input at image frequency $f_{IM} = f_{LO} - f_{IF} = f_{RF} - 2f_{IF}$, say from out-of-band interference that was not completely suppressed by filtering, then an interesting result occurs. Its output waveforms are at carrier frequency $f_{IM} - f_{RF} + f_{IF} = f_{RF} - 2f_{IF} - f_{RF} + f_{IF} = -f_{IF}$ and carrier frequency $f_{IM} + f_{RF} - f_{IF} = f_{RF} - 2f_{IF} + f_{RF} - f_{IF} = 2f_{RF} - 3f_{IF}$. Since the cosine is symmetric, the first cosine places the interference at IF—it has been mixed onto the same frequencies occupied by the desired signal, even though the original carrier frequency may have been far away from the carrier frequency of the desired signal. If $f_{RF} = 2f_{IF}$ or $f_{RF} = f_{IF}$, the second cosine would also be mixed onto the desired signal. But as long as $f_{RF} \gg f_{IF}$, filtering the mixer output will suppress this term. Since interference at frequency $f_{RF} - 2f_{IF}$ is "reflected" onto the desired signal by low-side mixing, the frequency $f_{RF} - 2f_{IF}$ is known as the image frequency for low-side mixing. The image frequency for high-side mixing is $f_{RF} + 2f_{IF}$.

Figure 14.7 illustrates these different mixing architectures with and without image frequencies. Even though only the carrier frequencies are portrayed, there typically is

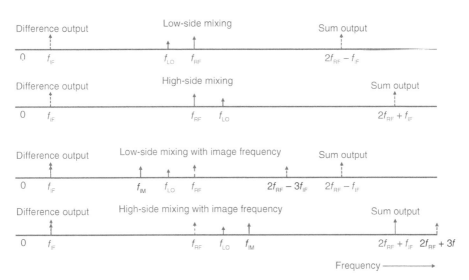

Figure 14.7. Illustration of Low-Side Mixing and High-Side Mixing without and with Image Frequencies

a complex envelope with nonzero bandwidth around the carrier frequency. The sum outputs are readily suppressed by filtering since they are at much higher frequencies than the desired different outputs. Inputs at the image frequency f_{IM} are mixed into the difference output, adding interference to the desired signal that has been mixed to f_{IF}.

Image frequencies do not result from any imperfection of the mixer; they are caused by inadequate suppression of out-of-band interference before mixing. One mitigation is to use RF filters with high selectivity so that expected levels of out-of-band interference are suppressed enough so that the images are small relative to thermal noise and in-band interference. But as discussed in Section 14.1.2, there are challenges to obtaining such RF filters. Another solution is to place a filter after the LNA, with stopband specifically selected to suppress image frequencies. This filter, called an image filter, can trade higher insertion loss for greater stopband loss with minimal effect on front-end performance since it is placed after the LNA, as shown in Section 14.2. Image rejection mixers can also be employed, as described in Reference 12.

There are other ways to mitigate image frequencies. One alternative is to select a high IF, which moves the image frequency far from RF, where the image can be attenuated more by RF filtering before the mixer. However, a high IF comes with a number of disadvantages, including the need for higher Q IF filters and potentially for higher sampling rates in the ADC. A second alternative is to choose between low-side or high-side mixing with associated IF values, so that the image frequency is centered in a part of the frequency band known to be "quiet"—that is, a band where transmitters are expected not to transmit much power into the satnav receive antenna. A third alternative is to use an IF of zero Hz, in which case $f_{RF} - 2f_{IF} = f_{RF}$ and there is no out-of-band image frequency.

Still another option is similar to that encountered in previous discussions of amplifiers and filters—use cascades rather than a single device. In the case of mixers, the result is known as a superheterodyne receiver, having two or more mixer stages, each downconverting to sequentially lower IFs. As discussed in Section 14.3.1, each stage uses appropriate IFs, filters, and amplifiers to address the issues discussed in this and previous subsections (e.g., interference dynamic range, insertion loss, group delay distortion, compression, image frequencies) to produce the desired signals with adequate quality at the final IF. Properly engineered cascading components reduce the need for any single component to be of exquisite quality. However, with cascaded components there are more components (with associated higher costs for assembly and more space consumed by the circuitry) and potentially higher power consumption.

The preceding discussion has assumed the ideal mixer equation (14.20) applies. In reality, the small-signal gain has the same types of nonlinearities discussed in Section 14.1.3, and input voltage levels must be commensurate with the dynamic range of the mixer. In addition, the mixer output contains versions of $v_{in}(t)$ at frequencies that are sums of integer multiples of the RF and integer multiples of the LO. Also, the LO input is not a pure sinusoid, but actually the desired sinusoid, plus noise, plus harmonics. Section 14.1.5 discusses the effect of these LO imperfections on mixer performance.

There are many mixer implementations, from one as simple as a single transistor to entire circuits that provide much better performance. As with other types of front-end components, cost, size, and power consumption must be taken into account when choosing a mixer. In addition, since mixer inputs and outputs must be more carefully conditioned when the mixer has less capability, the selected mixer characteristics affect the choice of other front-end components as well.

14.1.5 Oscillators and Frequency Synthesis

Reference oscillators produce a stable sinusoid used by multiple frequency synthesizers that develop different LO and clock frequencies used in the receiver. Reference oscillators drift, and also contribute phase noise (also known as jitter) that can degrade the performance of carrier tracking and code tracking. Frequency synthesizers are affected by this phase noise, but in addition frequency synthesizers produce spurious waveforms at frequencies other than the desired frequency, and these spurious waveforms can also affect performance of the receiver front end. These considerations are addressed in this subsection.

While an ideal oscillator would produce a pure sinusoid of constant frequency, real oscillators vary from this ideal. With aging, temperature changes, and changes in power supply voltage, the output frequency drifts slowly. This frequency drift is typically modeled as a deterministic variation over time, often using a Taylor series approximation over a finite time interval. Here, the Taylor series coefficients are random with specified distributions, or estimated during a calibration procedure. Since the deterministic variations of frequency, or frequency drift, occur on a relatively long time scale of minutes, hours, and days, they have little effect on receiver processing.

In addition, the frequency fluctuates rapidly and randomly due to noise as well as external stimuli such as vibration. This phase noise is characterized by the Allan deviation [13] and its square, the Allan variance. Since phase noise is well modeled as a random walk whose variance is not defined, the Allan deviation characterizes the phase variation over specific time scales so the variance is finite. Time scales of the order of a second are of most interest to receiver processing. The Allan deviation, $\sigma_y(\tau)$, is expressed as a dimensionless measure of deviation over a specific time interval τ, sometimes called the gate time.

The deviation of an oscillator's frequency, when the oscillator's nominal frequency is f_0 Hz, over a time scale of the order of τ, has a standard deviation of $\sigma_y(\tau)f_0$ Hz. Timekeeping accuracy based on this oscillator jitters over a time scale of τ seconds by $\sigma_y(\tau)\tau$ seconds. For example, if $\sigma_y(\tau) = 10^{-9}$ over a gate time of 3 seconds, the standard deviation of a 1575.42 MHz frequency source driven by this oscillator is approximately 1.6 Hz over a 3 second interval, and over this time interval there would be a high probability that the frequency remains within ± 5 Hz of its initial value. The standard deviation of a timekeeping circuit driven by this oscillator would be 3 ns over 3 seconds, and there would be high probability that the reported time is accurate within ± 10 ns after 3 seconds pass.

The deterministic variations dictate the drift of time keeping and frequency keeping over time intervals when the receiver's oscillator is not being disciplined by time and frequency from satnav signals. The time drift and frequency drift contribute to the initial time uncertainty and initial frequency uncertainty faced in initial synchronization processing, as discussed in Chapter 16. The time drift must also be accounted for in position calculations using signals from only three satellites, as discussed in Chapter 20. The random fluctuations (phase noise or jitter), which occur on time scales of fractions of a second or seconds, affect the receiver's ability to track signals as seen in Chapter 18. These random fluctuates include not only those characterized by Allan deviation, but also those introduced by vibrations, including acceleration stresses.

The reference oscillator produces a sinusoid at a typical frequency of multiple MHz (near 10 MHz is a common value). The frequency synthesizer, or LO, is driven by the reference oscillator. Typically, a phase-locked loop (PLL) is used to divide down the LO frequency to a common value of the order of tens or hundreds of kHz. This value is then multiplied by different values to provide the needed LO frequencies, clock frequencies, and other frequency and timing needs of the receiver circuitry. The LO PLL is phase locked to the reference oscillator frequency.

In practice, the LO is not the pure sine wave that is desired; there are phase noise and harmonics of the common LO frequency as well as harmonics of the reference oscillator frequency. The harmonics of the common LO frequency are typically well suppressed and not a cause for concern. However, the phase noise and any harmonics of the reference oscillator frequency can cause out-of-band interference to be mixed into the frequency band in a phenomenon known as reciprocal mixing. If the LO has a harmonic or phase noise at a frequency f_Δ above the intended LO frequency f_{LO}, then interference f_Δ above the intended carrier frequency f_{RF} is downconverted to f_{IF}, passing through the post-mixing filter and adding to the thermal noise and in-band interference. Suppressing reciprocal mixing involves a combination of filtering the input before mixing and ensuring adequately low harmonics and phase noise in the reference oscillator. Thus, there is an opportunity to trade quality of the preselection filtering and quality of the reference oscillator.

When interference has a carrier frequency offset from the RF by an amount approximately equal to the frequency of the reference oscillator or one of its harmonics, the interference is mixed to the IF by the LO harmonics. The interference thus is converted to IF and passes through the post-mixing filter, adding to other noise and in-band interference faced in subsequent receiver processing. Practical considerations and alternatives for oscillators are extensively addressed in Reference 14. A standard crystal oscillator (XO) is the least expensive and uses the least power, but has the greatest drift and Allan deviation. A typical specification for XO drift and aging over a wide temperature range could be ±5 parts per million (ppm), meaning that a 1575.42 MHz frequency derived from that source would be in the approximate range ±8 kHz—a large value.[7] Consequently, XOs are typically not used in satnav receivers except perhaps for crude timekeeping while the receiver is mostly powered down. A temperature-compensated crystal oscillator (TCXO) uses analog or digital circuitry to counteract the crystal's natural change in resonant frequency with temperature, typically obtaining on the order of a five times reduction in drift compared to an XO and a reduction in Allan deviation by

---

[7] Crystal drift over time, called aging, is also a potential issue but its effects are typically significantly less than those of temperature variations. Crystal frequencies also vary with changes in applied voltage, but once again variation with temperature dominates. Shock and vibration also influence phase noise in the oscillator output.

a factor of two. TCXOs are commonly used in low-cost satnav receivers, but their Allan deviation can limit the performance of signal tracking loops as described in Chapter 18. An oven-controlled crystal oscillator (OCXO) involves a small heater and an insulated package to keep the crystal at nearly a constant temperature, significantly reducing the frequency drift—perhaps by two orders of magnitude relative to a TCXO. OCXOs also have substantially smaller Allan deviation—also as much as two orders of magnitude less than a TCXO. Yet their cost and power consumption make them poorly suited for many consumer applications. While all crystals require warm-up time after power is applied, OCXO warm-up time is typically much longer than that for XOs or TCXOs, and OCXO performance is much more sensitive to warm up.

14.2 FRONT-END NOISE FIGURE

If all the front-end components were ideal, the output of the receiver front end would be a digital version of the input desired signal and interference, translated to the IF used for receiver processing. However, even in this case of ideal components, noise would be added due to the motion of atomic particles in the receiver front end electronics. As introduced in Section 14.1.3, ideal electronics at room temperature would introduce a power spectral density $N_{0,B} = k_B T$, where $k_B = 1.381 \times 10^{-23}$ J/°K (equivalently, watts/(°K-Hz)) is the Boltzmann constant and T is the temperature in degrees Kelvin. Standard room temperature is denoted T_0 and defined to be 290°K, so the resulting noise PSD for ideal electronics at room temperature is $N_{0,B} = -204.0$ dBW/Hz. If there were no interference, effective C/N_0 as described in Chapter 5 would be calculated using this noise PSD, with C defined at the input to the receiver.

But the components described in Section 14.1, along with others as simple as the cable that connects antenna to preamplifier, are not ideal. Their electronics produce more noise than predicted by $N_{0,B} = k_B T$, and some components introduce losses as well, while others introduce gains. This section describes how to account for the way these device characteristics degrade C/N_0 by representing their aggregate effect as an increase in noise PSD. This increase in noise PSD can then be discussed in different ways. The ratio $N_0/N_{0,B}$, for PSDs expressed in W/Hz, is called the front end's noise factor, a dimensionless quantity. When PSDs are expressed in dBW/Hz, the difference $N_0 - N_{0,B}$ is called the receiver noise figure, and is equal to the noise factor expressed in decibels. Equivalently, the higher PSD can be considered as resulting from ideal electronics exposed to a higher ambient temperature, denoted the effective temperature of the receiver front end, defined by $T_{\text{eff}} \stackrel{\Delta}{=} N_0/k_B$. The difference between the effective temperature and 290°K is called the effective noise temperature increase. This section describes how to calculate T_{eff} and thus N_0.

Besides the components in the receiver front end like those in Figure 14.1, there are interconnections formed by cables and wires. When they have low losses, their effects can be neglected, but longer cable runs can introduce losses of 1–10 dB or more and must be taken into account.

The model is based on a lumped element representation of the cascade of components. In this model, changes in carrier frequency are irrelevant, since the important consideration is the effect of components on desired signal power and on noise PSD, so only the desired signal PSD and noise PSD are addressed. A typical cascade of components might be antenna, cable, prefilter, LNA, cable, filter, amplifier, filter, mixer, filter.

Even though the components affect desired signal power and noise PSD, the model describes all effects in terms of changes to the noise PSD referred to the receiver input, with no change to the received power of the desired signal. The effective temperature of the receiver front end, T_{eff} is then greater than or equal to room temperature T_0, and the carrier power to noise PSD ratio at the output of the receiver front end is $\frac{C}{k_B T_{\text{eff}}} \leq \frac{C}{k_B T_0}$.

The terminology here follows the following convention: terms called a "factor" are power ratios not expressed in decibels, while terms called a "figure" are power ratios expressed in decibels. Using the formulation in Reference 15, distinct parametric models are provided for the receive antenna, for any passive component, and for any active component. The antenna is modeled by its effective temperature, passive components and active components are each described by a gain factor, G, and noise factor, F. The noise factor and gain factor used in the model are power ratios, and must be converted out of decibels.

The effective temperature for the receive antenna, denoted T_{ant}, is typically less than room temperature since the antenna's beam pattern typically encompasses the sky and satellites in space, which are very cold, approaching 0°K. However, within the beam pattern also may be structures, terrain, vegetation, vehicles, and even people, all of which may be warmer than room temperature. A typical value used for T_{ant} is 130°K, which accounts for the combined effects of cold sky, warmer surroundings, and antenna gain.

A specification for a passive component describes its loss figure. The loss factor L is the resulting power ratio, so if the input power of the desired signal is C_{in} W, the output power is $C_{\text{out}} = C_{\text{in}}/L$ W. For the passive component model, the noise factor is $F = L$ and the gain factor is $G = 1/L$. For an ideal conductor, $L = F = G = 1$. Typically the noise factor and gain factor are expressed in decibels as the noise figure $10 \log_{10}(F)$ dB and the gain figure $10 \log_{10}(G)$ dB. For example, if the insertion loss figure of a filter is 3 dB, $L = 2$, $F = 2$, and $G = 1/2$; the noise figure is 3 dB and the gain figure is -3 dB.

A specification for an active component describes its noise figure and its gain figure. The noise factor is the noise figure expressed as a power ratio, and the gain factor is the gain figure expressed as a power ratio. For example, if the noise figure is 6 dB and the gain figure is 20 dB, $F = 4$ and $G = 100$.

The effective temperature for a concatenated set of five components, numbered in sequence starting with the first component after the antenna, is then [15]

$$\begin{aligned} T_{\text{eff}} &= T_{\text{ant}} + \left[(F_1 - 1) + \frac{(F_2 - 1)}{G_1} + \frac{(F_3 - 1)}{G_1 G_2} + \frac{(F_4 - 1)}{G_1 G_2 G_3} + \frac{(F_5 - 1)}{G_1 G_2 G_3 G_4} \right] T_0 \\ &= T_{\text{ant}} + \left[(F_1 - 1) + \frac{1}{G_1} \left[(F_2 - 1) + \frac{1}{G_2} \left[(F_3 - 1) \right. \right. \right. \\ &\quad \left. \left. \left. + \frac{1}{G_3} \left[(F_4 - 1) + \frac{(F_5 - 1)}{G_4} \right] \right] \right] \right] T_0 \end{aligned} \quad (14.21)$$

Clearly the nesting of brackets can be continued based on the number of components. Denoting the effective noise factor for the entire circuit as F_{eff} allows (14.21) to be rewritten as

$$T_{\text{eff}} = T_{\text{ant}} + (F_{\text{eff}} - 1)T_0$$

where

$$F_{\text{eff}} = F_1 + \frac{1}{G_1}\left[(F_2 - 1) + \frac{1}{G_2}\left[(F_3 - 1) + \frac{1}{G_3}\left[(F_4 - 1) + \frac{(F_5 - 1)}{G_4}\right]\right]\right] \quad (14.22)$$

The review questions provide opportunities to explore this model and its implications.

Suppose the gain of the nth (active) device is very large. Then $1/G_n$ is very small, making the term $\frac{1}{G_n}[(F_{n+1} - 1) + \frac{(F_{n+2}-1)}{G_{n+1}}[\cdots]]$ so small that it can be neglected to good approximation, as long as $G_n \gg F_m, \forall m > n$. All subsequent components in the circuit have negligible effect on the effective noise factor, hence on the effective temperature of the circuit, and the effective noise factor can be approximated by the first n brackets.

This result shows that the receiver noise figure is dominated by losses of the passive components and the noise figures of the active components that include and precede the first high gain amplifier. Placing the preamplifier before as many other components as possible is critical for achieving a low noise figure. This strategy is consistent with active antennas, which co-locate the first amplifier with the antenna and before any cable connecting antenna to receiver. However, this strategy conflicts with the need to precede the first amplifier with a capable prefilter (with its associated insertion loss) in order to protect this amplifier from strong out-of-band interference.

14.3 FRONT-END ARCHITECTURES AND FREQUENCY PLANS

The components discussed and analyzed in Sections 14.1 and 14.2 can be assembled in various ways to perform the filtering, amplification, frequency translation, and digitization functions of the receiver front end. This section describes three common front-end architectures, along with their respective advantages and disadvantages. An extensive and clear discussion of alternative receiver architectures is found in Reference 16. Like much of the literature on this topic, it focuses on the general topic of communications receivers, which have two important differences from satnav receivers. First, communication receivers generally need to be tuned to different center frequencies, or channels, while satnav receivers generally do not need this capability since they use a small number of fixed carrier frequencies. Second, satnav receivers increasingly receive signals simultaneously on two or more carrier frequencies, while communication receivers often receive only one signal at a single carrier frequency at a time. Efficient architectures are still being developed for multifrequency receivers, and some related topics are explored in Review Question QT14.11.

The receiver's frequency plan describes the choices of frequencies used for the frequency source, mixers, sampling, and other clocks. These frequencies are carefully selected to address the considerations already described in mixing and filtering, subject to constraints of cost and power consumption. The different architectures described in this section involve differences in frequency plan and components. Section 14.3.1 addresses several variations of heterodyne and superheterodyne architectures. Section 14.3.2 describes a homodyne, or zero-IF architecture, while Section 14.3.3 shows how to use aliasing to achieve downconversion without a mixer by using a direct sampling architecture. The discussion focuses on key components and signal flow, not addressing ancillary aspects such as the reference oscillator and frequency synthesizer.

All of the architectures, as described, can be implemented with additional filtering and gain stages. Conversely, some of the filters and amplifiers that are described in some architectures could be omitted if a designer were to sacrifice come performance and robustness in order to use fewer parts. The invariants are the mixers and the LOs, while some of the amplifiers and filters are only representative.

14.3.1 Heterodyne and Superheterodyne Front Ends

A heterodyne architecture is shown in Figure 14.8, with nominal amplification and filtering, including image rejection (IR) filtering after the preamplifier. (This positioning allows a greater selectivity IR filter to be used, since any accompanying insertion loss is better accommodated after the preamplifier.) The IF filter also serves as an antialiasing filter before digitization. AGC and ADC details are discussed in Chapter 15. The sampling clock input is shown at frequency f_s, which must be greater than twice the frequency where the IF filter's upper edge transition band meets the stop band in order to avoid aliasing frequencies in the transition band. The IF amplifier can also be used for AGC; in that case there would be feedback from the ADC or its output to the IF amplifier, as discussed in Chapter 15.

Heterodyne architectures use a small number of components, and hence can be compactly packaged. However, the performance demands on each component are great, typically requiring more costly components and often consuming greater power. Also, the digitized waveform is centered at a digital IF, increasing the required sampling rate. The higher sampling rate at IF also increases digital processing and storage demands unless there is subsequent digital conversion to baseband, as discussed later in this subsection.

Figure 14.9 shows an extension of the heterodyne architecture, called the superheterodyne. Here, two or more mixing stages reduce the demands on the heterodyne's single stage. (The mixing inside the box with the dashed line could be replicated at a

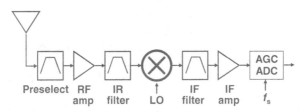

Figure 14.8. Heterodyne Architecture

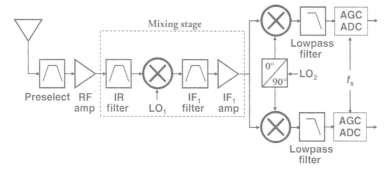

Figure 14.9. Superheterodyne Architecture

second, lower, LO, performing the frequency conversion in more steps.) The final stage uses in-phase/quadraphase (I/Q) mixing to preserve the output waveform independent of its phase relative to the phase of the second LO. As pointed out in the discussion of the heterodyne architecture, the last amplifier before the ADC can serve as an AGC, as long as the appropriate feedback is present. The sampling clock input is shown at frequency f_s, which must be greater than twice the frequency where the lowpass filter's transition band meets the stop band, in order to avoid aliasing frequencies in the transition band. Both ADCs should sample simultaneously.

The superheterodyne architecture uses more components, but places less demand than the heterodyne architecture on the characteristics of each component. Amplification and filtering are distributed across multiple devices, and this architecture is often selected for operation in the most stressing out-of-band interference conditions. The ADCs can operate at a slower rate consistent with the bandwidth of the waveform's complex envelope. Conversely, the multiple mixers generate multiple spurious frequencies that have to be accounted for through a combination of frequency plans that place them outside the frequency bands of interest, and filtering. Also, the additional components increase the physical size of the circuit, and increase power consumption. Finally, the final parallel chains must be carefully matched, including matching the ADC characteristics and the sample timing. In the superheterodyne architecture, the antialiasing filter bandwidths are half those in the heterodyne architecture, and the sampling rate of each ADC is half as well.

Digital IF architectures are a hybrid analog/digital implementation of a superheterodyne architecture that can be used as an adaption of Figure 14.9. Here, a single ADC would be placed after the IF_1 amplifier (which could serve as the AGC), and the second stage mixing and lowpass filtering would be implemented digitally to produce baseband I and Q waveforms for digital processing. The use of digital processing in the final stage enables very precise control of filtering and mixing, with no device imperfections introduced and no challenges matching the I and Q channels. In particular, the filtering can have excellent selectivity and perfectly linear phase. However, the ADC typically uses more bits of sampling than needed for correlation processing as described in Chapter 15, and must operate at a higher sampling rate to match the highest unfiltered frequency at IF. Also, the digital processing for filtering and mixing tends to consume more power than an

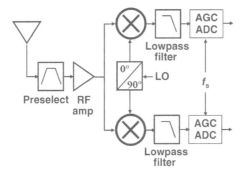

Figure 14.10. Homodyne Architecture

analog implementation. When this architecture is employed, the digital downconversion may be followed by decimation and resampling with a smaller number of bits, consistent with the results in Chapter 15, to reduce the complexity of subsequent processing.

14.3.2 Homodyne Front Ends

The superheterodyne architecture can be simplified, in concept, as shown in Figure 14.10, yielding an architecture known as homodyne, direct conversion, or zero IF. Here, the first stage mixer and supporting circuitry, in the box with dashed lines shown in Figure 14.9, have been removed, and the I/Q mixing converts directly from RF to baseband, using an LO that is equal to the RF. Digitizing rate considerations are the same as in the superheterodyne architecture.

In the homodyne architecture there is no image problem since the IF is zero, eliminating the need for an IR filter as discussed in Section 14.1.4. The number of filters and amplifiers has been reduced, making the circuit more compact and reducing its power consumption. The ADC sampling rate is minimal. However, the RF filter provides all the selectivity, meaning this architecture is typically less resistant to out-of-band interference than a superheterodyne receiver. AGCs are required, unless one-bit sampling is performed as discussed in Chapter 15. Further, any leakage of the LO into the RF path produces a DC term at the mixer outputs that can contaminate the AGC and ADC. In addition, all the gain is provided in the RF amplifier stage, resulting in a tendency for oscillation or ringing. The trend toward higher-quality components and the desire for smaller and simpler receiver front ends both favor the homodyne architecture for modern receivers.

14.3.3 Direct Sampling Front Ends

Direct sampling receivers, also known as direct digitization receivers use bandpass sampling [17] that relies on aliasing to provide frequency downconversion without any mixing. Figure 14.11 shows the direct sampling architecture—the epitome of simplicity. While a preselection filter could be included, it is not shown in order to emphasize the potentially small number of parts required with direct sampling. As discussed subsequently, the sampling rate must be chosen carefully based on the carrier frequency and

FRONT-END ARCHITECTURES AND FREQUENCY PLANS

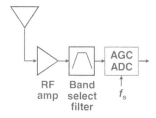

Figure 14.11. Direct Sampling Architecture

the bandwidth selected by the bandpass filter. The sampled waveform is at digital IF, and can either be processed at that frequency or digitally downconverted as discussed in Sections 14.3.1 and 14.3.2.

The direct sampling architecture's great advantage is its small number of components, yet the demands on each component are high. The RF amplifier must provide the dynamic range needed to accommodate out-of-band interference, although the dynamic range need can be reduced if a prefilter is added. The bandpass filter needs excellent selectivity to avoid aliasing waveforms in the stopband into the downconverted passband. The ADC needs to operate at very low jitter [18], since deviations in sampling times relative to the carrier frequency add noise to the quantized waveform. Just as in the homodyne architecture, all of the gain is provided in the RF amplifier stage, making it difficult to avoid ringing or oscillation, particularly when all components are in a small package. The direct-conversion architecture is the ultimate all-digital receiver, and the goal of many research projects.

To determine the sampling rate for a single-channel direct sampling receiver, follow the formulation in Reference 17. Denote the range of frequencies to be sampled as $f_L < f < f_H$, noting the strict equalities, since the endpoint frequencies will be aliased. The band select filter should attenuate all power at frequencies outside this range to low enough levels that the aliased versions contribute a negligible amount to the noise in the desired passband. Equivalently, the upper and lower frequencies should span, at a minimum, the transition bands and passband of the band select filter, to avoid aliasing in power from the transition bands. In addition, a guard band can be added. Define the resulting bandwidth to be downconverted as $\beta = f_H - f_L$.

An acceptable uniform sampling frequency f_s is any value within the range

$$\frac{2f_H}{n} \leq f_s \leq \frac{2f_L}{n-1} \tag{14.23}$$

where n is any integer given by

$$1 < n \leq \left\lfloor \frac{f_H}{\beta} \right\rfloor \tag{14.24}$$

where $\lfloor \ \rfloor$ indicates the next smaller integer, also called the floor.

The lowest center frequency, the resulting digital IF, is given by the smallest positive value of $(f_L + f_H)/2 - k2f_s$, for k an integer. Sampling rates for multiple-channel direct sampling receivers are selected to alias the different channels to adjacent and nonoverlapping bands at digital IF, suited for subsequent digital downconversion [19, 20].

14.4 SUMMARY

The art and science of designing satnav receiver front ends merits a book in its own right. This chapter has provided only an initial examination of the daunting set of interacting tradeoffs and decisions involving selection of architectures, frequencies, and components for designing a satnav receiver front end. In general, using many large, high-quality, and power-hungry components yields a front end with the best performance, but there may be no business case for such a design. Hence, there is a need to select architectures and components that merely meet the conditions under which the receiver needs to function adequately, recognizing that more stressing conditions can produce degraded operation. Unfortunately, these conditions are becoming more challenging, with the increasing presence of powerful out-of-band interference and the imminent availability of many satnav signals at many different frequencies.

Designing a receiver front end for the highest sensitivity in a weak signal environment such as that encountered under thick foliage is very different from designing a front end to maintain operation in a high interference environment. Designing for both conditions requires some compromises.

The receiver front end primarily should "do no harm." While antennas and filters can suppress interference and enhance the desired signal, for the most part the front end is designed simply to provide a digitized signal with acceptable degradation to the signal and its C/N_0. Ultimately, it is the signal processing that provides the exquisite performance, as long as the receiver front end has preserved adequate signal quality.

Antennas are often physically the largest single component, as well as the component that can most influence performance. For some classes of receivers, severe size and cost constraints limit antenna performance, and the rest of the receiver must make up for the resulting low C/N_0.

Filtering affects amplification and mixing. Recent publicity concerning strong out-of-band interference has increased awareness of the need for good filtering, along with strategies such as image rejection and the suppression of reciprocal mixing, in order to sustain operation with interference in adjacent frequency bands.

One of the few universal rules of thumb in front-end design is to ensure that only components with low losses and low noise figures should precede the first amplifier, and that the first amplifier should have high gain. This strategy yields a low effective noise figure for the front end, and results in high sensitivity. But implementing that strategy can conflict with the need for good preselection filtering to mitigate out-of-band interference and to protect the dynamic range of that first amplifier.

Dynamic range is an essential characteristic of the front end. It complements the emphasis on low noise for sensitivity, with a corresponding emphasis on handing large magnitude input waveforms. Since the power of satnav signals is less than that from receiver noise, the dynamic range is set by the interference power that must be accommodated, not by the signal power.

In general, there is a tension between obtaining low noise figure and accommodating higher levels of interference. As satnav receivers are exposed to greater levels of interference (from sources external to the receiver as well as from electronics in the same package as the receiver), there may be a trend toward designs and components that accommodate higher levels of interference, at the expense of higher noise figures.

Many different receiver architectures can be selected and adapted, each with different considerations influencing component count, component quality, size, and power consumption. Technology evolution yielding better ADC performance tends to favor earlier digitization and digital implementation of more front-end functions, such as receivers with digital IF, homodyne receivers, and direct-conversion receivers.

The remaining sections in Part III, which address receiver processing, assume a well-designed receiver front end that provides the signals with the power and quality needed for signal processing, without out-of-band interference, saturation of amplifiers and mixers, or presence of image frequencies or reciprocal mixing.

REVIEW QUESTIONS

Theoretical Questions

QT14.1. Use (14.4) to derive the approximation $G \cong \frac{29000}{\beta_\theta \beta_\phi}$ where β_θ and β_ϕ are, respectively, the azimuth and elevation beamwidths in degrees. Show for small beamwidths that $G \cong \frac{22689}{\beta_\theta \beta_\phi}$, for large beamwidths that $G \cong \frac{35640}{\beta_\theta \beta_\phi}$, and the result $G \cong \frac{29000}{\beta_\theta \beta_\phi}$ is their average.

QT14.2. Use expressions from Section 4.1 to show that the effective area of a receive antenna, in units of squared wavelengths, is given by

$$\frac{A_R}{\lambda^2} \cong \frac{1300}{\beta_\theta \beta_\phi}$$

where A_R is the effective area of the antenna, λ is the wavelength, β_θ and β_ϕ are, respectively, the azimuthal beamwidth and elevation beamwidth, each in degrees, of the antenna beam.

QT14.3. Derive the expression for input voltage at the saturation point (14.11).

QT14.4. Derive the input voltage at the fifth-order 1 dB compression point (14.12), and show there is a 1 dB compression point for a fifth-order model if $k_3^2 \geq 0.4348 k_1 k_5$.

QT14.5. Derive the expression for sinusoidal third-order 1 dB compression point in (14.16).

QT14.6. Define IP2 as the input voltage a of a sinusoidal input $v_{in}(t) = a\cos\left(2\pi f_0 t\right)$ applied to an amplifier characterized by a quadratic symmetric nonlinearity $v_{out}(t) = k_1 v_{in}(t) + k_2 v_{in}^2(t)$, where the magnitude of the output second harmonic is equal to the magnitude of the output first harmonic. Show that this

IP2 is $a'_{IP2} = \frac{2k_1}{k_2}$. Examine the frequency content of the output and discuss why IP2 is typically of less concern in receiver design than IP3.

QT14.7. Now consider the results of inputting the sum of two sinusoids into a nonlinear amplifier. The input is $v_{in}(t) = a_1 \cos(2\pi f_1 t) + a_2 \cos(2\pi f_2 t)$, where both input frequencies f_1 and f_2, are within or near the frequency range of interest so that they are not substantially attenuated before the amplifier.

a. If the amplifier is modeled by the quadratic symmetric nonlinearity in QT14.6, what frequencies occur in the output, and which are of greatest concern and why?

b. If the amplifier is modeled by the cubic odd-symmetric nonlinearity (14.13), what frequencies occur in the output, which are of greatest concern, and why?

QT14.8. Examine the effect of out-of-band interference, using the results of QT14.7.

a. Suppose there is out-of-band interference consisting of a sinusoid 50 MHz below the carrier frequency of the desired signal at 1575.42 MHz, and this interference is not completely suppressed by filtering. If the amplifier is characterized by a cubic odd-symmetric nonlinearity, what frequencies are produced at the output and are any of them of particular concern? Repeat for an interfering sinusoid 25 MHz below the carrier frequency.

b. Suppose there is out-of-band interference consisting of a pair of sinusoids, one 25 MHz below center frequency and one 50 MHz below the center frequency of the desired signal at 1575.42 MHz, and this interference is not completely suppressed by filtering. If the amplifier is characterized by a cubic odd-symmetric nonlinearity, what frequencies are produced at the output and are any of them of particular concern?

QT14.9. Consider cascaded amplifiers by assessing two important characteristics: 1 dB compression and noise figure.

a. Develop an approximate third-order nonlinear model for models of two amplifiers cascaded, when each amplifier's nonlinearity is modeled as a cubic odd-symmetric nonlinearity, assuming no delay in the amplifiers. Which amplifier's compression is more important—the first or the second?

b. Find the equivalent noise figure of the cascaded amplifiers. Which amplifier's noise figure is more important?

c. Summarize the resulting strategy for selecting the order of amplifiers in a cascade.

QT14.10. Consider cascaded passive components by assessing the noise figure of two cascaded passive components. Summarize the resulting strategy for selecting the order of passive components in a cascade.

QT14.11. Explore the frequency plan of a dual-frequency satnav receiver. Suppose the two carrier frequencies are denoted f_1 and f_2, with $f_1 > f_2$. Show that if $f_{LO} = (f_1 - f_2)/2$, a combination of low-side injection and high-side injection can be used with a single LO to produce the same IF for both carrier frequencies. Find numerical values of the LO and IF frequencies for three different pairs of carrier frequencies: L1 and L2, L1/E1 and L5/E5a, and E1 and E6.

Application Questions

QA14.1. Suppose a wideband receiver has precorrelation bandwidth of 30 MHz. For ideal electronics at room temperature, derive the total noise in the receiver front end. What received signal power would be 10 dB less than this total noise? Repeat for a narrow precorrelation bandwidth of 3 MHz.

QA14.2. For an amplifier modeled by (14.10) with $k_1 = 50$, $k_3 = -32 \times 10^4 \text{V}^{-2}$, and $k_5 = 4 \times 10^8 \text{V}^{-4}$, find the input voltage at saturation, the 1 dB compression points using (14.12), the 1 dB compression point using (14.14), the 1 dB compression point for sinusoidal inputs using (14.16), the third-order intercept point (14.17), and the alternate third-order intercept point (14.18).

QA14.3. Consider a receiver front end consisting of five components: antenna, cable and prefilter, LNA, cable and filter, mixer. Let the cable and prefilter loss be 0.5 dB, the LNA gain be 25 dB, the LNA noise figure be 1.9 dB, the loss of the second cable and filter is 1 dB, and the mixer noise figure is 9 dB.

   a. Find the noise figure and the effective noise power spectral density of the receiver front end noise for antenna temperature 130.0°K.

   b. Find the noise figure and the effective noise power spectral density of the receiver front end noise for antenna temperature 75.0°K.

   c. Find the noise figure and the effective noise power spectral density of the receiver front end noise for antenna temperature 130.0°K, if the mixer noise figure is increased by 6 dB to 15 dB.

   d. Comment on the results.

REFERENCES

1. C. J. Hegarty, D. Bobyn, J. Grabowski, and A. J. Van Dierendonck, "An Overview of the Effects of Out-of-Band Interference on GNSS Receivers," Proceedings of ION-GNSS-2011, Institute of Navigation, September 2011.

2. RF Design for GPS Receivers, u-blox Application Note GPS-X-0215, 2002, available at http://www.datasheetarchive.com/GPS-X-02015-A-datasheet.html, accessed 11 October 2013.

3. S. A. Maas, *The RF and Microwave Circuit Design Cookbook* (*Artech House Mobile Communications*), Artech House.

4. J. J. H. Wang, "Antennas for Global Navigation Satellite System (GNSS)," *Proceedings of the IEEE*, Vol. 100, No. 7, July 2012, pp. 2349–2355.
5. B. Rama Rao, W. Kunysz, R. Fante, and K. McDonald, *GPS/GNSS Antennas*, Artech House, 2013.
6. GPS Antennas: RF Design Considerations for u-blox GPS Receivers, u-blox Application Note GPS-X-08014, 2009, available at https://www.u-blox.com/sites/default/files/products/documents/GPS-Antenna_AppNote_(GPS-X-08014).pdf, accessed 20 September 2015.
7. J. G. Van Bladel, *Electromagnetic Fields*, 2nd edition, Wiley-IEEE Press, 2007.
8. J. S. McLean, "A Re-examination of the Fundamental Limits on the Radiation Q of Electrically Small Antennas," *IEEE Transactions on Antennas and Propagation*, Vol. 44, No. 5, 1996, pp. 672–675.
9. 2015 GPS World Antenna Survey, http://gpsworld.com/wp-content/uploads/2015/02/GPSWorld_2015_AntennaSurvey.pdf, accessed 20 September 2015.
10. J. W. Betz, "Effect of Linear Time-Invariant Distortions on RNSS Code Tracking Accuracy," Proceedings of the Institute of Navigation Conference on Global Navigation Satellite Systems 2002, ION-GNSS-2002, Institute of Navigation, September 2002.
11. National Space-Based Positioning, Navigation, and Timing Systems Engineering Forum (NPEF), "Assessment of Light Squared Terrestrial Broadband System Effects on GPS Receivers and GPS-dependent Applications," 28 June 2011.
12. T. Ihalainen and M. Multanen, "Radio Frequency Mixer with an Image Reject Filter for PMR446 Radio Receiver," available at https://www.ele.tut.fi/teaching/elc-3400/vastaanotinrakentelu/Ryhmien_raportit/pdf/05.pdf, accessed 16 October 2013.
13. D. Allan, "Statistics of Atomic Frequency Standards," *Proceedings of IEEE*, Vol. 54, No. 2, 1966, pp. 221–230.
14. "Fundamentals of Quartz Oscillators," Hewlett-Packard Application Note 200-2, available at http://literature.agilent.com/litweb/pdf/5965-7662E.pdf, accessed 14 October 2013.
15. F. van Diggelen, *A-GPS: Assisted GPS, GNSS, and SBAS*, Artech House, 2009.
16. Md. Maruf Hossain, "Design of RF Front End for Multi-Band Multi-System GNSS Receivers," Master's Thesis, University of Gävle, 31 January 2008.
17. R. G. Vaughan, N. L. Scott, and D. R. White, "The Theory of Bandpass Sampling," *IEEE Transactions on Signal Processing*, Vol. 39, 1991, pp. 1973–1984.
18. A. Brown and B. Wolt, "Digital L-Band Receiver Architecture with Direct RF Sampling," IEEE Position Location and Navigation Symposium, April 1994, pp. 209–216.
19. C.-H. Tseng and S.-C. Chou, "Direct Downconversion of Multiple RF Signals Using Bandpass Sampling," IEEE International Communications Conference, 2003.
20. D. M. Akos, M. Stockmaster, J. B. Y. Tsui, and J. Caschera, "Direct Bandpass Sampling of Multiple Distinct RF Signals," *IEEE Transactions on Communications*, Vol. 47, No. 7, July 1999, pp. 983–988.

15

ANALOG-TO-DIGITAL CONVERSION

Analog-to-digital conversion (ADC) provides the transition between the analog receiver front-end capabilities described in Chapter 14 and the digital receiver processing described in Chapters 16 through 20. While the receiver architectures introduced in Chapter 14 include ADC, many of these architectures produce digitized waveforms at IF. These waveforms are typically digitally downconverted to baseband, then decimated (resampled at a lower sampling frequency) and with a smaller number of bits per sample before digital receiver processing in order to reduce storage and processing demands.

> This chapter is based in part on instructional materials prepared by Dr. Alessandro P. Cerruti, The MITRE Corporation. These materials were prepared as part of the NavtechGPS course "Using Advanced GPS/GNSS Signals and Systems." The author thanks Dr. Cerruti for permission to use these materials in the preparation of this chapter, and also thanks Ms. Carolyn McDonald, President of NavtechGPS, for sponsoring the original development of these materials. Any errors or omissions, however, are the author's.

Engineering Satellite-Based Navigation and Timing: Global Navigation Satellite Systems, Signals, and Receivers, First Edition. John W. Betz.
© 2016 The Institute of Electrical and Electronics Engineers, Inc. Published 2016 by John Wiley & Sons, Inc.

This chapter addresses ADC for two different types of applications. The first application, linear ADC, is traditionally encountered in digital signal processing, where the ADC is configured so that a continuous-time, continuous amplitude waveform reconstructed from the ADC outputs closely matches the ADC input. The second and very different application is a digitizing correlator, where the ADC output is crosscorrelated with a digitized signal replica. In this situation the ADC is configured so the correlator output SNR approaches what would result with continuous-time, continuous-amplitude inputs. Very different sampling frequencies and amplitude quantizations are needed for these two distinct situations.

Section 15.1 provides an overview of ADC and associated automatic gain control (AGC), defining the conversion process and describing the role of gain control and approaches for implementing it. Section 15.2 reviews traditional guidelines for sampling frequency and quantitation steps used in conventional linear digital signal processing. Section 15.3 introduces the concept of a digitizing correlator and provides results for ADC design for a digitizing correlator, while Section 15.4 summarizes the chapter.

15.1 INTRODUCTION TO ANALOG-TO-DIGITAL CONVERSION AND AUTOMATIC GAIN CONTROL

In general, the ADC process involves bandlimiting, sampling, and quantization (BSQ). Bandlimiting attenuates power at frequencies not to be included in the sampled waveform, to prevent these frequencies from being aliased,[1] or folded, into the sampled waveform. Sampling turns a continuous-time waveform into a discrete-time waveform whose values are represented only at specific time epochs. Quantization replaces continuous amplitudes with a discrete set of amplitude values. AGC is used to adjust the amplitude of the continuous-time, continuous-amplitude waveform input to the ADC, to best use the amplitude values available.

Several simplifying assumptions are employed in this discussion of ADC. Bandlimiting is modeled by an ideal rectangular low-pass filter transfer function having a flat and unit-gain passband, transition bands with zero width, zero gain in the stopband, and linear phase over the passband. When approximating an actual filter transfer function in this way, it is important that the approximating ideal filter's passband be wide enough to include the actual filter's transition bands, rather than merely matching the actual filter's passband. Further, all the discussion of ADC in this book is limited to regularly spaced samples in time and uniform input quantization steps. The quantization is modeled as a memoryless, nonlinear, odd-symmetric transformation between input and output, where the input is continuous and the output takes on a discrete set of values set by the number of quantization levels, N_Q.

While it is more general to describe the number of quantization levels in an ADC, N_Q, it is common instead to describe the number of bits required to represent its outputs.

---

[1] See Reference 3 for a discussion of sampling and aliasing.

INTRODUCTION TO ANALOG-TO-DIGITAL CONVERSION AND AUTOMATIC GAIN CONTROL

Conventionally, N_Q is a power of two and in that case the number of bits used to represent the output is a positive integer Q given by

$$Q = \lg N_Q \tag{15.1}$$

where $\lg(\bullet)$ is a logarithm base 2. The input waveform can be positive-valued or negative-valued, and the output values, which are symmetric about zero, are represented as $-2^Q + 2k - 1 = -N_Q + 2k - 1, k = 1, 2, \ldots, N_Q$. Any symmetric and uniformly spaced set of integers can be selected, although this set is commonly used.

In some applications such as digitizing correlation described in Section 15.3, there can be useful performance benefits at modest implementation cost by adding a zero output to the ADC; the resulting number of output quantization levels is typically equal to unity plus a power of two. Even though it is not technically accurate, the colloquial description of the number of bits used to represent the output is then incremented by 0.5 from what would be used to represent the power of two. For example three-level quantization is often called 1.5 bit quantization, and five-level quantization is called 2.5 bit quantization. When the number of quantization levels is unity plus a power of two because zero is included as an ADC output, the number of bits used to represent the output is said to be

$$Q = \lg(N_Q - 1) + 0.5 \tag{15.2}$$

In this case, the output values are $-2^{Q-0.5} + 2(k-1) = -N_Q + 2k - 1, k = 1, 2, \ldots, N_Q$. The output step size used here is the same as for the case where N_Q is a power of two, but all output values can be multiplied by any nonzero constant with no loss of generality.

Figure 15.1 shows the input–output relationships for quantizers with Q of one bit, 1.5 bits, two bits, and 2.5 bits. The ADC input quantization step size, which defines the range of input values that produce a given output value, is given by α_q, which is finite in all cases except the one-bit ADC, where it is semi-infinite. The input values at which output levels change are called quantization set points; for the uniform quantization addressed here, only the single parameter α_q is needed to define all quantization set points, except for one-bit quantization where the only quantization set point is always zero.

This chapter deals only with the AGC/ADC configuration shown in Figure 15.2. The input is formed from a real baseband continuous-time, continuous-amplitude RF waveform, downconverted as described in Chapter 14, and represented as a complex envelope baseband signal having real and imaginary parts. This input is bandpass filtered for antialiasing, then adjusted in amplitude by the AGC, discussed below. The AGC is represented as a variable gain amplifier (VGA), but the feedback that adjusts this gain is not shown explicitly in this figure. The two legs in Figure 15.2 separately extract the inphase component as the real part, and the quadraphase component as the imaginary part. Both components are synchronously sampled, and the samples are then quantized. (Conceptually, the order of sampling and quantization can be reversed with no effect on overall function, or both can happen simultaneously. However, in practice, some ADC approaches, like successive approximation ADC, involve sampling first.) Note that each

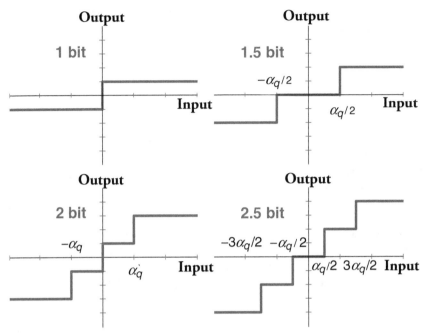

Figure 15.1. Input–Output Relationships for Quantizers with Small Numbers of Quantization Levels

ADC is operating on a real-valued waveform. Two samples, corresponding to the real and imaginary part of the input waveform, are taken at each sampling epoch. In an actual circuit implementation, the digitized real and imaginary parts are retained separately. However, the signal processing algorithms described in subsequent chapters employ the mathematical concept of a digitized complex envelope formed by multiplying the digitized quadraphase component by i and adding it to the digitized inphase component, as shown in Figure 15.2.

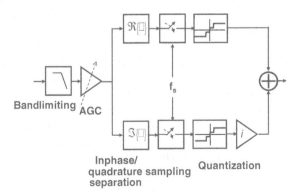

Figure 15.2. AGC/ADC Configuration for Baseband Quantization

AGC is used to adjust the level of the input waveform to the ADC in order to most effectively use the number of bits in the ADC. In some receiver designs the ADC may have enough bits so that the expected range of values of the input waveform (containing signal plus noise plus interference) can be quantized appropriately. In this case, no AGC is needed. In the opposite extreme, there is no need for AGC with a one-bit ADC since the only quantization set point is zero. In some cases of linear ADC (Section 15.2) and most cases of the digitizing correlator (Section 15.3) using more than one bit quantization, however, AGC is needed. While different AGC criteria are used for linear ADC and digitizing correlators, as described in Sections 15.2 and 15.3, AGC implementations can apply to either ADC application, and are discussed here.

Traditional AGC employs an analog amplifier with variable gain—a VGA. This approach is still used in lower-cost applications. It, however, lacks flexibility and exhibits limited dynamic range. The VGA's output power is compared to a desired value, and the difference is fed back to control the gain. Typically, a VGA is implemented [1] using the cascade of a variable attenuator and a fixed-gain amplifier. While conceptually the attenuator and amplifier can be placed in either sequence, it is common to precede the amplifier with the attenuator so that the amplifier's input maintains a consistent range of values that remain on the linear part of the gain curve as discussed in Section 14.1.3, resulting in low distortion.

A typical analog AGC is shown in Figure 15.3. The detector measures the amplifier's output power, using a linear (in magnitude) input–output relationship or a nonlinear relationship in magnitude, such as a square-law or logarithmic. Since the output power of the amplifier is intended to remain relatively constant, the detector need not have large dynamic range. The choice of nonlinearity affects the AGC's response to abrupt, large changes in input power—a square-law detector will respond more quickly than a logarithmic detector. The detector output is filtered or averaged, and then input to a differential amplifier whose other input is a reference voltage. The difference between the filtered detector output and the reference voltage determines the gain adjustment in the VGA. This design is a feedback control system, and parameters must be selected to balance responsiveness and stability.

Greater control and dynamic range can be obtained through the use of a digital attenuator before the fixed-gain amplifier. The digital attenuator, more properly called a

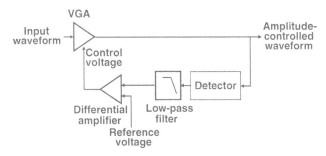

Figure 15.3. Analog AGC

digitally controlled attenuator, adjusts its attenuation based on a digital input determined from the ADC output. Control of the attenuation is then digital, known as digital gain control. The two primary differences between an analog AGC and AGC implemented with digital attenuator are:

- With the digital attenuator, a VGA is not needed, simplifying the amplifier design,
- With the digital attenuator, the feedback comes from the ADC output rather than the amplifier output, allowing more precise monitoring of AGC effects, and more sophisticated criteria for controlling gain.

Various approaches, such as counting the fraction of time the ADC outputs a value corresponding to the largest magnitude, or histograms of ADC output values, are used to determine the input to the digital attenuator and maintain the desired input levels to the ADC.

15.2 LINEAR ANALOG-TO-DIGITAL CONVERSION

Linear ADC's objective is to produce a digitized waveform that is the best replication of the input waveform, in the sense that an analog waveform reconstructed from the ADC output best matches the input waveform, approximating a linear transformation. The standard guidelines for ADC are well-known, found, for example, in Reference 2, and summarized in this section. These linear ADC guidelines apply for ADC applications where the digitized waveform will subsequently be subjected to linear signal processing (e.g., FFTs or digital filtering). Completely different guidelines, yielding very different results, are provided in Section 15.3 to design BSQ for input to a correlator.

Linear ADC guidelines permit separate selection of sampling frequency and amplitude quantization. The sampling frequency is based on the Nyquist criterion [2]: for the real low-pass waveforms being quantized here as shown in Figure 15.2, the sampling frequency must be greater than twice the highest frequency of interest in the input waveform. This highest frequency of interest is typically chosen to be equal to or greater than the lowest frequency in the stopband of the low-pass antialiasing. If the sampling frequency is lower than this, power in the transition band or passband can be aliased into the band of interest, adding noise and interference.

In most satnav applications, incommensurate sampling is used, meaning the sampling frequency is not an integer multiple of the spreading code chip rate or other frequencies in the received signal. Commensurate sampling should only be used with extreme caution and understanding; otherwise, samples can fall at spreading symbol transitions causing poor performance of the digital processing.

Once the sampling frequency is selected, the amplitude quantization can be determined separately. The number of quantization levels and the quantization step size are jointly established to achieve acceptably low quantization errors and large enough dynamic range. Since BSQ is not an injective function, information is lost and the original waveform can never be perfectly reconstructed from a finite number of amplitude quantization steps. However, the reconstruction error can be made acceptably small by using enough quantization levels with small enough quantization step size.

The error introduced by amplitude quantization is often modeled as contributing additive white Gaussian noise (AWGN) to the waveform. For the largest waveform magnitude that can be represented with the fixed number of bits, the SNR due to quantization noise is approximated by [3] $6Q + 1.8$ dB, where Q is the (integer) number of bits used in quantization.[2] This approximation becomes more accurate with increasing number of bits. As analyzed in Section 14.1.3, a satnav receiver's thermal noise power is always much larger than the power of any input desired signal. Thus, it is common practice in linear ADC to set the quantization step size 10 dB smaller than the receiver's thermal noise power spectral density, so the resulting quantization noise increases the total noise level in the waveform by only a small fraction of a decibel.

For linear ADC in satnav, the number of bits is established by the dynamic range to be supported, which is usually established by the largest interference power, I, at the ADC input. Typically, I is specified as the average received aggregate power for continuous interference. However, noise-like interference does not have a constant envelope, and the ADC should not saturate even when the instantaneous power fluctuates and is greater than the average. Often a 12 dB crest factor is used to accommodate envelope variation four times the average voltage, or 12 dB greater than the average power. In that case, the number of bits required is based on the expected input power of $10 \log_{10} I + 12$ dB. When, as described in Section 14.2, the thermal noise power is $N = \beta k_B T_{eff}$ with β the equivalent rectangular bandwidth of all preceding filters combined, k_B the Boltzmann constant, and T_{eff} the effective noise temperature of the front end, then the minimum number of bits for linear ADC is selected so

$$Q \geq \lceil (10 \log_{10}(I/N) + 20.2)/6 \rceil \quad (15.3)$$

where $\lceil \bullet \rceil$ is the ceiling function that indicates the smallest integer equal to or larger than the argument. The expression (15.3) accounts both for setting quantization step size 10 dB smaller than the receiver's thermal noise density and the 12-dB crest factor.

For example, a 12-bit ADC with $N_Q = 4096$, properly configured, can accommodate interference whose average power is approximately 50 dB greater than the total noise power in the receiver front end. If $k_B T_{eff}$ is -201.5 dBW/Hz and the precorrelation bandwidth is $\beta = 10$ MHz, N is -131.5 dBW, and average interference power of -79.5 dBW can be tolerated without distortion. As long as its maximum envelope is less than -67.5 dBW, digital filtering and other linear processing of the ADC output will comply with theoretical predictions.

Typically, in satnav receiver applications where linear digital processing is performed, linear ADC is implemented with enough bits to handle the full range of input

---

[2] Or, use the next smallest integer for noise and dynamic range calculations, establishing a small degree of conservatism.

waveform power levels without AGC. When that is not the case, there are three extreme strategies for AGC, none particularly satisfactory:

- Gain can be set to ensure that large-magnitude input waveforms are not distorted, meaning that when the input waveform power is high enough to saturate the ADC, the AGC gain is reduced. This reduction, however, means the quantization noise increases relative to the signal that is included in the waveform, degrading SNR for subsequent processing.
- Gain can be set to preserve the quantization noise level, making the ADC saturate at large input power levels. In this case the performance of subsequent linear processing is degraded, since the waveform's shape is not preserved when it takes on large magnitudes.
- The input can be blanked when large inputs occur. If the duty cycle of such inputs is low enough, then the degradation from blanking may be deemed acceptable.

15.3 PRECORRELATOR ANALOG-TO-DIGITAL CONVERSION—THE DIGITIZING CORRELATOR

BSQ for the digitizing correlator is based on a completely different criterion than what is used for linear digital signal processing. Instead of designing the BSQ to preserve the shape of the waveform being sampled, the criterion instead is to choose the sampling frequency, number of quantization levels, and quantization step size that degrade the correlator output SNR by only an acceptable amount.

Section 15.3.1 provides an intuitive discussion of the digitizing correlator. Section 15.3.2 provides a comprehensive set of numerical results for designing the BSQ for digitizing correlators, addressing different spreading modulations, and providing the resulting BSQ loss values for use in performance assessment. These results have been generated specifically for this book and are not available elsewhere. Section 15.3.3 addresses choosing the sampling frequency to avoid undesired effects from the aliasing of the replica.

15.3.1 Intuitive Motivation for the Digitizing Correlator

To understand the effects of BSQ in a digital correlator, consider Figure 15.4. The upper plot is a portion of a low-pass filtered biphase signal, $s(t)$, with a phase transition part way through, along with its one-bit quantization denoted $[\![s(t)]\!]_1$. The second plot shows only noise, $n(t)$, passed through the same low-pass filter as the signal in the upper plot. Also in the second plot is the one-bit quantization of the filtered noise, $[\![n(t)]\!]_1$. The third plot shows the same filtered noise with the filtered signal attenuated and added to produce $as(t) + n(t)$, with a selected to yield a C/N_0 of 40 dB-Hz. Also shown is the resulting one-bit quantization $[\![as(t) + n(t)]\!]_1$. There is no way the input analog waveforms could be accurately reconstructed from the quantized waveforms, but that is not the objective for the digitizing correlator. Instead, the ADC and correlator are

PRECORRELATOR ANALOG-TO-DIGITAL CONVERSION—THE DIGITIZING CORRELATOR 341

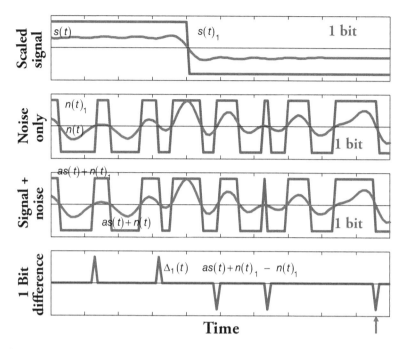

Figure 15.4. Example of Bandlimiting and Quantization for Weak Signal in Noise

considered as a unit, and the objective is to design BSQ to preserve correlator output SNR.

The one-bit quantizations of noise only and signal plus noise appear identical, and most of the time they are. The lowest plot shows the difference between the one-bit quantization of signal plus noise and the one-bit quantization of noise only, $\Delta_1(t) \stackrel{\Delta}{=} [\![as(t)+n(t)]\!]_1 - [\![n(t)]\!]_1$. Most of the time, $[\![as(t)+n(t)]\!]_1 = [\![n(t)]\!]_1$ and $\Delta_1(t) = 0$. But for a total time T_1 out of the correlation integration time T, the magnitude of the noise is very small and at those times, $\Delta_1(t) \cong [\![as(t)]\!]_1$. For example, near the end of the segments shown in Figure 15.4 (at the time indicated by the arrow), the noise transitions from positive to negative, causing the one-bit quantization to transition from +1 to −1. The signal, which takes on a negative but very small magnitude value at this time, pulls the signal plus noise value to negative slightly sooner than the noise-only value, causing their quantized values to differ and $\Delta_1(t) = [\![as(t)]\!]_1 = -1$ over a very small segment of time. Note that never does $\Delta_1(t) = -[\![as(t)]\!]_1$, since when the signal is positive-valued, $[\![as(t)+n(t)]\!]_1 \geq [\![n(t)]\!]_1$, and when the signal is negative-valued, $[\![as(t)+n(t)]\!]_1 \leq [\![n(t)]\!]_1$.

Since signal and noise are zero-mean and uncorrelated, the cross-correlation of quantized noise with quantized signal, $\frac{1}{T} \int_{-T/2}^{T/2} [\![n(t)]\!]_1 [\![s(t+\tau)]\!]_1 \, dt$, also has zero mean $\forall \tau$. However, at times when the magnitude of the noise is small, the cross-correlation of quantized signal-plus-noise with quantized signal $\frac{1}{T} \int_{-T/2}^{T/2} [\![as(t) + n(t)]\!]_1 [\![s(t+\tau)]\!]_1 \, dt$ has a small positive mean for small values of τ:

$$\frac{1}{T} \int_{-T/2}^{T/2} [\![as(t) + n(t)]\!]_1 [\![s(t)]\!]_1 \, dt = \frac{1}{T} \int_{-T/2}^{T/2} \Delta_1(t) [\![s(t)]\!]_1 \, dt + \frac{1}{T} \int_{-T/2}^{T/2} [\![n(t)]\!]_1 [\![s(t)]\!]_1 \, dt$$

so

$$E\left\{ \frac{1}{T} \int_{-T/2}^{T/2} [\![as(t) + n(t)]\!]_1 [\![s(t)]\!]_1 \, dt \right\} = \frac{T_1}{T} \qquad (15.4)$$

Recall from Chapter 5 that the correlator output SNR is mean^2/variance of the correlation function peak, so a larger T_1 yields a larger mean, producing a larger numerator in the expression for output SNR. While these positive values are evident for bandlimited and quantized waveforms, they remain after sampling only if the sample times occur at instances where the magnitude of the noise is small.

Based on the above, intuition can be developed to explain subsequent theoretical results. The discussion focuses on affecting the numerator of the expression for correlator output SNR. As the number of quantization levels increases, there are additional amplitude values where the presence of even very small signal levels can cause the quantized signal plus noise to differ from quantized noise only, with their difference having the same sign as the desired signal. Thus, increasing the number of quantization levels increases the numerator in the expression for correlator output SNR.

When the bandwidth and quantization remain unchanged, but the sampling rate increases, there is increasing chance that sample points will occur at times when the signal plus noise quantizes to a value different from noise only. Thus, both more quantization levels and faster sampling frequency for a given bandwidth produce more contributions to the mean of the correlation function, helping increase the correlation function SNR.

The most significant effect of precorrelation bandwidth on correlator output SNR comes from increasing the desired signal power as the bandwidth increases. Once the bandwidth is wide enough to pass most of the desired signal power, the correlator output SNR remains relatively insensitive to further increases in bandwidth, as long as the sampling rate increases proportional with the bandwidth increase. With increased bandwidth, the slope in time of the noise increases so the regions where the noise amplitude is small become narrower. If, however, the sampling rate increases proportionally with the bandwidth, then the fraction of samples that occur in one of the regions where noise amplitude is small remains approximately the same. Conversely, if the bandwidth is increased but the sampling rate does not increase proportionally, the output SNR decreases, since there is a reduction in the fraction of samples that occur at places where the magnitude of the noise is small enough for the signal to affect the quantized output.

15.3.2 Quantitative Guidelines for the Digitizing Correlator

A rigorous development of the theory behind the digitizing correlator guidelines is found in References 4 and 5. The results in white noise are expressed in terms of the loss in output SNR from cross-correlating the ADC output with a sampled replica of the desired signal, relative to the output SNR from an ideal cross-correlator whose inputs are the analog (continuous-time, continuous-amplitude) waveforms, or, equivalently, the waveforms at infinite bandwidth, sampled at infinite rates using an infinite number of quantization levels, with infinitesimal quantization step size. It is assumed here that the aggregate interference appears to be random; highly structured interference like a single sinusoid or single keyed signal may benefit from other techniques such as adaptive ADC [6].

The following tables of results are calculated using numerical evaluation of the expressions developed in References 4 and 5. They could also be replicated using computer simulations. In either case, they are developed for a desired spreading modulation in white noise, with a given bandlimiting, a given sampling frequency equal to or greater than the Nyquist rate associated with bandlimiting (but typically not commensurate with the spreading code chip rate), and a given number of output levels. For each case, the quantization step size is found that minimizes the loss in correlator output SNR for a quantizing correlator, relative to an ideal correlator with analog input waveforms. The associated BSQ losses are then determined. The receiver designer can then decide whether the associated BSQ losses are acceptable. If not, the BSQ values can be adjusted to balance implementation complexity (which tends to increase with wider bandlimiting, faster sampling, and larger number of quantization levels) against correlator output SNR loss. The results in References 4 and 5 are only given for quantization levels optimized for very wide precorrelation bandwidth and very high sampling frequency, while the quantization levels presented here and associated BSQ losses are optimized for each precorrelation bandwidth and number of quantization level.

For a given desired signal, precorrelation bandwidth, and sampling frequency, the optimal (in the sense that it minimizes the loss in output SNR, or equivalently, maximizes the output SNR for given conditions) quantization step size is described as a multiple of the standard deviation of the input waveform. Thus, the role of AGC in the quantizing correlator is to adjust the input level to the ADC so its standard deviation is approximately fixed over time, regardless of how the power in the input to the AGC may fluctuate. Setting the quantization set points to the indicated multiple of this standard deviation produces the optimum digitizing correlator performance for a given bandlimiting, sampling frequency, and number of quantization levels.

Chapter 5 shows how the correlator output SNR is related to the effective C/N$_0$, and how interference degrades the effective C/N$_0$ when correlating analog waveforms. The theory and numerical results can be extended for a digitizing correlator. The resulting expression is [7]

$$(C/N_0)_{\text{eff}} = \frac{C/\left(L_{sq}L_{\beta_R}^2\right)}{\dfrac{N_0}{L_{\beta_R}} + \sum_{k=1}^{K} \dfrac{I_k \kappa_{I_k s}}{\lambda_{k,sq}}} \tag{15.5}$$

In the numerator of (15.5), C is the power in the received signal at the receiver input, L_{sq} is the SNR loss in white noise due to sampling and quantization, L_{β_R} is the bandlimiting loss in the loss of desired signal power (having normalized PSD $\bar{\Phi}_s(f)$) due to an ideal precorrelation filter having rectangular bandpass with bandwidth $\pm \beta_r/2$

$$L_{\beta_r} = \left[\int_{-\beta_r/2}^{\beta_r/2} \bar{\Phi}_s(f) df \right]^{-1} \tag{15.6}$$

In the denominator of (15.5), there are K different types of interference, each having a different normalized PSD $\bar{\Phi}_{I,k}(f)$ and aggregate interference power at the receiver input I_k, κ_{sk} is the SSC between the desired signal and the kth type of interference as defined in (5.3), and $\lambda_{k,sq}$ is the interference power loss due to sampling and quantization. (If this interference power loss is less than unity, the sampling and quantization actually magnify the effect of the interference.)

The expression (15.5) simplifies to well-known cases. When the sampling frequency is infinite and the number of quantization steps is infinite with infinitesimal quantization set point, $L_{sq} = 1$ and $\lambda_{k,sq} = 1$, and the only loss is due to signal power lost outside the precorrelation filter bandwidth. In that case, (15.5) becomes equivalent to expressions in Chapter 5,

$$(C/N_0)_{\text{eff}}\big|_{\text{noSQloss}} = \frac{C/(L_{\beta_R}^2)}{\frac{N_0}{L_{\beta_R}} + \sum_{k=1}^{K} I_k \kappa_{l_k s}} \tag{15.7}$$

with $L_{\beta_r} = \eta^{-1}$. When there is BSQ but no interference and only signal plus white noise, (15.5) becomes

$$(C/N_0)_{\text{eff}}\big|_{\text{nointerference}} = \frac{C/(L_{sq} L_{\beta_R})}{N_0} = \frac{C/L_{bsq}}{N_0} \tag{15.8}$$

with the BSQ loss in white noise denoted by L_{bsq}.

Given the highly nonlinear quantization when the number of quantization levels is small, it is remarkable that the effects of different interference types add in the denominator of (15.5). Nonetheless, this is a very good approximation as long as the contribution of each interference type is less than the noise power. Consequently, (15.5) can be employed to represent the effects of different types of interference, each with different PSDs and power levels. Even with severe quantization, the SSCs and interference power losses due to sampling and quantization can be determined separately for each interference type and their combined effect is well-described by (15.5).

Each interference term in (15.5) can be rewritten $I_k \kappa_{l_k s}/\lambda_{k,sq} = I_k(\kappa_{l_k s})_{\text{eff}}$, where the effective SSC $(\kappa_{l_k s})_{\text{eff}} \stackrel{\Delta}{=} \kappa_{l_k s}/\lambda_{k,sq}$ includes the combined effect of BSQ on the interference effects. The interference loss $\lambda_{k,sq}$ (which describes the contribution to the correlation function variance) is different from the desired signal loss L_{bsq} (which describes the contribution to the mean correlation peak). Results examined over a range of applicable BSQ values, signal PSDs, and noise PSDs indicate that, in many cases, the effective SSC is typically close to the infinite bandwidth SSC with no sampling or quantization. Exceptions include when the sampling frequency introduces undesired aliasing of the correlation replica as discussed in Section 15.3.3, C/A-on-C/A interference discussed in Section 5.6, and one bit quantization that tends to fill in spectral nulls [6]. When the receiver encounters multiple interference types, it is often reasonable to use $(\kappa_{l_k s})_{eff} \approx \int_{-\beta_r/2}^{\beta_r/2} \bar{\Phi}_s(f) \bar{\Phi}_{l_k}(f) df$, avoiding the need to calculate, tabulate, and apply a large assortment of different values for $\lambda_{k,sq}$.

An extensive set of tabulated results is presented next for different spreading modulations received in white noise. These results are calculated numerically using software based on the theory in References 4 and 5. Since the column labels in the top row are symbols in order to preserve space, they are defined here. β_r is the two-sided precorrelation bandwidth, typically set by the low-pass antialiasing filter preceding ADC. The sampling frequency for each ADC in Figure 15.2 is described as proportional to the precorrelation bandwidth and given by $f_s = \phi_s \beta_r$, where ϕ_s is the sampling rate factor, with $\phi_s = 1$ corresponding to the Nyquist rate for the two-sided bandwidth β_r. Pairs of columns are provided for different numbers of quantization levels, N_Q, corresponding to one-bit, 1.5-bit, two-bit, 2.5-bit, and three-bit quantization. The quantization losses are so small for even larger quantization levels than those that there is little reason to compute them. For each number of quantization levels, the optimum step size α_q, defined in Figure 15.1, is given relative to a unit-variance input to minimize the BSQ loss for that precorrelation bandwidth and sampling rate factor. The associated BSQ loss is also provided. Results are calculated for specified transmit bandwidths, with the PSD modeled as zero outside that bandwidth. These values are calculated at C/N_0 values of approximately 40 dB-Hz, yet hold over a wide range of C/N_0 values, as long as the signal power is dominated by the white noise power.

Table 15.1 provides the results for a desired signal with BPSK-R(1) spreading modulation. An extensive discussion of the results in Table 15.1 provided here; the reader can develop corresponding interpretations and insights from subsequent tables.

Table 15.1 provides tradeoffs in BSQ design, showing that even with the same precorrelation bandwidth, there are multiple options for sampling and quantization to achieve the same target BSQ loss values. For example, to achieve approximately 1 dB of BSQ loss with $\beta_r = 2$ MHz, one could choose any of the following: $\phi_s = 2$, $N_Q = 3$, $\alpha_q = 1.3$; $\phi_s = 1.5$, $N_Q = 4$, $\alpha_q = 1.0$; $\phi_s = 1.5$, $N_Q = 5$, $\alpha_q = 0.9$; $\phi_s = 1$, $N_Q = 8$, $\alpha_q = 0.6$. The designer can then trade component complexity, power consumption, and other implementation considerations.

In some cases the same sampling frequency with a narrower precorrelation bandwidth yields a smaller BSQ loss than with a wider precorrelation bandwidth. Observe that the same sampling frequency is used in the following two cases, $\beta_r = 2$ MHz with $\phi_s = 2$, and $\beta_r = 4$ MHz with $\phi_s = 1$, yet with two-level quantization, the BSQ loss

TABLE 15.1. BSQ Designs (Optimal Quantization Step Size in White Noise) with Associated BSQ Loss for Digitizing Correlator and BPSK-R(1) Desired Signal, Transmit Bandwidth ±15.345 MHz

Precorrelation Bandwidth (MHz)	ϕ_s	$N_Q = 2$		$N_Q = 3$		$N_Q = 4$		$N_Q = 5$		$N_Q = 8$	
		α_q	L_{bsq} (dB)	α_q	L_{bsq} (dB)	α_q	L_{bsq} (dB)	α_q	L_{bsq} (dB)	α_q	L_{bsq} (dB)
±1.0	1.0	N/A	2.8	1.2	1.8	1.0	1.4	0.8	1.2	0.6	1.0
±1.0	1.5	N/A	2.1	1.3	1.3	1.0	1.0	0.9	0.9	0.6	0.7
±1.0	2.0	N/A	1.8	1.3	1.1	1.1	0.9	0.9	0.7	0.6	0.6
±1.0	3.0	N/A	1.6	1.4	0.9	1.1	0.7	1.0	0.6	0.6	0.5
±2.0	1.0	N/A	2.4	1.2	1.3	1.0	1.0	0.8	0.8	0.6	0.6
±2.0	1.5	N/A	1.8	1.3	1.0	1.0	0.7	0.9	0.6	0.6	0.4
±2.0	2.0	N/A	1.5	1.4	0.8	1.1	0.6	0.9	0.5	0.6	0.3
±2.0	3.0	N/A	1.4	1.4	0.7	1.2	0.5	1.0	0.4	0.7	0.3
±4.0	1.0	N/A	2.2	1.2	1.1	1.0	0.7	0.8	0.6	0.6	0.4
±4.0	1.5	N/A	1.6	1.3	0.8	1.0	0.5	0.9	0.4	0.6	0.2
±4.0	2.0	N/A	1.4	1.4	0.7	1.1	0.4	0.9	0.3	0.6	0.2
±4.0	3.0	N/A	1.2	1.4	0.5	1.2	0.3	1.0	0.3	0.7	0.2
±8.0	1.0	N/A	2.0	1.2	1.0	1.0	0.6	0.8	0.5	0.6	0.2
±8.0	1.5	N/A	1.5	1.3	0.7	1.0	0.4	0.9	0.3	0.6	0.2
±8.0	2.0	N/A	1.3	1.4	0.6	1.1	0.4	0.9	0.2	0.6	0.1
±8.0	3.0	N/A	1.2	1.4	0.5	1.2	0.3	1.0	0.2	0.7	0.1
±15.0	1.0	N/A	2.0	1.2	0.9	1.0	0.6	0.8	0.4	0.6	0.2
±15.0	1.5	N/A	1.5	1.3	0.7	1.0	0.4	0.9	0.3	0.6	0.1
±15.0	2.0	N/A	1.3	1.4	0.6	1.1	0.3	0.9	0.2	0.6	0.1
±15.0	3.0	N/A	1.1	1.4	0.4	1.2	0.2	1.0	0.2	0.7	0.1

is smaller with $\beta_r = 2$ MHz and $\phi_s = 2$. This result matches the intuition introduced in Section 15.3.1—the narrower bandwidth decreases the slope of the noise waveform, increasing the opportunity for samples to occur at time intervals when the noise magnitude is small enough to permit the signal to affect the quantized value.

The bandlimiting loss alone for $\beta_r = 2$ MHz is derived analytically using (15.6) to be 0.4 dB, showing the loss from sampling and quantization is only 0.1 dB at that bandwidth when the sampling rate factor is three (sampling at three times the Nyquist rate) and eight quantization levels are used. With 2 MHz precorrelation bandwidth and sampling factor of unity (Nyquist rate sampling), using three-level quantization rather than two-level quantization reduces the BSQ loss by 1 dB. Similarly, with 2 MHz precorrelation bandwidth and two-level quantization, increasing from Nyquist rate sampling to twice Nyquist rate sampling also reduces the BSQ loss by 1 dB.

Figures 15.5 and 15.6 plot the resulting BSQ loss values for different quantization set points with two different numbers of quantization levels, in each case for different precorrelation bandwidths and different sampling rate factors. Each line corresponds

PRECORRELATOR ANALOG-TO-DIGITAL CONVERSION—THE DIGITIZING CORRELATOR 347

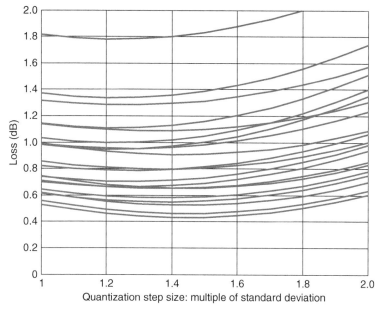

Figure 15.5. Dependence of BSQ Losses for BPSK-R(1) on Quantization Set Points for Different Precorrelation Bandwidths and Different Sampling Rate Factors, Three Quantization Levels

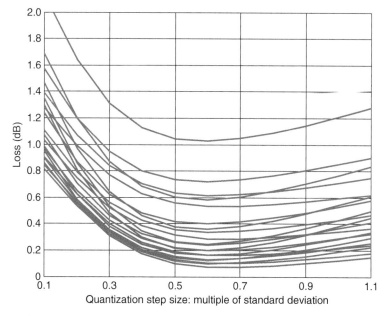

Figure 15.6. Dependence of BSQ Losses for BPSK-R(1) on Quantization Set Points for Different Precorrelation Bandwidths and Different Sampling Rate Factors, Eight Quantization Levels

to a pair of values of precorrelation bandwidth and sampling rate factor. The minima are broad, showing there is not extreme sensitivity either to AGC or to the quantization set point values. As a rule of thumb, ±20% deviations from the optimum set point (corresponding to ADC input power deviations of ±1.6 dB) produce only 0.1 dB of increase in BSQ loss. However, as the number of quantization levels increases, the losses become asymmetric, with significantly greater losses when the quantization set point is too small as shown in Figure 15.6. Thus, it is prudent to err on the side of designing AGC and quantization points to avoid having the quantization set point more than 20% smaller than optimum; it is preferable to err on the side of having the quantization set point slightly too large.

The losses due to BSQ can be deduced separately from the information in the tables. The bandlimiting loss can be computed analytically using (15.6) or is given to good approximation by the loss values in Table 15.1 which correspond to the highest sampling rate factor and number of quantization levels, so

$$L_{\beta_R} \approx L_{bsq}(\beta_r, \phi_s = 3, N_Q = 8) \tag{15.9}$$

where the BSQ loss is explicitly shown as a function of the BSQ parameters. For example, for BPSK-R(1) with 2 MHz precorrelation bandwidth, the bandlimiting loss estimated using (15.9) is 0.5 dB, while the analytically calculated value using (15.6) is 0.4 dB.

The remaining loss is due to the combination of sampling and quantization; when the number of quantization levels is large, this loss is dominated by sampling, so if we factor the sampling and quantization loss into separate loss terms, $L_{sq} = L_s L_q$, the sampling loss (with quantities expressed in power, not in decibels) is

$$L_s(\beta_r, \phi_s) \approx \frac{L_{bsq}(\beta_r, \phi_s, N_Q = 8)}{L_{bsq}(\beta_r, \phi_s = 3, N_Q = 8)} \tag{15.10}$$

For example, using Table 15.1 and ±1 MHz precorrelation bandwidth, the sampling loss is approximately 0.5 dB for a sampling rate factor of unity.

Finally, the quantization loss is the remaining loss

$$L_q(\beta_r, \phi_s, N_Q) \approx \frac{L_{bsq}(\beta_r, \phi_s, N_Q)}{L_{\beta_R} L_s(\beta_r, \phi_s)} \approx \frac{L_{bsq}(\beta_r, \phi_s, N_Q)}{L_{bsq}(\beta_r, \phi_s, N_Q = 8)} \tag{15.11}$$

For example, using Table 15.1 and ±1 MHz precorrelation bandwidth with unity sampling rate factor, the quantization loss is approximately 1.8 dB for $N_Q=2$ and 0.8 dB for $N_Q = 3$, showing the benefits of increasing the quantization level from one bit to 1.5 bits.

In contrast to linear ADC, increasing the sampling frequency above the Nyquist rate makes a significant difference when the number of quantization levels is small. A 50% increase in sampling rate factor over the Nyquist rate reduces the BSQ loss by 0.5 dB or more in many cases when the number of quantization levels is two or three.

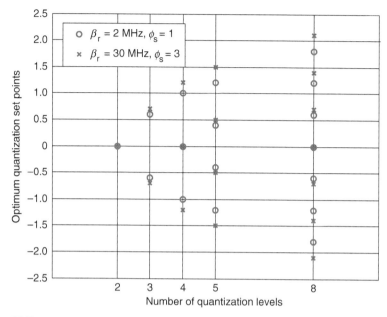

Figure 15.7. Optimum (Minimum Loss) Values Quantization Set Points at Different Numbers of Quantization Levels, for BPSK-R(1) in White Noise, at Extremes of Precorrelation Bandwidth (±1 MHz and ±15 MHz) and Sampling Rate

Figure 15.1 shows that, for odd-valued numbers of quantization levels, the smallest magnitude quantization set point has magnitude $\alpha_q/2$, while for even-valued numbers of quantization levels the smallest magnitude quantization set point has magnitude α_q. Figure 15.7 shows the optimum quantization set points (multipliers of the standard deviation) for different numbers of quantization levels, at both extremes of precorrelation bandwidth and sampling frequency. For Nyquist rate sampling, the optimum quantization set points do not change with different precorrelation bandwidths. For three times Nyquist rate sampling, the optimum quantization set points are slightly larger for wider precorrelation bandwidths. Since BSQ losses are insensitive to such small deviations from the optimum set points, particularly to increases from the optimum values, there would be little effect on BSQ losses if the optimum quantization set points for wide precorrelation bandwidths were used at smaller precorrelation bandwidths.

Table 15.2 shows the same results for a desired signal with BPSK-R(0.511) spreading modulation. In this case, to replicate GLONASS specifications for the L1OF and L2OF signals, the transmit bandwidth is ±0.511 MHz, so there is no need to use a precorrelation bandwidth greater than ±2.0 MHz. Because of the narrow transmit bandwidth, the bandlimiting loss is negligible.

Table 15.3 shows the BSQ results for a digitizing correlator with a desired signal having BOC(1,1) spreading modulation. For the same narrow precorrelation bandwidths, BSQ losses are as much as 0.8 dB greater than for a BPSK-R(1) spreading modulation.

TABLE 15.2. BSQ Designs (Optimal Quantization Step Size in White Noise) with Associated BSQ Loss for Digitizing Correlator and BPSK-R(0.511 MHz) Desired Signal, Transmit Bandwidth ±0.511 MHz

Precorrelation Bandwidth (MHz)	ϕ_s	$N_Q = 2$		$N_Q = 3$		$N_Q = 4$		$N_Q = 5$		$N_Q = 8$	
		α_q	L_{bsq} (dB)	α_q	L_{bsq} (dB)	α_q	L_{bsq} (dB)	α_q	L_{bsq} (dB)	α_q	L_{bsq} (dB)
±0.5	1.0	N/A	2.4	1.2	1.4	1.0	1.0	0.8	0.8	0.6	0.6
±0.5	1.5	N/A	1.7	1.3	0.9	1.0	0.6	0.9	0.5	0.6	0.3
±0.5	2.0	N/A	1.4	1.3	0.7	1.1	0.4	0.9	0.3	0.6	0.2
±0.5	3.0	N/A	1.2	1.4	0.5	1.1	0.3	1.0	0.2	0.6	0.1
±0.75	1.0	N/A	2.2	1.2	1.2	1.0	0.8	0.8	0.6	0.6	0.4
±0.75	1.5	N/A	1.5	1.3	0.7	1.0	0.4	0.9	0.3	0.6	0.1
±0.75	2.0	N/A	1.3	1.4	0.5	1.1	0.3	0.9	0.2	0.6	0.0
±0.75	3.0	N/A	1.1	1.4	0.4	1.2	0.2	1.0	0.1	0.7	0.0
±1.0	1.0	N/A	2.0	1.2	0.9	1.0	0.6	0.8	0.4	0.6	0.2
±1.0	1.5	N/A	1.3	1.4	0.5	1.0	0.3	0.9	0.1	0.6	0.0
±1.0	2.0	N/A	1.1	1.4	0.4	1.1	0.2	0.9	0.0	0.6	0.1
±1.0	3.0	N/A	0.9	1.4	0.2	1.2	0.0	1.0	0.0	0.7	0.1

TABLE 15.3. BSQ Designs (Optimal Quantization Step Size in White Noise) with Associated BSQ Loss for Digitizing Correlator and BOC(1,1) Desired Signal, Transmit Bandwidth ±15.345 MHz

Precorrelation Bandwidth (MHz)	ϕ_s	$N_Q = 2$		$N_Q = 3$		$N_Q = 4$		$N_Q = 5$		$N_Q = 8$	
		α_q	L_{bsq} (dB)	α_q	L_{bsq} (dB)	α_q	L_{bsq} (dB)	α_q	L_{bsq} (dB)	α_q	L_{bsq} (dB)
±2.0	1.0	N/A	3.2	1.2	2.2	1.0	1.8	0.8	1.6	0.6	1.4
±2.0	1.5	N/A	2.4	1.2	1.6	1.0	1.3	0.9	1.2	0.6	1.0
±2.0	2.0	N/A	2.1	1.3	1.3	1.1	1.1	0.9	1.0	0.6	0.9
±2.0	3.0	N/A	1.8	1.4	1.1	1.1	0.9	0.9	0.8	0.6	0.7
±4.0	1.0	N/A	2.6	1.2	1.5	1.0	1.1	0.8	1.0	0.6	0.8
±4.0	1.5	N/A	1.9	1.3	1.1	1.0	0.8	0.9	0.7	0.6	0.5
±4.0	2.0	N/A	1.6	1.4	0.9	1.1	0.7	0.9	0.5	0.6	0.4
±4.0	3.0	N/A	1.4	1.4	0.7	1.2	0.5	1.0	0.4	0.6	0.4
±8.0	1.0	N/A	2.2	1.2	1.2	1.0	0.8	0.8	0.6	0.6	0.4
±8.0	1.5	N/A	1.6	1.3	0.8	1.0	0.5	0.9	0.4	0.6	0.3
±8.0	2.0	N/A	1.4	1.4	0.7	1.1	0.4	0.9	0.3	0.6	0.2
±8.0	3.0	N/A	1.2	1.4	0.5	1.2	0.3	1.0	0.3	0.7	0.2
±15.0	1.0	N/A	2.1	1.2	1.0	1.0	0.6	0.8	0.5	0.6	0.3
±15.0	1.5	N/A	1.5	1.3	0.7	1.0	0.4	0.9	0.3	0.6	0.2
±15.0	2.0	N/A	1.3	1.4	0.6	1.1	0.3	0.9	0.2	0.6	0.1
±15.0	3.0	N/A	1.1	1.4	0.4	1.2	0.3	1.0	0.2	0.7	0.1

PRECORRELATOR ANALOG-TO-DIGITAL CONVERSION—THE DIGITIZING CORRELATOR

TABLE 15.4. BSQ Designs (Optimal Quantization Step Size in White Noise) with Associated BSQ Loss for Digitizing Correlator and TMBOC(6,1,4/33) Desired Signal, Transmit Bandwidth ±15.345 MHz

Precorrelation Bandwidth (MHz)	ϕ_s	$N_Q = 2$		$N_Q = 3$		$N_Q = 4$		$N_Q = 5$		$N_Q = 8$	
		α_q	L_{bsq} (dB)	α_q	L_{bsq} (dB)	α_q	L_{bsq} (dB)	α_q	L_{bsq} (dB)	α_q	L_{bsq} (dB)
±2.0	1.0	N/A	4.0	1.2	2.9	1.0	2.6	0.8	2.4	0.6	2.2
±2.0	1.5	N/A	3.2	1.2	2.3	1.0	2.1	0.9	1.9	0.6	1.8
±2.0	2.0	N/A	2.7	1.3	1.9	1.1	1.7	0.9	1.6	0.6	1.4
±2.0	3.0	N/A	2.2	1.4	1.5	1.1	1.3	0.9	1.2	0.6	1.1
±4.0	1.0	N/A	3.3	1.2	2.2	1.0	1.9	0.8	1.7	0.6	1.5
±4.0	1.5	N/A	2.3	1.2	1.5	1.0	1.2	0.9	1.0	0.6	0.9
±4.0	2.0	N/A	2.1	1.3	1.1	1.1	1.1	0.9	1.0	0.6	0.8
±4.0	3.0	N/A	1.8	1.4	1.2	1.1	0.9	1.0	0.8	0.6	0.7
±8.0	1.0	N/A	2.3	1.2	1.2	1.0	0.9	0.8	0.7	0.6	0.5
±8.0	1.5	N/A	1.7	1.3	0.9	1.0	0.6	0.9	0.5	0.6	0.3
±8.0	2.0	N/A	1.4	1.4	0.7	1.1	0.5	0.9	0.3	0.6	0.2
±8.0	3.0	N/A	1.2	1.4	0.5	1.2	0.4	1.0	0.3	0.6	0.2
±15.0	1.0	N/A	2.1	1.2	1.1	1.0	0.7	0.8	0.5	0.6	0.3
±15.0	1.5	N/A	1.5	1.3	0.7	1.0	0.5	0.9	0.3	0.6	0.2
±15.0	2.0	N/A	1.3	1.4	0.6	1.1	0.4	0.9	0.3	0.6	0.1
±15.0	3.0	N/A	1.2	1.4	0.5	1.2	0.3	1.0	0.2	0.7	0.1

Table 15.4 shows the BSQ results for a digitizing correlator with a desired signal having TMBOC(6,1,4/33) spreading modulation. For narrower precorrelation bandwidths that do not pass the BOC(6,1) component, BSQ losses are somewhat larger than for BOC(1,1), but at larger precorrelation bandwidths the losses are the same for TMBOC and BOC(1,1).

Table 15.5 shows the BSQ results for a digitizing correlator with a desired signal having CBOC− spreading modulation. The results are similar to those for TMBOC.

Table 15.6 shows the BSQ results for a digitizing correlator with a desired signal having CBOC+ spreading modulation. The BSQ losses are less than those for TMBOC and CBOC−, but greater than those for BOC(1,1).

Table 15.7 shows the BSQ results for a digitizing correlator with a desired signal having BPSK-R(5) spreading modulation. Not surprisingly, the design and BSQ losses are virtually identical to those for BPSK-R(1) with precorrelation bandwidths scaled by the factor of five that relates the spreading code chip rates.

Table 15.8 shows the BSQ results for a digitizing correlator with a desired signal having BPSK-R(5.11 MHz) spreading modulation. In this case, to replicate GLONASS specifications for the L1SF and L2SF signals, the transmit bandwidth is ±5.11 MHz, so there is no need to use a precorrelation bandwidth greater than ±10.0 MHz. Because of the narrow transmit bandwidth, the bandlimiting loss is negligible. The results are

TABLE 15.5. BSQ Designs (Optimal Quantization Step Size in White Noise) with Associated BSQ Loss for Digitizing Correlator and CBOC− Desired Signal, Transmit Bandwidth ±15.345 MHz

Precorrelation Bandwidth (MHz)	ϕ_s	$N_Q = 2$ α_q	L_{bsq} (dB)	$N_Q = 3$ α_q	L_{bsq} (dB)	$N_Q = 4$ α_q	L_{bsq} (dB)	$N_Q = 5$ α_q	L_{bsq} (dB)	$N_Q = 8$ α_q	L_{bsq} (dB)
±2.0	1.0	N/A	4.2	1.2	3.1	1.0	2.8	0.8	2.6	0.6	2.4
±2.0	1.5	N/A	3.4	1.2	2.5	1.0	2.3	0.9	2.1	0.6	2.0
±2.0	2.0	N/A	2.9	1.3	2.1	1.1	1.9	0.9	1.7	0.6	1.6
±2.0	3.0	N/A	2.3	1.4	1.6	1.1	1.4	0.9	1.3	0.6	1.2
±4.0	1.0	N/A	3.5	1.2	2.4	1.0	2.1	0.8	1.9	0.6	1.7
±4.0	1.5	N/A	2.5	1.2	1.6	1.0	1.3	0.9	1.1	0.6	1.0
±4.0	2.0	N/A	2.2	1.3	1.4	1.1	1.2	0.9	1.1	0.6	0.9
±4.0	3.0	N/A	1.9	1.4	1.2	1.1	1.0	1.0	0.9	0.6	0.8
±8.0	1.0	N/A	2.3	1.2	1.2	1.0	0.9	0.8	0.7	0.6	0.5
±8.0	1.5	N/A	1.7	1.3	0.9	1.0	0.6	0.9	0.5	0.6	0.3
±8.0	2.0	N/A	1.4	1.4	0.7	1.1	0.5	0.9	0.4	0.6	0.2
±8.0	3.0	N/A	1.3	1.4	0.6	1.2	0.4	1.0	0.3	0.6	0.2
±15.0	1.0	N/A	2.1	1.2	1.1	1.0	0.7	0.8	0.6	0.6	0.4
±15.0	1.5	N/A	1.5	1.3	0.7	1.0	0.5	0.9	0.3	0.6	0.2
±15.0	2.0	N/A	1.3	1.4	0.6	1.1	0.4	0.9	0.3	0.6	0.1
±15.0	3.0	N/A	1.2	1.4	0.5	1.2	0.3	1.0	0.2	0.7	0.1

TABLE 15.6. BSQ Designs (Optimal Quantization Step Size in White Noise) with Associated BSQ Loss for Digitizing Correlator and CBOC+ Desired Signal, Transmit Bandwidth ±15.345 MHz

Precorrelation Bandwidth (MHz)	ϕ_s	$N_Q = 2$ α_q	L_{bsq} (dB)	$N_Q = 3$ α_q	L_{bsq} (dB)	$N_Q = 4$ α_q	L_{bsq} (dB)	$N_Q = 5$ α_q	L_{bsq} (dB)	$N_Q = 8$ α_q	L_{bsq} (dB)
±2.0	1.0	N/A	3.8	1.2	2.7	1.0	2.4	0.8	2.2	0.6	2.0
±2.0	1.5	N/A	3.0	1.2	2.1	1.0	1.9	0.9	1.7	0.6	1.6
±2.0	2.0	N/A	2.5	1.3	1.8	1.1	1.5	0.9	1.4	0.6	1.3
±2.0	3.0	N/A	2.1	1.4	1.4	1.1	1.2	0.9	1.1	0.6	1.0
±4.0	1.0	N/A	3.1	1.2	2.0	1.0	1.7	0.8	1.5	0.6	1.3
±4.0	1.5	N/A	2.2	1.3	1.4	1.0	1.1	0.9	0.9	0.6	0.8
±4.0	2.0	N/A	2.0	1.3	1.2	1.1	1.0	0.9	0.8	0.6	0.7
±4.0	3.0	N/A	1.7	1.4	1.0	1.1	0.8	1.0	0.7	0.6	0.6
±8.0	1.0	N/A	2.2	1.2	1.2	1.0	0.8	0.8	0.7	0.6	0.4
±8.0	1.5	N/A	1.7	1.3	0.8	1.0	0.6	0.9	0.4	0.6	0.3
±8.0	2.0	N/A	1.4	1.4	0.7	1.1	0.4	0.9	0.3	0.6	0.2
±8.0	3.0	N/A	1.2	1.4	0.5	1.2	0.3	1.0	0.3	0.7	0.2
±15.0	1.0	N/A	2.1	1.2	1.1	1.0	0.7	0.8	0.5	0.6	0.3
±15.0	1.5	N/A	1.5	1.3	0.7	1.0	0.5	0.9	0.3	0.6	0.2
±15.0	2.0	N/A	1.3	1.4	0.6	1.1	0.4	0.9	0.2	0.6	0.1
±15.0	3.0	N/A	1.2	1.4	0.5	1.2	0.3	1.0	0.2	0.7	0.1

PRECORRELATOR ANALOG-TO-DIGITAL CONVERSION—THE DIGITIZING CORRELATOR 353

TABLE 15.7. BSQ Designs (Optimal Quantization Step Size in White Noise) with Associated BSQ Loss for Digitizing Correlator and BPSK-R(5) Desired Signal, Transmit Bandwidth ±15.345 MHz

Precorrelation Bandwidth (MHz)	ϕ_s	$N_Q = 2$		$N_Q = 3$		$N_Q = 4$		$N_Q = 5$		$N_Q = 8$	
		α_q	L_{bsq} (dB)	α_q	L_{bsq} (dB)	α_q	L_{bsq} (dB)	α_q	L_{bsq} (dB)	α_q	L_{bsq} (dB)
±5.0	1.0	N/A	2.7	1.2	1.7	1.0	1.3	0.8	1.1	0.6	0.9
±5.0	1.5	N/A	2.0	1.3	1.2	1.0	0.9	0.9	0.8	0.6	0.6
±5.0	2.0	N/A	1.7	1.3	1.0	1.1	0.7	0.9	0.6	0.6	0.5
±5.0	3.0	N/A	1.5	1.4	0.8	1.1	0.6	1.0	0.5	0.6	0.4
±10.0	1.0	N/A	2.3	1.2	1.2	1.0	0.8	0.8	0.7	0.6	0.5
±10.0	1.5	N/A	1.6	1.3	0.8	1.0	0.6	0.9	0.4	0.6	0.3
±10.0	2.0	N/A	1.4	1.4	0.7	1.1	0.5	0.9	0.3	0.6	0.2
±10.0	3.0	N/A	1.2	1.4	0.5	1.2	0.3	1.0	0.3	0.7	0.2
±15.0	1.0	N/A	2.1	1.2	1.1	1.0	0.7	0.8	0.5	0.6	0.3
±15.0	1.5	N/A	1.5	1.3	0.7	1.0	0.5	0.9	0.3	0.6	0.2
±15.0	2.0	N/A	1.3	1.4	0.6	1.1	0.4	0.9	0.3	0.6	0.1
±15.0	3.0	N/A	1.2	1.4	0.5	1.2	0.3	1.0	0.2	0.7	0.1

TABLE 15.8. BSQ Designs (Optimal Quantization Step Size in White Noise) with Associated BSQ Loss for Digitizing Correlator and BPSK-R(5.11 MHz) Desired Signal, Transmit Bandwidth ±5.11 MHz

Precorrelation Bandwidth (MHz)	ϕ_s	$N_Q = 2$		$N_Q = 3$		$N_Q = 4$		$N_Q = 5$		$N_Q = 8$	
		α_q	L_{bsq} (dB)	α_q	L_{bsq} (dB)	α_q	L_{bsq} (dB)	α_q	L_{bsq} (dB)	α_q	L_{bsq} (dB)
±5.0	1.0	N/A	2.4	1.2	1.4	1.0	1.0	0.8	0.8	0.6	0.6
±5.0	1.5	N/A	1.7	1.3	0.9	1.0	0.6	0.9	0.5	0.6	0.3
±5.0	2.0	N/A	1.4	1.3	0.7	1.1	0.4	0.9	0.3	0.6	0.2
±5.0	3.0	N/A	1.2	1.4	0.5	1.1	0.3	1.0	0.2	0.6	0.1
±7.5	1.0	N/A	2.2	1.2	1.1	1.0	0.8	0.8	0.6	0.6	0.4
±7.5	1.5	N/A	1.5	1.3	0.7	1.0	0.4	0.9	0.3	0.6	0.1
±7.5	2.0	N/A	1.3	1.4	0.5	1.1	0.3	0.9	0.2	0.6	0.1
±7.5	3.0	N/A	1.1	1.4	0.4	1.2	0.2	1.0	0.1	0.7	0.0
±10.0	1.0	N/A	2.0	1.2	0.9	1.0	0.6	0.8	0.4	0.6	0.2
±10.0	1.5	N/A	1.3	1.3	0.5	1.0	0.3	0.9	0.1	0.6	0.0
±10.0	2.0	N/A	1.1	1.4	0.4	1.1	0.2	0.9	0.1	0.6	0.1
±10.0	3.0	N/A	0.9	1.4	0.2	1.2	0.1	1.0	0.0	0.7	0.0

TABLE 15.9. BSQ Designs (Optimal Quantization Step Size in White Noise) with Associated BSQ Loss for Digitizing Correlator and BOC(5,2) Desired Signal, Transmit Bandwidth ±15.345 MHz

Precorrelation Bandwidth (MHz)	ϕ_s	$N_Q = 2$		$N_Q = 3$		$N_Q = 4$		$N_Q = 5$		$N_Q = 8$	
		α_q	L_{bsq} (dB)	α_q	L_{bsq} (dB)	α_q	L_{bsq} (dB)	α_q	L_{bsq} (dB)	α_q	L_{bsq} (dB)
±7.0	1.0	N/A	3.0	1.2	2.0	1.0	1.6	0.8	1.5	0.6	1.2
±7.0	1.5	N/A	2.4	1.2	1.6	1.0	1.4	0.8	1.3	0.6	1.1
±7.0	2.0	N/A	1.8	1.3	1.1	1.0	0.9	0.9	0.8	0.6	0.6
±7.0	3.0	N/A	1.5	1.4	0.9	1.1	0.7	0.9	0.6	0.6	0.5
±10.0	1.0	N/A	3.0	1.2	2.0	1.0	1.6	0.8	1.4	0.6	1.2
±10.0	1.5	N/A	2.1	1.2	1.3	1.0	1.0	0.9	0.9	0.6	0.7
±10.0	2.0	N/A	1.8	1.3	1.0	1.1	0.8	0.9	0.7	0.6	0.5
±10.0	3.0	N/A	1.5	1.4	0.8	1.1	0.6	0.9	0.5	0.6	0.4
±15.0	1.0	N/A	2.7	1.2	1.7	1.0	1.3	0.8	1.1	0.6	0.9
±15.0	1.5	N/A	1.9	1.3	1.0	1.0	0.7	0.9	0.6	0.6	0.4
±15.0	2.0	N/A	1.6	1.3	0.8	1.1	0.6	0.9	0.5	0.6	0.3
±15.0	3.0	N/A	1.3	1.4	0.6	1.1	0.4	1.0	0.3	0.6	0.2

identical to those in Table 15.2, when scaled by a factor of 10 representing the chip rate increase.

Table 15.9 shows the BSQ results for a digitizing correlator with a desired signal having BOC(5,2) spreading modulation. The benefits of increasing the sampling rate factor above unity and using three-level quantization rather than two-level quantization can be significant. With two-level quantization, using a precorrelation bandwidth of 14.0 MHz and sampling rate factor of 1.5 (for a sampling frequency of 21 MHz) provides 0.6 dB lower BSQ loss than using a precorrelation bandwidth of 20.0 MHz and sampling rate factor of unity (resulting in almost the same sampling frequency of 20 MHz). With a 14.0 MHz precorrelation bandwidth and sampling rate factor of unity, using three quantization levels instead of two reduces the BSQ loss by 1 dB.

Table 15.10 shows the BSQ results for a digitizing correlator with a desired signal having BOC(5,2.5) spreading modulation. The BSQ design and losses are very similar to those given in Table 15.9 for BOC(5,2).

Table 15.11 shows the BSQ results for a digitizing correlator with a desired signal having BPSK-R(10) spreading modulation. The BSQ design and losses for 20.0 MHz precorrelation bandwidth are very similar to those given in Table 15.1 for BPSK-R(1) and 2.0 MHz precorrelation bandwidth, because of the scaling by the factor of 10 in chip rate. Compared to BPSK-R(1), the BPSK-R(10) signal has slightly lower bandlimiting loss because the precorrelation bandwidths are a larger fraction of the transmit bandwidth for BPSK-R(10).

Table 15.12 shows the BSQ results for a digitizing correlator, with a desired signal having BOC(10,5) spreading modulation. The losses for 24.0 MHz precorrelation bandwidth are higher than for BPSK-R(10) because that bandwidth does not even include

TABLE 15.10. BSQ Designs (Optimal Quantization Step Size in White Noise) with Associated BSQ Loss for Digitizing Correlator and BOC(5,2.5) Desired Signal, Transmit Bandwidth ±15.345 MHz

Precorrelation Bandwidth (MHz)	ϕ_s	$N_Q = 2$ α_q	L_{bsq} (dB)	$N_Q = 3$ α_q	L_{bsq} (dB)	$N_Q = 4$ α_q	L_{bsq} (dB)	$N_Q = 5$ α_q	L_{bsq} (dB)	$N_Q = 8$ α_q	L_{bsq} (dB)
±7.5	1.0	N/A	3.0	1.2	2.0	1.0	1.6	0.8	1.5	0.6	1.2
±7.5	1.5	N/A	2.4	1.2	1.6	1.0	1.4	0.9	1.3	0.6	1.1
±7.5	2.0	N/A	1.8	1.3	1.1	1.0	0.9	0.9	0.8	0.6	0.6
±7.5	3.0	N/A	1.5	1.4	0.9	1.1	0.7	0.9	0.6	0.6	0.5
±10.0	1.0	N/A	3.0	1.2	2.0	1.0	1.6	0.8	1.4	0.6	1.2
±10.0	1.5	N/A	2.1	1.2	1.3	1.0	1.0	0.9	0.9	0.6	0.7
±10.0	2.0	N/A	1.8	1.3	1.0	1.1	0.8	0.9	0.7	0.6	0.5
±10.0	3.0	N/A	1.5	1.4	0.8	1.1	0.6	0.9	0.5	0.6	0.4
±15.0	1.0	N/A	2.7	1.2	1.7	1.0	1.3	0.8	1.1	0.6	0.9
±15.0	1.5	N/A	1.9	1.3	1.0	1.0	0.7	0.9	0.6	0.6	0.4
±15.0	2.0	N/A	1.6	1.3	0.8	1.1	0.6	0.9	0.4	0.6	0.3
±15.0	3.0	N/A	1.3	1.4	0.6	1.1	0.4	1.0	0.3	0.6	0.2

TABLE 15.11. BSQ Designs (Optimal Quantization Step Size in White Noise) with Associated BSQ Loss for Digitizing Correlator and BPSK-R(10) Desired Signal, Transmit Bandwidth ±15.345 MHz

Precorrelation Bandwidth (MHz)	ϕ_s	$N_Q = 2$ α_q	L_{bsq} (dB)	$N_Q = 3$ α_q	L_{bsq} (dB)	$N_Q = 4$ α_q	L_{bsq} (dB)	$N_Q = 5$ α_q	L_{bsq} (dB)	$N_Q = 8$ α_q	L_{bsq} (dB)
±10.0	1.0	N/A	2.5	1.2	1.5	1.0	1.1	0.8	1.0	0.6	0.7
±10.0	1.5	N/A	1.8	1.3	1.0	1.0	0.7	0.9	0.6	0.6	0.4
±10.0	2.0	N/A	1.6	1.3	0.8	1.1	0.6	0.9	0.5	0.6	0.3
±10.0	3.0	N/A	1.3	1.4	0.6	1.1	0.4	1.0	0.4	0.6	0.3
±12.0	1.0	N/A	2.5	1.2	1.5	1.0	1.1	0.8	0.9	0.6	0.7
±12.0	1.5	N/A	1.8	1.3	1.0	1.0	0.7	0.9	0.6	0.6	0.4
±12.0	2.0	N/A	1.5	1.4	0.8	1.1	0.6	0.9	0.4	0.6	0.3
±12.0	3.0	N/A	1.3	1.4	0.6	1.2	0.4	1.0	0.3	0.6	0.2
±15.0	1.0	N/A	2.3	1.2	1.3	1.0	0.9	0.8	0.7	0.6	0.5
±15.0	1.5	N/A	1.6	1.3	0.8	1.0	0.5	0.9	0.4	0.6	0.3
±15.0	2.0	N/A	1.4	1.4	0.7	1.1	0.4	0.9	0.3	0.6	0.2
±15.0	3.0	N/A	1.2	1.4	0.5	1.2	0.3	1.0	0.2	0.6	0.1

TABLE 15.12. BSQ Designs (Optimal Quantization Step Size in White Noise) with Associated BSQ Loss for Digitizing Correlator and BOC(10,5) Desired Signal, Transmit Bandwidth ±15.345 MHz

Precorrelation Bandwidth (MHz)	ϕ_s	$N_Q = 2$ α_q	L_{bsq} (dB)	$N_Q = 3$ α_q	L_{bsq} (dB)	$N_Q = 4$ α_q	L_{bsq} (dB)	$N_Q = 5$ α_q	L_{bsq} (dB)	$N_Q = 8$ α_q	L_{bsq} (dB)
±12.0	1.0	N/A	3.5	1.2	2.5	1.0	2.1	0.8	1.9	0.6	1.7
±12.0	1.5	N/A	2.6	1.2	1.7	1.0	1.4	0.8	1.3	0.6	1.2
±12.0	2.0	N/A	2.0	1.3	1.3	1.0	1.0	0.9	0.9	0.6	0.8
±12.0	3.0	N/A	1.7	1.4	1.0	1.1	0.8	0.9	0.7	0.6	0.6
±15.0	1.0	N/A	2.8	1.2	1.7	1.0	1.4	0.8	1.2	0.6	1.0
±15.0	1.5	N/A	2.0	1.2	1.2	1.0	0.9	0.9	0.8	0.6	0.7
±15.0	2.0	N/A	1.6	1.3	0.9	1.0	0.7	0.9	0.6	0.6	0.4
±15.0	3.0	N/A	1.3	1.4	0.6	1.1	0.4	0.9	0.3	0.6	0.2

the first nulls of the BOC(10,5) spectrum. With 30.0 MHz precorrelation bandwidth, however, the BOC(10,5) BSQ loss is within 0.5 dB of that for BPSK-R(10).

Table 15.13 shows the BSQ results for a digitizing correlator with a desired signal having $BOC_c(10,5)$ spreading modulation. Many of the losses for 24.0 MHz precorrelation bandwidth are higher than for BOC(10,5), in large part because of the $BOC_c(10,5)$'s 0.8 dB greater bandlimiting loss with the cosine-phasing that pushes power outside the main lobes, as discussed in Chapter 3.

Table 15.14 shows the BSQ results for a digitizing correlator with a desired signal having BOC(14,2) spreading modulation. A larger transmit bandwidth is used than

TABLE 15.13. BSQ Designs (Optimal Quantization Step Size in White Noise) with Associated BSQ Loss for Digitizing Correlator and $BOC_c(10,5)$ Desired Signal, Transmit Bandwidth ±15.345 MHz

Precorrelation Bandwidth (MHz)	ϕ_s	$N_Q = 2$ α_q	L_{bsq} (dB)	$N_Q = 3$ α_q	L_{bsq} (dB)	$N_Q = 4$ α_q	L_{bsq} (dB)	$N_Q = 5$ α_q	L_{bsq} (dB)	$N_Q = 8$ α_q	L_{bsq} (dB)
±12.0	1.0	N/A	5.3	1.2	4.3	1.0	3.9	0.8	3.7	0.6	3.5
±12.0	1.5	N/A	3.8	1.2	2.9	1.0	2.6	0.8	2.4	0.6	2.2
±12.0	2.0	N/A	3.1	1.2	2.2	1.0	1.9	0.8	1.8	0.6	1.6
±12.0	3.0	N/A	2.7	1.3	1.9	1.1	1.7	0.9	1.5	0.6	1.4
±15.0	1.0	N/A	3.3	1.2	2.3	1.0	1.9	0.8	1.7	0.6	1.5
±15.0	1.5	N/A	2.2	1.2	1.4	1.0	1.1	0.8	0.9	0.6	0.8
±15.0	2.0	N/A	1.8	1.3	1.1	1.0	0.8	0.9	0.7	0.6	0.6
±15.0	3.0	N/A	1.4	1.4	0.7	1.1	0.5	0.9	0.4	0.6	0.3

PRECORRELATOR ANALOG-TO-DIGITAL CONVERSION—THE DIGITIZING CORRELATOR

TABLE 15.14. BSQ Designs (Optimal Quantization Step Size in White Noise) with Associated BSQ Loss for Digitizing Correlator and BOC(14,2) Desired Signal, Transmit Bandwidth ±20.46 MHz

Precorrelation Bandwidth (MHz)	ϕ_s	$N_Q = 2$		$N_Q = 3$		$N_Q = 4$		$N_Q = 5$		$N_Q = 8$	
		α_q	L_{bsq} (dB)	α_q	L_{bsq} (dB)	α_q	L_{bsq} (dB)	α_q	L_{bsq} (dB)	α_q	L_{bsq} (dB)
±16.0	1.0	N/A	3.0	1.2	2.0	1.0	1.6	0.8	1.4	0.6	1.2
±16.0	1.5	N/A	2.3	1.2	1.5	1.0	1.2	0.8	1.1	0.6	0.9
±16.0	2.0	N/A	1.7	1.3	1.0	1.0	0.8	0.9	0.7	0.6	0.5
±16.0	3.0	N/A	1.4	1.3	0.7	1.1	0.5	0.9	0.5	0.6	0.4
±20.0	1.0	N/A	2.8	1.2	1.8	1.0	1.4	0.8	1.3	0.6	1.1
±20.0	1.5	N/A	2.2	1.2	1.4	1.0	1.2	0.8	1.0	0.6	0.9
±20.0	2.0	N/A	1.6	1.3	0.9	1.0	0.7	0.9	0.5	0.6	0.4
±20.0	3.0	N/A	1.3	1.4	0.6	1.1	0.4	0.9	0.4	0.6	0.3

for previous cases, in order to accommodate the wider null-to-null bandwidth of this signal.

Table 15.15 shows the BSQ results for a digitizing correlator with a desired signal having BOC(15,2.5) spreading modulation. With the slightly larger precorrelation bandwidth based on the higher spreading code rate, the BSQ designs and BSQ losses are very similar to those for BOC(14,2).

Table 15.16 shows the BSQ results for a digitizing correlator with a desired signal having $BOC_c(15,2.5)$ spreading modulation. The BSQ losses for the narrower precorrelation bandwidth and Nyquist rate sampling are larger than for sine-phased signal in Table 15.15.

TABLE 15.15. BSQ Designs (Optimal Quantization Step Size in White Noise) with Associated BSQ Loss for Digitizing Correlator and BOC(15,2.5) Desired Signal, Transmit Bandwidth ±20.46 MHz

Precorrelation Bandwidth (MHz)	ϕ_s	$N_Q = 2$		$N_Q = 3$		$N_Q = 4$		$N_Q = 5$		$N_Q = 8$	
		α_q	L_{bsq} (dB)	α_q	L_{bsq} (dB)	α_q	L_{bsq} (dB)	α_q	L_{bsq} (dB)	α_q	L_{bsq} (dB)
±17.0	1.0	N/A	3.1	1.2	2.1	1.0	1.7	0.8	1.5	0.6	1.3
±17.0	1.5	N/A	2.3	1.2	1.5	1.0	1.3	0.8	1.1	0.6	1.0
±17.0	2.0	N/A	1.7	1.3	1.0	1.0	0.8	0.9	0.7	0.6	0.6
±17.0	3.0	N/A	1.4	1.3	0.8	1.1	0.6	0.9	0.5	0.6	0.4
±20.0	1.0	N/A	2.9	1.2	1.8	1.0	1.4	0.8	1.3	0.6	1.1
±20.0	1.5	N/A	2.3	1.2	1.5	1.0	1.2	0.8	1.1	0.6	1.0
±20.0	2.0	N/A	1.6	1.3	0.9	1.0	0.6	0.9	0.5	0.6	0.4
±20.0	3.0	N/A	1.3	1.4	0.7	1.1	0.5	0.9	0.4	0.6	0.3

TABLE 15.16. BSQ Designs (Optimal Quantization Step Size in White Noise) with Associated BSQ Loss for Digitizing Correlator and $BOC_c(15,2.5)$ Desired Signal, Transmit Bandwidth ±20.46 MHz

Precorrelation Bandwidth (MHz)	ϕ_s	$N_Q = 2$		$N_Q = 3$		$N_Q = 4$		$N_Q = 5$		$N_Q = 8$	
		α_q	L_{bsq} (dB)	α_q	L_{bsq} (dB)	α_q	L_{bsq} (dB)	α_q	L_{bsq} (dB)	α_q	L_{bsq} (dB)
±17.0	1.0	N/A	3.6	1.2	2.5	1.0	2.2	0.8	2.0	0.6	1.8
±17.0	1.5	N/A	2.6	1.2	1.8	1.0	1.5	0.8	1.4	0.6	1.2
±17.0	2.0	N/A	2.0	1.2	1.2	1.0	1.0	0.9	0.9	0.6	0.7
±17.0	3.0	N/A	1.6	1.3	1.0	1.1	0.8	0.9	0.7	0.6	0.6
±20.0	1.0	N/A	3.1	1.2	2.0	1.0	1.7	0.8	1.5	0.6	1.3
±20.0	1.5	N/A	2.4	1.2	1.6	1.0	1.3	0.8	1.2	0.6	1.0
±20.0	2.0	N/A	1.7	1.3	0.9	1.0	0.7	0.9	0.6	0.6	0.4
±20.0	3.0	N/A	1.4	1.4	0.7	1.1	0.5	0.9	0.4	0.6	0.3

15.3.3 Replica Aliasing in the Digitizing Correlator

The constant envelope spreading modulations introduced in Chapter 3 and used in the different satnav systems in Part II all can be represented using two values, and thus with one-bit quantization.[3] When the desired signal uses one of these spreading modulations, the sampled one-bit representation of the replica is correlated against the digitized waveform from the receiver front end. Hardware implementation of the multiplications in the correlation is simplified with this one-bit replica, since the multiplications merely indicate whether to increment or decrement an accumulator, rather than requiring an actual multiplication. Consequently, there is a significant implementation advantage in using a one-bit replica.

However, there is a subtlety in correlating with a one-bit sampled replica—the ideal replica has infinite bandwidth, and it is not bandlimited before a sampled version is generated in the receiver. Consequently, the sampled replica contains aliased versions of itself. While aliasing actually involves the Fourier transform of the waveform, qualitative insights can be obtained by examining the aliased PSD, defined as

$$\overset{\leftrightarrow}{\Phi}_x(f) \overset{\Delta}{=} \sum_{k=-\infty}^{\infty} \Phi_x(f + kf_s)$$

$$= \Phi_x(f) + \sum_{\substack{k=-\infty \\ k \neq 0}}^{\infty} \Phi_x(f + kf_s) \qquad (15.12)$$

Here, the first term in the second line is the ideal PSD, and the second term is the aliasing; the sum of the ideal PSD and the aliasing yields the aliased PSD.

---

[3] Signals using CBOC require an extension discussed in Section 18.1.3.

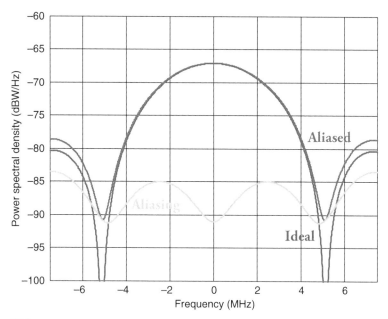

Figure 15.8. Aliased PSD for BPSK-R(5) Spreading Modulation, Precorrelation Bandwidth 15.0 MHz, Sampling Frequency 20.0 MHz

Consider first signals with BPSK-R spreading modulations, and assume the sampling frequency is greater than twice the spreading code chip rate. Then the largest peak outside the main lobe of the PSD has 5% of the value of the peak of the main lobe of the PSD, and the power in the first sidelobe is less than 3% of that in the main lobe. The entire power outside the main lobe is less than 10% of the signal power. Consequently, aliasing of the replica causes minor distortion, at most, of the correlator output. Figure 15.8 shows an example aliased PSD for the sampled one bit BPSK-R(5) spreading modulation. The aliased PSD shows very little change, relative to the ideal PSD, in the main lobes and even the sidelobes. The dominant effect of the aliasing is to fill in the spectral nulls, which tends to have negligible effect for a BPSK-R spreading modulation—the receiver would be more sensitive to narrowband interference at the null frequencies. Figure 15.9 shows similar results when the sampling frequency is reduced slightly to the Nyquist rate.

Now consider a BOC spreading modulation. One of the common motivations for using BOC is to obtain a deep spectral null at the carrier frequency, so that other signals having PSDs with peaks at the same carrier frequency produce little interference to reception of the BOC signal. Yet replica aliasing can fill in that null if the sampling frequency is not chosen carefully. Figure 15.10 shows the PSD for a BOC(5,2.5) signal, with sampling frequency 20 MHz. Spectral nulls are aliased onto the null in band center, preserving the receiver's isolation from interference in band center. Figure 15.11, however, shows the PSDs for the same signal with sampling frequency 15 MHz. In this case, sidelobes are aliased onto band center, filling in the ideal PSD's null. Interference

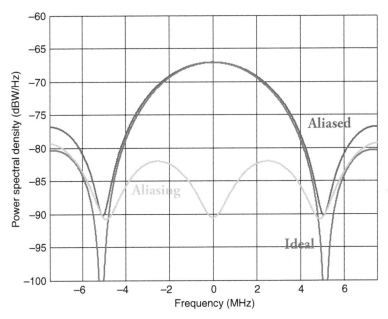

Figure 15.9. Aliased PSD for BPSK-R(5) Spreading Modulation, Precorrelation Bandwidth 15.0 MHz, Sampling Frequency 15.0 MHz

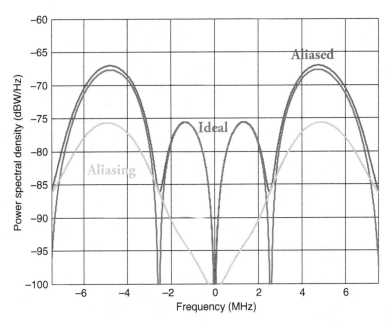

Figure 15.10. Aliased PSD for BOC(5,2.5) Spreading Modulation, Precorrelation Bandwidth 15.0 MHz, Sampling Frequency 20.0 MHz

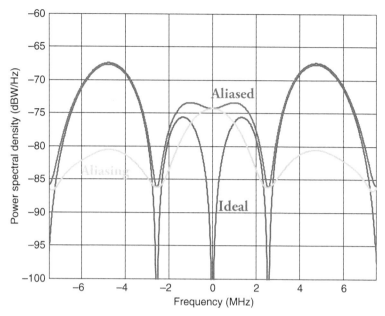

Figure 15.11. Aliased PSD for BOC(5,2.5) Spreading Modulation, Precorrelation Bandwidth 15.0 MHz, Sampling Frequency 15.0 MHz

at band center would produce much greater effect on reception of the BOC signal, due to this replica aliasing. The BOC(5,2.5) PSD has nulls near multiples of 10 MHz away from band center, and peaks 15 MHz away from band center as well as every 10 MHz further. For a BOC spreading modulation, choosing sampling frequencies that alias nulls onto band center, rather than those that alias spectral peaks onto band center, avoids filling in the spectral nulls in band center.

These heuristics are matched by calculations using the approach outlined in References 4 and 5, showing that for a BOC desired signal and low chip rate BPSK-R interferer in band center, when the sampling frequency aliases the BOC's spectral sidelobes onto its spectral null in band center, the interference power loss due to sampling and quantization, is much less than unity (taking on values less than 0.1 in some instances), meaning the sampling and quantization amplify the effect of the interference, just as predicted by the aliased PSDs.

The aliased replica has little effect on performance in white noise, but can significantly affect expected performance in nonwhite interference, if the aliasing fills in spectral nulls of the desired signal that were to be exploited for mitigating the interference. This replica aliasing does not affect the interference from a BOC interferer to a BPSK-R desired signal, but only the interference from a BPSK-R interferer to a BOC desired signal. Even then, it can be mitigated through proper choice of the sampling frequency.

15.4 SUMMARY

While the same structure of AGC and ADC is used for linear ADC (where subsequent processing relies on preservation of the input waveform's shape) and the digitizing correlator (where the objective is to preserve the correlator output SNR), very different BSQ strategies are used for these two different applications. BSQ design for linear ADC is well-known and consists of sampling at the Nyquist rate or higher, using enough bits to preserve dynamic range while keeping quantization noise much lower than other noise sources. Once the Nyquist rate is satisfied, there is little benefit to further increasing the sampling frequency.

The design of AGC and ADC for a digitizing correlator is based on completely different considerations. Performance with a small number of quantization levels can be close to ideal, as long as the quantization set points are established properly. Further, there can be significant benefits to sampling at higher than Nyquist rate when the number of quantization levels is small. An extensive set of BSQ design guidelines is provided for the digitizing correlator, accompanied by the associated processing loss values. A model is also provided for assessing the combination of BSQ and non-white interference on effective C/N_0. Even though the quantization may be highly nonlinear, superposition applies to the interference effects.

ADC is yet another receiver function where performance must be traded against implementation complexity. In general, wider bandlimiting, faster sampling frequencies, and more quantization levels provide better ADC performance, but at the cost of more costly components, greater power consumption, more hardware, and greater digital processing complexity. The tools provided in this chapter equip the reader for making these trades.

Even though the ADC output is digitized, much of the description and analysis of receiver processing in the remainder of this book is based on continuous-time, continuous amplitude representations of the waveforms, since this representation tends to be more compact. The use of digitized signals is implicit, and performance can be assessed using the tools provided in this chapter.

REVIEW QUESTIONS

Application Questions

QA15.1. If a receiver has ±2 MHz precorrelation bandwidth and must handle interference with average power as high as −100 dBW, how many bits of ADC is needed to achieve linear dynamic range and low quantization noise relative to thermal noise. Assume the receiver noise power spectral density is −201.5 dBW/Hz.

QA15.2. If a receiver has ±16 MHz precorrelation bandwidth and must handle interference with average power as high as −100 dBW, how many bits of ADC are needed to achieve linear dynamic range and low quantization noise relative to thermal noise. Assume the receiver noise power spectral density is −201.5 dBW/Hz.

QA15.3. If a receiver has ±2 MHz precorrelation bandwidth, identify three distinct sets of sampling and quantization parameters (number of bits, sampling rates) that would yield practical ADC designs for a digitizing correlator while introducing a quantization loss less than 2 dB. Provide results for desired signal having spreading modulation BPSK-R(1), BOC(1,1), CBOC−, CBOC+, and TMBOC.

QA15.4. For each of the following spreading modulations with precorrelation bandwidth set to the null-to-null bandwidth, explore sampling frequencies slightly greater than the Nyquist rate. Find one "good" sampling frequency greater than the that avoids replica aliasing that fills in the spectral null at the carrier frequency, and a "bad" sampling frequency that does fill in the spectral null: BOC(1,1) BOC(10,5), BOCc(10,5), BOC(14,2), BOC(15,2.5), BOCc(15,2.5).

REFERENCES

1. D. Whitlow, "Design and Operation of Automatic Gain Control Loops for Receivers in Modern Communications Systems," available at http://www.analog.com/static/imported-files/tutorials/42575412022953450461111812375Design_and_Operation_of_AGC_Loops.pdf, accessed 25 October 2013.
2. A. V. Oppenheim and R. W. Schafer, *Discrete-Time Signal Processing*, 3rd edition, Prentice Hall Signal Processing, 2010.
3. E. O. Brigham, *Fast Fourier Transform and Its Applications*, Prentice Hall, 1988.
4. C. J. Hegarty and A. P. Cerruti, "Results from an Analytical Model for GNSS Receiver Implementation Losses," Proceedings of the 23rd International Technical Meeting of the Satellite Division of the Institute of Navigation (ION GNSS 2010), Portland, OR, September 2010, pp. 2820–2834.
5. C. J. Hegarty, "Analytical Model for GNSS Receiver Implementation Losses", *NAVIGATION: Journal of The Institute of Navigation*, Vol. 58, No. 1, Spring 2011, pp. 29–44.
6. F. Amoroso, "Adaptive A/D Converter to Suppress CW Interference in DSPN Spread-Spectrum Communications," *IEEE Transactions on Communications*, Vol. 31, No. 10, pp. 1117–1123.
7. J. W. Betz and N. R. Shnidman, "Receiver Processing Losses with Bandlimiting and One-Bit Sampling," Proceedings of the Institute of Navigation Conference on Global Navigation Satellite Systems 2007, ION-GNSS-2007, Institute of Navigation, September 2007.

16

ACQUISITION

Acquisition is the process where the receiver detects the presence of desired satnav signals, obtains measurements and other information from the signals, and then computes and outputs an initial estimate of PVT with acceptable accuracy. Figure III.1 shows the digital signal processing stage in receiver processing, and Figure 16.1 provides a more detailed description of the different aspects of digital processing.

The inputs to the digital processing are the digitized baseband signals produced by the ADC as described in Chapter 15. Acquisition begins with initial synchronization, which produces three outputs: indication of whether a desired satnav signal has been detected, and if so a coarse estimate of the signal's time delay relative to the receiver clock, and a coarse estimate of the signal's frequency offset relative to the receiver's frequency reference. Based on the specific signal detected, the receiver determines which baseband replica to generate, using the signal's spreading code and spreading modulation from the ICD, with timing based on the coarse time delay estimate. Frequency adjustment uses the coarse frequency offset estimate to remove as much of the frequency offset (caused by Doppler shift and drift of the receiver's frequency reference) as possible from the desired signal in the received baseband waveform. The frequency-adjusted waveform is then crosscorrelated against the locally generated replica, producing measurements that are provided to discriminators. The discriminator outputs are smoothed

---

Engineering Satellite-Based Navigation and Timing: Global Navigation Satellite Systems, Signals, and Receivers, First Edition. John W. Betz.
© 2016 The Institute of Electrical and Electronics Engineers, Inc. Published 2016 by John Wiley & Sons, Inc.

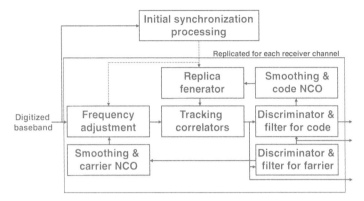

Figure 16.1. Digital Processing in a Conventional Receiver

and used to drive numerically controlled oscillators (NCOs) that generate timing for the replica generation and frequency adjustment. As indicated in Figure 16.1, there is a different receiver channel dedicated to each desired signal being processed, with separate frequency adjustment and tracking loops for each signal's complex envelope and carrier. Measurements from these channels are output to the digital processing function that calculates position, time, and velocity, and measurements may also be used to read and interpret the data message. Chapters 17, 18, 19, and 20 describe these other processes, while this chapter focuses on initial synchronization processing and other receiver processing steps unique to acquisition, as well as their performance.

While the term "acquisition" is sometimes used in slightly different ways, in this book it is defined as all the steps between when the receiver is powered on and completes any self-check, and when it outputs an initial estimate of PVT with specified accuracy. The time it takes to perform this sequence of steps is known as Time to First Fix (TTFF). The steps in acquisition include initial synchronization, signal tracking, synchronization to data message symbols or overlay code bits, synchronization to the data message, reading the data message, making measurements, and calculating PVT. All of these steps are addressed in this chapter, with the details for some of them provided in subsequent chapters of Part III.

Different variants of the process outlined in Figure 16.1 exist. Multiple initial synchronization circuits may be available to search in parallel for different signals. Instead of separate tracking loops in each channel with different channels dedicated to different signals, code and carrier tracking can be coupled or integrated, and joint tracking of multiple signals can be performed as well. Further, measurements from other sensors, such as inertial sensors, a compass, or an altimeter, can be used to aid signal tracking. These variations are introduced in Part IV of this book.

Figure 16.2 provides a more detailed description of the steps involved in acquisition for conventional receiver processing. The "acquisition circuit" performs initial synchronization; often this circuitry is complex enough that there is only one instance of it, shared sequentially for signals at different carrier frequencies and from different satellites. In that case, once initial synchronization is completed for the first signal, the

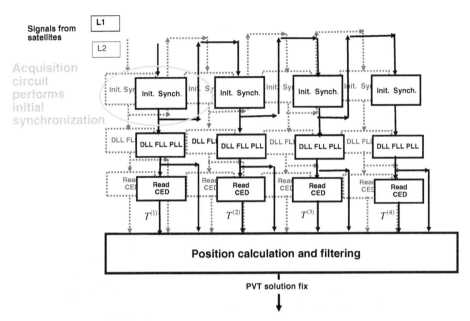

Figure 16.2. Sequence of Acquisition Processing Steps

estimates of its time delay and frequency offset can be used to reduce the uncertainty associated with time and frequency offset for the second and subsequent signals. If the receiver is a dual-frequency receiver as shown in the figure, initial synchronization is also performed on the second-frequency signal from each satellite. In that case, the search begins with an accurate estimate of the frequency offset, obtained by scaling the estimate from the signal at the other frequency, and the time of delay uncertainty is dominated by the unknown ionospheric delay difference at the two frequencies. Often, initial synchronization of signals after the first signal is less challenging, since the time and frequency uncertainties can be reduced by using information from the first signal. When signals on more than two frequencies from the same satellites are used by the receiver, after acquisition of the first two frequencies and estimation of the ionospheric delay as described in Section 20.2.4, tracking of subsequent signals from the same satellite can employ a hot start.

Initial synchronization provides information about what spreading code to use, along with coarse estimates of time delay and frequency offset, to tracking loops for that signal. A delay-locked loop (DLL), also known as a code tracking loop,[1] maintains track of the time delay of the baseband signal. A frequency-locked loop (FLL) tracks the residual carrier frequency. Often, a Costas loop or phase-locked loop (PLL), depending

---

[1] Actually, code tracking is a misnomer, since the entire baseband signal is being tracked, not only the spreading code, as described in Chapter 19. Calling it code tracking is so common and so succinct, however, that this terminology is used here as well.

upon the design of the signal as discussed in Chapter 18, is used to track the residual carrier phase. Once the first signal's parameters are being tracked, that signal's data message can be demodulated and interpreted. The critical data message contents needed to complete acquisition are time of transmission, along with the clock and ephemeris data (CED). This information enables the receiver to determine the satellite time when a specific point on the signal was transmitted, the position of the satellite at that time, and the relationship between satellite time and system time.

In parallel with beginning to track the first signal and read its CED, the receiver applies initial synchronization circuitry to other signals, each of which is then tracked and its CED is read. Ultimately, there is a sufficient set of delay measurements transformed to pseudorange measurements, carrier frequency measurements, and CED that the receiver's position and velocity can be calculated, along with time. At that point, acquisition is complete.

Acquisition processing and performance depend critically on the state of the receiver at the beginning of the acquisition process. Section 16.1 defines a set of commonly encountered initial conditions for acquisition. Initial synchronization is the critical part of acquisition, requiring the greatest processing and storage of any receiver function. Section 16.2 introduces the basics of initial synchronization and associated fundamental concepts. Section 16.3 describes several powerful algorithms for initial synchronization, including discussion of implementation considerations. Section 16.4 characterizes initial synchronization performance, including providing a simple way to estimate performance, while Section 16.5 addresses implementation of the entire acquisition process. Section 16.6 summarizes this chapter.

16.1 INITIAL CONDITIONS FOR ACQUISITION

While there are commonly employed names of different initial conditions for acquisition, there is considerable variability across the satnav community in defining the details of each. This section introduces the terminology used throughout this book. As described more fully in Section 16.2, the fundamental parameters that define the initial synchronization problem are initial time uncertainty (ITU) and initial frequency uncertainty (IFU). ITU is dictated by the combination of unknown system time and unknown propagation delay between each satellite and the receiver, resulting from the uncertainty in the satellite location, receiver location, ionospheric delay, and tropospheric delay. IFU is dictated by the combination of unknown offset of the receiver's reference frequency from the true frequency, and Doppler uncertainties caused by the combination of unknown satellite-to-receiver geometry and unknown receiver motion.

Table 16.1 summarizes different initial conditions for acquisition. While many of the characteristics vary continuously and do not fit into discrete categories, defining discrete categories is often useful for conceptual purposes and for the definition of performance requirements or capabilities. The information available a priori to the receiver defines three different issues: ITU, IFU, and what needs to be read from the data message. Limited precision of the almanac, unknown system time, and uncertain receiver location

TABLE 16.1. Representative Initial Conditions for Acquisition

Information Available A Priori to Receiver	Initial Condition				
	Cold Start	Warm Start	Hot Start	Assisted Start	Restart
Current almanac	No	Yes	Yes	Yes	Yes
System time	No	Tens of seconds	Within ±10 ms	Coarse time: Within ±2 s Fine time: Within ±10 ms	Within ±1 μs
Receiver location	No	Many tens s of km	Within ±5 km	Within ±3 km	Within ±10 m
Reference frequency	±3 ppm	±1 ppm	±0.5 ppm	±0.1 ppm	Within ±10 Hz
Receiver speed	Limited by maximum speed of platform	Limited by maximum speed of platform	Within ±30 m/s	Within ±30 m/s	Within ±1 m/s
Current clock and ephemeris data for each satellite to be tracked	No	No	Yes	Yes	Yes

INITIAL CONDITIONS FOR ACQUISITION

contribute to ITU. Coarse information in the almanac, uncertain receiver location, errors in reference frequency, and unknown receiver motion contribute to IFU. CED (and, eventually, almanac data) must be read from the data message if not available otherwise, along with the time that the signal was transmitted. The defined initial conditions range from starting with the least information available in the left-hand column to having the most information available in the right-hand column. Cold start occurs after a receiver has been completely powered off, has no information stored from previous signal reception, and may have moved anywhere in the world since it last received signals. It uses no a priori information except what is in the ICD, and the capabilities of the reference oscillator in the receiver. The receiver must search an IFU corresponding to all physically possible frequency offsets, and an ITU corresponding to the duration of the spreading code, which must be a repeating code to avoid a search over infinite time. After initial synchronization is complete, the receiver must then read all necessary information from the signals' data messages. Typically, the delay in reading the almanac is so long that, even after the first satellite is acquired in cold start, subsequent initial synchronization processing proceeds without almanac information, lengthening the time required for synchronizing to signals from subsequent satellites, compared to the situation where the almanac is already known. Warm start occurs when the receiver has not received signals for hours or days, has not moved far during that time, and has a coarse time estimate from an input or from a timekeeping device. Warm start benefits from prior knowledge of the almanac, approximate system time, and approximate receiver location, so initial synchronization can focus on satellites known to be above the horizon (although they may not be in view if sky visibility is impaired). This prior knowledge also limits the ITU, although typically it is still greater than the spreading code duration. The prior knowledge, combined with limits on receiver speed and improved reference frequency accuracy, determines the IFU. CED and time of transmission must still be read from each satellite signal. Hot start involves situations where the receiver has access to a received signal's time of arrival within hundreds of nanoseconds and frequency of arrival within tens of hertz. The most common hot start situation occurs when signal tracking is interrupted for tens of seconds. In hot start, the received signal falls within the tracking loop pull-in region (see Chapters 18 and 19), allowing tracking to commence without initial synchronization. Assisted start involves communications/navigation integration, where a communication link is employed to provide satnav information to the receiver. A civilian example of assisted start is in mobile telephone applications [1], where the telephone network provides almanac and ephemeris to the receiver, along with a frequency reference and approximate time (described in Chapter 24). A military example of communications/navigation integration involves deployment of a satnav-guided munition, where a datalink from the launching platform can provide the almanac, ephemeris, and approximate time. Two variants of assisted start are defined, coarse time and fine time, depending upon the uncertainty of system time provided to the receiver as discussed in Chapter 24. All other information has at least the same quality as that for hot start. Restart occurs after the receiver has very briefly lost access to the satnav signals—for example, a receiver that has been carried deep indoors where satellite signals were lost, then is quickly returned to near where it was previously tracking the signals.

If the receiver attempts to restart very far from where it had been previously tracking, its position change must be known by other means, such as some combination of an inertial sensor, pedometer, compass, or speedometer. The preferable situation in restart is for tracking loops to recommence tracking without any initial synchronization processing, as discussed in Section 16.6. The time between when a receiver begins restart and when PVT is produced is known as Time to Subsequent Fix (TTSF).

16.2 INITIAL SYNCHRONIZATION BASICS

As introduced in Section 16.1, initial synchronization searches over different signals, time (relative to receiver time) and frequency offset (relative to the receiver's frequency source). Often, the time dimension is called delay, since it is the delay or advance of the signal's time of arrival relative to a previous estimate or hypothesis. Also, the frequency offset dimension is often called Doppler, even though drift of the receiver's reference oscillator also contributes to the uncertainty in carrier frequency of the received signal.

Initial synchronization is essentially a process of guessing and checking—the processing hypothesizes a specific signal with its spreading modulation and spreading code, received at a particular time and frequency, and checks to see if it detects that signal at a given time delay and frequency offset. If it does detect that signal, the hypothesized time delay and frequency offset become the basis of the coarse estimates provided for tracking. If not, it searches another hypothesis.

Section 16.2.1 describes the process for searching over ITU and IFU, introducing the cross-ambiguity function (CAF) and its essential role in initial synchronization processing. Since the search space is segmented into a discrete set of delay and frequency offset hypotheses, Section 16.2.2 describes the considerations involved in choosing the dimensions of each hypothesis through sampling in time and frequency, demonstrating the advantages of broad correlation functions for initial synchronization. The discussion in this section focuses on signals having BPSK-R spreading modulations. Section 16.2.3 then describes techniques for broadening the correlation functions of signals with BOC spreading modulations.

16.2.1 Search Space and the Cross-Ambiguity Function

Figure 16.3 portrays the initial synchronization search space defined by ITU and IFU. If a specific desired signal is being received, it is being received at some time delay and frequency offset within these limits. For signals with repeating spreading codes, the ITU is usually limited by the spreading code duration, since once initial synchronization determines the delay relative to the spreading code duration, other processing, described in Section 16.4, is used to remove the time ambiguity. While the time and frequency values are continuous, all initial synchronization processing searches over a discrete set of values in time and frequency. A specific discrete-time delay and -frequency value is called a time-frequency cell in the ITU-IFU plane, and it actually encompasses a range of time delay and frequency offset values. Serial search algorithms guess and check one

INITIAL SYNCHRONIZATION BASICS

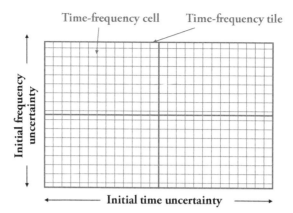

Figure 16.3. Initial Synchronization Search Space

time-frequency cell at a time, as described in Section 16.3.1. With modern algorithms and processing, however, many time-frequency cells can be guessed and checked in parallel. Such a set of time-frequency cells is called a time-frequency tile. Although sometimes the number of cells in a tile is called the number of correlators, such a metric is usually an oversimplification, since the implementation complexity for increasing the size of a tile in the frequency dimension can be very different, and often less, than increasing the size of a tile in the time dimension, as seen in Section 16.3.

Signal processing for initial synchronization is based on the CAF [2] defined as

$$\Xi(\tau,\phi) = \frac{1}{T} \int_{-T/2}^{T/2} x(t)\, y^*(t+\tau)\, e^{i2\pi\phi t} dt \qquad (16.1)$$

The CAF is closely related to a correlation function; the CAF includes a frequency translation in the integrand. While correlation is a function of the delay τ between one input $x(t)$ and the other input $y(t)$, the CAF is a function of the delay τ between the input $x(t)$ and the frequency shifted other input $y(t)$ where the frequency shift is ϕ. In particular, the correlation between $x(t)$ and $y(t)$ is $\Xi(\tau, 0)$.

If the first input is either noise only or noise plus a signal that has a known complex envelope $s(t)$ with unknown amplitude, delay, phase shift, and frequency shift, then that input can be modeled as either $x(t) = n(t)$ or $x(t) = \sqrt{\zeta C} e^{-i\theta} s(t+D) e^{-i2\pi F t} + n(t)$, where $s(t)$ is known but received with unknown amplitude $\sqrt{\zeta C} > 0$, ζ is the fraction of total signal power C in the signal component being processed, θ is the unknown carrier phase, and $n(t)$ is additive white Gaussian noise uncorrelated with $s(t)$. Initial synchronization determines whether the input contains signal or only noise, and estimates the delay, D, and the residual carrier frequency after downconversion, F, if the signal is present.

When $x(t) = \sqrt{\zeta C} e^{-i\theta} s(t+D) e^{-i2\pi Ft} + n(t)$ and $y(t) = s(t)$ in (16.1),[2]

$$\Xi(\tau, \phi) = \frac{1}{T} \int_{-T/2}^{T/2} \left[\sqrt{\zeta C} e^{-i\theta} s(t+D) e^{-i2\pi Ft} + n(t) \right] s(t+\tau) e^{i2\pi \phi t} dt$$

$$= \frac{\sqrt{\zeta C} e^{-i\theta}}{T} \int_{-T/2}^{T/2} s(t+D) s(t+\tau) e^{-i2\pi(F-\phi)t} dt \qquad (16.2)$$

$$+ \frac{1}{T} \int_{-T/2}^{T/2} n(t) s(t+\tau) e^{i2\pi \phi t} dt$$

A sufficient statistic for detecting the presence of a known signal with unknown amplitude and carrier phase, time delay, and frequency shift in white noise is the magnitude or magnitude-squared of the CAF: $|\Xi(\tau, \phi)|$ or $|\Xi(\tau, \phi)|^2$ [3]. Further, the maximum likelihood estimates of the delay and frequency shift are $(\hat{D}, \hat{F}) = \arg\max |\Xi(\tau, \phi)|$ [3].

When the frequency shift estimate is correct so that $\hat{F} = F$, the CAF is the sum of the signal's correlation function and a noise-like term:

$$\Xi(\tau, F) = \frac{\sqrt{\zeta C}}{T} \int_{-T/2}^{T/2} e^{-i\theta} s(t+D) s(t+\tau) dt + \frac{1}{T} \int_{-T/2}^{T/2} n(t) s(t+\tau) e^{i2\pi Ft} dt$$

$$= \frac{\sqrt{\zeta C} e^{-i\theta}}{T} \int_{-T/2}^{T/2} s(t+D) s(t+\tau) dt + \frac{1}{T} \int_{-T/2}^{T/2} n'(t) s(t+\tau) dt \qquad (16.3)$$

where $n'(t) \stackrel{\Delta}{=} n(t) e^{i2\pi Ft}$.

Initial synchronization involves computing the CAF between the input waveform that may contain the desired signal (with unknown amplitude, delay, and frequency shift) with additive noise and interference, and the replica generated in the receiver having the same spreading modulation and spreading code as the desired signal. The CAF peak occurs at delay D and frequency offset F. If it is determined that a signal is present, the coordinates of the peak in the ITU-IFU plane (i.e., the time-frequency cell where the peak is located) provide the coarse estimates of time delay and frequency (these estimates are typically refined by interpolation after the detection decision). An example of the CAF magnitude is shown in Figure 16.4. The magnitude of the CAF peak is large enough that it stands out clearly from the values in other cells, which are due to noise.

---

[2] Here and in the remainder of this discussion, formulations are provided for biphase signals, allowing the complex conjugation in the CAF to be neglected.

INITIAL SYNCHRONIZATION BASICS

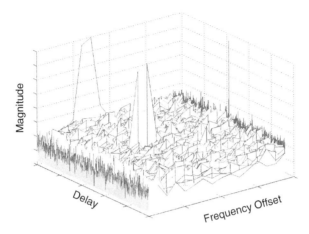

Figure 16.4. Example Cross-Ambiguity Function Magnitude

The signal model used to develop (16.2) explicitly accounts for shift of the carrier frequency, but it ignores time companding of the complex envelope, as discussed in Section 3.2.1. It is important to ensure that initial synchronization signal processing parameters are consistent with this approximation, or that processing accounts for deviations from this approximation as discussed in Section 16.3.5.

Other models and assumptions used in the above development are also approximations. The input waveform rarely consists of a single desired signal with additive white Gaussian noise. Instead, typically there are multiple satnav signals, along with interference that may not be constant envelope, is rarely white, and may not be Gaussian. Under these conditions, the CAF defined in (16.1) is no longer an optimal detector of signal presence or an optimal estimator of delay and Doppler. In some cases (i.e., strong narrowband interference or pulsed interference) the input waveform can be pre-processed to mitigate this mismatch between model and reality and obtain better performance, as discussed in Chapter 21. However, in other situations such as broadband but nonwhite interference, or interference from known satnav signals, theoretically optimum processing may be impractical and suboptimal performance from straightforward CAF computations is accepted.

16.2.2 Sampling Considerations

Since modern receivers perform initial synchronization processing on digitized signals, and since the CAF is a generalization of crosscorrelation, the design of AGC and ADC for a digitizing correlator and the resulting BSQ losses, all provided in Section 15.3, are relevant to initial synchronization processing as well. However, since the BSQ losses described in Section 15.3 apply when a sample falls at the peak of the correlation function or CAF, sample placement introduces an additional loss in initial synchronization processing. The CAF peak may occur at a delay and frequency value different from the discrete time and frequency values associated with the time-frequency cells where the CAF is computed, since the location of that peak is not known even approximately before initial synchronization. The result is an additional loss in output SNR, known as scalloping loss, that must be considered. One way to reduce scalloping loss is to compute the CAF with very dense samples in time and frequency, producing very small time-frequency cells, one of which must be very close to the CAF peak. However, the complexity of implementing initial synchronization processing increases with finer sampling in time and frequency space, requiring additional processing hardware and additional storage, with other implications such as power consumption. Since initial synchronization often dominates the processing and storage capabilities of a receiver, there are significant benefits to reducing the complexity of initial synchronization processing through sparser sampling in time and frequency. The result is another of the performance/complexity trades encountered in satnav receiver design.

Sometimes the input waveform does not contain the desired signal. This occurs if there is no satellite transmitting a signal with this carrier frequency and spreading modulation and spreading code, or if the receiver is searching for signals from a satellite on the other side of the Earth, or if the satellite signal is blocked by terrain, structures, or foliage so that its received power is negligible. In this case only the second term in the lower expression in (16.2) is present. The resulting CAF is then zero-mean, with variance determined by the PSD of the noise and interference, and the integration time used in the CAF.

The CAF output SNR is defined under the assumption that the carrier phase of the signal is known and used to phase rotate the CAF so that all signal power is in its real part. The imaginary part of the CAF then can be disregarded, and the output SNR is the squared mean of the real part of the phase-rotated peak divided by its variance:

$$\rho \stackrel{\Delta}{=} \frac{[E\{\Re\{e^{i\theta}\Xi(D,F)\}\}]^2}{Var\{\Re\{e^{i\theta}\Xi(D,F)\}\}} \tag{16.4}$$

As shown in (16.3), when the frequency offset is properly estimated, the CAF becomes a correlation. The CAF output SNR is thus the same as the correlator output SNR discussed in Chapter 5.

Review Question QT16.1 shows that the mean of the CAF is $E\{e^{i\theta}\Xi(\tau,\phi)\} \cong \sqrt{\zeta C}\bar{R}_s(\tau-D)e^{i2\pi F(\tau-D)}\text{sinc}\,[\pi(\phi-F)T]$, so the mean-squared of the CAF at the correct delay is

$$|E\{e^{i\theta}\Xi(\tau,\phi)\}|^2 = \zeta C[\bar{R}_s(\tau-D)]^2\text{sinc}^2\,[\pi(\phi-F)T] \tag{16.5}$$

When the frequency estimate is correct so $\phi = F$, the shape of the mean-squared CAF peak in the delay dimension is defined by $[\bar{R}_s(\tau-D)]^2$, so it depends entirely on the shape of the desired signal's autocorrelation function. Conversely, when the delay estimate is correct so $\tau = D$, the shape of the CAF peak mean-squared in the frequency offset dimension is defined by $\text{sinc}^2\,[\pi(\phi-F)T]$, which has nothing to do with the signal design, but only the CAF integration time T.

Review Question QA16.1 shows that, for a signal with BPSK-R(f_c) spreading modulation, the two-sided half-power width of the CAF peak in delay is approximately

INITIAL SYNCHRONIZATION BASICS

$\pm 0.29/f_c$. Similarly, the same Review Question shows that the two-sided half power width of the CAF peak in frequency offset is approximately $\pm 0.44/T$, for CAF integration time T.

When the CAF is computed using discrete-time values and at discrete frequency offsets, (16.1) can be written as

$$\Xi[k\Delta_\tau, m\Delta_\phi] = \frac{1}{L} \sum_{\ell=-L/2}^{L/2} x[\ell\Delta_\tau] y[\ell\Delta_\tau + k\Delta_\tau] e^{i2\pi m\ell\Delta_\phi\Delta_\tau} \qquad (16.6)$$

where L, assumed even, is the number of discrete products being summed (the correlation integration time dividing by the temporal sample spacing), Δ_τ is a discrete step in delay, and Δ_ϕ is a discrete step in carrier frequency. The sampled input waveform is $x[\ell\Delta_\tau] = \sqrt{\zeta C} e^{-i\theta} s[\ell\Delta_\tau + D] e^{-i2\pi Fn\Delta_\tau} + n[\ell\Delta_\tau]$, and the replica is $y[\ell\Delta_\tau] = s[\ell\Delta_\tau]$, with these quantities all evaluated at the discrete time and discrete frequency values indicated. When the replica contains a segment of the desired signal no longer than the duration of the data message bit or the overlay code bit, there is no opportunity that the signal in the input waveform has a phase transition that would degrade performance.

Typically, a commensurate sampling rate is avoided, since samples could repeatedly occur at or near points where spreading symbols are zero. Instead, noncommensurate sampling with f_c/Δ_τ not an integer, with f_c the spreading code ship rate, is used so that samples fall at different points on different spreading symbols.

The discrete-time, discrete-frequency version of (16.5) is

$$[\langle E\{e^{i\theta}\Xi[k\Delta_\tau, m\Delta_\phi]\}\rangle_t]^2 = \zeta C[\bar{R}_s(k\Delta_\tau - D)]^2 \mathrm{sinc}^2[\pi(m\Delta_\phi - F)T] \qquad (16.7)$$

Denote the delay and frequency offset indices closest to the true values by k^p and m^p, so the largest value of (16.7) is $[\langle E\{e^{i\theta}\Xi[k^p\Delta_\phi, m^p\Delta_\phi]\}\rangle_t]^2 = \zeta C[\bar{R}_s(k^p\Delta_\tau - D)]^2 \mathrm{sinc}^2[\pi(m^p\Delta_\phi - F)T]$. Substituting $\tau = D$ and $\phi = F$ in (16.5) shows that the mean-squared CAF peak has value C. The ratio between that value and (16.7) yields what is often called the scalloping loss

$$L_{\mathrm{scalloping}} = 1/[\bar{R}_s(k^p\Delta_\tau - D)\mathrm{sinc}[\pi(m^p\Delta_\phi - F)T]]^2 \qquad (16.8)$$

quantifying the loss in signal power due to sampling at points in delay and frequency offset that are different from the location of the CAF peak.

The continuous function shown in Figure 16.5 illustrates scalloping loss in the delay dimension when the frequency offset is perfectly matched, for a BPSK-R($1/T_c$) signal. In each plot, the solid lines show the correlation function, and the "lollypops" are sample points whose interpolations are dashed curves, using cubic spline interpolation. The left column is sampled at approximately one sample per spreading code chip, $\Delta_\tau \approx T_c$, and the right column is sampled at approximately two samples per spreading code chip, $\Delta_\tau \approx T_c/2$. Different rows have different sample phasings, determined by chance during initial synchronization. The upper row in each column has the fortunate situation where $k^p\Delta_\tau = D$, so a sample falls directly at the peak of the correlation

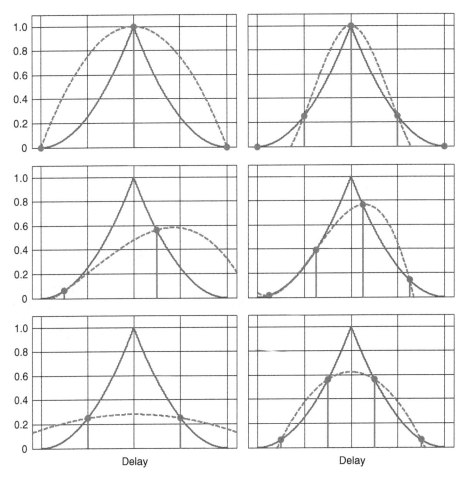

Figure 16.5. Examples of Sampled and Interpolated Correlation Functions

function, yielding $L_{\text{scalloping}} = 1$, or 0 dB, corresponding to no scalloping loss. The lowest row shows the worst possible sample placement for each sample spacing, where the largest samples fall on the shoulders of the correlation function and have the lowest possible amplitudes for that sample spacing. The middle row shows intermediate sample placement. Interpolations of the samples are also shown using a dashed line; interpolation uses a cubic spline interpolator constrained to pass through the sample points.

The narrower the sample spacing (i.e., the higher the sampling rate), the larger the sample values under worst-case sample placement and the lower the scalloping loss, which occurs in addition to the BSQ losses analyzed in Section 15.3. The results in Figure 16.5 are commonly interpreted to imply that sampling for initial synchronization should use approximately two or more samples per spreading code chip (half chip sample spacing). With approximately two samples per spreading code chip, scalloping loss of $20 \log_{10}(1/0.75) = 2.5$ dB is often assumed in performance prediction. Based on the figure, scalloping loss of 6 dB would be assumed in performance prediction with approximately one sample per spreading code chip (half chip sample spacing).

This analysis appears to overstate scalloping loss in delay for several reasons. At the worst-case sample placement, there are two opportunities to detect the peak, improving the probability of detection. Furthermore, the correlation function peak may shift in delay somewhat during initial synchronization processing due to uncompensated code Doppler (Section 16.3.5). The result is a broadened correlation peak less sensitive to scalloping loss. Interpolating the samples can provide improved detection performance. For example, the interpolation in the lowest rows in Figure 16.5 reduces the loss of signal power by approximately 0.5 dB in each column. Further, when noise is present, sampling and any subsequent interpolation affect the CAF variance as well as the mean of the interpolated CAF peak, so only examining the effect on the mean peak power ignores the effect on correlation function variance.

Computer simulations have indicated that, when the CAF output SNR is 15 dB and infinite precision is used, sample spacing of approximately half a spreading code chip, followed by interpolation, yields output SNRs within 0.5 dB of ideal, while sample spacing approaching a spreading code chip duration, followed by interpolation, yields output SNRs within 2 dB of ideal. These preliminary results, calculated over a range of sample placements relative to the correlation peak, suggest that the scalloping loss predicted by (16.8) may be excessively conservative.

Figure 16.6 shows similar effects in the frequency offset dimension when the delay is perfectly matched, for any spreading modulation. The curve is the function $\text{sinc}^2\left[\pi\left(\phi - F\right)T\right]$ from (16.5), plotted in terms of the normalized frequency offset $(\phi - F)T$—the larger the CAF integration time T, the greater the loss for a given frequency offset $(\phi - F)$, for $|(\phi - F)T| < 1$. Samples in the left column are spaced in frequency by $1/T$, while in the right column samples are spaced in frequency by $0.5/T$. Different rows have different sample phasing, determined by chance during initial synchronization. The upper row in each column has the fortunate situation where $m^p \Delta_\phi = F$, so a sample falls directly at the peak of the correlation function, yielding no scalloping loss. The lowest row shows the worst possible sample placement for each sample spacing, where the largest samples fall on the shoulders of the sinc-squared function and have the lowest possible amplitudes for that sample spacing. The middle row shows intermediate sample placement. Interpolations of the samples are also shown using a dashed line; interpolation uses a cubic spline interpolator constrained to pass through the sample points. Scalloping loss is usually interpreted as the reciprocal of the largest sample.

The worst case scalloping loss in frequency offset with the wider sample spacing in Figure 16.6 is 3.9 dB, which is generally considered excessive. It is common to space samples in the frequency domain by $0.5/T$, so the worst-case loss of signal power is 0.9 dB. As was observed in Figure 16.5, interpolation over frequency recovers 0.5 dB of signal power for the worst sample placement in both columns of Figure 16.6.

Scalloping loss considerations motivate small time-frequency cells, produced by narrow spacing of time samples and frequency samples. Implementation considerations, however, motivate wider sample spacing in both dimensions, as shown here. First, consider sample spacing in the time domain, when the CAF integration time T and the

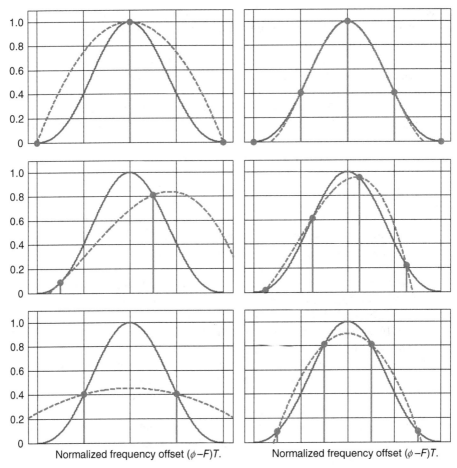

Figure 16.6. Examples of Sampled and Interpolated CAF Peak in Frequency Offset Dimension

ITU are fixed. For a specific value of m, directly implementing (16.6) involves $L+1$ operations, each involving the multiplication $x[\ell \Delta_\tau] y[\ell \Delta_\tau + k\Delta_\tau] e^{i2\pi m \ell \Delta_\phi \Delta_\tau}$ and an accumulation, for each k. The number of terms in the sum is proportional to the number of samples, which is proportional to the sample rate in time—if the sample rate is doubled the value of L doubles to keep the CAF integration time the same. Also, the number of values of k needed to search the ITU is proportional to the sample rate—as the sample rate doubles, the time width of each time-frequency cell is halved and the number of cells needed to search the ITU doubles. Consequently, for direct implementation of (16.6), the number of operations required with a time-domain implementation, while searching a given ITU with a given CAF integration time, is proportional to the reciprocal of the square of the time sample spacing [4].

If the computation of (16.6) for a fixed value of m is instead implemented using fast Fourier transform (FFT) techniques as described in Section 16.3.4, the number of

INITIAL SYNCHRONIZATION BASICS

operations still increases more than linearly with increases in the sample rate, since the number of multiply-adds in FFT processing is proportional to $N \lg(N)$, where N is the number of points in the FFT and $\lg(\cdot)$ is the base 2 logarithm.

The relationship of computational complexity to sample spacing in frequency depends on algorithm choice. Searching a given ITU using direct implementation of (16.6) involves computing $\frac{1}{L}\sum_{\ell=-L/2}^{L/2} x[\ell\Delta_\tau] y[\ell\Delta_\tau + k\Delta_\tau] e^{i2\pi m\ell \Delta_\phi \Delta_\tau}$ for each value of m. In this case the number of operations is proportional to the reciprocal of frequency sample spacing. For other algorithms, such as those described in Sections 16.3.3 and 16.3.4, the number of operations is much less than linear with the reciprocal of frequency sample spacing.

Besides increasing the number of operations, reducing the sample spacing also increases the needed storage. Depending upon the specific implementation of initial synchronization processing, the sampled input waveform may need to be stored, and the sampled CAF is also stored, along with intermediate quantities in the computations. Potentially, hundreds of thousands or millions of values must be stored and accessed rapidly. The resulting storage can dominate the size of the circuitry used for a receiver, and the high-rate input/output can affect processor architecture and contribute significantly to power consumption. Consequently, from an implementation perspective there is a strong motivation to space samples as widely as possible both in time and frequency.

Yet another factor that contributes to implementation complexity is the length of the primary spreading code. Since the ITU is often dictated by the duration of the spreading code, the number of time-frequency cells goes down with reduction in the spreading code length. CAF storage is also typically proportional to the length of the spreading code. Consequently initial synchronization complexity is lower for signals with short spreading codes, like GPS C/A code, and the GLONASS signals, L1OF and L2OF.

The receiver designer can control some aspects of this complexity. Operational steps that help keep the receiver's initial conditions in the right-hand columns of Table 16.1 are beneficial. So is use of high-quality timekeeping devices and frequency sources to reduce the ITU and IFU. Avoiding the need to acquire at low C/N_0 values allows smaller CAF integration times. Efficient algorithms, like those described in Section 16.3, are also essential.

In many cases, search over delay dominates the operations count and storage. The results provided previously all describe the spacing of time samples relative to the spreading code chip duration for BPSK-R signals. If the choice is made to process a BPSK-R signal with 10 times the spreading code chip rate, keeping the same scalloping loss requires samples with one-tenth the spacing, potentially increasing the operations count by a factor of 100 and storage by a factor of 10. This observation helps explain why, in the original GPS signal design, the C/A (recall that "C/A" stands for coarse acquisition) code signal was intended for acquisition. Only after acquisition would receivers hand over to tracking of the P(Y) code signal that has the much higher spreading code chip rate.

Many modern signals described in Part II have faster spreading code chip rates and narrower correlation function peaks, as well as longer spreading codes, compared to the GLONASS L1OF and GPS C/A code signals. While these characteristics provide better performance in a number of ways, they also increase the complexity of initial synchronization processing. Advances in digital circuitry can offset the increased complexity, but for a given circuit technology, initial synchronization processing is still simpler for these original GPS and GLONASS signals.

16.2.3 Broadening BOC Correlation Functions for Initial Synchronization

Section 16.2.2's discussion of delay sample spacing emphasizes signals with BPSK-R spreading modulations. This section addresses BOC spreading modulations and their variations.

As seen in Part II, variants of the BOC(1,1) spreading modulation are used for several modern signals. Comparing correlation functions for BPSK-R(1) in Figure 7.3 and BOC(1,1) in Figure 7.4 shows that the BOC(1,1) correlation function peak is approximately one-third the width of that for BPSK-R(1). The implication is that, for the same scalloping loss, the sampling rate would have to be increased by a factor of three, increasing the operations count by as much as a factor of nine, and storage by a factor of three, compared to initial synchronization of a BPSK-R(1) spreading modulation. The widths of correlation peaks for signals with higher subcarrier frequencies, such as GPS M-code signal, Galileo PRS signals, and some BeiDou signals, are even narrower, requiring greater initial synchronization implementation complexity.

Fortunately, there are ways to broaden the correlation function peak for initial synchronization of BOC spreading modulations. Figure 16.7 shows the concept of sideband processing BOC signals [5]. Parallel processing chains downconvert each sideband to baseband, transforming a BOC(f_s,f_c) signal to a pair of BPSK-R(f_c) signals. A CAF is computed for each downconverted sideband using a BPSK-R(f_c) replica. Since the relative carrier phase between sidebands is unknown, the squared magnitudes or the magnitudes of each sideband CAF are summed. The result can be treated just like the result from initial synchronization of any signal with BPSK-R spreading modulation.

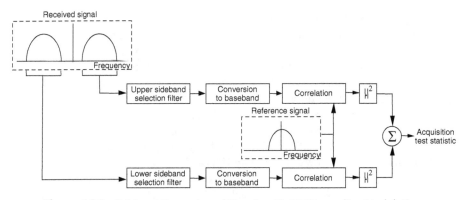

Figure 16.7. Sideband Processing of Signals with BOC Spreading Modulations

INITIAL SYNCHRONIZATION BASICS

While sideband processing doubles the number of CAFs that must be computed, each CAF computation is simplified by the use of sample spacing corresponding to the correlation peak width of the BPSK-R signal rather than that of the BOC signal. For example, performing initial synchronization processing on BPSK-R(1) rather than BOC(1,1) allows time sample spacing to be increased by a factor of three for the same scalloping loss. The reduction of up to a factor of nine in CAF computations outweighs the need to compute two CAFs, providing an overall reduction in computational complexity by up to a factor of 4.5. Storage of the CAF is also reduced by a factor of three, providing additional simplification.

Other ways to broaden the BOC correlation function have also been identified. One approach, proposed in Reference 6 for signal tracking, can be adapted to initial synchronization. This approach, called IBOC/QBOC processing, also computes two separate CAFs, one using a $BOC_s(1,1)$ replica and the other using a BOC_c replica, both employing the same spreading code. (In Reference 6 the $BOC_s(1,1)$ is called IBOC and the BOC_c is called QBOC, reflecting the in-phase/quadrature phase 90° relationship between the subcarriers in the two replicas.) The magnitudes or magnitudes squared of both CAFs are summed to yield the CAF employed for subsequent processing.

Initial exploration of IBOC/QBOC processing for initial synchronization used a wide bandwidth signal, yielding a correlation function with too narrow a peak to be advantageous. Figure 16.8 plots the resulting squared magnitude for a wide bandwidth

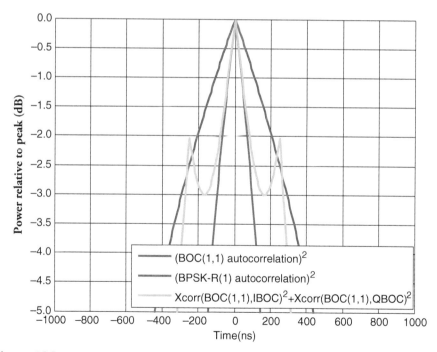

Figure 16.8. Correlation Peaks Computed Using Wideband, Sideband, and IBOC/QBOC Processing with ±12 MHz Bandwidth BOC(1,1) Signal

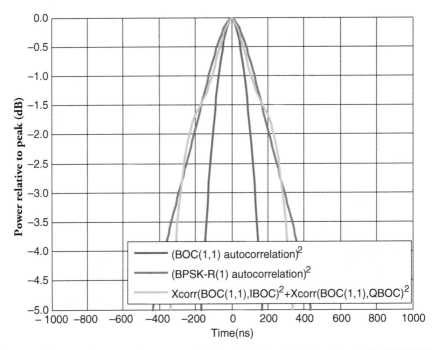

Figure 16.9. Correlation Peaks Computed Using Wideband, Sideband, and IBOC/QBOC Processing with ±3.069 MHz Bandwidth BOC(1,1) Signal

BOC(1,1) signal, comparing it to the squared magnitudes from sideband processing (corresponding to BPSK-R(1)) and from direct processing of the wideband BOC(1,1) signal. While the peak from IBOC/QBOC processing is broader at the base than the BOC(1,1) peak, it is still too narrow near the top of the peak, where the scalloping loss would be acceptable.

If the BOC(1,1) signal's precorrelation bandwidth is narrowed, however, the conclusion changes. Figure 16.9 shows the same correlation peaks computed from a narrower bandwidth signal. The IBOC/QBOC correlation function peak is now as broad as that from sideband processing, indicating the potential utility of IBOC/QBOC processing to increase the sample spacing for initial synchronization of BOC signals.

Initial synchronization of signals having MBOC spectrum, with a blend of BOC(1,1) and BOC(6,1) symbols, must also be addressed. The solution is to process these signals, whether implemented using TMBOC or CBOC, as if they are BOC(1,1) signals, neglecting the BOC(6,1) component for initial synchronization. The fraction of power in the BOC(6,1) components is small enough, especially for highly bandlimited signals typically employed for initial synchronization, that appropriate processing (including blanking of the BOC(6,1) spreading symbols in TMBOC, as described in Section 18.1.3) affects initial synchronization processing performance by a negligible amount. Any additional complexity for using the BOC(6,1) components in MBOC signals for initial synchronization appears to be unwarranted.

16.3 INITIAL SYNCHRONIZATION COMPUTATION

This section outlines a variety of algorithms for initial synchronization, ranging from direct implementation of the CAF to more sophisticated and complex techniques. Section 16.3.1 starts with serial search, while Section 16.3.2 introduces use of the FFT for CAF parallel search in the time domain. Section 16.3.3 describes an approximate approach for parallel search of a time-frequency tile using a mixed time-domain/frequency-domain algorithm, and Section 16.3.4 shows its implementation as a pure frequency-domain calculation. Section 16.3.5 addresses noncoherent integration and the resulting need for code Doppler compensation.

16.3.1 Serial Search

In early GPS receivers, each correlator was a precious commodity involving considerable hardware and consuming considerable power. For initial synchronization, each correlator would compute the CAF at one specific time-frequency cell, sequentially stepping through different time-frequency cells in a serial search of ITU and IFU. Often, the same correlator hardware would be shared between initial synchronization and tracking.

In a serial search, for each value of m, (16.6) is evaluated over sequential values of k until the ITU is searched, then the search is repeated for different m. If multiple correlators are available, they can be used in parallel, searching different values of k simultaneously while stepping serially through values of m.

When $x[\ell \Delta_\tau] = \sqrt{\zeta C} e^{-i\theta} s[\ell \Delta_\tau + D] e^{-i2\pi F n \Delta_\tau} + n[\ell \Delta_\tau]$ and $y[\ell \Delta_\tau] = s[\ell \Delta_\tau]$, the term $x[\ell \Delta_\tau] y[\ell \Delta_\tau + k \Delta_\tau] e^{i2\pi m \ell \Delta_\phi \Delta_\tau}$ within the summation in (16.6) can be written as the product of $\{ \sqrt{\zeta C} e^{-i\theta} s[\ell \Delta_\tau + D] e^{-i2\pi F n \Delta_\tau} + n[\ell \Delta_\tau] \} e^{i2\pi m \ell \Delta_\phi \Delta_\tau}$ and $s[\ell \Delta_\tau + k \Delta_\tau]$. When the signal is a constant-envelope biphase signal, this second term takes on values of ± 1, so directly implementing the product in the time domain does not require a multiplication, but merely a sign change of the first term whenever the second term takes on the value -1, resulting in incrementing or decrementing an accumulator. The resulting hardware implementation complexity can be much less than with an actual multiplication.

16.3.2 Parallel Time Search

It is well-known [7] that correlation can be implemented using the discrete Fourier transform (DFT), often implemented using an FFT. Implementing (16.6) using the DFT involves first selecting the range of delay values to search. Let the range of delay values be given by $-K/2 \leq k \leq K/2$, where K is assumed even and the true delay must be in the range $-K\Delta_\tau/2 \leq D \leq K\Delta_\tau/2$. A set of $K+L+1$ samples $x[m,j]|_{j=0}^{K+L} = \{\alpha e^{-i\theta} s[\ell\Delta_\tau + D]e^{-i2\pi Fn\Delta_\tau} + n[\ell\Delta_\tau]\}e^{i2\pi m\ell\Delta_\phi\Delta_\tau}\big|_{\ell=-L/2-K/2}^{L/2+K/2}$ is used for a given frequency offset index m, with a zero-padded signal replica of the same length $y[j]|_{j=0}^{K+L} = s[-L\Delta_\tau/2], \ldots, s[L\Delta_\tau/2], \underbrace{0, 0, \ldots, 0}_{K}$. Representing the DFT of $x[j]|_{j=0}^{K+L}$ by $X[q]|_{q=0}^{K+L} \triangleq \sum_{j=0}^{K+L} x[j]e^{-i2\pi qj/(K+L+1)}$ and the DFT of $y[j]|_{j=0}^{K+L}$ by $Y[q]|_{q=0}^{K+L} \triangleq \sum_{j=0}^{K+L} y[j]e^{-i2\pi qj/(K+L+1)}$, compute the conjugate product of the DFTs $Z[m,q] \triangleq X[m,q]Y^*[q]$, and the correlation is

$$z[m,k] = \frac{1}{K+L+1} \sum_{q=0}^{K+L} Z^*[m,q] e^{-i2\pi qj/(K+L+1)}, \quad k = -K/2, \ldots, K/2 \quad (16.9)$$

which is also a DFT.

When the number of samples $K + L + 1$ is chosen to be a power of two,[3] the FFT can be used to simplify these three DFT computations. In addition, since $Y[q]|_{q=0}^{K+L}$ does not depend on m, it does not need to be recomputed to search different frequency offsets indexed by m, meaning that only two DFTs need to be computed for the second and subsequent values of m.

Whereas the computation of (16.6), as described in Section 16.3.1, produces a single value at a particular time-frequency cell indexed by k and m, the computation (16.9) produces a set of values over the time delay index range $k = -K/2, \ldots, K/2$ and for a particular frequency offset index m. The result is a time-frequency tile with only one frequency row but many time columns, all computed in parallel.

The number of multiplies and additions involved in computing (16.9) is far less than the direct implementation of (16.6) for all values of k using serial search. FFT-based parallel time search uses many fewer arithmetic operations than serial search.

16.3.3 Parallel Time-Frequency Search Using a Mixed Time-Domain/Frequency-Domain Algorithm

The theme of expanding from serial search over time delay to parallel search over time delay can be extended to parallel search over both time delay and frequency offset via a clever approximation [8] described here. Begin with the definition of the CAF

---

[3] If this value is not a power of two, it can be increased to the next largest value of two through additional zero padding in order to take advantage of FFT computational simplicity.

in (16.1),[4] but partition the integral into a sum of N integrals over contiguous time intervals each $\Lambda = T/N$ long:

$$\Xi(\tau,\phi) = \frac{1}{T}\int_{-T/2}^{T/2} x(t)y^*(t+\tau)e^{i2\pi\phi t}dt$$

$$= \frac{1}{T}\sum_{n=0}^{N-1}\int_{-T/2+n\Lambda}^{-T/2+(n+1)\Lambda} x(t)y^*(t+\tau)e^{i2\pi\phi t}dt \quad (16.10)$$

Now approximate the time variable in the complex exponential by its midpoint value. The complex exponential is no longer a function of the variable of integration, allowing the complex exponential to be removed from the integral, enabling several simplifications:

$$\Xi(\tau,\phi) \approx \frac{1}{T}\sum_{n=0}^{N-1}\int_{-T/2+n\Lambda}^{-T/2+(n+1)\Lambda} x(t)y^*(t+\tau)e^{i2\pi\phi(-T/2+n\Lambda+\Lambda/2)}dt$$

$$= \frac{1}{T}e^{i\pi\phi(-T+\Lambda)}\sum_{n=0}^{N-1} e^{i2\pi\phi n\Lambda}\int_{-T/2+n\Lambda}^{-T/2+(n+1)\Lambda} x(t)y^*(t+\tau)dt \quad (16.11)$$

$$= \frac{1}{T}e^{i\pi\phi(-T+\Lambda)}\sum_{n=0}^{N-1} e^{i2\pi\phi n\Lambda}\chi_n(\tau)$$

where $\chi_n(\tau) \triangleq \int_{-T/2+n\Lambda}^{-T/2+(n+1)\Lambda} x(t)y^*(t+\tau)dt$ are called short-time correlations. The approximation in the first line of (16.11) is known as a staircase approximation, since the linear "ramp" in carrier phase versus time has been approximated by a set of discrete "steps" in carrier phase.

The short-time correlations can be computed using sampled waveforms, yielding the discrete-time short-time correlations

$$\chi_n(k\Delta_\tau)\big|_{k=-K/2}^{K/2} = \frac{1}{L}\sum_{\ell=-L/2}^{L/2} x[\ell\Delta_\tau]y^*[\ell\Delta_\tau + k\Delta_\tau]\bigg|_{k=-K/2}^{K/2} \quad (16.12)$$

---

[4] A similar development can be based on the discrete CAF (16-6), but using (16-1) simplifies the notation.

The resulting CAF approximation, computed at discrete time delay and frequency offset indices, can be written

$$\Xi[k\Delta_\tau, m\Delta_\phi]\Big|_{k=-K/2,m=0}^{k=K/2,m=M-1} \approx \frac{1}{T} e^{i\pi m \Delta_\phi(-T+\Lambda)} \sum_{n=0}^{N-1} \chi_n(k\Delta_\tau) e^{i2\pi m \Delta_\tau n\Lambda} \Big|_{k=-K/2,m=0}^{k=K/2,m=M-1}$$

$$= \frac{1}{T} e^{i\pi m \Delta_\phi(-T+\Lambda)} X_m(k\Delta_\tau) \Big|_{k=-K/2,m=0}^{k=K/2,m=M-1} \quad (16.13)$$

where $X_m(k\Delta_\tau) \stackrel{\Delta}{=} \sum_{n=0}^{N-1} \chi_n(k\Delta_\tau) e^{i2\pi m \Delta_\phi n\Lambda}$ is the inverse DFT (IDFT) of the short-time correlations, and the magnitude-squared of the CAF is $|\Xi[k\Delta_\tau, m\Delta_\phi]|^2 \approx |\frac{1}{T} X_m(k\Delta_\tau)|^2$.

In practice, it is beneficial to zero-pad the short-time correlations $\chi_n(\tau)$ before taking their inverse DFT in (16.13). Typically, N zeroes are added, providing the ability to interpolate by a factor of two and thus obtain frequency offset resolution on the order of $\Delta_\phi/2$. In addition, the staircase approximation to the carrier phase ramp is poorer for larger values of m, so often only the values $m = 0, 1, \ldots, \lfloor M/2 \rfloor$ are retained.

Figure 16.10 shows this mixed time-domain/frequency-domain initial synchronization algorithm on a small scale, computing a time-frequency tile with five delay values and five frequency offset values. The four sampled short-time correlations are computed, each at five delay indices, in the time domain, with each short-time correlation function

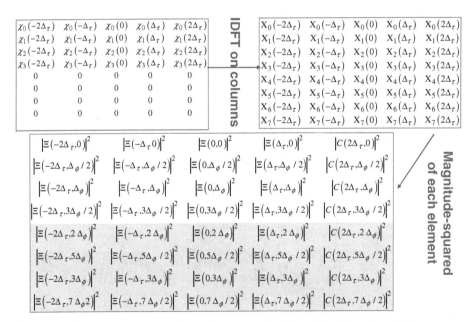

Figure 16.10. Small-Scale Example Implementation of the Mixed Time-Domain/Frequency-Domain Algorithm

a row in a matrix. After additional rows of zeroes are added for zero padding, the IDFT is computed down each column. The squared-magnitude of each element approximates the magnitude-squared of the CAF at that delay and frequency offset index; values with larger frequency offset indices are discarded because of the approximation losses.

The number of columns in the matrices corresponds to the different delays being searched, and the number of retained rows corresponds to the frequency offsets being searched, defining the time-frequency tile. Typically, the number of short-time correlations is a modest power of two, in the range 16–64, while the number of delay values may be in the thousands.

This algorithm has remarkable implications. The short-time correlations can be computed once and stored, avoiding the need to compute correlations for each frequency offset index. The search over frequency offset reuses the short-time correlations, employing modest-sized FFTs. Computing the short-time correlations typically dominates the computational burden; the IDFTs for computing the frequency offset dimension can represent less than 10% of the computational burden. Thus, the frequency search is very efficient in terms of operations. In a hardware embodiment of this algorithm, such an application-specific integrated circuit (ASIC) or field-programmable gate array (FPGA), it is often determined that time-domain computation of the short-time correlations is the most hardware-efficient implementation. This hardware can take advantage of the fact that the replica can be represented with one bit in most cases, replacing multiplications with accumulator increments or decrements, as discussed in Section 16.3.1.

Computing (16.13) over a range of indices in time delay and frequency offset produces a rectangular time-frequency tile, but implementation considerations can limit the size of the tile. As shown in Figure 16.3, if the tile is smaller than the ITU and IFU, different tiles can be computed by delaying or advancing, and frequency shifting, the input waveform, denoting this delayed/advanced and frequency shifted input $x(t)$ in the short-time correlations.

In 2002 and 2003, an ASIC was designed, produced, and tested to use this algorithm in order to demonstrate the ability to perform direct acquisition of the GPS M-code signal [9]. This ASIC implemented sideband processing of the BOC(10,5) signal, using hardware shared at twice real time to compute CAFs from each sideband. The ASIC, built using a 180-nm CMOS process, has dimensions of 9.40 × 9.36 mm and contains 21.5 million transistors with 1.3 million placeable cells, and a maximum clock speed of 80 MHz, with the CAF stored in off-board memory. Its demonstrated performance closely matched theoretical predictions. With subsequent reductions in lithography to smaller features, and support for higher clock rates, a more modern version of this ASIC would have much smaller dimensions.

16.3.4 Parallel Time-Frequency Search Using a Frequency-Domain/Frequency-Domain Algorithm

The final algorithm for CAF computation described here is a blend of those described in Sections 16.3.2 and 16.3.3. In this implementation, the discrete-time short-time correlations (16.12) are computed using the frequency-domain approach described in Section 16.3.2. The resulting short-time correlation functions are combined to approximate the CAF using (16.13). When implementation considerations are driven by the count of mathematical operations, such as a software receiver implementation with a digital signal processor or general purpose processor, this algorithm may be preferred.

16.3.5 Noncoherent Integration and Code Doppler Compensation

The CAF output SNR determines performance of detection and affects the accuracy of time delay and frequency offset estimates. For given received signal power, noise density, and received interference, there are only a few ways to improve CAF output SNR, which is the same as the correlator output SNR (5.6). One alternative is to reduce losses, including BSQ losses and scalloping loss, but doing this has limited benefit and can significantly increase implementation complexity. Another alternative is to increase the CAF integration time T.

However, a number of considerations must be taken into account when increasing T. The CAF integration time is typically limited by signal characteristics, such as duration of a modulating bit (corresponding to a data message symbol or overlay code bit). CAF integrations that straddle unknown bit transitions do not provide the desired enhancement of output SNR. Data wiping, described in Section 16.5, can sometimes be used to overcome the limitation due to modulating bits.

Implementation complexity also can limit CAF integration time. Not only does the number of arithmetic or logical operations involved in computing the CAF at each time-frequency cell increase at least linearly with T, but the CAF must be evaluated at more time-frequency cells to search the same IFU, since the number of frequency indices needed to cover the IFU is inversely proportional to T. Further, a longer CAF integration time also increases storage, since a longer segment of the input waveform or replica must be stored, and storage of the CAF increases due to the larger number of frequency indices required to cover a given IFU.

In addition, there are physical constraints on the CAF integration time. While the CAF accommodates Doppler shift of the carrier, it does not account for time companding of the signal's complex envelope as discussed in Section 3.2.1, so T must be limited accordingly. Also, any line of sight accelerations between satellite and receiver, caused by motion of satellite or receiver, also can diminish the benefit from longer CAF integration time since these dynamics cause the CAF peak to change location in frequency offset, just as code Doppler causes the peak to change location in delay. Further, phase noise on the received signal or from the oscillator in the receiver also limits performance as T increases. These physical constraints can limit CAF integration time to many tens of milliseconds, even if implementation complexity considerations discussed in previous paragraphs are not limiting factors.

Noncoherent integration is an additional way to improve initial synchronization performance without encountering these limitations. Consider a sequence of CAFs computed over contiguous time intervals T seconds apart,

$$\Xi_n(\tau,\phi) = \int_{-T/2+nT}^{T/2+nT} x(t)y^*(t+\tau)e^{i2\pi\phi t}dt, \quad n=0,1,\ldots,N_T-1 \quad (16.14)$$

INITIAL SYNCHRONIZATION COMPUTATION

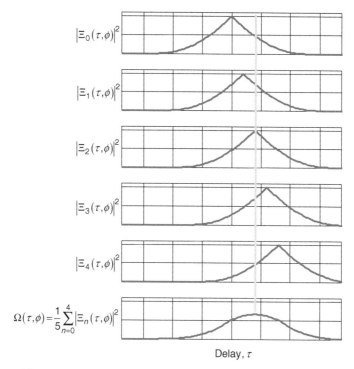

Figure 16.11. Example of Uncompensated Code Doppler Introducing 1.8 dB Loss

The noncoherent integration of these CAFs is

$$\Omega(\tau, \phi) = \frac{1}{N_T} \sum_{n=0}^{N_T-1} |\Xi_n(\tau, \phi)|^2 \qquad (16.15)$$

where either the magnitude or the squared magnitude can be used. The performance benefits of noncoherent integration are assessed in Section 16.5.

The amount of received waveform needed for noncoherent integration is TN_T. The code Doppler can change considerably over this time, limiting performance unless it is addressed. Figure 16.11 shows the effect of code Doppler on correlation functions where the carrier Doppler[5] is perfectly matched. Five-squared correlation functions for a BPSK-R signal are summed to produce the noncoherent integration in the bottom trace. Clearly, the peak locations change in each correlation function, producing a noncoherent integration whose peak is smoothed and diminished by 1.8 dB. Even though the code Doppler is much smaller than the carrier Doppler, neglecting the code Doppler degrades performance.

---

[5] In this subsection frequency offset is called Doppler, following common usage.

To quantify the effect of code Doppler, denote by Υ the ratio of carrier frequency to spreading code chip rate for any BPSK-R signal. A complete table of Υ values for different signals is provided in Section 19.3. When the maximum magnitude of the carrier frequency offset is $\psi_{\max}$, the maximum code Doppler is $\psi_{\max}/\Upsilon$. With no compensation for code Doppler, the maximum number of CAFs noncoherently integrated before the peak of the final correlation function is shifted by one-quarter spreading code chip duration relative to the peak of the first correlation function is

$$N_{T,\max} = \frac{\Upsilon}{4T\psi_{\max}} \tag{16.16}$$

When more than this number of noncoherent integrations is used, the full benefit of noncoherent integration may not be obtained. The result can be a significant limitation on the number of useful noncoherent integrations, particularly for signals with larger spreading code chip rates and lower carrier frequencies, since they have smaller values of Υ.

Fortunately, code Doppler can be partially compensated, enabling a larger number of noncoherent integrations. The ability to compensate arises from inherent compensation for carrier Doppler in the CAF. When the CAF is computed for a particular frequency offset index m, this computation corresponds to the hypothesis that the signal has carrier Doppler $m\Delta\phi$. If that is the carrier Doppler, then the code Doppler for that frequency cell must be $m\Delta\phi/\Upsilon$. However, the actual carrier Doppler may not be $m\Delta\phi$, since the time-frequency cell includes a range of Doppler values. When frequency cells are spaced by $0.5/T$, the frequency offsets in a given cell span $m\Delta\phi \pm 0.25/T$. Consequently, the worst case magnitude error in carrier Doppler is $0.25/T$, and the largest magnitude error in residual code Doppler is $1/(4\Upsilon T)$. The maximum number of noncoherent integrations before the last correlation function's peak is shifted by one-quarter spreading code chip duration from the first peak is obtained by finding the time $N_{\max}T$ it takes for the largest magnitude error in residual code Doppler to be one-fourth of a cycle:

$$\frac{N_{T,\max}T}{4\Upsilon T} = \frac{1}{4}$$
or
$$N_{T,\max} = \Upsilon \tag{16.17}$$

This result states that, with code Doppler compensation and CAF frequency bin spacing by $0.5/T$, a conservative upper limit on the number of noncoherent integrations (where almost full benefit from noncoherent integrations is obtained) is the ratio between carrier frequency and spreading code chip rate, for initial synchronization of signals with BPSK-R spreading modulation. (This includes sideband processing of BOC spreading modulations, and any other approach for broadening the correlation function that produces correlation functions as wide as the corresponding BPSK-R signal.) With code Doppler compensation, many more noncoherent integrations can be used, as seen by comparing (16.17) with (16.16).

INITIAL SYNCHRONIZATION COMPUTATION

The criteria (16.16) and (16.17) are both conservative, only allowing the correlation function peak to shift by one-quarter of the spreading code chip duration over the entire set of noncoherent integrations, and assuming worst case code Doppler. Some authors adopt similar rules of thumb, but allow up to four times the code Doppler, leading to proportionally longer integration times. Also, using narrower carrier Doppler spacing in the CAF enables more precise code Doppler compensation, allowing proportionally more noncoherent integrations than indicated in (16.17). This is yet another case of obtaining performance at the cost of implementation complexity.

There are many ways to implement code Doppler compensation, but receiver manufacturers tend to keep them proprietary. In principle, each squared-magnitude CAF must be shifted in delay before the summation indicated in (16.15), with the same shift for all cells having the same frequency index (that is, for each row of the time-frequency tile). For the time-frequency cell at location $(k\Delta\tau, m\Delta\phi)$, noncoherent integration with code Doppler compensation is then given by

$$\Omega(k\Delta_\tau, m\Delta_\phi) = \frac{1}{N_T} \sum_{n=0}^{N_T-1} |\Xi_n(k\Delta_\tau + (n - N_T/2)m\Delta_\phi/\Upsilon, m\Delta_\phi)|^2 \qquad (16.18)$$

Typically, the shift is not an integer number of samples, so some form of interpolation of the discrete-time samples is needed. One published method is found in Reference 10. This patent describes the use of delay lines to interpolate between delay samples for code Doppler compensation, as well as for interpolation of the CAF after noncoherent integration in order to reduce scalloping loss as discussed in Section 16.2.1.

When sideband processing, or an equivalent way of broadening the CAF peak, is employed, two separate CAFs, denoted $\Xi_{n,1}(\tau, \phi)$ and $\Xi_{n,2}(\tau, \phi)$, are computed using the same T seconds of data. The CAF from that time interval is the noncoherent integration of these two CAFs:

$$|\Xi_n(\tau, \phi)|^2 = [|\Xi_{n,1}(\tau, \phi)|^2 + |\Xi_{n,1}(\tau, \phi)|^2]/2 \qquad (16.19)$$

When initial synchronization processing is performed on pilot and data components, the phase relationship between these components is not known a priori, so when wideband processing is used, two separate CAFs, denoted $\Xi_{n,P}(\tau, \phi)$ and $\Xi_{n,D}(\tau, \phi)$, are computed using the same T seconds of data. The CAF from that time interval is the noncoherent integration of these two CAFs:[6]

$$|\Xi_n(\tau, \phi)|^2 = [|\Xi_{n,P}(\tau, \phi)|^2 + |\Xi_{n,D}(\tau, \phi)|^2]/2 \qquad (16.20)$$

---

[6] The expression assumes equal power pilot and data, so equal weighting of the two CAFs. When pilot and data power are not equal, the two CAFs must be weighted appropriately. The benefits of computing and combining both CAFs diminish significantly with unequal power, so the only case considered here is equal power. For example, the benefit from noncoherent integration of the 25% power data component with the 75% power pilot component is shown to have minimal benefit in Reference 15 for L1C.

When initial synchronization processing is performed on pilot and data components using sideband processing, four separate CAFs, denoted $\Xi_{n,P,1}(\tau,\phi)$, $\Xi_{n,P,2}(\tau,\phi)$, $\Xi_{n,D,1}(\tau,\phi)$, and $\Xi_{n,D,2}(\tau,\phi)$, are computed using the same T seconds of data. The CAF from that time interval is the noncoherent integration of these four CAFs:

$$|\Xi_n(\tau,\phi)|^2 = [|\Xi_{n,P,1}(\tau,\phi)|^2 + |\Xi_{n,P,2}(\tau,\phi)|^2 + |\Xi_{n,D,1}(\tau,\phi)|^2 + |\Xi_{n,D,2}(\tau,\phi)|^2]/4$$

(16.21)

While use of pilot and data components and separate sidebands increases the number of noncoherent integrations, the signal power is divided among these different CAFs, diminishing each CAF output SNR as seen in Section 16.4.

16.4 INITIAL SYNCHRONIZATION PERFORMANCE

Initial synchronization performance is primarily assessed in terms of two parameters, the minimum effective C/N_0 at which initial synchronization can be achieved, and the time it takes for the receiver to complete initial synchronization, called time to initial synchronization (TTIS). Minimum effective C/N_0 may be expressed equivalently in terms of the maximum tolerable power of interference or jamming, or the maximum tolerable I/S or J/S level (for a particular value of S), as described in Chapter 5. TTIS is typically longer at lower effective C/N_0, as seen below.

Initial synchronization performance is typically evaluated under the assumption that noise and interference are uncorrelated with the desired signal, and that assumption is used here initially as well. The effect of strong interfering signals with partly correlated spreading codes is addressed subsequently.

For CAFs computed using noncoherent integration as defined in (16.15), initial synchronization detection performance is assessed using the generalized Marcum Q-function [11, 12]. Two parameters dictate the calculation of this function, the CAF (or correlator) output SNR, and the total number of degrees of freedom used in noncoherent integration, N_{nci}, which is the number of CAF squared magnitudes being summed in noncoherent integration. The CAF output SNR is in turn calculated from the CAF integration time T and the effective C/N_0, where the signal power in the numerator is the fraction of desired signal power provided to that correlator, adjusted for processing losses.

The Neyman–Pearson criterion [11] is employed in assessing detection performance. The detection threshold—the value that the squared magnitude correlation function peak must exceed for a detection to be declared—is determined from the desired false detection probability computed using the generalized Marcum Q-function with SNR of 0 (or $-\infty$ in dB) and the number of degrees of freedom being used. This threshold setting is then used to determine the detection probability for a peak at a specific CAF output SNR (based on the CAF integration time T and the effective C/N_0) and the same number of degrees of freedom.

INITIAL SYNCHRONIZATION PERFORMANCE

TABLE 16.2. Parameters for Calculating Initial Synchronization Performance

	Wideband Processing		Processing Both Sidebands	
Signal Component(s) Used for Initial Synchronization	CAF Output SNR, ρ	Number of Degrees of Freedom per CAF Integration Time, N_{perT}	CAF Output SNR, ρ	Number of Degrees of Freedom per CAF Integration Time, N_{perT}
Only one, having fraction ζ of total signal power	$\zeta 2T(C/N_0)_{eff}$	1	$\zeta T(C/N_0)_{eff}$	2
Pilot and data each having 50% signal power	$T(C/N_0)_{eff}$	2	$0.5T(C/N_0)_{eff}$	4

As a shortcut or supplement to rigorous calculation of the generalized Marcum Q-function, the following approach can be used to predict initial synchronization performance. Table 16.2 shows how to relate CAF output SNR and number of degrees of freedom for different initial synchronization situations. Initial synchronization may use only one signal component, or may separately process two components in parallel, combining both resulting CAFs using noncoherent integration. Both cases are addressed in Table 16.2, with the case of using two components specific to a 50%/50% power division between pilot and data component. Different columns are provided for wideband processing, where the CAF is computed from the entire signal bandwidth, and for sideband processing of a BOC signal as outlined in Figure 16.7. CAF output SNR is defined for each CAF; for wideband processing, the CAF output SNR is twice that for processing each sideband, since the signal power provided to each sideband correlator is half the wideband power. Also, for separate processing of pilot and data each having 50% signal power, the CAF output SNR is half that for a single component containing 100% of the power. Each CAF computed in parallel contributes one degree of freedom per CAF integration time. Two CAFs are computed in parallel when processing both sidebands, doubling the number of degrees of freedom per integration time. Separate processing of pilot and data components also doubles the number of degrees of freedom per integration time.

A high detection probability is usually desired for initial synchronization, with 0.95 a typical value. If this probability is not high, there is a significant chance of a missed detection, motivating tiles to be searched again if no detection is reported after initial search of the ITU and IFU space. Selecting the false detection probability depends upon what logic is used after a detection report. With the larger number of tracking loops available in modern receivers, an increasingly common approach is to hand each detection report (with its associated estimate of delay and frequency offset) to a tracking loop for verification processing. If tracking is successfully initiated, the detection report is verified; if tracking is not sustained, the detection report is considered

TABLE 16.3. Number of Noncoherent Integrations Needed for Detection Probability of 0.95

CAF Output SNR, ρ, After Implementation Loss (dB)	Total Number of Noncoherent Integrations, N_{nci}	
	Per-Decision False Detection Probability 10^{-3}	Per-Decision False Detection Probability 10^{-6}
10	4	7
9	6	10
8	7	13
7	10	18
6	14	25
5	20	35
4	28	49
3	40	71
2	58	103
1	85	152
0	126	226
−1	188	338
−2	284	511
−3	431	778
−4	659	1193

a false detection. With this approach, the rate of false detections (that is, the total number of false detections per unit time) should not exceed the availability of tracking loops for verification processing. If there are v_{cells} in a time-frequency tile, the expected number of false detections is approximately $p_{fa} v_{cells}$, where p_{fa} is the per-decision false detection probability. Typically, processing is assumed instantaneous in modern receivers, and in that case the time to compute the tile and produce detection reports is TN_T. Then, the expected rate of false detections is $p_{fd} v_{cells}/(TN_T)$. The false detection probability must be established to keep this rate from being unmanageably high. Typically, the per-decision false detection probability is between 10^{-3} and 10^{-6}.

Table 16.3 shows the total number of noncoherent integrations needed to obtain a detection probability of 0.95, for different CAF output SNRs and two different false detection probabilities. Over the range of values portrayed, for a given CAF output SNR, N_{nci} roughly doubles to reduce the false detection probability from 10^{-3} to 10^{-6}. Also, doubling N_{nci} allows initial synchronization at approximately 2 dB lower SNR for SNRs greater than 0 dB, and approximately 1.7 dB lower SNR for CAF output SNRs less than 0 dB.

To assess the TTIS, recognize that $N_{nci} = N_T N_{perT}$, so when the processing is assumed instantaneous,

$$\text{TTIS} = T(N_{nci}/N_{perT}) + T_{validation} \qquad (16.22)$$

where $T_{validation}$ is the time needed to validate the detection report. The expression (16.22) is actually a lower bound, since time to initial synchronization is longer on average due to the need for dealing with missed detections.

TABLE 16.4. Example of Desensitizing Initial Synchronization Performance to Address Sidelobe Detections from Strong Interfering Signal Having Effective C/N_0 of 50 dB-Hz

Number of Noncoherent Integrations	Sidelobe Level (dB)	Minimum Detectable Effective C/N_0 (dB-Hz)
100	$-\infty$	29.1
100	-32	29.6
100	-24	31.6
1000	$-\infty$	23.4
1000	-32	24.7
1000	-24	28.4

The previous results in this section apply when the received waveform consists of desired signal plus uncorrelated noise and interference. If there are additional signals present at high enough power level, with large enough crosscorrelations to the desired signal, however, the false detections will be larger than expected from the preceding analysis. Suppose, for example,[7] it is desired that initial synchronization operate when the effective C/N_0 is 28 dB-Hz, in order to work with weak signals resulting from a combination of low receive antenna gain and excess attenuation, such as due to foliage or building materials. Suppose the crosscorrelation between desired and interfering signal is 24 dB lower than the desired signal's CAF peak (i.e., CAF sidelobe values of -24 dB), and that an interfering signal is received with effective C/N_0 of 50 dB-Hz, due to a combination of receive antenna gain and absence of excess attenuation. Then the crosscorrelation between desired signal replica and interfering signal produces CAF sidelobes having effective C/N_0 of 26 dB-Hz, significantly increasing the false detection probability when noise and sidelobes combine.

One way to deal with this situation is to accept the false detections and reject them through verification processing. In that case, the detection threshold can be established as described previously, but many verifications may be needed. Sometime a tracking loop may even lock on such a crosscorrelation, risking the use of faulty measurements in PVT calculation. Receivers can even use almanac information, when available, to predict when such problems might occur and avoid using signals from those satellites.

Another option is to increase the detection threshold, desensitizing the acquisition processing to avoid false detections from crosscorrelations. In this case, the detection threshold is established using the generalized Marcum Q-function with output SNR based on sidelobes as well as noise, rather than based on noise-only conditions. Table 16.4 shows numerical results, calculated analytically, for desensitizing initial synchronization for three different sidelobe levels, including perfectly low sidelobes corresponding to a noise-only situation. The results are shown for an interfering signal having effective C/N_0 of 50 dB-Hz, 1 ms CAF integration times, detection probability of 0.95, and per-decision false detection probability of 10^{-6}. No processing loss is included.

The results in Table 16.4 show that the lower sidelobe level of -32 dB causes modest desensitization of less than 1.5 dB even with a large number of noncoherent integrations.

---

[7] The numerical values in this example correspond roughly to the GPS C/A code signal.

The higher sidelobe level of −24 dB, however, introduces 5 dB of desensitization with the larger number of noncoherent integrations. Similar results can be found elsewhere—in Reference 13, for example.

16.5 OTHER ASPECTS OF ACQUISITION

Since the focus of preceding sections has been on initial synchronization processing, this section addresses the overall acquisition process, describing many operations and functions needed, as well as considerations specific to individual signal designs.

Determining the detection threshold is implicit to the discussion of initial synchronization, and in particular to the initial synchronization performance assessed in Section 16.4. Different receivers employ different approaches, but most are based on a constant false alarm rate (CFAR) approach that estimates or infers the CAF variance and then sets the threshold to produce the desired per-decision false detection probability. When modern algorithms are used to compute a large time-frequency tile in parallel, the CAF variance can be estimated from a region of the CAF where no peaks exist, or by using censoring approaches to exclude possible peaks from contributing to the variance estimate.

Selecting the CAF integration time involves the tradeoffs discussed earlier concerning performance and implementation complexity. Another factor, however, is the effect on false detection rate. Longer CAF integration times require more time-frequency cells for searching the same IFU with the same scalloping loss, providing more opportunities for false detections. If the false detection rate is to be kept the same, the per-decision false detection must be decreased with larger CAF integration times, resulting in higher detection thresholds that in turn yield lower detection probabilities. Consequently, when the false detection rate must be maintained, the performance benefits of increasing the CAF integration time may not be as large as expected.

The receiver typically does not have access to the effective C/N_0 before initial synchronization, so does not know a priori how many noncoherent integrations are needed. Numerous strategies exist for selecting the number of noncoherent integrations for initial synchronization. With serial search, it is common to start with a small number of noncoherent integrations, allowing a high false detection probability. Additional serial searches with continued noncoherent integrations are performed on the initial detection reports; surviving detection reports are validated using tracking loops.

With highly parallel computation of large time-frequency tiles, other approaches are favored. One approach is to assume the minimum effective C/N_0 and use the largest number of noncoherent integrations, employing a detection threshold that provides an acceptable false detection rate. Alternatively, a small number of noncoherent integrations can be used with low false detection probability, yielding a detection only if the effective C/N_0 turns out to be high. If initial synchronization does not occur, it is assumed that the effective C/N_0 is low, and the tile is re-searched with the maximum number of noncoherent detections.

There can be multiple causes for an absence of detection report in a time-frequency tile: it is possible that there is a missed detection, or that the signal being searched for is present in an adjacent tile, or that the signal is not present (e.g., in cold start when the satellite may be on the other side of the world), or that the effective C/N_0 may be too low due to blockage or to severe attenuation by foliage or building material. There is no theory to guide the search strategy—deciding whether to re-search the tile for the same signal, to search adjacent tiles for the same signal, or to re-search for a different signal are all plausible steps. The receiver designer must select a favored approach.

While the CAF integration time often is selected to match the duration of the spreading code, in order to avoid integrating over bit boundaries due to the navigation symbols or overlay code bits, many variations are possible. These variations are tied to the characteristics of individual signals. For GPS C/A code, with 20 repetitions of the spreading code within a data message bit, using 1 ms CAF integration times guarantees no bit alignment loss, but at the price of small CAF integration times. It is shown in Reference 1 that when the CAF integration time is set to an odd positive integer less than 20 and not a divisor of 20, some benefit can be obtained at the expense of a predictable worst-case loss due to occasional straddling of bit boundaries.

For signals like GPS L5 and Galileo E5a and E5b, where the spreading code has short duration and the overlay codes are of medium duration, initial synchronization can use the overlaid spreading code for a longer CAF integration time. Such approaches are explored in Reference 14.

If time is known a priori to within the duration of a data message symbols or overlay code bit (from assisted acquisition or a very warm start), and if these bit values are known a priori (as they are for overlay code bits), the receiver can phase-rotate the segments of the received time series where data message symbols or overlay code bits invert the spreading code. This process is called data wiping. The result contains no phase inversions of the signal, and enables CAF integration times longer than the duration of a data message symbol or overlay code bit. When these bit values are not known in advance, they can be estimated from the signal or hypothesized, then used for data wiping. This process is known as quasi-coherent integration. There is a chance of estimation errors that introduce losses in processing gain. Longer CAF integration times are most needed under low effective C/N_0 conditions, but it is under these same conditions that bit estimation errors are more frequent, producing greater losses in processing gain from quasi-coherent integration. Often, when all considerations of performance are considered, the net benefits of quasi-coherent integration are found not to justify the implementation complexity.

A series of papers [15–17] explores various alternatives for initial synchronization of the GPS L1C signal. Except under specialized conditions, the results show little net performance benefit to employing an approach other than the straightforward one of performing initial synchronization only on the L1C pilot, blanking the BOC(6,1) symbols, using CAF integration time of 10 ms corresponding to the spreading code duration, and noncoherently integrating over sequential 10 ms spreading code durations using code Doppler compensation.

Since initial synchronization processing complexity increases with spreading code length, receiver designers typically avoid performing initial synchronization on the L2C signal's pilot component with its very long L2CL spreading code. Instead, the receiver may acquire GPS C/A code signal on L1 and then hand over to tracking of the L2C signal. Or, if only the L2C signal is being used, initial synchronization is typically performed only using the L2C data component with its shorter L2CM spreading code.

Some variations for initial synchronization processing of the Galileo E1 OS signal are assessed in Review Question QA16.4. Using both the pilot and data components can provide benefits at the expense of doubling the complexity of initial synchronization processing. Alternatively, since the duration of overlay codes is only 100 ms, it is conceivable that 100 ms CAF integration times can be used on the pilot component only, hypothesizing and searching each phase of the overlay code.

When the signal uses a non-repeating spreading code, as is the case for signals whose spreading codes are produced using cryptographic algorithms, initial synchronization yields unambiguous alignment to the received signal. In this case, initial synchronization provides the receiver with time of transmission. There is also unique alignment to data message symbols, as well as to the data message itself. However, for signals with repeating spreading codes, time ambiguities remain after initial synchronization, requiring further processing in order to synchronize to boundaries of the data message symbol or overlay code bit, to the pilot component's overlay code phase, and to the data message. Each of these is discussed below.

When the replica used in initial synchronization processing has duration equal to that of a data message symbol or overlay code bit, initial synchronization inherently yields alignment to the data message symbol or overlay code bit. However, if the spreading code repeats multiple times in a data message symbol duration or overlay code bit and the replica employed in initial synchronization is only one duration of the spreading code, then initial synchronization reduces the ITU from a continuum to a finite number of values greater than unity, but the start time of a data message symbol is still ambiguous. For example, if the spreading code repeats N_p times within a data message symbol duration or overlay code bit, there are still N_p possible times when each data message symbol begins. A similar situation occurs with the C/A code signal, where after initial synchronization with a 1 ms duration spreading code replica, there are 20 possible start times for a data message symbol. Similarly, with the GPS L5 signal initial synchronization with a 1 ms duration spreading code replica leaves 10 possible start times for a data message symbol and 20 possible overlay code phases for the pilot component.

Figure 16.12 illustrates this situation. The time alignment of spreading waveform repetitions, in the top row of the figure, is known from initial synchronization. The receiver knows the sequence of overlay code bits $\{p_n\}$ from the signal's ICD, but does not know how they are aligned to the spreading waveform, so, any of the N_p alignments of overlay code bits and data message bits shown in the figure are possible. The receiver also knows that the data message bits $\{d_k\}$ are aligned to the overlay code, but does not know the values of the data message bits or the carrier phase variation, which is assumed to be slowly changing so that it can be approximated as taking on a constant value over each data message bit duration.

Histogramming is one way to resolve such ambiguities and determine the data message bit boundaries. A received segment of signal, with noise and interference, is obtained, with duration an integer number $K + 1$ data bits of time, so that the segment duration is $(K + 1) N_a N_p T_c$, where the duration of a spreading waveform is $N_a T_c$ with N_a

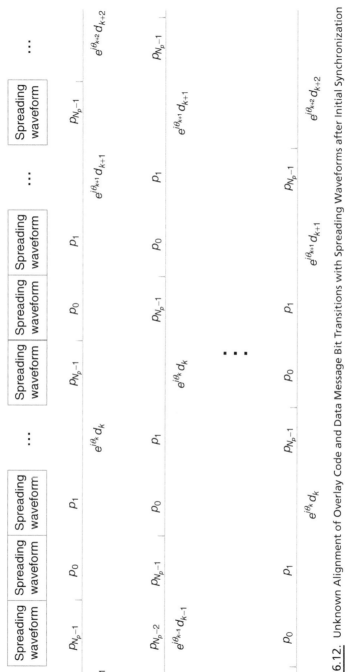

Figure 16.12. Unknown Alignment of Overlay Code and Data Message Bit Transitions with Spreading Waveforms after Initial Synchronization

the length of the spreading code and T_c the duration of a spreading symbol, and N_p the length of the overlay code. A replica with same length of repeating spreading waveforms is generated in the receiver, aligned with the spreading waveforms in the received segment based on initial synchronization. The dot product between each spreading waveform in the replica and the received segment is computed, wiping off the spreading waveform and producing the sequence of complex values $\beta_1, \beta_2, \ldots, \beta_{(K+1)N_p}$. New data bit transitions occurs every N_p of these values, but it is unknown between which two of these values the data bit transitions occur, and thus where the overlay code value begins at p_0 again.

Now compute the test values

$$\alpha_0 \stackrel{\Delta}{=} \frac{1}{K} \sum_{k=0}^{K-1} \left| \sum_{n=0}^{N_p-1} p_n \beta_{kN_p+n} \right|$$

$$\alpha_1 \stackrel{\Delta}{=} \frac{1}{K} \sum_{k=0}^{K-1} \left| \sum_{n=0}^{N_p-1} p_n \beta_{kN_p+1+n} \right| \qquad (16.23)$$

$$\vdots$$

$$\alpha_{N_p-1} \stackrel{\Delta}{=} \frac{1}{K} \sum_{k=0}^{K-1} \left| \sum_{n=0}^{N_p-1} p_n \beta_{kN_p+N_p-1+n} \right|$$

Each of these real values corresponds to a different hypothesis concerning the phasing of the overlay code and associated data bit transitions. For one of these hypotheses, the correct hypothesis, the overlay code bits are aligned with those in the complex values, and the inner sums run over the same data bit value. In all other test values, the inner sums include two adjacent data bit values, and when these data bit values change (which occurs half the time, on average), terms in the inner sum cancel each other. In the incorrect hypotheses, misaligned overlay code bits can also lead to partial cancellations in the inner sum. When these test values (16.23) are compared, colloquially called histogramming, the index of the largest one typically indicates where the new data bit value occurs.

A simple application of histogramming occurs with the GPS C/A code signal, where $N_a = 1023$, $N_p = 20$, and $p_m = 1, m = 0, 1, \ldots, 19$. Even for these cases where the overlay code values are all unity, as long as the histogramming process includes a sufficient number of bit durations, correct alignment to the data bits is obtained.

After synchronization to the data message symbol or overlay code bit, the receiver still needs to synchronize to the beginning of each data word or message. In all signal designs, messages contain a preamble—a distinct bit sequence that the receiver can search for and use to obtain alignment to the data word or message. The receiver is then able to read time, clock corrections, and ephemeris from the message in order to calculate PVT. In other cases, such as the GPS L1C signal, the overlay code's length is the same as that of the data message, so synchronization to the overlay code achieves alignment to the data message, avoiding any need to perform additional synchronization steps.

The preceding discussion demonstrates one of the benefits of the L1C design, where the length of the pilot overlay code matches that of the data message. For this signal, synchronization to the overlay code provides alignment to the data message, avoiding the time and complexity of processing for data message synchronization.

Acquisition of other signals follows the same general approach as described earlier in this section. The GLONASS L1OF and L2OF signals are processed in a similar way to C/A code signals; the meander sequence is a simple overlay code that reduces, but does not eliminate, the challenge of data bit synchronization after initial synchronization is complete. The well-designed overlay codes for the BeiDou B1I and B2I signals simplify data bit synchronization for these signals. With the similarity of QZSS signal designs to GPS signals designs, common acquisition approaches can be used. Like the GPS C/A code signal, IRNSS SPS signals have a spreading code that repeats 20 times within a data message bit with no overlay code, and consequently require histogramming or a similar approach to achieve bit synchronization. Until more detail is available for other signals, such as BeiDou Phase 3 signals and GLONASS CDMA signals, detailed discussion of acquisition approaches for these signals is premature.

Once data bit synchronization is complete for any of these signals, synchronization words or preambles are processed to obtain alignment to the data message, enabling the data message to be read while signal tracking, described in Chapters 18 and 19, provides the measurements needed for PVT calculation as described in Chapter 20.

TTFF—the time between completing power up and first output of PVT—is typically the most important performance parameter for acquisition. Reducing TTFF is one of the key motivations behind signal designs, receiver processing, and augmentations.

16.6 SUMMARY

Acquisition is the most demanding and complicated receiver function, in part since it subsumes all other receiver functions in order to produce the measurements and data message information needed to compute PVT.

Initial conditions are critical in determining what receiver functions are needed for acquisition, as well as the resulting acquisition performance. This chapter identifies five distinct initial conditions and provides explicit definitions of each. While different references may employ somewhat different specific definitions, the underlying concepts are virtually universal and essential for clearly discussing and describing acquisition. All initial conditions except restart involve initial synchronization. Restart uses enough tracking loops spaced over delay and frequency to resume tracking without initial synchronization.

Much of this chapter focuses on initial synchronization, typically the receiver processing function that is most complex to implement. Initial synchronization processing is based on computing the CAF; a peak, or large magnitude of the CAF, signifies presence of a signal and the coordinates of the peak reveal the approximate time delay and frequency offset of the signal. Various algorithms are described for computation of the CAF, ranging from sequential search of time delay and frequency offset that are implemented in the time domain by a single correlator, to sophisticated techniques for computing in parallel an entire tile of time-frequency space using digital signal processing involving time-domain and frequency-domain techniques equivalent to hundreds of thousands or millions of correlators operating in parallel. Other tradeoffs in initial synchronization implementation involve choosing the CAF integration time and sampling rate, both of which influence performance as well as implementation complexity. Increased sampling rates can drastically increase implementation complexity of initial synchronization, favoring wide correlation function peaks for initial synchronization. Techniques are described for widening the correlation function peaks to simplify implementation of initial synchronization for signals with BOC spreading modulations.

Even with inexpensive digital processing, initial synchronization complexity is likely to remain a dominant consideration in receiver design for two significant reasons:

- The processing load, storage, and power consumption for this function are dominant even in many modern receiver designs. Fortunately, initial synchronization is performed only intermittently so the energy used in initial synchronization may not be a significant factor in the overall energy budget, but circuit size and storage considerations can dominate the size and complexity of receiver circuitry.
- The human factors aspect remains important. When a user activates a satnav receiver, he or she wants instant PVT. Delays in acquisition are a frustration and an important discriminator among receivers, so there is great motivation to speed acquisition time. As initial synchronization circuits become more capable and advanced algorithms simplify implementation, there can be advantages in using the increased capability to speed up acquisition (e.g., searching in parallel for multiple satnav signals), rather than to reduce circuit size and power consumption.

Noncoherent integration is a way to improve the sensitivity of initial synchronization processing while avoiding some implementation challenges, and code Doppler processing is described to enable more noncoherent integrations for greater sensitivity.

Performance of initial synchronization is typically described in terms of the effective C/N_0 at for initial synchronization, as well as the time needed to search a given ITU and IFU at some value of effective C/N_0. Since computing detection and false detection probabilities involves detailed numerical calculations, precomputed tables are provided to enable performance assessment over a range of typically useful conditions, and for different processing strategies applied to different signal designs.

Ultimately, acquisition involves not only initial synchronization, but all the steps outlined in Figure 16.2. While initial synchronization identifies time alignment of the spreading code, additional steps can be needed to achieve time alignment with data message symbols and overlay code periods, and also with data words and messages. These steps are described as well.

The receiver designer must make many decisions in implanting acquisition processing. Some decisions involve generic tradeoffs between performance and implementation complexity, and between speed of acquisition and ability to acquire under stressing conditions. Choosing sampling rates, CAF integration times, and noncoherent integration strategies are all part of these tradeoffs. Other decisions involve the design of the signal being acquired, for example, initial synchronization using only the pilot component, only the data component, or both. This chapter contains tools needed to perform these trades and make these design choices.

REVIEW QUESTIONS

Theoretical Questions

QT16.1. For a CAF $\Xi(\tau,\phi) = \frac{1}{T} \int_{-T/2}^{T/2} [\sqrt{\zeta C} s(t+D) e^{-i2\pi Ft} + n(t)] s(t+\tau) e^{-i2\pi\phi t} dt$
with $s(t)$ a real-valued cyclostationary signal with delay D and residual carrier frequency F, $n(t)$ zero-mean additive noise uncorrelated with the signal, and correlation integration time T, show that the mean of the CAF at $\tau = D$ is $E\{\Xi(\tau,\phi)\} \cong \sqrt{\zeta C} \bar{R}_s(\tau - D) \operatorname{sinc}[\pi(\phi - F)T]$.

Application Questions

QA16.1. Show that for a signal with BPSK-R(f_c) spreading modulation, the two-sided half-power width of the CAF peak in delay is approximately $\pm 0.29/f_c$. Show also that the two-sided half power width of the CAF peak in frequency offset is approximately $\pm 0.44/T$, for CAF integration time T. For what spreading modulations does the latter result apply to?

QA16.2. Consider design of an initial synch processing circuit for a receiver that is stationary. For each of the following GPS signals (C/A, L1C, L5):
  a. Describe the ITU and IFU in units of s and Hz, respectively.
  b. Determine the dimensions of a time-frequency cell.
  c. Identify the number of cells in the ITU/IFU window.
  Make the coherent integration time and ITU each equal to the period of the spreading code, and make the IFU large enough to cover satellite Doppler in a cold start condition in addition to a frequency reference having ± 1 part per million uncertainty.

QA16.3. Consider the DirAc design described in Section 16.3.3. Estimate the number of transistors needed to perform equivalent initial synchronization processing on a BOC(1,1) signal using the same sideband processing architecture, assuming half the transistors are used for processing and half for storage. Estimate the area of the chip if a 45-nm process were used with the same architecture as DirAc.

QA16.4. Consider a situation where initial synchronization processing uses 1 second of data, the detection threshold is established to provide a 10^{-6} probability of false detection per decision, the detection probability is 0.95, the processing loss is 3 dB, and the signal has pilot and data components.
  a. For wideband processing of both components having a 50%/50% power split, with a 4-ms CAF integration time, find the C/N_0 needed.
  b. For sideband processing of both components having a 50%/50% power split, with a 4-ms CAF integration time, find the C/N_0 needed.
  c. For wideband processing of only the 50% power pilot component with a 100-ms CAF integration time, find the C/N_0 needed.

d. For sideband processing of only the 50% power pilot component with a 100-ms CAF integration time, find the C/N_0 needed.

e. For wideband processing of only the 75% power pilot component with a 10-ms CAF integration time, find the C/N_0 needed.

f. For sideband processing of only the 75% power pilot component with a 10-ms CAF integration time, find the C/N_0 needed.

REFERENCES

1. F. van Diggelen, *A-GPS: Assisted GPS, GNSS, and SBAS*, Artech House, 2009.
2. L. E. Franks, *Signal Theory*, Prentice-Hall, 1969.
3. A. D. Whalen, *Detection of Signals in Noise*, Academic Press, 1971.
4. P. Fishman and J. W. Betz, "Predicting Performance of Direct Acquisition for the M Code Signal," *Proceedings of Institute of Navigation National Technical Meeting 2000, ION-NTM-2000*, Institute of Navigation, January 2000.
5. J. W. Betz, "Binary Offset Carrier Modulations for Radio Navigation," *NAVIGATION: The Journal of the Institute of Navigation*, Vol. 48, No. 4, 2001–2002, pp. 227–246.
6. P. Ward, "A Design Technique to Remove the Correlation Ambiguity in Binary Offset Carrier (BOC) Spread Spectrum Signals," *Proceedings of ION NTM 2004*, January 2004.
7. E. O. Brigham, *The Fast Fourier Transform*, Prentice-Hall, 1974.
8. S. Stein, "Algorithms for Ambiguity Function Processing," *IEEE Transactions on Acoustics, Speech, and Signal Processing*, Vol. ASSP-29, No. 3, 1981, pp. 588–599.
9. J. W. Betz, J. D. Fite, and P. T. Capozza, "DirAc: An Integrated Circuit for Direct Acquisition of the M-Code Signal," *Proceedings of the Institute of Navigation Conference on Global Navigation Satellite Systems 2004, ION-GNSS-2004*, Institute of Navigation, September 2004.
10. J. D. Fite and J. W. Betz, Interpolation Processing for Enhanced Signal Acquisition, United States Patent US 7995676 B2, August 9, 2011.
11. R. N. McDonough and A. D. Whalen, *Detection of Signals in Noise*, 2nd edition, Academic Press, 1995.
12. D. A. Shnidman, "The Calculation of the Probability of Detection and the Generalized Marcum Q-Function," *IEEE Transactions on Information Theory*, Vol. IT-35, 1989, pp. 389–400.
13. J.A. Avila-Rodriguez, V. Heiries, T. Pany, and B. Eissfeller, "Theory on Acquisition Algorithms for Indoor Positioning," *Proceedings of 12th Saint Petersburg International Conference on Integrated Navigation Systems*, Saint Petersburg, Russia, May 23–25, 2005, available at http://www.researchgate.net/publication/228855044_Theory_on_Acquisition_algorithms_for_indoor_positioning, accessed 22 September 2015.
14. C. Hegarty, M. Tran, and A. J. Van Dierendonck, "Acquisition Algorithms for the GPS L5 Signal," *Proceedings of the 16th International Technical Meeting of the Satellite Division of The Institute of Navigation (ION GPS/GNSS 2003)*, Portland, OR, September 2003, pp. 165–177.
15. K. C. Seals, W. R. Michalson, P. F. Swaszek, and R. J. Hartnett, "Analysis of Coherent Combining for GPS L1C Acquisition," *Proceedings of the 25th International Technical*

Meeting of The Satellite Division of the Institute of Navigation (ION GNSS 2012), Nashville, TN, September 2012, pp. 384–393.

16. K. C. Seals, W. R. Michalson, P. F. Swaszek, and R. J. Hartnett, "Analysis of L1C Acquisition by Combining Pilot and Data Components over Multiple Code Periods," *Proceedings of the 2013 International Technical Meeting of The Institute of Navigation*, San Diego, CA, January 2013, pp. 526–535.

17. K. C. Seals and W. R. Michalson, "Semi-Coherent and Differentially Coherent Integration for GPS L1C Acquisition," *Proceedings of the ION 2013 Pacific PNT Meeting*, Honolulu, HI, April 2013, pp. 123–130.

17

DISCRETE-UPDATE TRACKING LOOPS

Tracking involves maintaining estimates of one or more time-varying parameters using measurements that are contaminated by disturbances such as noise and interference, differentiating between actual changes in parameter values and jitter caused by the disturbances. Tracking is integral to many applications including radar and sonar, where tracking is used to estimate location and motion of targets, and digital communications, where tracking is used to maintain receiver synchronization to the received signals despite relative motion between the transmitter and receiver.

> This chapter is based in part on instructional materials prepared by Dr. Alessandro P. Cerruti, The MITRE Corporation. These materials were prepared as part of the NavtechGPS course "Using Advanced GPS/GNSS Signals and Systems." The author thanks Dr. Cerruti for permission to use these materials in the preparation of this chapter, and also thanks Ms. Carolyn McDonald, President of NavtechGPS, for sponsoring the original development of these materials. Any errors or omissions, however, are the author's.

Engineering Satellite-Based Navigation and Timing: Global Navigation Satellite Systems, Signals, and Receivers, First Edition. John W. Betz.
© 2016 The Institute of Electrical and Electronics Engineers, Inc. Published 2016 by John Wiley & Sons, Inc.

DISCRETE-UPDATE TRACKING LOOPS

In satnav, the carrier frequency, carrier phase, and delay of the complex envelope all vary with time, and tracking loops are often used to estimate the current values of these parameters. Disturbances include additive noise and interference, multipath, and ionospheric effects, as well as phase noise in the received signal and in the replica generated by the receiver.

In the conventional satnav receiver processing flow portrayed in Figure 16.1, each receiver channel employs at least two tracking loops: one to track parameters associated with the signal's carrier frequency, and one to track parameters associated with the delay of the signal's complex envelope, commonly known as code tracking. Since many details of carrier tracking and code tracking are different, carrier tracking and code tracking are addressed separately in Chapters 18 and 19, respectively. However, the fundamental design of tracking loops is common to carrier tracking and code tracking, so it is addressed in this chapter using a general formulation.

While some tracking algorithms involve mere smoothing or filtering of the data, the tracking loops employed in conventional satnav receivers are actually feedback control systems that use parameter estimates to generate an estimate of the received signal (called a replica), perform signal processing to estimate the error between the replica and the received signal, then use smoothed errors to refine and update the parameter estimates before repeating the process.

GPS receivers originally implemented tracking loops using analog circuitry that provided continuous updating of the loop. Modern satnav receivers implement tracking loops digitally, either in digital hardware or in software running on a digital signal processor or general purpose processor. Often, however, the design of the tracking loops has reverted to historical techniques for continuous-update loops. These analog loop designs are then mapped, with certain approximations and limitations, into designs for the discrete-update loops that are actually implemented, as documented in References 1 and 2. (Note that while common terminology refers to analog tracking loops and digital tracking loops, the distinction is more precisely stated as continuous-update tracking loops and discrete-update tracking loops.) Due to the analog to digital mapping, loop parameters such as the loop noise bandwidth and closed loop pole locations can deviate from their intended values, resulting in different performance than desired and potentially even loop instabilities. These discrepancies are particularly poor in loops having large bandwidth-update time products [3,4].

This chapter uses a different approach, drawing upon the techniques pioneered in References 4 and 5 to design and characterize tracking loops that are fundamentally designed for discrete updates. The result is loops with improved performance over a wider range of parameter values, and more precisely designed behavior.

Section 17.1 provides the basic formulation for a discrete-update tracking loop using a generic representation that applies to all the tracking loops addressed in Chapters 18 and 19. Equations are provided for the operation of the discrete-update tracking loops. Section 17.2 provides the algorithms for designing the discrete-update tracking loop parameters and numerical results. Section 17.3 discusses and describes considerations in tracking loop design, including selection of loop bandwidth, loop order, and damping factors, as well as effects of receiver oscillator phase noise, loop pull-in, and loop loss of lock. Section 17.4 provides a summary of this chapter.

17.1 DISCRETE-UPDATE TRACKING LOOP FORMULATION

Figure 17.1 shows the flow of a discrete-update tracking loop. The formulation here is predicated on instantaneous processing in each block of the diagram, with synchronous outputs from each block every T seconds, where T is the loop update time and $1/T$ is the loop update rate. The input waveform to the loop is the sum of signal, noise, and interference, typically at or near baseband and after appropriate BSQ as discussed in Chapter 15. The first step in the tracking loop is signal processing, which also makes use of a signal replica generated within the tracking loop. The signal processing outputs measurements at the loop update rate. Specific descriptions of the replica, the signal processing, and the measurements are provided for different tracking loops in Chapters 18 and 19, but in this chapter the descriptions are generic. The measurements are used in a discriminator to produce an error signal that is smoothed, or filtered, then used to calculate an updated estimate of the parameter rate (the first derivative with respect to time of the parameter being tracked) and the parameter itself. The updated parameter rate and parameter are then used to produce an updated replica of the signal, which is used in signal processing for the next loop update.

While the parameter itself is part of the derivation and description of tracking loops, some loops only produce and use the parameter rate, while others produce and use the parameter itself as well. The description in this chapter employs both for generality.

For a more rigorous description of tracking loop operation, the following notation is used for the nth update of the loop:

- The true value of the parameter being tracked is p_n (it could be a vector of parameters, but the formulation here is restricted to a single parameter),
- The vector of input signal samples is $s_n(p_n)$, explicitly showing the dependence of the signal on the parameter being tracked,
- The vector of noise plus interference is w_n, so the segment of input waveform used in the signal processing is $s_n(p_n) + w_n$,

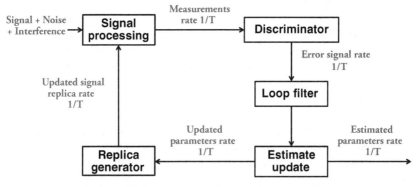

Figure 17.1. Discrete-Update Tracking Loop Flow

DISCRETE-UPDATE TRACKING LOOP FORMULATION

- The replica vector is $\hat{\mathbf{s}}_n = \mathbf{s}_n(\hat{p}_n, \hat{\dot{p}}_n)$, an estimate of the signal that is computed using information from the ICD, along with the estimate of the parameter being tracked $\hat{p}_n$ and the estimate of the parameter rate $\hat{\dot{p}}_n$,
- The measurements produced by the signal processing are $\{\mu_1(\mathbf{s}_n(p_n), \hat{\mathbf{s}}_n), \ldots, \mu_M(\mathbf{s}_n(p_n), \hat{\mathbf{s}}_n)\}$, which depend on the input signal and the signal replica,
- The error in the parameter estimate is the difference between the true value and the estimated value, $\tilde{p}_n = K_0(p_n - \hat{p}_n)$, where K_0 is a constant of proportionality that is the discriminator gain,
- Filtering that smooths the error values, as described by the expression $\hat{\dot{p}}_{n+1}T =$

$$\hat{p}_{n+1} - \hat{p}_n = K_0 K_1 \tilde{p}_n + K_0 K_2 \sum_{i=1}^{n} \tilde{p}_i + K_0 K_3 \sum_{i=1}^{n} \sum_{j=1}^{i} \tilde{p}_j + K_0 K_4 \sum_{i=1}^{n} \sum_{j=1}^{i} \sum_{k=1}^{j} \tilde{p}_k + \ldots,$$

where the loop gain values K_1, K_2, K_3, K_4 are used to design the loop characteristics as described subsequently.

The formulation provided here assumes the length of the vector of input signal plus noise plus interference corresponds to T seconds, as does the length of the signal replica, so that T is not only the loop update rate but also the duration of the input waveform used in the signal processing.

Representing the replica as dependent on the parameter being tracked is directly relevant in satnav receivers. In carrier tracking, it is known a priori that the carrier is a sinusoid, and the parameters are either carrier frequency or carrier phase. In code tracking, the form of the complex envelope (spreading symbols, spreading code bits, overlay code bits) is known a priori from the ICD; the unknown parameter is the delay of the complex envelope. Thus, the only differences between $\mathbf{s}_n(p_n)$ and $\hat{\mathbf{s}}_n = \mathbf{s}_n(\hat{p}_n, \hat{\dot{p}}_n)$, other than signal imperfections introduced by the satellite or the transmission channel, are due to differences between the true parameter value and its estimate.

Since the parameter being tracked varies over time and the input to the loop represents T seconds of input waveform, the convention is to estimate the parameter and its rate at the midpoint of the T seconds.

Figure 17.2 shows the tracking loop flow using this notation, more fully defining the tracking loop algorithm. The discriminator output equation

$$\tilde{p}_n = K_0(p_n - \hat{p}_n) \tag{17.1}$$

applies for small errors, producing an output proportional to the error. As will be seen in Chapters 18 and 19, the proportionality does not hold as the error gets larger. More generally, the discriminator response is known as an S-curve and given by $S(\tilde{p}_n)$. The range of input errors that includes the origin and for which $S(\tilde{p}_n) \approx K_0 \tilde{p}_n$ is known as the linear region of the S-curve with $-E/2 \leq \tilde{p}_n \leq E/2$. Often, the discriminator is normalized so that $K_0 = 1$.

Figure 17.3 shows an example of an S-curve. It passes through the origin, has a positive slope at the origin, and is approximately linear for small-magnitude errors. With larger errors, the slope approaches zero, and in some cases beyond that point the slope of S-curves can change sign, sometimes causing the S-curve to have multiple zero crossings.

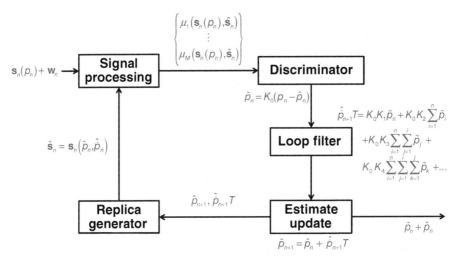

Figure 17.2. Tracking Loop Algorithm

Design of the tracking loop involves the following steps. The loop update time, T, must be selected. For the specific parameter to be tracked, the signal processing is defined, along with the appropriate replica. Based on the parameter being tracked and the discriminator that is selected, specific measurements are identified. (The parameters, signal processing, replica, measurements, and discriminators for specific tracking loops used in satnav receivers are described in Chapters 18 and 19.) Loop design then involves determining the loop gain parameters K_1, K_2, K_3, K_4. The number of nonzero parameters is called the tracking loop order; loop orders as high as four are rare, and loops of order higher than four are almost never used and thus not treated here. The loop gain parameters are selected based on the loop update time and loop order as well as two other loop characteristics: the loop bandwidth and the loop damping.

As with most other aspects of receiver design, selection of the loop bandwidth and the loop damping involves numerous tradeoffs, with no single set of values best for all applications and conditions. As will be shown, smaller loop bandwidths provide

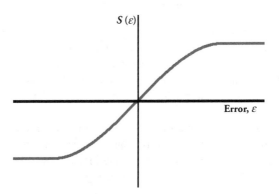

Figure 17.3. Example S-Curve

DISCRETE-UPDATE TRACKING LOOP FORMULATION

more smoothing of noise and interference effects, while also making the loop more sensitive to local oscillator imperfections and less able to accommodate dynamics (large time derivatives of the parameters being tracked). Higher-order loops can better handle dynamics, but are more sensitive to effects of noise and interference. Choices of loop damping affect how rapidly and accurately the loop responds to dynamics.

The tracking loops described here use the "immediate update" formulation from Reference 4, under the assumption that calculations in loop operation are performed essentially instantaneously, so that the outputs from each block in Figure 17.2 are produced with negligible delay. If the computation time is long relative to the loop update time so that updates are delayed by one or more loop update times, the same general formulation provided here can be used, with the parameter values for delayed-update loops found in Reference 4. Also, for immediate update loops, there is negligible difference between loops that estimate only parameter rate and those that estimate both parameter and parameter rate, so only the latter are addressed here.

Define the output of the loop as the parameter estimate $\hat{p}_n$, and the input as the true parameter value p_n. For small errors, the loop behavior can be modeled as a linear system having the loop transfer function, in terms of the z-transform [4]

$$\hat{P}_z(z) = H_N(z) P_z(z)$$

$$H_N(z) = \frac{\hat{P}_z(z)}{P_z(z)} = \frac{D_N(z) - (z-1)^N}{D_N(z)} \qquad (17.2)$$

$$D_N(z) \stackrel{\Delta}{=} (z-1)^N K_0 + (z-1)^{N-1} K_0 K_1 + z(z-1)^{N-2} K_0 K_2$$
$$+ z^2 (z-1)^{N-3} K_0 K_3 + \cdots + z^{N-1} K_0 K_N$$

where $\hat{P}_z(z)$ is the z-transform of the parameter estimate, $P_z(z)$ is the z-transform of the true parameter value, and N is the order of the loop. It is desired that the loop behaves like a low-pass filter, rejecting higher frequency disturbances while preserving lower frequency dynamics of the parameter. All of the numerical values in Reference 4 are based on the discriminator gain, K_0, taking on a value of unity. If K_0 is other than unity, as long as it is nonzero, in all of the remaining expressions in this chapter, replace K_1 by $K_0 K_1$, K_2 by $K_0 K_2$, K_3 by $K_0 K_3$, and K_4 by $K_0 K_4$.

Tracking loop transfer functions for different order loops are obtained from (17.2),

$$H_1(z) = \frac{K_1}{z - 1 + K_1}$$

$$H_2(z) = \frac{(z-1)K_1 + zK_2}{(z-1)^2 + (z-1)K_1 + zK_2}$$

$$H_3(z) = \frac{(z-1)^2 K_1 + z(z-1)K_2 + z^2 K_3}{(z-1)^3 + (z-1)^2 K_1 + z(z-1)K_2 + z^2 K_3} \qquad (17.3)$$

$$H_4(z) = \frac{(z-1)^3 K_1 + z(z-1)^2 K_2 + z^2(z-1)K_3 + z^3 K_4}{(z-1)^4 + (z-1)^3 K_1 + z(z-1)^2 K_2 + z^2(z-1)K_3 + z^3 K_4}$$

Loop parameter values need to be chosen carefully in order to yield useful tracking loop designs; good loop parameter values are provided in Section 17.2.

Loop frequency responses can be found by substituting $z = e^{i2\pi fT}$ into (17.3), and are commonly expressed in terms of the normalized frequency fT. The one-sided bandwidth of the loop is given by the equivalent rectangular bandwidth of the closed loop transfer function, discussed in Appendix Section A.6. Here, the DC gain (the gain at zero frequency) is constrained to unity, so the equivalent rectangular bandwidth of the loop transfer function (the rectangular transfer function having the same unit DC gain as the actual transfer function and passing the same white noise power as the actual loop transfer function) is

$$B_L = \frac{1}{2} \int_{-0.5/T}^{0.5/T} |H_N(z = e^{i2\pi fT})|^2 df \qquad (17.4)$$

Reference 4 derives the equivalent rectangular loop normalized bandwidths for different loop orders,

$$(B_L T)_{N=1} = \frac{K_1}{2(2 - K_1)}$$

$$(B_L T)_{N=2} = \frac{2K_1^2 + 2K_2 + K_1 K_2}{2K_1(4 - 2K_1 - K_2)}$$

$$(B_L T)_{N=3} = \frac{4K_1^2 K_2 - 4K_1 K_3 + 4K_2^2 + 2K_1 K_2^2 + 4K_1^2 K_3 + 4K_2 K_3 + 3K_1 K_2 K_3 + K_3^2 + K_1 K_3^2}{2(K_1 K_2 - K_3 + K_1 K_3)(8 - 4K_1 - 2K_2 - K_3)}$$

$$(B_L T)_{N=4} = \frac{\begin{pmatrix} 8K_1^2 K_2 K_3 - 8K_1 K_3^2 - 8K_1^2 K_4 + 8K_2^2 K_3 - 8K_1 K_2 K_4 - 8K_3 K_4 \\ +4K_1 K_2^2 K_3 + 8K_1^2 K_3^2 + 8K_2 K_3^2 + 4K_1^2 K_2 K_4 + 8K_2^2 K_4 \\ -20K_1 K_3 K_4 - 8K_4^2 + 6K_1 K_2 K_3^2 + 2K_3^3 + 14K_1^2 K_3 K_4 \\ +14K_2 K_3 K_4 + 4K_1 K_2^2 K_4 - 10K_1 K_4^2 + 11K_1 K_2 K_3 K_4 + 2K_1 K_3^3 \\ +5K_3^2 K_4 + 7K_1^2 K_4^2 + 6K_2 K_4^2 + 5K_1 K_3^2 K_4 + 5K_1 K_2 K_4^2 \\ +4K_3 K_4^2 + 4K_1 K_3 K_4^2 + K_4^3 + K_1 K_4^3 \end{pmatrix}}{\begin{array}{c} 2(K_1 K_2 K_3 - K_3^2 - K_1^2 K_4 + K_1 K_3^2 + K_1 K_2 K_4 - 2K_3 K_4 \\ +2K_1 K_3 K_4 - K_4^2 + K_1 K_4^2)(16 - 8K_1 - 4K_2 - 2K_3 - K_4) \end{array}} \qquad (17.5)$$

17.2 DISCRETE-UPDATE TRACKING LOOP DESIGN

The loop parameters $\{K_n\}_{n=1}^{N}$ can be designed using an iterative procedure described in Reference 4, with results summarized in this section. For the results provided here, the relative magnitudes of all roots are equal, so in the notation of Reference 4, $\lambda_n \equiv 1, \forall\, n$. Also, the damping factor is set to either supercritically damped with $\eta_n^2 = 0, \forall n$, or standard underdamped, with $\eta_n^2 = -1, \forall n$, in the notation of Reference 4. Behavior of loops with different damping factors is explored in Section 17.3, with supercritically damped loops returning to equilibrium as rapidly as possible without oscillating, while standard underdamped loops return to equilibrium more rapidly, but oscillate.

DISCRETE-UPDATE TRACKING LOOP DESIGN

The designer selects the loop order, the normalized loop bandwidth, and either supercritical response or standard underdamping. An iterative numerical approach is then used to solve for the loop parameters that provide the desired loop characteristics.

First-order loops are simple and useful in practice when either the parameter being tracked does not undergo large and rapid variations with respect to the loop update time, or when the loop bandwidth can be made large enough to accommodate these parameter variations. In satnav receivers, the absence of large and rapid parameter variations occurs when the receiver dynamics are low (e.g., surveying), or when the loop is aided (see Section 19.3 and Chapter 25) by other measurements that are used to remove most dynamics so the loop bandwidth need not account for them. Large loop bandwidths and short loop update times, both of which enhance the loop's need to accommodate parameter variations, can be used when the effective C/N_0 is large. With a first-order loop, there is only one pole and damping is not a consideration.

Table 17.1 provides first-order loop parameter values for different normalized loop bandwidths, up to the limit of 0.5. Table 17.1 also provides parameter values for

TABLE 17.1. Parameters for First-Order and Second-Order Discrete-Update Tracking Loops

	First-Order Loop	Second-Order Loop			
		Supercritical Response		Standard Underdamped Response	
$B_L T$	K_1	K_1	K_2	K_1	K_2
0.005	0.0198	0.0158	6.31×10^{-5}	0.0132	8.75×10^{-5}
0.01	0.0392	0.0313	2.49×10^{-4}	0.0261	3.45×10^{-4}
0.02	0.0769	0.0612	9.68×10^{-4}	0.0511	1.34×10^{-3}
0.03	0.113	0.0899	2.12×10^{-3}	0.0750	2.92×10^{-3}
0.04	0.148	0.117	3.67×10^{-3}	0.0979	5.04×10^{-3}
0.05	0.182	0.144	5.58×10^{-3}	0.120	7.66×10^{-3}
0.06	0.214	0.169	7.82×10^{-3}	0.141	0.0107
0.07	0.246	0.193	0.0104	0.161	0.0142
0.08	0.276	0.217	0.0132	0.181	0.0180
0.09	0.305	0.239	0.0163	0.200	0.0222
0.1	0.333	0.261	0.0196	0.218	0.0267
0.2	0.571	0.438	0.0626	0.368	0.0834
0.3	0.750	0.564	0.116	0.476	0.151
0.4	0.889	0.658	0.172	0.558	0.221
0.5	1.00	0.728	0.229	0.621	0.290
0.6		0.783	0.285	0.672	0.356
0.7		0.826	0.339	0.713	0.418
0.8		0.860	0.392	0.748	0.476
0.9		0.887	0.441	0.776	0.531
1.0		0.910	0.489	0.801	0.581
2.0		0.910	0.489	0.801	0.581
2.5		1.00	1.00	1.00	1.00

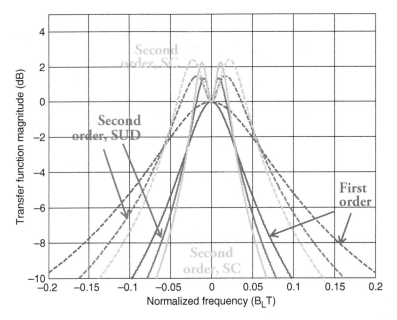

Figure 17.4. Example First-Order and Second-Order Transfer Function Magnitudes, with Supercritical (SC) and Standard Underdamped (SUD) Second-Order Loops, Solid Lines $B_L T = 0.1$, and Dashed Lines $B_L T = 0.3$

second-order loops, both with supercritical response and standard underdamped responses. As shown in Section 17.3, second-order loops can track a parameter having nonzero first derivative with respect to time. Since the loop transfer function has a pair of poles, damping is controlled by placement of poles—equal and real for supercritical response, and conjugate complex-valued having equal real and imaginary parts, for standard-underdamped response.

Figure 17.4 shows the magnitudes of first-order and second-order closed loop transfer functions for two different loop bandwidths. Transfer functions for first-order loops achieve a maximum magnitude at zero hertz, and roll off more slowly than the second-order loops. For both second-order loops, the real-valued zero in the transfer function pulls down the magnitude at zero hertz, making the magnitude transfer function peak at nonzero frequencies. Transfer functions for supercritically damped second-order loops have a higher maximum magnitude, and roll off more quickly, than for standard underdamped second-order loops.

Table 17.2 provides the parameters for third-order loops, for both supercritical response and standard underdamped responses. As shown in Section 17.3, third-order loops can track a parameter having nonzero second derivative with respect to time.

Figure 17.5 shows the magnitudes of third-order closed loop transfer functions for two different loop bandwidths. Compared to the second-order loops portrayed in Figure 17.4, transfer functions for third-order loops roll off more rapidly, with supercritically

TABLE 17.2. Parameters for Third-Order Discrete-Update Tracking Loops

	Supercritical Response			Standard Underdamped Response		
$B_L T$	K_1	K_2	K_3	K_1	K_2	K_3
0.005	0.0144	6.94×10^{-5}	1.12×10^{-7}	0.0129	7.44×10^{-5}	1.61×10^{-7}
0.01	0.0285	2.73×10^{-4}	8.78×10^{-7}	0.0255	2.93×10^{-4}	1.26×10^{-6}
0.02	0.0557	1.06×10^{-3}	6.76×10^{-6}	0.0500	1.13×10^{-3}	9.72×10^{-6}
0.03	0.0817	2.31×10^{-3}	2.20×10^{-5}	0.0734	2.47×10^{-3}	3.16×10^{-5}
0.04	0.107	3.99×10^{-3}	5.03×10^{-5}	0.0958	4.25×10^{-3}	7.20×10^{-5}
0.05	0.131	6.05×10^{-3}	9.48×10^{-5}	0.117	6.45×10^{-3}	1.36×10^{-4}
0.06	0.154	8.46×10^{-3}	1.58×10^{-4}	0.138	9.00×10^{-3}	2.26×10^{-4}
0.07	0.176	0.0112	2.43×10^{-4}	0.158	0.0119	3.46×10^{-4}
0.08	0.197	0.0142	3.50×10^{-4}	0.177	0.0151	4.98×10^{-4}
0.09	0.217	0.0175	4.83×10^{-4}	0.196	0.0185	6.85×10^{-4}
0.1	0.237	0.0210	6.40×10^{-4}	0.213	0.0223	9.07×10^{-4}
0.2	0.398	0.0652	3.78×10^{-3}	0.360	0.0684	5.27×10^{-3}
0.3	0.514	0.117	9.76×10^{-3}	0.466	0.122	0.0134
0.4	0.600	0.171	0.0182	0.546	0.177	0.0247
0.5	0.666	0.224	0.0286	0.609	0.229	0.0384
0.6	0.717	0.273	0.0406	0.658	0.279	0.0538
0.7	0.759	0.320	0.0537	0.698	0.325	0.0705
0.8	0.792	0.363	0.0677	0.731	0.367	0.0880
0.9	0.820	0.403	0.0823	0.759	0.406	0.106
1.0	0.843	0.440	0.0974	0.783	0.442	0.124
2.0	0.950	0.693	0.253	0.903	0.683	0.302
3.0	0.981	0.825	0.395	0.947	0.808	0.451
4.0	0.992	0.900	0.519	0.968	0.881	0.574
5.0	0.997	0.944	0.629	0.980	0.927	0.676
6.0	0.999	0.971	0.727	0.987	0.957	0.764
7.0	1.00	0.987	0.814	0.992	0.978	0.840
8.0	1.00	0.996	0.893	0.995	0.992	0.908
9.0	1.00	1.00	0.966	0.997	1.00	0.969
10.0	1.00	1.00	1.00	1.00	1.00	1.00

damped loops having larger maximum magnitude, and even faster roll off, than for standard underdamped loops.

Table 17.3 provides the parameters for fourth-order loops, for both supercritical response and standard underdamped responses. Fourth-order loops can track a parameter having nonzero third derivative with respect to time.

Figure 17.6 shows the magnitudes of fourth-order closed loop transfer functions for two different loop bandwidths. Compared to the third-order loops portrayed in Figure 17.4, transfer functions for fourth-order loops roll off slightly more rapidly, with supercritically damped loops having larger maximum magnitude, and even faster roll off, than for standard underdamped loops.

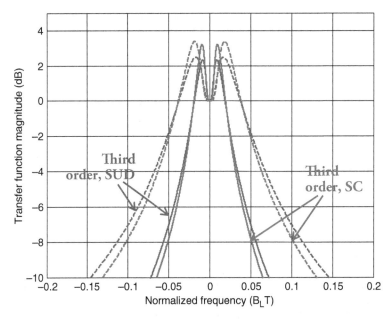

Figure 17.5. Example Transfer Function Magnitudes for Supercritical (SC) and Standard Underdamped (SUD) Third-Order Loops, Solid Lines $B_L T = 0.1$, and Dashed Lines $B_L T = 0.3$

17.3 TRACKING LOOP CHARACTERIZATION

Loop bandwidth, loop update rate, loop order, and loop damping must be selected to provide good performance under the varying conditions where the loop will operate. Numerous tradeoffs involved in making these selections are discussed in this section. Section 17.3.1 describes how loop design affects performance in noise. Parameter dynamics occur when there are nonzero time derivatives of the parameter being tracked; Section 17.3.2 describes how different loop designs affect the ability to handle dynamics. Receiver oscillator imperfections also affect tracking loop performance, and Section 17.3.3 analyzes this effect. Section 17.3.4 describes considerations of initiating tracking with loop pull-in, and also loss of tracking loop lock.

17.3.1 Noise Effects

As long as the small-error model applies consistent with (17.1), the variance of tracking error due to noise is proportional to the equivalent rectangular bandwidth of the closed loop transfer function. Loop order and loop damping have no effect on the variance of noise-induced parameter tracking errors when the small-error model applies. Specific relationships between input noise and interference and the variance of the parameter estimates vary for different parameters and discriminators, and are provided in Chapters 18 and 19.

TABLE 17.3. Parameters for Fourth-Order Discrete-Update Tracking Loops

$B_L T$	Supercritical Response				Standard Underdamped Response			
	K_1	K_2	K_3	K_4	K_1	K_2	K_3	K_4
0.005	0.0136	6.99×10^{-5}	1.60×10^{-7}	1.37×10^{-10}	0.0117	6.91×10^{-5}	2.04×10^{-7}	3.01×10^{-10}
0.01	0.0269	2.75×10^{-4}	1.25×10^{-6}	2.14×10^{-9}	0.0232	2.72×10^{-4}	1.59×10^{-6}	4.69×10^{-9}
0.02	0.527	1.06×10^{-3}	9.62×10^{-6}	3.27×10^{-8}	0.0454	1.05×10^{-3}	1.22×10^{-5}	7.15×10^{-8}
0.03	0.774	2.32×10^{-3}	3.12×10^{-5}	1.58×10^{-7}	0.0668	2.29×10^{-3}	3.96×10^{-5}	3.45×10^{-7}
0.04	0.101	4.00×10^{-3}	7.11×10^{-5}	4.77×10^{-7}	0.0873	3.94×10^{-3}	9.00×10^{-5}	1.04×10^{-6}
0.05	0.124	6.06×10^{-3}	1.34×10^{-4}	1.11×10^{-6}	0.107	5.97×10^{-3}	1.69×10^{-4}	2.42×10^{-6}
0.06	0.146	8.47×10^{-3}	2.23×10^{-4}	2.21×10^{-6}	0.126	8.33×10^{-3}	2.80×10^{-4}	4.79×10^{-6}
0.07	0.166	0.0112	3.41×10^{-4}	3.92×10^{-6}	0.144	0.0110	4.28×10^{-4}	8.48×10^{-6}
0.08	0.187	0.0142	4.90×10^{-4}	6.41×10^{-6}	0.162	0.0139	6.14×10^{-4}	1.38×10^{-5}
0.09	0.206	0.0174	6.73×10^{-4}	9.84×10^{-6}	0.179	0.0171	8.42×10^{-4}	2.12×10^{-5}
0.1	0.225	0.0209	8.91×10^{-4}	1.44×10^{-5}	0.195	0.0205	1.11×10^{-3}	3.10×10^{-5}
0.2	0.378	0.0644	5.14×10^{-3}	1.57×10^{-4}	0.330	0.0628	6.29×10^{-3}	3.31×10^{-4}
0.3	0.489	0.115	0.0130	5.67×10^{-4}	0.430	0.112	0.0156	1.18×10^{-3}
0.4	0.571	0.167	0.0238	1.32×10^{-3}	0.506	0.161	0.0282	2.70×10^{-3}
0.5	0.635	0.217	0.0368	2.46×10^{-3}	0.565	0.209	0.0430	4.95×10^{-3}
0.6	0.685	0.263	0.0513	3.97×10^{-3}	0.613	0.253	0.0592	7.88×10^{-3}
0.7	0.726	0.307	0.0669	5.83×10^{-3}	0.652	0.294	0.0763	0.0114
0.8	0.759	0.347	0.0832	8.03×10^{-3}	0.685	0.332	0.0939	0.0156
0.9	0.787	0.384	0.0998	0.010	0.712	0.367	0.112	0.0202
1.0	0.810	0.418	0.117	0.0133	0.736	0.400	0.129	0.0253
2.0	0.924	0.648	0.275	0.0506	0.736	0.400	0.129	0.0253
3.0	0.961	0.767	0.401	0.0957	0.911	0.731	0.401	0.160
4.0	0.978	0.837	0.500	0.142	0.937	0.800	0.487	0.228
5.0	0.986	0.881	0.578	0.188	0.952	0.846	0.552	0.292
6.0	0.991	0.912	0.641	0.232	0.962	0.879	0.603	0.350
7.0	0.994	0.933	0.693	0.274	0.969	0.903	0.645	0.404
8.0	0.996	0.949	0.737	0.315	0.975	0.921	0.679	0.454
9.0	0.997	0.960	0.774	0.354	0.979	0.936	0.707	0.501
10.0	0.998	0.969	0.805	0.391	0.982	0.947	0.731	0.544
20.0	1.00	0.997	0.959	0.693	0.994	0.994	0.855	0.856
25.0	1.00	0.999	0.986	0.812	0.996	1.00	0.883	0.964

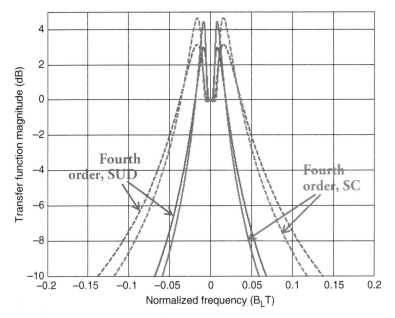

Figure 17.6. Example Transfer Function Magnitudes for Supercritical (SC) and Standard Underdamped (SUD) Fourth-Order Loops, Solid Lines $B_L T = 0.1$, and Dashed Lines $B_L T = 0.3$

17.3.2 Dynamics Effects

Loop bandwidth, loop update rate, loop order, and loop damping all affect loop behavior when the parameter being tracked changes over time. Typically, effects are considered for different kinds of parameter variations, or dynamics, characterized by different time derivatives. The following examples illustrate the effects, and are followed by more general results. No noise is added to the loop simulations shown here, so the only stress on the loop is the parameter dynamics.

Figure 17.7 shows the error in estimating the parameter value being tracked over a number of loop updates, in response to a unit change in parameter value at the first loop update shown, for first-order and second-order loops having two different bandwidths. The first-order loops converge directly to the new value, yielding zero parameter error, and the wider bandwidth loop responds approximately twice as fast as the loop with half the bandwidth. The second-order loops all overshoot the desired value, introducing errors of opposite sign, before settling to the correct parameter value. Here again, the wider bandwidth loop settles faster than the narrower bandwidth loop, better handling the dynamics. The standard underdamped loop overshoots more but settles to the correct value faster than does the supercritically damped loop. The narrower bandwidth second-order loop with supercritical response takes the longest to converge to zero error.

Figure 17.8 shows the estimated parameter rate for the same situation, corresponding to the slope of the curves in Figure 17.7. The second-order loops have larger magnitude rates of change, corresponding to the overshoot in Figure 17.7.

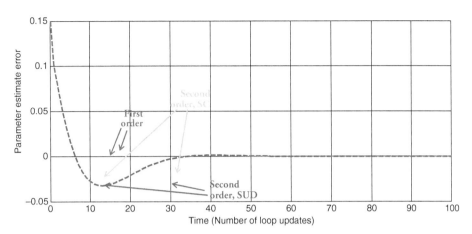

Figure 17.7. Parameter Estimate Errors of First-Order and Second-Order Loops to a Unit Step Change in Parameter Value, with Supercritical (SC) and Standard Underdamped (SUD) Second-Order Loops, Solid Lines $B_L T = 0.1$, and Dashed Lines $B_L T = 0.3$

Figure 17.9 shows the error in estimating the parameter being tracked over a number of loop updates, in response to a unit change in parameter value at the first loop update shown, for third-order loops having the same two bandwidths as in the previous two figures. Compared to the second-order loop responses in Figure 17.7, the third-order loops respond just as quickly (response time is set by the bandwidth), but overshoot more and take more loop updates to settle to zero error. Again, the standard underdamped loops respond more quickly to the unit step change in parameter values than the supercritically damped loops, and wider bandwidth loops also respond more quickly.

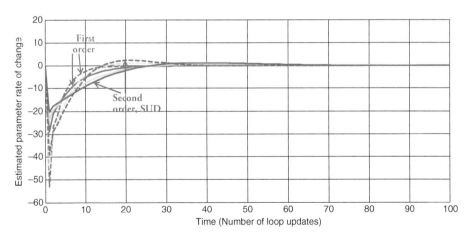

Figure 17.8. Estimated Parameter Rates of First-Order and Second-Order Loops Subjected to a Unit Step Change in Parameter Value, with Supercritical (SC) and Standard Underdamped (SUD) Second-Order Loops, Solid Lines $B_L T = 0.1$, and Dashed Lines $B_L T = 0.3$

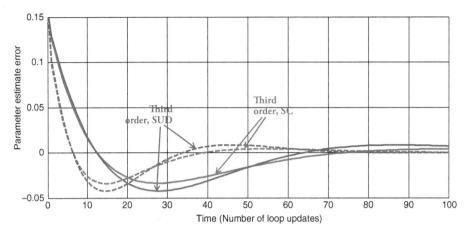

Figure 17.9. Parameter Estimate Errors of Supercritical (SC) and Standard Underdamped (SUD) Third-Order Loops to a Unit Step Change in Parameter Value, Solid Lines $B_L T = 0.1$, and Dashed Lines $B_L T = 0.3$

Figure 17.10 shows the estimated parameter rate for the same situation, corresponding to the slope of the curves in Figure 17.9. The supercritical and standard underdamped responses are almost indistinguishable in this plot, while the effect of different loop bandwidths is evident. The third-order loops have smaller magnitude rates of change than second-order loops shown in Figure 17.8.

Now consider tracking a parameter that varies over time, beginning at the initial loop update shown, corresponding to a constant rate of change in the parameter value of -10 per loop update. Figure 17.11 shows the resulting parameter estimation error of

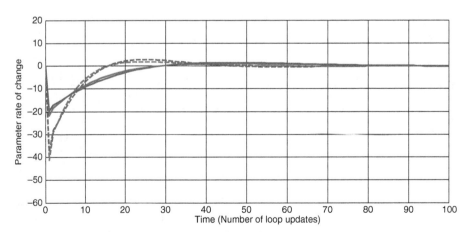

Figure 17.10. Estimated Parameter Rates of Supercritical (SC) and Standard Underdamped (SUD) Third-Order Loops Subjected to a Unit Step Change in Parameter Value, Solid Lines $B_L T = 0.1$, and Dashed Lines $B_L T = 0.3$

TRACKING LOOP CHARACTERIZATION

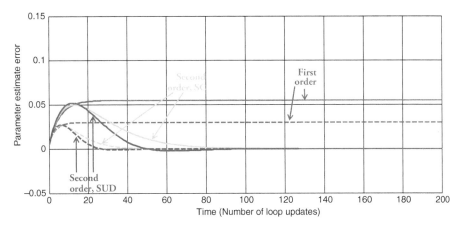

Figure 17.11. Parameter Estimate Errors of First-Order and Second-Order Loops to a Constant Rate of Change in Parameter Value, with Supercritical (SC) and Standard Underdamped (SUD) Second-Order Loops, Solid Lines $B_L T = 0.1$, and Dashed Lines $B_L T = 0.3$

first-order and second-order loops having different loop bandwidths. The first-order loops cannot follow this constant rate of change, and converge to nonzero parameter errors, with the wider bandwidth loop having a smaller asymptotic error. The second-order loops are able to track this first derivative in the parameter value, initially overshooting and then converging to zero asymptotic error, with the wider bandwidth loops converging faster, and the standard underdamped loops converging faster than the supercritically damped loops.

The associated estimates of parameter rates are shown in Figure 17.12. All loops converge to the correct estimate of parameter rate of change, but, as seen in Figure

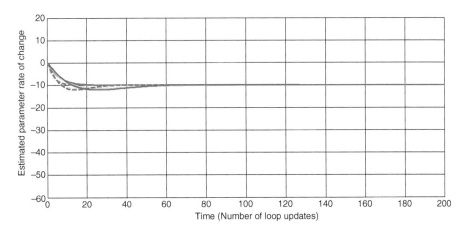

Figure 17.12. Estimated Parameter Rates of First-Order and Second-Order Loops Subjected to a Constant Rate of Change in Parameter Value, with Supercritical (SC) and Standard Underdamped (SUD) Second-Order Loops, Solid Lines $B_L T = 0.1$, and Dashed Lines $B_L T = 0.3$

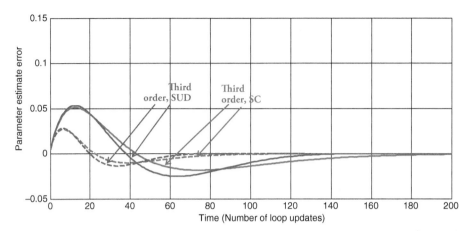

Figure 17.13. Parameter Estimate Errors of Supercritical (SC) and Standard Underdamped (SUD) Third-Order Loops to a Constant Rate of Change in Parameter Value, Solid Lines $B_L T = 0.1$, and Dashed Lines $B_L T = 0.3$.

17.11, the first-order loops converge to a fixed and nonzero parameter error because they cannot "catch up" with the linearly changing parameter value.

Figure 17.13 shows the parameter estimation error of third-order loops having different loop bandwidths to this constant rate of change. The third-order loops are also able to track this first derivative in the parameter value, but take longer to converge than second-order loops having the same bandwidth. As with the second-order loops, the wider bandwidth loops converge faster, and the standard underdamped loops overshoot by more but then converge faster than the supercritically damped sloops.

The associated estimates of parameter rates are shown in Figure 17.14. The third-order loops converge to the correct estimate of parameter rate of change.

Now consider a second-order rate of change in the parameter value. Figure 17.15 shows the parameter estimation error of first-order and second-order loops having different loop bandwidths to this quadratic rate of change. The first-order loops cannot track this constant change in the rate of change, and the error grows linearly with time, although the wider bandwidth loop has smaller error. The second-order loops are also unable to track this second derivative in the parameter value, but converge to a constant error, with the wider bandwidth loops having smaller error, and the supercritical loops having smaller asymptotic errors than the standard underdamped loops.

The associated estimates of parameter values are shown in Figure 17.16. The first-order loops underestimate the magnitude of the parameter rate of change, while the second-order and third-order loops estimate the parameter rate of change accurately, although the second-order loops converge to a fixed error in parameter value because they cannot "catch up" with the quadratically changing parameter value.

Figure 17.17 shows the parameter estimation error of third-order loops having different loop bandwidths to this quadratic rate of change. The third-order loops are able to track this second derivative in the parameter value, converge to zero parameter

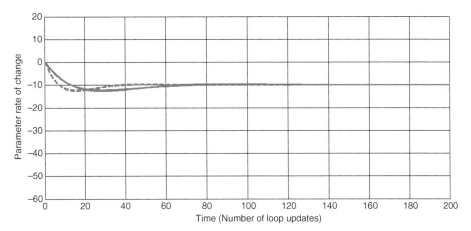

Figure 17.14. Estimated Parameter Rates of Supercritical (SC) and Standard Underdamped (SUD) Third-Order Loops Subjected to a Constant Rate of Change in Parameter Value, Solid Lines $B_L T = 0.1$, and Dashed Lines $B_L T = 0.3$

estimate error. As seen in other cases, the wider bandwidth loops converge faster, and the standard underdamped loops converge faster than the supercritically damped loops.

The associated estimates of parameter rates are shown in Figure 17.18. The third-order loops converge to the correct estimate of parameter rate of change.

This ability to track dynamics with different order derivatives is consistent with standard results in linear control systems [6]. The asymptotic errors provided in Table 17.4 are derived using these standard results.

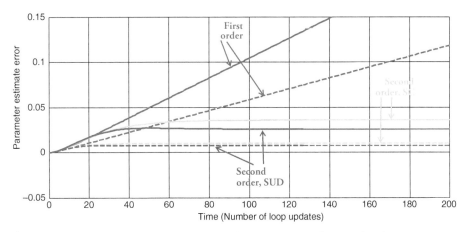

Figure 17.15. Parameter Estimate Errors of First-Order and Second-Order Loops to a Quadratic Rate of Change in Parameter Value, with Supercritical (SC) and Standard Underdamped (SUD) Second-Order Loops, Solid Lines $B_L T = 0.1$, and Dashed Lines $B_L T = 0.3$

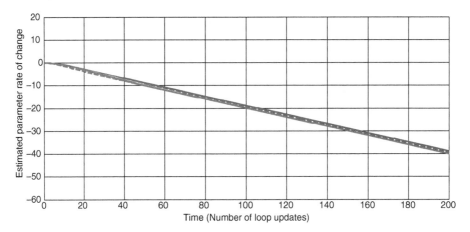

Figure 17.16. Estimated Parameter Rates of Third-Order Loops Subjected to a Quadratic Rate of Change in Parameter Value, with Supercritical (SC) and Standard Underdamped (SUD) Second-Order Loops, Solid Lines $B_L T = 0.1$, and Dashed Lines $B_L T = 0.3$

17.3.3 Receiver Oscillator Noise Effects

Receiver oscillator phase noise, discussed in Section 14.1.5, also degrades the parameter estimate produced by tracking loops. Since the oscillator phase noise enters through the replica into the signal processing, the transfer function relating parameter estimation errors and oscillator phase noise is given by $(1 - H_N[z])$. If the oscillator phase noise

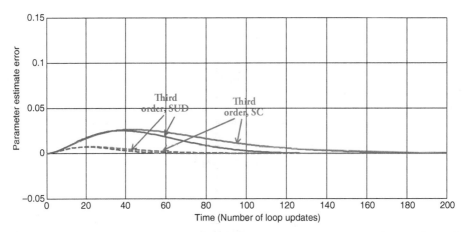

Figure 17.17. Parameter Estimate Errors of Supercritical (SC) and Standard Underdamped (SUD) Third-Order Loops to a Quadratic Rate of Change in Parameter Value, Solid Lines $B_L T = 0.1$, and Dashed Lines $B_L T = 0.3$

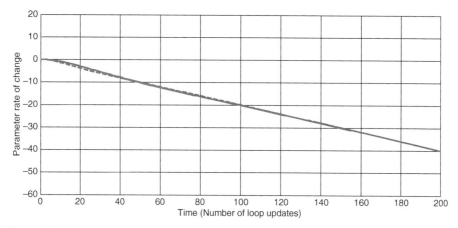

Figure 17.18. Estimated Parameter Rates of Supercritical (SC) and Standard Underdamped (SUD) Third-Order Loops Subjected to a Quadratic Rate of Change in Parameter Value, Solid Lines $B_L T = 0.1$, and Dashed Lines $B_L T = 0.3$

has PSD $\Phi_\phi[z]$, the variance contributed to the parameter estimate by the phase noise is given by

$$\sigma_\phi^2 = \int_0^\infty |1 - H_N(e^{i2\pi fT})|^2 \Phi_\phi(e^{i2\pi fT}) df \qquad (17.6)$$

This result is striking. The loop transfer function performs low-pass filtering of input noise and interference, so narrower loop bandwidths reduce their contribution to the parameter error variance. In contrast, however, the variance contributed to the parameter estimate by the oscillator phase noise is a high-pass filtered version of the phase noise PSD, so the narrower the loop bandwidth, the greater the contribution of the oscillator phase noise to the parameter estimation variance.

17.3.4 Loop Closure and Loss of Lock

Initiating tracking loop operation is commonly called loop closure or loop pull-in. The loop will only close if the initial estimate that begins loop operation is on the linear portion

TABLE 17.4. Tracking Loop Final Errors in Response to Input Dynamics

Input Dynamics	First-Order Loop	Second-Order Loop	Third-Order Loop
Unit step	Zero	Zero	Zero
Linear change with time	Nonzero constant	Zero	Zero
Quadratic Change with Time	Infinity	Nonzero constant	Zero

of the S-curve. Faster and more reliable loop closure occurs when the initial value is near the center of the linear portion of the S-curve. Loops with smaller loop update times, wider loop bandwidths, and lower order all exhibit faster and more reliable pull-in.

Loss of lock occurs when a loop has been tracking the parameter, but the error in the discriminator takes on a value outside the linear region of the S-curve, $-E/2 \leq \tilde{p}_n \leq E/2$. If the loop order is not high enough to track higher derivatives of the parameter variation, the loop update rate can be made small enough so that the differences between sequential parameter values remain well within the linear range $[-E/2, E/2]$.

When the loop is most stressed by random effects due to a combination of input noise and interference, receiver oscillator phase noise, and random variation of the true parameter value caused by dynamics, all assumed uncorrelated, the parameter error is characterized by the variance

$$\sigma_p^2 = \sigma_{\text{noise+interference}}^2 + \sigma_{\text{phase noise}}^2 + (\Delta p)^2 \qquad (17.7)$$

where the variances on the right hand are caused, respectively, by noise and interference and phase noise, and Δp is the unmodeled change in parameter value between loop updates.

A rule of thumb for loop loss of lock is that the loop is likely to maintain lock as long as the following loop lock criterion is satisfied:

$$3\sigma_p < E/4 \qquad (17.8)$$

where, as discussed in Section 17.1, the linear part of the S-curve is $[-E/2, E/2]$.

When the loop is most stressed by dynamics, higher-order loops, wider loop bandwidths, and smaller loop update times can all help avoid loss of lock. When the loop is most stressed by oscillator phase noise, wider loop bandwidths help avoid loss of lock. When the loop is most stressed by input noise and interference and weak signals, narrow loop bandwidths and longer loop update times help avoid loss of lock.

In practice, loop pull-in and loss of lock cannot be characterized analytically and need to be explored through hardware experiment or computer simulation. Loss of lock, in particular, is difficult to characterize since, near the loss of lock threshold, loss of lock remains a low probability event that needs to be evaluated through Monte Carlo experiments; each experiment can represent hundreds or thousands of seconds of tracking time. Even establishing a criterion for loss of lock involves careful thought, since loops can temporarily lose lock, then regain lock subsequently.

17.4 SUMMARY

The treatment of tracking loop design and performance in this chapter differs from those in other satnav references in two important ways:

- A general formulation is provided that applies to carrier and code tracking loops of different types, with specifics for each tracking loop described in subsequent chapters.

- Design procedure and performance characterization are based on discrete-update loops typically employed in modern satnav receivers, rather than the older approach of designing an analog loop and then transforming the design to a discrete-update algorithm.

While the treatment of loop design and characterization in this chapter are kept general in order to apply to all tracking loops in a satnav receiver, specifics of signal processing and discriminator design are addressed specifically for each type of loop considered in the subsequent two chapters.

A number of observations apply to the design of any tracking loop.

- Smaller loop update times allow the loop to better accommodate dynamics, but can cause more sensitivity to noise and interference, and to weak signals.
- Narrower loop bandwidths better suppress the effects of noise and interference and weak signals, but larger loop bandwidths enable better tracking of dynamics, better loop pull-in performance, and better suppression of oscillator phase noise.
- Underdamped loops may be slightly less stable in noise and interference and weak signals, but can provide more accurate tracking of parameter transients.
- Higher loop orders better accommodate dynamics, allowing the loop bandwidth to be narrowed for better performance in noise and interference and weak signals; lower loop orders provide tighter tracking of parameter dynamics as well as better pull-in performance, and also can accommodate more noise and interference and weaker signals before losing lock due to signal-to-noise ratio stresses. A rule of thumb is that third-order or fourth-order loop loss of lock occurs at 1–2 dB higher signal-to-noise ratio than loss of lock in a first-order or second-order loop.

More specifics are found in Chapters 18 and 19, which address carrier tracking loops and code tracking loops, respectively.

REFERENCES

1. B. W. Parkinson, J. J. Spilker, Jr., P. Axelrad, and P. Enge, *Global Positioning System: Theory and Applications*, American Institute of Aeronautics and Astronautics, 1996.
2. E. Kaplan and C. Hegarty, *Understanding GPS: Principles and Applications*, 2nd edition, Artech House, 2006.
3. W. Lindsey and C. Chie, "A Survey of Digital Phase Locked Loops," *Proceedings of the IEEE*, Vol. 69, 1981, pp. 410–431.
4. S. A. Stephens and J. B. Thomas, "Controlled-Root Formulation for Digital Phase-Locked Loops," *Aerospace and Electronic Systems, IEEE Transactions on*, Vol. 31, No. 1, 1995, pp. 78–95.
5. J. B. Thomas, "An Analysis of Digital Phase-Locked Loops," *JPL Publication*, Vol. 89-2, 1989.
6. R. T. Stefani, B. Shahian, C. Savant, and G. Hostetter, *Design of Feedback Control Systems*, 4th edition, Oxford University Press, 2001.

18

CARRIER TRACKING AND DATA DEMODULATION

Carrier tracking produces estimates of the carrier frequency or the carrier phase, which are used as measurements and also to remove, or wipe off, any residual carrier frequency and, in some cases, carrier phase, from the signal being processed. Figure 18.1 shows the digital processing in a receiver channel, highlighting the carrier tracking loop. This chapter describes the signal processing used for carrier tracking, then introduces three fundamentally different carrier tracking loops that employ different measurements and different discriminators, and track different parameters. It also provides analytical models for predicting the performance of these different tracking loops, along with discussion of design considerations. Data message demodulation is also addressed since it uses the same measurements employed for carrier tracking.

Section 18.1 describes signal processing for carrier tracking. With the details of tracking loop design addressed in Chapter 17, and signal processing for carrier tracking addressed in Section 18.1, specific carrier tracking loops are then readily addressed. Section 18.2 describes the measurements, discriminator design, performance, and design considerations for frequency-locked loops (FLL) that track the carrier frequency but not the carrier phase. Section 18.3 addresses measurements, discriminator design, performance, and design considerations for Costas loops that track carrier phase even when the carrier phase is also biphase modulated by data message bits or overlay code bits.

Engineering Satellite-Based Navigation and Timing: Global Navigation Satellite Systems, Signals, and Receivers, First Edition. John W. Betz.
© 2016 The Institute of Electrical and Electronics Engineers, Inc. Published 2016 by John Wiley & Sons, Inc.

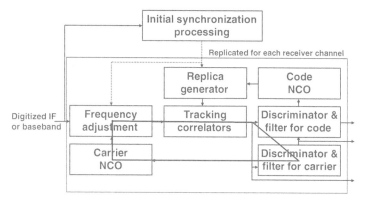

Figure 18.1. Digital Processing in a Receiver Channel, with Highlighted Carrier Tracking Loop

Section 18.4 addresses measurements, discriminator design, performance, and design considerations for phase-locked loops (PLLs) that track carrier phase when the carrier phase has no unknown additional modulation. Section 18.5 describes demodulation of data message bits and performance of data message demodulation and decoding. Section 18.6 summarizes the chapter.

18.1 SIGNAL PROCESSING FOR CARRIER TRACKING

Signal processing for carrier tracking can be applied to either the data component or the pilot component, if one exists, or both, depending upon signal design and receiver processing strategy. Section 18.1.1 describes signal processing of data components and pilot components, reemphasizing the central role that correlation processing plays in satnav receivers. Section 18.1.2 outlines the extraction of data components and pilot components from a composite signal containing multiple components or signals combined onto the same RF carrier as described in Section 3.3.12. Section 18.1.3 outlines specialized strategies for correlation processing of signals whose spreading modulations have MBOC spectra.

18.1.1 Signal Processing of Pilot Components and Data Components

Signal processing is based on the assumption that time of arrival of the signal relative to the receiver time, and the received carrier frequency relative to the frequency reference in the receiver, are known approximately, either from acquisition or from prior estimates of these parameters from code and carrier tracking loops.

The received signal components are modeled as analytic signals, with the data component analytical signal at RF given by $\sqrt{C_d}\gamma_d(t-D)\exp[i2\pi f_0(t-D)]$, and the pilot component analytical signal at RF given by $\sqrt{C_p}\gamma_p(t-D)\exp[i2\pi f_0(t-D)]$, using the same notation employed in Figures 3.23 and 3.25. In these expressions, C_d is the power in the received data component, C_p is the power in the received pilot component, the delay D and carrier frequency f_0 are each slowly varying functions of time, reflecting the time-varying geometry between satellite and receiver. Mixing in the receiver front end, using an estimate of the received carrier frequency $\hat{f}_0$, performs frequency translation of these components,

$$\sqrt{C_d}\gamma_d(t-D)\exp[i2\pi f_0(t-D)]\exp[-i2\pi\hat{f}_0 t] = \sqrt{C_d}\gamma_d(t-D)\exp[i2\pi(Ft-f_0 D)]$$
$$\sqrt{C_p}\gamma_p(t-D)\exp[i2\pi f_0(t-D)]\exp[-i2\pi\hat{f}_0 t] = \sqrt{C_p}\gamma_p(t-D)\exp[i2\pi(Ft-f_0 D)]$$
(18.1)

where the residual frequency after downconversion is $F \triangleq f_0 - \hat{f}_0$.

Frequency adjustment, shown in Figure 18.1, frequency translates the components in (18.1) by the estimated residual frequency from acquisition or prior estimates from tracking, $\hat{F}$, yielding

$$\sqrt{C_d}\gamma_d(t-D)\exp[i2\pi(Ft-f_0 D)]\exp[i2\pi\hat{F}t] = \sqrt{C_d}\gamma_d(t-D)\exp[i2\pi(\tilde{F}t-f_0 D)]$$
$$\sqrt{C_p}\gamma_p(t-D)\exp[i2\pi(Ft-f_0 D)]\exp[i2\pi\hat{F}t] = \sqrt{C_p}\gamma_p(t-D)\exp[i2\pi(\tilde{F}t-f_0 D)]$$
(18.2)

where $\tilde{F} \triangleq F - \hat{F}$ is the residual carrier frequency after downconversion and frequency adjustment. Denote the time-varying phase of the resulting components by $\theta(t) \triangleq 2\pi(\tilde{F}t - f_0 D)$. Carrier tracking then estimates the residual frequency

$$\tilde{F} = \frac{1}{2\pi}\frac{d\theta(t)}{dt}$$
(18.3)

or the phase directly, $\theta(t)$.

Invoking the notation in Figures 3.23 and 3.25 yields the two received components

$$s_d(t) = \sqrt{C_d}e^{i\theta(t)}\sigma_d(t-D(t))\beta(t-D(t))$$
$$= \sqrt{C_d}e^{i\theta(t)}\sum_{n=-\infty}^{\infty} b'_n g_d(t-D(t)-nT_c - T_c/2)$$
$$\times \sum_{k=-\infty}^{\infty} d_k \text{rect}_{T_d}(t-D(t)-kT_d - T_d/2)$$
(18.4)

$$s_p(t) = \sqrt{C_p}e^{i\theta(t)}\left[\sum_{n=-\infty}^{\infty} a_n g_p(t-D(t)-nT_c - T_c/2)\right]$$
$$\times \left[\sum_{m=-\infty}^{\infty} q_m \text{rect}_{T_q}(t-D(t)-mT_q - T_q/2)\right]$$

where the (typically slow) time variation of the delay is made explicit using the notation $D(t)$.

It is assumed that the receiver is approximately synchronized to the overlaid spreading waveform, so the receiver can generate the replica at some delay[1] τ chosen to be offset from the estimated time delay, $\hat{D}(t)$, $\sum_{n=-\infty}^{\infty} b'_n g_d(t + \tau - \hat{D}(t) - nT_c - T_c/2)$, where the overlaid data spreading code b'_n and data spreading symbols $g_d(t)$ are both known from the ICD, and the approximate time delay is known from acquisition or prior tracking estimates. The replica generator then produces this waveform, and the data component signal processing calculates the measurement

$$\mu_d(j;\tau) = \frac{1}{T} \int_{jT-T/2}^{jT+T/2} s_d(t - D(t)) \sum_{n=-\infty}^{\infty} b'_n g_d(t + \tau - \hat{D}(t) - nT_c - T_c/2) dt$$

$$= \frac{\sqrt{C_d}}{T} \int_{jT-T/2}^{jT+T/2} e^{i\theta(t)} \left[\sum_{n=-\infty}^{\infty} b'_n g_d(t - D(t) - nT_c - T_c/2) \right]$$

$$\times \left[\sum_{n=-\infty}^{\infty} b'_n g_d(t + \tau - \hat{D}(t) - nT_c - T_c/2) \right] \quad (18.5)$$

$$\times \left[\sum_{k=-\infty}^{\infty} d_k \text{rect}_{T_d}(t - D(t) - kT_d - T_d/2) \right] dt$$

$$\cong e^{i\theta[j]} \frac{\sqrt{C_d}}{T} \bar{R}_d(\tilde{D}(t) - \tau) \int_{jT-T/2}^{jT+T/2} \sum_{k=-\infty}^{\infty} d_k \text{rect}_{T_d}(t - D(t) - kT_d - T_d/2) dt$$

where the integration limits are established so no data bit transitions occur within the limits, $\tilde{D}(t) \stackrel{\Delta}{=} D(t) - \hat{D}(t)$, and the final line involves two different approximations. One is that the phase $\theta(t)$ is so slowly varying over the time interval T that it can be approximated by a single value over the time interval, and thus represented as a discrete sequence $\theta[k]$ over all time. The other approximation is that $\left[\sum_{n=-\infty}^{\infty} b'_n g_d(t - D(t) - nT_c - T_c/2) \right] \left[\sum_{n=-\infty}^{\infty} b'_n g_d(t + \tau - \hat{D}(t) - nT_c - T_c/2) \right] \cong \bar{R}_d(\tilde{D}(t) - \tau)$, assuming the spreading code satisfies the ideal long spreading code model.

Since $\hat{D}(t) \cong D(t)$ and the data message bit duration is known from the ICD, the correlation integration time in (18.5) can be set either to the data message bit duration or to a fraction of the data message bit duration such that the integral limits are always aligned in time with the end of a data message bit, meaning that data message bit transitions only occur between integrals in (18.5).

---

[1] Note that the term "delay" is used consistently; an advance is represented by a negative-valued delay.

The signal processing in (18.5) is equivalent to crosscorrelating the received waveform, after downconversion and frequency adjustment, to the replica of the overlaid data spreading waveform, at a delay offset by τ from the estimated delay. When the delay offset is $\tau = 0$, this operation is called wiping the data spreading waveform, since it produces a result that is independent of the spreading code and the spreading modulation, as well as the data overlay code.

Since the signal processing operates not only on received signal, but also on additive noise and interference, the signal processing for the data component then yields the prompt measurements

$$Z_P[j] \cong \mu_d(j; \tilde{D}(t)) + \text{Noise terms}$$
$$= \sqrt{C_d} \bar{R}_d(\tilde{D}[j]) e^{i\theta_d[j]} + \text{Noise terms} \quad (18.6)$$

where $\theta_d[j] = \theta[j] + d_k \pi$, with $d_k = \{0, 1\}$ depending on the data message bit value. In (18.6), the time-varying delay is approximated by a constant value at the midpoint of the loop update interval.

The measurement (18.6) is termed "prompt" since the replica is intended to be aligned in time with the received signal; this measurement is sometimes called the prompt correlator output or prompt correlator tap. This signal processing of the data component is used in subsequent sections to yield measurements employed for carrier tracking using the data component and also for data message demodulation.

Signal processing of the pilot component in (18.4) involves two different cases. First consider when initial synchronization has provided a delay estimate that indicates the delay of the pilot spreading code, but is still ambiguous with respect to the phase of the pilot overlay code. In that case, the receiver can generate the replica at some delay τ offset from the time-varying approximate time delay, $\hat{D}(t)$, $\sum_{n=-\infty}^{\infty} a_n g_p(t + \tau - \hat{D}(t) - nT_c - T_c/2)$ where the pilot spreading code a_n and pilot spreading symbols $g_p(t)$ are both known from the ICD, and the approximate time delay is known from initial synchronization or code tracking. The replica generator then produces this waveform, and the pilot component signal processing calculates the measurement

$$\mu_{p,o}(j; \tau) = \frac{1}{T} \int_{jT-T/2}^{jT+T/2} s_p(t - D(t)) \sum_{n=-\infty}^{\infty} a_n g_p(t + \tau - \hat{D}(t) - nT_c - T_c/2) dt$$

$$= \frac{\sqrt{C_p}}{T} \int_{jT-T/2}^{jT+T/2} e^{i\theta(t)} \left[\sum_{n=-\infty}^{\infty} a_n g_p(t - D(t) - nT_c - T_c/2) \right]$$

$$\times \left[\sum_{n=-\infty}^{\infty} a_n g_p(t + \tau - \hat{D}(t) - nT_c - T_c/2) \right] \quad (18.7)$$

$$\times \left[\sum_{m=-\infty}^{\infty} q_m \text{rect}_{T_q}(t - D(t) - mT_q - T_q/2) \right] dt$$

$$\cong e^{i\theta[j]} \frac{\sqrt{C_p}}{T} \bar{R}_p(\tilde{D}(t) - \tau) \int_{jT-T/2}^{jT+T/2} \sum_{m=-\infty}^{\infty} q_m \text{rect}_{T_q}(t - D(t) - mT_q - T_q/2) dt$$

where the integration limits are established so that no pilot overlay code bit transitions occur within the limits, and the approximation in the final line arises from the same two approximations discussed after (18.5).

Since $\hat{D}(t) \cong D(t)$ and the pilot bit duration is known from the ICD, the correlation integration time in (18.7) can be set either to the pilot overlay code bit duration or a fraction of the pilot overlay code bit duration such that the integral limits are always almost aligned in time with the end of a pilot overlay code bit, meaning that pilot overlay code bit transitions only occur between integrals in (18.7).

The signal processing in (18.7) is equivalent to that for data component processing previously described, except in this case the received signal is crosscorrelated, after downconversion and frequency adjustment, to the replica of the pilot waveform, at zero delay. When the delay offset is $\tau = 0$, this operation is sometimes referred to as wiping the pilot spreading code, since it produces a result that is independent of the pilot spreading code.

Since the signal processing operates not only on received signal, but also on additive noise and interference, the signal processing for the pilot component with synchronization to the pilot spreading code but not to the pilot overlay code then yields the prompt measurements

$$Z_P[j] \cong \mu_{p,o}(j; \tilde{D}(t)) + \text{Noise terms}$$
$$= \sqrt{C_p} e^{i\theta_p[j]} \bar{R}_p(\tilde{D}[j]) + \text{Noise terms} \quad (18.8)$$

where $\theta_p[j] = \theta[j] + q_m \pi$, with $q_m = \{0, 1\}$ depending on the overlay code bit value. In (18.8), the time-varying delay is approximated by a constant value at the midpoint of the loop update interval.

Like (18.6), this measurement is termed "prompt" since the replica is intended to be aligned in time with the received signal; this measurement is sometimes called the prompt correlator output. This signal processing of the pilot component is used in subsequent sections to yield measurements employed for carrier tracking using the pilot component and also for estimating pilot overlay code bits to be used for pilot code synchronization. Since it is exactly analogous to data component processing, it is not addressed specifically any further.

The other case for signal processing of the pilot component in (18.4) is when the receiver is synchronized to the pilot overlay code as well as the pilot spreading code. In that case, the receiver can generate the replica at some delay τ offset from the approximate time-varying time delay, $\hat{D}(t)$, $\left[\sum_{n=-\infty}^{\infty} a_n g_p(t - \hat{D}(t) - nT_c - T_c/2)\right] \left[\sum_{m=-\infty}^{\infty} q_m \text{rect}_{T_q}(t + \tau - \hat{D}(t) - mT_q - T_q/2)\right]$ where the pilot spreading code a_n, the pilot spreading symbols $g_p(t)$, and pilot overlay code bits q_m are all known from the ICD, and the approximate time delay $\hat{D}(t)$ is known from initial synchronization or code tracking. The replica generator then produces this waveform, and the pilot component signal processing calculates the measurement

$$\begin{aligned}\mu_p(j;\tau) &= \frac{1}{T}\int_{jT-T/2}^{jT+T/2} s_p(t-D(t))\left[\sum_{n=-\infty}^{\infty} a_n g_p(t+\tau-\hat{D}(t)-nT_c-T_c/2)\right] \\ &\quad \times \left[\sum_{m=-\infty}^{\infty} q_m \mathrm{rect}_{T_q}(t+\tau-\hat{D}(t)-mT_q-T_q/2)\right] dt \\ &= \frac{\sqrt{C_p}}{T}\int_{jT-T/2}^{jT+T/2} e^{i\theta(t)}\left[\sum_{n=-\infty}^{\infty} a_n g_p(t-D(t)-nT_c-T_c/2)\right] \\ &\quad \times \left[\sum_{n=-\infty}^{\infty} a_n g_p(t+\tau-\hat{D}(t)-nT_c-T_c/2)\right] \\ &\quad \times \left[\sum_{m=-\infty}^{\infty} q_m \mathrm{rect}_{T_q}(t-D(t)-mT_q-T_q/2)\right] \\ &\quad \times \left[\sum_{m=-\infty}^{\infty} q_m \mathrm{rect}_{T_q}(t+\tau-\hat{D}(t)-mT_q-T_q/2)\right] dt \\ &\cong e^{i\theta[j]} \frac{\sqrt{C_p}}{T} \bar{R}_p(\tilde{D}(t)-\tau)\end{aligned} \quad (18.9)$$

where the integration limits can be set without concern for pilot overlay code bit transitions, and the approximation in the final line arises from three different approximations. One is that the phase $\theta(t)$ is so slowly varying over the time interval T that it can be approximated by a single value over the time interval, and is thus represented as a discrete sequence $\theta[k]$ over all time. The second approximation is that $\left[\sum_{n=-\infty}^{\infty} a_n g_p(t-D(t)-nT_c-T_c/2)\right]\left[\sum_{n=-\infty}^{\infty} a_n g_p(t+\tau-\hat{D}(t)-nT_c-T_c/2)\right] \cong \bar{R}_p(\tilde{D}(t)-\tau)$ which implies the ideal long spreading code model and assumes that $\hat{D}(t) \cong D(t)$. The third approximation is that $\left[\sum_{m=-\infty}^{\infty} q_m \mathrm{rect}_{T_q}(t-D(t)-mT_q-T_q/2)\right]\left[\sum_{m=-\infty}^{\infty} q_m \mathrm{rect}_{T_q}(t+\tau-\hat{D}(t)-mT_q-T_q/2)\right]$ almost everywhere over the time interval T, with a negligible fraction of time that this product takes on a value of -1.

Since $\hat{D}(t) \cong D(t)$ and the pilot bit duration is known from the ICD, the correlation integration time in (18.9) can be set either to the pilot bit duration or a fraction of the pilot bit duration such that the integral limits are always aligned in time with the end of a pilot overlay code bit, meaning that pilot overlay code bit transitions only occur between integrals in (18.9).

The signal processing in (18.9) is equivalent to that for data component processing previously described, except in this case the received signal is crosscorrelated, after downconversion and frequency adjustment, to the replica of the overlaid pilot waveform, at zero delay. When the delay offset is $\tau = 0$, this operation is sometimes referred to as wiping the overlaid pilot spreading code, since it produces a result that is independent of the pilot spreading code, the pilot spreading symbols, and the overlay code.

Since the signal processing operates not only on received signal, but also on additive noise and interference, the signal processing for the pilot component with synchronization to both the pilot spreading code and to the pilot overlay code then yields the prompt measurements

$$Z_P[j] \cong \mu_p(j; \tilde{D}(t)) + \text{Noise terms}$$
$$= \sqrt{C_p} e^{i\theta_p[j]} \bar{R}_p(\tilde{D}[j]) + \text{Noise terms} \quad (18.10)$$

where $\theta_p[j] = \theta[j]$, which is the residual carrier phase with no phase contributions from data message bits or overlay code bits. In (18.10), the time-varying delay is approximated by a constant value at the midpoint of the loop update interval.

Observe that (18.10) is a slowly varying sinusoid in noise, resulting from wiping the overlaid pilot spreading code. This sinusoid's phase is the residual carrier phase, and this unique characteristic is exploited in PLL processing as described in Section 18.4.

18.1.2 Signal Processing for Data Components and Pilot Components from a Composite Signal

The preceding discussion describes processing of a signal's data component and pilot component as if each were isolated, except for additive noise and interference. Section 3.3.12 introduces transmitter combining of multiple signals and components, used by all current satnav systems to modulate more than one signal or component onto the same RF carrier. This subsection discusses how receive processing needs to account for this signal combining, and in this subsection the residual carrier frequency is neglected for notational simplicity.

For phase multiplexing, represent the composite received analytical signal after downconversion and frequency adjustment

$$[x(t) + iy(t)] \exp[i(2\pi Ft + \theta)] \quad (18.11)$$

where $x(t)$ and $y(t)$ are each biphase signals and uncorrelated. If $x(t)$ is the desired signal or component, consider the crosscorrelation between (18.11) and the desired signal

$$\frac{1}{T} \int_{-T/2}^{T/2} [x(t) + iy(t)] x(t + \tau) \exp(i 2\pi Ft + \theta) \, dt \quad (18.12)$$

The mean of (18.12) is $e^{i\theta}R_x(\tau)\operatorname{sinc}(\pi FT) \cong e^{i\theta}R_x(\tau)$, where the approximation holds when the residual frequency F is small relative to $1/T$. The biphase signal in phase quadrature, $y(t)$, has no effect on the mean output of correlation processing. While $y(t)$ does contribute to the variance that affects correlator output SNR, the contribution is very small in practice compared to noise and other sources of interference. Consequently, when two uncorrelated signals or components are phase multiplexed, signal processing can proceed as if only the desired signal or component is present.

Next, consider when the desired signal is code division multiplexed with another signal, using simple summation, interplexing, or any other technique that produces a composite received analytical signal after downconversion and frequency adjustment of the form

$$[x(t) + y(t) + iz(t) + iw(t)]\exp[i(2\pi Ft + \theta)] \qquad (18.13)$$

where $w(t)$, $x(t)$, $y(t)$, and $z(t)$ are each biphase signals and mutually uncorrelated. The same conclusions apply as with phase multiplexing.

The next type of signal combining considered here is time division multiplexing of two keyed signals having the same duration spreading symbols

$$[x(t) + y(t)]\exp[i(2\pi Ft + \theta)] \qquad (18.14)$$

where $x(t) = \sum\limits_{k=-\infty}^{\infty} a_k g_1(t - kT_c)$ and $y(t) = \sum\limits_{k=-\infty}^{\infty} b_k g_2(t - kT_c)$ where $g_1(t) = \begin{cases} p_1(t), & 0 \leq t \leq T_c \\ 0, & T_c \leq t \leq 2T_c \end{cases}$ and $g_2(t) = \begin{cases} 0, & 0 \leq t \leq T_c \\ p_2(t), & T_c \leq t \leq 2T_c \end{cases}$, and the spreading codes a_k and b_k are uncorrelated.

Let $x(t)$ be the desired signal or component, and let the ideal long spreading code model apply to $x(t)$, so that $R_x(\tau) = \frac{1}{2T_c}\int\limits_{-\infty}^{\infty} g_1(t)g_1(t+\tau)\,dt$. Consider the crosscorrelation between (18.14) and the desired signal

$$\frac{1}{T}\int\limits_{-T/2}^{T/2} [x(t) + y(t)]x(t+\tau)\exp(i2\pi Ft + \theta)\,dt \qquad (18.15)$$

The mean of (18.12) is $e^{i\theta}R_x(\tau)\operatorname{sinc}(\pi FT) \cong e^{i\theta}R_x(\tau)$, where the approximation holds when the residual frequency F is small. For time division multiplexing, it is essential that the replica be set to zero during the times the desired signal is zero. With this blanking of the replica during the times the other component or signal is present, the signal processing proceeds as desired. Performance analysis merely takes into account the fraction of time that the signal is blanked, in accounting for the correlation integration time.

Finally, consider an AltBOC signal, as introduced in Section 3.1. The Galileo E5 signal, which employs an AltBOC design described in Section 10.3.5, is addressed specifically here. From (10.4) and (10.5), the E5 signal is given by

$$x_{E5}(t) = [s_{E5a-d}(t) + is_{E5a-p}(t)]\psi_1^*(t) + [s_{E5b-d}(t) + is_{E5b-p}(t)]\psi_1(t) \\ + s_{E5b-d}(t)s_{E5b-p}(t)[s_{E5a-p}(t) + is_{E5a-d}(t)]\psi_2^*(t) \\ + s_{E5a-d}(t)s_{E5a-p}(t)[s_{E5b-p}(t) + is_{E5b-d}(t)]\psi_2(t) \quad (18.16)$$

The receiver designer can choose among four options for processing the E5 signal, or any AltBOC signal:

- Process entire E5 signal as a wideband 8-phase signal,
- Process both E5a and E5b, but separately,
- Process only the E5a component,
- Process only the E5b component.

Processing the entire E5 signal provides benefits of accessing all the signal power, as well as access to the widest bandwidth satnav signal with opportunities for interference mitigation and large RMS bandwidth for benefits in code tracking and multipath mitigation, as seen in Chapters 19 and 22. Yet there are also implementation considerations to evaluate before making this choice. The large fractional bandwidth, typically defined as the null-to-null bandwidth divided by the center frequency, implies the risk of significantly different Doppler effects over the signal bandwidth, potentially limiting integration times if there is expected to be significant satellite-to-receiver dynamics. Dispersive effects due to the ionosphere may be a limiting factor [1], and dispersion due to receiver front-end hardware must be evaluated, possibly requiring more expensive front-end hardware. The needed wide receiver bandwidth can expose the receiver to powerful interference, in-band and adjacent-band, for which the receiver front end must provide adequate dynamic range. The wide bandwidth receiver and associated high sampling rate both drive power consumption of the front end and the digital processing.

Initial synchronization of the entire wideband signal would be challenging, so presumably initial synchronization would be performed using sideband processing because of the narrow correlation function with wideband processing, and data messages for E5a and E5b would be demodulated separately, so only tracking is required of the wideband signal. If the data message bits can be demodulated with a low enough error rate, as discussed in Section 18.5, the replica (18.15) can be produced and correlated with the received waveform that included the E5 signal, noise, and interference. While such processing is straightforward to express mathematically, exquisite attention to implementation details is required for this very wideband signal. Typically, the signal would be filtered to a bandwidth no less than ±25 MHz around its center frequency of 1191.795 MHz. After processing in the receiver front end and ADC, the digitized frequency translated waveform would be frequency adjusted and correlated against the replica as described in Section 18.1.1.

Coherently processing each E5a and E5b signal separately, then combining measurements in a single receiver channel, is a compromise both in terms of benefits and in terms of implementation considerations, but still makes use of the entire signal. Processing only the E5a or E5b signal reduces implementation to that of a BPSK-R(10) signal as shown below, but neglects more than half the signal power, as well as the wider signal bandwidth.

Processing the E5a component from an input waveform that includes the E5 signal, noise, and interference proceeds as follows. If the E5 signal has already been frequency translated using a 1191.795 MHz center frequency, then, except for Doppler shift, $x_{E5}(t)$ in (18.16) is available. If no frequency translation has been performed, then the analytic signal is $x_{E5}(t)e^{if_0 t}$, where $f_0 = 1191.795$ MHz, not accounting for Doppler shifts, and downconverting by 1191.795 MHz yields $x_{E5}(t)$.

Multiplying $x_{E5}(t)$ by $\psi_1(t)$ yields

$$x_{E5}(t)\psi_1(t) = [s_{E5a-d}(t) + is_{E5a-p}(t)] + [s_{E5b-d}(t) + is_{E5b-p}(t)]\psi_1^2(t)$$
$$+ s_{E5b-d}(t)s_{E5b-p}(t)[s_{E5a-p}(t) + is_{E5a-d}(t)]\psi_2^*(t)\psi_1(t) \quad (18.17)$$
$$+ s_{E5a-d}(t)s_{E5a-p}(t)[s_{E5b-p}(t) + is_{E5b-d}(t)]\psi_2(t)\psi_1(t)$$

The first term is the desired baseband E5a signal, while the other terms are noise like, containing high frequencies relative to the spreading modulation. These noise-like terms do not correlate with either E5a component, and contribute negligible noise at the correlator output compared to thermal noise and interference. Recall that $\psi_1(t)$ is a discrete approximation to a complex sinusoid of frequency -15×1.023 MHz, so the processing to recover the baseband E5a signal from E5 at RF is equivalent to downconverting by 1150×1.023 MHz, and processing to recover the baseband E5a signal from E5 at baseband is equivalent to upconverting by 15×1.023 MHz. The result can then be further processed as described in the preceding subsections.

Processing the E5b component is analogous to what was described for the E5a signal, except $x_{E5}(t)$ is multiplied by $\psi_1^*(t)$, and the processing to recover the baseband E5b signal from E5 at RF is equivalent to downconverting by 1180×1.023 MHz, and processing to recover the baseband E5b signal from E5 at baseband is equivalent to downconverting by 15×1.023 MHz.

Figure 18.2 shows all of these options for processing the E5 signal in a single signal flow diagram, summarizing the above discussion.

18.1.3 Processing Signals Having MBOC Spectra

As described in Sections 3.1, 7.3, and 10.3, the GPS L1C signal and Galileo E1 OS signal use different spreading modulations that each have the MBOC spectrum. Each signal's spreading modulation is a different blend of BOC(1,1) and BOC(6,1) spreading modulations, requiring receiver designers to adopt different processing approaches while maintaining efficient implementation and good performance. This subsection outlines some of the considerations for processing these signals.

SIGNAL PROCESSING FOR CARRIER TRACKING

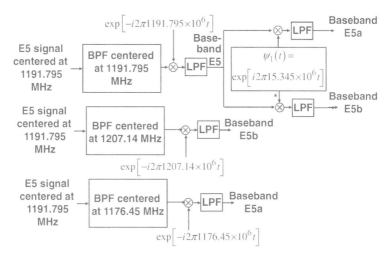

Figure 18.2. Options for Processing the Galileo E5 Signal

The L1C signal's pilot and data components use spreading codes designed to be orthogonal, and are transmitted on the same RF carrier using code division multiplexing. Section 18.1.2 describes processing to recover individual components for this case. The data component, with 25% of the total power, uses a BOC(1,1) spreading modulation, so its signal processing is as outlined in Section 18.1.1.

Processing the L1C pilot component, with 75% of the total power and its TMBOC spreading modulation, however, merits special consideration. Since the pilot component contains the preponderance of signal power, receivers are expected to perform initial synchronization and tracking using only the pilot component, since processing the data component and combining the resulting measurements with those from the pilot component is typically not worth the complexity.

While the BOC(6,1) symbols in the pilot component provide better code tracking performance, using them also complicates the receiver due to the need for wide precorrelation bandwidth and associated sampling rates. Thus, simpler receivers may not make use of the BOC(6,1) symbols. The change in pilot power from not using the BOC(6,1) symbols is approximately 0.4 dB for MBOC signals bandlimited to ± 15.345 MHz, so the difference in effective C/N_0 is insignificant. During initial synchronization, when a wide correlation function is desirable as discussed in Section 16.1, not processing the BOC(6,1) symbols is essential; typically, the input waveform would be bandlimited to allow a lower sampling rate, removing the higher frequency BOC(6,1) symbols in the process. Or, if sideband processing is used, the BOC(6,1) symbols cannot be used anyway. Also, narrower bandwidth receiver designs designed to track the input waveform with precorrelation bandwidths less than ± 7 MHz filter out the BOC(6,1) component.

Figure 18.3 shows examples of the waveforms. The upper trace shows a segment of an ideal TMBOC waveform. When the waveform is filtered with a narrow precorrelation bandwidth, as in the second trace, the BOC(6,1) symbols are eliminated. A full replica

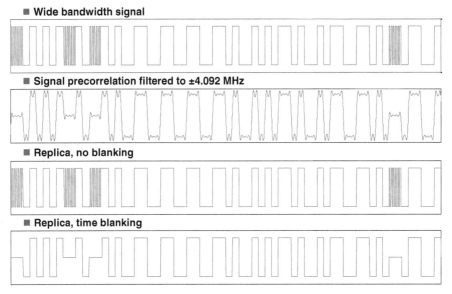

Figure 18.3. Examples of TMBOC Waveforms

of the TMBOC waveform, shown in the third trace, contains the entire waveform. The blanked replica in the fourth trace sets all BOC(6,1) symbols to zero and consequently would "ignore" BOC(6,1) symbols even if they are not filtered out.

If the signal is precorrelation filtered as in the second trace in Figure 18.3, but a full replica is employed as shown in the third trace in Figure 18.3, the correlation function remains that of a BOC(1,1) just as with a blanked replica. The difference, however, is that the BOC(6,1) symbols in the replica only correlate against noise and interference that remain at the times where the BOC(6,1) symbols had been filtered out, degrading the correlator output SNR by a factor of 29/33, or 0.56 dB,[2] since signal is present only 29 of every 33 symbols, but noise and interference are present in all 33 of every 33 symbols. Consequently, when the BOC(6,1) symbols in TMBOC are not used, blanking should be employed by setting all the BOC(6,1) symbols in the replica to zero. Signal processing then proceeds using this replica and the same approaches described earlier in this section.

For the Galileo E1 OS signal with the CBOC spreading modulation described in Sections 3.1 and 10.3, different considerations are involved. The pilot and data component each have 50% of the total signal power, making it more advantageous to process both components in acquisition and tracking. Further, since the BOC(6,1) spreading symbols are continuously applied rather than time multiplexed, if it is desired to process the CBOC signals as if they are BOC(1,1) signals, the receiver processing can use a BOC(1,1) replica rather than a CBOC replica. Some minor distortions may occur in the resulting correlation measurements unless the received signal is bandlimited

---

[2] This value is smaller when computed for bandlimited signal, making the output SNR degradation from not using the BOC(6,1) less than 0.4 dB for signals bandlimited to ± 15.345 MHz.

SIGNAL PROCESSING FOR CARRIER TRACKING

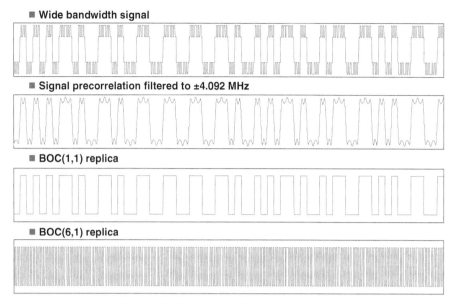

Figure 18.4. Examples of CBOC+ Waveforms

to remove its BOC(6,1) components, since if the BOC(6,1) component in the received signal is not removed by low-pass filtering, then crosscorrelating the received CBOC signal against a BOC(1,1) replica produces a somewhat asymmetric correlation function. Many techniques are being developed for processing signals having CBOC spreading modulations; some of these are described in Reference 2.

Figure 18.4 shows CBOC waveforms analogous to those for TMBOC in Figure 18.3. While CBOC+ is shown, the results are analogous for CBOC−. Precorrelation filtering with bandwidth less than ±6 MHz removes the BOC(6,1) part of the waveform, leaving BOC(1,1) which is readily processed using a BOC(1,1) replica, producing a symmetric correlation function.

As discussed in Chapter 16, there can be significant simplifications when correlation processing is implemented in hardware, if the correlator uses a one-bit replica. When a CBOC replica is used in correlation processing, however, the four-level waveform cannot be represented using one bit. Using dual one-bit correlators, however, is an attractive option as shown in Figure 18.5. The single correlator shown in the upper portion employs a one-bit BOC(1,1) replica, and is all that would be used if the CBOC signal were processed as having a BOC(1,1) spreading modulation. However, if the full CBOC spreading modulation is to be exploited using one-bit correlators, the lower correlator is added. It employs a one-bit replica consisting of pure BOC(6,1), with the same spreading code used for the BOC(1,1) replica. The two correlator outputs are weighed and then either added (for CBOC+) or subtracted (for CBOC−). The weighing typically reflects the relative amplitude or power between BOC(1,1) and BOC(6,1) used in producing the CBOC waveform. This implementation preserves the use of one-bit correlators for CBOC, although two correlators are needed instead of one for TMBOC or BOC(1,1).

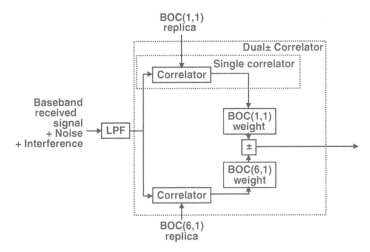

Figure 18.5. Dual One-Bit Correlators for Signal Processing of Signals with CBOC Spreading Modulation

18.2 FREQUENCY-LOCKED LOOPS

The first carrier tracking loop considered is one that tracks the carrier frequency, the FLL. It is the only loop considered in this book where the discriminator output depends on measurements taken at two different times; all other discriminators addressed in this book use measurements only computed at the current time.

The measurements used for the FLL at time jT are $Z_{\text{PROMPT}}[j]$ and $Z_{\text{PROMPT}}[j-1]$, using (18.6), (18.8), or (18.10) as appropriate. Operation of an FLL is degraded when there is a transition of a data message bit or overlay code bit between these two measurements. If the receiver channel is not synchronized to the data bit or overlay code bit (typically this happens only for GPS C/A code or GLONASS L1OF or L2OF, where initial synchronization does not provide bit synchronization), then typically a value of T (the correlation integration time and loop update time) smaller than the bit duration is used, and the occasional bit transitions merely add noise to the loop. Once bit synchronization is achieved, either the data bit value can be estimated and wiped, or T is set to be smaller than half the bit duration, and the loop is updated only when adjacent measurements occur during the same bit value.

When there is no bit transition between adjacent measurements, the FLL discriminator generates the error

$$\tilde{F}[j] = \frac{\arctan2[\Im\{Z_{\text{PROMPT}}[j]Z^*_{\text{PROMPT}}[j-1]\}, \Re\{Z_{\text{PROMPT}}[j]Z^*_{\text{PROMPT}}[j-1]\}]}{2\pi T}$$

(18.18)

where arctan2(y, x) is the four-quadrant arctangent, defined to have the range $(-\pi, \pi]$, producing a unique output for all angles over the unit circle. The function arctan 2(y, x) can be defined[3] in terms of the two-quadrant arctangent as

$$\arctan 2(y, x) \stackrel{\Delta}{=} \begin{cases} \arctan(y/x), & x > 0 \\ \arctan(y/x) + \pi, & y \geq 0, x < 0 \\ \arctan(y/x) - \pi, & y < 0, x < 0 \\ \pi/2, & y > 0, x = 0 \\ -\pi/2, & y < 0, x = 0 \\ \text{undefined}, & y = 0, x = 0 \end{cases} \quad (18.19)$$

where arctan (y/x) is the two-quadrant arctangent defined to have the range $(-\pi/2, \pi/2]$.

To better understand (18.18), express the prompt measurement in terms of its magnitude and phase, $Z_{\text{PROMPT}}[j] = |Z_{\text{PROMPT}}[j]| e^{i\varphi[j]}$. From (18.19) the magnitude does not affect the four-quadrant arctangent, so with no loss of generality,

$$\begin{aligned} \tilde{F}[j] &= \frac{\arctan 2 \left[\sin (\varphi[j] - \varphi[j-1]), \cos (\varphi[j] - \varphi[j-1]) \right]}{2\pi T} \\ &= \frac{\text{PV}_{\arctan 2} \{\varphi[j] - \varphi[j-1]\}}{2\pi T} \end{aligned} \quad (18.20)$$

where $\text{PV}_{\arctan 2} \{\bullet\}$ denotes the principal value for the arctan2 function.

As long as there is no transition of a data message bit or a pilot code bit between the two measurements, this result is merely the discrete-time approximation to (18.3), the definition of residual frequency. Hence, the parameter estimated by the FLL is the residual frequency. Observe that (18.20) is not sensitive to a fixed carrier phase applied to the received signal, as expected from a derivative, showing that it is not a carrier phase estimator, but only a carrier frequency estimator.

When there is no noise or interference, so that the input consists only of the desired signal, if $\tilde{F}[j]$ is the residual carrier frequency after downconversion and frequency adjustment, then $\varphi[j] = \varphi[j-1] + 2\pi T \tilde{F}[j]$, and (18.20) yields $\tilde{F}[j] = \frac{\text{PV}_{\arctan 2}\{2\pi T \tilde{F}[j]\}}{2\pi T}$. The FLL S-curve is then

$$S_{\text{PLL}}(\tilde{F}[j]) = \frac{\text{PV}_{\arctan 2}\{2\pi T \tilde{F}[j]\}}{2\pi T} \quad (18.21)$$

---

[3] One might think that the four-quadrant arctangent should be denoted arctan 4(y, x) and the two-quadrant arctangent denoted arctan 2(y/x), but the two-quadrant arctangent has been labeled arctan(y/x) or atan(y/x) in computer programming for decades, and the four-quadrant arctangent has been denoted arctan 2(y, x) or atan2(y, x) for almost as long. Changing this notation for this book would be confusing at the very least. Actually, no distinction in the function name is needed since the presence of two arguments versus one argument is sufficient, but the different function names are used as well.

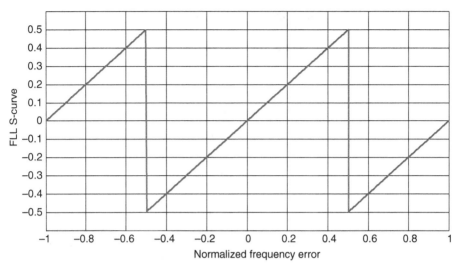

Figure 18.6. FLL S-Curve

Figure 18.6 shows this FLL S-curve, in terms of the normalized frequency error $T\tilde{F}[j]$. This S-curve has the desirable properties introduced in Section 17.1: it passes through the origin with a unity slope, and is linear for small errors around the origin. Also, its slope does not change with received signal power, meaning the loop bandwidth and other loop design parameters do not vary with SNR.

As long as $|\tilde{F}[j]| < \frac{1}{2T}$, the loop continues to track the correct frequency and to correct errors. But if the residual carrier frequency's magnitude is larger than $\frac{1}{2T}$, the principal value computation in (18.21) produces an incorrect estimate of the residual carrier frequency, and the loop may lose lock or produce grossly incorrect estimates. Such an event can occur if noise and interference or receiver oscillator phase noise corrupt the signal processing measurements excessively, or if the frequency value changes too much between loop updates due to dynamics. If the errors are dominated by dynamics or oscillator phase noise, reducing the value of T extends the linear portion of the S-curve. However, if the errors are dominated by noise and interference, reducing T also increases the jitter, as shown below, and actually can degrade loss of lock performance. If the loop bandwidth can be reduced without unacceptably degrading loop response, ability to accommodate dynamics in the loop, and receiver oscillator phase noise effects, it can mitigate the effects of noise and interference.

In early years of satnav receiver development, implementing four-quadrant arctangent calculations was expensive in terms of computation and memory. Approximations to (18.21) were developed and are still discussed in many texts. Review Question QT18.1 examines two of these, and demonstrates their deficiencies relative to the FLL S-curve based on the four-quadrant arctangent. In modern receivers, there is rarely any reason to use any FLL discriminator other than one based on the four-quadrant arctangent. If the calculations in (18.20) are too complicated to implement through direct evaluation of the four-quadrant arctangent, they can be implemented using table lookup.

FREQUENCY-LOCKED LOOPS

High fidelity characterization of FLL performance requires measurements using computer simulations or hardware implementations. However, when errors are small relative to the linear region of the FLL S-curve, the mean-squared FLL error can be expressed by

$$\sigma^2_{FLL} = \sigma^2_{FLL,noise+interference} + \sigma^2_{FLL,phase\,noise} + (\Delta F)^2 \quad (Hz^2) \quad (18.22)$$

where $\sigma^2_{FLL,noise+interference}$ is the variance due to noise and interference, $\sigma^2_{FLL,phase\,noise}$ is the variance due to phase noise in the receiver oscillator, and ΔF is the frequency change during a loop update expected due to receiver dynamics along the line of sight relative to the satellite transmitting the signal. Typically, phase noise has little effect on FLL performance, and is often neglected.

The variance of the FLL jitter due to noise and interference is approximately [3]

$$\sigma^2_{FLL,noise+interference} = \frac{\alpha B_L}{(\pi T)^2 \zeta \left(\frac{C}{N_0}\right)_{eff}} \left[1 + \frac{1}{T\zeta \left(\frac{C}{N_0}\right)_{eff}}\right] \quad (Hz^2) \quad (18.23)$$

where $\left(\frac{C}{N_0}\right)_{eff}$ is the effective carrier power to noise PSD ratio for the entire signal, ζ is the fraction of total signal power in the component being tracked, $\alpha = 1$ when errors are very small relative to the linear region of the FLL S-curve, and $\alpha = 2$ when errors are not very small. Typically, it is prudent to be conservative and use $\alpha = 2$. The term in square brackets in (18.23) is called the squaring loss, and directly involves the correlator output SNR as defined in Sections 5.3 and 5.4. The FLL variance due to noise and interference increases with decreasing effective C/N_0, with approximately inverse linear dependence until the squaring loss becomes large. This variance is also proportional to the loop bandwidth, and inversely proportional to the square of the correlation integration time in the signal processing.

Loop pull-in requires an initial frequency estimate on the linear portion of the S-curve that intersects the origin, and preferably near the center. When FLL tracking begins after initial synchronization or a restart, the value of T may be reduced in order to widen the pull-in region and speed loop updates, both of which improve pull-in and convergence. A good rule of thumb would be to set T initially to same value or less than the crossambiguity integration time used for initial synchronization. Once stable frequency tracking has been achieved, the value of T may be increased, but typically no larger than half the duration of the data message bit if the data component is being tracked, and no larger than half the pilot code bit duration if pilot code synchronization has not yet been achieved.

Loop loss of lock can be characterized very approximately using the rule of thumb introduced in Section 17.3.4. Applied to FLL tracking, this rule of thumb for the loop to remain locked is

$$3\sigma_{FLL} < \frac{1}{4T} \quad (18.24)$$

If FLL variance is dominated by noise and interference, the loop-lock criterion (18.24) becomes approximately

$$\frac{30 B_L}{\zeta \left(\frac{C}{N_0}\right)_{\text{eff}}} \left[1 + \frac{1}{T\zeta \left(\frac{C}{N_0}\right)_{\text{eff}}}\right] < 1 \qquad (18.25)$$

This result says that, when noise and interference dominate FLL loss of lock, reducing loop bandwidth has great benefit, but changing T makes little difference until the correlator output SNR is small. As long as the correlator output SNR remains more than 8 dB, any change in T affects the linear region of the S-curve proportional to the change in standard deviation of the FLL jitter, having little effect on loss of lock performance.

Figure 18.7 plots (18.23) for different parameter values using $\alpha = 2$, with vertical lines at the point where the loop-lock criterion (18.25) is no longer satisfied. The effect of different correlation integration times and loop bandwidths is clear, with smaller jitter and lower loss of lock thresholds when the loop bandwidth is smaller and correlation integration time is longer. Unlike most other places in this book where the effective C/N_0 is that for the entire signal, here the effective C/N_0 is for the component whose frequency is being tracked. Observe that, while T decreases by a factor of 10, the loop-lock point decreases by less than 4 dB—hardly more than a factor of two. In many applications, loss of lock is more important than frequency tracking accuracy, since as long as the FLL remains locked, the receiver channel continues to operate properly with small errors due to noise and interference.

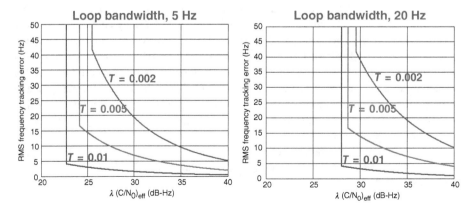

Figure 18.7. FLL Tracking Error Caused by Noise and Interference

18.3 COSTAS LOOPS

A Costas loop tracks the carrier phase of a signal that is biphase modulated by an unknown bit stream. Costas loops are used for carrier tracking of pilot components before synchronization to the pilot code, and for carrier tracking of data components. Costas loops are sometimes called PLLs, but in this book the term PLL is reserved to the coherent PLLs described in Section 18.4.

The measurement used for the Costas loop at time jT is $Z_{PROMPT}[j]$, using (18.6) or (18.8) as appropriate. Carrier phase tracking of a pilot component after pilot code synchronization should not use a Costas loop, but rather a PLL as described in Section 18.4, since there is no unknown biphase modulation and a PLL outperforms a Costas loop.

The Costas loop discriminator generates the error

$$\tilde{\theta}[j] = \arctan\left[\frac{\Im\{Z_{PROMPT}[j]\}}{\Re\{Z_{PROMPT}[j]\}}\right] \tag{18.26}$$

To better understand (18.26), express the prompt measurement in terms of its magnitude and phase, $Z_{PROMPT}[j] = |Z_{PROMPT}[j]|e^{i\varphi[j]}$. From (18.26) the magnitude does not affect the two-quadrant arctangent, so with no loss of generality,

$$\tilde{\theta}[j] = \arctan\left[\frac{\sin(\varphi[j])}{\cos(\varphi[j])}\right] = PV_{\arctan}\{\varphi[j]\} \tag{18.27}$$

where $PV_{\arctan}\{\bullet\}$ denotes the principal value for the two-quadrant arctangent function. Observe that $PV_{\arctan}\{\varphi[j]\} = PV_{\arctan}\{\varphi[j] + \pi\}$, confirming that biphase modulation of the signal or component does not affect the phase estimate.

When there is no noise or interference, so the input consists only of the desired signal, if $\tilde{\theta}[j]$ is the residual carrier phase after removal of the current carrier phase estimate, $\varphi[j] = \tilde{\theta}[j]$ or $\varphi[j] = \tilde{\theta}[j] + \pi$, and (18.27) yields $\tilde{\theta}[j] = PV_{\arctan}\{\tilde{\theta}[j]\}$. The Costas loop S-curve is then

$$S_{Costas}(\tilde{\theta}[j]) = PV_{\arctan}\{\tilde{\theta}[j]\} \tag{18.28}$$

Figure 18.8 shows this Costas loop S-curve, in terms of the carrier phase error $\tilde{\theta}[j]$. This S-curve has the desirable properties introduced in Section 17.1: it passes through the origin with a unity slope, and is linear for small errors around the origin. Also, its slope does not change with received signal power, meaning the loop bandwidth and other loop design parameters do not vary with SNR.

As long as $|\tilde{\theta}[j]| < \pi/2$, the loop continues to track the correct carrier phase and to correct errors. But if the carrier phase error's magnitude is larger than $\pi/2$, the principal value computation in (18.28) produces an incorrect estimate of the carrier phase, and the loop may lose lock or produce grossly incorrect estimates. Such an event can occur if noise and interference or receiver oscillator phase noise corrupt the signal processing measurements excessively, or if the carrier phase changes too much between loop updates due to dynamics. If the errors are dominated by dynamics, reducing T

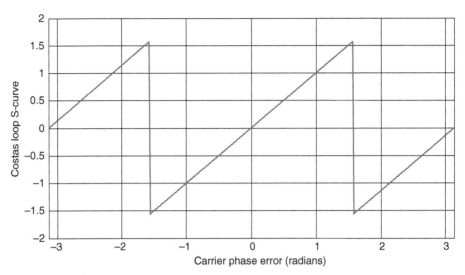

Figure 18.8. Costas Loop S-Curve

increases the update rate, thus reducing the carrier phase changes between updates. If the errors are dominated by receiver oscillator phase noise, increasing the loop bandwidth, as discussed in Section 17.3, reduces the effect while increasing the effect of noise and interference. If the loop bandwidth can be reduced without unacceptably degrading loop response and the ability to accommodate dynamics in the loop, doing so will mitigate effects of noise and interference.

In early years of satnav receiver development, when the costs of implementing two-quadrant arctangent calculations were expensive in terms of computation or memory, approximations to (18.28) were developed and are still discussed in many texts. Review Question QT18.2 examines two of these, and demonstrates their deficiencies relative to the Costas loop S-curve based on the two-quadrant arctangent. In modern receivers, there is rarely any reason to use any Costas loop discriminator other than one based on the two-quadrant arctangent. If the calculations in (18.20) are too complicated to implement through direct implementation of the two-quadrant arctangent, they can be implemented using table lookup.

High fidelity characterization of Costas loop performance requires measurements using computer simulations or hardware implementations. However, when errors are small relative to the linear region of the Costas loop S-curve, the mean-squared Costas loop error can be expressed by

$$\sigma^2_{Costas} = \sigma^2_{Costas,noise + interference} + \sigma^2_{Costas,phase\ noise} + (\Delta\theta)^2 \quad (radian^2) \quad (18.29)$$

where $\sigma^2_{Costas,noise+interference}$ is the variance due to noise and interference, $\sigma^2_{Costas,phase\ noise}$ is the variance due to phase noise in the receiver oscillator, and $\Delta\theta$ is the change in carrier phase during a loop update due to receiver dynamics along the line of sight relative to the satellite transmitting the signal.

The variance of the Costas loop jitter due to noise and interference is approximately [3]

$$\sigma^2_{\text{Costas,noise + interference}} = \frac{B_L}{\zeta \left(\frac{C}{N_0}\right)_{\text{eff}}} \left[1 + \frac{1}{2T\zeta \left(\frac{C}{N_0}\right)_{\text{eff}}}\right] \quad (18.30)$$

$$= \frac{1}{\rho_L}\left[1 + \frac{1}{\rho}\right] \quad (\text{radian}^2)$$

where

$$\rho_L \triangleq \zeta \left(\frac{C}{N_0}\right)_{\text{eff}} \frac{1}{B_L} \quad (18.31)$$

is the loop SNR, and $\rho = 2T\zeta(C/N_0)_{\text{eff}}$ is the correlator output SNR as defined in Sections 5.3 and 5.4, with ζ the fraction of total signal power in the component being tracked.

The term in square brackets in (18.30) is called the squaring loss, and directly involves the correlator output SNR. The Costas loop variance due to noise and interference increases with decreasing effective C/N_0, with approximately inverse linear dependence until the squaring loss becomes large. This variance is also proportional to the loop bandwidth, and dependent on the correlation integration time in the signal processing only for the squaring loss. When the correlator output SNR is large enough that squaring loss is approximately unity, the approximation $\sigma^2_{\text{Costas,noise+interference}} \cong 1/\rho_L$ (radian^2) can be employed.

Loop pull-in requires an initial estimate of carrier phase on the linear portion of the S-curve that intersects the origin, and preferably near the center. Loop loss of lock can be characterized very approximately using the rule of thumb introduced in Section 17.3.4. Applied to Costas loop tracking this rule of thumb for the loop to stay in lock is

$$3\sigma_{\text{Costas}} < \frac{\pi}{4} \quad (18.32)$$

This says that the RMS carrier phase tracking error should remain less than $\pi/12$ radians, or 15°, to ensure that the loop remains locked.

If Costas loop variance is dominated by noise and interference, the loop-lock criterion (18.32) becomes approximately

$$\rho_L > 14.6 \left[1 + \frac{1}{\rho}\right] \quad (18.33)$$

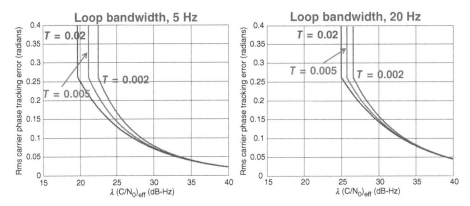

Figure 18.9. Costas Loop Tracking Error Caused by Noise and Interference

This result says that, when noise and interference dominate Costas loop loss of lock, reducing loop bandwidth has great benefit, but changing T makes very little difference until the correlator output SNR is small. As long as the correlator output SNR remains more than approximately 3 dB, any change in T affects the linear region of the S-curve proportional to the change in standard deviation of the Costas loop jitter, having little effect on loss of lock performance.

An alternative, simpler, and more conservative criterion for retaining lock in a Costas loop is that the loop SNR remains greater than 16 dB [Dr. C. Cahn, personal communication, 1999]. Expressed like (18.33), this criterion is $\rho_L > 40$, which is more than a factor of two more conservative than (18.33). This difference between loss of lock criteria is explained by the fact that loss of lock is a random event, and different criteria may be based upon different probabilities of loss of lock as well as durations of time over which lock must be maintained.

Figure 18.9 plots (18.30) for different parameter values, with vertical lines at the point where the loop-lock criterion (18.33) is no longer satisfied. The effect of different correlation integration times and loop bandwidths is clear, with smaller jitter and lower loss of lock thresholds when the loop bandwidth is smaller and correlation integration time is longer. For Costas loops, the correlation integration time only affects the squaring loss, so there is little performance difference with modest changes in value of T as long as the correlator output SNR is large enough. Unlike most other places in this book where the effective C/N_0 is that for the entire signal, here the effective C/N_0 is for the component whose phase is being tracked. Often, loss of lock is more important than carrier phase tracking accuracy, since as long as the Costas loop remains locked, the errors are small.

18.4 PHASE-LOCKED LOOPS

PLLs track the carrier phase of a signal with no unknown phase modulation. PLLs are used for carrier tracking of pilot components after synchronization to the pilot code, and for carrier tracking if data message bits are known so that their modulation of the signal can be removed, or wiped. PLLs are different from Costas loops, in that PLLs cannot accommodate unknown biphase modulation as Costas loops can. However, PLL coherent processing yields better performance than Costas loop processing.

The measurement used for the PLL at time jT is $Z_{\text{PROMPT}}[j]$, using (18.10). The PLL discriminator generates the error estimate

$$\tilde{\theta}[j] = \arctan 2[\Im\{Z_{\text{PROMPT}}[j]\}, \Re\{Z_{\text{PROMPT}}[j]\}] \tag{18.34}$$

To better understand (18.34), express the prompt measurement in terms of its magnitude and phase, $Z_{\text{PROMPT}}[j] = |Z_{\text{PROMPT}}[j]|e^{i\varphi[j]}$. From (18.19) the magnitude does not affect the four-quadrant arctangent, so with no loss of generality,

$$\tilde{\theta}[j] = \arctan 2\left[\sin(\varphi[j]), \cos(\varphi[j])\right] = PV_{\arctan 2}\{\varphi[j]\} \tag{18.35}$$

where $PV_{\arctan 2}\{\bullet\}$ denotes the principal value for the four-quadrant arctangent function. Observe that $PV_{\arctan 2}\{\varphi[j]\} \neq PV_{\arctan 2}\{\varphi[j] + \pi\}$, so unknown biphase modulation of the signal or component would affect the phase estimate.

When there is no noise or interference so that the input consists only of the desired signal, if $\tilde{\theta}[j]$ is the residual carrier phase after removal of the current carrier phase estimate, $\varphi[j] = \tilde{\theta}[j]$, and (18.35) yields $\tilde{\theta}[j] = PV_{\arctan 2}\{\tilde{\theta}[j]\}$. The PLL S-curve is then

$$S_{PLL}(\tilde{\theta}[j]) = PV_{\arctan 2}\{\tilde{\theta}[j]\} \tag{18.36}$$

Figure 18.10 shows this PLL S-curve, in terms of the carrier phase error $\tilde{\theta}[j]$. This S-curve has the desirable properties introduced in Section 17.1: it passes through the

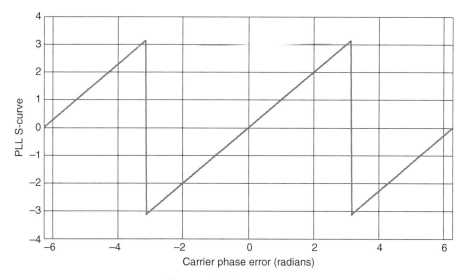

Figure 18.10. PLL S-Curve

origin with a unity slope, and is linear for small errors around the origin. Also, its slope does not change with received signal power, meaning the loop bandwidth and other loop design parameters do not vary with SNR.

As long as $|\tilde{\theta}[j]| < \pi$, the loop continues to track the correct carrier phase and to smooth jitter. But if the carrier phase error's magnitude is larger than π, the principal value computation in (18.35) produces an incorrect estimate of the carrier phase, and the loop may lose lock or produce grossly incorrect estimates. Such an event can occur if noise and interference or receiver oscillator phase noise corrupt the signal processing measurements excessively, or if the carrier phase changes too much between loop updates due to dynamics. If the errors are dominated by dynamics, reducing T increases the update rate and thus reduces the carrier phase changes between updates. If the errors are dominated by receiver oscillator phase noise, increasing the loop bandwidth, as discussed in Section 17.3, reduces the effect while increasing the effect of noise and interference. If the loop bandwidth can be reduced without degrading loop response and the ability to accommodate dynamics in the loop, doing so will mitigate effects of noise and interference.

In early years of satnav receiver development, when the costs of implementing four-quadrant arctangent calculations were expensive in terms of computation or memory, approximations to (18.34) were developed and are still discussed in many texts. Review Question QT18.3 examines two of these, and demonstrates their deficiencies relative to the PLL S-curve based on the four-quadrant arctangent. In modern receivers, there is rarely any reason to use any PLL discriminator other than one based on the four-quadrant arctangent. If the calculations in (18.34) are too complicated to implement through direct implementation of the four-quadrant arctangent, they can be implemented using table lookup.

High fidelity characterization of PLL performance requires measurements using computer simulations or hardware implementations. However, when errors are small relative to the linear region of the PLL S-curve, the mean-squared PLL error can be expressed by

$$\sigma_{\text{PLL}}^2 \cong \sigma_{\text{PLL,noise+interference}}^2 + \sigma_{\text{PLL,phase noise}}^2 + (\Delta\theta)^2 \quad (\text{radian}^2) \quad (18.37)$$

where $\sigma_{\text{PLL,noise+interference}}^2$ is the variance due to noise and interference, $\sigma_{\text{PLL,phase noise}}^2$ is the variance due to phase noise in the receiver oscillator, and $\Delta\theta$ is the change in carrier phase during a loop update due to receiver dynamics along the line of sight relative to the satellite transmitting the signal.

The variance of the PLL loop jitter due to noise and interference is approximately [3]

$$\sigma_{\text{PLL,noise + interference}}^2 \cong \frac{B_L}{\zeta \left(\dfrac{C}{N_0}\right)_{\text{eff}}} \quad (18.38)$$

$$= \frac{1}{\rho_L} \quad (\text{radian}^2)$$

where ρ_L is the loop SNR defined in (18.31), and ζ is the fraction of total signal power in the component being tracked.

Unlike the analogous expressions in the previous sections, (18.38) contains no squaring loss term. This is because the PLL involves coherent processing over multiple loop updates, and it makes no difference whether the coherent processing is performed in the signal processing or in the loop updating. Like the FLL and Costas loop, the PLL variance is also proportional to the loop bandwidth.

Loop pull-in requires an initial estimate of carrier phase on the linear portion of the S-curve that intersects the origin, and preferably near the center. Observe the PLL's S-curve linear region has twice the extent of the Costas loop's S-curve linear region, providing advantages in loop pull-in and loss of lock.

Loop loss of lock can be characterized very approximately using the rule of thumb introduced in Section 17.3.4. Applied to PLL tracking, this rule of thumb for the loop to stay in lock is

$$3\sigma_{\text{PLL}} < \frac{\pi}{2} \qquad (18.39)$$

This says that the RMS carrier phase tracking error should remain less than $\pi/6$ radians, or 30°, to ensure that the loop remains locked.

If PLL variance is dominated by noise and interference, substituting (18.38) into the loop-lock criterion (18.39) yields

$$\rho_L > 3.6 \qquad (18.40)$$

This result says that, when noise and interference dominate PLL loss of lock, reducing loop bandwidth has great benefit, but changing T makes no difference within the validity of these approximations and rules of thumb.

An alternative, simpler, and more conservative criterion for retaining PLL lock is that the loop SNR remains greater than 10 dB [4]. Expressed like (18.40), this criterion is $\rho_L > 10$, almost a factor of three more conservative than (18.40). The differences between these criteria involves different criteria for loss of lock, including the probability of loss of lock as well as the duration of time over which lock must be maintained. Observe that both criteria consistently indicate the PLL loop-lock threshold is approximately 6 dB less than the Costas loop loop-lock threshold.

Figure 18.11 plots (18.38) for different parameter values, with vertical lines at the point where the loop-lock criterion (18.40) is no longer satisfied. The effect of different loop bandwidths is clear, with smaller jitter and lower loss of lock thresholds when the loop bandwidth is smaller. In contrast to analogous results in previous sections, however, the results are independent of correlation integration time T. Often, loss of lock is more important than carrier phase tracking accuracy, since as long as the PLL remains locked, errors are small.

18.5 DATA MESSAGE DEMODULATION

Design, implementation, and performance of data message demodulation involve techniques from digital communication, and are addressed more completely in References

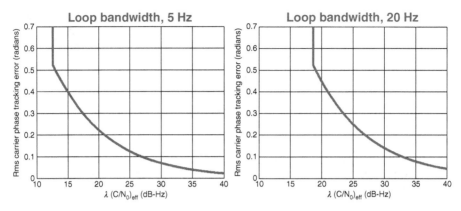

Figure 18.11. PLL Tracking Error Caused by Noise and Interference

4 and 5. For completeness, however, this book outlines demodulation techniques and provides results for data message demodulation performance, when the data message bits are biphase modulated onto the data component or the entire signal.

Data message demodulation is closely related to carrier tracking, since carrier tracking is a prerequisite for data demodulation, and also since data demodulation uses prompt correlator outputs, just as does carrier tracking.

Figure 18.12 shows the processing in a receiver channel when the receiver is already tracking the data component. In this case, the prompt correlator output (18.6) provides the measurements needed to read data symbol values. In demodulating the data message, the quantities $Z_{PROMPT}[j]$ are known as soft decisions, and are not really decisions at all since they are raw measurements undergoing little quantization. In contrast, hard decisions are one-bit quantized estimates of the binary value of the data message symbol value. In either case, if there is forward error coding applied to the data message bits, decoding uses either the soft decisions or the hard decisions. Modern satnav receivers

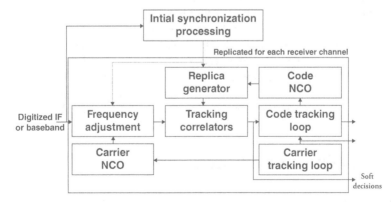

Figure 18.12. Extraction of Soft Decisions When Data Component Is Being Tracked

DATA MESSAGE DEMODULATION

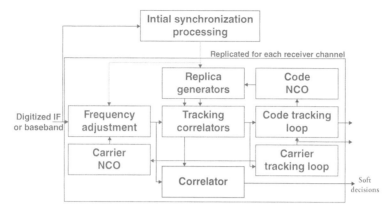

Figure 18.13. Extraction of Soft Decisions When Pilot Component Is Being Tracked, with Differences from Figure 18.12 Noted in Red

typically use soft decisions because of the improved performance, since the additional implementation complexity associated with processing soft decisions is modest with low satnav data rates and modern processing.

When only the pilot component is being tracked, obtaining soft or hard decisions from the data component involves adding a replica generator to produce the overlaid data spreading waveform. The overlaid data spreading waveform is then wiped from the data component, to enable data symbol demodulation. An additional correlator is needed to produce $Z_{PROMPT}[j]$ from the data component. Figure 18.13 shows the corresponding modifications to a receiver channel for data demodulation when only the pilot component is being tracked. Note that in this approach the data component is not being tracked, but only used for data message demodulation.

Three different types of data message demodulation are addressed in this section. Section 18.5.1 describes demodulation performance of biphase keyed bits with no forward error control, while Section 18.5.2 describes demodulation performance when data symbols are encoded using half-rate convolutional coding with constraint length 7 are biphase modulated onto the data component. Section 18.5.3 describes demodulation performance when data symbols encoded using half-rate LDPC coding are biphase modulated onto the data component, for the particular coding parameters used in the GPS L1C signal, and also describes the special techniques for data symbol combining available when reading the L1C data message.

Performance results provided here are all based on assumptions of an AWGN channel. Some results for urban channels are available in Reference 6.

18.5.1 Coherent Demodulation of Uncoded Binary Phase Shift Key Symbols

The original GPS and GLONASS signals essentially employ no forward error control encoding. If carrier phase tracking is being used with either a Costas loop or a PLL, a reference carrier phase $\hat{\theta}_{\text{ref}}[j]$ is established using the current carrier phase estimate. If the data component is being tracked or if the pilot component is being tracked but the pilot and data component share the same carrier phase, the reference carrier phase is the current output from the carrier phase tracking loop. If the pilot is being tracked and the data component is in phase quadrature to the pilot component, the reference carrier phase is adjusted by this quadrature offset.

The carrier phase of the data component with overlaid spreading waveform wiped is estimated from the prompt tap of the data component: $\phi[j] = \arctan 2[\Im\{Z_{\text{PROMPT}}[j]\}, \Re\{Z_{\text{PROMPT}}[j]\}]$. A hard decision for the data bit value $\hat{b}[j]$ is formed by

$$\hat{b}[j] = \begin{cases} 0, & |\hat{\theta}_{\text{ref}}[j] - \phi[j]| \leq \pi/2 \\ 1, & \text{otherwise} \end{cases} \quad (18.41)$$

deciding whether the estimated phase of the data component is closer to reference phase or closer to the reference phase rotated by π. Equivalently, the hard decision is

$$\hat{b}[j] = \begin{cases} 0, & \Im\{e^{-i\hat{\theta}_{\text{ref}}[j]} Z_{\text{PROMPT}}[j]\} < 0 \\ 1, & \text{otherwise} \end{cases} \quad (18.42)$$

This approach is used for reading data message bits from the GPS or QZSS C/A code signal, and for the GLONASS L1OF and L2OF signals.

Assessing data demodulation performance in real-world situations involving burst noise, multipath, and fading requires extensive computer simulations. The performance results provided here are developed analytically and apply to the ideal AWGN channel characterized by a time-invariant effective C/N_0.

For uncoded data bits modulated onto the signal using coherent biphase-shift keying (BPSK), the probability of bit error is given by [5]

$$P_{b,\text{BPSK}} = \frac{1}{2}\text{erfc}\left[\sqrt{\left(\frac{E_b}{N_0}\right)_{\text{eff}}}\right] \quad (18.43)$$

where

$$\left(\frac{E_b}{N_0}\right)_{\text{eff}} = \zeta T_b \left(\frac{C}{N_0}\right)_{\text{eff}} \quad (18.44)$$

with ζ the fraction of total signal power used in the data component, T_b the duration (in seconds) of the uncoded bit in the data message, $\left(\frac{C}{N_0}\right)_{\text{eff}}$ (in units of hertz) the effective total signal power to noise PSD ratio at the correlator output in the receiver, and $\text{erfc}(x) \triangleq \frac{2}{\sqrt{\pi}} \int_x^\infty e^{-t^2} dt$ is the complementary error function. Figure 18.14 shows the

DATA MESSAGE DEMODULATION

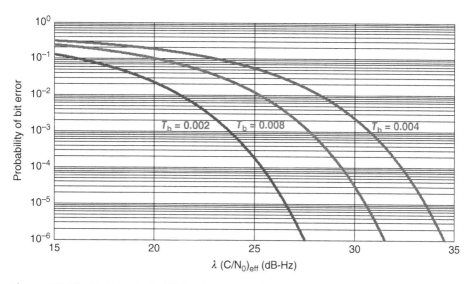

Figure 18.14. Bit Error Probabilities for Uncoded Binary Phase-Shift Keying in an AWGN Channel

results for some representative bit rates, including the 50 bits per second used by C/A code, L1OF, and L2OF. All the performance expressions provided here are based on perfect carrier tracking; near the loop-lock threshold, a more conservative calculation would reduce the $\left(\frac{C}{N_0}\right)_{\text{eff}}$ by perhaps 1 dB to account for carrier tracking imperfections.

When the carrier is tracked with an FLL, absolute phase measurements cannot be made. As an alternative, differential demodulation can be performed by comparing the carrier phase of sequential data message symbols and noting a change any time the magnitude of the phase change $|\phi[j] - \phi[j-1]|$, where $\phi[j] = \arctan 2[\Im\{Z_{\text{PROMPT}}[j]\}, \Re\{Z_{\text{PROMPT}}[j]\}]$, is greater than $\pi/2$. Since no phase reference is available, the receiver cannot determine whether the phase changes from 0 to 1 or from 1 to 0, but a record of the phase changes can be established, leading to two possible bit sequences that are complements of each other. The correct bit sequence can be established by comparing with message preambles or a synchronization pattern, or by performing CRCs. The bit error performance of differential demodulation with no error control coding is given by [5]

$$P_{b,\text{diff demod}} = 0.5 \exp\left[-\left(\frac{E_b}{N_0}\right)_{\text{eff}}\right] \qquad (18.45)$$

Figure 18.15 shows the performance for differential demodulation with no error control coding. In an AWGN channel, compared to coherent demodulation, approximately 2 dB larger effective C/N_0 is needed for the same probability of bit error using differential demodulation.

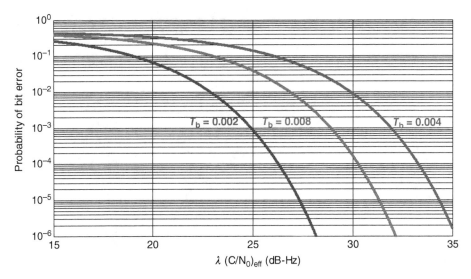

Figure 18.15. Bit Error Probabilities for Uncoded Binary Phase-Shift Keying with Differential Demodulation in an AWGN Channel

When the probability of bit error, P_b is small (e.g., less than 0.1) and bit errors are independent, the probability of an error in a message that is N bits long is, as shown in Figure 18.16,

$$P_{\text{message}} \cong 1 - (1 - P_b)^N \qquad (18.46)$$

This approximation is best for coherent demodulation with uncoded bits, but less accurate for differential demodulation and decoding with forward error control since bit errors are not independent in this case.

18.5.2 Demodulation with Half-Rate Convolutional Coding

For half-rate convolutional coding with constraint length 7, and lossless Viterbi decoding, a tight upper bound on the decoded bit error rate [7, pp. 148] is found in terms of the Hamming distance

$$D \stackrel{\Delta}{=} \exp\left[-\frac{1}{2}\left(\frac{E_b}{N_0}\right)_{\text{eff}}\right] \qquad (18.47)$$

and the bit error probability is then upper bounded by the series

$$P_{b,\text{conv}} \leq \frac{1}{2}(36D^{10} + 211D^{12} + 1404D^{14} + 11633D^{16} + 77433D^{18} + \ldots) \qquad (18.48)$$

Figure 18.17 shows the probability of bit error for different information bit durations. This coding provides bit error probability of 10^{-5} at an effective C/N₀ more than 4 dB

DATA MESSAGE DEMODULATION

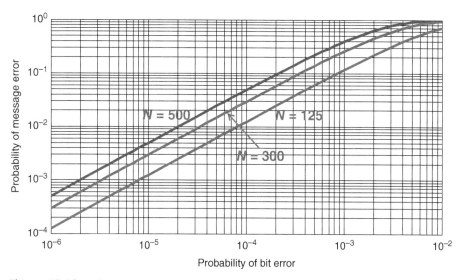

Figure 18.16. Relationship between Message Error Rate and Bit Error Rate for Message Length N Bits

lower than with uncoded coherent biphase keying. Thus, even if a signal design applies only 50% power to the data component, reducing the data component effective C/N_0 by 3 dB relative to 100% data component, using the half-rate convolutional coding with the 50% power data component provides even lower bit error probability at the same composite signal effective C/N_0.

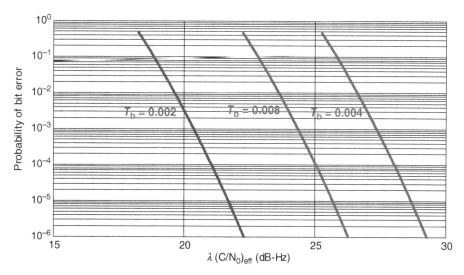

Figure 18.17. Bit Error Probabilities for Coherent Binary Phase-Shift Keying with Half-Rate Convolutional Coding, Constraint Length 7

18.5.3 Demodulation of the L1C Data Message

The L1C signal, described in Section 7.3, employs two unique features in its message encoding that warrant special attention, LDPC coding and the opportunity for data symbol combining in decoding the clock and ephemeris data (CED). Each of these features is described here.

Although LDPC codes were discovered in 1960, their rediscovery in the 1990s, coupled with increases in processing capability, motivated recent interest in them. LDPC codes, along with turbo codes, have the potential to operate close to the Shannon limit [8] on coding effectiveness, when very long blocks of data are encoded. Even though latency considerations in satnav prohibit encoding of very long blocks, the additional error protection provided by LDPC codes enables use of a 25% power data component in L1C.

LDPC codes are block codes with a very sparse parity matrix [9]. Decoding is performed iteratively. There are no available analytical predictions of LDPC code effectiveness, so computer simulations must be used to assess performance.

Figure 18.18 shows such simulation results [10] for the two different lengths of bits in the L1C message subframes. The CED, encoded using LDPC(1200, 600), has a 10^{-5} bit error probability at an effective C/N_0 approximately 2.5 dB lower than with the half-rate convolutional coding used in other new and modernized signals. Codeword error probabilities are also shown for the 600 bit codewords and the 274 bit codewords.

The second innovation in the L1C signal's design involves the opportunity for a receiver to trade latency against the effective C/N_0 at which the CED can be demodulated. To understand this innovation, first consider data message demodulation for all other new and modernized signals. Once the receiver channel is synchronized to the signal's data message symbols, it recovers soft decisions from the data message symbols modulated onto the data component, and decodes the forward error control encoding. If decoding is reported to be error free, the CRC is then used to check for uncorrected errors. If the CRC passes, the decoded data message is almost certainly correct, and can be used in calculating position and time. If the decoding and CRC are not successful, the receiver discards the symbols, waits for the next message, and attempts to decode the CED from that message. If the signal's effective C/N_0 is even slightly below the data message threshold, however, many messages would need to be processed before a chance occurrence of no bit errors in the decoded message.

In contrast, the L1C data message structure is designed so that the CED is LDPC encoded separately from other information in the data message. Even after interleaving with Subframe 3 encoded symbols (which contain system information but are not essential for initial positioning), symbols containing the CED information in Subframe 2 occur at fixed and known locations in the data message modulated onto the L1C data component.

Once the receiver channel is synchronized to the L1C spreading waveform and pilot overlay code, it is perfectly synchronized to the L1C data message as well, since the overlay code has the same length as the data message. The receiver recovers soft decisions from the Subframe 2 symbols modulated onto the L1C data component, and

DATA MESSAGE DEMODULATION

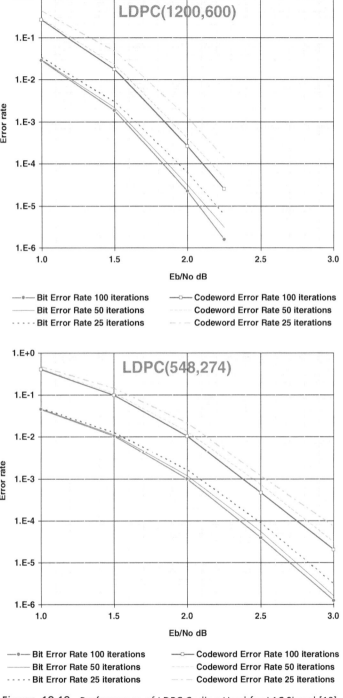

Figure 18.18. Performance of LDPC Coding Used for L1C Signal [10]

performs LDPC decoding. Decoding and CRC check proceed as for the other signals, as long as decoding and CRC pass.

However, there is an opportunity for an L1C receiver to also use a more sophisticated and capable approach if there are uncorrectable errors in the data message. This approach is called here data symbol combining, and was originally termed code combining [11]. Let $e^{-i\hat{\theta}_{\text{ref}}[j]}\mathbf{Z}_{\text{PROMPT}}[j]$ be a vector containing all the phase-corrected soft decision corresponding to one of the Subframe 2 message symbols at time index j, and suppose that the decoding yields an uncorrected error. As long as the CED has not been updated between adjacent messages (such updates occur rarely and at scheduled times), then the same CED symbols are repeated 1800 symbols later. The receiver stores all the phase-corrected soft decisions, and waits 18 seconds for repetition of Subframe 2 in the next data message. Each vector of phase-corrected soft decisions is then summed, $(e^{-i\hat{\theta}_{\text{ref}}[j]}\mathbf{Z}_{\text{PROMPT}}[j] + e^{-i\hat{\theta}_{\text{ref}}[j+1800]}\mathbf{Z}_{\text{PROMPT}}[j + 1800])/2$, and all of these summed soft decisions then go through LDPC decoding and CRC. If an undetected error still remains, the process could be continued a third time to form $(e^{-i\hat{\theta}_{\text{ref}}[j]}\mathbf{Z}_{\text{PROMPT}}[j]+ e^{-i\hat{\theta}_{\text{ref}}[j+1800]}\mathbf{Z}_{\text{PROMPT}}[j + 1800] + e^{-i\hat{\theta}_{\text{ref}}[j+3600]}\mathbf{Z}_{\text{PROMPT}}[j + 3600])/3$. As long as the carrier phase of the pilot component is tracked, the signal power sums coherently and the noise power sums noncoherently, enhancing $(E_b/N_0)_{\text{eff}}$ by 3 dB when two identical data symbols are combined, and by 4.8 dB when three data symbols are combined. Data symbol combining thus allows data message demodulation at lower C/N_0 values.

Data symbol combining cannot be used on the Subframe 3 contents, which typically changes every message. But the signal can be used for positioning and timing once the CED is read from the message and the time of transmission is known. If needed, the Subframe 3 information can be read from other signals being received at higher effective C/N_0.

18.6 SUMMARY

This chapter has provided the specific details of signal processing, measurements, and discriminator design needed to draw on the general loop design theory developed in Chapter 17, and apply it to FLLs, Costas loops, and PLLs. Analytical expressions for tracking errors have been provided for each type of loop, along with rules of thumb regarding loss of lock. The signal processing defined in Section 18.1 is general enough to also apply in code tracking, as seen in Chapter 19.

All the measurements used in conventional carrier tracking and code tracking involve correlation functions evaluated at specific delay values; these correlation functions, evaluated at specific delay values, are known as correlator taps. Carrier tracking uses the correlation function evaluated at what is estimated to be zero delay, often known as the prompt tap or punctual tap. Code tracking, described in Chapter 19, makes use of correlation functions evaluated at other delay values.

Since measurements for carrier tracking are formed by wiping off the spreading modulation, carrier tracking does not depend on the spreading modulation except for how it affects effective C/N_0. Signals having very different spreading modulations provide the same carrier tracking performance, as long as the loop parameters are the same and the effective C/N_0 is the same.

Receiver design involves choosing the type of carrier tracking as well as the associated loop parameters. In general, FLLs are less sensitive to dynamics and can remain locked under more stressing conditions than Costas loops or PLLs. Some receivers use FLLs after initial synchronization to begin tracking, handing over to carrier phase tracking after the FLL has converged.

There is a much clearer choice between Costas loops and PLLs when carrier phase tracking is needed. If there is unknown biphase data modulation of the signal or component being tracked, a Costas loop is needed. Even though a Costas loop would function with a pilot component having no unknown phase modulation, a PLL has smaller phase tracking errors, twice as long a linear region of the S-curve for loop pull-in and for accommodating dynamics, and approximately 6 dB lower effective C/N_0 loop-lock threshold than a Costas loop under the same conditions. If the signal has a 50% power pilot component, using a PLL on the pilot component provides approximately 3 dB lower effective C/N_0 loop-lock threshold than would be obtained with a Costas loop tracking a 100% power data component.

Choosing carrier tracking parameters involves careful trades, as well as good characterization of conditions where the receiver must operate. Smaller values of T, the correlation integration time, and loop update time, provide more robust operation when dynamics are large and effective C/N_0 is moderate to high. Large loop bandwidths also reduce degradation from receiver oscillator noise, as discussed in Section 17.3.3. Yet when low effective C/N_0 is the dominant stress, smaller loop bandwidths provide better accuracy and loss of lock performance, and larger values of T are beneficial for FLLs and Costas loops. The tradeoffs are well defined here, but the designer must make the choices.

REVIEW QUESTIONS

Theoretical Questions

QT18.1. Derive and evaluate the following approximation to the four-quadrant FLL discriminator:
$$\tilde{F}[j] \approx -\frac{\Re\{Z_{\text{PROMPT}}[j-1]\}\Im\{Z_{\text{PROMPT}}[j]\} - \Im\{Z_{\text{PROMPT}}[j-1]\}\Re\{Z_{\text{PROMPT}}[j]\}}{2\pi T}.$$

The quantity $(2\pi T)^{-1}$ can be computed in advance, so then the calculation involves a total only of three multiplications and a subtraction.

a. Starting with
$$Z_{\text{PROMPT}}[j-1]Z^*_{\text{PROMPT}}[j] = |Z_{\text{PROMPT}}[j-1]Z^*_{\text{PROMPT}}[j]|e^{i[\varphi(t)-\varphi(0)]},$$
$\tilde{F}[j] = \frac{\varphi[j]-\varphi[j-1]}{2\pi T}$, show that $\tilde{F}[j] \cong \frac{\sin(\varphi[j]-\varphi[j-1])}{2\pi T}$, with the approximation more accurate when $|\varphi(t) - \varphi(0)|$ is small.

b. Now assume that $|Z_{\text{PROMPT}}[j-1]Z^*_{\text{PROMPT}}[j]| \cong 1$ (this can be achieved by normalizing the prompt values using estimates of the received signal power). Show that the result from (a) can be approximated as $\tilde{F}[j] \cong \frac{\Im\{Z_{\text{PROMPT}}[j-1]Z^*_{\text{PROMPT}}[j]\}}{2\pi T}$.

c. Then show that
$$\tilde{F}[j] \cong -\frac{\Re\{Z_{\text{PROMPT}}[j-1]\}\Im\{Z_{\text{PROMPT}}[j]\} - \Im\{Z_{\text{PROMPT}}[j-1]\}\Re\{Z_{\text{PROMPT}}[j]\}}{2\pi T}$$

d. Plot the S-curve for the four-quadrant arctan discriminator and for the approximation derived here, and comment on the pull-in region and loss of lock characteristics.

e. Now suppose the estimate of received signal power is in error, and the actual $|Z_{\text{PROMPT}}[j-1]Z^*_{\text{PROMPT}}[j]|$ is twice as large as estimated. Plot the discriminator curves as in (d) and comment on the results.

QT18.2. Derive and evaluate the following approximations to the two-quadrant Costas discriminator.

a. Start with the two-quadrant arctangent discriminator $\tilde{\theta}[j] = \tan^{-1}\left[\frac{-\Im\{Z_{\text{PROMPT}}[j]\}}{\Re\{Z_{\text{PROMPT}}[j]\}}\right]$, where $Z_{\text{PROMPT}}[j] = \begin{cases} |Z_{\text{PROMPT}}[j]|e^{i[-\theta[j]]} \\ |Z_{\text{PROMPT}}[j]|e^{i[-\theta[j]+\pi]} \end{cases}$, and use trigonometric small-angle approximations to obtain the ratio discriminator $\tilde{\theta}[j] \approx \frac{-\Im\{Z_{\text{PROMPT}}[j]\}}{\Re\{Z_{\text{PROMPT}}[j]\}}$.

b. Now show that the ratio discriminator provides the same estimate regardless of the received signal power—it is "self-normalizing."

c. The "original discriminator" is $\tilde{\theta}[j] \approx -\frac{1}{2}\Im\{Z_{\text{PROMPT}}[j]\}$. Show that the original discriminator is sensitive to the estimate of power—it is not "self-normalizing."

d. Plot the S-curve for the two-quadrant arctan discriminator, the ratio discriminator, and the original discriminator, and comment on the pull-in region and loss of lock characteristics.

e. Now suppose the estimate of received signal power is in error, and the actual $|Z_{\text{PROMPT}}[j-1]Z^*_{\text{PROMPT}}[j]|$ is twice as large as estimated. Plot the discriminator curves as in (d) and comment on the results.

QT18.3. Evaluate the approximate PLL discriminator is $\tilde{\theta}[j] \cong -\Im\{Z_{\text{PROMPT}}[j]\}$.

a. Show the approximate PLL discriminator yields an estimate of the carrier phase error for small values of carrier phase error, when $|Z_{\text{PROMPT}}[j]| \cong 1$ (this can be achieved by normalizing the prompt values using estimates of the received signal power).

b. Now show that the approximate PLL discriminator provides an estimate proportional to the received signal power—it is not "self-normalizing."

c. Plot the S-curve for the four-quadrant arctan discriminator and the approximate PLL discriminator, and comment on the pull-in region and loss of lock characteristics.

d. Now suppose the estimate of received signal power is in error, and the actual $|Z_{PROMPT}[j]|$ is twice as large as estimated. Plot the discriminator curves as in (c) and comment on the results.

Application Questions

QA18.1. Compare loop-lock thresholds, in terms of total signal effective C/N_0, for phase tracking from different signals: C/A code signal, L1C, and E1OS. Use the criteria that the Costas loop loop-lock threshold is loop SNR of 16 dB and the PLL loop-lock threshold is loop SNR of 10 dB, and assume 10 Hz loop bandwidths.

QA18.2. Compare data demodulation performance, in terms of total signal effective C/N_0 threshold, for different signals: C/A code signal, L1C, and E1OS. Assume perfect carrier tracking and the BER threshold for C/A code signal is an E_b/N_0 of 9.6 dB, the BER threshold for L1C is an E_b/N_0 of -0.3 dB (assume one round of data symbol combining), and the BER threshold for E1 OS is an E_b/N_0 of 4.8 dB.

REFERENCES

1. G. X. Gao, S. Datta-Barua, T. Walter, and P. Enge, "Ionosphere Effects for Wideband GNSS Signals," *Proceedings of the 63rd Annual Meeting of The Institute of Navigation* 2007, Cambridge, MA, April 2007, pp. 147–155.
2. O. Julien, C. Macabiau, and E. Bertrand, "Analysis of Galileo E1 OS unbiased BOC/CBOC tracking Techniques for Mass Market Applications," Proceedings of 2010 5th ESA Workshop on Satellite Navigation Technologies and European Workshop on GNSS Signals and Signal Processing (NAVITEC), 2010, December 2010.
3. E. Kaplan and C. Hegarty, *Understanding GPS: Principles and Applications*, 2nd edition, Artech House, 2006.
4. B. Sklar, *Digital Communications: Fundamentals and Applications*, 2nd edition, Prentice-Hall, 2001.
5. J. Proakis and M. Salehi, *Digital Communications*, 5th edition, McGraw-Hill, 2007.
6. M. Roudier, A. J. Garcia-Pena, O. Julien, T. Grelier, L. Ries, C. Poulliat, M-L. Boucheret, and D. Kubrak, "Demodulation Performance Assessment of New GNSS Signals in Urban Environments," *Proceedings of the 27th International Technical Meeting of The Satellite Division of the Institute of Navigation (ION GNSS+ 2014)*, Tampa, Florida, September 2014, pp. 3411–3429.
7. M. K. Simon, J. K. Omura, R. A. Scholtz, and B. K. Levitt, *Spread Spectrum Communications, Volume II*, Computer Science Press, 1985.
8. T. M. Cover and J. A. Thomas, *Elements of Information Theory*. John Wiley & Sons, New York, 2006.
9. T. J. Richardson and R. L. Urbanke, Efficient Encoding of Low-Density Parity-Check Codes, *IEEE Transactions on Information Theory*, Vol. 47, No. 2, 2001, pp. 638–656.

10. J. W. Betz, M. A. Blanco, C. R. Cahn, P. A. Dafesh, C. J. Hegarty, K. W. Hudnut, V. Kasemsri, R. Keegan, K. Kovach, L. S. Lenahan, H. H. Ma, J. J. Rushanan, D. Sklar, T. A. Stansell, C. C. Wang, and S. K. Yi, "Description of the L1C Signal," *Proceedings of the Institute of Navigation Conference on Global Navigation Satellite Systems 2006*, ION-GNSS-2006, Institute of Navigation, September 2006.

11. D. Chase, Code Combining–A Maximum-Likelihood Decoding Approach for Combining an Arbitrary Number of Noisy Packets, *IEEE Transactions on Communications*, Vol. 33, No. 5, 1985, pp. 385–393.

19

CODE TRACKING

Code tracking processes the signal's complex envelope to produce estimates of the time-varying time of arrival, relative to receiver time, of the satnav signal. (Even though code tracking involves the spreading waveform that is composed of the spreading modulation as well as the spreading code, it is conventionally called code tracking and that terminology is employed here as well.) The resulting time of arrival estimate is used to produce the pseudorange estimate as described in Chapter 20, and is also used to generate the replica in tracking loops. As seen in Chapter 18, this time of arrival estimate also allows the receiver to align the spreading waveform with the received signal, enabling wipe off of the spreading waveform before carrier tracking. Figure 19.1 shows the digital processing in a receiver channel, highlighting the code tracking loop. This chapter describes the signal processing used for code tracking, then introduces different discriminators. It also provides analytical models for predicting the performance of these different tracking loops, along with discussion of design considerations.

The description of code tracking follows directly from tracking loop design addressed in Chapter 17, and signal processing for carrier tracking described in Section 18.1. Section 19.1 describes signal processing for a code tracking loop, also known as a delay-locked loop. Section 19.2 describes the resulting measurements and design of different discriminators for code tracking. Section 19.3 introduces carrier-aided code

Engineering Satellite-Based Navigation and Timing: Global Navigation Satellite Systems, Signals, and Receivers, First Edition. John W. Betz.
© 2016 The Institute of Electrical and Electronics Engineers, Inc. Published 2016 by John Wiley & Sons, Inc.

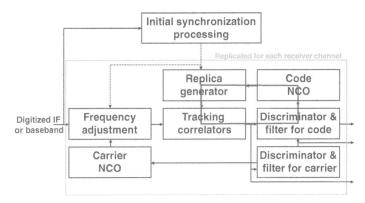

Figure 19.1. Digital Processing in a Receiver Channel, with Highlighted Code Tracking Loop

tracking, which uses measurements from carrier tracking to enhance the performance of code tracking. Section 19.4 provides an analytical model for code tracking loop errors in white noise, and Section 19.5 describes the generalization of this analytical model to include the effects of nonwhite interference. Tracking of signals whose spreading modulations can introduce ambiguous tracking points is addressed in Section 19.6. Section 19.7 provides a summary of the chapter.

19.1 SIGNAL PROCESSING FOR CODE TRACKING

Signal processing for code tracking is based on the same signal processing approaches described in Section 18.1. When code tracking is performed on a data component, the signal contribution to the measurement from (18.5) is of the form

$$\mu_d(j;\tau) \cong e^{i\theta[j]} \frac{\sqrt{C_d}}{T} \bar{R}_d(\tilde{D}(t) - \tau) \int_{jT-T/2}^{jT+T/2} \sum_{k=-\infty}^{\infty} d_k \text{rect}_{T_d}(t - D(t) - kT_d - T_d/2) dt$$

(19.1)

where the integration limits are established so no data bit transitions occur within the limits. When code tracking is performed on a pilot component without synchronization to the pilot overlay code, the signal contribution to the measurement from (18.7) is of the form

$$\mu_{p,o}(j;\tau) \cong e^{i\theta[j]} \frac{\sqrt{C_p}}{T} \bar{R}_p(\tilde{D}(t) - \tau) \int_{jT-T/2}^{jT+T/2} \sum_{m=-\infty}^{\infty} q_m \text{rect}_{T_q}(t - D(t) - mT_q - T_q/2) dt$$

(19.2)

where the integration limits are established so that no pilot overlay code bit transitions occur within the limits. When code tracking is performed on a pilot component after the receiver is synchronized to the pilot overlay code so that this code can be wiped, the signal contribution to the measurement from (18.9) is of the form

$$\mu_p(j;\tau) \cong e^{i\theta[j]} \frac{\sqrt{C_p}}{T} \bar{R}_p(\tilde{D}(t) - \tau) \tag{19.3}$$

Conventional code tracking loops use two specific measurements based on the above expressions, calculated at delay values that are symmetrically offset by $\pm\Delta/2$ from the estimated true delay. The measurement computed at $\tau = \tilde{D}(t) - \Delta/2$ is called the "early measurement" or early correlator tap, while the measurement computed at $\tau = \tilde{D}(t) + \Delta/2$ is known as the "late measurement" or late correlator tap. The quantity Δ is called the early–late spacing used in the code tracking loop. In some cases code tracking loops also use the prompt measurements introduced in Section 18.1.

Thus, for code tracking of the data component, the early and late measurements are, from (19.1),

$$\begin{aligned}
Z_E[j] &\cong \mu_d(j; \tilde{D}(t) - \Delta/2) + \text{Noise terms} \\
&= \sqrt{C_d}\bar{R}_d(j; \tilde{D}[j] - \Delta/2)e^{i\theta_d[j]} + \text{Noise terms} \\
Z_L[j] &\cong \mu_d(\tilde{D}(t) + \Delta/2) + \text{Noise terms} \\
&= \sqrt{C_d}\bar{R}_d(j; \tilde{D}[j] + \Delta/2)e^{i\theta_d[j]} + \text{Noise terms}
\end{aligned} \tag{19.4}$$

where $\theta_d[j] = \theta[j] + d_k\pi$, with $d_k = \{0, 1\}$ depending on the data message bit value.

Similarly, for code tracking of the pilot component before synchronization to the pilot code bits, the early and late measurements are, from (19.2),

$$\begin{aligned}
Z_E[j] &\cong \mu_{p,o}(j; \tilde{D}(t) - \Delta/2) + \text{Noise terms} \\
&= \sqrt{C_p}e^{i\theta_p[j]}\bar{R}_p(\tilde{D}[j] - \Delta/2) + \text{Noise terms} \\
Z_L[j] &\simeq \mu_{p,o}(j; \tilde{D}(t) + \Delta/2) + \text{Noise terms} \\
&= \sqrt{C_p}e^{i\theta_p[j]}\bar{R}_p(\tilde{D}[j] + \Delta/2) + \text{Noise terms}
\end{aligned} \tag{19.5}$$

where $\theta_p[j] = \theta[j] + q_m\pi$, with $q_m = \{0, 1\}$ depending on the overlay code bit value. Here, the correlation integration time is selected to ensure an overlay code bit transition does not happen within an integration.

Finally, for code tracking of the pilot component after synchronization to the pilot code bits, the early and late measurements are, from (19.3),

$$\begin{aligned}
Z_E[j] &\cong \mu_p(j; \tilde{D}(t) - \Delta/2) + \text{Noise terms} \\
&= \sqrt{C_p}e^{i\theta_p[j]}\bar{R}_p(\tilde{D}[j] - \Delta/2) + \text{Noise terms} \\
Z_L[j] &\cong \mu_p(j; \tilde{D}(t) + \Delta/2) + \text{Noise terms} \\
&= \sqrt{C_p}e^{i\theta_p[j]}\bar{R}_p(\tilde{D}[j] + \Delta/2) + \text{Noise terms}
\end{aligned} \tag{19.6}$$

where $\theta_p[j] = \theta[j]$, which is the residual carrier phase with no phase contributions from data message bits or overlay code bits.

In the first two sets of measurements, (19.4) and (19.5), even if the carrier phase is estimated perfectly by the carrier tracking loop, the phase of the signal contribution to the measurement is unknown unless the data bit or overlay code bit, respectively, is known.

These models assume the carrier wipe off has been mostly effective, but some residual slowly time-varying carrier phase remains, as represented by the complex exponentials. Two types of discriminators are considered: coherent discriminators that rely on a nearly perfect estimate of $\theta(t)$ and any phase modulation from data message bits or overlay code bits to remove the resulting carrier phase rotation, and noncoherent discriminators that produce a measurement independent of $\theta(t)$ and any phase modulation from data message bits or overlay code bits.

Measuring the delay $D[j]$ is equivalent to measuring the time of arrival of the signal; it is the objective of code tracking to produce an accurate estimate of $D[j]$. While $D[j]$ is unknown, the receiver has an initial estimate for each update, denoted $\hat{D}[j]$. When code tracking commences, the first estimate of $D[j]$ is provided by initial synchronization. Subsequently, the most recent estimate of $D[j]$ from the code tracking loop is used as the initial estimate for the next update. It will be shown that these estimation errors are typically much smaller than a spreading code chip duration.

Coherent signal processing is illustrated in Figure 19.2. Here the replica being correlated with the received waveform is a generic spreading waveform that can apply

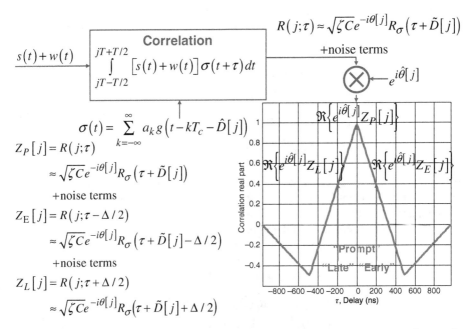

Figure 19.2. Coherent Signal Processing with No Delay Error; No Noise or Interference in Correlation Function, ζ Is the Fraction of Total Signal Power in the Component Being Tracked

to any of the cases addressed earlier in this section, and the correlator output is also expressed generically as a function of delay τ. The replica used at time jT need only be generated over the time interval $-jT/2 - \Delta/2 \leq t \leq jT/2 + \Delta/2$. The correlator output is the correlation function of the spreading waveform, $R_\sigma(\tau)$, scaled in amplitude by the square root of the received signal power, and rotated in phase by the combination of residual carrier phase and any phase modulation from data bit modulation or overlay code modulation that has not been removed. Coherent signal processing relies on an accurate estimate of this phase rotation, $\hat{\theta}[j]$ so that it is wiped off, rotating the correlation function of the spreading waveform into the real part of the correlator output. This phase-wiped correlator output is evaluated at delay values corresponding to early, late, and prompt estimates, so the correlation need not be computed at all delay values, but only at these three specific delays. The example correlation function shown in Figure 19.2 corresponds to a noise-free BOC(1,1) spreading modulation. If the delay estimate $\hat{D}[j]$ is perfect so there is no error, the prompt tap is computed at the peak of the correlation function, while the early and late taps are each located on different "shoulders" of the correlation function peak, having the same amplitude so that a line connecting them is horizontal.

Figure 19.3 shows similar results, but with an erroneously early initial delay estimate, so that the prompt tap is computed slightly to the left of the correlation function peak, and the line joining the early and late prompt taps has positive slope. As in Figure 19.2, the correlation function shown is ideal with no effects of noise or interference.

Figure 19.4 shows similar results, but when the initial delay estimate is erroneously late, so the prompt tap is computed slightly to the right of the correlation function peak,

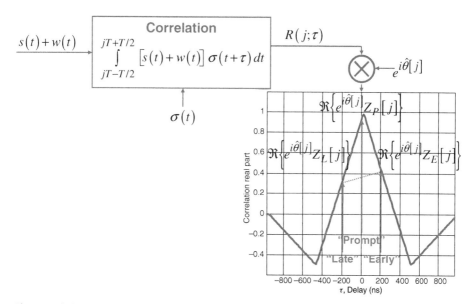

Figure 19.3. Coherent Signal Processing with Early Delay Error; No Noise or Interference in Correlation Function

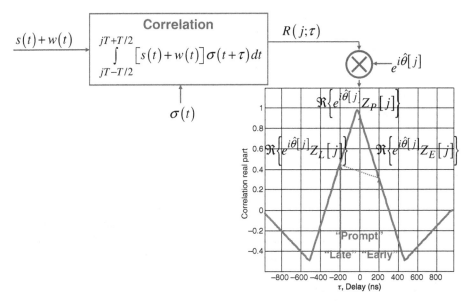

Figure 19.4. Coherent Signal Processing with Late Delay Error; No Noise or Interference in Correlation Function

and the line joining the early and late prompt taps has negative slope. While the fact that the delay estimate was erroneous might be gleaned by recognizing the incorrect amplitude of the prompt peak, such a measurement would be subject to errors in gain calibration, signal fading, and other distortions of the received signal amplitude, and by itself would not indicate whether the estimate is too early or too late. In contrast, comparing the amplitude of early and late measurements is insensitive to many factors (although it is sensitive to distortions in the shape of the correlation function peak), and the slope of the comparison also indicates whether the estimate is early or late.

Figure 19.5 shows a correlation function free of noise and interference, and a correlation function distorted by noise and interference, when the initial delay estimate has no error. Even though the initial delay estimate is correct, noise and interference distort the shape of the correlation function so the amplitudes of the early and late taps differ, erroneously indicating the delay estimate is late. The correlation function computed in the next time interval would be distorted by different interference and noise waveforms also causing imbalance in early and late prompt amplitudes. These errors, often referred to as code tracking jitter, are typically modeled as zero-mean Gaussian, and uncorrelated over sequential correlation integration times.

The coherent signal processing portrayed in the previous figures relies upon knowledge of the carrier phase so the receiver can derotate, or wipe, the carrier phase from the correlator tap values so that all signal-related information is in their real parts. Figure 19.6 shows noncoherent signal processing that does not rely on carrier phase at all. Instead, the magnitude-squared (or, as a variation, the magnitudes themselves could be used) of the early and late taps are compared.

SIGNAL PROCESSING FOR CODE TRACKING

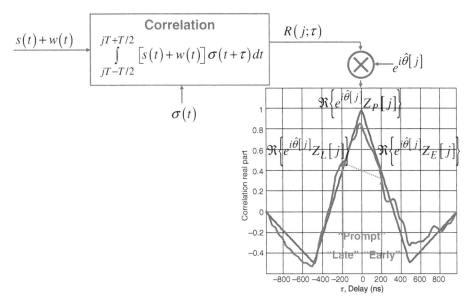

Figure 19.5. Coherent Signal Processing with No Delay Error; Noise and Interference Effects

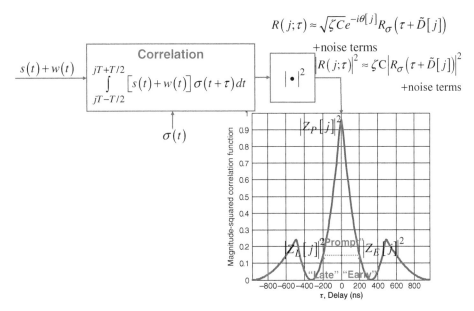

Figure 19.6. Noncoherent Signal Processing with No Delay Error; No Noise or Interference in Correlation Function

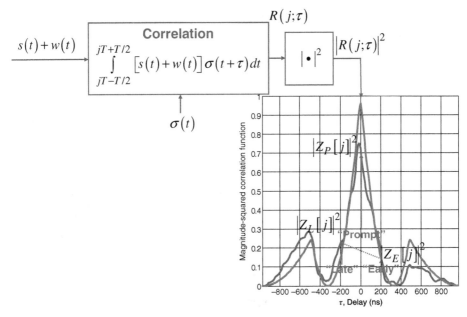

Figure 19.7. Noncoherent Signal Processing with Noise and Interference Effects

Figure 19.7 shows noncoherent signal processing analogous to what is shown in Figure 19.5 for coherent signal processing. Here again, the noise and interference distort the correlation function, causing the early and late tap amplitudes to be mismatched even though the initial delay estimate is correct.

19.2 DISCRIMINATORS FOR CODE TRACKING

Returning now to coherent processing, define a coherent code tracking discriminator function by the difference between the real part of the early and late prompt taps, after the carrier phase rotation is removed. The discriminator function is $\Re\{e^{i\theta}Z_E\} - \Re\{e^{i\theta}Z_L\}$, and it compares the amplitudes of the early and late taps. As introduced in Chapter 8, it is common to plot the discriminator response as a function of error in the initial estimate of delay, computed with no noise or interference,

$$S_{\text{CELP}}(\tilde{D}) = R_\sigma(\tilde{D} - \Delta/2) - R_\sigma(\tilde{D} + \Delta/2) \tag{19.7}$$

where $R_\sigma(\tau)$ is the correlation function of the spreading modulation computed using an ideal long code model for the spreading code, and $\tilde{D}$ is the error in estimating the time of arrival.

The discriminator function is commonly called an S-curve because of its shape. It translates the vector of measurements resulting from signal processing into a scalar

DISCRIMINATORS FOR CODE TRACKING

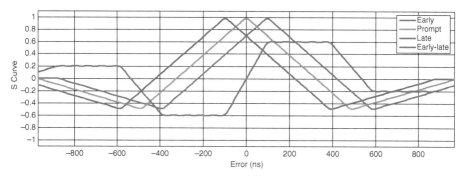

Figure 19.8. CELP S-Curve for BOC(1,1) with Precorrelation Bandwidth ±15.345 MHz and Early–Late Spacing $\Delta = 0.2/(1.023 \times 10^6 \text{ Hz}) = 195.5$ ns

value that is filtered in the tracking loop. The specific discriminator defined in (19.7) applies to coherent early–late processing, or CELP.

Figure 19.8 illustrates the formation of a CELP S-curve for a signal with BOC(1,1) spreading modulation, computed using (19.7). As discussed in Section 17.1, this S-curve has the desirable characteristics of passing through the origin (no bias), positive slope (producing correct feedback sign in the tracking loop), and linearity in the region of the origin (so the discriminator output value is proportional to the error). However, the positive slope and linearity exist only for small errors; once the error is large enough, these characteristics are lost. As discussed in Sections 19.3 and 19.4, code tracking performance can be modeled analytically for small errors corresponding to the linear segment of the S-curve near the origin, but once the errors are large enough to move onto the nonlinear part of the S-curve, the loop usually loses lock, standard analytic models for performance no longer hold, and only computer simulations can be used to characterize performance.

Figure 19.9 shows a BOC(1,1) S-curve for CELP with narrower early–late spacing, again computed using (19.7). It has the same overall shape as the S-curve, Figure 19.10,

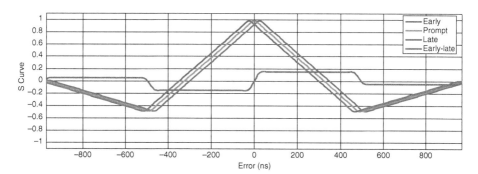

Figure 19.9. CELP S-Curve for BOC(1,1) with Precorrelation Bandwidth ±15.345 MHz and Early–Late Spacing $\Delta = 0.05/(1.023 \times 10^6 \text{ Hz}) = 48.9$ ns

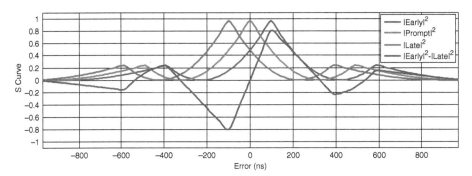

Figure 19.10. PELP S-Curve for BOC(1,1) with) with Precorrelation Bandwidth ±15.345 MHz and Early–Late Spacing $\Delta = 0.2/(1.023 \times 10^6 \text{ Hz}) = 195.5$ ns

but its linear region is narrower and its slope is smaller over the linear region. In the discussion of code tracking performance that follows, the narrower early–late spacing will be shown to provide better accuracy in noise and interference (as well as multipath, as discussed in Chapter 21), but the narrower linear region makes it more challenging to begin the code tracking process (i.e., "pull-in the code tracking loop") and to maintain lock in the presence of disturbances such as dynamics or rapidly fluctuating multipath.

Appendix 19.A shows that the correlation function of a bandlimited signal can be approximated near its peak by $R_x(\tau) = \sqrt{\zeta C} \tilde{R}_x(\tau) \cong \sqrt{\zeta C} \eta (1 - 2\pi^2 \beta_{rms}^2 \tau^2)$, where $\sqrt{C}$ is the amplitude of the received signal, ζ is the fraction of total signal power in the component being tracked, $\eta = \int_{-\beta_r/2}^{\beta_r/2} \Phi_s(f) df$ is the fraction of signal power passed by the precorrelation bandwidth, and β_{rms} is the RMS bandwidth of the signal defined in (19.21). Review Question QT19.1 shows that as long as $|\tilde{D}|$ and Δ are small, the CELP S-curve in (19.7) is approximated by $S(\tilde{D}) \cong 4\pi^2 \sqrt{\zeta C} \eta \beta_{rms}^2 \tilde{D}\Delta$, with slope proportional to the received signal amplitude, the early–late spacing, and to the squared RMS bandwidth. Larger slope corresponds to higher code tracking accuracy.

A discriminator function for noncoherent processing is $|Z_E|^2 - |Z_L|^2$, yielding the S-curve for noncoherent processing known as power early–late processing, or PELP, defined by

$$S(\tilde{D}) = |R_s(\tilde{D} - \Delta/2)|^2 - |R_s(\tilde{D} + \Delta/2)|^2 \qquad (19.8)$$

and shown in Figure 19.10. The linear region's shape, which is most important for code tracking, is similar to that for CELP. Review Question QT19.2 shows that, for small $|\tilde{D}|$ and Δ, the S-curve in (19.8) is approximated by $S(\tilde{D}) \cong \zeta C \eta^2 4\pi^2 \beta_{rms}^2 \tilde{D}\Delta [2 - \pi^2 \beta_{rms}^2 \Delta^2]$, with small-error slope proportional to the received signal power. The second term in square brackets, however, indicates that the slope of a PELP S-curve can decrease with increasing early–late spacing, producing lower code tracking accuracy.

DISCRIMINATORS FOR CODE TRACKING

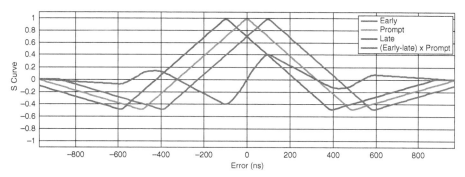

Figure 19.11. Coherent Dot Product S-Curve Based for BOC(1,1) with Precorrelation Bandwidth ±15.345 MHz and Early–Late Spacing $\Delta = 0.2/(1.023 \times 10^6 \text{ Hz}) = 195.5$ ns

A commonly employed discriminator is the coherent dot product $(\Re\{e^{i\theta}Z_E\} - \Re\{e^{i\theta}Z_L\})\Re\{e^{i\theta}Z_P\}$ yielding the S-curve

$$S(\tilde{D}) = (R_s(\tilde{D} - \Delta/2) - R_s(\tilde{D} + \Delta/2))R_s(\tilde{D}) \qquad (19.9)$$

and shown in Figure 19.11. Review Question QT19.3 shows that, for small $|\tilde{D}|$ and Δ, the S-curve in (19.9) is approximated by $S(\tilde{D}) \cong \zeta C \eta^2 4\pi^2 \beta_{rms}^2 \tilde{D}\Delta$, with small-error slope proportional to the received signal power.

All the S-curves described above are sensitive to the received signal power—the approximations provided above show that the S-curve slope for small errors is proportional to the received signal amplitude in some cases, and to the received signal power in other cases. Consequently, the slope of the S-curve changes with received signal power. The loop gain parameters would be designed based on a nominal received signal power, but if the received signal power changes, the loop gain parameters change, affecting the loop bandwidth and other loop characteristics discussed in Chapter 17. Thus, variations in received signal power away from nominal cause the loop behavior to vary from what was carefully designed.

Normalized discriminators avoid this undesirable behavior by attempting to make the S-curve insensitive to variations of the received signal amplitude. Prompt-normalized CELP and PELP discriminator functions are respectively $(\Re\{e^{i\theta}Z_E\} - \Re\{e^{i\theta}Z_L\})/\Re\{e^{i\theta}Z_P\}$ and $(|Z_E|^2 - |Z_L|^2)/|Z_P|^2$, yielding the S-curves

$$S(\tilde{D}) = \frac{R_s(\tilde{D} - \Delta/2) - R_s(\tilde{D} + \Delta/2)}{R_s(\varepsilon)}$$

$$S(\tilde{D}) = \frac{|R_s(\tilde{D} - \Delta/2)|^2 - |R_s(\tilde{D} + \Delta/2)|^2}{|R_s(\tilde{D})|^2} \qquad (19.10)$$

In the early days of satnav receiver development, there was great incentive to minimize the number of tracking correlators to reduce hardware cost and power. Rather than using an additional correlator to calculate the prompt correlator output for normalization,

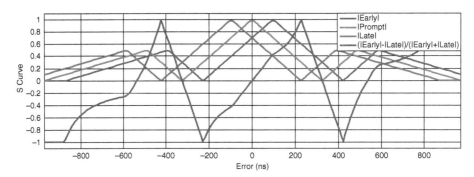

Figure 19.12. Normalized NELP S-Curve Based on (19.11) for BOC(1,1) with Precorrelation Bandwidth ±15.345 MHz and Early–Late Spacing $\Delta = 0.2/(1.023 \times 10^6 \text{ Hz}) = 195.5$ ns

as discussed above, designers normalized using the average of the early and late tap values. The resulting normalized noncoherent early–late processing (NELP) discriminator is $(|Z_E| - |Z_L|)/(|Z_E| + |Z_L|)$ producing the S-curve

$$S(\tilde{D}) = \frac{|R_s(\tilde{D} - \Delta/2)| - |R_s(\tilde{D} + \Delta/2)|}{|R_s(\tilde{D} - \Delta/2)| + |R_s(\tilde{D} + \Delta/2)|} \quad (19.11)$$

plotted in Figure 19.12.

In practice, several complicating factors occur with normalized discriminator functions. One is that the tap values used for normalization can fluctuate considerably due to noise and interference, as seen in Figures 19.5 and 19.7. Dividing by these highly fluctuating values increases code tracking jitter. In addition, when the term used for normalization involves the magnitude or the magnitude-squared of correlator taps, the correlator taps are biased estimators of the signal power, since the noise and interference power also contribute to the estimate of signal power. This bias is small and typically not a problem when the received signal power is high relative to the noise and interference power, but if the received signal is weak or the noise and interference power is large, the denominators in the normalized discriminator functions are larger than they should be, making the loop gain parameters smaller than intended.

Excessive fluctuation in the denominator values can be mitigated by replacing the instantaneous values by time-averaged values, using a sliding window filter or a fading-memory filter to balance smoothing against using more current values. When the coherent prompt value is averaged, an unbiased estimate results. To avoid biased estimators of magnitude or magnitude-squared correlator tap values, the noise and interference power can be estimated separately from an additional correlation tap far from the correlation peak, and this separate estimate can remove the bias from the magnitude or magnitude-squared of early, late, or prompt taps.

S-curves take on different shapes with different early–late spacings. Figure 19.13 shows PELP S-curves for BOC(1,1) with different precorrelation bandwidth and different early–late spacings. The linear part of the S-curve is wider with larger early–late

DISCRIMINATORS FOR CODE TRACKING

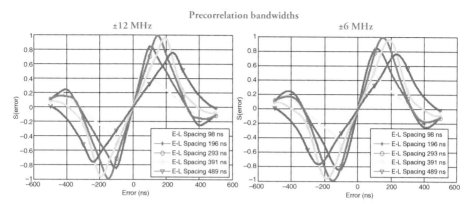

Figure 19.13. PELP S-Curves for BOC(1,1) with Precorrelation Bandwidth ±12 MHz and ±6 MHz, and Different Early–Late Spacings

spacings (making loop pull-in easier since a larger error can be accommodated), but the slope of the S-curve is larger for smaller early–late spacings (providing better code tracking accuracy in noise and interference, as shown in Section 19.4).

Figure 19.14 shows PELP S-curves for BPSK-R(1) and TMBOC(6,1,4/33), using the same precorrelation bandwidths and early–late spacings. For the same early–late spacing, TMBOC's S-curves have larger slope than those of BOC(1,1). However, as the early–late spacing increases to approach 200 ns, TMBOC's S-curve has reduced slope at the origin, becoming horizontal at early–late spacing of 244 ns. In order to avoid this undesirable situation, which also occurs for CBOC, either the early–late spacing should be kept below approximately 150 ns, or the BOC(6,1) components should not be processed so that the signal is processed like a BOC(1,1), as discussed in Section 18.1.

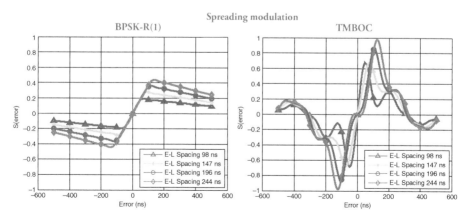

Figure 19.14. PELP S-Curves for BPSK-R(1) and TMBOC(6,1,4/33) with Precorrelation Bandwidth ±6 MHz and Different Early–Late Spacings

19.3 CARRIER-AIDED CODE TRACKING

Chapter 17's discussion of loop response to input dynamics showed that higher-order loops and larger loop bandwidths better accommodate dynamic stresses. Yet lower-order loops and smaller loop bandwidths provide better performance in noise.

One way to improve code tracking performance in the presence of dynamic stress and noise is carrier-aided code tracking. This aiding exploits the high accuracy of carrier tracking to provide an accurate estimate of the dynamics along the line of sight between receiver and satellite. While Chapter 24 discusses general approaches for using other sensors, such as inertial sensors, for this purpose, carrier-aided code tracking is a particularly simple and effective approach for reducing the need to track dynamics in the code tracking loop. Unlike other sensors, carrier aiding measures dynamics already projected onto the line of sight between receiver and satellite, and uses measurements already available in the receiver.

Figure 19.15 shows carrier-aided code tracking. The carrier loop output (its measurement of residual received carrier frequency) is scaled by a factor reflecting the ratio between carrier frequency and spreading code chip rate, then added to the spreading code chip rate in the code loop. The replica is then generated using the accurate estimate of chip rate change due to line of sight Doppler, removing the effect of these dynamics from the measurements output from signal processing. As long as the carrier loop maintains lock, the carrier-aided code tracking loop uses measurements that do not include effects of line of sight dynamics, and consequently the code tracking loop must only accommodate jitter from noise and interference, as well as time-varying ionospheric effects. Consequently, lower-order code tracking loops with narrower loop bandwidths and longer update times can be employed for better code tracking performance. Table 19.1 provides scale factors for signals described in Part II of this book.

Carrier-aided code tracking makes code tracking rely on carrier tracking; when carrier tracking loses lock the code tracking loop bandwidth must be widened to accommodate line of sight dynamics.

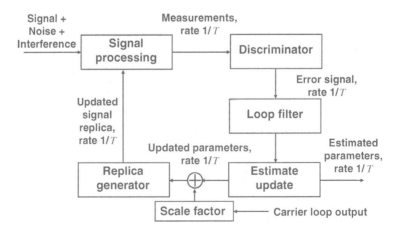

Figure 19.15. Code Tracking Loop with Carrier Aiding

TABLE 19.1. Scale Factors for Carrier Aiding of Code Tracking Loops

Signals	Scale Factor
IRNSS S SPS	1/2436
IRNSS S RS	1/1218
GLONASS L1OC	1/1565
GLONASS L1SC	1/626
BeiDou B1-A	1/770
GPS C/A, GPS L1C, SBAS L1, Galileo E1 OS, BeiDou B1-C, QZSS L1C, QZSS L1-C/A, QZSS L1S	1/1540
GPS L1 P(Y)	1/154
GPS L1 M	1/308
Galileo E1 PRS	1/616
BeiDou B1	1/763
Galileo E6 CS, Galileo E6 PRS, QZSS L6	1/250
BeiDou B3	1/124
BeiDou B3-A	1/496
GLONASS L2OC	1/1220
GLONASS L2SC	1/488
GPS L2C, QZSS L2C	1/1200
GPS L2 P(Y)	1/120
GPS L2 M	1/240
GLONASS L3OC	1/117.5
Galileo E5b, BeiDou B2b	1/118
GPS L5, SBAS L5, Galileo E5a, BeiDou B2a, QZSS L5	1/115
IRNSS L5 SPS	1/1150
IRNSS L5 RS	1/575

19.4 CODE TRACKING PERFORMANCE IN WHITE NOISE

Code tracking performance depends upon the shape of the signal's correlation function (equivalently, the shape of its PSD), and the resulting shape of the S-curve. Appendix 19.A shows how the sharpness of the signal's correlation function peak is related to its PSD and to the RMS bandwidth. This section shows that the RMS bandwidth is a key parameter in analytical models for code tracking accuracy.

Code tracking errors are unbiased when the S-curve passes through the origin and is linear in the region of the origin. In practice, imperfections in received signals or receiver hardware can introduce biases. Reference 1 shows how to predict these biases when the imperfections can be modeled as a signal passed through a linear time-invariant distortion. Some receivers estimate and remove biases resulting from receiver hardware imperfections either during the manufacturing process or even adaptively during operation.

Code tracking errors are typically modeled as random, including the effect of receiver dynamics (which, to the extent they are not modeled, introduce additional errors into the delay estimate), the effects of noise and interference, and timing jitter in the clock reference used to generate the replica—usually a negligible effect. Typically, these error sources are modeled as independent, and the overall variance of code tracking errors is the sum of the variances from each of these error sources.

When there is no interference, code tracking errors caused by additive white noise can be predicted analytically, as long as the errors are small enough to stay on the linear portion of the S-curve. (Effects of interference are included in Section 19.4.) Modeling the precorrelation filter as an ideal rectangular band-pass filter with passband $\pm\beta_r/2$, the variance of code tracking error, in units of seconds-squared, for CELP is [2]

$$\sigma^2_{\text{CELP,white}} = \frac{B_L \int_{-\beta_r/2}^{\beta_r/2} \bar{\Phi}_s(f) \sin^2(\pi f \Delta) df}{(2\pi)^2 \frac{\zeta C}{N_0} \left(\int_{-\beta_r/2}^{\beta_r/2} f \bar{\Phi}_s(f) \sin(\pi f \Delta) df \right)^2} \qquad (19.12)$$

Here, B_L is the one-sided bandwidth of the code tracking loop selected as part of the loop design procedure described in Chapter 17, ζ is the fraction of total signal power in the component being tracked, $\bar{\Phi}_s(f)$ is the normalized PSD (unit area) of the component being tracked, Δ is the early–late spacing, C is the power of the desired signal, and N_0 is the PSD of the (complex-valued) white noise. This is a very general expression, holding for any spreading modulation. In fact, the signal could be a random waveform with no defined spreading symbols or chip rate, as long as its replica can be generated in the receiver. While the variance is inversely proportional to C/N_0, the shape of the signal's PSD affects the code tracking error as well. Observe that the code tracking error for coherent processing is independent of correlation integration time.

For NELP, that uses the magnitude or magnitude-squared of the discriminator taps, the corresponding expression is a scaled version of the variance for CELP (again in units of seconds-squared) [3]:

$$\sigma^2_{\text{NELP,white}} = \sigma^2_{\text{CELP,white}} \left[1 + \frac{\int_{-\beta_r/2}^{\beta_r/2} \bar{\Phi}_s(f) \cos^2(\pi f \Delta) df}{T \frac{\zeta C}{N_0} \left(\int_{-\beta_r/2}^{\beta_r/2} \bar{\Phi}_s(f) \cos(\pi f \Delta) df \right)^2} \right] \qquad (19.13)$$

The term in square brackets is often called the squaring loss, reflecting the increase in variance due to noncoherent processing. When the correlator output SNR $2T\zeta C/N_0$ is large, the squaring loss is approximately unity, and NELP performance is close to that of CELP. At low output SNR values, the squaring loss term can be substantially larger than unity, magnifying NELP code tracking errors relative to those of CELP.

CODE TRACKING PERFORMANCE IN WHITE NOISE

Expressing code tracking variance in units of squared meters involves multiplying (19.12) and (19.13) by the squared speed of light, c^2.

When the early–late spacing is small, the preceding expressions simplify to (again in units of seconds-squared) [2, 3]

$$\sigma^2_{\text{CELP,white},\Delta\to 0} = \frac{B_L}{(2\pi)^2 \frac{\eta \zeta C}{N_0} \beta^2_{\text{rms}}}$$

$$\sigma^2_{\text{NELP,white},\Delta\to 0} = \sigma^2_{\text{CELP,white},\Delta\to 0}\left[1 + \frac{1}{T \frac{\eta \zeta C}{N_0}}\right] \tag{19.14}$$

$$\text{where } \eta = \int_{-\beta_r/2}^{\beta_r/2} \bar{\Phi}_s(f)df, \quad \beta^2_{\text{rms}} = \frac{1}{\eta}\int_{-\beta_r/2}^{\beta_r/2} f^2 \bar{\Phi}_s(f)df$$

The variance of CELP processing with vanishing early–late spacing equals the value of the Cramér–Rao lower bound [2], showing that CELP with small early–late spacing is the minimum variance unbiased estimator under these conditions. Appendix 19.A provides insight and numerical results for the RMS bandwidth, β_{rms}, that is fundamental to (19.14). The squaring loss term in square brackets affects the variance of NELP, and is $[1 + 2/\rho]$, where $\rho = 2T\eta\zeta C/N_0$ is the correlator output SNR in white noise defined in Chapter 5.

Figure 19.16 shows the squaring loss term, which is always greater than unity. The value is very close to unity for C/N_0 greater than 30 dB-Hz with typical correlation integration times. At lower values of C/N_0, the squaring loss increases. Ironically, while an initial implication is that coherent discriminators would be preferable at these lower C/N_0 values, carrier phase tracking is also increasingly unreliable in this case.

Simpler approximate expressions for code tracking error have been developed for the special case of NELP and BPSK_R spreading modulations [4], in terms of the dimensionless parameters $\hat{\Delta} = \Delta/T_c$, $\hat{\beta}_r = \beta_r T_c$,

$$\frac{\sigma^2_{\text{NELP}}}{T_c^2} \cong \begin{cases} \dfrac{B_L}{2\frac{\zeta C}{N_0}}\hat{\Delta}\left[1 + \dfrac{2}{T\frac{\zeta C}{N_0}(2-\hat{\Delta})}\right], & \pi \le \hat{\Delta}\hat{\beta}_r \\[2ex] \dfrac{B_L}{2\frac{\zeta C}{N_0}}\left[\dfrac{1}{\hat{\beta}_r} + \dfrac{\hat{\beta}_r}{\pi - 1}\left(\hat{\Delta} - \dfrac{1}{\hat{\beta}_r}\right)^2\right]\left[1 + \dfrac{2}{T\frac{\zeta C}{N_0}(2-\hat{\Delta})}\right], & 1 \le \hat{\Delta}\hat{\beta}_r < \pi \\[2ex] \dfrac{B_L}{2\frac{\zeta C}{N_0}}\left(\dfrac{1}{\hat{\beta}_r}\right)\left[1 + \dfrac{2}{T\frac{\zeta C}{N_0}}\right], & \hat{\Delta}\hat{\beta}_r < 1 \end{cases}$$

$$\tag{19.15}$$

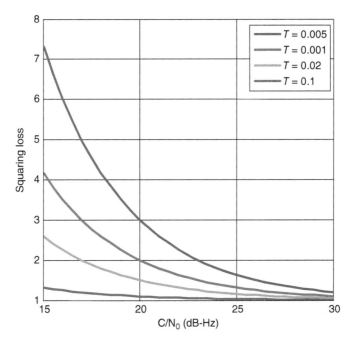

Figure 19.16. Squaring Loss for Different Correlator Integration Times In Units of Seconds

Equivalent expressions for CELP are obtained from (19.15) by setting the squaring loss terms in square brackets to unity.

The expression in the top line of (19.15) was published in the last century [5], and has been widely misapplied to any spreading modulation, any combination of precorrelation bandwidth and early–late spacing, and even when interference is present. Readers are cautioned to be on the lookout for misuse of that expression, and instead to recognize when the different and more general expressions provided in this section are needed.

Once code tracking errors (resulting from the combination of tracking jitter caused by noise, and dynamic stress errors caused by receiver dynamics) are no longer small enough to stay on the linear part of the S-curve, the above analytical expressions no longer apply. The full nonlinear behavior of code tracking loops, including pull-in and loss of lock, can only be assessed using hardware measurements or computer simulations. Even then the behavior is still a probabilistic event and must be characterized carefully.

In general, the linear region of the S-curve corresponds to code tracking errors with magnitude less than $\Delta/2$. Pull-in is easiest when early–late spacing is larger, providing a longer linear portion of the S-curve. Receivers can adaptively adjust early–late spacing, making the spacing larger to ease pull-in or to sustain lock during significant receiver dynamics, then using smaller early–late spacing for higher accuracy in noise. However, there is rarely any benefit to setting the early–late spacing smaller than the reciprocal of the precorrelation bandwidth.

Higher-order tracking loops tend to lose lock sooner when noise is the dominant stress. Loss of lock at 1–2 dB higher SNR than for first-order loops is often reported.

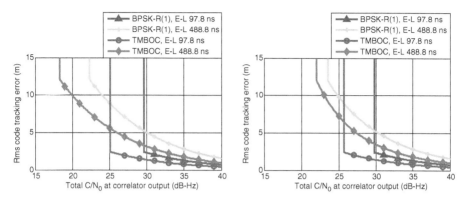

Figure 19.17. Code Tracking Errors Due to Noise, Precorrelation Bandwidth ±4 MHz, with Code Tracking Loop Bandwidth 1 Hz; CELP on Left, and NELP on Right

From Section 17.3, a simple approximate rule of thumb for a code tracking loop-lock criterion is then

$$\sigma_{\text{code tracking}} < \frac{\Delta}{12} \qquad (19.16)$$

where $\sigma_{\text{code tracking}}$ is the standard deviation of code tracking error from all sources root-sum squared together, including dynamic stress as well as jitter induced by noise and interference.

The remainder of this section examines show code tracking performance when noise is the dominant source of errors, examining the effect of coherent and noncoherent processing, different spreading modulations, precorrelation bandwidths, early–late spacings, and correlation integration times. All the results are shown for C/N_0 in the signal or component being tracked.

Figure 19.17 shows CELP and NELP code tracking errors for two different spreading modulations, and two different early–late spacings. While TMBOC is used here, the narrow precorrelation bandwidth makes it equivalent to BOC(1,1). The error curves are given infinite slope at the point where the loop-lock criterion (19.16) is satisfied. At the precorrelation bandwidth shown here, the BOC(6,1) component in TMBOC is filtered out, making the results of TMBOC equivalent to those for BOC(1,1). BPSK-R(1)'s smaller RMS bandwidth produces larger code tracking errors, as does the larger early–late spacing. While CELP and NELP errors are similar at larger values of C/N_0, the effect of squaring loss on NELP is evident in the larger errors at lower C/N_0 values, with the loop-lock criterion occurring at several dB larger C/N_0 in some cases. Although the narrower early–late spacings provide smaller errors, loop lock for this relatively narrow precorrelation bandwidth is maintained at lower C/N_0 values with the larger early–late spacing, which provides a larger linear region of the S-curve. Typically, carrier-aided code tracking would be needed to sustain tracking with a 1 Hz code tracking loop bandwidth, unless the receiver is stationary as in a surveying application, and it can still be challenging to maintain carrier track below 20 dB-Hz in order to aid the code tracking. As indicated earlier, CELP can only be used if there is no unknown phase modulation,

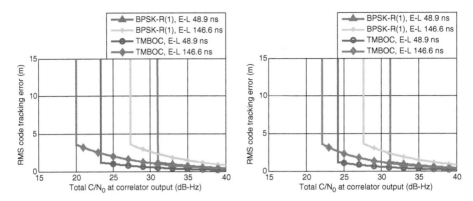

Figure 19.18. Code Tracking Errors Due to Noise, Precorrelation Bandwidth ±12 MHz, with Code Tracking Loop Bandwidth 1 Hz; CELP on Left, and NELP on Right

and if carrier phase tracking is accurate. Ultimately, the loop-lock thresholds for carrier tracking and code tracking must be compared to determine which is the limiting factor.

Figure 19.18 shows comparable results with a much wider precorrelation bandwidth. The performance of TMBOC is greatly improved since the BOC(6,1) component is included, and narrower early–late spacing performs much better as well. While wider early–late spacing still produces lower loss of thresholds for BPSK-R(1), the loop-lock thresholds for TMBOC are comparable or better for narrower early–late spacing with TMBOC, since the improved accuracy with noise counterbalances the smaller linear region of the S-curve. Of course, if other factors like dynamics are significant, the wider early–late spacing would provide better loop-lock performance.

The effect of different correlation integration times on NELP code tracking accuracy is explored in Figures 19.19 and 19.20. As seen previously, with the narrower precorrelation bandwidth, narrow early–late spacing produces smaller code tracking errors

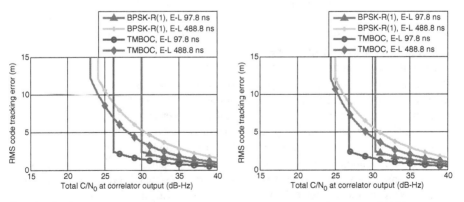

Figure 19.19. NELP Code Tracking Errors Due to Noise, Precorrelation Bandwidth ±4 MHz, with Code Tracking Loop Bandwidth 1 Hz; Correlation Integration Time 10 ms on Left and 5 ms on Right

CODE TRACKING PERFORMANCE IN WHITE NOISE

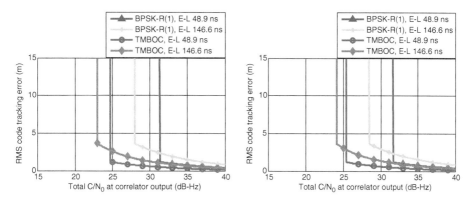

Figure 19.20. NELP Code Tracking Errors Due to Noise, Precorrelation Bandwidth ±12 MHz, with Code Tracking Loop Bandwidth 1 Hz; Correlation Integration Time 10 ms on Left and 5 ms on Right

but higher loop-lock thresholds due to noise. With NELP and larger early–late spacing, BPSK-R(1) and TMBOC provide similar performance. However, with narrower early–late spacing, TMBOC outperforms BPSK-R(1), and with wide precorrelation bandwidth TMBOC with narrow early–late spacing outperforms the other alternatives shown.

Finding the best early–late spacing involves trading code tracking errors, which can be smaller with smaller early–late spacing, against loop-lock threshold, which may be smaller with somewhat larger early–late spacings. Figure 19.21 shows the results of selecting specific precorrelation bandwidths, then finding a value of early–late spacing for each spreading modulation and precorrelation bandwidth that provides a low loop-lock threshold and good tracking accuracy. At all precorrelation bandwidths, TMBOC provides lower loop-lock threshold and smaller code tracking errors for the well-chosen early–late spacings.

The effect of different early–late spacings is portrayed in Figure 19.22 for wide precorrelation bandwidth. Here, the early–late spacing must be less than approximately 150 ns for TMBOC, as discussed previously. TMBOC's larger RMS bandwidth translates

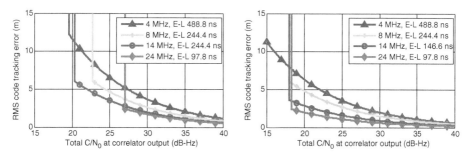

Figure 19.21. CELP Code Tracking Errors Due to Noise with Different Precorrelation Bandwidths and Early–Late Spacings, with Code Tracking Loop Bandwidth 1 Hz; BPSK-R(1) on Left and TMBOC on Right

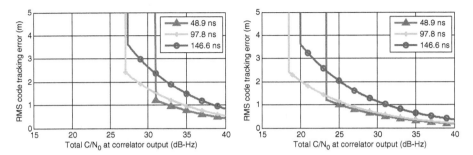

Figure 19.22. CELP Code Tracking Errors Due to Noise with Different Early–Late Spacings, Precorrelation Bandwidth ±12 MHz, with Code Tracking Loop Bandwidth 1 Hz; BPSK-R(1) on Left and TMBOC on Right

to providing the same code tracking accuracy at approximately 4 dB lower C/N_0, and correspondingly lower loop-lock thresholds. For this precorrelation bandwidth, code tracking accuracy is no better with 48.9 ns early–late spacing than with 97.8 ns, and the loop-lock threshold is several dB lower with 97.8 ns early–late spacing compared to 48.9 ns early–late spacing. This result exemplifies a very useful rule of thumb that, for code tracking in noise, choose early–late spacing that is twice the reciprocal of the precorrelation bandwidth, unless the S-curve requires a smaller early–late spacing for stable loop performance. This result typically comes close to providing the smallest code tracking error in noise. Lower loop-lock thresholds, however, may be obtained with larger early–late spacings.

Similar results are observed with the narrower precorrelation bandwidth used in Figure 19.23, although the code tracking errors are much larger. Here, the narrower precorrelation bandwidth filters out the BOC(6,1) component of TMBOC, allowing use of larger early–late spacings than in Figure 19.21. The same early–late spacing rule of thumb is illustrated—the reciprocal of the precorrelation bandwidth is 125 ns, and the early–late spacing of 244.4 ns provides code tracking errors as small as the early–late

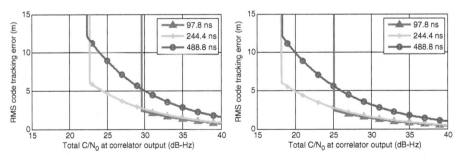

Figure 19.23. CELP Code Tracking Errors Due to Noise with Different Early–Late Spacings, Precorrelation Bandwidth ±4 MHz, with Code Tracking Loop Bandwidth 1 Hz; BPSK-R(1) on Left and TMBOC on Right

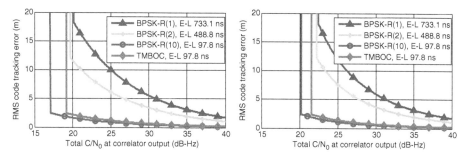

Figure 19.24. NELP Code Tracking Errors Due to Noise for Different Spreading Modulations with Precorrelation Bandwidth ±12 MHz, 100 ms Correlation Integration Time on Left, 10 ms Correlation Integration Time on Right, Code Tracking Loop Bandwidth 1 Hz

spacing of 97.8 ns, and much lower loop-lock threshold, with lower errors and the same loop-lock threshold as a wider early–late spacing.

Figure 19.24 shows the performance of different spreading modulations, all with the same precorrelation bandwidth and narrow early–late spacing, correlation integration time, and loop bandwidth. Early–late spacings were chosen to provide low loop-lock thresholds. The result is that TMBOC and BPSK-R(10) provide comparable code tracking errors, with BPSK-R(10) offering several dB lower loop-lock threshold.

The analytical expressions presented in this section can be applied to trade different processing approaches. Tracking performance for some specific spreading modulations has also been evaluated in Reference 6.

19.5 CODE TRACKING PERFORMANCE IN WHITE NOISE AND INTERFERENCE

When the receiver input includes not only the desired signal, but also continuous interference having a normalized PSD $\bar{\Phi}_I(f)$ and interference power I, the small-error variance is affected by the interference as well as the white noise. The variance of code tracking error for CELP, in seconds-squared, is given by

$$\sigma^2_{\text{CELP}} \cong \frac{B_L}{\left[2\pi\left(\dfrac{\zeta C}{N_0}\right)\displaystyle\int_{-\beta_r/2}^{\beta_r/2} f\bar{\Phi}_s(f)\sin(\pi f\Delta)df\right]^2}$$

$$\times \left[\int_{-\beta_r/2}^{\beta_r/2}\left[1+\left(\frac{I}{N_0}\right)\bar{\Phi}_I(f)\right]\bar{\Phi}_s(f)\sin^2(\pi f\Delta)df\right] \quad (\text{s}^2) \quad (19.17)$$

$$= \frac{B_L}{\left[2\pi\left(\frac{\zeta C}{N_0}\right)\int_{-\beta_r/2}^{\beta_r/2} f\bar{\Phi}_s(f)\sin(\pi f\Delta)df\right]^2}$$

$$\times \left[\int_{-\beta_r/2}^{\beta_r/2} \bar{\Phi}_s(f)\sin^2(\pi f\Delta)df + \frac{I}{N_0}\int_{-\beta_r/2}^{\beta_r/2} \bar{\Phi}_I(f)\bar{\Phi}_s(f)\sin^2(\pi f\Delta)df\right] \quad (s^2)$$

When the early–late spacing is vanishingly small, (19.17) simplifies to, in units of seconds-squared,

$$\sigma^2_{\text{CELP},\Delta\to 0} \cong \left(\frac{1}{2\pi}\right)^2 \frac{B_L}{(\zeta C/N_0)} \left(\frac{1}{\eta\beta_{\text{rms}}}\right)^2 \left(1 + \frac{I}{N_0}\frac{\chi_{is}}{\beta_{\text{rms}}^2}\right)$$

$$\eta \triangleq \int_{-\beta_r/2}^{\beta_r/2} \bar{\Phi}_s(f)df, \quad \beta_{\text{rms}} \triangleq \left(\frac{1}{\eta}\int_{-\beta_r/2}^{\beta_r/2} f^2\bar{\Phi}_s(f)df\right)^{1/2}, \quad (19.18)$$

$$\chi_{is} \triangleq \frac{1}{\eta}\int_{-\beta_r/2}^{\beta_r/2} f^2\bar{\Phi}_I(f)\bar{\Phi}_s(f)df$$

This expression shows the dependence on the signal's RMS bandwidth, along with dependence on a code tracking spectral separation coefficient χ_{is}. The term $\left(\frac{1}{\eta\beta_{\text{rms}}}\right)\left(1 + \frac{I}{N_0}\frac{\chi_{is}}{\beta_{\text{rms}}^2}\right)^{1/2}$ is a spectrum-specific factor that captures how the PSDs of desired signal and interference affect code tracking standard deviation.

When NELP is used instead of CELP, the variance of code tracking error in interference and noise becomes

$$\sigma^2_{\text{NELP}} \cong \sigma^2_{\text{CELP}}\left[1 + \frac{\int_{-\beta_r/2}^{\beta_r/2}\Phi_s(f)\cos^2(\pi f\Delta)df + \frac{I}{N_0}\int_{-\beta_r/2}^{\beta_r/2}\Phi_I(f)\Phi_s(f)\cos^2(\pi f\Delta)df}{\left(T\frac{\zeta C}{N_0}\left(\int_{-\beta_r/2}^{\beta_r/2}\Phi_s(f)\cos(\pi f\Delta)df\right)\right)^2}\right] \quad (s^2)$$

(19.19)

For vanishingly small early–late spacing, (19.19) becomes

$$\sigma^2_{\text{NELP},\Delta \to 0} \approx \sigma^2_{\text{CELP},\Delta \to 0} \left[1 + \frac{1 + \frac{I}{N_0}\kappa_{ts}}{T\eta \frac{\zeta C}{N_0}} \right] \quad (s^2) \quad (19.20)$$

In (19.19) and (19.20), the terms in square brackets are the squaring loss, similar to those for white noise in (19.13) and (19.14). However, there is an additional term in the numerator of the squaring loss term that involves the PSDs of desired signal and interference, and simplifies to the code tracking SSC when early–late spacing is small.

In some cases, the implications of these results are consistent with intuition, yet in other cases the results may differ from commonly held expectations. Figure 19.25 shows the effect of matched spectrum interference (interference having the same PSD as that of the desired signal) on the spectrum-specific factor for code tracking using CELP, with vanishing early–late spacing. At the smaller values of C_i/N_0, the differences are merely due to different RMS bandwidths of the different spreading modulations. At higher interference power levels, however, the different code tracking SSCs produce different results, consistent with what might be expected intuitively for the different spreading modulations.

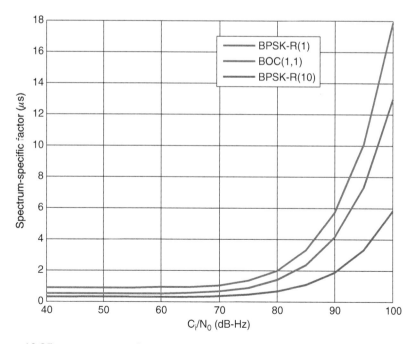

Figure 19.25. Spectrum-Specific Factor for Matched Spectrum Interference and Different Spreading Modulations, 24 MHz Precorrelation Bandwidth

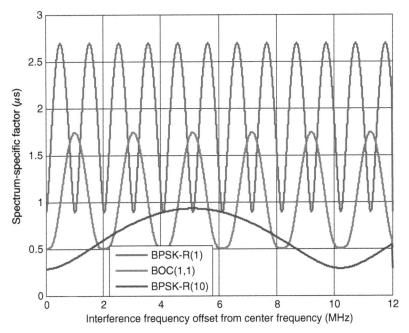

Figure 19.26. Effect of Narrowband Interference, at C_i/N_0 of 80 dB-Hz, on Spectrum-Specific Factor for CELP with Vanishing Early–Late Spacing

More surprising results occur in Figure 19.26. Here, the effect of narrowband interference at a fixed power is assessed for different interference center frequencies, with vanishing early–late spacing for CELP. In contrast to results for effective C/N_0 in Section 5.6, narrowband interference at frequencies far outside the null-to-null bandwidth of the desired signals PSD can introduce code tracking errors similar to those from narrowband interference within the null-to-null bandwidth. This result occurs because of the frequency-squared weighting within the integral for the code tracking SSC. Another interesting observation from Figure 19.26 is that, for a BPSK-R spreading modulation, narrowband interference most degrades code tracking error when the interference is centered midway between the spectral peak and the first null of the PSD. Other results, not shown here, indicate that, when early–late spacing is not vanishingly small, the errors due to narrowband interference are largest for frequencies near band center, with peak values falling off gradually at larger offsets from center frequency. However, for these wider early–late spacings all code tracking errors are larger than with vanishing early–late spacing.

19.6 AMBIGUOUS CODE TRACKING

The previous two sections examined small-error models of code tracking where the errors are zero mean with standard deviations well within the linear portion of the S-curve.

AMBIGUOUS CODE TRACKING

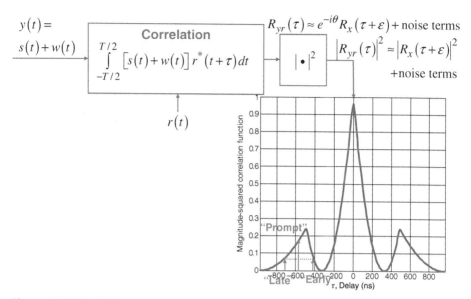

Figure 19.27. Falsely Locked Noncoherent Signal Processing with No Delay Error, and No Noise or Interference Effects

BOC spreading modulations have correlation functions with multiple peaks, leading to S-curves that have additional zero crossings. Compare Figure 19.27 with Figure 19.6. In both cases, the early and late taps have the same power, so the discriminator outputs a zero error indicating the initial estimate of delay was perfect. But while Figure 19.6 shows the prompt tap on the correlator peak, Figure 19.27 shows the prompt tap on a sidelobe of the correlation function, located at a delay where early and late prompt taps are balanced, so the tracking loop can remain locked and tracking. However, this lock point has a bias of approximately 550 ns in the delay estimate, corresponding to a consistent pseudorange error of approximately 165 m. Figure 19.10 shows the corresponding S-curve, with a zero crossing having positive slope at delays of approximately ±550 ns, yielding a stable tracking point.

A tracking loop can get into a false lock situation in one of the two ways: either a large error in delay estimation from initial synchronization processing, or a large error after the code tracking loop has already been pulled in and locked at the correct point. In the former case, if such a delay error arises from initial synchronization processing, the loop could pull in at this incorrect delay, so it is critical for initial synch processing to provide a delay estimate accurate enough to begin code tracking on the correct segment of the discriminator S-curve. At correlator output SNRs large enough for reliable initial synchronization, good techniques for delay estimation yield estimation errors having standard deviations of approximately 10% of a spreading code chip duration, so for initial synchronization of BOC(1,1) using sideband processing, the three-sigma error should be of the order of 300 ns, meaning the probability of a 550 ns error would be very small. In addition, code tracking can commence with wider early–late spacing to accommodate larger initial errors.

Once the code tracking loop is pulled in and locked, an error resulting from noise and interference could also occur, but typically only at low correlator output SNR. If the correlator output SNR is low enough at the true peak for loss of lock to occur, it is many dB lower at false peaks—likely not high enough to sustain code tracking most of the time. For example, Figure 19.27 shows the signal power in the sidelobe is approximately 20% that of the main peak, so if the SNR at the true peak is already marginal, the 7 dB lower SNR of the sidelobe may not sustain tracking. High line of sight dynamics also could cause the tracking point to shift to a stable but erroneous zero crossing as well, but typically not when carrier aiding or inertial aiding has removed most of the dynamics.

Different spreading modulations with different correlation functions have different levels of susceptibility to ambiguous tracking points. BOC spreading modulations with higher BOC ratios (ratio of subcarrier frequency to spreading code frequency) have more sidelobes, increasing the number of false lock points in the S-curve. Higher subcarrier frequencies produce correlation function sidelobes spaced more closely in delay to the main peak, increasing the likelihood of false lock. These correlation function sidelobes also have larger amplitudes relative to the main peak have higher SNR, enabling false tracking at higher levels of noise and interference. Cosine-phased BOC modulations also exhibit larger ratios of sidelobe amplitude to peak amplitude after even modest bandlimiting. Consequently, BOC signals with lower chance of false lock have small BOC ratios, lower subcarrier frequencies, and use sine phasing.

Coherent discriminators preserve negative-going correlation function peaks, eliminating the potential problem for BOC(1,1) (see the lack of false lock points in Figure 19.8, e.g., since the slope is negative) and reducing the problem for other BOC modulations. Multiple approaches described below have been developed to address false lock of BOC spreading modulations, most following the paradigm illustrated in Figure 19.28. Two separate sets of measurements are employed: higher accuracy but possibly ambiguous; and lower accuracy and unambiguous. The measurements can be used directly in

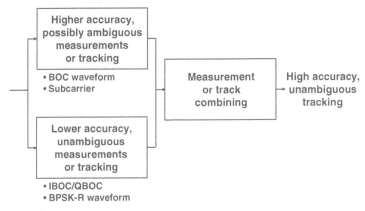

Figure 19.28. Processing Multiple Correlation Functions to Address False Lock Points

AMBIGUOUS CODE TRACKING

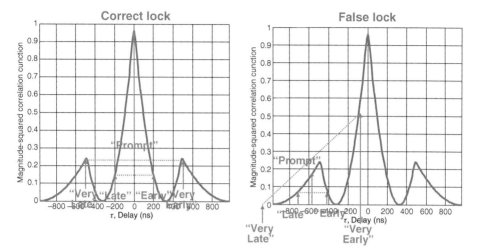

Figure 19.29. Very Early/Very Late Processing Against Ambiguous Track Points

tracking, or separate tracking loops can be employed using each set of measurements. Ultimately, the lower accuracy but unambiguous measurements are used to detect false lock and adjust the lock point, with higher accuracy measurements used to drive code tracking. There has been no comprehensive and rigorous evaluation and comparison of the different approaches.

The original approach, known as "bump jumping" or very early/very late processing [7], is illustrated in Figure 19.29. Two additional correlator taps are employed on the same wideband correlation function, spaced from the prompt correlator tap so they are placed at the correlation function's first sidelobes. When the correct correlation function peak is being tracked, the very early and very late prompt values are balanced, but when ambiguous tracking occurs, the very early and very late tap values are grossly different. The NCO value can then be reset to the correct tracking point, since the sign of the very early/very late difference shows which direction to move, and knowledge of the correlation function indicates how far to change the initial delay estimate. In the context of Figure 19.28, the very early and very late taps are lower accuracy but unambiguous, while the early and late taps provide higher accuracy but potentially ambiguous measurements.

Another proposed approach is to use the noncoherent combination of correlator outputs produced using sine-phased and cosine-phased replicas of the spreading signal, each generated using the same spreading code [8]. The measurements are the summed magnitude-squared correlator taps formed using sine-phased and cosine-phased replicas in two different correlators. As discussed in Chapter 16, this approach (termed IBOC/QBOC by its developer) broadens the correlation function. Figure 19.30 shows the IBOC and QBOC spreading symbols for BOC(1,1), while Figure 19.31 shows how the IBOC/QBOC correlation function is employed. Adding very early and very late correlator taps to the IBOC/QBOC correlation function at delays corresponding to its sidelobes provides false lock detection as described previously.

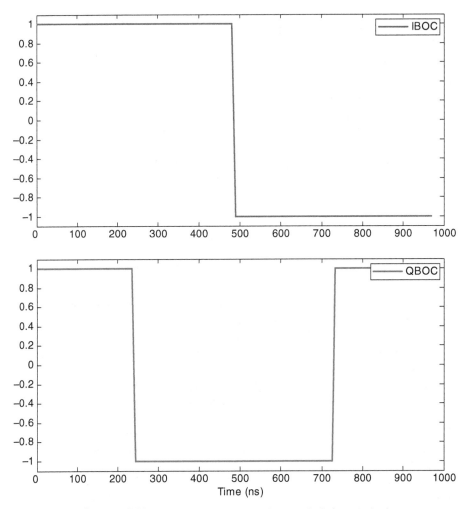

Figure 19.30. IBOC and QBOC Spreading Symbols for BOC(1,1)

Sideband processing, described in Chapter 16, can be employed to produce a wider correlation function used to compute very early and very late tap values, as shown in Figure 19.32.

A different approach, the multiple estimator loop [9] illustrated in Figure 19.33, uses three parallel sets of signal processing and discriminators whose outputs are combined into a composite error signal, in an extension of the basic tracking loop processing flow portrayed in Figure 19.1. Here, the carrier signal processing is analogous to that described in Chapter 18, wiping off the spreading modulation and spreading code, and producing measurements that are the most accurate yet most ambiguous. Subcarrier signal processing is also analogous to that described in Chapter 18, but operating on the subcarrier rather than the carrier. This processing wipes off the carrier and the spreading

AMBIGUOUS CODE TRACKING

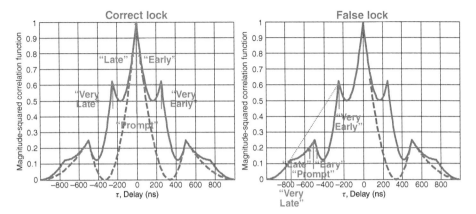

Figure 19.31. Use of IBOC/QBOC Squared Correlation Function (Solid Blue Curve) to Detect and Correct False Lock Points for BOC(1,1) Spreading Modulations; Corresponding BOC(1,1) Squared Correlation Shown as Dashed Red Curve

code, leaving the subcarrier which is less accurate and less ambiguous than the carrier, being at a much lower frequency, but still ambiguous. Sideband signal processing wipes off the carrier and the subcarrier, leaving a BPSK-R sideband signal to be tracked as described in this chapter. These measurements are less accurate but unambiguous. Weighted combining of the three resulting error signals produces a composite error signal that is loop filtered and used to produce estimates of carrier phase and delay and their rates. The replica generator produces the different replicas needed for all three sets of signal processing. It is claimed that the multiple estimator loop, properly implemented, provides the accuracy of conventional BOC code tracking while eliminating ambiguous tracking.

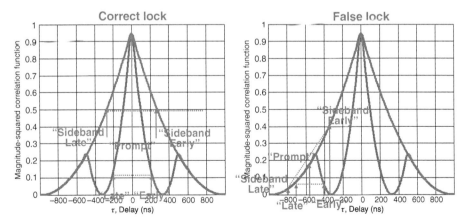

Figure 19.32. Very Early/Very Late Taps Using Sideband Processing (Solid Blue Curve) to Detect and Correct False Lock Points for BOC(1,1) Spreading Modulations; Corresponding BOC(1,1) Squared Correlation Shown as Red Curve

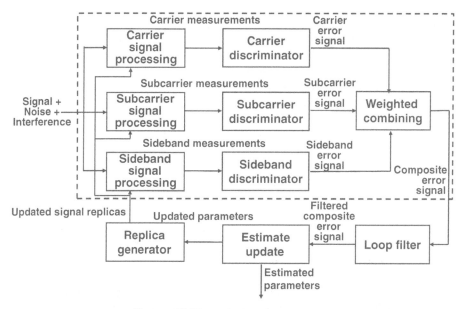

Figure 19.33. Multiple Estimator Loop

Many other approaches to addressing false tracking points of BOC spreading modulations can be found in the literature, including References 10–12. Many of these exploit the theoretical shape of ideal correlation functions, and their ability to perform in real environments with real hardware remains to be demonstrated. Ultimately, as long as the BOC parameter is small, even simple techniques like that in Reference 7 may suffice, since at high values of C/N_0 false lock points can be readily detected and corrected, while at low C/N_0 the false lock points, which have lower C/N_0 than the valid lock point and may not support tracking under the stressed conditions that caused loss of lock at the valid lock point. Chapter 3 of Reference 13 provides additional information concerning tracking of signals with ambiguous lock points.

19.7 SUMMARY

Signal processing for code tracking may involve wiping overlay code bits or data message bits when possible and needed for longer correlation integration times, followed by correlating the resulting signal and noise and interference against the locally generated replica of the spreading time series. If overlay code bits or data bits are not wiped, the correlation integration time intervals must be chosen not to straddle a bit transition opportunity. The correlations are computed at several delay values, or taps, producing the measurements.

A set of discriminators with associated S-curves can be developed to produce an output that depends on delay estimation error. Some are coherent discriminators, needing a good estimate of the carrier phase, while others are noncoherent estimators designed to be independent of carrier phase. The S-curves have linear segments with positive slopes passing through the origin, but become nonlinear for larger error values. It is undesirable for a discriminator response to vary with changes in received signal power, so normalized discriminators are beneficial for the best code tracking over a range of received signal powers. Some normalization factors are biased, providing erroneously large estimates of signal power in high noise and interference environments.

Code tracking loops respond predictably as long as errors are small enough to stay on the linear part of the S-curve. Then, analytical models of code tracking accuracy are useful. General expressions are presented for code tracking error in white noise, along with simpler versions for small early–late spacing and for BPSK-R spreading modulations. With larger errors, analytical approaches do not exist, in general, to predict code tracking loop performance, and Monte Carlo computer simulations or hardware experiments must be employed. In general, signals having larger RMS bandwidths provide smaller code tracking errors, as do code tracking filters with narrower loop bandwidths.

Carrier-aided code tracking removes line of sight dynamics from the measurements, allowing lower-order code tracking loops with smaller loop bandwidths, providing better performance against noise and interference.

Some S-curves have multiple zero crossings with positive slope, and the loop can achieve stable tracking at any of these points. In applications where such ambiguous tracking is deemed likely, a variety of receiver processing approaches are available to mitigate the problem, all using a blend of lower accuracy and unambiguous measurements along with higher accuracy but ambiguous measurements.

APPENDIX 19.A RMS BANDWIDTH

The signal $s(t)$ has normalized PSD $\bar{\Phi}_s(f)$, with unit area over all frequencies. The corresponding normalized autocorrelation function of the bandlimited received signal is $\bar{R}_s(\tau) = \int_{-\beta_r/2}^{\beta_r/2} \bar{\Phi}_s(f) e^{i2\pi f \tau} df$, with $\bar{R}_s(0) = \eta = \int_{-\beta_r/2}^{\beta_r/2} \bar{\Phi}_s(f) df$. Expand the bandlimited correlation function in a polynomial series near the peak $\tau = 0$ as $\bar{R}_s(\tau) = \eta(1 + a_1\tau + a_2\tau^2 + \cdots)$. Since the correlation function is symmetric, all the odd terms in the polynomial expansion are zero. Assuming it exists,[1] the second derivative of the correlation function is $\frac{d^2\bar{R}_s(\tau)}{d\tau^2} = \eta(2a_2 + \text{Higher order terms involving } \tau)$. Evaluating the result at the peak and rearranging yields $a_2 = \frac{1}{2\eta} \frac{d^2\bar{R}_s(\tau)}{d\tau^2}\big|_{\tau=0}$. Using the Fourier relationship between correlation function and PSD, $\frac{d^2}{d\tau^2}\bar{R}_s(\tau) = -(2\pi)^2 \int_{-\beta_r/2}^{\beta_r/2} f^2 \bar{\Phi}_s(f) e^{i2\pi f \tau} df = -(2\pi)^2 \int_{-\beta_r/2}^{\beta_r/2} f^2 \bar{\Phi}_s(f) \cos(2\pi f \tau) df$. Defining the RMS bandwidth by

$$\beta_{\text{rms}} \stackrel{\Delta}{=} \left[\frac{1}{\eta} \int_{-\beta_r/2}^{\beta_r/2} f^2 \bar{\Phi}_s(f) df \right]^{1/2} \tag{19.21}$$

---

[1] For all the satnav spreading modulations addressed in this book, this second derivative exists anytime the signal has finite bandwidth; consequently, it exists for all practical purposes.

TABLE 19.A.1. RMS Bandwidths for Different Spreading Modulations and Precorrelation Bandwidths

Spreading Modulation	Precorrelation Bandwidth (MHz)	RMS Bandwidth (MHz)
BPSK-R(1)	±12	1.13
BPSK-R(1)	±6	0.80
BOC(1,1)	±12	1.98
BOC(1,1)	±6	1.42
TMBOC(6,1,4/33)	±12	2.69
TMBOC(6,1,4/33)	±6	1.84
BPSK-R(10)	±12	3.48
BOC(10,5)	±12	9.06

yields the coefficient of the quadratic term in the polynomial expansion of the correlation function $a_2 = -2\pi^2 \beta_{rms}^2$. Since in (19.21), the division by η within the brackets keeps the area of $\bar{\Phi}_s(f)/\eta$ unity over any precorrelation bandwidth β_r, the RMS bandwidth as defined is affected only by the shape of the bandlimited PSD, and not by the loss in area caused by the bandlimiting. Also, observe the RMS bandwidth, as defined in (19.21), is a two-sided bandwidth, which is the convention. Finally, note the RMS bandwidth obtains its name from the RMS operation in its definition (19.21); elsewhere in the literature it is sometimes called the Gabor bandwidth.

Over infinite precorrelation bandwidth, the spreading modulations encountered in this book have infinite RMS bandwidth, corresponding to a perfectly triangular peak. With finite precorrelation bandwidth, however, different spreading modulations have distinctly different RMS bandwidth values. Table 19.A.1 shows RMS bandwidths for some representative spreading modulations and precorrelation bandwidths. Observe the legacy spreading modulations BPSK-R(1) and BPSK-R(10) have considerably smaller RMS bandwidths than their more modern counterparts BOC(1,1) and BOC(10,5), respectively, and that TMBOC's RMS bandwidth is significantly greater than that of BOC(1,1).

A correlation function can be approximated near its peak by the two-term polynomial $\bar{R}_s(\tau) \cong \eta(1 + a_2\tau^2)$. Figures 19.A.1–19.A.3 compare the approximation with

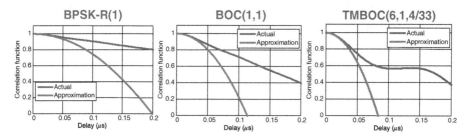

Figure 19.A.1. Comparison of Polynomial Approximations and Exact Correlation Functions for Narrowband Spreading Modulations, ±12 MHz Precorrelation Bandwidths

RMS BANDWIDTH

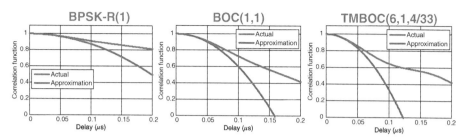

Figure 19.A.2. Comparison of Polynomial Approximations and Exact Correlation Functions for Narrowband Spreading Modulations, ±6 MHz Precorrelation Bandwidths

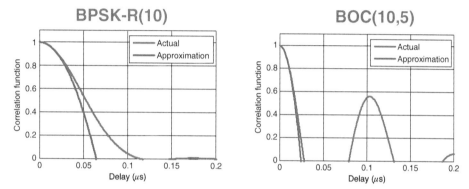

Figure 19.A.3. Comparison of Polynomial Approximations and Exact Correlation Functions for Wideband Spreading Modulations, ±12 MHz Precorrelation Bandwidths

the exact value for different spreading modulations and precorrelation bandwidths. Table 19.A.2 lists the largest delay value where the approximation has less than 10% error relative to the exact correlation function. The approximations apply over larger ranges of delay when signals are more bandlimited, and also for narrower bandwidth spreading modulations.

TABLE 19.A.2. Delay Ranges Over Which Two-Term Polynomial Approximation Is Within 10% of Exact Correlation Function Value

Spreading Modulation	Precorrelation Bandwidth (MHz)	Maximum Delay Magnitude (ns) for Approximation Within 10%
BPSK-R(1)	±12	82
BPSK-R(1)	±6	128
BOC(1,1)	±12	55
BOC(1,1)	±6	87
TMBOC(6,1,4/33)	±12	46
TMBOC(6,1,4/33)	±6	67
BPSK-R(10)	±12	41
BOC(10,5)	±12	18

REVIEW QUESTIONS

Theoretical Questions

QT19.1. Show that as long as $|\tilde{D}|$ and Δ are small, the CELP S-curve in (19.7) is approximated by $S(\tilde{D}) \cong 4\pi^2 \sqrt{C\eta\beta_{rms}^2} \tilde{D}\Delta$.

QT19.2. Show that, for small $|\tilde{D}|$ and Δ, the S-curve in (19.8) is approximated by $S(\tilde{D}) \cong C\eta^2 4\pi^2 \beta_{rms}^2 \tilde{D}\Delta[2 - \pi^2 \beta_{rms}^2 \Delta^2]$.

QT19.3. Show that, for small $|\tilde{D}|$ and Δ, the S-curve in (19.9) is approximated by $S(\tilde{D}) \cong C\eta^2 4\pi^2 \beta_{rms}^2 \tilde{D}\Delta$.

Application Questions

QA19.1. Given a signal with RMS bandwidth of 1 MHz received in white noise, and a receiver using wide precorrelation bandwidth and narrow early–late spacing, with errors dominated by noise.

   a. If the code tracking loop bandwidth is 1 Hz, what is the lowest C/N_0 where the code tracking error is 1 m RMS?

   b. If the RMS bandwidth instead is 2 MHz, how much lower C/N_0 will provide the same code tracking error of 1 m RMS?

   c. If the RMS bandwidth is 2 MHz, how much can the code tracking loop bandwidth be increased while keeping the same code tracking accuracy as in (a)?

REFERENCES

1. J. W. Betz, "Effect of Linear Time-Invariant Distortions on RNSS Code Tracking Accuracy," *Proceedings of the Institute of Navigation Conference on Global Navigation Satellite Systems 2002, ION-GNSS-2002*, Institute of Navigation, September 2002.
2. J. W. Betz and K. R. Kolodziejski, "Generalized Theory of Code Tracking with an Early-Late Discriminator, Part 1: Lower Bound and Coherent Processing," *IEEE Transactions on Aerospace and Electronic Systems*, Vol. 45, No. 4, 2009, pp. 1538–1550.
3. J. W. Betz and K. R. Kolodziejski, "Generalized Theory of Code Tracking with an Early-Late Discriminator, Part 2: Noncoherent Processing and Numerical Results," *IEEE Transactions on Aerospace and Electronic Systems*, Vol. 45, No. 4, 2009, pp. 1551–1564.
4. J. W. Betz and K. R. Kolodziejski, "Extended Theory of Early-Late Code-Tracking for a Bandlimited GPS Receiver," *NAVIGATION: The Journal of the Institute of Navigation*, Vol. 47, No. 3, 2000, pp. 211–226.
5. A. J. Van Dierendonck, P. Fenton, and T. Ford, "Theory and Performance of Narrow Correlator Spacing in a GPS Receiver," *Navigation: Journal of The Institute of Navigation*, Vol. 39, No. 3, 1992, pp. 265–283.
6. M. Soellner and P. Erhard, Comparison of AWGN code tracking accuracy for Alternative-BOC, Complex-LOC and Complex-BOC modulation options in Galileo E5-Band, *Proceedings of the European Navigation Conference ENC-GNSS 2003*, Graz, Austria, April 2003.

7. P. Fine and W. Wilson, "Tracking Algorithm for GPS Offset Carrier Signals," *Proceedings of the 1999 National Technical Meeting of The Institute of Navigation*, ION NTM 1999, January 1999. pp. 671–676.
8. Ward, P., "A Design Technique to Remove the Correlation Ambiguity in Binary Offset Carrier (BOC) Spread Spectrum Signals (Revised Version)," *Proceedings of U.S. Institute of Navigation NTM, (San Diego, CA, January 26–28,* 2004. pp. 886–896.
9. M. S. Hodgart, R. M. Weiler, and M. Unwin, "A Triple Estimating Receiver of Multiplexed Binary Offset Carrier (MBOC) Modulated Signals," *Proceedings of the 21st International Technical Meeting of the Satellite Division of The Institute of Navigation (ION GNSS 2008)*, Savannah, GA, September 2008, pp. 877–886.
10. O. Julien, E. Cannon, G. Lachapelle, C. Mongrédien, and C. Macabiau, "A New Unambiguous BOC(n,n) Signal Tracking Technique," Proceedings of GNSS'2004, *European Navigation Conference*, Rotterdam, Netherlands, May 2004.
11. P. B., Anantharamu, D. Borio, and G. Lachapelle, "Pre-Filtering, Side-Peak Rejection and Mapping: Several Solutions for Unambiguous BOC Tracking," *Proceedings of the 22nd International Technical Meeting of The Satellite Division of the Institute of Navigation (ION GNSS 2009)*, Savannah, GA, September 2009, pp. 3142–3155.
12. O. Julien, C. Macabiau, M. E. Cannon, and G. Lachapelle, "ASPeCT: Unambiguous Sine-BOC(n,n) Acquisition/Tracking Technique for Navigation Applications," *IEEE Transactions on Aerospace and Electronic Systems*, Vol. 43, No. 1, 2007, pp. 150–162.
13. S. Jin, *Global Navigation Satellite Systems: Signal, Theory and Applications*, ISBN: 978-953-307-843-4, InTech, DOI: 10.5772/29453. Available at: http://www.intechopen.com/books/global-navigation-satellite-systems-signal-theory-and-applications, accessed 25 October 2014.

20

POSITION, VELOCITY, AND TIME CALCULATION

This chapter serves as the culmination of the first three parts of this book. Part I describes the design and analysis of signals and systems for satnav. Part II portrays satnav systems being designed, deployed, and operated, and the technical details of their signals. Preceding chapters in Part III provide descriptions of hardware and techniques, and their performance, all leading to measurements of carrier frequency, carrier phase, and signal time of arrival, as well as reading the signals' data messages. This chapter describes how to use those measurements, along with information from the data messages, to calculate PVT.

> This chapter is based in part on instructional materials prepared by Professor Michael Braasch, Thomas Professor of Electrical Engineering and Computer Science, Ohio University. These materials were prepared as part of the NavtechGPS course "Using Advanced GPS/GNSS Signals and Systems." The author thanks Professor Braasch for permission to use these materials in the preparation of this chapter, and also thanks Ms. Carolyn McDonald, President of NavtechGPS, for sponsoring the original development of these materials. Any errors or omissions, however, are the author's.

Engineering Satellite-Based Navigation and Timing: Global Navigation Satellite Systems, Signals, and Receivers, First Edition. John W. Betz.
© 2016 The Institute of Electrical and Electronics Engineers, Inc. Published 2016 by John Wiley & Sons, Inc.

This chapter focuses on standalone receiver operation. Augmentations and differential satnav, as well as assisted satnav, which use multiple spatially separate receivers communicating with the user receiver, provide extensions to standalone receiver operation and are addressed in Chapters 23 and 24, respectively.

Satnav, like many navigation techniques, relies on time. The development in this section addresses many types of time. System time is the time basis to which a satnav system's time is referenced. Each satellite's clock unavoidably deviates from its system time, and these deviations are also addressed. The receiver makes measurements referenced to receiver time, which is offset by an initially unknown amount from system time. As multiple satnav systems become operational, receivers must deal with multiple system times. Other times involved in this chapter include the time of transmission—the system time at which a point on the signal was transmitted, the time of arrival—the receiver time at which the point on the signal was received, and the time of propagation—the time it takes a signal to travel from the satellite transmit antenna to the receive antenna. All of these different times are involved in calculating PVT.

This chapter focuses on standard point positioning (SPP) for PVT, where pseudoranges (and perhaps pseudorange rates) calculated from code tracking (actually tracking of the signals' complex envelopes) are used with information read from the signals' data messages to calculate PVT for a standalone receiver. Section 20.1 describes the formation of pseudorange measurements from delay measurements produced by code tracking techniques in Chapter 19, and carrier phase measurements using carrier phase tracking described in Chapter 18. The initial pseudorange measurements contain errors from multiple sources, and the magnitude of these errors would make the pseudorange measurements virtually useless. Section 20.2 provides algorithms that correct and mitigate these errors, producing corrected and smoothed pseudorange measurements. Section 20.3 provides details of SPP—how to calculate position and time using these enhanced measurements and information from each signal's data message. With multiple operational satnav systems, a receiver can use signals from multiple systems' satellites. Section 20.4 relates the approach for SPP using multiple systems' satellites. Section 20.5 then describes algorithms for calculating receiver velocity. Regardless of the number of satellites available, some receivers will always be disadvantaged, seeking to calculate PVT without having enough measurements from different satellites to use conventional techniques for PVT solutions, or without being able to read each signal's data message. Section 20.6 outlines approaches for disadvantaged receivers to calculate position and time. Section 20.7 provides an introduction to PPP, an emerging set of techniques that can provide positioning accuracies of the order of decimeters or centimeters—one or two orders of magnitude better than typical SPP techniques—without differential techniques. Section 20.8 describes approaches for using redundant measurements to check the integrity of the information obtained from each satellite, while Section 20.9 summarizes the chapter.

20.1 FORMING MEASUREMENTS

Suppose the receiver knows the true time of transmission of the beginning of a data message for the kth signal, in system time. As long as the receiver counts the integer and fractional number of spreading code durations between the beginning of this data message and a particular point on the received signal, the time of transmission at that point of the signal is known in system time.

Denote the time of transmission for this particular point on the signal by $\tau_t^{(k)}$, where the superscript (k) is placed in parentheses to designate the satellite but avoid misinterpretation as an exponent. Ideally, the receiver measures the time of arrival of the same point on the signal relative to the same system time reference, denoted $\tau_{a,m}^{(k)}$, where the subscript a indicates "arrival," the subscript m refers to the signal on carrier frequency f_m that was used to obtain the measurement. If the true speed of propagation is c, then the true range from the receiver (actually, the phase center of the receive antenna) to the satellite (actually, the phase center of the transmit antenna) is

$$r^{(k)} = \left(\tau_{a,m}^{(k)} - \tau_t^{(k)}\right)c \tag{20.1}$$

where $\tau_{a,m}^{(k)} - \tau_t^{(k)}$ is the propagation delay.

Since many of the assumptions leading to (20.1) do not apply in practice, things get more complicated. The satellite clock that dictates time of transmission typically differs from system time by an offset $\Delta_t^{(k)}$ at the time of transmission, yielding an uncorrected time of transmission $\hat{\tau}_t^{(k)} = \Delta_t^{(k)} + \tau_t^{(k)}$. The propagation delay includes unknown contributions due to delays from ionospheric $\Delta_{iono,m}^{(k)}$ and tropospheric propagation $\Delta_{tropo}^{(k)}$. (Note ionospheric delay is specific to carrier frequency f_m while tropospheric delay is the same for any signal at any L band carrier frequency from that satellite.) The receiver clock is offset from system time by an unknown amount, Δ_r which is the same for all receiver channels tracking signals, so is not indexed by satellite or by carrier frequency, yielding an uncorrected time of arrival $\hat{\tau}_a^{(k)} = \Delta_r + \tau_a^{(k)}$.

In practice, the term $\Delta_t^{(k)}$ actually varies not only by carrier frequency but also by specific signal modulated onto that carrier. To address these imperfections, the satellite broadcasts a group delay correction that describes how to adjust pseudorange values computed from signals at different carrier frequencies due to slightly different delays at different frequencies within the satellite. Furthermore, when multiple signals are transmitted on a given carrier frequency, that satellite broadcasts inter-signal corrections so the receiver can correct the delay of each signal relative to the delay of the signal for which group delay corrections are provided. Each signal's ICD should be referenced to understand satellite group delay correction and inter-signal corrections, which are not discussed further here.

The true pseudorange is based on satellite time aligned with system time yielding

$$\begin{aligned}\tilde{r}^{(k)} &= \left(\hat{\tau}_a^{(k)} - \tau_t^{(k)}\right)c \\ &= \left(\tau_a^{(k)} - \tau_t^{(k)}\right)c + \Delta_r c\end{aligned} \tag{20.2}$$

where in the last line the first term is the geometric range, $r^{(k)}$, and the second term is the unknown range bias caused by the unknown time offset between receiver time and system time, turning geometric range into pseudorange. This range bias is the same for all satellites and frequencies, and thus not labeled with a satellite index or a frequency index. The corresponding pseudodelay is $\tau_a^{(k)} - \tau_t^{(k)} + \Delta_r$.

The delay at the receiver is measured relative to the receiver clock, and also includes ionospheric and tropospheric delay, so the measured delay using code tracking, relative to system time, would be $D_m^{(k)} = \Delta_r + \tau_a^{(k)} + \Delta_{iono,m}^{(k)} + \Delta_{tropo}^{(k)}$. However, the delay estimate for the signal at that frequency also has jitter due to noise, interference, and multipath, denoted $\tilde{D}_m^{(k)}$, so that the delay estimate is $\hat{D}_m^{(k)} = D_m^{(k)} + \tilde{D}_m^{(k)} = \Delta_r + \tau_a^{(k)} + \Delta_{iono,m}^{(k)} + \Delta_{tropo}^{(k)} + \tilde{D}_m^{(k)}$.

The estimated pseudorange, before correcting for satellite clock errors, ionospheric delay, tropospheric delay, and jitter due to noise and interference, is then

$$\begin{aligned}
\widehat{r}_m^{(k)} &= \left(\hat{D}_m^{(k)} - \hat{\tau}_t^{(k)}\right)c \\
&= \left(\Delta_r + \tau_a^{(k)} + \Delta_{iono,m}^{(k)} + \Delta_{tropo}^{(k)} + \tilde{D}_m^{(k)} - \Delta_t^{(k)} - \tau_t^{(k)}\right)c \\
&= \left(\tau_a^{(k)} - \tau_t^{(k)}\right)c + \Delta_r c + \left(\Delta_{iono,m}^{(k)} + \Delta_{tropo}^{(k)} - \Delta_t^{(k)} + \tilde{D}_m^{(k)}\right)c \\
&= \widehat{r}^{(k)} + \widetilde{r}_m^{(k)}
\end{aligned} \qquad (20.3)$$

where $\widetilde{r}_m^{(k)} \triangleq (\Delta_{iono,m}^{(k)} + \Delta_{tropo}^{(k)} - \Delta_t^{(k)} + \tilde{D}_m^{(k)})c$ is the pseudorange error from the signal at frequency f_m.

When carrier phase measurements are also available, the true carrier phase at the receiver is the fraction of a cycle relative to $N_m^{(k)}$, the integer number of wavelengths between receive antenna phase center and transmit antenna phase center, so $\theta_m^{(k)} = r^{(k)}/\lambda_m - N_m^{(k)}$. Defining $\Phi_m^{(k)} \triangleq \theta_m^{(k)} \lambda_m$ yields $\Phi_m^{(k)} = r^{(k)} - N_m^{(k)} \lambda_m$. The estimated carrier phase then contains similar terms to those for estimated pseudorange, so

$$\widehat{\Phi}_m^{(k)} = r^{(k)} + \left(\Delta_{tropo}^{(k)} - \Delta_{iono,m}^{(k)}\right)c + \left(\Delta_r - \Delta_t^{(k)}\right)c - N_m^{(k)} \lambda_m + \widetilde{\Phi}_m^{(k)} \qquad (20.4)$$

A fundamental difference between (20.4) and (20.3) is the sign of the ionospheric delay, resulting from the dispersive phenomenon described in Section 3.2.2. In addition, $N_m^{(k)}$ is unknown a priori, and the estimation error, $\widetilde{\Phi}_m^{(k)} \triangleq \frac{\tilde{\theta}_m^{(k)}}{2\pi} \lambda_m$ typically has variance much smaller than the variance of the corresponding estimation error from code tracking, $\tilde{D}_m^{(k)} c$.

While in practice, four or more pseudorange estimates like (20.3) (or carrier phase estimates like (20.4) if $N_m^{(k)}$ were known) could be used to solve for receiver location, the errors can be very large—up to tens or hundreds of kilometers—resulting in correspondingly large position and time errors. Errors due to satellite clock offset, troposphere, code tracking, and ionosphere must be reduced before PVT can be calculated accurately. Section 20.2 describes how to reduce these errors, but they are not eliminated completely.

20.2 REDUCING PSEUDORANGE ERRORS

This section describes techniques for reducing each contributor to pseudorange error. Many of these pseudorange corrections require the receiver to use time or receiver position in the correction. This can seem contradictory since the receiver is trying to correct the pseudorange errors in order to solve for time and position. However, the corrections do not require precise time or position, so the receiver can use an estimate, or a value calculated without all corrections applied, in the correction calculation. The use of carrier phase measurements for positioning is deferred to Section 20.7 and Chapter 23.

Section 20.2.1 addresses correcting time of transmission, which should be performed first since it corrects for potentially the largest errors. Section 20.2.2 describes corrections for tropospheric errors. Section 20.2.3 relates techniques for reducing pseudorange jitter caused by noise and interference and rapidly fluctuating multipath, while Section 20.2.4 describes corrections for ionospheric errors. The approaches described in Sections 20.2.1, 20.2.2, and 20.2.4 also adapt to correcting carrier phase measurements.

20.2.1 Correcting Time of Transmission

While conceptually all satellite clocks are perfectly synchronized to system time, in practice, each satellite's clock drifts from system time. Much of this drift is smooth and predictable, and each satellite's data message provides the correction between satellite time and system time. When the offset gets large enough, the satellite is taken out of service, the clock is reset, and then the satellite is brought back into service.

The ICD describing each signal describes how clock corrections are represented in that signal's data message, and how to correct satellite time to system time. The following example applies to GPS signals defined in Reference 1, and similar representations are employed for most other signals and systems. The estimated satellite clock error is

$$\widehat{\Delta_t^{(k)}} = a_{f0}^{(k)} + a_{f1}^{(k)}(t - t_{oc}) + a_{f2}^{(k)}(t - t_{oc})^2 + \Delta t_{rel}^{(k)} \tag{20.5}$$

representing a quadratic curve fit to the clock offset for any time t different from the instant in time t_{oc} at which the clock parameters apply (for GPS signals the quantity $t - t_{oc}$ must be adjusted modulo time of the week per the algorithms provided in the GPS IS). In some cases, more modern signals may only use a linear curve fit for $t - t_{oc}$ less than days or weeks, reflecting better quality satellite clocks that can be adequately modeled by only a linear approximation. In (20.5), $\Delta t_{rel}^{(k)} \stackrel{\Delta}{=} F \cdot e \cdot \sqrt{A} \sin(E^{(k)})$ is a relativistic correction due to any eccentricity of the kth satellite's orbit (the relativistic effect due to uniform motion in a circular orbit is accounted for by purposely setting the satellite clock frequency slightly below the frequency seen by an Earth-bound observer; for GPS the clock frequency is 0.0046 Hz below 10.23 MHz), with $F = -4.442807633 \times 10^{-10} \, s/\sqrt{m}$, e the eccentricity of the orbit, A the semi-major axis of the orbit, and $E^{(k)}$ the satellite's eccentric anomaly, indicating the satellite's location in its orbit, all obtained from the signal's data message. An estimate of t can be used in (20.5), since the correction changes very little over time differences of many seconds. Without the relativistic correction $\Delta t_{\text{rel}}^{(k)}$, time estimates could have errors of tens of nanoseconds and position errors could be more than 10 m for satellites in nominally circular orbits, and much larger for satellites in orbits having greater eccentricity. The clock corrected pseudorange estimate at the mth frequency is then

$$\left(\hat{r}_m^{(k)}\right)_{\text{clock corrected}} = \hat{r}_m^{(k)} + \widehat{\Delta_t^{(k)}} c \qquad (20.6)$$

Any residual satellite clock error is typically of the order of a nanosecond or less, so deemed negligible for SPP. It is then included in the residual error term $\tilde{D}_m^{(k)}$. The remaining pseudorange error is then $(\Delta_{\text{tropo}}^{(k)} + \tilde{D}_m^{(k)} + \Delta_{\text{iono},m}^{(k)})c$.

20.2.2 Tropospheric Propagation Delay Correction

As discussed in Chapter 6, the troposphere extends from ground level up to approximately 50 km altitude, although the bulk of the tropospheric effect is due to the lower 20 km or so. The troposphere is nondispersive at GPS frequencies and primarily causes refraction, or ray bending, that changes the perceived line-of-sight propagation speed of the signal. Tropospheric delay errors can cause pseudorange errors of a few meters for high elevation angle satellites up to tens of meters for low elevation angle satellites, whose signals have a longer path length through the troposphere. Fortunately, most of the tropospheric delay is due to the dry-gas component which is well-modeled by the ideal gas law. The zenith dry component for a receiver at sea level is approximately 2.5 m. The water-vapor component, also known simply as the wet component, is not easily modeled. It varies from approximately 1% of the dry component in hot desert regions to approximately 10% of the dry component in tropical regions, depending upon weather conditions, and is also highly sensitive to receiver altitude, since much of the effect is below 10 km altitude.

The receiver designer can employ a tropospheric correction model of his or her own choice, based on considerations of accuracy and complexity, including the availability of meteorological data. In general, the form of the model is (Zenith dry + Zenith wet) × (Elevation mapping function) × (Receiver altitude factor). Some useful tropospheric models are discussed in Reference 2, and a widely used model for tropospheric delay, with good accuracy and moderate complexity, is provided in Reference 3.

As an example, a simple model that compensates for the dry component is

$$\widehat{\Delta_{\text{tropo}}^{(k)}} c = \frac{2.4224}{0.026 + \sin(\eta^{(k)})} e^{-0.13345 h_r} \quad (\text{m}) \qquad (20.7)$$

where $\eta^{(k)}$ is the elevation angle from receiver to the kth satellite, and h_r is the receiver altitude in kilometers. Figure 20.1 shows the result, with predicted delays as large as 22 m for low elevation satellites and a receiver at sea level, to delay of 62 cm for a

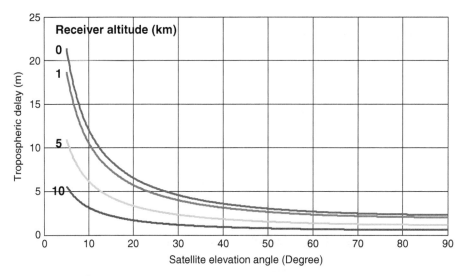

Figure 20.1. Predicted Dry Component of Tropospheric Error

satellite at zenith and receiver at aircraft altitude. Correcting for the tropospheric error removes an additional term from the pseudorange estimate (20.3), leaving

$$\left(\hat{r}_m^{(k)}\right)_{\substack{\text{clock corrected} \\ \text{tropo corrected}}} = \hat{r}_m^{(k)} + \left(\widehat{\Delta_t^{(k)}} - \widehat{\Delta_{\text{tropo}}^{(k)}}\right) c \qquad (20.8)$$

Typically, the residual tropospheric error is small relative to other errors in SPP. The residual tropospheric error is then included in the residual error term $\tilde{D}_m^{(k)}$, and the remaining error is then $\left(\Delta_{\text{iono},m}^{(k)} + \tilde{D}_m^{(k)}\right) c$.

20.2.3 Reducing Jitter in the Delay Estimate

Before pseudorange estimates are used to develop an ionosphere-free pseudorange estimate, and either before or after satellite clock correction and troposphere correction, there is a further opportunity to reduce pseudorange errors that are due to noise, interference, and rapidly fluctuating multipath. While multipath is addressed in Chapter 22; if it fluctuates rapidly relative to the reciprocal of the code tracking loop's bandwidth, its effects are jitter—rapidly fluctuating errors—that are smoothed by the loop filter and can also be addressed using techniques discussed in this subsection. If the multipath causes errors that remain invariant over times comparable to or longer than the reciprocal of the code tracking loop bandwidth, then techniques described in Chapter 22 must be employed to reduce the effects.

As discussed in Chapters 17 and 19, the first steps to reducing jitter in the delay estimate involve good design of the code tracking loop, including use of carrier aiding to remove dynamics, and the selection of as small a loop bandwidth as possible, given other constraints.

Comparing carrier tracking errors due to noise in Chapter 18 with code tracking errors due to noise in Chapter 19 provides important insights. RMS carrier tracking errors are typically calculated in units of radians, and may be several tenths of a radian. A 0.2 radian carrier phase error is less than 3% of a wavelength, which is less than 1 cm over the range of carrier frequencies from 1176 MHz to 1600 MHz. In contrast, RMS code tracking errors may be of the order of 1 m or more, or more than a factor of 100 larger.

However, the accuracy of carrier phase tracking is accompanied by carrier phase ambiguities over wavelengths of the order of 20 cm. Even if the correct number of cycles can be found initially between the receiver and the satellite, any carrier phase tracking loss of lock (known as a cycle slip) introduces ambiguities that need to be resolved. In contrast, while code tracking jitter may be two orders of magnitude greater than carrier tracking jitter, if there are any code tracking ambiguities at all, they typically occur over ranges of tens or hundreds of meters—two or more orders of magnitude larger than carrier tracking ambiguities.

Some of the advances in receiver processing accuracy result from ways of combining carrier phase measurements with code tracking measurements in order to exploit carrier phase accuracy with code tracking's resistance to ambiguities. One example already examined is carrier-aided code tracking described in Section 19.3. Carrier-smoothed code (CSC) measurements are described here. Even more sophisticated approaches are found in Section 20.6 and in Chapter 23.

The Hatch filter [4, 5] produces CSC measurements by combining delta carrier phase measurements with pseudorange measurements. The algorithm, derived in Review Question QT20.1, is as follows

$$\langle \hat{D} \rangle [1] = \hat{D}[1]$$

$$\langle \hat{D} \rangle [\ell] = \frac{1}{\ell} \hat{D}[\ell] + \frac{\ell-1}{\ell} \left(\langle \hat{D} \rangle [\ell-1] + \frac{\hat{\theta}[\ell] - \hat{\theta}[\ell-1]}{2\pi f_m} \right), \quad 1 < \ell \leq L$$

$$\langle \hat{D} \rangle [\ell] = \frac{1}{L} \hat{D}[\ell] + \frac{L-1}{L} \left(\langle \hat{D} \rangle [\ell-1] + \frac{\hat{\theta}[\ell] - \hat{\theta}[\ell-1]}{2\pi f_m} \right), \quad L < \ell \quad (20.9)$$

where $\langle \hat{D} \rangle [\ell]$ is the CSC at time index ℓ counted from the beginning of signal tracking or over the time that carrier phase tracking has been continuous, with no cycle slips or losses of lock, $\hat{D}[\ell]$ is the delay estimate from the code tracking loop at time index ℓ, and $\hat{\theta}[\ell]$ is the carrier phase estimate at time index ℓ.

As analyzed in Review Question QT20.1, the Hatch filter aligns carrier phase and code delay, using the higher accuracy changes in carrier phase to steer the delay so that averaging removes jitter without smearing the valid changes due to line-of-sight velocity. Each time the carrier phase tracking loses lock or slips cycles, the delta carrier phase term $\hat{\theta}[\ell] - \hat{\theta}[\ell-1]$ then includes an unknown integer multiple of 2π, which biases the smoothed delay. To avoid corruption of subsequent smoothed information by this bias, the filter must begin with a new initialization, producing delay estimates with larger errors until L measurements have been smoothed.

The variance of jitter in the delay estimate is reduced by 2ℓ for $1 < \ell \leq L$, and by $2L$ for $L > \ell$, motivating large values of L. However, if single-frequency measurements are used, code-carrier divergence caused by hanging ionospheric effects over time and pierce point, as described in Section 3.3, limits the size of L. The value of LT, where T is the correlation integration time used to update the estimate of delay, is restricted, typically ranging between 10 seconds and 100 seconds. If ionosphere-free measurements are used from dual-frequency receivers as described in Section 20.2.4, the variance of each measurement is larger but the first-order code-carrier divergence is reduced and much larger values of LT can be used.

Another variation of CSC is found in Reference 6. The CSC algorithm involves first taking the difference between the code measurement (20.3) and the carrier measurement (20.4), to yield $2\Delta_{\text{iono},m}^{(k)} c + N_m^{(k)} \lambda_m + \tilde{D}_m^{(k)} c - \tilde{\Phi}_m^{(k)}$. The resulting sequential carrier-minus-code differences are then smoothed or averaged over tens or hundreds of seconds, resulting in $2\Delta_{\text{iono},m}^{(k)} c + N_m^{(k)} \lambda_m + \langle \tilde{D}_m^{(k)} c - \tilde{\Phi}_m^{(k)} \rangle_L$, where L is the number of differences that are smoothed, as long as the ionospheric delay is constant over the smoothing time, and there are no carrier cycle slips so that $N_m^{(k)}$ remains the same. Finally, the most recent carrier measurement, corrected for satellite clock offset and tropospheric delay (actually, these corrections can be applied afterward), is added to the smoothed result, yielding $r^{(k)} + \Delta_{\text{iono},m}^{(k)} c + \langle \tilde{D}_m^{(k)} c - \tilde{\Phi}_m^{(k)} \rangle_L + \tilde{\Phi}_m^{(k)}$. The ionospheric error remains, the variance of the code tracking measurement has been reduced by L, and error terms related to carrier phase jitter have been added. Recognizing that the variance of the code tracking measurement can easily be two orders of magnitude greater than the variance of the carrier phase measurement, it is clear that the CSC can provide a much more accurate pseudorange measurement.

The result of these error reduction steps is an improved set of pseudorange measurements, corrected for satellite clock offset from system time, for tropospheric delay, and with jitter smoothed. These improved measurements at frequency f_1 denoted

$$\widehat{\hat{r}_m^{(k)}} = \left(\hat{r}_m^{(k)}\right)_{\substack{\text{clock-corrected} \\ \text{tropo-corrected} \\ \text{carrier-smoothed}}} \tag{20.10}$$

are used in the PVT calculations described in the remainder of this chapter.

20.2.4 Ionospheric Propagation Delay Correction

As discussed in Sections 3.2 and 6.1, the ionosphere, which extends from about 50 km to 1000 km above the Earth's surface, introduces an unknown additional delay in signal propagation time. Approximately 99% of this delay is accounted for by a simple model of the delay as inversely proportional to the signal's center frequency. However, the constant of proportionality depends on ionospheric conditions that vary with time and location. Receivers tracking signals on only a single frequency can use an ionospheric model to reduce the ionospheric error. Each signal's data message provides an ionospheric model that can be used as described in that signal's ICD. These models only account for general trends, and residual errors can be many meters even after the model is applied. Alternatively, augmentation signals or other sources may provide higher fidelity sophisticated models that account for ionospheric variations based on time of day and location. The associated documentation must be used to employ these models; such a discussion is beyond the scope of this chapter. Receivers tracking signals on two frequencies, however, can significantly reduce ionospheric errors without the use of models, and this approach is addressed here in greater detail. A dual-frequency receiver has two pseudorange estimates of the form (20.8), corrected for satellite clock error and tropospheric error, at carrier frequencies f_{m_1} and f_{m_2}. The pair of frequencies 1575.42 MHz and 1176.42 MHz (L1 and L5 for GPS and QZSS, E1 and E5a for Galileo) may be most common in future. The objective is to provide an "ionosphere-free pseudorange estimate." (Actually, this approach removes the first-order effects of ionosphere, which under normal ionospheric conditions leaves only centimeter-level errors.) The ionosphere-free pseudorange estimate is defined to be

$$\left(\hat{r}^{(k)}\right)_{\substack{\text{clock-corrected}\\\text{tropo-corrected}\\\text{carrier-smoothed}\\\text{iono-free}}} \stackrel{\Delta}{=} \frac{\omega_{m_1,m_2}\left(\hat{r}^{(k)}_{m_1}\right)_{\substack{\text{clock-corrected}\\\text{tropo-corrected}\\\text{carrier-smoothed}}} - \left(\hat{r}^{(k)}_{m_2}\right)_{\substack{\text{clock-corrected}\\\text{tropo-corrected}\\\text{carrier-smoothed}}}}{\omega_{m_1,m_2} - 1} \quad (20.11)$$

where $\omega_{m_1,m_2} \stackrel{\Delta}{=} f_{m_1}^2/f_{m_2}^2$ and without loss of generality require $f_{m_1} > f_{m_2}$ so that $\omega_{m_1,m_2} > 1$. Review Question QT20.2 shows that, when there are no pseudorange errors, the ionosphere-free pseudorange is the true pseudorange. This review question further shows that, when there are pseudorange estimation errors at each frequency, with variance at frequency f_m denoted $(\sigma_m^{(k)})^2$, and the errors are zero mean and uncorrelated, the variance of the ionosphere-free pseudorange error is

$$\left(\sigma_{\text{iono-free}}^{(k)}\right)^2 = \frac{1}{(\omega_{m_1,m_2} - 1)^2}\left[\omega_{m_1,m_2}^2\left(\sigma_{m_1}^{(k)}\right)^2 + \left(\sigma_{m_2}^{(k)}\right)^2\right] \quad (20.12)$$

This result shows three important characteristics of ionosphere-free pseudorange accuracy:

- The ionosphere-free pseudorange error is a weighted combination of the pseudorange errors at each frequency,
- The errors of the higher-frequency pseudorange estimate have greater effect on the ionosphere-free pseudorange variance than do the errors of the lower-frequency pseudorange estimate,
- The variance of the ionosphere-free pseudorange decreases significantly as the squared frequency ratio ω_{m_1,m_2} increases.

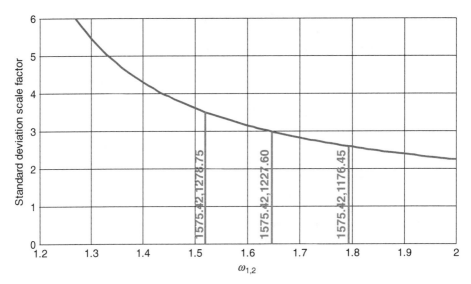

Figure 20.2. Scaling of Single-Frequency Pseudorange Standard Deviations in Ionosphere-Free Standard Deviation

The same review question shows further that when, in addition, the pseudorange error variances are the same at both frequencies so that $(\sigma^{(k)})^2 \triangleq (\sigma_{m_1}^{(k)})^2 = (\sigma_{m_2}^{(k)})^2$, then the variance of the ionosphere-free pseudorange error is

$$\left(\sigma_{\text{iono-free}}^{(k)}\right)^2 = \frac{\omega_{m_1,m_2}^2 + 1}{(\omega_{m_1,m_2} - 1)^2}(\sigma^{(k)})^2 \tag{20.13}$$

In this case, the standard deviation of ionosphere-free pseudorange error is greater than the standard deviation of either single-frequency pseudorange error by the scale factor $\sqrt{\omega_{m_1,m_2}^2 + 1}/(\omega_{m_1,m_2} - 1)$. Whereas this scale factor would be $\sqrt{2}$ for the unweighted difference between two uncorrelated quantities having the same variance, Figure 20.2 shows that this scale factor exceeds 2.5 for any of the typical pairs of L band satnav frequencies, and exceeds 3.5 for Galileo E1/E6.

Removing the dual-frequency estimate of the ionospheric error leaves the corrected pseudorange estimate

$$(\hat{r}^{(k)})_{\substack{\text{clock-corrected} \\ \text{tropo-corrected} \\ \text{carrier-smoothed} \\ \text{iono-free}}} = r^{(k)} + c\tilde{D}^{(k)} \tag{20.14}$$

Observe the carrier frequency is not shown explicitly in (20.14) since the correction for ionospheric delay makes that distinction unnecessary, also, while the same symbol is used for the error term in (20.14) as in earlier expressions, in fact its value is different and its variance is larger due to the ionospheric correction.

Since the errors in ionosphere-free pseudorange are magnifications of the errors in pseudorange errors at each frequency, there are great benefits in reducing those pseudorange errors. Good design of code tracking loops is beneficial, as are other approaches, including the smoothing described in Section 20.2.3.

20.3 STANDARD POINT POSITIONING

This section addresses SPP, using pseudorange estimates with clock and ephemeris data (CED) from each signal's data message. The approach is the same regardless of which corrections in Section 20.2 are employed. At a minimum, however, the receiver typically corrects the satellite clock, applies a model for the troposphere, and either uses a model for the ionosphere to reduce ionospheric errors, or employs dual-frequency processing and uses the ionosphere-free pseudorange estimate. Smoothing of the pseudorange measurements may or may not be employed, but is highly beneficial.

The pseudorange estimate to the kth satellite, after any or all corrections and smoothing described in Section 20.2, is simply denoted

$$\widehat{\widehat{r}}_k \stackrel{\Delta}{=} \widehat{r}_k + c\tilde{D}_k \tag{20.15}$$

where the notation is simplified to indicate the satellite index as a subscript, and no explicit frequency dependence. The receiver clock offset from GPS system time is Δ_r. The resulting geometry is portrayed in Figure 20.3. The true pseudorange is the sum of the range and the range bias, which is $\Delta_r c$ and may be positive or negative, so that the pseudorange may be longer than or shorter than the true range.

Section 20.3.1 describes determination of the satellite position, while Section 20.3.2 sets up the system of equations for finding receiver position and clock offset. An algorithm for solving this system of equations is developed and described in Section 20.3.3.

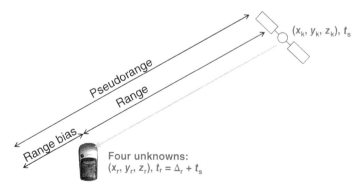

Figure 20.3. Relationship between Range, Range Bias, and Pseudorange

20.3.1 Determining Satellite Position at Time of Arrival

The satellite position is obtained from the CED in the signal's data message, and processed per the algorithm in each signal's ICD to obtain the satellite's position in ECEF coordinates at the time of transmission. Although there are errors in the ephemeris, those errors are typically small and included in pseudorange errors.

However, a subtle but important refinement is needed. The satellite position at time of transmission is in ECEF coordinates, and the desired receiver location is also in ECEF coordinates. But this coordinate frame rotates in an inertial coordinate frame during the time the signals propagate to the receiver. Section 2.4 shows the propagation delay can approach 0.1 second between satellites in higher altitude medium Earth orbits and terrestrial receivers. Delays can exceed 0.1 second for terrestrial receivers and satellites in geosynchronous orbits, and for space-based receivers. Since the circumference of the Earth at the equator is approximately 40,075 km at the equator, a point on the equator moves approximately 464 m/s due to Earth rotation, or more than 46 m during 0.1 second of propagation time. Clearly the error introduced by this Sagnac effect depends upon the latitude of the receiver, with no error introduced for a receiver at either pole.

Removing the Sagnac effect involves using a rotation matrix to transform each satellite's ECEF coordinates at time of transmission into the ECEF coordinates at time of arrival. During the time the signal from the kth satellite propagates to the receiver, $\tau_{\text{prop}}^{(k)}$, the Earth has rotated by an angle $\phi_{\text{Earth}}^{(k)} = \tau_{\text{prop}}^{(k)} \Omega_e$, where Ω_e is the Earth rotation rate, with value provided in the signal's ICD. The ECEF coordinates of the kth satellite at time of transmission $[x_k^{\text{TOT}} \; y_k^{\text{TOT}} \; z_k^{\text{TOT}}]^T$, are rotated into ECEF coordinates at time of arrival $[x_k^{\text{TOA}} \; y_k^{\text{TOA}} \; z_k^{\text{TOA}}]^T$ by

$$\begin{bmatrix} x_k^{\text{TOA}} \\ y_k^{\text{TOA}} \\ z_k^{\text{TOA}} \end{bmatrix} = \begin{bmatrix} \cos \phi_{\text{Earth}}^{(k)} & \sin \phi_{\text{Earth}}^{(k)} & 0 \\ -\sin \phi_{\text{Earth}}^{(k)} & \cos \phi_{\text{Earth}}^{(k)} & 0 \\ 0 & 0 & 1 \end{bmatrix} \begin{bmatrix} x_k^{\text{TOT}} \\ y_k^{\text{TOT}} \\ z_k^{\text{TOT}} \end{bmatrix} \qquad (20.16)$$

and the latter are used in calculations described in the remainder of this section.

Before discussing how to obtain $\tau_{\text{prop}}^{(k)}$, consider the accuracy that is needed in this parameter value. For every kilometer of error in the geometric range between receiver and satellite, the propagation time error is one-third of a microsecond. A point on the equator has rotated a fraction of 1 cm during this time. Consequently, errors of even tens of kilometers in geometric range or multiple microseconds in propagation time are not significant in compensating for the Sagnac effect in SPP.

Consequently, there are several approaches that can be used for obtaining $\tau_{\text{prop}}^{(k)}$ to perform the coordinate conversion in (20.16). The most basic approach is to solve for receiver location, as described in the remainder of this section, without removing the Sagnac effect. Use the resulting estimate of receiver position to perform the coordinate rotation (20.16), and solve again for receiver location using the corrected satellite position. Alternately, if the receiver has current ephemeris or even almanac information for the satellite, and a recent estimate of receiver location (with accuracy typically within a few kilometers), the resulting estimate of geometric range is sufficient to compute $\tau_{prop}^{(k)}$ with adequate accuracy.

20.3.2 System of Equations for Finding Receiver Position and Clock Offset

Suppose $K \geq 4$ satellites are being tracked. The Pythagorean theorem yields the system of equations

$$
\begin{aligned}
r_1 &= \sqrt{(x_r - x_1)^2 + (y_r - y_1)^2 + (z_r - z_1)^2} + \Delta_r c \\
r_2 &= \sqrt{(x_r - x_2)^2 + (y_r - y_2)^2 + (z_r - z_2)^2} + \Delta_r c \\
r_3 &= \sqrt{(x_r - x_3)^2 + (y_r - y_3)^2 + (z_r - z_3)^2} + \Delta_r c \\
&\vdots \\
r_K &= \sqrt{(x_r - x_K)^2 + (y_r - y_K)^2 + (z_r - z_K)^2} + \Delta_r c
\end{aligned}
\quad (20.17)
$$

Here, the satellite position is $[x_k \ y_k \ z_k]^T$ may be computed with or without removal of the Sagnac effect, as discussed at the end of Section 20.3.1.

After 20 chapters and hundreds of pages and hundreds of equations, the first equation in the book, (1.1), has resurfaced. At this point the reader has learned how to obtain all the needed quantities in order to obtain a navigation solution.

20.3.3 Solving the System of Equations

The conventional way to solve this nonlinear system of equations (there are other approaches, such as in Reference 7) is to expand each equation in a Taylor series, as discussed in Appendix Section A8. The result of expanding the kth equation around the values $(x_r^{(\ell)}, y_r^{(\ell)}, z_r^{(\ell)}, \Delta_r^{(\ell)})$, where ℓ is an iteration index to be explained subsequently, is

$$
\begin{aligned}
r_k = r_r^{(\ell)} &+ \frac{x_r^{(\ell)} - x_k}{r_r^{(\ell)} - \Delta_r^{(\ell)} c}(x_r - x_r^{(\ell)}) + \frac{y_r^{(\ell)} - y_k}{r_r^{(\ell)} - \Delta_r^{(\ell)} c}(y_r - y_r^{(\ell)}) + \frac{z_r^{(\ell)} - z_k}{r_r^{(\ell)} - \Delta_r^{(\ell)} c}(z_r - z_r^{(\ell)}) \\
&+ (\Delta_r - \Delta_r^{(\ell)}) c
\end{aligned}
\quad (20.18)
$$

where $r_r^{(\ell)} \triangleq \sqrt{(x_r^{(\ell)} - x_k)^2 + (y_r^{(\ell)} - y_k)^2 + (z_r^{(\ell)} - z_k)^2} + \Delta_r^{(\ell)} c$ is the pseudorange calculated for the ℓth iteration. The known quantities in (20.18) are the pseudorange estimate (actually, the pseudorange estimate is known, as introduced below) and the three coordinates of satellite position. The unknown quantities are the three coordinates of receiver position and the clock bias.

If the pseudorange were known exactly, the quantity $\delta r_k^{(\ell)} = r_k - r_r^{(\ell)}$ would be known and the quantities $\delta x_r^{(\ell)} \triangleq x_r - x_r^{(\ell)}$, $\delta y_r^{(\ell)} \triangleq y_r - y_r^{(\ell)}$, $\delta z_r^{(\ell)} \triangleq z_r - z_r^{(\ell)}$, $\delta \Delta_r^{(\ell)} \triangleq \Delta_r - \Delta_r^{(\ell)}$ would be unknown. Then (20.18) could be written

$$\delta r_k^{(\ell)} = \frac{x_r^{(\ell)} - x_k}{r_r^{(\ell)} - \Delta_r^{(\ell)} c} \delta x_r^{(\ell)} + \frac{y_r^{(\ell)} - y_k}{r_r^{(\ell)} - \Delta_r^{(\ell)} c} \delta y_r^{(\ell)} + \frac{z_r^{(\ell)} - z_k}{r_r^{(\ell)} - \Delta_r^{(\ell)} c} \delta z_r^{(\ell)} + \delta \Delta_r^{(\ell)} c \quad (20.19)$$

This equation is linear in the four unknowns. Define the $\delta \hat{r}_k^{(\ell)} \triangleq \hat{r}_k - r_r^{(\ell)}$ where the first quantity on the left hand is from (20.15). Since the pseudorange is not known exactly, $\delta \hat{r}_k^{(\ell)}$ is substituted into (20.19). Admitting a resulting error term on the right-hand side yields

$$\delta \hat{r}_k^{(\ell)} = \frac{x_r^{(\ell)} - x_k}{r_r^{(\ell)} - \Delta_r^{(\ell)} c} \delta x_r^{(\ell)} + \frac{y_r^{(\ell)} - y_k}{r_r^{(\ell)} - \Delta_r^{(\ell)} c} \delta y_r^{(\ell)} + \frac{z_r^{(\ell)} - z_k}{r_r^{(\ell)} - \Delta_r^{(\ell)} c} \delta z_r^{(\ell)} + \delta \Delta_r^{(\ell)} c + \varepsilon_k^{(\ell)}$$

(20.20)

The system of nonlinear equations (20.17) is then replaced by the system of linear equations known as the SPP equations

$$\begin{bmatrix} \delta \hat{r}_1^{(\ell)} \\ \delta \hat{r}_2^{(\ell)} \\ \delta \hat{r}_3^{(\ell)} \\ \vdots \\ \delta \hat{r}_K^{(\ell)} \end{bmatrix} = \begin{bmatrix} \mathbf{u}_1^{(\ell)} & 1 \\ \mathbf{u}_2^{(\ell)} & 1 \\ \mathbf{u}_3^{(\ell)} & 1 \\ \vdots & \vdots \\ \mathbf{u}_K^{(\ell)} & 1 \end{bmatrix} \begin{bmatrix} \delta x_r^{(\ell)} \\ \delta y_r^{(\ell)} \\ \delta z_r^{(\ell)} \\ c \delta \Delta_r^{(\ell)} \end{bmatrix} + \begin{bmatrix} \varepsilon_1^{(\ell)} \\ \varepsilon_2^{(\ell)} \\ \varepsilon_3^{(\ell)} \\ \vdots \\ \varepsilon_K^{(\ell)} \end{bmatrix} \quad (20.21)$$

where $\mathbf{u}_k^{(\ell)} \triangleq \begin{bmatrix} \frac{x_r^{(\ell)} - x_k}{r_r^{(\ell)} - \Delta_r^{(\ell)} c} & \frac{y_r^{(\ell)} - y_k}{r_r^{(\ell)} - \Delta_r^{(\ell)} c} & \frac{z_r^{(\ell)} - z_k}{r_r^{(\ell)} - \Delta_r^{(\ell)} c} \end{bmatrix}$ is the direction vector between the estimate of the receiver location at the ℓth iteration and the kth satellite. Represent (20.21) in compact form as

$$\mathbf{v}^{(\ell)} = \mathbf{H}^{(\ell)} \boldsymbol{\beta}^{(\ell)} + \boldsymbol{\varepsilon}^{(\ell)} \quad (20.22)$$

The matrix $\mathbf{H}^{(\ell)}$ is often called the measurement matrix or the geometry matrix.

The algorithm for solving (20.22) is as follows. For the K satellites, find each satellite's location from its CED and the user algorithm in the ICD. Make the K pseudorange measurements and reduce errors using approaches described in Section 20.2.

For the first iteration with $\ell = 0$, select initial values $\mathbf{u}_k^{(0)} \triangleq \begin{bmatrix} \frac{x_r^{(0)} - x_k}{r_r^{(0)} - \Delta_r^{(0)} c} & \frac{y_r^{(0)} - y_k}{r_r^{(0)} - \Delta_r^{(0)} c} & \frac{z_r^{(0)} - z_k}{r_r^{(0)} - \Delta_r^{(0)} c} \end{bmatrix}$, which could be previous values if they are available, or a vector of all zeroes (corresponding to the center of the Earth) with no clock offset. Compute the pseudorange, $r_r^{(0)}$, using the initial values, and then use the measured pseudoranges to compute the initial delta pseudoranges $\delta\hat{r}_k^{(0)} \triangleq \hat{r}_k - r_r^{(0)}$, forming $\mathbf{v}^{(0)}$. Use the initial values to calculate all the elements of $\mathbf{H}^{(0)}$, yielding $\mathbf{v}^{(0)} = \mathbf{H}^{(0)}\boldsymbol{\beta}^{(0)} + \boldsymbol{\varepsilon}^{(0)}$ with $\boldsymbol{\beta}^{(0)}$ and $\boldsymbol{\varepsilon}^{(0)}$ unknown.

If $K = 4$ and $\mathbf{H}^{(0)}$ is invertible,[1] then $\boldsymbol{\beta}^{(0)} = (\mathbf{H}^{(0)})^{-1}\mathbf{v}^{(0)}$. If $K > 4$ then the system of equations is overdetermined. The ordinary least squares (OLS) solution is (see Appendix Section A.7) is found from the left pseudoinverse of $\mathbf{H}^{(0)}$, defined for a matrix $\mathbf{A}$ as $\mathbf{A}^{\dagger} \triangleq (\mathbf{A}^T\mathbf{A})^{-1}\mathbf{A}^T$, yielding $\hat{\boldsymbol{\beta}}_{OLS}^{(0)} = (\mathbf{H}^{(0)})^{\dagger}\mathbf{v}^{(0)}$. This OLS solution minimizes the mean-squared error $(\boldsymbol{\varepsilon}^{(0)})^T\boldsymbol{\varepsilon}^{(0)}$.

If the measurement errors in (20.21) do not all have same variance, and the variances are known, then using the weighted least squares (WLS) solution (Appendix Section A.7), denoted $\hat{\boldsymbol{\beta}}_{WLS}^{(0)}$, can provide smaller errors. Sometimes receivers model measurements from lower elevation satellites as having larger variance errors, and apply such weighting. However, WLS solutions should be used with care. If the prior estimates of the variances are incorrect, the weighting can magnify, rather than mitigate, estimation errors. In addition, some error contributions (e.g., residual errors from using tropospheric and ionospheric correction models) are correlated, invalidating the model used to derive the WLS solution.

But the algorithm is not yet complete for two reasons. First, the solution is only for the delta values $\delta x^{(0)}, \delta y^{(0)}, \delta z^{(0)}, \delta\Delta^{(0)}$. The updated receiver position and time offset is found from $\mathbf{u}^{(1)} = \mathbf{u}^{(0)} + \hat{\boldsymbol{\beta}}_{OLS}^{(0)}$ or $\mathbf{u}^{(1)} = \mathbf{u}^{(0)} + \hat{\boldsymbol{\beta}}_{WLS}^{(0)}$. Secondly, the updated solution is only based on a linearization of the actual equations, so is only a step toward the correct solution.

The algorithm then must be iterated for values of $\ell = 1, 2, \ldots$ Using the updated receiver position and time offset $\mathbf{u}^{(\ell)}$, compute the updated pseudorange, $r_r^{(\ell)}$, using the initial values, and then use the measured pseudoranges to compute the updated delta pseudoranges $\delta\hat{r}_k^{(\ell)} \triangleq \hat{r}_k - r_r^{(\ell)}$, forming $\mathbf{v}^{(\ell)}$. Use the initial values to calculate all the elements of $\mathbf{H}^{(\ell)}$, so $\mathbf{v}^{(\ell)} = \mathbf{H}^{(\ell)}\boldsymbol{\beta}^{(\ell)} + \boldsymbol{\varepsilon}^{(\ell)}$ with $\boldsymbol{\beta}^{(\ell)}$ and $\boldsymbol{\varepsilon}^{(\ell)}$ unknown. Solve this system of linear equations as described previously using either the inverse of $\mathbf{H}^{(\ell)}$, or the corresponding OLS or WLS solution if the system is overdetermined, yielding $\hat{\boldsymbol{\beta}}^{(\ell)} = (\mathbf{H}^{(\ell)})^{-1}\mathbf{v}^{(\ell)}$, $\hat{\boldsymbol{\beta}}_{OLS}^{(\ell)} = (\mathbf{H}^{(\ell)})^{\dagger}\mathbf{v}^{(\ell)}$, or $\hat{\boldsymbol{\beta}}_{WLS}^{(\ell)}$. The updated receiver position and time offset is then $\mathbf{u}^{(\ell+1)} = \mathbf{u}^{(\ell)} + \boldsymbol{\beta}^{(\ell)}$.

These iterations employ the same measurements, but increasingly more accurate estimates of receiver position. The iterations should continue until the solution has converged, which is determined by assessing when $\mathbf{u}^{(\ell+1)}$ and $\mathbf{u}^{(\ell)}$ are almost equal, typically using a vector norm [8]. In most cases, fewer than 10 iterations are needed if

---

[1] If $\mathbf{H}^{(0)}$ is a square matrix, it would not be invertible if it were rank deficient. In practice, this would occur only with specific and rare satellite geometries, such as all four satellites being coplanar. Such situations almost never occur in practice, and if they do, satellite motion promptly resolves the situation.

the algorithm was initialized with all zeroes, and only a single iteration may be needed if the initialization was based on a recent solution.

It is important to recognize that, while Δ_r is commonly called the receiver clock offset from satnav system time, it actually includes more than receiver clock offset. This term accounts for all of the common biases in the signals being processed, and thus includes all delays between when the signals arrive at the receive antenna phase center, and when the parameter estimates are produced. This distinction makes no difference in finding position and velocity, but if the receiver is used to produce precise timing estimates, all contributions to common biases, other than receiver clock offset, need to be estimated and removed so that they do not bias the receiver's precise estimate of satnav system time.

20.4 BLENDING SOLUTIONS FROM MULTIPLE SATNAV SYSTEMS

As more satnav systems and their signals are available, receivers increasingly will take advantage of them by using measurements based on signals from at least two systems. Ideally, all satnav systems would use the same system time, the same geodetic coordinate system, and the same values for physical constants, making their measurements interchangeable in solving for position and time.

Unfortunately, this is not the case. At a minimum, each satnav system uses a different system time. Of the systems described in Part II, currently only QZSS steers its clock bias term to that of GPS, so that GPS and QZSS signals can be treated as if they have the same system time. In other cases, system time offsets may be measured by one or both systems and provided in data messages, but receivers must wait for that data message to arrive, suffering additional latency as a result. Alternatively, if there are enough satellites in view, the receiver can solve for the system time offsets as described below.

Even though every satnav system uses a different geodetic coordinate system, all of their coordinate systems are converging to ITRS, and typically differ only at the centimeter level, making them interchangeable for SPP. Finally, receivers using signals from multiple satnav systems also need to carefully employ the correct physical constants listed in each signal's ICD. Although values for some of these constants have been refined since WGS-84 was defined, GPS cannot update the values since doing so would not be interoperable with legacy receivers. Consequently, other systems may use slightly different values of physical constants than GPS, and the values defined for each system need to be applied in processing the measurements from that system.

When the receiver has not been provided the time offsets between different system times, then an extended version of the system of equations (20.22) is needed.

When there are K_1 satellites from the first constellation and K_2 satellites from the second constellation, the left-hand side is

$$\mathbf{v}^{(\ell)} = \begin{bmatrix} \delta \hat{r}_1^{(\ell)} & \cdots & \delta \hat{r}_{K_1}^{(\ell)} & \delta \hat{r}_{(K_1+1)}^{(\ell)} & \cdots & \delta \hat{r}_{(K_1+K_2)}^{(\ell)} \end{bmatrix}^T \quad (20.23)$$

the geometry matrix is

$$\mathbf{H}^{(\ell)} = \begin{bmatrix} \mathbf{u}_1^{(\ell)} & 1 & 0 \\ \vdots & \vdots & \vdots \\ \mathbf{u}_{K_1}^{(\ell)} & 1 & 0 \\ \mathbf{u}_{K_1+1}^{(\ell)} & 0 & 1 \\ \vdots & \vdots & \vdots \\ \mathbf{u}_{K_1+K_2}^{(\ell)} & 0 & 1 \end{bmatrix} \quad (20.24)$$

and the update vector is

$$\boldsymbol{\beta}^{(\ell)} = \begin{bmatrix} \delta x_r^{(\ell)} & \delta y_r^{(\ell)} & \delta z_r^{(\ell)} & c\,\delta\Delta_{r,1}^{(\ell)} & c\,\delta\Delta_{r,2}^{(\ell)} \end{bmatrix}^T \quad (20.25)$$

where $\delta\Delta_{r,1}^{(\ell)}$ is the unknown offset between receiver time and the first system's time, and $\delta\Delta_{r,2}^{(\ell)}$ is the unknown offset between receiver time and the second system's time.

This system of equations can be solved using the same approaches described in Section 20.3. Using one measurement from one satellite in a second system provides one more measurement but also introduces an additional unknown, providing no net benefit.

Using measurements from at least two satellites in a second system potentially provides a benefit, under the following conditions:

- If the number of measurements from satellites in the first system is $1 < K_1 < 4$, then using K_2 measurements from satellites in the second system, with $K_2 > 4 - K_1$, provides enough measurements to solve for receiver position even if $1 < K_2 < 4$.
- If $K_1 > 3$ so there is a sufficient number of measurements from satellites in the first system, and the measurement errors from the second system have comparable variance to those in the first system, then positioning accuracy may be improved from use of at least two measurements from satellites in the second system. The extra state theorem [9], however, shows that DOP is nondecreasing when an additional unknown (in this case, time offset to the second satnav system) is added, so a net improvement in accuracy is not guaranteed.
- If $K_1 > 3$, and the measurement error variances from the second system are larger than those in the first system, but all variances are known, then a WLS solution may still provide a benefit when $K_2 > 1$, subject to previously expressed cautions about using WLS, and also subject to the extra state theorem.

20.5 VELOCITY CALCULATION

In general, satnav enables a PVT solution, but the earlier discussion has addressed only position and time. This section completes the discussion of obtaining a PVT solution when position and time offset have already been determined using the approach in Section 20.3 or 20.4. Section 20.5.1 describes an algorithm based on delta pseudoranges, while Section 20.5.2 describes an algorithm based on delta ranges or pseudorange rates.

20.5.1 Using Delta Pseudoranges for Velocity Calculation

If receiver dynamics are modest and the receiver produces position updates frequently, the line-of-sight velocity to each satellite can be estimated using sequential pseudorange updates. From (20.15), denote here explicitly the time at which the pseudorange estimate is made by $\hat{r}_k[nT] \triangleq r_k[nT] + c\tilde{D}_k[nT]$. The delta pseudorange estimate is the difference between two sequential pseudorange estimates,

$$\hat{r}_k[(n+1)T] - \hat{r}_k[nT] = r_k[(n+1)T] - r_k[nT] + c\left(\tilde{D}_k[(n+1)T] - \tilde{D}_k[nT]\right) \tag{20.26}$$

Observe that if the uncorrected pseudorange (20.3) is used instead, as long as T is small enough that the tropospheric error, clock offset, and ionospheric error are approximately the same for the sequential updates, these errors cancel in the delta pseudorange. However, computing the delta pseudorange magnifies the pseudorange error due to jitter, since the variance of $\tilde{D}_k[(n+1)T] - \tilde{D}_k[nT]$ is the sum of the variances of $\tilde{D}_k[(n+1)T]$ and $\tilde{D}_k[nT]$, when they are uncorrelated.

The result is a linear system of equations [10]

$$\begin{bmatrix} \hat{r}_k[(n+1)T] - \hat{r}_k[nT] \\ \hat{r}_k[(n+1)T] - \hat{r}_k[nT] \\ \vdots \\ \hat{r}_k[(n+1)T] - \hat{r}_k[nT] \end{bmatrix} = T \begin{bmatrix} \mathbf{u}_1 \\ \mathbf{u}_2 \\ \vdots \\ \mathbf{u}_K \end{bmatrix} \begin{bmatrix} \dot{x}_r \\ \dot{y}_r \\ \dot{z}_r \end{bmatrix} + T \begin{bmatrix} \varepsilon_1 \\ \varepsilon_2 \\ \vdots \\ \varepsilon_K \end{bmatrix} \tag{20.27}$$

where the direction vectors are $\mathbf{u}_k \triangleq \begin{bmatrix} \frac{x_r - x_k}{r_r - \Delta_r c} & \frac{y_r - y_k}{r_r - \Delta_r c} & \frac{z_r - z_k}{r_r - \Delta_r c} \end{bmatrix}$ and $[\dot{x}_r \ \dot{y}_r \ \dot{z}_r]^T$ is the unknown receiver velocity vector in ECEF coordinates. Although the same notation is used for the error vector in (20.27) as in (20.21), they are very different. Note that (20.27) has important similarities and differences with the SPP equations, (20.21). In both cases, each row of the geometry matrix contains a direction vector between the receiver and kth satellite, but in the calculation or velocity using delta pseudoranges, (20.27), there is no fourth column. For velocity calculation, the direction vectors are known since position has already been determined, so there is no need for an iterative solution as there was in (20.21). Further, there are only three unknowns, so as few as three measurements are needed, unless a unique satellite geometry causes the rank of the resulting measurement matrix to be less than three. Thus, (20.27) can be solved using the matrix inverse if only three measurements are used, or OLS or WLS solutions if the system of equations is overdetermined. While this delta pseudorange approach has the advantage of only having three unknowns to solve for, it has the disadvantage of using pseudorange measurements. Not only are these pseudorange measurements far noisier than phase measurements, but the differencing further increases the error.

Obtaining the system of equations (20.27) involves a number of assumptions and approximations that must be recognized. Implicit to the development is that the rates of change of satellite system clock offset and receiver clock offset are negligible. Otherwise the different satellite system clock offset must be accounted for, and the rate of change of receiver clock offset is an additional variable that must be found as part of the solution, increasing the order of the system of equations. Further, the receiver's position must be close to constant. All of these assumptions and approximations best apply when T is small, but this is the case when pseudorange jitter most degrades the solution.

20.5.2 Using Delta Ranges or Pseudorange Rates for Velocity Calculation

In this case, the receiver position has already been determined using an approach consistent with Section 20.3 or 20.4. Carrier tracking produces estimates of the Doppler shift from each satellite, biased by any frequency offset in the receiver's oscillator. As described in Section 18.1, an FLL provides an estimate of the kth satellite's received carrier frequency over some time interval T, denoted $\hat{f}_m^{(k)}$.

Using measurements from the signal at carrier frequency f_m, the estimated delta range to this satellite is then $\lambda_m(\hat{f}_m^{(k)} + f_\Delta)T$, where λ_m is the wavelength corresponding to the carrier frequency, and f_Δ is the total frequency translation applied to the signal in the receiver from RF to the carrier tracking loop. Generally, the receiver's oscillator is offset by an amount f_r, so that the estimated frequency of arrival $\hat{f}_m^{(k)} + f_\Delta$ is biased by that f_r. An estimate of the average line-of-sight velocity to the satellite over time interval T is then the delta range divided by T, or $\hat{v}^{(k)} \triangleq \lambda_1(\hat{f}_1^{(k)} + f_\Delta)$.

Alternatively, if a Costas loop or PLL is used, the receiver can count the number of carrier cycles N_θ over a time interval T, (where N_θ need not be an integer) and find the pseudorange rate $\lambda_0 N_\theta/T$, which also is biased by f_r. In this case, the estimate of line-of-sight velocity is $\hat{v}^{(k)} \triangleq \lambda_1(N_\theta/T + f_\Delta)$.

The resulting system of equations is [10]

$$\begin{bmatrix} \hat{v}^{(1)} \\ \hat{v}^{(2)} \\ \hat{v}^{(3)} \\ \vdots \\ \hat{v}^{(K)} \end{bmatrix} = \begin{bmatrix} \mathbf{u}_1 & 1 \\ \mathbf{u}_2 & 1 \\ \mathbf{u}_3 & 1 \\ \vdots & \vdots \\ \mathbf{u}_K & 1 \end{bmatrix} \begin{bmatrix} \dot{x}_r \\ \dot{y}_r \\ \dot{z}_r \\ \lambda_1 f_r \end{bmatrix} + \begin{bmatrix} \varepsilon_1^{(\ell)} \\ \varepsilon_2^{(\ell)} \\ \varepsilon_3^{(\ell)} \\ \vdots \\ \varepsilon_K^{(\ell)} \end{bmatrix} \qquad (20.28)$$

where the direction vectors are the same as in (20.27). Even though the errors in (20.28) use the same notation as in (20.21) and (20.27), they are not related in any way, but using the same variables prevents an explosion of notation.

The linear system of equations (20.28) can be solved directly for the velocity vector and frequency offset using the same approaches (matrix inverse if $K = 4$ and the matrix is invertible, and for $K > 4$ either the OLS or WLS solutions) as for SPP, but without any need for iterations. Compared to using delta pseudoranges described in Section 20.5.1, the disadvantage of using delta ranges is the need for an additional measurement. However, using delta ranges employs much higher accuracy measurements (typically FLL measurements have an order of magnitude smaller variance than delta pseudoranges, and phase measurements have two orders of magnitude smaller variances than delta pseudoranges), providing more accurate estimates of receiver velocity. Further, using delta ranges does not assume the receiver is stationary, so is more generally applicable. It should be clear that, unlike the solution approach described in Section 20.3.3 for SPP, neither of these approaches for finding velocity needs an iterative solution.

20.6 WORKING WITH DISADVANTAGED RECEIVERS

In spite of the increasing number of satnav satellites that are operating, receivers that have a limited view of the sky still may not receive signals from four or more satellites. Further, if there is not sufficient effective C/N_0 or time available, receivers cannot read the signal's data messages to obtain time of transmission, satellite time offsets, and ephemeris. The discussion in previous sections would lead to the conclusion that PVT cannot be obtained under these conditions. This section outlines ways that receivers can provide PVT in spite of having less information than previously presumed.

Section 20.6.1 describes ways to obtain position and time using measurements from fewer than four satellites. Section 20.6.2 outlines ways to use measurements even if the signal's data messages are not available.

20.6.1 Calculating Position and Time with Fewer than Four Satellites

Section 20.3 demonstrates that, in general, at least four unknown quantities need to be found to solve for receiver position and clock offset, meaning that at least four independent measurements are needed. If fewer than four satellites are available, measurements of pseudorange and pseudorange rates can both be made, leading to four independent measurements with fewer satellites as described in Reference 9. However, unless the receiver's time offset and frequency offset are known in advance, both of these are unknowns and need to be included in the solution. In this case, a minimum of three satellites is needed to obtain six measurements (pseudorange and delta range from three satellites provides six independent measurements needed to solve for receiver position in three dimensions, receiver clock offset, and receiver oscillator frequency offset). Further, DOP is very large when using velocity measurements for position calculation, making this approach of little practical use with modern satnav.

If the receiver has good enough quality in its clock and frequency reference, and has recently been calculating PVT solutions, it can develop a model of its clock offset from system time and its frequency offset from system frequency over time, using the same type of polynomial model employed for the system clock in (20.5). These models can be used to predict the offset values, reducing the number of unknowns to three and thus further reducing the number of needed measurements. This technique is known as clock hold.

In many cases, the receiver's altitude may be known to sufficient accuracy a priori, either from a recent measurement of position, from a separate sensor such as an altimeter, or because the receiver is known to be on the surface of the Earth and terrain elevation is known via a data base. In that case, the following altitude hold technique can be used. Place a virtual satellite at the center of the Earth, and set its range to the receiver equal to the (known) receiver's distance away from the center of the Earth, denoted $\widehat{r}_4$. There is no clock offset in this "measurement." The resulting linearized system of equations for measurements from three satellites is somewhat different from (20.21).

$$\begin{bmatrix} \delta \widehat{r}_1^{(\ell)} \\ \delta \widehat{r}_2^{(\ell)} \\ \delta \widehat{r}_3^{(\ell)} \\ 0 \end{bmatrix} = \begin{bmatrix} \mathbf{u}_1^{(\ell)} & 1 \\ \mathbf{u}_2^{(\ell)} & 1 \\ \mathbf{u}_3^{(\ell)} & 1 \\ \frac{x_r^{(\ell)}}{\widehat{r}_4} & \frac{y_r^{(\ell)}}{\widehat{r}_4} & \frac{z_r^{(\ell)}}{\widehat{r}_4} & 0 \end{bmatrix} \begin{bmatrix} \delta x_r^{(\ell)} \\ \delta y_r^{(\ell)} \\ \delta z_r^{(\ell)} \\ c\delta\Delta_r^{(\ell)} \end{bmatrix} + \begin{bmatrix} \varepsilon_1^{(\ell)} \\ \varepsilon_2^{(\ell)} \\ \varepsilon_3^{(\ell)} \\ \varepsilon_4^{(\ell)} \end{bmatrix} \qquad (20.29)$$

where the $\mathbf{u}_k^{(\ell)}$ are the same as in (20.21). The fourth equation reflects the fact that the pseudorange to the virtual satellite does not change with iterative updates in the receiver's location and clock offset, and involves range, not pseudorange. The same iterative solution approach discussed for (20.21) and (20.22) applies.

Next, consider timing receivers whose locations are surveyed-in and known to very high accuracy. They only need to solve for the clock offset. Review Question QT20.3 shows that a minimum of a single measurement is needed, yielding the solution is $\widehat{\Delta}_r = (\widehat{r}_1 - \widehat{r}_1)/c$ when only one measurement is available. When K measurements are available, the OLS solution is $\widehat{\Delta}_r = \frac{1}{K} \sum_{k=1}^{K} (\widehat{r}_k - \widehat{r}_k)/c$ where $\widehat{r}_k$ is the known true range between the receiver and the kth satellite.

20.6.2 Solutions without Reading the Signals' Data Messages

While the previous subsection addresses disadvantaged receivers that do not have access to the normally expected number of satellite measurements, this subsection addresses situations where the receiver does not or cannot read the signals' data message. In some cases, the desire for a rapid initial fix means that the receiver does not want to incur the latency in waiting for CED and time of transmission to appear in each signal's data message. Sometimes the signal quality may be insufficient for reading CED and time of transmission.

The receiver can obtain CED in other ways. In assisted satnav, described in Chapter 25, the cellular infrastructure provides CED to a receiver over the communications network. Receivers can also store and reuse, over the validity interval, CED read from previous data messages. Further advances in comm/nav integration, where communications and navigation functions are integrated to mutual benefit, will enable CED to be available from other sources as well. An additional trend is toward long-term CED, also called ephemeris extension [9], where third parties develop models for ephemeris and clock correction that remain useful for days; long-term CED is provided to receivers over communications networks when a connection is present. Thus, there are viable ways to obtain CED without reading it from the signal's data message.

Obtaining time of transmission without reading the data message, however, requires further attention. Assume that the receiver has CED and thus can correct satellite time to system time. If the spreading code or overlay code never repeated and was tied to time of transmission, then the receiver could determine time of transmission after synchronizing to the spreading code and overlay code. Since civil signals use repeating spreading codes and overlay codes, their time of transmission is ambiguous even after initial synchronization and synchronization to the overlay code.

A technique called coarse-time navigation [9] has been developed for position calculation without reading time of transmission from the data message. It applies to situations where, through comm/nav integration, a stationary receiver is provided with CED, line-of-sight velocity to each satellite, and coarse time—the offset between time of week and time of transmission. The algorithm requires the coarse time to be provided within a time accuracy that depends on the design of the signal being used, as discussed below. The coarse-time navigation algorithm solves for the ambiguity resulting from repeating spreading code and repeating overlay code, then estimates the time of transmission, using an extension of (20.21).

In the coarse-time algorithm, the navigation problem is five-dimensional (three dimensions of position, bias between receiver clock and satnav system clock, and time offset between time of week and coarse time). Altitude hold can be used to reduce the problem to four dimensions. Since the algorithm was developed for the C/A code signal, it is first described here in this context, then followed by discussion of its extension to other signal designs.

For coarse-time navigation using the C/A code signal, time of transmission must be provided to the receiver with an accuracy of a few seconds or better. If the receiver position is known a priori to within 150 km, then the time of flight for each signal is known within half a millisecond of time (i.e., about 150 km), then the 1 millisecond ambiguities can be observed, with the exception of cases where the unknown receiver clock bias happens to add close to half a millisecond of offset. This difficulty is overcome by an algorithm that eliminates the clock bias through single differencing before assigning integer millisecond values [9].

When the time of transmission has been determined, but the broadcast time of week has not yet been decoded, the absolute time can be determined modulo the C/A code signal data bit duration, which is 20 ms, once the C/A code signal data message bit boundaries are determined as discussed in Section 16.5. If the five-dimensional navigation solution is calculated with enough accuracy, then the correct time can be determined. Specifically, the time must be solved to better than 10 ms to determine the correct time modulo the 20 ms data message bit duration. Since GPS satellite range rates are up to 800 m/s, the positioning accuracy corresponds to better than 800 m/s × 0.01 s = 8 m. This may seem like modest accuracy, but in urban environments where large errors commonly occur due to multipath and shadowing (see Chapter 22), this level of accuracy is considered acceptable. Also, if the receiver clock offset is within 10 ms of GPS time, then unambiguous time can also be determined. However, local oscillator drift, as well as time provided by the cellular system, may differ from GPS time by more than 1 or 2 seconds. In that case, receivers must make do with coarse time until the broadcast time of week is decoded.

New and modernized signals, with overlay codes and longer duration spreading codes, provide the opportunity for improved coarse-time navigation. The allowable uncertainty in a priori position increases, relative to that for C/A code signals, proportional to increased duration of the spreading codes (e.g., 600 km for the Galileo E1 OS signal with 4 ms spreading code duration, 1500 km for the GPS L1C signal with 10 ms duration). Overlay codes having duration longer than 20 ms enable immediate determination of the time of the week with larger positioning errors and larger time uncertainties. For example, with Galileo's 100 ms duration overlay codes, the receiver can determine the correct time if the position solution is accurate to within 40 m, which is usually the case even in urban environments. And the 18 second duration overlay code on the GPS L1C signal enables receivers to determine the correct time of week after obtaining the coarse-time fix, whenever time at the receiver is within 9 seconds of GPS time, rather than the 10 ms needed for C/A code signals.

20.7 PRECISE POINT POSITIONING

Precise point positioning (PPP) is defined here as the set of techniques that provide centimeter or decimeter positioning accuracy with a single receiver—that is, without the use of differential satnav discussed in Chapter 23. PPP inevitably requires reliance on carrier phase measurements along with the use of more precise CED than is available in the data messages of standard signals. PPP typically involves dual-frequency measurements to remove ionospheric effects, although augmentations may provide accurate enough ionospheric models to allow adequate correction of single-frequency measurements. Higher accuracy PPP also requires precise information about other error sources that are significant at the centimeter level, such as real time meteorology for correcting tropospheric delay and Earth-crustal motion. Other error sources such as transmit and receive antenna phase variations, carrier phase windup, ocean loading, Earth tide, and receiver oscillator variations, must also be accounted for in order to obtain best accuracy. These error sources are described in detail in Reference 11.

Some PPP approaches require stationary receivers (static PPP), while others permit moving receivers (dynamic PPP). PPP performed in real time requires continuous access to current error models. PPP performed with post processing can use final models developed in non-real time that are typically more accurate. The discussion here relies heavily on References 12–14.

The measurement equations used in PPP, for measurements to the kth satellite at the mth frequency, are an application of the pseudorange equations (20.3) and (20.4), rewritten here as a pair of equations

$$\widehat{r}_m^{(k)} = r^{(k)} + \left(\Delta_{tropo}^{(k)} + \Delta_{iono,m}^{(k)}\right)c + \left(\Delta_r - \Delta_t^{(k)}\right)c + \tilde{D}_m^{(k)}c$$
$$\widehat{\Phi}_m^{(k)} = r^{(k)} + \left(\Delta_{tropo}^{(k)} - \Delta_{iono,m}^{(k)}\right)c + \left(\Delta_r - \Delta_t^{(k)}\right)c - N_m^{(k)}\lambda_m + \widetilde{\Phi}_m^{(k)} \quad (20.30)$$

where $r^{(k)}$ is the geometric range between the receiver and the kth satellite, Δ_r is the receiver clock offset from system time so that the true pseudorange is $\widehat{r}^{(k)} = r^{(k)} + \Delta_t^{(k)}c$, $\Delta_t^{(k)}$ is the offset of the kth satellite's clock from system time, $\Delta_{tropo}^{(k)}$ is the delay of the signal through the troposphere, $\Delta_{iono,m}^{(k)}$ is the ionospheric effect for the signal at the mth frequency (delaying the code and advancing the carrier as discussed in Section 3.3), c is the speed of light propagation in vacuum, $\tilde{D}_m^{(k)}$ is the error in estimating code time of arrival (units of seconds) that accounts for effects of multipath and noise and interference, $N_m^{(k)}$ is the integer number of wavelengths between receive antenna phase center and transmit antenna phase center, and $\widetilde{\Phi}_m^{(k)}$ is the error in estimating carrier phase, converted to units of meters, that accounts for effects of multipath and noise and interference.

The first equation in (20.30) is called the code measurement equation and is the same as (20.3), with quantities reordered. The second equation in (20.30) is called the carrier phase measurement equation.

The traditional mode of PPP calculation corrects for the satellite clock offset from system time as described in Section 20.2.1, employs a high-precision tropospheric model to essentially remove the tropospheric error, then forms ionosphere-free measurements, similar to the dual-frequency ionospheric correction described in Section 20.2.4, by calculating

$$\hat{r}_{iono-free}^{(k)} = \frac{f_{m_1}^2 \hat{r}_{m_1}^{(k)} - f_{m_2}^2 \hat{r}_{m_2}^{(k)}}{f_{m_1}^2 - f_{m_2}^2} = r^{(k)} + \Delta_r c + \tilde{D}_{iono-free}^{(k)} c$$

$$\hat{\Phi}_{iono-free}^{(k)} = \frac{f_{m_1}^2 \hat{\Phi}_{m_1}^{(k)} - f_{m_2}^2 \hat{\Phi}_{m_2}^{(k)}}{f_{m_1}^2 - f_{m_2}^2} = r^{(k)} + \Delta_r c + \lambda_{iono-free} N_{iono-free}^{(k)} + \tilde{\Phi}_{iono-free}^{(k)} \quad (20.31)$$

where $\tilde{D}_{iono-free}^{(k)} \triangleq \frac{f_{m_1}^2 \tilde{D}_{m_1}^{(k)} - f_{m_2}^2 \tilde{D}_{m_2}^{(k)}}{f_{m_1}^2 - f_{m_2}^2}$, $\lambda_{iono-free} \triangleq \frac{c}{f_{m_1} - f_{m_2}}$, $N_{iono-free}^{(k)} \triangleq \frac{f_{m_1} N_{m_1}^{(k)} - f_{m_2} N_{m_2}^{(k)}}{f_{m_1} + f_{m_2}}$ and

$\tilde{\Phi}_{iono-free}^{(k)} \triangleq \frac{f_{m_1}^2 \tilde{\theta}_{m_1}^{(k)} - f_{m_2}^2 \tilde{\theta}_{m_2}^{(k)}}{2\pi \left(f_{m_1}^2 - f_{m_2}^2\right)}$.

Solving the traditional mode PPP equations is a complicated process. Typically, the ionosphere-free code equation is used to solve for the receiver position and receiver clock offset. These values are used to initialize the ionosphere-free carrier phase equation.

Solving for the ionosphere-free ambiguity involves complicated algorithms, repeated over time. As long as there are no cycle slips in carrier tracking, the ionosphere-free carrier phase ambiguity must change consistently with satellite and receiver motion. Observing this evolution over time leads to a consistent and unique ionosphere-free ambiguity parameter. Further discussion of this processing for carrier phase ambiguity resolution is provided in Chapter 23, and Reference 11 provides an excellent detailed description and comparison of ambiguity resolution techniques for PPP.

The use of highly accurate carrier phase measurements is essential in PPP; the challenge is estimating the ambiguity of carrier phase measurements. Although many ambiguity resolution techniques have been published and implemented in receivers, ambiguity resolution is still the subject of active research. There are benefits to resolving ambiguities known to be integers, since within the code tracking error there is a countable and typically modest number of potential ambiguity values. However, the ambiguity term $N_{iono-free}^{(k)}$ is a real number, not constrained to the integers, making its resolution more difficult. PPP convergence depends on several factors including the number and geometry of visible satellites, user environment and dynamics, observation quality and rate. It is asserted [13] that receiver and satellite hardware variations can also be significant contributors to long convergence times. As these different factors interplay, the period of time (from session start) required for the solution to reach a pre-defined precision level will vary, but often takes tens of minutes.

20.8 INTEGRITY MONITORING: RECEIVER AUTONOMOUS INTEGRITY MONITORING AND FAULT DETECTION AND EXCLUSION

Section 20.6.1 examines ways to cope with what might normally be considered an inadequate number of measurements. In contrast, this section considers how to exploit more than four measurements. One way to benefit from more than four measurements is described in Section 20.3.3, where the additional measurements in the overdetermined system of equations are used to reduce, essentially through averaging, measurement errors in each equation by finding the OLS solution or WLS solution.

While overdetermined solutions can provide greater accuracy, they can also be used to check the consistency of information in individual equations to confirm its validity. Receiver Autonomous Integrity Monitoring (RAIM) uses five equations to detect if one of them significantly deviates from the others. Fault Detection and Exclusion (FDE) uses six or more equations to not only detect an invalid equation but also exclude it while providing valid PVT using the others. Information that is potentially invalid includes not only the measurements, but also the CED in the signals' data messages. In the aviation community, such crosschecking of information is part of what is known as integrity monitoring.

Chapter 7 of Reference 15, along with its references, provide a comprehensive treatment of integrity monitoring and fault exclusion that can isolate and remove erroneous information from the calculation of position and time. This section provides an overview of the underlying concepts.

Integrity monitoring has been applied initially to GPS, whose operation is so reliable that at most one fault is expected. If other less reliable systems are to be assessed, techniques may need to be revised in order to anticipate more than one simultaneous fault; in this case typically more measurements are needed. Integrity checking involves statistical tests on different solutions, and thus provides better results when measurement errors are known to be small, when the satellite geometry provides low DOP, and when faults are large.

Different approaches have been developed that check consistency in different domains, such as position, range, or parity space. The solution separation method [16] is useful to introduce the general concept. Suppose $K > 4$ in the SPP system of equations (20.22). Use all equations to solve for the receiver position and receiver clock offset. Now use each different set of $K - 1$ measurements to solve for receiver position and receiver clock offset. Compare all of these solution vectors using a selected vector norm (checking in position space), or, alternatively, compute the range to each satellite using these different solutions (checking in range space).

Each solution is expected to have different errors, and the solutions using fewer measurements would be expected to be less accurate. But all of the solution vectors or ranges to satellites should be "close" to each other, within statistical limits that can be calculated rigorously under certain assumptions.

If all of the solution vectors or ranges are "close" to each other, all equations are considered valid and the full solution obtained with K equations is used. If one or more solutions found when excluding one equation is determined to differ from the others to an extent judged statistically significant, then a fault is assumed present. If the number of equations is large enough, the specific faulty equation can be excluded and a useful solution is found with the remaining equations. Otherwise, a fault is declared and no solution is reported.

While RAIM was developed for the GPS constellation, the concept has been extended to multi-constellation processing, and called Advanced RAIM (ARAIM) [17]. It applied the solution separation method using a second frequency and more satellites from multiple constellations, and will apply for vertical guidance as well as the horizontal separation for which RAIM is used.

Although integrity monitoring was developed for aviation and other safety of life applications, wider applications are likely. With more processing capability available and more signals available from more satellites, even consumer receivers may use similar techniques to benefit from many measurements on many signals. The primary application would be not to find faults with the signals as transmitted, but rather to identify and exclude measurements that are significantly contaminated by local propagation errors, such as shadowing and multipath, associated with environments where consumer receivers are often employed.

20.9 SUMMARY

Many techniques are available for PVT calculation. They use different measurements, different information provided by satellites or other sources, and different algorithms.

They operate under different constraints, involve different complexities and different time durations to obtain a solution. They provide different accuracies.

Pseudorange measurements are essential for PVT calculation. The delay estimates produced by code tracking require further processing to develop the pseudorange measurements with small enough errors to provide useful PVT. These resulting corrected and smoothed pseudorange measurements from multiple satellites can be used to form a system of linear equations that is readily solved using iterative application of standard linear algebra techniques to find the receiver position and the offset between receiver time and satnav system time for SPP. Extensions of this technique can be used to calculate position and time using measurements from different satnav systems having different system times. Delta pseudorange or delta range measurements can be used to solve for receiver velocity. Disadvantaged receivers may not have access to the four pseudorange measurements typically stated as necessary for finding receiver position, and several approaches are outlined for obtaining a position solution using fewer than the four measurements needed for SPP. If disadvantaged receivers cannot read CED and time of transmission from the signal, approaches have also been developed to still estimate receiver position under some constraints. PPP is a rapidly emerging set of techniques that provide much greater positioning accuracy than SPP, but without the use of differential techniques. PPP uses carrier phase measurements in addition to pseudorange measurements, precise ephemeris and clock corrections, and either dual-frequency measurements or precision corrections from augmentation systems to remove ionospheric errors. RAIM and FDE, developed for aviation applications, use measurements beyond the minimum needed for PVT solution to detect, isolate, and exclude flawed inputs to PVT calculations. Variants of these techniques may appear in many types of receivers to benefit from the many measurements that will be available from the many satnav satellites being launched.

Much of the existing literature, understandably, has focused on PVT calculation using the GPS C/A code signal, and using measurements only from GPS satellites. Like the rest of this book, this chapter has provided results not tied to any one signal design, so that concepts described here can be widely applied to other signals and systems. The themes of this chapter: accuracy, speed of solution, disadvantaged receivers, exploiting signals from multiple systems on multiple carrier frequencies, and using information provided over communications networks, will be continuing themes for satnav in the years to come. Expect an emergence of literature that provides further advances that address these themes.

REVIEW QUESTIONS

Theoretical Questions

QT20.1. Derive the Hatch filter using the formulation provided in Reference 5, and assess its performance.

    a. Suppose at time index ℓ the actual propagation pseudodelay is $D[\ell]$, the pseudodelay estimate is $\hat{D}[\ell]$, the true carrier phase is $\theta[\ell]$ radians, and the carrier phase estimate at the same time is $\hat{\theta}[\ell]$ radians. Between time index $n < \ell$ and time index ℓ, express the change in path length between two separate perfect pseudorange measurements and also between changes in carrier phase measurements.

b. Show as a result that $D[\ell] = D[n] + (\theta[\ell] - \theta[n])\frac{1}{2\pi f_0}$ for any value of ℓ and n, where f_0 is the carrier frequency.

c. When estimates of pseudodelay and phase are used, however, the equality is not exact, and the equation can be written $\hat{D}_n[\ell] = \hat{D}[n] + (\hat{\theta}[\ell] - \hat{\theta}[n])\frac{1}{2\pi f_0}$, or $D_n[\ell] = \hat{D}_n[\ell] + \varepsilon_{\ell,n}$ or $D_n[\ell] = \hat{D}[n] + (\hat{\theta}[\ell] - \hat{\theta}[n])\frac{1}{2\pi f_0} + \varepsilon_{\ell,n}$ where $\varepsilon_{\ell,n}$ is the error term.

If there were no motion between satellite and receiver, the pseudodelay estimates could merely be averaged. But since there is relative motion, averaging the pseudodelay measurements would average out the effects of motion. However, the effect of motion can be estimated using the much more accurate phase differences, averaging out the errors but not the motion. Write a system of equations for estimates of the expressions derived in (b), for $n = 1, 2, \ldots, \ell$.

d. Now obtain the Hatch filter. Sum each of these equations and divide by ℓ to form an estimate of the this pseudodelay, denoted $\langle \hat{D}[\ell] \rangle$, as the average of the separate estimates of $D_n[\ell]$: $\langle \hat{D}[\ell] \rangle \overset{\Delta}{=} \frac{1}{\ell} \sum_{n=1}^{\ell} \hat{D}_n[\ell]$, and write expressions for $\ell = 1$, $\ell = 2$, $2 < \ell < L$ where all previous values are used, and $\ell \geq L$ where only the previous L values are used. Comment on the need to find the number of wavelengths in the path length.

e. Represent the error in estimating the pseudodelay $\hat{D}[n]$ as $\tilde{r}_n$ and the error in estimating the carrier phase $\hat{\theta}[n]$ as $\tilde{\theta}_n$. Suppose initially each of these errors have the same variance σ_r^2 and σ_θ^2 respectively. Then express the error in $\langle \hat{D}[\ell] \rangle$, denoted $\tilde{d}_\ell$, in terms of the pseudodelay estimate error and carrier phase error.

f. Find the mean of the errors in $\langle \hat{D}[\ell] \rangle$ in terms of the mean of the error in pseudodelay and carrier phase estimation errors.

g. Assuming phase estimation errors and pseudodelay estimation errors are uncorrelated, and all estimation errors at different times are also uncorrelated, find the means of the squared errors in $\langle \hat{D}[\ell] \rangle$.

h. In general, the variance is related to these moments by $Var\{x\} = E\{x^2\} - (E\{x\})^2$. If it is assumed that the pseudodelay and phase estimation errors are each zero mean, find the variance of the error in $\langle \hat{D}[\ell] \rangle$, denoted σ_ℓ^2, for $k = 1, 2$ and $\ell \geq L$. Assume that for $\ell \geq L$, the variance is approximately the same at sequential times, so $\sigma_\ell^2 \cong \sigma_{\ell-1}^2$. Comment on how the error in the Hatch filtered output is reduced as long as the carrier phase variance is much less than the pseudodelay variance.

i. Now suppose the filter has converged and $\ell > L$, and there is a carrier phase tracking cycle slip of N wavelengths. Analyze the effect of a cycle slip on the error in the Hatch filter estimate.

REVIEW QUESTIONS   533

j. Now examine code-carrier divergence in Hatch filtering using single-frequency measurements, addressing the steady state case $\ell \geq L$. Neglect all sources of error in the pseudodelay and phase estimates except the ionospheric delay. The pseudodelay estimate at index ℓ is $\hat{D}[\ell] = D[\ell] + \Delta_{iono}[\ell]$ while the carrier phase estimate at the same time index is $\hat{\theta}[\ell] = \theta[\ell] - \Delta_{iono}[\ell](2\pi f_0)$, neglecting integer ambiguities in the phase measurements. Substitute these expressions into the Hatch filter equation, and show that the resulting bias as time index ℓ is $b_\ell = \frac{L-1}{L} b_{\ell-1} + \frac{2-L}{L} \Delta_{iono}[\ell] + \frac{L-1}{L} \Delta_{iono}[\ell-1]$ so that $b_\ell = -\Delta_{iono}[\ell] + \frac{2}{\ell} \sum_{n=1}^{\ell} \Delta_{iono}[n]$.

QT20.2. Examine properties of the iono-free pseudorange estimate $\hat{r}_{k,iono-free} \overset{\Delta}{=} \frac{\omega_{1,2}(\hat{r}_{k,1})_{\substack{\text{clock corrected} \\ \text{tropo corrected}}} - (\hat{r}_{k,2})_{\substack{\text{clock corrected} \\ \text{tropo corrected}}}}{\omega_{1,2} - 1}$, where $(\hat{r}_{k,1})_{\substack{\text{clock corrected} \\ \text{tropo corrected}}} = r_{k,1} + \left(\Delta_{iono,1}^{(k)} + \tilde{D}_1^{(k)}\right) c$.

a. Show when there are no pseudorange errors the result is the true pseudorange, to first order. Here, $(\hat{r}_{k,m})_{\substack{\text{clock corrected} \\ \text{tropo corrected}}} = r_k + c\Delta_{iono,m}^{(k)}$, and

$$\hat{r}_{k,iono-free} \overset{\Delta}{=} \frac{\omega_{1,2}(\hat{r}_{k,1})_{\substack{\text{clock corrected} \\ \text{tropo corrected}}} - (\hat{r}_{k,2})_{\substack{\text{clock corrected} \\ \text{tropo corrected}}}}{\omega_{1,2} - 1}$$

$$= \frac{\omega_{1,2}\left(r_k + c\Delta_{iono,1}^{(k)}\right) - \left(r_k + c\Delta_{iono,2}^{(k)}\right)}{\omega_{1,2} - 1}$$

b. Show that when there are pseudorange estimation errors at each frequency and these errors are zero mean and uncorrelated, the variance of the iono-free pseudorange error is

$$\sigma_{\hat{r}_{k,iono-free}}^2 = \frac{1}{(\omega_{1,2} - 1)^2} \left[\omega_{1,2}^2 \left(\sigma_1^{(k)}\right)^2 + \left(\sigma_2^{(k)}\right)^2\right]$$

where $\left(\sigma_1^{(k)}\right)^2$ and $\left(\sigma_2^{(k)}\right)^2$ are, respectively, the variances of the pseudorandom estimation errors to the satellite at the two frequencies.

c. Show that when, in addition, the pseudorange error variances are the same at both frequencies so that $\left(\sigma^{(k)}\right)^2 \overset{\Delta}{=} \left(\sigma_1^{(k)}\right)^2 = \left(\sigma_2^{(k)}\right)^2$, then the variance of the ionosphere-free pseudorange error is

$$\sigma_{\hat{r}_{k,iono-free}}^2 = \frac{\omega_{1,2}^2 + 1}{(\omega_{1,2} - 1)^2} (\sigma^{(k)})^2.$$

QT20.3. Consider a stationary timing receiver whose set up procedure includes surveying in the receiver position, so its location is known subsequently. The true range to each satellite $\overline{r}_k$ is known, and only the time offset between receiver time and system time is unknown.

a. Write the equation that relates pseudorange estimate $\widehat{r}_k$, known range to the satellite, $\overline{r}_k$, and unknown time offset between receiver time and system time Δ_r for the kth satellite, and find the expression for the estimate of the time offset.

b. Now suppose there are measurements from K satellites. Write the resulting system of equations, find the ordinary least squares solution, and provide an interpretation of the result.

REFERENCES

1. IS-GPS-200, available at http://www.gps.gov/technical/icwg/, accessed 25 March 2014.
2. B. Hofmann-Wellenhof, H. Lichtenegger, and J. Collins, *GPS Theory and Practice*, 4th edition, Springer-Verlag, Vienna, New York, 1998
3. RTCA, Inc., "Minimum Operational Performance Standards for Global Positioning System/Wide Area Augmentation system Airborne Equipment," RTCA DO-229D, December 2008.
4. R. R. Hatch, "The synergism of GPS code and carrier measurements," Proceedings of the Third International Geodetic Symposium on Satellite Doppler Positioning, New Mexico, 1982, pp. 1213–1232.
5. R. Abhinav, "Code Phase Smoothing using Carrier Phase Measurements," Aerospace Engineering Department, IIT Bombay, Roll No.:110010019, April 2013, available at http://abhinavr110010019.files.wordpress.com/2013/05/slpreport12.pdf, accessed 25 March 2014.
6. G. McGraw, "How can dual frequency code and carrier measurements be optimally combined to enhance position solution accuracy?," in GNSS Solutions, InsideGNSS Magazine, July/August 2006, pp. 17–19.
7. J. L. Leva, "An Alternative Closed-Form Solution to the GPS Pseudo-Range Equations," *IEEE Transactions On Aerospace and Electronic Systems*, Vol. 32, No. 4, 1996, pp. 1430–1439.
8. G. H. Golub and C. F. Van Loan, *Matrix Computations*, 3rd edition, The Johns Hopkins University Press, 1996.
9. F. van Diggelen, *A-GPS: Assisted GPS, GNSS, and SBAS*, Artech House, 2009.
10. P. Misra and P. Enge, *Global Positioning System: Signals, Measurements, and Performance*, Revised 2nd edition, Ganga-Jamuna Press, 2006.
11. J. Shi, "*Precise Point Positioning Integer Ambiguity Resolution with Decoupled Clocks*," PhD Dissertation, University of Calgary, 2012.
12. P. Héroux, Y. Gao, J. Kouba, F. Lahaye, Y. Mireault, P. Collins, K. Macleod, P. Tétreault, and K. Chen, "Products and Applications for Precise Point Positioning-Moving Towards Real-Time," ION GNSS 17th International Technical Meeting of the Satellite Division, Long Beach, CA, September 21–24, 2004.

13. M. Elsobeiey and A. El-Rabbany, "An Efficient Precise Point Positioning Model for Near Real-Time Applications," International Technical Meeting (ITM) of The Institute of Navigation San Diego, CA, January 28–30, 2013.
14. P. Collins, S. Bisnath F. Lahaye, and P. Héroux, "Undifferenced GPS Ambiguity Resolution Using the Decoupled Clock Model and Ambiguity Datum Fixing," *NAVIGATION: Journal of The Institute of Navigation*, Vol. 57, No. 2, 2010, pp. 123–135.
15. E. Kaplan and C. Hegarty, *Understanding GPS: Principles and Applications*, 2nd edition, Artech House, 2006.
16. R. G. Brown and P. McBurney, "Self-Contained GPS Integrity Check Using Maximum Solution Separation," *NAVIGATION: Journal of the Institute of navigation*, Vol. 35, No. 1, 1988, pp. 41–54.
17. J. Blanch, T. Walter, P. Enge, Y. Lee, B. Pervan, M. Rippl, and A. Spletter, "Advanced RAIM user Algorithm Description: Integrity Support Message Processing, Fault Detection, Exclusion, and Protection Level Calculation," Proceedings of the 25th International Technical Meeting of The Satellite Division of the Institute of Navigation (ION GNSS 2012), Nashville, TN, September 2012, pp. 2828–2849.

PART IV

SPECIALIZED TOPICS

This final part of the book discusses topics that either integrate across those addressed in multiple earlier chapters, or extend concepts introduced in previous chapters. These topics reflect areas where research and development continue, providing ample opportunity for readers of this book to make further contributions.

The style and level of detail in Part IV differ from those in earlier parts. Each chapter in Part IV represents an overview, or survey, of issues and of work that is being performed. Emphasis is on describing concepts in an organized structure for the topic, rather than providing detailed techniques and their performance. The intent is to provide relevant concepts and terminology, equipping the reader to pursue further detail in the literature. Thus, the chapters in this part are higher level and shorter, and do not include review questions.

Chapter 21 addresses the critical topic of interference. While some chapters in Part I and many chapters in Part III consider aspects of interference, its effects, and how to mitigate it, this chapter provides an integrated treatment of this topic. It draws upon discussion, analysis, and techniques introduced in the first 20 chapters to describe interference and its effects on different receiver functions, and to outline approaches for mitigating interference and its effects on receiver performance.

Chapter 22 addresses the problems of multipath and shadowing. It describes how multipath occurs, how multipath is characterized in satnav, and how the effects of multipath are compounded by shadowing. Effects of multipath on measurements produced by code tracking and carrier tracking are explained. Techniques for mitigating multipath and its effects on receiver performance, and for dealing with shadowing, are outlined.

Satnav augmentations have become essential ways of extending the performance and reliability of satnav, integrating communications and navigation (comm/nav integration) to improve the performance of a satnav receiver. Two chapters describe different types of satnav augmentations.

Chapter 23 describes satnav augmentations that use differential techniques for improving accuracy and for enhancing the integrity of received signals. A general architecture is introduced for differential augmentations, followed by description of different types of differential systems and the associated aspects of the receiver processing.

Engineering Satellite-Based Navigation and Timing: Global Navigation Satellite Systems, Signals, and Receivers, First Edition. John W. Betz.
© 2016 The Institute of Electrical and Electronics Engineers, Inc. Published 2016 by John Wiley & Sons, Inc.

Real-time kinematic (RTK) navigation is introduced for achieving decimeter accuracy in real time.

Another type of satnav augmentation, called assisted satnav, is introduced in Chapter 24. While the augmentations described in Chapter 23 enhance accuracy and integrity, assisted satnav enhances availability. The augmentations described in Chapter 23 target professional users (e.g., surveying, machine control, agriculture) and safety of life applications (aviation, maritime), while those in Chapter 24 primarily target consumer applications including smartphones and other uses of location-based services. The contrast between Chapter 23's emphasis on accuracy and integrity, and Chapter 24's emphasis on user friendliness and availability in challenging environments, is marked. Chapter 24 introduces key concepts in assisted satnav and describes architectures and approaches for implementing the associated processing.

While Chapters 23 and 24 describe ways to exploit and expand the integration of communications with navigation, Chapter 25 describes integration of multiple sensors providing PNT information to enhance the robustness of satnav processing in a distinctly different way than the assisted satnav processing introduced in Chapter 24. After an introduction to Kalman filtering as the most commonly used processing approach for integrated receiver processing, the concepts of loose coupling and tight coupling between other sensors and satnav are introduced. A complementary type of integrated processing, involving combined processing of all signals in a vector-locked loop (VLL) rather than in separate loops as described in Part III, is described. While there are practical limitations and disadvantages to a VLL, it has great promise as receiver processing power increases, and merits careful consideration for receiver design in challenged environments. The epitome of integrated processing combines concepts involved in tight coupling with those of vector processing, yielding ultratight coupling. Various architectural approaches are summarized, offering a range of performance and associated complexities.

Part IV thus describes key aspects of the present and future of satnav receiver techniques, building on the conventional design approaches introduced in Part III. All of the concepts described in Part IV were developed initially to extract even better performance from the legacy GPS C/A code signal, and all can be extended and applied to exploit the features and capabilities of newer satnav signals currently being introduced.

The first 15 years of this century have emphasized the design and development of new and modernized satnav systems and signals. As these advances transition from concept to operational reality, there is tremendous opportunity to exploit them using applications and extensions of the techniques described in Part III and Part IV of this book.

21

INTERFERENCE

Radio frequency interference is the bane of satnav receivers, degrading performance or preventing operation. The increasing use of L band transmissions for many applications including consumer communications contributes to increased interference environments encountered by satnav receivers. The interference situation is exacerbated by recent increases in the use of "personal privacy jammers" that, although typically illegal, are used to counter the use of satnav for tracking an individual, vehicle, or item by employers, owners, investigators, and law enforcement. Since interference is likely to be an increasing challenge, coping with interference should be an increasing focus of receiver design in future.

Section 21.1 outlines different ways that interference is characterized, introducing terminology encountered in discussions of interference, and categorizing different types of interference that have different effects on receiver operation. Section 21.2 summarizes the effect of interference on receiver operation, drawing upon topics introduced in earlier chapters and applying them to interference. Section 21.3 outlines strategies for dealing with interference, including ways to mitigate its effect on receiver operation as well as ways that a receiver can characterize the interference environment. Section 21.4 summarizes this chapter.

Engineering Satellite-Based Navigation and Timing: Global Navigation Satellite Systems, Signals, and Receivers, First Edition. John W. Betz.
© 2016 The Institute of Electrical and Electronics Engineers, Inc. Published 2016 by John Wiley & Sons, Inc.

21.1 INTERFERENCE CHARACTERISTICS

There are many ways to categorize the different types of interference, but none of them are unique or uniformly useful. Sometimes interference is distinguished by the intent associated with its generation. Benign interference results from accidental transmissions, from legitimate transmissions by systems that share spectrum or operate in adjacent spectrum, or from spurious emissions (such as harmonics) from legitimate transmissions. In contrast, hostile interference, often called jamming, is attributed to transmissions that are intended to degrade or deny receiver operation. Such distinctions can be useful for regulatory or legal purposes, but make no difference to a receiver, whose response to interference does not change with the intent of the transmission.

There is a fundamental difference between uncorrelated interference, whose cross-correlation function with the desired signal is zero (or approximately zero) at all delays, and correlated interference, like multipath. Multipath is addressed separately in Chapter 22, and this chapter's discussion is limited to uncorrelated interference.

Another distinction often encountered in discussion of satnav interference involves the difference between intrasystem, intersystem, external, and device interference. Intrasystem interference to reception of a desired signal is produced by all other satnav signals transmitted by satellites from the same system that transmits the desired signal. Intersystem interference to reception of a signal results from transmissions of other satnav systems' signals, while external interference is caused by transmissions from other than satnav systems. Device interference is caused by emissions from other electronics in the device that hosts the receiver.

Intrasystem interference is carefully analyzed and controlled by each satnav system's designers. Intersystem interference is also carefully controlled under procedures of the International Telecommunications Union Radiocommunications Sector (ITU-R) [1]. External interference should also be controlled under ITU-R procedures as well as regulations in each nation, but variations in spectrum allocations in different nations, combined with limited enforcement of spectrum protection regulations, can result in anomalous levels of external interference. Device interference involves very low power emissions very close to the satnav receiver, and must be characterized and controlled by the device manufacturer—particularly by receiver manufacturers of multifunction devices that include satnav receivers along with other RF transmitters and high speed digital circuitry.

An important technical distinction is made between in-band interference, which occupies frequencies within the receiver's precorrelation bandwidth, and out-of-band interference, whose power is at frequencies outside the receiver's precorrelation bandwidth. Even out-of-band interference still can affect analog electronics in the receiver front end. Further, receiver processing imperfections, such as local oscillator spurs or insufficient antialiasing filtering, as discussed in Chapter 14, can transform out-of-band interference to in-band interference that reduces effective C/N_0.

The technical characteristics of interference can be described in many different domains including temporal, spectral, locational, directional, and polarization. Temporal characteristics include whether the interference power is relatively constant over time, or highly time varying or intermittent, and whether the interference amplitude has a Gaussian distribution, is constant envelope, or has some other distribution over one or more time scales. Spectral characteristics describe the frequency content of the interference, and whether that frequency content is constant or time varying. Common examples of interference spectral content include narrowband interference—concentrated in a limited frequency range—or wideband interference whose power is approximately the same at all frequencies over a bandwidth not less than precorrelation bandwidth. If the frequency content is time varying, narrowband interference can be a swept tone with a specific profile of carrier frequency over time, or some other time-frequency profile. Another example of interference spectral content is matched spectrum interference, whose PSD is the same as that of the desired signal. Also, interference can be geographically widespread or highly localized, and may arrive from a specific direction or appear omnidirectional. Finally, interference can have circular polarization with either left-hand or right-hand sense, elliptically polarized with a specific sense, or linearly polarized.

The interference environment also varies by frequency band. In upper L band (1559–1610 MHz), radionavigation satellite service is universally designated a co-primary service by the ITU-R, meaning it is protected from all other interference sources. In lower L band (1164–1215 MHz and 1215–1300 MHz), a wider variety of systems is permitted to share the spectrum including high power terrestrial transmitters. Satnav receivers are more likely to encounter significant interference in lower L band.

21.2 EFFECTS OF INTERFERENCE ON RECEIVER OPERATION

Figure 21.1 illustrates different ways that interference can affect a receiver. High power interference, even if out-of-band, can affect the receiver front end components, as

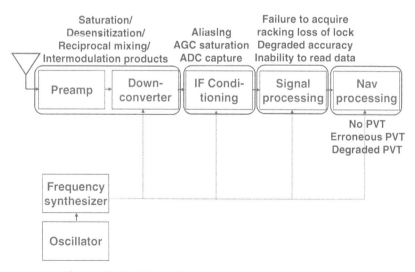

Figure 21.1. Effects of Interference on Receiver Processing

described in Chapter 14. If the interference power is large enough, it causes nonlinear effects such as saturation or desensitization that distort the input waveforms, degrading subsequent processing. Distortion of the desired signal causes decorrelation with the replica used in correlation processing for acquisition, tracking, and data demodulation, degrading or disrupting the processing. When the interference is also distorted by nonlinear behavior of receiver components, the performance of any subsequent receiver processing intended to mitigate the interference (such as filtering) is also degraded. Out-of-band interference can also be translated to in-band interference by reciprocal mixing, image frequencies, or aliasing during ADC. As indicated in Figure 21.1, interference can also saturate the AGC or capture the ADC, severely degrading subsequent receiver processing.

If the receiver front end responds linearly to the input combination of signal, noise, and interference, then the receiver processing has an opportunity to continue operation, although it may be degraded by the interference. If the interference is continuous and can be characterized by a PSD, its effect on acquisition, carrier tracking, and data message demodulation can be assessed using SSCs, as described in Chapters 5, 16, and 18. Interference affects code tracking differently from these other receiver functions, and the code tracking SSC can be used to assess interference in this case, as described in Chapter 19.

Ultimately, interference affects navigation processing. Depending upon the interference characteristics and the receiver design, interference may cause the PVT output to be absent, erroneous, or degraded. Providing acceptable performance in the presence of interference requires careful selection and design of every part of the receiver hardware, algorithms, and processing for specified interference characteristics.

21.3 DEALING WITH INTERFERENCE

As described in this section, dealing with interference involves both characterization (providing the receiver with an assessment of the interference conditions) and mitigation (processing to reduce the effects of interference on receiver performance). Section 21.3.1 describes an approach for estimating received interference power levels, and approaches for detecting and characterizing interference in the amplitude and frequency domains. Section 21.3.2 addresses approaches for estimating effective C/N_0 within a receiver, while Section 21.3.3 outlines approaches for mitigating interference.

21.3.1 Interference Characterization

It is valuable for a receiver to estimate the interference environment in order to adapt its processing accordingly. The most critical point in receiver processing for assessing the interference environment is before acquisition, so that processing parameters for initial synchronization and subsequent steps in acquisition can be selected to provide the needed sensitivity without excessive time to first fix. Since the receiver is not yet tracking the signal, it cannot estimate the received signal power before acquisition. It can, however, estimate the interference level using a "J/N meter" [2, 3].

DEALING WITH INTERFERENCE

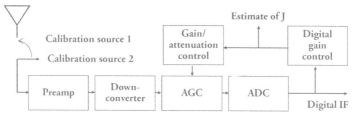

Figure 21.2. J/N Meter

Figure 21.2 shows the architecture for the J/N meter included in a receiver front end using digital gain control as described in Chapter 15. If added to a receiver front end during the design stage, there is very little impact. The digital gain control is proportional to the power level at the receiver front end, and can be calibrated and monitored as follows.

Two calibration sources (or, equivalently, a single calibration source capable of two different levels) are applied at the receive antenna output. These sources can be applied only at the factory, or can be built into the receiver. Factory calibration provides less precise calibration due to aging of components and to temperature changes, but has less implementation impact than including calibration sources in the receiver hardware. If calibration sources are included in the receiver hardware, calibration can be repeated, providing greater ability to react to aging and temperature effects. When the antenna is connected under benign conditions, and no calibration source is applied, the digital gain control value indicates the expected level with no interference present. The two different calibration source levels are set to be different from each other, both larger than the expected level with no interference present but within the dynamic range of the receiver front end. When the digital gain control value is measured for both calibration source levels, these two values, along with the value when the antenna is connected under benign conditions, define an affine transformation from which the input interference level, not including the noise power with no interference present, can be estimated.

Another way to detect and characterize interference is to compute a histogram of ADC outputs, as illustrated in Figure 21.3. This approach typically requires at least a three-bit ADC in order to obtain eight levels of granularity. When the input noise and interference has a Gaussian distribution, a histogram like that on the left is obtained—unimodal with small tails. However, if there is constant-envelope interference present with power exceeding that of the thermal noise, the shape of the histogram changes to bimodal when the interference power is high, and to unimodal with a much broader peak when the interference power is only slightly larger than the thermal noise power.

Standard tests for Gaussianity can be applied to detect the presence of constant-envelope interference having power above that of thermal noise. One simple approach is to estimate the kurtosis from a series of ADC outputs, and determine if it deviates significantly from that from Gaussian noise. Since interference is typically a concern only when its power significantly exceeds that of the thermal noise, relatively crude statistical tests can be adequate. Such statistical tests can be performed on ADC outputs without

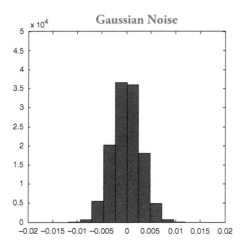
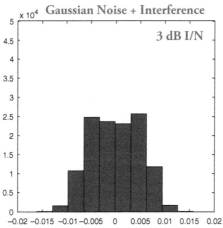
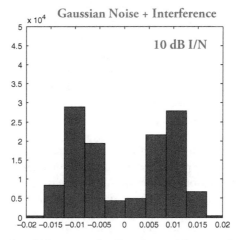

Figure 21.3. Examples of Histograms for Gaussian and Constant-Envelope Interference

actually computing a histogram, but this approach is still known as ADC histogramming. Typically, a few hundred ADC samples, each separated by a few milliseconds, would be used in the statistical test.

It must be noted that histogramming is sensitive to constant-envelope interference, whether narrowband or broadband, but not to Gaussian interference, since its histogram would have the same shape as thermal noise. Many types of man-made interference have nearly constant envelopes, however, making this type of interference detection of practical use. Another limitation of this approach is that it is primarily useful for merely detecting the presence of interference, rather than estimating its power, although some sense of the interference power may be inferred from the separation between bimodal peaks if there are enough ADC bits.

Yet another technique for interference characterization involves spectrum analysis, which is increasingly being implemented in receivers that can be purchased today. Even small FFTs can be used to determine if there are spectral peaks caused by narrowband interference in the receiver front end, or else a narrowband digital filter can be swept over the precorrelation bandwidth to detect and characterize narrowband interference.

The three techniques described in this section—J/N meter, ADC histogramming, and spectrum analysis—are all complementary and can be implemented together if affordable. ADC histogramming may add the least implementation complexity to a receiver, but is only sensitive to constant-envelope interference and does not estimate the interference power. Spectrum analysis may be the most complex to implement, and is only sensitive to partial-band or narrowband interference. The J/N meter provides an estimate of interference power, but does not characterize the interference PSD or amplitude distribution. For example, while knowing the specific frequency of narrowband interference is important in understanding its effect on effective C/N_0 and code tracking accuracy, the J/N meter reports the same value for narrowband interference anywhere in the receiver's precorrelation bandwidth.

Any or all of these approaches can be used to identify situations where acquisition processing parameters should be established to provide higher sensitivity due to elevated interference conditions. However, since all of these approaches only characterize interference, they are not helpful in detecting weak signal conditions caused by foliage penetration or building penetration, and thus will not instigate high sensitivity acquisition processing under weak signal conditions.

21.3.2 Estimating Effective C/N_0

Once the receiver is tracking a signal, it is in a much better position to estimate the effective C/N_0. It is important that this estimate be consistent with the definition and discussion found in Chapter 5. Work exploring the theoretical and practical aspects of this estimation is found in References 4 and 5.

The estimate of received desired signal power at the correlator output, denoted C, is proportional to the magnitude of the correlation function peak. Estimating the effective noise PSD is the tricky part. If the input were always white, the total input power could be estimated, then divided by the precorrelation bandwidth and combined with the estimate of received desired signal power. However, this precorrelation C/N_0 approach yields fallacious results when the interference is not white over the precorrelation bandwidth.

Alternatively, the sample variance of the correlation function amplitude can be measured at the correlation function peak or at a different correlation function delay, then used with the estimate of received desired signal power based on the definition of effective C/N_0 in Chapter 5. This postcorrelation C/N_0 approach provides valid estimates of C/N_0 over a wide range of conditions, and should be employed whenever practicable.

When a receiver's reported C/N_0 is used for testing, it is prudent to characterize how the receiver estimates C/N_0, and in particular to determine whether the receiver uses a precorrelation approach or postcorrelation approach. A signal generator that produces a relatively narrowband (of the order of 10 kHz bandwidth) Gaussian waveform with adjustable center frequency and power is very useful as a probe waveform for this purpose. Initially, the center frequency of this probe waveform should be set to the frequency where the desired signal's PSD is maximum. The power should be increased from very small until the reported C/N_0 degrades by approximately 6 dB. Next, using this same power level, sweep the center frequency of the probe waveform in 1 MHz steps, both below and above the starting center frequency, until the reported C/N_0 returns to the same value as with no interference. This frequency sweep establishes the precorrelation bandwidth of the receiver. Finally, place the center frequency of the probe waveform near the edge of the precorrelation bandwidth, starting at a very low power, and increase the power to measure reported C/N_0 versus different values of probe signal C/N_0, using an estimate of the thermal noise PSD.

The postcorrelation C/N_0 is also computed analytically since it is the same as the effective C/N_0, and the precorrelation C/N_0 can be computed analytically using the received probe signal power divided by the precorrelation bandwidth to compute the precorrelation noise density. The measurements and analytical computations can be plotted on the same axes, then compared.

Figure 21.4 shows results from Reference 4 that implement the above procedure for two receivers of C/A code signals. The receiver labeled "Receiver A" has a measured precorrelation bandwidth of approximately ±4 MHz, while the receiver labeled

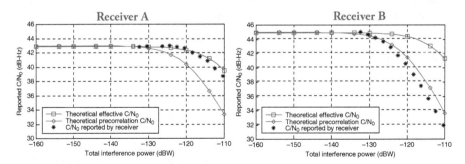

Figure 21.4. Examples of Comparing Reported C/N_0 to Analytical Predictions of C/N_0 Measurement Approaches

"Receiver B" has a measured precorrelation bandwidth of approximately ±8 MHz. For Receiver A, the reported C/N_0 values are closely aligned with the analytical prediction of effective C/N_0, indicating that the receiver properly reports effective C/N_0. For Receiver B, the reported C/N_0 values are closely aligned with the analytical prediction of precorrelation C/N_0, indicating that the receiver does not properly report effective C/N_0.

Experience has shown unexpected variation in such receiver characterizations. The same receiver model with different software versions may employ different algorithms for reporting C/N_0. Even the same receiver and software however, can exhibit different characteristics under different conditions. Receivers have been found that report C/N_0 using effective C/N_0 when a spare tracking channel is available but use precorrelation C/N_0 if all tracking channels are occupied. Presumably, the spare tracking channel is used for variance estimation when it is available, and when it is not the receiver reverts to dividing the input power by the precorrelation bandwidth.

21.3.3 Mitigating Interference and Its Effects

Operating through interference conditions successfully involves a combination of steps:

- Reduce the interference power incident on the receive antenna, without affecting received power of desired signals,
- For a given interference power incident on the receive antenna, reduce the ratio of interference power to desired signal power at the output of the receive antenna,
- Design the receiver front end to reduce interference effects and provide adequate dynamic range to avoid nonlinear behavior,
- Using baseband processing in the receiver, further reduce the ratio of interference power to desired signal power into the measurement processing in the receiver,
- Adapt signal processing parameters and techniques for better performance in acquisition, carrier tracking, code tracking, and data message demodulation.

Each step is discussed below in further detail.

Increasing the separation between the receive antenna and sources of interference is beneficial, as can be positioning the receive antenna to increase blockage losses by removing direct line of sight between the receive antenna and the interference source. If the interference source is near the Earth's surface, reducing the height of the receive antenna may increase the propagation loss, as discussed in Chapter 4, and increase blockage losses for the interference.

One interesting observation involves the benefits of moving away from interference. Assume propagation loss between the interference source and the receiver can be modeled as following a power law, so that received interference power can be represented by $P_I \propto r_{I,rx}^{-n}$, with $r_{I,rx}$ the distance between interference source and receiver. Here, n is a positive integer, the propagation exponent, taking on a value of 2 for free-space propagation, and four for the approximate model of two-ray model approximation derived in Section 4.3. Then the change or received interference power with a change in $r_{I,rx}$ is proportional to $-nr_{I,rx}^{-(n+1)}$, so the change in P_I, denoted ΔP_I, has the following proportionality to a change in $r_{I,rx}$ denoted $\Delta r_{I,rx}$: $\Delta P_I \prec -nr_{I,rx}^{-(n+1)}(\Delta r_{I,rx})$. Several insights can be inferred from this simple result. One is the confirmation that increasing $r_{I,rx}$ reduces P_I. Another is that increasing $r_{I,rx}$ by a given amount decreases P_I more when the propagation exponent is larger. Finally, when $r_{I,rx}$ is smaller, a given change in $r_{I,rx}$ produces a greater reduction P_I. This latter observation indicates that increasing the distance from an interference source by a given amount is more beneficial when that source is nearby than when the source is distant.

When everything possible has been done to reduce the interference power incident on the receive antenna, the next step is to reduce the interference power output by the antenna. Since many receive antennas have backlobes with low gain, raising the receive antenna's height can be beneficial, although such a step must be balanced against the possibility of lower interference propagation loss as described earlier. Receive antennas with reduced gain toward the horizon can be beneficial as well, to the extent that adequate gain is still available at elevations where satellite signals must be received. If the interference does not have RHCP, decreasing the antenna's response to interference having other polarizations is also beneficial.

Adaptive antenna processing can be very effective, especially when facing a small number of interference sources that are highly directional. The design of antenna arrays, processing hardware, and processing algorithms for this purpose merits entire chapters—far beyond the scope of this chapter. One useful resource is Chapter 4 of Reference 6 and its references.

Interference not rejected by the receive antenna passes to the receiver front end. Chapter 14 describes approaches for designing the front end to reduce effects of out-of-band interference, and to keep out-of-band and in-band interference from degrading receiver performance. Adaptive AGC techniques [7] can mitigate constant-envelope interference as well.

Baseband processing in the receiver provides the next opportunity to discriminate against interference. This processing is particularly successful in mitigating interference whose temporal or spectral characteristics differ from those of the received signal and noise. Pulse blanking has been studied extensively, particularly for receivers of the GPS L5 signal [8], and the approaches developed and examined for this application can be adapted more broadly. Constant-envelope interference can be mitigated through algorithms that estimate and remove such interference, such as Reference 9. Similarly, narrowband interference excision [10] can reduce by substantial amounts the effects of interference if its PSD is instantaneously narrowband, in the sense that it occupies a small fraction of the signal's null-to-null bandwidth during the time period when the spectrum is being estimated. An excellent survey of interference mitigation techniques is found in Reference 11.

In general, it is preferable to mitigate interference as early in receiver processing as possible—before it impinges on the receive antenna, before it is output by the receive antenna, before it enters baseband processing, or before correlation processing. Ultimately, though some interference will be present in what receiver processing deals with, and the processing for acquisition, carrier tracking, code tracking, and data message demodulation must cope with this residual interference. One way to improve all of these processing steps is to reduce processing loss through wider precorrelation filters, faster sampling rates, and sampling with more bits, but each of these steps is accompanied by implementation complexity and other undesired consequences.

As described in Chapter 16, a number of approaches are available to enhance acquisition processing under conditions of low effective C/N_0. Coherent integration times can be extended, and the number of noncoherent integrations can be increased. Toward this end, steps that reduce ITU and IFU enable initial synchronization resources to be concentrated on a smaller time-frequency tile. Code Doppler compensation is an enabler for these steps.

Carrier and code tracking can be made more robust under interference conditions as described in Chapters 17 through 19. Lower-order tracking loops, narrower loop filters, and the use of coherent discriminators all help tracking loops deal with interference, but can introduce greater fragility to dynamics, oscillator imperfections, and other challenges.

Some of the topics addressed in subsequent chapters also provide benefits in operating through interference. Integrated receiver processing, addressed in Chapter 24, includes vector locked loops and coupling sensor-integrated processing with signal tracking, all enabling tracking to operate under more challenging jamming conditions, as well as the opportunity to "coast" through conditions that cause temporary tracking loss of lock to some or all signals. Assisted satnav, described in Chapter 25, was developed for satnav reception in weak-signal conditions, but provides the same benefits when low effective C/N_0 is caused by high levels of effective N_0.

Finally, PVT calculation can benefit from using more signals that provide more opportunity for signals to be received at high elevation angle where received signal power tends to be higher due to reduced path loss and typically higher receive antenna gain. In addition, using measurements from more signals reduces the DOP, and hence the PVT errors.

21.4 SUMMARY

Interference will remain the preeminent challenge of satnav. Characterizing interference by its source is useful from a regulatory perspective, but makes little difference from the perspective of designing receivers to mitigate or accommodate interference. Technical characteristics of interference, however, are useful in assessing effects of interference and implementing receiver design steps that mitigate or accommodate interference.

Interference can degrade or deny every phase of receiver operation. Every aspect of receiver processing must be designed and assessed to ensure proper performance, for carefully specified interference conditions. Overdesigning against interference is costly in terms of components, receiver complexity, power consumption, and other aspects of performance like sensitivity. However, underdesigning against interference causes the receiver to provide inadequate performance.

A key aspect of receiver processing for interference involves the receiver's ability to characterize the interference environment in order to adapt or incorporate the appropriate receiver processing. The J/N meter, ADC histogramming, and spectrum analysis all provide tools for characterizing the interference environment even before signal tracking commences. Proper reporting of signal C/N_0 is also critical, but some receivers provide misleading results that can either underestimate or overestimate the effective signal C/N_0. Part III of this book, along with Chapters 24 and 25, provide design approaches, both at the receiver level and at the system level, needed to counter interference and its effects. Almost every step for dealing with interference is accompanied by some disadvantages, so numerous tradeoffs are needed to achieve adequate performance, with minimal detrimental effects, for the interference environments where the receiver is being designed to operate.

REFERENCES

1. "ITU Radiocommunications Sector," http://www.itu.int/en/ITU-R/pages/default.aspx, accessed 24 August 2014.
2. P. W. Ward, "RFI Situational Awareness in GNSS Receivers: Design Techniques and Advantages," Proceedings of the 63rd Annual Meeting of The Institute of Navigation, Cambridge, MA, April 2007, pp. 189–197.
3. P. W. Ward, "Simple Techniques for RFI Situational Awareness and Characterization in GNSS Receivers," Proceedings of the 2008 National Technical Meeting of The Institute of Navigation, January 2008, pp. 154–163.
4. J. W. Betz, "Effect of Partial-Band Interference on Receiver Estimation of C/N_0: Theory," *Proceedings of Institute of Navigation National Technical Meeting 2001*, ION-NTM-2001, Institute of Navigation, January 2001.
5. J. T. Ross, J. L. Leva, and S. Yoder, "Effect of Partial-Band Interference on Receiver Estimation of C/NO: Measurements," Proceedings of Institute of Navigation National Technical Meeting 2001, ION-NTM-2001, Institute of Navigation, January 2001.
6. B. Rama Rao, W. Kunysz, R. Fante, and K. McDonald, *GPS/GNSS Antennas*, Artech House, 2013.
7. F. Amoroso, "Adaptive A/D converter to suppress CW interference in DSPN spread spectrum communications," Military Communications Conference, MILCOM 1983. IEEE, 2009, pp. 720–728.
8. C. Hegarty, A.J. Van Dierendonck, D. Bobyn, M. Tran, and J. Grabowski, "Suppression of Pulsed Interference through Blanking," Proceedings of the IAIN World Congress and the 56th Annual Meeting of The Institute of Navigation, San Diego, CA, June 2000, pp. 399–408.
9. P. Henttu, "A New Interference Suppression Algorithm Against Broadband Constant Envelope Interference," IEEE Conference on Military Communications, MILCOM 2000. 21st Century Military Communications Conference Proceedings (Volume:2), October 2000.
10. J. Raasakka and M. Orejas, "Analysis of Notch Filtering Methods for Narrowband Interference Mitigation," Proceedings of IEEE/ION PLANS 2014, Monterey, CA, May 2014, pp. 1282–1292.
11. T. Lotz, Adaptive Analog-to-Digital Conversion and Pre-correlation Interference Mitigation Techniques in a GNSS Receiver, Doctoral Thesis, Technische Universitat Kaiserslautern, 2008.

22

MULTIPATH

Multipath is interference from delayed versions of a signal that are received in addition to the same signal arriving directly from the transmitter. These delayed versions of the direct-path signal are caused by reflections or diffractions in the propagation channel. In satnav, each delayed version is typically modeled as a replica of the direct-path signal, but with different amplitude, delay, and carrier phase. (Even though the Doppler shift of the reflected signal can differ from that of the direct-path signal, this Doppler shift difference is typically assumed to be negligible.)

In contrast to the uncorrelated interference addressed in Chapter 21, multipath interference is correlated with the desired signal. Consequently its effects can be different from uncorrelated interference, and some techniques for mitigating it also differ from those described in Chapter 21. Like uncorrelated interference, multipath introduces errors in receiver measurements of time of arrival and carrier phase, thus causing errors in the estimation of PVT. Unlike uncorrelated interference, however, multipath can have the same or similar characteristics over multiple correlation integration times, introducing constant or slowly varying errors rather than the rapidly fluctuating jitter caused by noise and interference.

As analyzed in Chapter 6, multipath-induced errors can often be the largest source of PVT error with new and modernized satnav systems, and are typically the dominant

---

Engineering Satellite-Based Navigation and Timing: Global Navigation Satellite Systems, Signals, and Receivers, First Edition. John W. Betz.
© 2016 The Institute of Electrical and Electronics Engineers, Inc. Published 2016 by John Wiley & Sons, Inc.

errors in differential satnav. Consequently, there is considerable motivation to understand multipath and its effects, and to design satnav systems with multipath mitigation in mind.

In practice, multipath can occur in combination with shadowing, where the direct-path signal is attenuated by an obstacle such as terrain, buildings, or foliage. In this case, the received power of multipath signals can be greater than the received power of the shadowed direct path. Sometimes the direct-path signal is sufficiently attenuated that its received power is below the level the receiver needs to process a signal. In this case the receiver processing only has access to the multipath arrivals. This situation—of severe direct-path shadowing plus multipath—provides a very different and severe challenge to receiver processing that is rarely addressed in the literature.

This chapter describes the origin, characteristics, and effects of multipath, as well as ways to mitigate multipath and its effects, including the combination of multipath and shadowing. Section 22.1 provides models of multipath channels and characterizes different multipath environments, while Section 22.2 shows the effects of multipath on receiver processing and its performance. Section 22.3 addresses multipath mitigation at the system level and by receiver processing. Section 22.4 summarizes this chapter. Some of the contents and development in this chapter are drawn from those in Section 6.2 of Reference 1. An extensive survey of multipath mitigation techniques is also available in Chapter 17 of Reference 2.

22.1 MULTIPATH CHARACTERISTICS

Figure 22.1 shows the type of situation that gives rise to satnav signal multipath and shadowing. The direct path from satellite to receiver may involve line of sight propagation, or there may be additional attenuation or shadowing from foliage or other obstructions, as shown in the figure. Regardless of whether shadowing occurs or not, signal reflections also occur from structures, terrain, vehicles, or other conductive or partially conductive objects in the environment. These signal reflections always arrive at the receiver later than the direct path since the reflected paths are longer than the direct path.[1] When there is no shadowing of the direct path, the received power in each signal reflection is less than the received power in the direct-path signal.

Many models of multipath have been proposed and applied in digital communications and in satnav. The challenge is to find a model simple enough to be useful, while adequately capturing the essence of the immensely complicated real-world multipath environment so that results using the model are of practical use.

In most cases of practical interest, the power levels and materials involved in satnav multipath produce phenomena that can be modeled as causing a linear response. Thus, in principle, satnav multipath must be modeled as a linear time-varying channel, as discussed in References 3 and 4. The time-varying effects are caused by a combination of satellite motion (so even the direct path is time varying), motion of reflectors

---

[1] While multipath always arrives at the receiver later than the direct path, multipath can induce delay estimates that are either earlier than or later than the true delay, due to carrier phase effects, as shown subsequently.

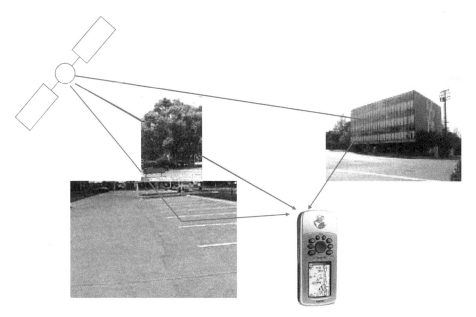

Figure 22.1. Representative Situation Involving Shadowing and Multipath

in the channel (e.g., vehicles), and receiver motion in some cases. The need for accurate measurements in satnav can require higher-fidelity models than typically used in communications applications.

For simplicity, the satnav multipath channel is often modeled as a linear time-invariant channel over a limited time interval. Even though diffuse scattering can occur and would be modeled by a continuous impulse response, in most cases the channel is modeled as tapped delay line having impulse response

$$h(\tau) = \alpha_0 e^{-i\theta_0} \delta(\tau - \tau_0) + \sum_{n=1}^{N} \alpha_n e^{-i\theta_n} \delta(\tau - \tau_n) \quad (22.1)$$

where α_0^2 is the received power of the direct path, θ_0 is the received carrier phase, and τ_0 is the propagation delay in the direct path. The quantity α_n^2/α_0^2 is called the multipath-to-direct-path ratio (MDR), $\theta_n - \theta_0$ is the relative carrier phase, and $\tau_n - \tau_0$ is the excess delay, each associated with the nth discrete multipath.

Since the wavelengths are short relative to path lengths and motion of satellites, reflectors, and receiver, the carrier phases θ_n are often modeled as IID and uniformly distributed on $[0, 2\pi)$. In this case, the multipath channel impulse response (22.1) can be characterized by the power-delay profile (PDP)

$$\begin{aligned} P(\tau) &= P_0 \delta(\tau - \tau_0) + P_1 \delta(\tau - \tau_1) + P_2 \delta(\tau - \tau_2) + \cdots + P_N \delta(\tau - \tau_N) \\ &= P_0 [\delta(\tau - \tau_0) + \overline{P}_1 \delta(\tau - \tau_1) + \overline{P}_2 \delta(\tau - \tau_2) + \cdots + \overline{P}_N \delta(\tau - \tau_N)] \end{aligned} \quad (22.2)$$

Figure 22.2. Example PDP

where $P_n \stackrel{\Delta}{=} \alpha_n^2$ and the MDR of the nth discrete multipath is $\overline{P}_n = P_n/P_0$. Figure 22.2 shows an example PDP based on the model (22.2). Paths with relatively small excess delays (up to a few hundred nanoseconds) are called near echoes, while those with larger excess delays are called far echoes. Near echoes typically arise from objects near the receiver,[2] are more common, and tend to have larger MDRs. In contrast, far echoes are often associated with more distant buildings or terrain. As will be seen, receiver processing often can more readily mitigate the effects of far echoes, while near echoes are more challenging to mitigate in receiver processing.

The PDP of the discrete channel model (22.2) is parameterized by the number of paths, the range of delay spreads, and the MDRs for each delay. This model is only an approximation—in reality there typically is a continuity of echoes at different relative power levels, changing over time.

Multipath channels are often characterized by the first two moments of their PDP. The mean delay of a multipath channel, relative to the direct-path delay, is

$$\bar{\tau} \stackrel{\Delta}{=} \frac{\int_0^\infty (\tau + \tau_0) P(\tau + \tau_0) d\tau}{\int_0^\infty P(\tau + \tau_0) d\tau} \qquad (22.3)$$

and the so-called[3] RMS delay spread of a multipath channel [6] is

$$\sigma_\tau \stackrel{\Delta}{=} \left[\frac{\int_0^\infty (\tau + \tau_0 - \bar{\tau})^2 P(\tau + \tau_0) d\tau}{\int_0^\infty P(\tau + \tau_0) d\tau} \right]^{1/2} \qquad (22.4)$$

The mean delay provides a measure of the multipath's excess delay, while the RMS delay spread is related to the number of multipaths, their excess delay, and their MDR.

---

[2] This is not always the case. As an example, the GPS satellite SVN 49 has a hardware flaw causing an internal reflection of the RF signals, producing multipath-type phenomena [5].
[3] Observe this is actually an expression for the standard deviation, not the root mean squared.

MULTIPATH CHARACTERISTICS

For satnav, where measurement accuracy is the dominant consideration, the true RMS delay spread of a multipath channel may be defined as

$$\sqrt{\overline{\tau^2}} \stackrel{\Delta}{=} \left[\frac{\int\limits_0^\infty (\tau + \tau_0)^2 P(\tau + \tau_0) d\tau}{\int\limits_0^\infty P(\tau + \tau_0) d\tau} \right]^{1/2} \tag{22.5}$$

and is likely more relevant to satnav considerations.

For a simple one-multipath channel with $N = 1$, $\bar{\tau} \stackrel{\Delta}{=} \overline{P}_1 \tau_1/(1 + \overline{P}_1)$, which is upper bounded by $\overline{P}_1 \tau_1$, $\sigma_\tau = \sqrt{\overline{P}_1} \tau_1/(1 + \overline{P}_1)$, which is upper bounded by $\sqrt{\overline{P}_1} \tau_1$, and $\sqrt{\overline{\tau^2}} \stackrel{\Delta}{=} \tau_1 \sqrt{\overline{P}_1/(1 + \overline{P}_1)}$ which is also upper bounded by $\sqrt{\overline{P}_1} \tau_1$ and takes on similar values to σ_τ when MDR is small. As long as the MDR < 1, the delay spread for this channel is greater than the mean delay, and the standard deviation of delay spread and true RMS delay spread are roughly interchangeable.

Extensive efforts have been made to model and measure multipath characteristics. An extensive measurement campaign for L band satcom channels is documented in Reference 7, and may still be the most comprehensive relevant model available. Simulations using ray tracing have also been used to generate representative PDPs based on detailed models of structures and their materials. In Reference 8, PDPs were obtained through high-fidelity simulations for receivers indoors and outdoors, as the receiver is moved different distances from a starting point. The indoor PDPs have shorter delay spreads, since single bounces from within a room have small additional path lengths, and multiple bounces are typically attenuated significantly by scattering from multiple surfaces. Occasionally, the indoor data in Reference 8 shows a spike at longer delay spread, presumably caused by a multipath entering through a window. In contrast, the outdoor PDPs have longer delay spreads, caused by reflections from buildings and other structures far from the receiver.

The results in Reference 9 show L band delay spreads below 50 ns in multiple indoor areas, with delay spreads less than 20 ns in some indoor areas. Other measurements [10] confirm that even in outdoor rural and urban conditions, paths having significant MDR values typically occur at excess delays less than several hundred nanoseconds. These characterizations of multipath conditions provide important context for the multipath mitigation examined in subsequent sections of this chapter.

Table 22.1 contrasts multipath models commonly used in analytical, simulation, and even field testing of multipath mitigation approaches compared with characteristics of multipath channels most relevant to actual satnav applications. Clearly, the mismatch between common models and real-world conditions can be significant, and only more sophisticated models, like those in Reference 7 or detailed ray-tracing simulations, have much prospect for assessing performance in real-world conditions.

Several items in the table merit additional discussion. Modeling MDR as constant with excess delay is equivalent to assuming that the multipath as occurring from a specular reflection—a reflection from a smooth surface where there is essentially no change in received power with different ranges between receiver and reflector, as long as the

TABLE 22.1. Comparison of Multipath Models and Real-World Conditions

Common Model	Common Real-World Conditions
Static, time invariant	Time varying
Single multipath	Many multipaths
Discrete path	Diffuse and discrete paths
MDR constant with excess delay	MDR varies with excess delay
Direct path present, MDR < 1	Shadowing; some MDRs>1 or no direct path present
No noise or interference	Noise and interference also present

path length from the transmitter to reflector is much larger than the path length from reflector to receiver. Also, early emphasis on multipath in satnav arose in surveying and aviation, which are applications where received signals are relatively strong and interference is often negligible. The result was common use of multipath models that did not include noise and other interference, but emphasized only the degradation due to multipath. There are very few attempts in the literature to evaluate receiver performance under conditions of combined noise, interference, and multipath, although such combined conditions are of increasing practical importance.

Even these simple common models have demonstrated utility as diagnostic tools and to assess relative performance. Their limitations should be realized, however, so that results using these models are not interpreted as representing real-world performance in multipath. The static one-multipath model has been used extensively as a simple model, and is also used extensively in the remainder of this chapter. For this model, let $N = 1$ in (22.1), so when the transmitted signal is $x(t)$, the received signal is (assuming no noise or other interference)

$$r(t) = \alpha_0 e^{-i\theta_0} x(t - \tau_0) + \alpha_1 e^{-i\theta_1} x(t - \tau_1) \tag{22.6}$$

22.2 MULTIPATH EFFECTS

To understand the effect of multipath on receiver processing, consider the crosscorrelation of $r(t)$ in (22.6) with a replica of the transmitted signal, producing a composite correlation function whose stationary part is $\alpha_0 e^{-i\theta_0} R_x(\tau - \tau_0) + \alpha_1 e^{-i\theta_1} R_x(\tau - \tau_0 - \tau_1)$. The magnitude-squared version of this correlation function make code tracking insensitive to carrier phase tracking,

$$\begin{aligned}&\left|\alpha_0 e^{-i\theta_0} R_x(\tau - \tau_0) + \alpha_1 e^{-i\theta_1} R_x(\tau - \tau_0 - \tau_1)\right|^2 \\ &= P_0 \big[R_x^2(\tau - \tau_0) + \overline{P}_1 R_x^2(\tau - \tau_0 - \tau_1) \\ &\quad + 2\sqrt{\overline{P}_1} \cos(\theta_1 - \theta_0) R_x(\tau - \tau_0) R_x(\tau - \tau_0 - \tau_1)\big]\end{aligned} \tag{22.7}$$

The shape of this magnitude-squared correlation function is dictated by three parameters, the MDR $\overline{P}_1 \stackrel{\Delta}{=} \alpha_1^2/\alpha_0^2$, the relative carrier phase $\tilde{\theta}_1 \stackrel{\Delta}{=} \theta_1 - \theta_0$, and the excess delay

$\tilde{\tau}_1 \overset{\Delta}{=} \tau_1 - \tau_0$. These three parameters characterize the effect of the one-multipath channel on the shape of the magnitude correlation function.

If there were no multipath, the mean estimate of time of arrival from a code tracking loop would be the true time of arrival of the direct path, τ_0. With multipath, the mean estimate of time of arrival from a code tracking loop is a function of the three multipath parameters, $\hat{D}(\overline{P}_1, \tilde{\theta}_1, \tilde{\tau}_1)$, and the mean error is then $\varepsilon(\overline{P}_1, \tilde{\theta}_1, \tilde{\tau}_1) = \hat{D}(\overline{P}_1, \tilde{\theta}_1, \tilde{\tau}_1) - \tau_0$. Unlike code tracking errors due to noise and uncorrelated interference, code tracking errors caused by a time-invariant multipath channel have a bias error that does not diminish with averaging or smoothing of the measurements.

Figure 22.3 shows squared correlation functions for different relative carrier phases, along with the mean estimate of delay for NELP with different early–late spacings, for two different spreading modulations. Multipath causes the magnitude-squared correlation function to be asymmetric, so that NELP produces bias errors whose magnitude increases with larger early–late spacing. It is common, but incorrect, to think that the sign of the error is always positive, since excess delay is always positive. As shown in the lower right case with relative phase of π, when the relative phase causes the last term in brackets in (22.7) to be negative, the magnitude-squared correlation function can be asymmetric toward negative delay, causing negative bias errors. This case also shows that some relative phase values cause the terms in (22.7) to partially cancel, reducing the magnitude of the correlation peak and consequently reducing effective C/N_0 at the correlator output. These results also show that the magnitude of the multipath-induced bias errors is smaller for BPSK-R(10) than for BPSK-R(1). The importance of spreading modulation design will become more evident in Section 22.3.

Consider next BPSK-R(1) code tracking using NELP. Suppose the static one-multipath model is applied for many different excess delays and relative phases, with an MDR of −6 dB for any excess delay and relative phase. The resulting multipath-induced bias errors (in units of meters) are shown in Figure 22.4, for a phase-equalized precorrelation filter with magnitude transfer function of a four-pole Butterworth filter having −3 dB points ±12 MHz away from the center frequency. For any given multipath delay, the bias errors can be positive or negative, depending upon the relative phase. All errors at each multipath delay fit within an envelope of values, independent of relative phase.

This depiction of multipath errors leads to two common ways of describing performance for the static one-multipath model, the multipath error envelope and the average worst-case multipath error. The multipath error envelope is defined as the extreme values of the multipath error, over all relative phase values

$$E(\tilde{\tau}_1, \overline{P}_1) \overset{\Delta}{=} \left\{ \max_{\tilde{\theta}_1} \varepsilon(\overline{P}_1, \tilde{\theta}_1, \tilde{\tau}_1), \min_{\tilde{\theta}_1} \varepsilon(\overline{P}_1, \tilde{\theta}, \tilde{\tau}_1) \right\} \qquad (22.8)$$

The average worst-case multipath error provides a running average of the envelope:

$$\bar{E}(\tilde{\tau}_1, \overline{P}_1) \overset{\Delta}{=} \frac{1}{\tilde{\tau}_1} \int_0^{\tilde{\tau}_1} \frac{1}{2} \left[\left| \max_{\tilde{\theta}_1} \varepsilon(\overline{P}_1, \tilde{\theta}_1, \tilde{\tau}_1) \right| + \left| \min_{\tilde{\theta}_1} \varepsilon(\overline{P}_1, \tilde{\theta}_1, \tilde{\tau}_1) \right| \right] du \qquad (22.9)$$

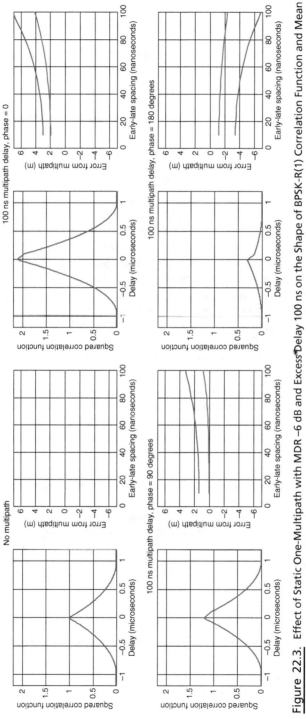

Figure 22.3. Effect of Static One-Multipath with MDR −6 dB and Excess Delay 100 ns on the Shape of BPSK-R(1) Correlation Function and Mean Error Using NELP with Different Early-Late Spacings of BPSK-R(1) (Blue Lines) and of BPSK-R(10) (Red Lines) Spreading Modulations

MULTIPATH EFFECTS

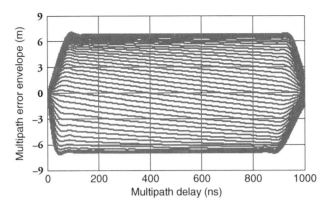

Figure 22.4. Code Tracking Bias Error for NELP with Early–Late Spacing of 97.8 ns and BPSK-R(1) and MDR of −6 dB

These two quantities portray the same information, but in slightly different ways. The multipath error envelope shows the range of possible errors at each excess delay. The average worst-case multipath error shows the multipath error envelope's average magnitude error, averaged over all values up to a given excess delay. It thus provides a sense for the error that might be expected if the excess delay were uniformly distributed between zero and a given excess delay. Since the multipath error is typically less than the envelope, as shown in Figure 22.4, (22.8) and (22.9) provide pessimistic, or worst-case, error values for a given static multipath channel model.

Figure 22.5 shows these quantities for the same conditions as Figure 22.4. The average of the worst-case error provides little additional insight for this spreading modulation, but is more useful for BOC and other spreading modulations examined in Section 22.3.

While the previous discussion has addressed how static multipath introduces bias errors in code tracking, static multipath also introduces bias errors in carrier phase tracking. Consider the composite correlation function from the static one-multipath model derived earlier, $\alpha_0 e^{-i\theta_0}[R_x(\tau - \tau_0) + \sqrt{\overline{P_1}} e^{-i\overline{\theta}_1} R_x(\tau - \tau_0 - \tau_1)]$. Then the carrier

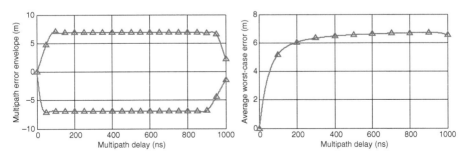

Figure 22.5. Multipath Error Envelope and Average Worst-Case Multipath Error for Same Conditions as Figure 22.4

phase angle of the term in brackets represents the carrier phase error from this composite correlation function $\tan^{-1}\left[\dfrac{-\sqrt{P_1}\sin(\tilde{\theta}_1)R_x(\tau-\tau_0-\tau_1)}{R_x(\tau-\tau_0)+\sqrt{P_1}\cos(\tilde{\theta}_1)R_x(\tau-\tau_0-\tau_1)}\right]$

As long as the MDR is less than unity, the largest phase error occurs when $R_x(\tau - \tau_0) \cong R_x(\tau - \tau_0 - \tau_1)$. In that case, the carrier phase angle error from this composite correlation function is given by

$$\tan^{-1}\left[\frac{-\sqrt{P_1}\sin(\tilde{\theta}_1)}{1+\sqrt{P_1}\cos(\tilde{\theta}_1)}\right] \tag{22.10}$$

which has a maximum magnitude of $\pi/2$ when the MDR approaches unity and the multipath carrier phase is 180° out of phase of the carrier phase in the direct path. Consequently, ranging errors caused by multipath to carrier phase tracking are less than half a wavelength—much smaller than corresponding code tracking errors.

While the preceding discussion is based on a static multipath model, the channel is actually time varying; what is important is how fast it is changing. The channel may be almost static under some situations, such as a stationary receiver in a parking lot where the dominant sources of multipath are reflections from pavement or buildings, causing bias errors as discussed. If the rate of change of channel fluctuation is of the order of the reciprocal of the loop tracking bandwidth, the loop's parameter estimates are biased, but any subsequent filtering of position and velocity estimates with longer averaging times can partially average out the errors. If the rate of channel variation is slower than the correlation integration time but fast relative to the reciprocal of the loop bandwidth (often due to carrier phase variations), then the multipath contributes to jitter, similar to that caused by noise and interference as described in Chapters 18 and 19, that is smoothed by the tracking loop. If the multipath is rapidly fluctuating, at a rate faster than the correlation integration time in the receiver processing, then the fluctuating multipath can reduce coherence between the received signal and multipath, and the replica, degrading correlator output SNR. In this case, it would be prudent to reduce the correlation integration time to a value where the multipath is approximately time invariant over the correlation integration time.

22.3 MULTIPATH MITIGATION

The deleterious effects of multipath motivate significant effort to mitigate these effects on receiver performance. This section describes layered approaches for mitigating multipath at the system level in Section 22.3.1, through choices of spreading modulations and early–late processing parameters in Section 22.3.2, and using advanced techniques introduced in Section 22.3.3.

22.3.1 System-Level Multipath Mitigation

Multipath mitigation at the system level emphasizes preventing multipath from entering the receiver. For fixed-site installations, nearby reflective structures that could produce multipath can be removed, or modified (through shaping or coating with special materials) to reduce signal reflections. Antennas can be placed further from these reflectors in order to reduce reception of multipaths, or closer to reflectors so that the excess delays are small enough to cause acceptably small errors. Antenna gain patterns can also be shaped to reduce gain in the directions where multipath may arrive, particularly low-elevation angles where satnav signal reception may not be needed. In addition, since some multipaths, depending upon the geometry and the material of the reflecting surface, can undergo polarization reversal, antennas designed to reject LHCP signals can be beneficial. Choke ring antennas [11] are intended to provide low gain at low-elevation angles and to reject LHCP signals, as are some other types of antennas. Beamforming antenna arrays are also useful for reducing gain toward low-elevation reflectors while maintaining good gain toward satellites.

For antennas in fixed installations, multipath effects from fixed structures reoccur with repeating satellite ground tracks. These effects can be measured and removed at least in part after repeated observations. Such techniques are particularly useful with GPS, since its ground tracks repeat daily as opposed to longer repeat times for other global systems.

22.3.2 Multipath Mitigation for Code Tracking Using Early–Late Processing

This subsection illustrates ways to adopt the code tracking described in Chapter 19 for better performance in multipath. The fundamental observation, illustrated in Figure 22.6, is that using narrow early spacing reduces multipath errors significantly, just as smaller early–late spacing reduces code tracking errors in noise and interference.

Narrow early–late spacing is only beneficial, however, if the precorrelation bandwidth is correspondingly wide. Figure 22.7 shows the effect of different precorrelation bandwidths with narrow early–late spacing. Clearly, it is the combination of wide precorrelation bandwidth and narrow early–late spacing that provides the desired effect. A rule of thumb is that multipath mitigation benefits little from narrowing the early–late spacing to less than the reciprocal of the two-sided precorrelation bandwidth.

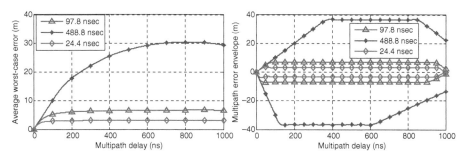

Figure 22.6. Multipath Error Envelopes for NELP of BPSK-R(1), Static One-Multipath Model with MDR of −6 dB, Precorrelation Filter Phase Equalized with Magnitude of Four-Pole Butterworth Filter Having −3 dB Points at ±12 MHz

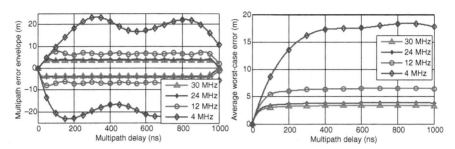

Figure 22.7. Multipath Error Envelopes for NELP of BPSK-R(1) with Early–Late Spacing of 48.9 ns, Static One-Multipath Model with MDR of −6 dB, Precorrelation Filters Phase Equalized with Magnitude of Four-Pole Butterworth Filter Having Narrow −3 dB Points at Indicated Bandwidths

Yet another way to affect multipath errors is by choice of spreading modulation. As more signals with different spreading modulations become available, the receiver designer will be able to choose which signals to process, as well as how to process the selected signals. Figure 22.8 shows results for four different spreading modulations, all with the same relatively wide precorrelation bandwidth and narrow early–late processing. The oscillatory multipath error envelopes associated with BOC(1,1) and TMBOC are difficult to compare with the error envelopes of the BPSK-R spreading modulations, and the average worst-case multipath error provides a clearer interpretation.

Figure 22.8 provides other useful information concerning multipath mitigation. With early–late processing, BOC(1,1) is no better than BPSK-R(1) for multipath delays less than 500 ns—the near echoes that are most common, often have largest MDR, and typically produce the most detrimental effects. Further, the results for TMBOC are significantly better than for BOC(1,1). This result explains the decision to change

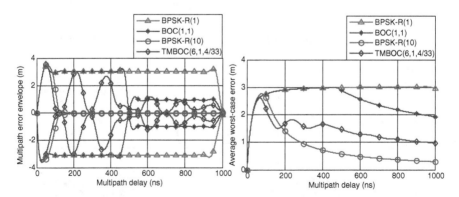

Figure 22.8. Multipath Error Envelopes for NELP of Different Spreading Modulations with Early–Late Spacing of 24.4 ns, Static One-Multipath Model with MDR of −6 dB, Precorrelation Filter Phase Equalized with Magnitude of Four-Pole Butterworth Filter Having −3 dB Points at ±12 MHz

the GPS L1C spreading modulation from BOC(1,1) to MBOC [12], and also informs receiver designers: for better performance with early–late processing of GPS L1C in multipath, use the BOC(6,1) component as well as the BOC(1,1) spreading modulation, as discussed in Section 18.1.3 and Reference 13. (The same conclusions apply for Galileo E1 OS, and using its CBOC spreading modulation instead of processing E1 OS as BOC(1,1).) For excess delays shorter than 150 ns, the multipath error is smaller for TMBOC than for BPSK-R(10), even though BPSK-R(10) has much larger bandwidth.

While these results are shown for a static one-multipath model, the same conclusions apply qualitatively for more complicated multipath conditions—multipath errors with early–late processing are smaller with wide precorrelation bandwidths, narrow early–late spacings, and superior spreading modulations like TMBOC and BPSK-R(10). The benefits of these modern spreading modulations with larger RMS bandwidths are being substantiated by field measurements, such as in Reference 14.

Other spreading modulations that provide excellent performance with early–late processing, without requiring very wide signal bandwidths, have also been identified [15, 16].

Another approach that can improve the performance of code tracking, as described in Chapter 19, applies in fluctuating multipath conditions. Reducing the code tracking loop bandwidth can smooth these fluctuating multipath errors, providing better accuracy.

22.3.3 Advanced Multipath Mitigation

Many other multipath mitigation approaches have been developed. Sophisticated techniques have also been devised for obtaining better multipath performance for the C/A code signal with its BPSK-R(1) spreading modulation. The most widely used approaches employ modified replica waveforms in the correlators used for the code tracking discriminator. Similar approaches have been developed and patented multiple times. References 17 and 18 provide representative descriptions of these approaches.

The basic concept for these modified replica multipath mitigation approaches is to use a replica formed by passing the signal through a filter having noncausal impulse response known as a double delta, $h(t) = -\delta(t + \Delta_1/2) + 2\delta(t) - \delta(t - \Delta_1/2)$. Figure 22.9 shows segments of a conventional BPSK-R(1) replica and the double-delta replica.

Crosscorrelating the received signal with the double-delta replica produces the time-averaged correlation function $-R_x(\tau + \Delta_1/2) + 2R_x(\tau) - R_x(\tau - \Delta_1/2)$, portrayed in Figure 22.10 for BPSK-R(1) spreading modulation and two different values of Δ_1. Early–late processing can then be applied, using early–late spacing $\Delta_2 < \Delta_1$. Observe that the double-delta correlation peak's amplitude is reduced significantly, which indicates lower correlator output SNR than with a conventional replica.

The resulting double-delta S-curves are displayed in Figure 22.11. The linear portions of the S-curves are very short relative to those for conventional BPSK-R(1) early–late S-curves, raising potential challenges in pulling in lock and maintaining lock with dynamics or other challenging conditions.

Figure 22.12 shows the resulting multipath error envelopes, for the same conditions and some of the same waveforms as with the conventional replica in Figure 22.8. The error envelopes are much smaller with double-delta processing.

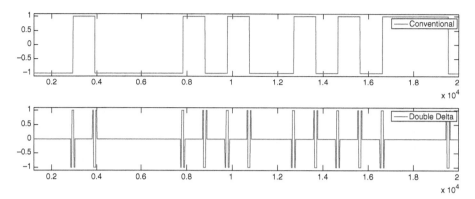

Figure 22.9. Comparison of Conventional and Double-Delta Replicas for Segment of BPSK-R(1) Waveform

Echoes appear in Figure 22.12 with the BPSK-R spreading modulations as well as the BOC(1,1) spreading modulation. An intuitive explanation can be obtained by recognizing that the double-delta impulse response is an approximation to taking the second derivative of the waveform. The result is equivalent to highpass filtering of the signal, producing a filtered PSD with a notch at band center and spectral peaks on both sides. Its shape is similar to that of a BOC, and so it is not surprising that the resulting multipath error envelope has echoes just as with BOC spreading modulations.

Interestingly, a variant of the double-delta replica can remove the echoes for BPSK-R spreading modulations. This variant places the same impulses at every spreading symbol edge, even where transitions in value do not occur. This echo erasing double-delta replica is portrayed in Figure 22.13.

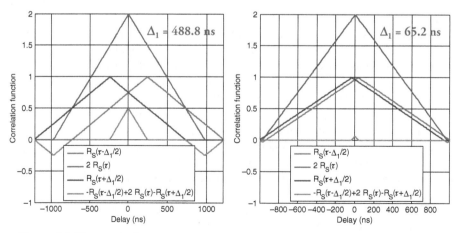

Figure 22.10. Result of Correlating BPSK-R(1) Waveform with Its Double-Delta Replica

MULTIPATH MITIGATION

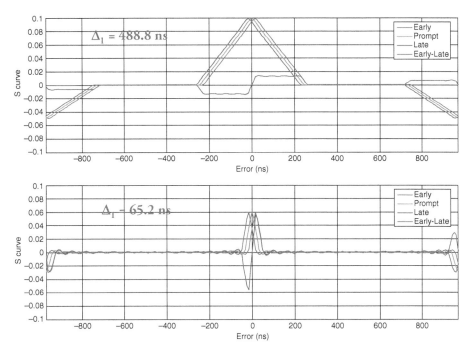

Figure 22.11. Double-Delta S-Curves for CELP with Early–Late Spacing 32.6 ns

While these double-delta approaches provide superior multipath mitigation performance for C/A code signals, they have been applied primarily in surveying applications where their behavior in significant noise and interference, and with receiver dynamics, has not been reported. Some performance analysis is available in Reference 19. Extending them to spreading modulations other than BPSK-R(1) has not been reported. It should also be noted that, because of the reduced output SNR, it is important also to

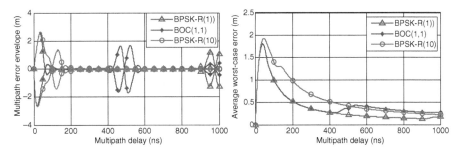

Figure 22.12. Multipath Error Envelopes for NELP with Double-Delta Processing of Different Spreading Modulations with Δ_1 65.2 ns and Early–Late Spacing of 48.9 ns, Static One-Multipath Model with MDR of −6 dB, Precorrelation Filter Phase Equalized with Magnitude of Four-Pole Butterworth Filter Having −3 dB Points at ±12 MHz

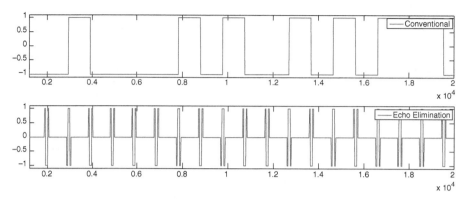

Figure 22.13. Comparison of Conventional and Double-Delta Echo Erasing Replicas for Segment of BPSK-R(1) Waveform

use a conventional prompt correlator with the original signal as replica, so that its output can be used for carrier tracking and data symbol demodulation.

Other advanced multipath mitigation approaches have been published and implemented in some cases. Correlation function shape matching [20] and blind deconvolution approaches (see, e.g., Reference 21) are among those that have been proposed.

Much more recent work has emphasized model-based approaches, where signal processing estimates parameters that describe the channel impulse response. Nuisance parameters associated with the multipaths and their amplitudes are then discarded while the estimated time of arrival of the direct path is retained and used. An example of this approach is described in Reference 22. In general, the parameter estimation algorithms are based on an assumption of a time-invariant channel, and require coherent processing of relatively long (of the order of a second) data records to obtain the needed parameter estimation accuracy. Channel variations over the processing time can distort or degrade the parameter estimates. Techniques that require coherent processing over such long data records are typically restricted to applications where the receiver is stationary and the channel does not vary rapidly. Evidence indicates that these advanced techniques perform better with wider precorrelation bandwidths and spreading modulations having larger RMS bandwidths.

While more work remains to be performed to develop practical multipath mitigation techniques that fully exploit modern spreading modulations, theory has been developed to inform this work. The Cramér–Rao lower bound on pseudorange accuracy in multipath has been shown to depend on RMS bandwidth [23]. Limits on precorrelation bandwidths and integration times, susceptibility to interference, the rate of channel fluctuations, and the ability to maintain tracking lock, are among the practical considerations that must be accounted for in developing and assessing these algorithms.

The issue of addressing the combined problem of shadowing and multipath also remains to be more fully addressed by further research efforts. With the increasing number of satellites available from multiple constellations, there is increasing opportunity to make measurements on enough additional satellites to use RAIM techniques, introduced in Section 20.8, in order to identify and exclude those measurements with large excess delay that result from tracking of multipaths whose corresponding direct paths are shadowed [24, 25].

22.4 SUMMARY

Unlike uncorrelated interference, whose power needs to be significantly greater than the power of the received satnav signal before causing errors, multipath can contribute significant errors even when its power is less than that of the desired signal and much less than the input noise power. Multipath is a dominant error source in modern satnav, and the combination of multipath and shadowing is even more detrimental to accuracy of positioning and timing.

Real-world multipath channels are often amazingly complex and variable, and extensive efforts have been made to characterize them using measurements and simulations. Near echoes that have excess delay of a few hundred nanoseconds are typically the most common and most severe multipaths that introduce the greatest error. In spite of the real-world complexity, a very simple static one-multipath model is used extensively for characterizing multipath performance. It is not useful for predicting performance in most real-world conditions, but is useful for diagnosis and comparison of techniques for receiver processing in multipath. The multipath error envelope and average worst-case multipath error, calculated using this model, are often used to characterize receiver processing performance in multipath.

Multipath introduces code tracking errors and carrier tracking errors; the former being much larger. Consequently multipath mitigation efforts concentrate on improving code tracking performance.

Receiver processing for multipath mitigation benefits from wide precorrelation bandwidth, and most such techniques also benefit from selection and design of spreading modulations having larger RMS bandwidths. Early–late discriminators exhibit reduced multipath errors with narrow early–late spacing, as long as the precorrelation bandwidth is correspondingly wide.

Specialized receiver processing techniques based on signal processing for parameter estimation have been developed that go beyond variations of early–late processing. Various types of approaches have been pursued, with recent research emphasizing model-based signal processing. The robustness of such specialized techniques to noise, interference, and dynamics merits additional attention.

Research in multipath mitigation remains an active research area. New and modernized satnav systems, including more capable spreading modulations, multiple signals at the same and different carrier frequencies, more satellites in view, and the availability of inexpensive signal processing hardware, all provide opportunities for further research and development of more capable techniques that can be applied in practice for many different applications.

It should be noted that, while multipath introduces degradation in navigation and timing applications, multipath signals are exploited for scientific purposes in remote sensing [26]. Characteristics of the Earth's surface including soil moisture content, and ocean waves conditions, as well as ice and snow coverage, can be estimated using signal processing approaches based on bistatic radar.

REFERENCES

1. E. Kaplan and C. Hegarty, *Understanding GPS: Principles and Applications*, 2nd edition, Artech House, 2006.
2. S. Jin, *Global Navigation Satellite Systems: Signal, Theory and Applications*, InTech, 2012.
3. J. Proakis and M. Salehi, *Digital Communications*, 5th edition, McGraw-Hill, 2007.
4. F. Hlawatsch and G. Matz, *Wireless Communications over Rapidly Time-Varying Channels*, Amsterdam, The Netherlands: Academic Press, 2011.
5. C. D. Goldstein, Request for Feedback on GPS IIR-20 (SVN-49) Mitigation Options, Slide Presentation dated 5 march 2010, available at http://www.insidegnss.com/assets/misc/GPSW_SVN-49_information_Briefing_(Mar_2010)_Final.pdf, accessed 12 October 2014.
6. A. Goldsmith, *Wireless Communications*, Cambridge University Press, 2005.
7. A. Jahn, H. Bischl, and G. Heiss, "Channel Characterisation for Spread Spectrum Satellite Communications," Proceedings of the 4th IEEE International Symposium on Spread Spectrum Techniques and Applications (Volume:3), 1996.
8. M. O'Donnell, T. Watson, J. Fisher, S. Simpson, G. Brodin, E. Bryant, and D. Walsh, "A Study of Galileo Performance - GPS Interoperability and Discriminators for Urban and Indoor Environments," Proceedings of the 15th International Technical Meeting of the Satellite Division of The Institute of Navigation (ION GPS 2002), September 2002, pp. 2160–2172.
9. R. Davies, M. Bensebti, M. A. Beach, and J. P. McGeehan, "Wireless Propagation Measurements in Indoor Multipath Environments at 1.7 GHz and 60 GHz for Small Cell Systems," 41st IEEE Vehicular Technology Conference, 1991.
10. A. Steingass and A. Lehner, "Measuring the Navigation Multipath Channel A Statistical Analysis," Proceedings of the 17th International Technical Meeting of the Satellite Division of The Institute of Navigation (ION GNSS 2004), September 2004, pp. 1157–1164.
11. B. Rama Rao, W. Kunysz, R. Fante, and K. McDonald, *GPS/GNSS Antennas*, Artech House, 2013.
12. United States and the European Union Announce Final Design for GPS-Galileo Common Civil Signal, Joint U.S.-EC Press Release on GPS-Galileo Common Civil Signal Design, July 2007, available at http://www.gps.gov/policy/cooperation/europe/2007/MBOC-agreement/, accessed 12 October 2014.
13. J. W. Betz, C. J. Hegarty, and J. J. Rushanan, "Time-Multiplexed Binary Offset Carrier Signaling and Processing—Group II," U.S. Patent Application No. 12/755,933.
14. C. Lee, Y.-H. Chen, G. Wong, S. Lo, and P. Enge, "Multipath Benefits of BOC vs. BPSK Modulated Signals Using On-Air Measurements," Proceedings of the 2013 International Technical Meeting of The Institute of Navigation, January 2013, pp. 742–751.
15. C. J. Hegarty, M. Tran, and J. W. Betz, "Multipath Performance of the New GNSS Signals," *Proceedings of Institute of Navigation National Technical Meeting 2004*, ION-NTM-2004, Institute of Navigation, January 2004.
16. C. J. Hegarty, J. W. Betz, and A. Saidi, "Binary Coded Symbol Modulations for GNSS," *Proceedings of the Institute of Navigation Annual Meeting 2004*, ION-AM-2004, Institute of Navigation, June 2004.

17. L. Garin, F. van Diggelen, and J.-M. Rousseau, "Strobe & Edge Correlator Multipath Mitigation for Code," Proceedings of the 9th International Technical Meeting of the Satellite Division of The Institute of Navigation (ION GPS 1996), September 1996, pp. 657–664.
18. G. A. McGraw and M. S. Braasch, GNSS Multipath Mitigation Using Gated and High Resolution Correlator Concepts, ION NTM 1999, 25–27 January 1999.
19. T. N. Morrissey, K. W. Shallberg, and B. Townsend, "Code Tracking Errors for Double Delta Discriminators with Narrow Correlator Spacings and Bandlimited Receivers," Proceedings of the 2006 National Technical Meeting of The Institute of Navigation, Monterey, CA, January 2006, pp. 914–926.
20. B. R. Townsend, P. C. Fenton, K. J. Van Dierendonck, and D. J. R. van Nee, "Performance Evaluation of the Multipath Estimating Delay Lock Loop", *NAVIGATION: Journal of The Institute of Navigation*, Vol. 42, No. 3, 1995, pp. 503–514.
21. N. Sokhandan, A. Broumandan, V. Dehghanian, and G. Lachapelle, "GNSS Multipath Error Reduction in Harsh Environments," Proceedings of the 24th International Technical Meeting of The Satellite Division of the Institute of Navigation (ION GNSS 2011), September 2011, pp. 2885–2895.
22. L. R. Weill, "Multipath Mitigation using Modernized GPS Signals: How Good Can it Get?," Proceedings of the 15th International Technical Meeting of the Satellite Division of The Institute of Navigation (ION GPS 2002), Portland, OR, September 2002, pp. 493–505.
23. M. Sahmoudi and M.G. Amin, "Fast Iterative Maximum-Likelihood Algorithm (FIMLA) for Multipath Mitigation in the Next Generation of GNSS Receivers," *IEEE Transactions on Wireless Communications*, Vol. 7, No. 11, 2008.
24. T. Iwase, N. Suzuki, and Y. Watanabe, "Estimation and Exclusion of Multipath Range Error for Robust Positioning," *GPS Solutions*, Vol. 17, No. 1, 2013, pp. 53–62.
25. Z. Jiang, P. D. Groves, W. Y. Ochieng, S. Feng, C. D. Milner, and P. G. Mattos, "Multi-Constellation GNSS Multipath Mitigation Using Consistency Checking," Proceedings of the 24th International Technical Meeting of the Satellite Division of the Institute of Navigation (ION GNSS 2011), September 2011, pp. 3889–3902.
26. U.S. Department of Commerce, National Oceanic & Atmospheric Administration, NOAA Research, "Remote Sensing Using GPS," http://www.esrl.noaa.gov/psd/psd3/multi/remote/, accessed 19 May 2015.

23

AUGMENTATIONS USING DIFFERENTIAL SATNAV

Differential satnav augmentations use reference stations, additional processing, and communications networks to enhance the accuracy and integrity a receiver can obtain with satnav signals. Absolute differential systems enable a user receiver to correct its measurements, improving (compared to stand-alone receiver processing) accuracy in an absolute coordinate system such as ECEF. In contrast, relative differential systems allow a user receiver, sometimes called a rover receiver, to accurately measure its position relative to a reference coordinate system. In some applications the relationship between the reference coordinate system and an absolute coordinate system is well known, allowing the user receiver to calculate its position and velocity in the absolute coordinate system using the relative position. In other cases, only the relative position is important (e.g., guiding a vehicle repeatedly in the same trajectory, or bringing two objects to a common point), and absolute position and velocity are not calculated.

Besides providing accuracy improvement, reference stations in differential satnav systems can also inspect received satellite signals for anomalies, and provide the resulting integrity information (assurance that the inspected signals do not contain anomalies, or alerts that an anomaly exists) to user receivers. Integrity information is primarily needed in critical transportation applications such as aviation safety of life, and maritime use in harbors and coastal waterways.

Section 23.1 provides an overview of differential satnav, reviewing the motivation, architectures, and alternative approaches that can be employed. Different categories of differential satnav, based on geometry of the reference stations and use of different measurements, are introduced. Differential systems based on code tracking measurements are discussed in Section 23.2. These are the simplest approaches to implement, but do not provide the exquisite accuracy of techniques based on carrier phase measurements, which Section 23.3 explores in greater detail. Section 23.4 summarizes this chapter.

23.1 OVERVIEW OF DIFFERENTIAL SATNAV

Chapter 6 describes the dominant sources of error in satnav, showing they arise from imperfections in the space segment, the control segment, propagation, and the user receiver. Typically, the resulting errors are projected onto the line of sight distance and velocity between receiver and satellite. SISRE refers to line of sight measurement errors associated with the space and control segments, while UEE accounts for line of sight measurement errors introduced by propagation and receiver processing. UERE refers to the combined contributions of SISRE and UEE to line of sight measurement errors. When line of sight measurements are transformed into estimates of user PVT, the geometry of the satellites whose signals are used by the receiver determines how the line of sight measurement errors are transformed into PVT errors, typically represented by a scalar multiplier called DOP.

Differential satnav exploits the spatial correlation of errors—the fact that some errors are common over different receiver locations. Some contributors to a satnav signal's SISRE are common to any receiver using the signal. Reference stations can estimate the SISRE, or some of the contributors to SISRE, and communicate the estimated errors to user receivers. In addition, some contributors to UEE, namely those introduced by propagation through the ionosphere and troposphere, can be estimated and communicated to a user receiver. To the extent that the user receiver experiences the same errors, it can remove these estimated errors from its measurements, improving the accuracy of the resulting PVT solution. An alternative to estimating errors is for the reference station to make independent measurements of satellite clock corrections and ephemeris, and of ionospheric and tropospheric conditions, and provide these measurements to user receivers.

Satnav systems are constrained by the number and locations of monitoring stations, the number of bits in the data message, the available satellite upload times, and large service volume, all of which contribute to SISRE due to estimation, prediction, curve fitting, and latency. In contrast, differential system designs can relax these constraints, and consequently can provide significantly smaller SISRE than experienced by stand-alone receivers using only the broadcast satnav signals.

A generic architecture for differential satnav is provided in Figure 23.1. In contrast to stand-alone satnav depicted in Figure 1.4, differential satnav adds one or more reference stations. These reference stations are independently surveyed, so their positions (actually the positions of the reference stations' receive antenna phase centers) are known independent of satnav measurements. When the receiver computes its position

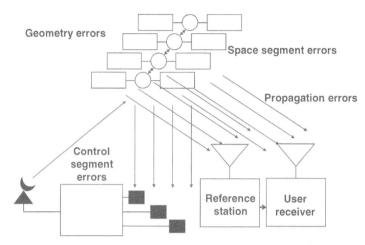

Figure 23.1. Generic Architecture for Differential Satnav

from the satnav signals, errors between the computed position and the a priori knowledge of position are ascribed to SISRE and propagation effects. Observations over time and from multiple reference stations help resolve the specific contributors to the error (i.e., clock corrections vs. ephemeris vs. propagation).

Satellite clock errors produce the same code and carrier errors observed by receivers in any location, and thus are highly spatially correlated. Satellite clock errors also tend to vary slowly, and thus exhibit high temporal correlation as well. For example, the Allan deviation of GPS satellites ranges from less than 10^{-11} for a 1 second observation time to approaching 10^{-13} for a 1000 second observation time [1]. In contrast, ephemeris errors exhibit less spatial correlation, since they produce code and carrier errors that are projected onto the line of sight between satellite and receiver, and thus vary with receiver location.

In some differential systems, the user receiver obtains information directly from a single reference station. The distance between the user receiver and the reference station in this case is called the baseline. In other differential systems becoming increasingly common, measurements from multiple reference stations are combined to calculate the corrections to a user receiver. These networked differential systems require a much lower density of reference stations, but involve more complicated interconnection of reference stations and more sophisticated processing to form the corrections. With networked systems, the errors remaining after correction tend to have the same statistics over the entire service region (the region where reference stations are located), rather than being strongly dependent upon the user receiver's distance from the nearest reference station.

For single-baseline positioning, the separation between reference receiver and user receiver, called the baseline length, is an important parameter. Short-baseline differential systems have baseline lengths typically less than 20 km, and sometimes an order of magnitude smaller than that. Long-baseline differential systems employ baseline lengths that may exceed 500 km. Intervisibility, the ability for the reference station and user receiver both to track common satellites, is essential for obtaining full performance from differential satnav. Longer baselines increase the fraction of time that intervisibility does not exist, and long-baseline systems also have generally larger errors because many error sources are less spatially correlated over the longer baseline.

Networked differential systems that service receivers over a large region, like a continent, are known as wide-area differential systems. Those that provide service over a small area, with distances between reference stations and user receivers on the order of kilometers or a few tens of kilometers are called local-area differential systems. Regional-area differential systems fall in between these two, with service areas extending many tens or even hundreds of kilometers. Regional-area differential systems can either be networked or can employ a dense set of reference stations with the user receiver establishing a baseline to the nearest one. Networked systems avoid sensitivity to spatial correlation of errors by estimating the errors at the user receiver's location, rather than merely employing the errors at the nearest reference station.

Code-based differential systems employ measurements based on code tracking, whereas carrier-based differential systems employ carrier phase measurements and must resolve integer wavelength ambiguities as described in Section 23.3. Typically, wide-area differential systems are networked and code-based, while regional-area and local-area differential systems can be either code based or carrier based. Some authors reserve the term differential satnav for code-based differential systems, calling carrier-based systems relative satnav. However, this book uses the term differential satnav for systems using either type of measurements.

Code-based wide-area differential satnav systems typically provide positioning accuracy on the order of a meter or slightly less. In contrast, carrier-based local differential satnav systems can provide centimeter-level accuracy. Positioning using real-time communication of carrier phase measurements from local or networked regional reference stations is called real-time kinematic (RTK) navigation, and can produce millimeter or centimeter accuracies.

The distinction between differential satnav and precise point positioning (PPP), described in Section 20.7, is becoming increasingly blurred. Strictly speaking, differential satnav supplies information that the user receiver employs to correct its measurements, while PPP provides information that replaces or augments information from the satnav signals. Historically, PPP has been used for postprocessing in nonreal time, but increasingly supports real-time processing as communications networks or broadcast satnav signals provide the information in real time. Finally, PPP relies on dual-frequency carrier-based processing, while some versions of differential satnav employ code-based processing, others employ carrier-based processing, and single-frequency differential satnav is common.

Different communications links are employed between user receivers and the reference stations or central processing site. Line of sight radio communications using VHF or UHF band data radios can be used over short baselines. Line of sight low power spread spectrum radios in unlicensed bands are also employed for short baselines. Commercial wireless networks are increasingly used for regional-area systems. Nationwide differential GPS (NDGPS) services are operational in multiple countries [2], and use low-frequency (30–300 kHz) or medium-frequency (300–3000 kHz) bands, with carrier frequencies near 300 kHz. These frequency bands support operation beyond line of sight. NDGPS systems were developed primarily to eliminate errors from GPS Selective Availability, serving maritime applications. The US NDGPS program is described in Reference 3.

Satellite communications are also used extensively in differential satnav. Commercial enterprises like StarFire [4] and OmniStar [5] use satellite communications (LEO satellites providing mobile satellite services or GEO communications satellites) to link short-baseline differential systems, as well as regional-area differential systems. Satellite-based augmentation systems (SBASs), described in Chapter 8, employ dedicated leased transponders on GEO communications satellites.

Some differential systems provide relative information, with respect to a local coordinate system such as a reference station. Others provide absolute information, with respect to an ECEF coordinate system. Differential information can be provided from reference stations to user receivers in different forms. Corrections to position measurements can be provided directly for local-area differential systems. Instead, corrections to pseudorange measurements can be provided. Also, corrections can be provided to the clock corrections and ephemeris information supplied in each satnav signal's data message. In addition, ionospheric and tropospheric conditions can be measured and provided for each satellite signal as received at reference station, or at various pierce points in the atmosphere. The conditions at a given pierce point can be used directly by the user receiver, or the receiver can interpolate the provided data in order to estimate conditions at pierce points associated with its measurements. Replacement clock and ephemeris data can also be provided by the differential system, along with the coordinates of reference stations.

The radio signals used for differential augmentations primarily serve as communications function. However, in some cases these signals can also be used by a receiver to obtain measurements that enhance geometry and improve signal availability, as is the case for some SBASs as discussed in Chapter 8.

The different architectures and options described in this section all vary in terms of performance, coverage area, communications capacity needed, and processing complexity in the differential infrastructure and the user receiver. While different choices are preferable in different situations, all can provide marked performance improvements relative to stand-alone satnav receivers. The next two sections provide more detail on specific approaches.

23.2 CODE-BASED DIFFERENTIAL SYSTEMS

The simplest differential systems are code based with a single reference station in a short-baseline configuration. The reference station computes and broadcasts a position correction, which the user receiver uses to correct its position calculation, or a pseudorange correction to each satellite in view, which the user receiver adds or subtracts from its pseudorange measurement before computing PVT as described in Chapter 20. Since the baseline is short, many sources of error (except for those resulting from multipath and noise) are highly correlated. The result can be meter-level accuracies with a single-frequency receiver. The best error reduction is achieved when both receivers are tracking the same signals from the same satellites using the same discriminator designs and tracking loop designs, and when receivers are at the same altitude. If receivers are at significantly different altitudes, an altitude correction for tropospheric errors can be applied, in a fashion similar to that described in Section 20.2.2. If differential corrections are only transmitted occasionally, then a polynomial representation of each time-varying pseudorange can be provided for each satellite, along with the relevant time epoch.

When single-frequency processing is used with carrier-smoothed code tracking, code minus carrier divergence cancels under ideal conditions: that is, when the ionospheric effects are the same at the reference station and user receiver (implying short baseline) and the code minus carrier smoothing filters use identical parameters and have converged. If either receiver experiences a cycle slip, however, the tracking loops must be reset and then converge again before error cancellation reoccurs. The restriction to use identical code minus carrier filters forces compromises between the reference receiver, which has no dynamics and would benefit from a very narrow filter bandwidth, and the user receiver, whose filter bandwidth must be wide enough to accommodate its dynamics [6, 7].

Several advantages accrue from the use of dual-frequency receivers and ionosphere-free measurements [6]. Most importantly, longer baselines can be employed since the ionospheric effects need not be the same at both receivers. Also, code minus carrier divergence does not occur in either receiver, allowing different smoothing parameters to be used in the reference receiver and user receiver, so the loop filters can be designed separately for each receiver.

If a networked regional-area differential system is employed, the user receiver can use various approaches. One approach is to compute a single set of pseudorange corrections using a weighted combination (typically with weighting based on baseline length) of corrections from the different stations, then apply the resulting corrections to the receiver's pseudorange measurements in order to compute PVT. An alternate approach is to use the correction from each reference station to compute a separate PVT solution, then compute a final PVT solution as the weighted sum of the separate PVT solutions. Sample statistics can also be computed from the separate PVT solutions and used to determine the weights applied to the separate pseudorange corrections.

For wide-area differential systems, pseudorange errors are not well correlated spatially, and it becomes preferable for the differential system to compute and provide estimated clock corrections and ephemeris, as well as ionospheric models, instead of pseudorange corrections. Reference stations typically make dual-frequency measurements allowing them to estimate the TEC at the pierce point corresponding to each satellite, and also to make ionosphere-free pseudorange measurements that can employ carrier aiding or carrier smoothing. When all reference stations are synchronized, they can report pseudorange estimates with a common time base to a central processing station, which can use the pseudorange equations introduced in Chapter 20, but rewritten to solve for location and time offset between satellite time and system time. The satellite positions and velocities can be represented and communicated in a selected coordinate system, such as ECEF. Tropospheric errors are usually corrected using models, due to low correlation over long baselines. Since this approach computes and communicates fundamental characteristics of each satellite, it allows variability in receiver altitude, signals being tracked, and receiver processing. SBASs, described in Chapter 8, are wide-area differential systems that provide meter-level accuracies with single-frequency user receivers over areas the size of a continent.

23.3 CARRIER-BASED DIFFERENTIAL SYSTEMS

The distance between receiver and satellite (actually the distance between phase centers of the receive antenna and transmit antenna), or the distance between a user receiver and base station (actually the distance between the phase centers of their receive antennas) can be expressed as the sum of an integer number of wavelengths of a carrier frequency and a fractional wavelength. Chapter 18 showed that while carrier phase can be measured to a small fraction of a cycle, the S-curve is periodic. Consequently, the integer number of wavelengths in the receiver-to-satellite distance, typically denoted by N, is ambiguous.

In contrast, the combination of spreading code and either data message or overlay code makes code tracking unambiguous. Yet Chapters 19 and 22 show that, in units of meters, code tracking errors are much larger than carrier phase estimation errors.

Like PPP introduced in Chapter 20, carrier-based differential systems attempt to exploit the accuracy of carrier phase measurements while using code tracking and other processing to resolve the carrier phase ambiguities. The focus here is on RTK techniques that accurately measure the baseline vector between a reference station and a user receiver in real time. Intervisibility is essential in RTK, since the processing involves making differences between measurements by different receivers on the same signals from the same satellites.

Section 23.3.1 introduces double differencing—the most common current approach to carrier-based differential processing used in RTK. Section 23.3.2 provides more detail concerning double differencing and its variants, the effects on errors, and the challenge of ambiguity resolution. Section 23.3.3 overviews ambiguity resolution processing.

23.3.1 Introduction to Double Differencing

Providing centimeter accuracy using measurements from satellites many thousands of kilometers away requires significant attention to detail, and, as a result, the complicated equations and notation used in Section 23.3.2. Although these details are needed in practice to achieve full accuracy, the resulting clutter obscures the underlying concepts. This section describes the essence of double difference processing without that clutter.

Figure 23.2 shows the simplified geometry, based on the exposition in Reference 8. Receiver 1 is the reference station, with known location, and Receiver 2 is the user receiver. The objective is to find the baseline vector $\mathbf{b}_{2,1}$ from Receiver 2 to Receiver 1. Since Receiver 1 is at a known location and the satellite ephemerides are known, the unit vectors from Receiver 1 to the two satellites, $\mathbf{u}_{1,1}$ and $\mathbf{u}_{1,2}$, are known.

The range from receivers to satellites is so large relative to the baseline separating the two receivers that the unit vectors are approximated as the same from either receiver to the same satellite, so $\mathbf{u}_{2,1} \cong \mathbf{u}_{1,1}$ and $\mathbf{u}_{2,2} \cong \mathbf{u}_{1,2}$. Denoting the range between the nth receiver and the kth satellite by $r^{(n,k)}$ enables the difference in path lengths between the

CARRIER-BASED DIFFERENTIAL SYSTEMS

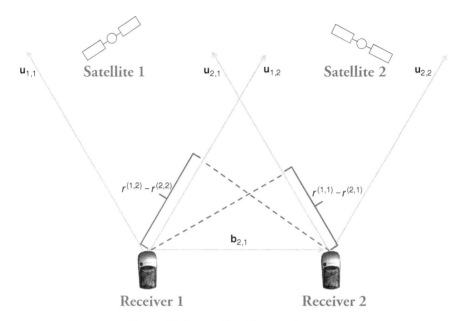

Figure 23.2. Double Difference Geometry

two receivers and a given satellite to be expressed as the vector dot product between the unit vector toward the satellite and the baseline vector.

$$r^{(1,1)} - r^{(2,1)} = \mathbf{u}_{1,1}^T \mathbf{b}_{2,1}$$
$$r^{(1,2)} - r^{(2,2)} = \mathbf{u}_{1,2}^T \mathbf{b}_{2,1}$$
(23.1)

Denoting a measurement of the range between the nth receiver and the kth satellite by $\hat{r}^{(n,k)}$,

$$\hat{r}^{(1,1)} - \hat{r}^{(2,1)} = \mathbf{u}_{1,1}^T \mathbf{b}_{2,1} + \varepsilon_{1,1}$$
$$\hat{r}^{(1,2)} - \hat{r}^{(2,2)} = \mathbf{u}_{1,2}^T \mathbf{b}_{2,1} + \varepsilon_{1,2}$$
(23.2)

where $\varepsilon_{1,1}$ and $\varepsilon_{2,1}$ are error terms modeled as random and zero mean.

The measurement double difference is then $(\hat{r}^{(1,1)} - \hat{r}^{(2,1)}) - (\hat{r}^{(1,2)} - \hat{r}^{(2,2)}) = (\mathbf{u}_{1,1} - \mathbf{u}_{1,2})^T \mathbf{b}_{2,1} + \varepsilon_{1,1} - \varepsilon_{1,2}$. When the same information is available for K satellites, the resulting system of double difference equations is

$$\begin{bmatrix} (\hat{r}^{(1,1)} - \hat{r}^{(2,1)}) - (\hat{r}^{(1,2)} - \hat{r}^{(2,2)}) \\ (\hat{r}^{(1,2)} - \hat{r}^{(2,2)}) - (\hat{r}^{(1,3)} - \hat{r}^{(2,3)}) \\ \vdots \\ (\hat{r}^{(1,K-1)} - \hat{r}^{(2,K-1)}) - (\hat{r}^{(1,K)} - \hat{r}^{(2,K)}) \end{bmatrix} = \begin{bmatrix} \mathbf{u}_{1,1}^T - \mathbf{u}_{1,2}^T \\ \mathbf{u}_{1,2}^T - \mathbf{u}_{1,3}^T \\ \vdots \\ \mathbf{u}_{1,K-1}^T - \mathbf{u}_{1,K}^T \end{bmatrix} \mathbf{b}_{2,1} + \begin{bmatrix} \varepsilon_{1,1} - \varepsilon_{1,2} \\ \varepsilon_{1,2} - \varepsilon_{1,2} \\ \vdots \\ \varepsilon_{1,K-1} - \varepsilon_{1,K} \end{bmatrix}$$
(23.3)

If $K = 4$, and the matrix is full rank, the estimate of $\hat{\mathbf{b}}_{2,1}$ is found in the standard way:

$$\mathbf{b}_{2,1} \cong \begin{bmatrix} \mathbf{u}_{1,1}^T - \mathbf{u}_{1,2}^T \\ \mathbf{u}_{1,2}^T - \mathbf{u}_{1,3}^T \\ \mathbf{u}_{1,3}^T - \mathbf{u}_{1,4}^T \end{bmatrix}^{-1} \begin{bmatrix} (\hat{r}^{(1,1)} - \hat{r}^{(2,1)}) - (\hat{r}^{(1,2)} - \hat{r}^{(2,2)}) \\ (r^{(1,2)} - r^{(2,2)}) - (\hat{r}^{(1,3)} - \hat{r}^{(2,3)}) \\ (\hat{r}^{(1,3)} - \hat{r}^{(2,3)}) - (\hat{r}^{(1,4)} - \hat{r}^{(2,4)}) \end{bmatrix} \quad (23.4)$$

If $K > 4$, $\hat{\mathbf{b}}_{2,1}$ is found by OLS or WLS solution of the overdetermined system of equations (23.3) as described in Appendix A.7.

23.3.2 Measurements for Carrier-Based Differential Systems

Generalizing (20.30), the code tracking and carrier phase measurements yield

$$\hat{r}_m^{(n,k)} = r^{(n,k)} + \left(\Delta_{\text{tropo}}^{(n,k)} + \Delta_{\text{iono},m}^{(n,k)}\right)c + \left(\Delta_{c,r}^{(n)} - \Delta_{c,s}^{(k)}\right)c + \left(\Delta_{h,r}^{(n)} - \Delta_{h,s}^{(k)}\right)c + p_{r,m}^{(n,k)} + \tilde{r}_m^{(n,k)}$$

$$\hat{\Phi}_m^{(n,k)} = r^{(n,k)} + \left(\Delta_{\text{tropo}}^{(n,k)} - \Delta_{\text{iono},m}^{(n,k)}\right)c + \left(\Delta_{c,r}^{(n)} - \Delta_{c,s}^{(k)}\right)c$$
$$+ \left(\Delta_{h,r}^{(n)} - \Delta_{h,s}^{(k)}\right)c - N_m^{(n,k)}\lambda_m + p_{\Phi,m}^{(n,k)} + \tilde{\Phi}_m^{(n,k)} \quad (23.5)$$

where

- $\hat{r}_m^{(n,k)}$ is the pseudorange measurement observed by the nth receiver processing a signal at the mth carrier frequency from the kth satellite,
- $r^{(n,k)}$ is the true distance between the nth receiver antenna phase center and the kth transmit antenna phase center,
- $\Delta_{\text{tropo}}^{(n,k)}$ is the tropospheric delay at the nth receiver for signals from the kth satellite,
- $\Delta_{\text{iono},m}^{(k)}$ is the ionospheric delay at the nth receiver for signals from the kth satellite at the mth carrier frequency,
- $\Delta_{c,r}^{(n)}$ is the offset between nth receiver time and system time,
- $\Delta_{c,s}^{(k)}$ is the offset between kth satellite time and system time,
- $\Delta_{h,r}^{(n)}$ is the hardware delay bias for the nth receiver,
- $\Delta_{h,s}^{(k)}$ is the hardware delay bias for the kth satellite,
- $p_{r,m}^{(n,k)}$ is the code tracking multipath error at the nth receiver for signals from the kth satellite at the mth carrier frequency,
- $\tilde{r}_m^{(n,k)}$ is the code tracking measurement noise at the nth receiver for signals from the kth satellite at the mth carrier frequency,
- $\hat{\Phi}_m^{(n,k)}$ is the product of the carrier phase measurement observed by the nth receiver processing in cycles and the wavelength of the signal's carrier frequency λ_m for the kth satellite at the mth carrier frequency,
- $N_m^{(n,k)}$ is the integer ambiguity and the uncalibrated phase delays originating in the satellite, for the kth satellite and nth receiver at the mth carrier frequency,

CARRIER-BASED DIFFERENTIAL SYSTEMS

- $p_{\Phi,m}^{(n,k)}$ is the carrier tracking multipath error at the nth receiver for signals from the kth satellite at the mth carrier frequency,
- $\tilde{\Phi}_m^{(n,k)}$ is the carrier tracking measurement noise at the nth receiver for signals from the kth satellite at the mth carrier frequency.

In general, many of these quantities are time-varying, and asynchronous in practice. For simplicity it is assumed here that all measurements are made simultaneously. Further, the processing described here does not exploit measurements made over time, although some other processing approaches do so, so time dependence is suppressed. Even when time-varying characteristics are taken into account, however, the integer ambiguity is not time-varying as long as there are no cycle slips in carrier phase tracking.

Between-receiver single differencing (BRSD) of measurements from a given satellite is defined by

$$\hat{r}_{BRSD,m}^{(n_1,n_2,k)} \triangleq \hat{r}_m^{(n_1,k)} - \hat{r}_m^{(n_2,k)}$$

$$= r^{(n_1,k)} - r^{(n_2,k)} + \left(\Delta_{tropo}^{(n_1,k)} - \Delta_{tropo}^{(n_2,k)}\right)c + \left(\Delta_{iono,m}^{(n_2,k)} - \Delta_{iono,m}^{(n_1,k)}\right)c$$

$$+ \left(\Delta_{c,r}^{(n_1)} - \Delta_{c,r}^{(n_2)}\right)c + \left(\Delta_{h,r}^{(n_1)} - \Delta_{h,r}^{(n_2)}\right)c + \varepsilon_{BRSD,r,m}^{(n_1,n_2,k)}$$

$$\hat{\Phi}_{BRSD,m}^{(n_1,n_2,k)} \triangleq \hat{\Phi}_m^{(n_1,k)} - \hat{\Phi}_m^{(n_2,k)}$$

$$= r^{(n_1,k)} - r^{(n_2,k)} + \left(\Delta_{tropo}^{(n_1,k)} - \Delta_{tropo}^{(n_2,k)}\right)c + \left(\Delta_{iono,m}^{(n_2,k)} - \Delta_{iono,m}^{(n_1,k)}\right)c$$

$$+ \left(\Delta_{c,r}^{(n_1)} - \Delta_{c,r}^{(n_2)}\right)c + \left(\Delta_{h,r}^{(n_1)} - \Delta_{h,r}^{(n_2)}\right)c$$

$$- \left(N_m^{(n_1,k)} - N_m^{(n_2,k)}\right)\lambda_m + \varepsilon_{BRSD,\Phi,m}^{(n_1,n_2,k)} \tag{23.6}$$

where the measurement errors are $\varepsilon_{BRSD,r,m}^{(n_1,n_2,k)} \triangleq p_{r,m}^{(n_1,k)} - p_{r,m}^{(n_2,k)} + \tilde{r}_m^{(n_1,k)} - \tilde{r}_m^{(n_2,k)}$, and $\varepsilon_{BRSD,\Phi,m}^{(n_1,n_2,k)} \triangleq p_{\Phi,m}^{(n_1,k)} - p_{\Phi,m}^{(n_2,k)} + \tilde{\Phi}_m^{(n_1,k)} - \tilde{\Phi}_m^{(n_2,k)}$ for code delay and carrier phase measurements, respectively. This differencing removes the satellite clock error and the satellite hardware delay bias.

In contrast, between-satellite single differencing (BSSD) of measurements by a given receiver involves computing

$$\hat{r}_{BSSD,m}^{(n,k_1,k_2)} \triangleq \hat{r}_m^{(n,k_1)} - \hat{r}_m^{(n,k_2)}$$

$$= r^{(n,k_1)} - r^{(n,k_2)} + \left(\Delta_{tropo}^{(n,k_1)} - \Delta_{tropo}^{(n,k_2)}\right)c + \left(\Delta_{iono,m}^{(n,k_2)} - \Delta_{iono,m}^{(n,k_1)}\right)c$$

$$+ \left(\Delta_{c,s}^{(k_2)} - \Delta_{c,s}^{(k_1)}\right)c + \left(\Delta_{h,s}^{(k_2)} - \Delta_{h,s}^{(k_1)}\right)c + \varepsilon_{BSSD,r,m}^{(n,k_1,k_2)}$$

$$\hat{\Phi}_{BSSD,m}^{(n,k_1,k_2)} \triangleq \hat{\Phi}_m^{(n,k_1)} - \hat{\Phi}_m^{(n,k_2)}$$

$$= r^{(n,k_1)} - r^{(n,k_2)} + \left(\Delta_{tropo}^{(n,k_1)} - \Delta_{tropo}^{(n,k_2)}\right)c + \left(\Delta_{iono,m}^{(n,k_2)} - \Delta_{iono,m}^{(n,k_1)}\right)c$$

$$+ \left(\Delta_{c,s}^{(k_2)} - \Delta_{c,s}^{(k_1)}\right)c + \left(\Delta_{h,s}^{(k_2)} - \Delta_{h,s}^{(k_1)}\right)c$$

$$- \left(N_m^{(n,k_1)} - N_m^{(n,k_2)}\right)\lambda_m + \varepsilon_{BSSD,\Phi,m}^{(n,k_1,k_2)} \tag{23.7}$$

where the measurement errors are $\varepsilon_{BSSD,r,m}^{(n_1,n_2,k)} \triangleq p_{r,m}^{(n,k_1)} - p_{r,m}^{(n,k_2)} + \tilde{r}_m^{(n,k_1)} - \tilde{r}_m^{(n,k_2)}$ and $\varepsilon_{BSSD,\Phi,m}^{(n,k_1,k_2)} \triangleq p_{\Phi,m}^{(n,k_1)} + \tilde{\Phi}_m^{(n,k_1)} - p_{\Phi,m}^{(n,k_2)} - \tilde{\Phi}_m^{(n,k_2)}$. This differencing removes the receiver clock error and receiver hardware delay bias.

Now consider double differencing of the single differences for two satellites and the same pair of receivers:

$$\begin{aligned}
\hat{r}_{DD,m}^{(n_1,n_2,k_1,k_2)} &\triangleq \hat{r}_{BSSD,m}^{(n_1,k_1,k_2)} - \hat{r}_{BSSD,m}^{(n_2,k_1,k_2)} = \hat{r}_{BRSD,m}^{(n_1,n_2,k_1)} - \hat{r}_{BRSD,m}^{(n_1,n_2,k_2)} \\
&= (\mathbf{u}_{n_1,k_1} - \mathbf{u}_{n_1,k_2})^T \mathbf{b}_{n_2,n_1} + \left(\Delta_{tropo}^{(n_1,k_1)} - \Delta_{tropo}^{(n_1,k_2)}\right)c - \left(\Delta_{tropo}^{(n_2,k_1)} - \Delta_{tropo}^{(n_2,k_2)}\right)c \\
&\quad + \left(\Delta_{iono,m}^{(n_1,k_2)} - \Delta_{iono,m}^{(n_1,k_1)}\right)c \\
&\quad - \left(\Delta_{iono,m}^{(n_2,k_2)} - \Delta_{iono,m}^{(n_2,k_1)}\right)c + \varepsilon_{BSSD,r,m}^{(n_1,k_1,k_2)} - \varepsilon_{BSSD,r,m}^{(n_2,k_1,k_2)} \\
\hat{\Phi}_{DD,m}^{(n_1,n_2,k_1,k_2)} &\triangleq \hat{\Phi}_{BSSD,m}^{(n_1,k_1,k_2)} - \hat{\Phi}_{BSSD,m}^{(n_2,k_1,k_2)} = \hat{\Phi}_{BRSD,m}^{(n_1,n_2,k_1)} - \hat{\Phi}_{BRSD,m}^{(n_1,n_2,k_2)} \\
&= (\mathbf{u}_{n_1,k_1} - \mathbf{u}_{n_1,k_2})^T \mathbf{b}_{n_2,n_1} + \left(\Delta_{tropo}^{(n_1,k_1)} - \Delta_{tropo}^{(n_1,k_2)}\right)c - \left(\Delta_{tropo}^{(n_2,k_1)} - \Delta_{tropo}^{(n_2,k_2)}\right)c \\
&\quad + \left(\Delta_{iono,m}^{(n_1,k_2)} - \Delta_{iono,m}^{(n_1,k_1)}\right)c \\
&\quad - \left(\Delta_{iono,m}^{(n_2,k_2)} - \Delta_{iono,m}^{(n_2,k_1)}\right)c - N_{DD,m}^{(n_1,n_2,k_1,k_2)} \lambda_m + \varepsilon_{BSSD,\Phi,m}^{(n_1,k_1,k_2)} - \varepsilon_{BSSD,\Phi,m}^{(n_2,k_1,k_2)}
\end{aligned} \quad (23.8)$$

where (23.1) is used to replace $(r^{(n_1,k_1)} - r^{(n_2,k_1)}) - (r^{(n_1,k_2)} - r^{(n_2,k_2)}) = (r^{(n_1,k_1)} - r^{(n_1,k_2)}) - (r^{(n_2,k_1)} - r^{(n_2,k_2)})$ with $(\mathbf{u}_{n_1,k_1} - \mathbf{u}_{n_1,k_2})^T \mathbf{b}_{n_2,n_1}$ using $\mathbf{u}_{n_1,k}^T \cong \mathbf{u}_{n_2,k}^T$ as in Section 23.3.1, and the double difference integer ambiguity is $N_{DD,m}^{(n_1,n_2,k_1,k_2)} \triangleq N_m^{(n_1,k_1)} - N_m^{(n_1,k_2)} - N_m^{(n_2,k_1)} + N_m^{(n_2,k_2)}$.

This double differencing removes the receiver and satellite clock errors, and also the satellite and receiver hardware biases, leaving ionospheric errors, tropospheric errors, and measurement errors due to multipath, noise, and interference.

The pseudorange and carrier phase expressions in (23.8) are similar, but have two important distinctions:

- The carrier phase expression includes the double difference integer ambiguity, which is unknown,
- The variance of the double difference carrier phase measurement errors, $\varepsilon_{BSSD,\Phi,m}^{(n_1,k_1,k_2)} - \varepsilon_{BSSD,\Phi,m}^{(n_2,k_1,k_2)}$, is much less than the variance of the double difference pseudorange errors, $\varepsilon_{BSSD,r,m}^{(n_1,k_1,k_2)} - \varepsilon_{BSSD,r,m}^{(n_2,k_1,k_2)}$, due to the smaller errors (when expressed in meters) of the carrier phase tracking jitter and carrier phase multipath errors, compared to their counterparts for pseudorange.

The remaining challenges involve removing or reducing tropospheric and ionospheric errors, then establishing a system of equations that can be solved for $\mathbf{b}_{n_2,n_1}$, exploiting the accuracy of carrier measurements while resolving the ambiguity associated with $N_{DD,m}^{(n_1,n_2,k_1,k_2)} \lambda_m$.

CARRIER-BASED DIFFERENTIAL SYSTEMS

In (23.8), the wavelength associated with the carrier phase ambiguity corresponds to the carrier frequency f_m. The wavelength ranges from approximately 19 cm at the highest frequency in upper L band to approximately 26 cm at the lowest frequency in lower L band. Consequently, for every meter of code tracking error, there are between 3.9 and 5.4 ambiguous wavelengths for each satellite, yielding hundreds of thousands or millions of different ambiguous values that $N_{DD,m}^{(n_1,n_2,k_1,k_2)}$ could take on if the overall code tracking uncertainty is ± 2 m for each satellite.

Reducing the number of ambiguous values is essential; this is accomplished through a combination of reducing code tracking errors and increasing the effective wavelength of the carrier phase measurements. Steps that reduce code tracking jitter and code tracking multipath error are highly beneficial. As described in Chapter 19, code tracking jitter is reduced through narrow tracking loop bandwidths with carrier-aided code tracking, use of narrow early-late spacing in code tracking discriminator design, and using signals whose spreading modulations have large RMS bandwidth. Carrier-smoothed code, introduced in Chapter 20, is also highly beneficial, and divergence-free carrier-smoothed code processing [6] can be used to further reduce jitter. Many of these same steps also help reduce code tracking multipath error.

In a short-baseline differential configuration (e.g., with baseline length less than approximately 20 km), the troposphere and ionosphere errors are often approximately the same on the paths from a given satellite to both receivers, so the double difference equations in (23.8) simplify to

$$\hat{r}_{DD,m}^{(n_1,n_2,k_1,k_2)} \cong \left(\mathbf{u}_{n_1,k_1} - \mathbf{u}_{n_1,k_2}\right)^T \mathbf{b}_{n_2,n_1} + \varepsilon_{BSSD,r,m}^{(n_1,k_1,k_2)} - \varepsilon_{BSSD,r,m}^{(n_2,k_1,k_2)}$$

$$\hat{\Phi}_{DD,m}^{(n_1,n_2,k_1,k_2)} \cong \left(\mathbf{u}_{n_1,k_1} - \mathbf{u}_{n_1,k_2}\right)^T \mathbf{b}_{n_2,n_1} - N_{DD,m}^{(n_1,n_2,k_1,k_2)} \lambda_m$$
$$+ \varepsilon_{BSSD,\Phi,m}^{(n_1,k_1,k_2)} - \varepsilon_{BSSD,\Phi,m}^{(n_2,k_1,k_2)} \quad (23.9)$$

These can be solved for $\mathbf{b}_{n_2,n_1}$ using the approach outlined in Section 23.3.3.

Dual-frequency receivers using carrier frequencies f_{m_1} and f_{m_2}, with $f_{m_1} > f_{m_2}$, produce dual-frequency pseudorange measurements $\hat{r}_{m_1}^{(n,k)}$ and $\hat{r}_{m_2}^{(n,k)}$, and carrier phase measurements $\hat{\Phi}_{m_1}^{(n,k)}$ and $\hat{\Phi}_{m_2}^{(n,k)}$ from the nth receiver to the kth satellite. Various linear combinations of the measurements can be formed from two frequencies, all of the form

$$\hat{r}_{LC}^{(n,k)} \triangleq \alpha_1 \hat{r}_{m_1}^{(n,k)} + \alpha_2 \hat{r}_{m_2}^{(n,k)}$$
$$\hat{\Phi}_{LC}^{(n,k)} \triangleq \beta_1 \hat{\Phi}_{m_1}^{(n,k)} + \beta_2 \hat{\Phi}_{m_2}^{(n,k)} \quad (23.10)$$

The constants are constrained to $\alpha_1 + \alpha_2 \triangleq 1$ and $\beta_1 + \beta_2 \triangleq 1$ in order to yield the correct line of sight ranges when there are no errors.

When measurement errors are modeled as random with zero mean and uncorrelated at the different carrier frequencies with the same variance, the resulting measurement errors due to random effects in the linearly combined measurements have variance

$$\sigma_{r_{LC}}^2 = \left(\alpha_1^2 + \alpha_2^2\right)\sigma_{r_m}^2$$
$$\sigma_{\Phi_{LC}}^2 = \left(\beta_1^2 + \beta_2^2\right)\sigma_{\Phi_m}^2 \quad (23.11)$$

The condition

$$(\alpha_1 + \beta_1)f_{m_2}^2 + (\alpha_2 + \beta_2)f_{m_1}^2 = 0 \quad (23.12)$$

is required for divergence-free carrier smoothing [6] of code minus carrier measurements.

Many different linear combinations can be used, as described in References 6 and 9. One that is commonly used is the combination of narrowlane pseudorange measurements and widelane carrier phase measurements, with $\alpha_1 = \frac{f_{m_1}}{f_{m_1}+f_{m_2}}$ and $\alpha_2 = \frac{f_{m_2}}{f_{m_1}+f_{m_2}}$. In this case, when the error variances are equal at both carrier frequencies and denoted $\sigma_{r_m}^2$, the variance of the narrowlane pseudorange measurement, from (23.11), is

$$\sigma_{r_{NL}}^2 = \frac{f_{m_1}^2 + f_{m_2}^2}{(f_{m_1}+f_{m_2})^2}\sigma_{r_m}^2 \quad (23.13)$$

The fraction in (23.13) is less than unity, indicating that the narrowlane pseudorange measurement has smaller error variance than the variance of the pseudorange measurement at either carrier frequency.

The corresponding widelane carrier phase measurement employs $\beta_1 = \frac{f_{m_1}}{f_{m_1}-f_{m_2}}$ and $\beta_2 = \frac{-f_{m_2}}{f_{m_1}-f_{m_2}}$, so the measurement is

$$\hat{\Phi}_{WL}^{(n,k)} \triangleq \frac{f_{m_1}}{f_{m_1}-f_{m_2}}\hat{\Phi}_{m_1}^{(n,k)} - \frac{f_{m_2}}{f_{m_1}-f_{m_2}}\hat{\Phi}_{m_2}^{(n,k)}$$

$$= r^{(n,k)} + \Delta_{\text{tropo}}^{(n,k)}c + \left(\Delta_r^{(n)} - \Delta_t^{(k)}\right)c + \left(\tau_r^{(n)} + \tau^{(k)}\right)c - \frac{f_{m_1}}{f_{m_1}-f_{m_2}}\Delta_{\text{iono},m_1}^{(n,k)}c$$

$$+ \frac{f_{m_2}}{f_{m_1}-f_{m_2}}\Delta_{\text{iono},m_2}^{(n,k)}c$$

$$+ \frac{f_{m_1}}{f_{m_1}-f_{m_2}}\left[p_{\Phi,m_1}^{(n,k)} - N_{m_1}^{(n,k)}\lambda_{m_1} + \tilde{\Phi}_{m_1}^{(n,k)}\right] - \frac{f_{m_2}}{f_{m_1}-f_{m_2}}\left[p_{\Phi,m_2}^{(n,k)} - N_{m_2}^{(n,k)}\lambda_{m_2} + \tilde{\Phi}_{m_2}^{(n,k)}\right]$$

$$\cong r^{(n,k)} + \Delta_{\text{tropo}}^{(n,k)}c + \left(\Delta_r^{(n)} - \Delta_t^{(k)}\right)c + \left(\tau_r^{(n)} + \tau^{(k)}\right)c + \frac{f_{m_1}}{f_{m_2}}\Delta_{\text{iono},m_1}^{(n,k)}c - N_{WL}^{(n,k)}\lambda_{WL}c$$

$$+ \frac{f_{m_1}}{f_{m_1}-f_{m_2}}\left[p_{\Phi,m_1}^{(n,k)} + \tilde{\Phi}_{m_1}^{(n,k)}\right] - \frac{f_{m_2}}{f_{m_1}-f_{m_2}}\left[p_{\Phi,m_2}^{(n,k)} + \tilde{\Phi}_{m_2}^{(n,k)}\right] \quad (23.14)$$

where the last step uses the approximation $\Delta_{\text{iono},m_2}^{(n,k)} \cong \frac{f_{m_1}^2}{f_{m_2}^2} \Delta_{\text{iono},m_1}^{(n,k)}$, $N_{\text{WL}}^{(n,k)} \triangleq N_{m_1}^{(n,k)} - N_{m_2}^{(n,k)}$ is an integer, and the widelane wavelength is based on the difference frequency: $\lambda_{\text{WL}} = \frac{c}{f_{m_1} - f_{m_2}}$. The widelane wavelength is much larger than the wavelength associated with either carrier frequency, and increases with closely spaced carrier frequencies.

When the carrier phase error variances are equal at both carrier frequencies and denoted $\sigma_{\Phi_m}^2$, the variance of the widelane carrier phase measurement is

$$\sigma_{\Phi_{\text{WL}}}^2 = \frac{f_{m_1}^2 + f_{m_2}^2}{(f_{m_1} - f_{m_2})^2} \sigma_{\Phi_m}^2 \tag{23.15}$$

The fraction in (23.15) is greater than unity, indicating that the widelane carrier phase measurement has larger error variance than the variance of the pseudorange measurement at either carrier frequency, and increases with more closely spaced carrier frequencies.

This combination of the narrowlane pseudorange measurement and widelane carrier phase measurement provides advantages in ambiguity resolution due to the smaller pseudorange errors and larger widelane wavelength. Its disadvantages are larger carrier phase tracking error variance, and larger ionospheric errors except in short-baseline differential configuration where the ionospheric errors approximately cancel.

Ionosphere-free carrier phase measurements $\hat{\Phi}_{\text{IF}}^{(n,k)}$ are formed by a linear combination of dual-frequency measurements using (23.10) with $\beta_1 = \frac{f_{m_1}^2}{f_{m_1}^2 - f_{m_2}^2}$ and $\beta_2 = \frac{-f_{m_2}^2}{f_{m_1}^2 - f_{m_2}^2}$, eliminating the ionospheric errors in the carrier phase measurement from each receiver to each satellite. These measurements can be used for divergence-free carrier smoothing with the ionosphere-free pseudorange measurements described in Section 20.2.4, $\hat{r}_{\text{IF}}^{(n,k)}$ using (23.10), with $\alpha_1 = \frac{f_{m_1}^2}{f_{m_1}^2 - f_{m_2}^2}$ and $\alpha_2 = \frac{-f_{m_2}^2}{f_{m_1}^2 - f_{m_2}^2}$.

The resulting carrier phase measurement variance is greater than that for either carrier frequency but smaller than for widelane measurements (23.15). The most significant disadvantage of ionosphere-free carrier phase measurements, however, is that the effective wavelength is less than a centimeter for typical L band carrier frequencies, making ambiguity resolution very difficult.

Once the linear combinations are formed from dual-frequency measurements to each satellite from each receiver, the between-satellite single differences are, per (23.7),

$$\begin{aligned}
\hat{r}_{\text{BSSD,LC}}^{(n,k_1,k_2)} &\triangleq \hat{r}_{\text{LC}}^{(n,k_1)} - \hat{r}_{\text{LC}}^{(n,k_2)} \\
&= \alpha_1 \left(\hat{r}_{m_1}^{(n,k_1)} - \hat{r}_{m_1}^{(n,k_2)} \right) + \alpha_2 \left(\hat{r}_{m_2}^{(n,k_1)} - \hat{r}_{m_2}^{(n,k_2)} \right) \\
\hat{\Phi}_{\text{BSSD,LC}}^{(n,k_1,k_2)} &\triangleq \hat{\Phi}_{\text{LC}}^{(n,k_1)} - \hat{\Phi}_{\text{LC}}^{(n,k_2)} \\
&= \beta_1 \left(\hat{\Phi}_{m_1}^{(n,k_1)} - \hat{\Phi}_{m_1}^{(n,k_2)} \right) + \beta_2 \left(\hat{\Phi}_{m_2}^{(n,k_1)} - \hat{\Phi}_{m_2}^{(n,k_2)} \right)
\end{aligned} \tag{23.16}$$

The double differences are then, from (23.8),

$$\hat{r}_{DD,LC}^{(n_1,n_2,k_1,k_2)} = \hat{r}_{BSSD,LC}^{(n_1,k_1,k_2)} - \hat{r}_{BSSD,LC}^{(n_2,k_1,k_2)}$$
$$\hat{\Phi}_{DD,LC}^{(n_1,n_2,k_1,k_2)} = \hat{\Phi}_{BSSD,LC}^{(n_1,k_1,k_2)} - \hat{\Phi}_{BSSD,LC}^{(n_2,k_1,k_2)} \quad (23.17)$$

23.3.3 Ambiguity Resolution

If code tracking errors were as small as carrier phase tracking errors, three or more pseudorange double differences using four or more satellites, based on the double difference pseudorange expression in (23.8), or based on any of the corresponding double difference of pseudorange linear combinations (23.17), could be used to solve for the baseline vector as outlined in Section 23.3.1. Even with ionosphere-free pseudorange measurements and a highly accurate troposphere model, however, the resulting errors would be on the order of meters—far too large for centimeter-level accuracy needed for applications like surveying, machine control, and precision agriculture.

Carrier measurements are essential, motivating ambiguity resolution, which uses carrier and code measurements to estimate the total number of wavelengths between receiver and satellite. Once the ambiguity is resolved, the baseline is then known to the accuracy of carrier phase estimates.

In some cases, the ambiguous quantity is a real-valued ambiguity. This situation can occur from unobservable phase delays in the satellite and receiver that must be included with the integer ambiguity, or from differencing of measurements that produce real-valued ambiguous quantities. In other cases, these phase delays are known through calibration or removed through differencing of measurements. Then, the ambiguity is an integer-valued ambiguity. Integer solutions provide much greater accuracy [10]. Another advantage of double differencing in (23.8) is that the ambiguity parameter is an integer because the noninteger terms in the GPS carrier phase observation, due to clock and hardware delays in the transmitter and receiver, are eliminated.

Ambiguity resolution consists of the following steps, using a combination of code tracking estimates and carrier phase measurements, which can be over time, from different receivers, and to different satellites. These measurements are used to establish the ambiguity resolution observation model: an overdetermined system of equations with unknowns consisting of the unknown ambiguous quantities as well as other unknown parameters such as ionospheric and tropospheric characteristics, receiver positions, and clock parameters. The following four ambiguity resolution steps are described in Reference 11:

1. Using estimated variances and covariances for these unknowns, define a least-squares problem and solve it for real-valued unknowns. This is the float solution, and completes the ambiguity resolution for real-valued ambiguities. If the ambiguities are actually integer-valued, the following three steps are taken.
2. Find the set of integer values—the integer solution—that best approximates the float solution. The easiest approach is to round the float solution. A better

approach is to test different possible integer values over an integer solution search space, with each vector of possible integer values in the search space known as a grid point.
3. Test the integer solution to ensure there is high enough confidence in it. If not, gather more data in the ambiguity resolution observation model and begin again.
4. Reestimate the unknown parameters using the integer solution, arriving at the fixed solution.

The ambiguity resolution observation model is given by Reference 10

$$\mathbf{y} \cong \mathbf{Aa} + \mathbf{Bb} \qquad (23.18)$$

where

- $\mathbf{y}$ is a j-dimensional column vector of pseudorange and carrier phase measurements, such as (23.8) or one of the linear combinations (23.17),
- $\mathbf{a}$ is an ℓ-dimensional column vector of ambiguous unknowns, $\mathbf{b}$ is a p-dimensional column vector of other unknown parameters, with $\mathbf{A}$ and $\mathbf{B}$ design matrices having dimensions $j \times \ell$ and $j \times p$, respectively, all obtained from (23.8) or one of the linear combinations (23.17). Assuming the design matrices are not rank deficient, this linear system is overdetermined when $j > (\ell + p)$. The covariance matrix of the measurements is $\mathbf{Q_y}$, assumed to be positive definite. The WLS solution is then given by $\arg\min_{\mathbf{a},\mathbf{b}} \|\mathbf{y} - \mathbf{Aa} - \mathbf{Bb}\|^2_{\mathbf{Q_y}}$, where $\|\bullet\|^2_{\mathbf{Q_y}} \stackrel{\Delta}{=} (\bullet)^T \mathbf{Q_y}^{-1}(\bullet)$.

As discussed previously, the entries in $\mathbf{a}$ are either real-valued (in the case of real-valued ambiguities) or constrained to the integers (for integer ambiguities, such as in double difference processing (23.8)), while the entries in $\mathbf{b}$ are real-valued.

Details of the Least-squares AMBiguity Decorrelation Adjustment (LAMBDA) approach for carrying out this computation are found in References 10 and 11, and toolboxes for performing the computations can be found through links provided in Reference 11. It is highly desirable to use the integer solution, and to reduce the search space as much as possible, yet ensure that the correct integer solution is within that search space. The LAMBDA algorithm reshapes the search space and decorrelates the ambiguities from different satellites to assist in this process.

Accurately estimating the ambiguous quantities benefits from increased measurement accuracies, using measurements from more satellites, using measurements from the same satellites but separated by longer time intervals, improved algorithms, and greater computational complexity.

Clearly there are advantages to using longer wavelengths—even using signals with carrier frequency of 1176.45 MHz rather than 1575.42 MHz can reduce the number of grid points with six satellites by almost a factor of seven. Carrier phase wide-laning, as in (23.14), provides much longer wavelengths and thus greater benefits in this respect. Also, better code tracking accuracy that reduces jitter and multipath errors can provide significant reductions in the search space as well, motivating the use of signals with improved spreading modulations, and the employment of advanced receiver processing techniques for enhanced code tracking accuracy, as discussed in Chapter 19.

The objective is "on the fly" ambiguity resolution to enable real-time centimeter-accuracy positioning relative to the reference receiver. More satellites, more constellations, signal designs that enable more accurate pseudorange measurements and more robust carrier tracking, increasing prevalence of dual-frequency signals, enhanced algorithms, and greater processing power all contribute toward this objective.

23.4 SUMMARY

While the accuracy of satnav systems continues to improve, there are many challenges to consistently achieving positioning accuracies much below meter-level using a stand-alone receiver. Decimeter-level accuracies are readily obtained using PPP or code-based differential satnav. Centimeter-level accuracies typically involve differential satnav, usually carrier based.

Consistent with the intent of Part IV of this book, this chapter introduces basic concepts and terminology associated with differential satnav. Greater detail is found in Chapter 8 of Reference 8, Chapter 7 of References 12, 9, and in Reference 13.

Short-baseline systems can be most readily implemented to provide centimeter accuracies, since longer baselines require modeling of an increasingly large number of different parameters in order to calibrate or estimate the many phenomena that can introduce centimeter-level errors. Carrier phase differential systems are increasingly used, and the opportunity remains for better ambiguity resolution, especially exploiting multiple signals on multiple carrier frequencies from multiple systems.

While differential satnav is very common, its long-term future remains a question. The accuracy provided by differential satnav can be matched in many cases by PPP, which is becoming increasingly popular given the increasingly widespread provision of PPP data by some emerging satnav systems. While integrity information from differential satnav systems is essential today, the availability of more signals from more satellites and more systems enables greater use of receiver-based processing for integrity checking. If these transitions away from differential satnav to PPP and receiver-based integrity processing do occur, they will be gradual, making differential satnav still important for many years.

REFERENCES

1. X. Chen, H. Landau, F. Zhang, M. Nitschke, M. Glocker, A. Kipka, U. Weinbach, and D. Salazar, "Towards a Precise Multi-GNSS. Positioning System Enhanced for the Asia–Pacific Region," Chapter 25 in *China Satellite Navigation Conference (CSNC) 2013 Proceeding*, Edited by J. Sun, W. Jiao, H. Wu, and C. Shi, Springer Science & Business Media, 2013.
2. World DGPS Database for DXers, http://www.ndblist.info/datamodes/worldDGPSdatabase.pdf, accessed 25 October 2014.

REFERENCES

3. "United States Departement of Transportation, Office of the Assistant Secretary for Research and Technology, Positioning, Navigation and Timing & Spectrum Management, Major Initiatives," http://www.rita.dot.gov/pnt/major_initiatives/nationwide_differential_gps_major _initiative.html, accessed 25 October 2014.
4. "Progressive Engineer Feature, Nothing Runs Like a Precision Framing System," http://www.progressiveengineer.com/PEWebBackissues2005/PEWeb%2060%20Mar05-2/ Deere.htm, accessed 25 October 2014.
5. OmniSTAR, http://www.omnistar.com/, accessed 25 October 2014.
6. G. McGraw, "GNSS Solutions: How Can Code and Carrier Measurements Be Optimally Combined to Enhance Position Solution Accuracy?" *Inside GNSS*, July/August 2006, pp. 17–19.
7. G. McGraw, "Generalized Divergence-Free Carrier Smoothing and Dual Frequency Differential GPS Architecture Implementing the Same," U.S. Patent 7,570,204 B1, August 2009.
8. E. Kaplan and C. Hegarty, *Understanding GPS: Principles and Applications*, 2nd edition, Artech House, 2006.
9. N. Zinas, GPS Network Real Time Kinematic Tutorial, Tekmon Geomatics LLP, March 2011, available at http://tekmon.gr/tekmon-research/gps-network-rtk-tutorial, accessed 15 December 2014.
10. P. J. Teunissen, G. P. J. De Jonge, and C. C. J. M. Tiberius, "Performance of the LAMBDA Method for Fast GPS Ambiguity Resolution", *NAVIGATION: Journal of The Institute of Navigation*, Vol. 44, No. 3, 1997, pp. 373–383.
11. P. J. G. Teunissen, "GNSS Integer Ambiguity Validation: Overview of Theory and Methods," *Proceedings of the Institute of Navigation Pacific PNT 2013 Conference, 2013.*
12. P. Misra and P. Enge, *Global Positioning System: Signals, Measurements, and Performance*, Revised 2nd edition, Ganga-Jamuna Press, 2006.
13. A. Leick, *GPS Satellite Surveying*, 3rd edition, John Wiley & Sons, 1984.

24

ASSISTED SATNAV

Assisted satnav is another form of augmentation, distinct from those discussed in Chapter 23. While the differential augmentations described in Chapter 23 are primarily intended for professional and safety applications where improved accuracy and integrity are needed, assisted satnav is primarily intended for consumer applications where improved availability in challenged environments and faster TTFF are needed. While the differential augmentations described in Chapter 23 provide enhancements to the receiver processing described in Part III, assisted satnav is not based on code or carrier tracking. Instead, it is based on block processing using CAF computations like those used for initial synchronization, as described in Chapter 16. In addition, the assisted techniques described in this chapter emphasize low-cost hardware and low power consumption—considerations that are typically not considered as critical in applications of the techniques described in Chapter 23.

Reference 1 provides an informative and thorough description of assisted satnav. The discussion in this chapter draws from this book and its references to summarize the key concepts, most of which integrate and apply material already addressed in earlier parts of this book.

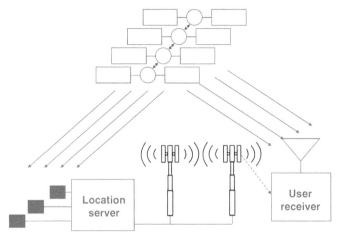

Figure 24.1. Architecture of Assisted Satnav

Assisted satnav's targeted application is consumer devices with cellular communications' connectivity, allowing exploitation of several key advantages:

- Extensive existing infrastructure,
- Reliable communication links from the infrastructure to devices, with ample capacity,
- Ability to amortize investments over millions of users.

The concept of assisted satnav is elegant but simple. Figure 24.1 shows the overall architecture. The reference stations on the lower left provide satnav measurements and broadcast data messages to the location server. While some reference stations may be specifically purposed for assisted satnav, there is also the opportunity to use reference stations provided by other networks such as the International GNSS Service (IGS). The location server computes and provides the essential information for assisted satnav. It gathers the ephemeris and other information received and provided by the reference stations. It may compute long-term ephemeris[1] that predicts ephemeris (and clock corrections) for days, with meter-level accuracy adequate for consumer devices. It also evaluates satellite health and other status information. Inputs to the location server from cellular base stations are used to obtain a coarse estimate of receiver location (in cellular telephony, the handset is called a mobile station (MS), and the user receiver is within the MS). The location server then provides information to the user receiver in order to assist its use of satnav. The user receiver then makes pseudorange measurements. Calculation of receiver position can take place either in the user receiver (MS-based positioning) or in the location server (network-based positioning). For MS-based positioning, the location server provides to the user receiver, through the cellular network, the ephemeris

---

[1] Also known as long-term orbits or ephemeris extension.

and system status, ionospheric corrections, and other information needed for position calculation; the receiver performs the position calculation and may provide the results through the cellular network to the location server or elsewhere. For network-based positioning, the receiver provides to the location server the pseudorange measurements, which performs the position calculation and provides the user receiver's location to the user receiver through the cellular network.

The remainder of this chapter provides further details concerning implementation of the approach outlined in Figure 24.1. Section 24.1 discusses reducing ITU and IFU, while Section 24.2 discusses the provision of clock corrections, ephemeris, and broadcast data message bits. The CAF computation fundamental to assisted satnav is addressed in Section 24.3, while computing pseudoranges and receiver position are described in Section 24.4. Section 24.5 summarizes this chapter.

24.1 REDUCING IFU AND ITU

As discussed in Chapter 16, the computational complexity and storage involved in computing CAFs can be reduced through reducing ITU and IFU. The amount of reduction depends upon the computational approach used, with alternative approaches described in Section 16.3.

Since IFU is dominated by the combination of frequency drift in the reference oscillator and unknown Doppler shift from relative motion between satellite and receiver, reducing IFU must address both of these factors. The user receiver is in a consumer device that also performs communications, so the receiver's reference oscillator is frequency-aligned to the communications signals, typically of the order of 0.1 ppm with remaining uncertainties due to drift in the base station frequency reference, and to Doppler shift due to receiver motion. The same reference oscillator is used for communications and for satnav; once the oscillator is aligned for communications, its frequency drift is typically reduced by an order of magnitude from frequency uncertainty of the order of ± 1 ppm due to oscillator stability, to on the order of ± 0.1 ppm.

Since satellite ephemeris is provided by the location server, the accuracy of predicting the received Doppler shift of the received satnav signal is based on the uncertainty of the receiver's location and velocity. Typically, the receiver's location can be established by the cellular infrastructure (using a combination of received signal strength and time of arrival measurements) to less than the spacing between base stations. Base station spacing can range from several kilometers in rural areas to fractions of a kilometer in urban areas. The resulting frequency uncertainty is typically less than ± 0.01 ppm [1].

Receiver velocity is typically limited to ground vehicle speeds, with unknown direction, producing frequency uncertainty of the order of ± 0.3 ppm or less. The resulting total frequency uncertainty is of the order of ± 0.4 ppm or less, or less than approximately ± 600 Hz for signals with center frequencies near 1600 MHz, and less than approximately ± 500 Hz for signals with center frequencies near 1200 MHz. Modern architectures for initial synchronization processing can typically address these IFU values in a single time–frequency tile.

The dominant factors contributing to ITU are uncertainties in satellite location, receiver location, and delay in transferring time over the communication network. Combined satellite and receiver location uncertainties contribute less than ±0.1 ms to ITU, so ITU is typically dominated by network time uncertainty. Different communication formats enable different accuracies of time uncertainty. The current trend seems to be toward asynchronous formats that neither require nor provide precise timing to receivers.

Two levels of time uncertainty are typically considered in assisted satnav, coarse time and fine time [1]. Coarse time involves uncertainties of up to several seconds—much greater than the spreading code duration, data message bit duration, and pilot code bit duration. Fine time involves uncertainties of a fraction of a millisecond for C/A code signals, so time is known within the C/A code signal spreading code duration. For other signals whose spreading code durations are longer than the 1 ms for C/A code signal, the definition of fine time is expanded proportionally, as long as the receiver's ITU is within a spreading code duration.

24.2 PROVISION OF CLOCK CORRECTIONS, EPHEMERIS, AND DATA MESSAGE BITS

The location server has access to the broadcast data message bits provided by all the reference stations. Typically, many reference stations read the broadcast data message bits from each satellite, and the location server computes the majority vote of the multiple inputs from each satellite, virtually eliminating any bit errors at a single reference station.

The resulting set of broadcast data message bits for each satellite can also be provided to each cellular base station for network-based positioning and also to each user receiver for MS-based positioning, and if fine time is available. These broadcast data message bits include clock corrections and ephemeris. For some signals, such as GPS C/A code and L1C, most of the broadcast message bits can be predicted, and provided to the user receiver in advance for data wiping, as discussed in Section 24.3.

Since the validity intervals for clock corrections and ephemeris provided by the broadcast data messages are only several hours at most, using this data requires updated information each time a different satellite comes into view, and every few hours for every satellite. The maximum update interval depends on the validity interval of the clock corrections and ephemeris for the signal being used. Another alternative is for the reference station to compute and provide long-term ephemeris (which also includes clock drift models) that can be used for days, with gradual degradation of accuracy as the number of days increases.

Chapter 8 in Reference 1 provides considerable detail concerning long-term ephemeris, and only the highlights are provided here. Typically, the long-term ephemeris is calculated over validity intervals whose length ranges between the validity interval used in broadcast ephemeris and twice that long. The long-term ephemeris is typically represented in the same format as used in broadcast ephemeris. Thus, the extended Keplerian model introduced in Chapter 2 is commonly employed. However, a sequence of parameter sets corresponding to adjacent future validity intervals is computed by the location server, drawing upon extended monitoring of satellite orbits and sophisticated models that include second-order and third-order effects on satellite motion and clock behavior. The product of the number of validity intervals and the length of each validity interval yields the total time interval where the long-term ephemeris can be used.

Even when long-term ephemeris is provided to the user receiver, it typically is used only until broadcast ephemeris becomes available. The accuracy of the long-term ephemeris can be compared to the broadcast clock corrections and ephemeris. If this comparison shows that the long-term ephemeris is no longer accurate, then it is purged and replaced by updated long-term ephemeris.

24.3 BLOCK PROCESSING

Block processing computes an instantaneous estimate of the signal's time of arrival and frequency of arrival using CAF processing explained in Chapter 16. The IFU is established based on the considerations described in Section 24.1. If fine time is available, the time uncertainty is less than the spreading code duration, and is used to determine ITU. If the receiver uses coarse time, then the ITU corresponds to the spreading code duration.

The size of each time–frequency cell is established based on the considerations discussed in Chapter 16. Hardware and power considerations dictate, to a large extent, which of the algorithmic approaches outlined in Section 16.3 are used.

Chapter 16 also introduces the considerations involved in selecting correlation integration times and the number of noncoherent integrations. If a data component is being processed and fine time is available along with predicted broadcast data message bits, data wiping can be used to allow selection of any desired correlation integration time without the limits imposed by data bit boundaries. Also, if a pilot component is being processed and fine time is available, the receiver can wipe the pilot code bits and select a correlation integration time accordingly. Correlation integration times approaching 100 ms or more are considered when receiver motion is negligible. Combined with ample noncoherent integrations with appropriate code, Doppler compensation usable CAF peaks can be obtained at effective C/N_0 values as low as 10 dB-Hz.

Since tracking loops are not used to smooth the estimates, there is additional motivation to accurately locate the CAF peak in the delay/frequency offset plane. Various steps, including interpolation or recomputation of the CAF in smaller time–frequency cells in the vicinity of a detected peak, can be used to enhance the accuracy of the estimated location of the CAF peak.

24.4 COMPUTING PSEUDORANGES AND POSITION

The coordinates of the CAF peak computed as part of block processing yield an estimate of the time of arrival of the signal, modulo the spreading code duration. The satellite's position at the time it transmitted the same epoch is also needed. Conventional tracking of the signal and reading of the broadcast data message can be used if C/N_0 levels are high enough to enable data message demodulation, but adds latency for reading the data message.

Operation at lower C/N_0 values or with shorter TTIF requires a different approach. When fine time is available, the receiver can compute the satellite's position within a few meters, which is typically accurate enough for consumer applications when other error sources, such as multipath, typically dominate. Position calculation follows the approach as described in Section 20.3.

If only coarse time is available, however, useful results can be obtained as long as the receiver is approximately stationary. In that case, the predicted pseudoranges each contain an error involving the time error and line of sight velocity (or pseudorange rate) between satellite and receiver.[2]

A coarse-time algorithm [1], which includes the pseudorange rate and adds an unknown to represent the coarse-time error, can provide useful results, with the penalty of needing measurements from an additional satellite to accommodate the additional state. Such algorithms suffer from time ambiguities related to the duration of the spreading code (termed the "millisecond ambiguity" in Reference 1 since the development is tied to C/A code signals having 1 ms spreading code duration), and approaches are available [1] to deal with this ambiguity. The resulting system of equations and iterative solution approach are extensions to those described in Chapter 20.

Since consumer devices are often used in environments that present a limited view of the sky, there can be significant benefits to using altitude hold or clock hold approaches, described in Chapter 20, to reduce the number of measurements needed.

Pseudorange measurements from block processing typically undergo less smoothing than those produced by code tracking loops, and hence usually have larger errors. The position estimates resulting from assisted satnav processing can be smoothed in the position domain, using a Kalman filter (see Section 25.1) or other filtering approach, in order to compensate, in part, for the limited smoothing in the measurement domain.

24.5 SUMMARY

Augmenting satnav with terrestrial infrastructure is one of the ways that satnav use has been extended to operation indoors and in other challenging applications. The assisted satnav augmentations introduced in this chapter are used in hundreds of millions of consumer devices, exploiting sophisticated infrastructure in order to provide fast first fix and operation under extremely challenging conditions.

While assisted satnav employs many of the receiver processing approaches described in Part III, the overall approach is distinctly different from what is used in conventional receivers. Receiver signal processing is based on CAF computations that produce a peak whose coordinates can be estimated to obtain delay and frequency offset measurements. Assistance is used to reduce the size of the time–frequency space to be searched, as well as to provide information normally obtained from the broadcast data messages. Clever approaches have been developed to calculate receiver positon, under

---

[2] Actually, it also includes the rate of change of the satellite clock, but this value is typically so small relative to the effect of satellite motion that it can be neglected.

some conditions even without accurate knowledge of the signal's time of transmission or other information usually considered essential.

All of the pioneering work on assisted satnav was performed to assist the use of GPS C/A code signals. The characteristics of new and modernized signals provide both opportunities and challenges in assisted satnav. Presumably, engineers are working hard at current providers of A-GPS services, developing ways to deal with these new and modernized signals, and to exploit their characteristics.

REFERENCE

1. F. van Diggelen, *A-GPS: Assisted GPS, GNSS, and SBAS*, Artech House, 2009.

25

INTEGRATED RECEIVER PROCESSING

While conventional receiver processing described in Part III involves many complicated aspects, even more complicated and capable processing schemes are being developed, based on integrated receiver processing. There are two important distinctions between conventional receiver processing and integrated receiver processing:

- Conventional receiver processing uses only satnav signals, while integrated receiver processing can include inputs from other sensors.
- Conventional receiver processing separately tracks individual satnav signals; even for an individual signal, there are separate code and carrier tracking loops, possibly with carrier-aiding of the code tracking. In contrast, some types of integrated receiver processing jointly track multiple signals.

Adding inputs from other sensors, such as inertial, barometric, or magnetic, can provide dramatic benefits. Recall from Chapters 17, 18, and 19 that tracking loops attempt to smooth out parameter jitter due to noise, while needing to accommodate time variations in parameter values due to dynamics. These two objectives are in conflict, forcing tradeoffs to be made in tracking loop design. When external sensors are used to measure receiver dynamics, integrated tracking loops need not accommodate these

---

Engineering Satellite-Based Navigation and Timing: Global Navigation Satellite Systems, Signals, and Receivers, First Edition. John W. Betz.
© 2016 The Institute of Electrical and Electronics Engineers, Inc. Published 2016 by John Wiley & Sons, Inc.

dynamics. Tracking loops can then be designed to perform more smoothing of noise, needing only to accommodate the much smaller residual dynamics. The result is tracking that can operate at lower C/N_0. In addition, inputs from other sensors can be used to help maintain PVT outputs even when satnav signals are temporarily unavailable, so signal tracking can be restored more rapidly when signals become available again.

Integrated receiver processing of multiple signals can also provide significant performance benefits. Joint processing of multiple satnav signals can enable receivers to maintain track and provide useful PVT measurements under more challenging conditions, such as lower C/N_0, than conventional receiver processing. Receiver performance then improves with the number of satellite signals being tracked. In addition, cross-channel aiding occurs, where tracking of signals with good C/N_0 helps the receiver track other signals having marginal C/N_0. Also, if tracking of one or several signals is lost temporarily due to inadequate C/N_0, it is quickly restored when better C/N_0 returns, as long as other signals continue to be tracked.

This chapter describes the key concepts of integrated receiver processing. Section 25.1 introduces the concept of Kalman filtering, which, along with its variants, is the widely employed algorithm used for integrated processing. Section 25.2 introduces how loose and tight coupling of satnav processing with measurements from other sensors can enhance performance. Section 25.3 then describes vector tracking of all satnav signals and the benefits that can be obtained. Section 25.4 then introduces the combination of vector processing with using measurements from other sensors in what is called ultratight coupling, or deep integration. The chapter is summarized in Section 25.5.

25.1 KALMAN FILTER OVERVIEW

A Kalman filter (KF) is a multiple-input, multiple-output recursive linear digital filter. It was introduced early in the age of digital computing [1] and has become the preeminent practical algorithm for solving complex problems in digital filtering. Its ability to accommodate multiple inputs and multiple outputs allows it to address multivariable problems in an integrated way. Since it is recursive, it only depends upon current and most recent information and calculations and does not require storage of the entire history. Although it inherently is based on linear behavior, it can be extended to accommodate nonlinear characteristics. Finally, it enables designers to provide a priori models of the behavior of the quantities being filtered, as well as models of the errors involved in the processing, providing superior performance through the use of this a priori information. Under certain conditions, it is an optimal linear estimator.

Thorough treatment of the KF requires a book in and of itself, and there are many useful such references available. The classic textbook, Reference 2 clearly and rigorously introduces the KF from a perspective of statistical estimation. Reference 3 provides excellent theoretical and practical insights into KF design and applications, with emphasis on GPS applications; so does the more recent Reference 4. A wonderful overview of the KF is provided by Reference 5, and the material in this section draws heavily on this reference.

KALMAN FILTER OVERVIEW

A KF[1] inherently represents time-varying quantities that are described in discrete time, with time index k. The set of time-varying characteristics to be estimated (e.g., the position and velocity of a receive antenna phase center, or the pseudoranges to a set of satellites, the price of a particular stock, or the coordinates of a particular point on a speaker's lip in a video) are called states and are denoted x_k. The description of how values of the states evolve over time is called a process. A process model, developed by the filter designer, provides an a priori description of how the states evolve over time. This process model is constrained to be linear, so the updated state values are linear combinations of the preceding state values. Typically, there is uncertainty or randomness in the process model, represented by additive Gaussian white noise with a specific covariance matrix. The resulting process model is

$$\mathbf{x}_{k+1} = \mathbf{F}_k \mathbf{x}_k + \mathbf{G}_k \mathbf{w}_k \tag{25.1}$$

where $\mathbf{F}_k$ is the process matrix, $\mathbf{w}_k$ is the process noise having (diagonal) covariance matrix $\mathbf{Q}_k$, and $\mathbf{G}_k$ is the noise input matrix, all at time index k. The state vector $\mathbf{x}_k$ has covariance matrix $\mathbf{P}_k$.

The states typically are not directly observable. Instead what may be observed is a set of measurements $\mathbf{z}_k$, which ideally are linear combinations of the states defined by the filter designer. Typically, there is also uncertainty in the measurements, often from noisy observations. This uncertainty is represented by additive Gaussian white noise with a specific covariance matrix. The resulting measurement model is

$$\mathbf{z}_{k+1} = \mathbf{H}_k \mathbf{x}_k + \mathbf{v}_k \tag{25.2}$$

where $\mathbf{H}_k$ is the measurement matrix and $\mathbf{v}_k$ is the measurement noise having (diagonal) covariance matrix $\mathbf{R}_k$, all at time index k. The initial state value, $\mathbf{x}_0$, is modeled as a Gaussian (but not necessarily white) random variable having mean $\mathbf{m}_0$ and covariance matrix $\mathbf{S}_0$. It is assumed that $\mathbf{x}_0$, $\mathbf{w}_k$, and $\mathbf{v}_k$ are uncorrelated, and that all covariance matrices are positive definite. The measurement noise and process noise are only concepts, in that they are not found explicitly in the KF equations; only their covariances appear.

The time index notion must be extended to explicitly account for how the iterative algorithm updates quantities at each time step. The notation $\mathbf{x}_{k/k-1}$ and $\mathbf{P}_{k/k-1}$ is used to represent optimal estimates of the state and its covariance matrix at time index k, based on all information available up to and including time index $k-1$. In contrast, $\mathbf{x}_{k/k}$ and $\mathbf{P}_{k/k}$ represent optimal estimates of the state and its covariance matrix at time index k, based on all information available up to and including time index k.

KF initialization involves setting the initial state to $\mathbf{x}_{0/-1} = \mathbf{m}_0$ and the initial state covariance matrix to $\mathbf{P}_{0/-1} = \mathbf{S}_0$. Then for every time index, the KF involves iterative

---

[1] Only the discrete time version is addressed here.

operation of two steps: updating (also called correction) and prediction. First, the Kalman gain at the current time index is computed

$$\mathbf{K}_k \triangleq \mathbf{P}_{k/k-1}\mathbf{H}_k^T \left[\mathbf{H}_k\mathbf{P}_{k/k-1}\mathbf{H}_k^T + \mathbf{R}_k\right]^{-1} \quad (25.3)$$

The predicted measurement at time index k is $\mathbf{H}_k\mathbf{x}_{k/k-1}$, so the prediction errors, or innovations, are $\mathbf{z}_k - \mathbf{H}_k\mathbf{x}_{k/k-1}$. The updated estimate of the state at this time is then

$$\mathbf{x}_{k/k} = \mathbf{x}_{k/k-1} + \mathbf{K}_k[\mathbf{z}_k - \mathbf{H}_k\mathbf{x}_{k/k-1}] \quad (25.4)$$

where $\mathbf{K}_k[\mathbf{z}_k - \mathbf{H}_k\mathbf{x}_{k/k-1}]$ is the correction to the previous state estimate.

The optimal estimate of the state covariance is then updated to be

$$\mathbf{P}_{k/k} = \mathbf{P}_{k/k-1} - \mathbf{K}_k\mathbf{H}_k\mathbf{P}_{k/k-1} = [\mathbf{I} - \mathbf{K}_k\mathbf{H}_k]\mathbf{P}_{k/k-1} \quad (25.5)$$

where $-\mathbf{K}_k\mathbf{H}_k\mathbf{P}_{k/k-1}$ is the correction to the previous state covariance estimate. Since the term being subtracted in (25.5) is positive definite, the covariances of the state estimates decrease over time, if the KF is well-designed.

The next state is then predicted using

$$\mathbf{x}_{k+1/k} = \mathbf{F}_k\mathbf{x}_{k/k} \quad (25.6)$$

and its covariance is predicted to be

$$\mathbf{P}_{k+1/k} = \mathbf{F}_k\mathbf{P}_{k/k}\mathbf{F}_k^T + \mathbf{G}_k\mathbf{Q}_k\mathbf{G}_k^T \quad (25.7)$$

At the next iteration, the Kalman gain is computed, the new measurement is used to update the predicted next state, the state covariance is also updated, and then the state and its covariance are then predicted for the subsequent time index.

Figure 25.1 portrays KF operation, with initialization inputs, measurement inputs, and state estimate outputs indicated.

Many practical issues need to be taken into account in KF design and application. Matrix inversions can be implemented using efficient algorithms rather than brute force implementations. Numerical precision must be considered carefully to avoid poorly conditioned covariance matrices, such as asymmetry or loss of positive definiteness. Computational complexity is an issue, since it increases linearly with the filter update rate and more than linearly with the number of states. Care must be taken in dealing with unsynchronized inputs, and in ensuring that all quantities are in common coordinate frames.

If different states change at very different rates, a single large KF can be approximated by the combination of several KFs, which then can be updated at different rates. Both the decomposition into several smaller KFs and the opportunity to update some KFs more slowly can help reduce the computational complexity of KF implementation.

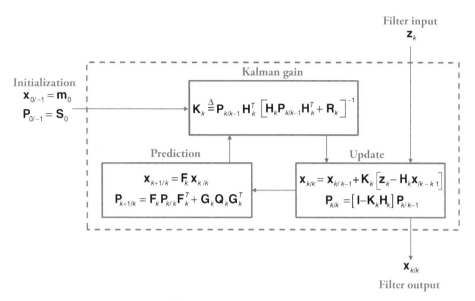

Figure 25.1. KF Processing

Uncertainty in the process model or using a reduced number of states can be accommodated with larger values in the process noise covariance matrix. The consequence, however, is larger errors in the state estimates, since (25.7) shows that the process noise is a lower bound on the state covariance.

The linearity of a KF is an obstacle to addressing many physical situations, including most associated with satnav, where significant nonlinearities occur. An extended Kalman filter (EKF) can be used for such situations. Here, the process matrix and measurement matrix are approximated using a Taylor's Series expansion. These approximations are updated at each iteration. The appearance of the resulting equations and resulting computations is very similar to those of the (linear) KF. An EKF does not provide an optimal solution to the nonlinear filtering problem but can provide a useful and practical approximation to an optimal solution.

25.2 LOOSELY AND TIGHTLY COUPLED SENSOR-INTEGRATED SATNAV PROCESSING

Now consider the use of other sensors in the development of PVT and perhaps attitude, or orientation. The original and primary sensor to be used is an inertial measurement unit (IMU) [4], which typically consists of three accelerometers (motion sensors) and three gyroscopes (rotation sensors), oriented so their measuring axes are mutually orthogonal,

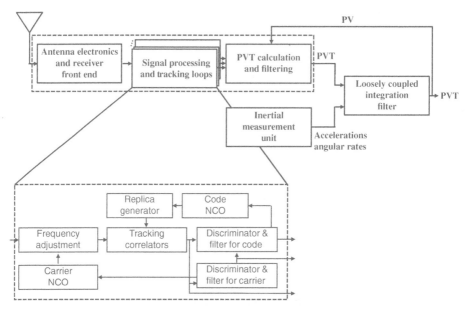

Figure 25.2. Loosely Coupled Integration

in order to measure acceleration in three dimensions.[2] IMUs may also include other sensors, such as three-axis magnetometers, to assist in determining orientation. IMUs output acceleration, or velocity changes, and orientation changes, each in three dimensions.

IMUs can be gyrostabilized (maintaining a fixed orientation relative to an external coordinate system, typically using gimbals or other types of bearings) or strapdown (maintaining a fixed orientation relative to the platform or apparatus on which they are mounted). Strapdown is increasingly used in modern IMUs, avoiding the mechanical complexity of gimbals and accepting the additional computational complexity of converting between the platform coordinate system and the external coordinate system.

Other sensors that can be integrated with satnav processing include barometers or other altimeters, magnetometers, pedometers, odometers, and even optical systems that measure feature location or optical flow. Each of these provides measurements of position, motion, or orientation that can be represented in a common coordinate system and used in conjunction with satnav measurements.

An IMU and a satnav receiver have complementary characteristics. The IMU's measurements are almost noise-free, but drift over time. In contrast, satnav measurements are noisy, but their accuracy does not change over time. Figure 25.2 shows an initial

---

[2] There is a clear distinction between an IMU and an inertial navigation system (INS). An INS consists of an IMU and a navigation computer that processes the IMU measurements. The navigation computer integrates acceleration over time to determine velocity change, and integrates velocity change over time to determine position change, using orientation rate information to convert the results into a standard coordinate system. The effects of gravity and gravity variations must be taken into account in order to accurately calculate velocity changes and position changes.

approach, known as loose coupling, for integrating an IMU with satnav [6] to take advantage of their complementary characteristics. The upper leg indicates a conventional satnav receiver described in Part III, with receive antenna and receiver functions within the dashed box, divided between three different functional segments. Signal processing and tracking loops involve conventional carrier tracking loops and code tracking loops, consistent with what was described in Part III. Although carrier aiding of the code tracking is not shown, it could also be employed. Acquisition processing is not shown explicitly, since it is not emphasized in integrated receiver processing. PVT calculation and filtering include reading the data message and computing PVT, which may be smoothed using a KF or other filtering approach. The IMU in the lower leg produces accelerations and angular rates. PVT from the satnav receiver, along with estimated accelerations and angular rates from the IMU, feed the integration filter. When the integration filter is a KF, it uses models of the IMU errors and the satnav receiver errors to blend these measurements in a way that benefits from the advantages of each. States of the KF can include three-dimensional position and velocity, along with receiver clock offset and its rate of change. Clock corrections, satellite ephemeris, and relative geometry between the satnav antenna phase center and IMU origin (this relative geometry is known as a lever arm) can also be included in the state vector. While the integration filter's PVT output is fed back to PVT calculation and filtering, there is no feedback to the signal processing and tracking loops.

More expensive and more capable systems now use tight coupling, depicted in Figure 25.3. While the difference in the figures appears to be minor, the differences in the underlying processing and resulting performance are significant. With tight coupling,

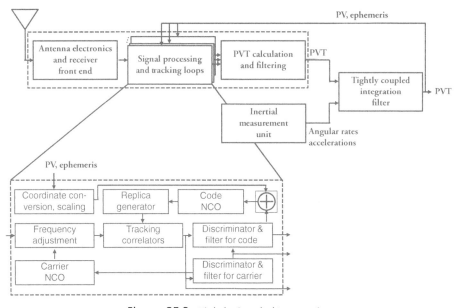

Figure 25.3. Tightly Coupled Integration

the integration filter's position and velocity outputs are provided to signal processing and filtering, converted into line of sight velocity to each satellite, then scaled by the ratio of spreading code chip rate to the speed of light. The resulting Doppler estimates are used in replica generation for each code tracking loop, removing much of the receiver's dynamics from the measurements. This inertial aiding is analogous to the carrier aiding of code tracking described in Section 19.3—the inertial aiding of the tracking loops reduces the need for tracking loops to accommodate receiver dynamics. The tracking loops then only need to account for jitter, imperfections in the integration filter's position and velocity outputs, and any unremoved dynamics, such as due to satellite motion. In applications where receiver dynamics are a significant stress that drives tracking loop design, the result is the ability to use lower order tracking loops and tighter loop bandwidths, providing better accuracy with lower loss of lock thresholds.

Aiding of the carrier loop can be performed in analogous ways. However, since the carrier wavelength is much smaller than that of the spreading code chip rate, much greater precision is needed for carrier loop aiding to be effective. Very precise coordinate conversions are needed, along with accurate position and velocity estimates, to avoid inducing cycle slips.

Integration filter processing with tight coupling is portrayed in Figure 25.4. Measurements from the IMU are used to estimate position and velocity, then converted into satnav receiver coordinates. The difference between these values and position and velocity from the satnav receiver yields the measurement errors. The measurement errors are input to a KF, whose output is estimates of the errors, which are then used to correct the position and velocity estimates from the IMU. A minimum of eight states is used in the KF: position error in three dimensions, velocity error in three dimensions, receiver clock offset error, and receiver clock offset error rate of change. Other satnav states that can be used, depending on the quality satnav receiver and upon the application needs, include satellite ephemeris errors, along with contributors to line of sight biases including satellite clock, ionosphere, troposphere, and multipath errors. Additional states related to IMU errors can include coordinate frame conversion and lever arm errors, biases in accelerometers and gyroscopes, scale factor errors in accelerometers and gyroscopes, along with errors associated with gravity variation. In alternative approaches, the integration filter uses as states measurement errors, IMU errors, and clock errors. More details concerning the KF design are provided in Chapter 9 of Reference 7 and in Reference 8.

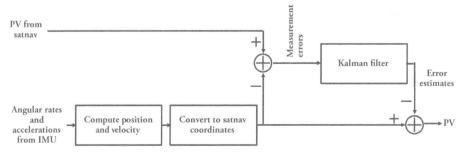

Figure 25.4. Integration Filter Processing in Tightly Coupled Integration

Many details need to be addressed carefully in implementing tightly coupled integration. All measurements must be converted to a common coordinate system—often centered at the receive antenna phase center. Even flexing of the structure holding the IMU and satnav receive antenna can introduce errors.

However, the benefits of tight coupling can be significant. The IMU aiding can allow use of low-order tracking loops having bandwidths up to an order of magnitude narrower than would have been needed to accommodate the full receiver dynamics. The result can be greater tracking accuracy along with tracking loss of lock thresholds that can handle on the order of 10 dB greater interference power.

25.3 STANDALONE VECTOR TRACKING

Conventional receivers, as described in Part III of this book, track each signal separately, with no information flow between tracking loops on different signal channels, and not necessarily any information flow even between code and carrier tracking loops for a given signal, except perhaps for carrier aiding of the code tracking. Even in loosely coupled and tightly coupled processing described in Section 25.2, separate tracking loops are used for each signal.

Consider now an alternate receiver processing architecture, known as vector tracking, or a vector-locked loop. Many different descriptions and variants of this processing are found in the literature, including References 9 and 10. An extensive and thorough treatment is found in Reference 11.

Figure 25.5 and the following discussion provide a generic description of vector tracking. The signal processing in each channel is open-loop, with filtered discriminator results output, rather than fed back to the NCOs. Separate signal processing and discriminators, with filtering, are employed for each signal being tracked. All the filtered discriminator outputs are provided to a vector integration filter, typically an EKF. Loop closure involves feeding back each predicted pseudorange and pseudorange rate from the vector integration filter to the NCOs used to generate replicas for that signal. The figure shows vector tracking for code and carrier associated with each signal. Some simplification can be obtained by using this approach only for code tracking, but retaining standalone carrier tracking loops.

States in the EKF can include position and velocity errors in three dimensions, receiver clock offset and its rate of change, as well as carrier phase and frequency errors. If signals at different carrier frequencies are tracked, the TEC or an equivalent ionospheric delay parameter in the direction of each satellite is added to the state vector [12].

Observe that the feedback to each signal replica generator is based on tracking of all signals. Essentially, vector tracking extends the concept of carrier-aided code tracking to aid the tracking of each signal with the tracking of all other signals. The performance benefit from this cross-channel aiding increases with the number of signals being tracked. Specific numerical values of the advantage depend upon many implementation details, as well as the conditions under which the comparison is assessed. An example of the type of reported improvement is approximately 2 dB lower tracking thresholds with five

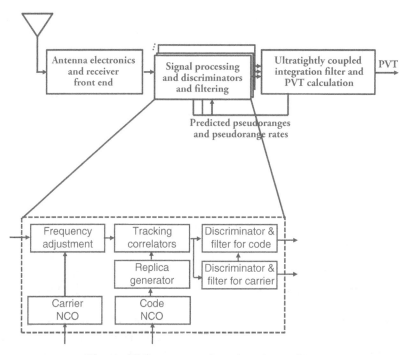

Figure 25.5. Vector Code and Carrier Tracking

signals tracked, to approximately 6 dB lower thresholds with 11 signals tracked [11]. Another benefit of vector tracking occurs under marginal conditions when tracking of signals having lower C/N_0 is lost, but tracking continues of other signals having higher C/N_0. The KF continues to propagate error estimates for the signals whose tracking has been lost, and as long as the outage duration is not too long, tracking of these signals can be restored quickly after their C/N_0 improves without resorting to initial synchronization processing, due to the cross-channel aiding.

Error models in the EKF typically do not include multipath errors in pseudorange or carrier cycle slips. If the measurement noise variance is not large enough to account for these errors when they occur, then the erroneous measurements have more weight than they should in the EKF, and the errors can degrade tracking of all signals, contaminating the state estimates. On the other hand, if the measurement noise variance is set large enough to account for these errors, when these errors do not occur, the reported measurement accuracy is much larger than it actually is, and the associated measurements do not have enough weight in the EKF. Essentially, the measurement variance should be small when multipath errors and carrier cycle slips do not occur, and much larger when they do occur. Approaches have been developed to inspect measurements for the presence of these errors, rejecting (or significantly deweighting) measurements judged to have grossly large estimation errors, while retaining small measurement noise variance otherwise.

Vector tracking requires an EKF with many states that are updated many times per second, imposing a significant computational burden. Satnav measurements are tied to the timing of each signal, and thus the measurements are asynchronous with each other, and must be adjusted for these timing offsets. Accurate state modeling and coordinate conversions are needed to provide accurate representations of the line of sight errors to each satellite.

25.4 ULTRATIGHTLY COUPLED SENSOR-INTEGRATED SATNAV PROCESSING

Ultratight coupling (UTC) [13], also known as deep integration [14], combines the attributes of tight coupling and vector tracking. As portrayed in Figure 25.6, the vector tracking architecture described in Section 25.3 is augmented with an IMU. Joint tracking of all signals and the IMU inputs is used to aid tracking of each signal.

As described in Reference 15, the integration filter in Figure 25.6 combines processing that integrates satnav measurements with IMU measurements, as in tightly coupled processing, with the joint signal tracking introduced in vector tracking. The result is then aiding of each tracking loop by a combination of IMU measurements and measurements from all other signals being tracked.

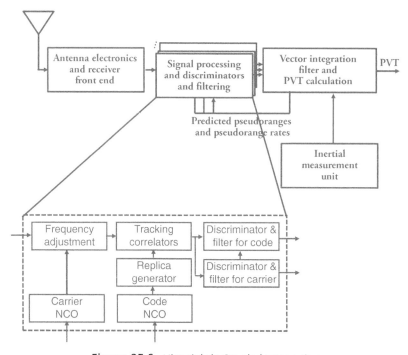

Figure 25.6. Ultratightly Coupled Integration

UTC can provide the combination of benefits from tightly coupled processing and vector integration. The use of the IMU measurements removes most receiver dynamics from satnav signals before they are tracked, allowing the equivalent of very narrow tracking loop bandwidths that provide lower loss of lock thresholds in noise and interference. The cross-channel aiding advantage of vector tracking further reduces loss of lock thresholds.

Obtaining reliable comparisons between UTC loss of lock thresholds and those for separate use of tightly coupled and vector processing approaches is very difficult. The results are highly sensitive to implementation details, to the quality of the components, to the number of signals being tracked, and to the scenario—especially the receiver dynamics. However, it is reasonable to project that the benefits of vector tracking combine with those of tight coupling. In that case, one would project that UTC provides 2–6 dB lower tracking thresholds than tightly coupled processing.

In addition, the IMU measurements can extend the length of time that signal dropouts can be accommodated without use of initial synchronization processing when adequate C/N_0 is restored.

It is common to employ an EKF in the UTC integration filter, with states that are errors in the quantities of interest, as described for tightly coupled processing. This approach is used since the error propagation is closer to linear than the propagation of the actual quantities, improving the approximations inherent to an EKF. An alternative is to use as states the actual quantities of interest, including receiver position and velocity, and employ a process model that is explicitly nonlinear rather than approximating the process model in an EKF. Reference 16 describes such an approach.

Just as UTC combines the functions and benefits of tightly coupled integration and vector tracking, its implementation considerations also include the combination of those discussed in Sections 25.2 and 25.3. Challenges to practical use of UTC include implementation complexity and the ability to handle unmodeled errors such as those induced by multipath and carrier cycle slips.

25.5 SUMMARY

This chapter has presented a variety of techniques for integrated receiver processing, with emphasis on their benefits and implementation considerations. The filtering inherent to this integrated processing is typically based on state-space models and often uses variants of Kalman filtering, as described in Section 25.1. Sections 25.2, 25.3, and 25.4 introduce a progression of increasingly complicated and increasingly capable approaches for integrated receiver processing. Each of these techniques, loosely coupled processing, tightly coupled processing, vector processing, and ultratightly coupled processing, builds upon the preceding concepts. They also represent the past state of the art (loose coupling), technology commonly fielded today (tight coupling), and approaches that are still being matured and are waiting for enabling technology to make them ready for common fielding (vector tracking and UTC).

Several aspects of integrated processing merit discussion.

- The primary benefits of integrated processing are robustness in the form of lower tracking thresholds, rather than accuracy. Very high accuracy approaches employ PPP (Chapter 20) or RTK (Chapter 23), but these techniques require relatively high C/N_0 to avoid loss of lock.
- Assisted satnav (Chapter 24) and integrated processing each provide functionality in low C/N_0 conditions, but using diametrically opposite approaches. Assisted satnav uses complex and sophisticated infrastructure with relatively simple receiver processing that eschews tracking loops and merely calculates CAFs. In contrast, integrated processing relies upon no infrastructure but employs very complicated and sophisticated receiver processing with exquisite tracking loops. Each approach has its own applications.
- The discussion of integrated receiver processing in this chapter has focused on tracking rather than the other essential receiver functions of acquisition and data message demodulation. While integrated processing assists in reacquisition and hot starts, it does not address the challenges of cold start or warm start acquisition. Further, while better carrier tracking may provide a minor improvement in data message demodulation threshold, integrated processing does not enable data message demodulation below the threshold established by E_b/N_0 and error correction coding. The benefits of integrated processing need to be assessed in the context of overall operational concepts that include all receiver functions.

The pilot components in modern signal designs, continuing advances in power efficiency of signal processing, and refinement of algorithms all will enable greater and more effective use of integrated processing techniques. Certainly integrated processing techniques are destined to be widely used in many applications. Many extensions are also emerging, including the use of signals at multiple carrier frequencies from multiple satnav systems, and the employment of multiple sensors in addition to IMUs.

REFERENCES

1. R. E. Kalman, "A New Approach to Linear Filtering and Prediction Problems," *Journal of Basic Engineering, the Transactions of the American Society of Mechanical Engineers*, Series D, Vol. 83, No. 1, 1960, pp. 35–45.
2. A. Gelb, Ed., *Applied Optimal Estimation*, MIT Press, 1974.
3. R. G. Brown and P. Y. C. Hwang, *Introduction to Random Signals and Applied Kalman Filtering (with Matlab Exercises and Solutions)*, 3rd edition, John Wiley & Sons, Inc., New York, 1997.
4. M. S. Grewal, A. P. Andrews, and C. G. Bartone, *Global Navigation Satellite Systems, Inertial Navigation, and Integration*, 3rd edition, John Wiley & Sons, 2012.

5. L. J. Levy, "The Kalman Filter: Navigation's Integration Workhorse," *GPS World*, Vol. 8, No. 9, 1997, pp. 65–71.
6. J. T. Nielson, G. W. Swearingen, and A. J. Witsmeer, "GPS Aided Inertial Navigation," *IEEE AES Magazine*, Vol. 1, No. 3, 1986, pp. 20–26.
7. E. D. Kaplan and C. Hegarty, *Understanding GPS: Principles and Applications*, 2nd edition, Artech House, 2006.
8. J. Zhang, T. Zhang, X. Jiang, and S. Wang, "Tightly Coupled GPS/INS Integrated Navigation Algorithm Based on Kalman Filter," Proceedings of the 2012 Second IEEE International Conference on Business Computing and Global Informatization, 2012, pp. 588–591.
9. S. Kiesel, C. Asher, D. Gramm, and G. F. Trommer, "GNSS Receiver with Vector Based FLL-Assisted PLL Carrier Tracking Loop," in Proceedings of the 21st International Technical Meeting of the Satellite Division of The Institute of Navigation (ION GNSS 2008), September 2008.
10. M. Lashley, D. M. Bevly, and J. Y. Hung, "A Valid Comparison of Vector and Scalar Tracking Loops," Proceedings of IEEE/ION PLANS 2010, May 2010, pp. 464–474.
11. S. Bhattacharyya, Performance and Integrity Analysis of the Vector Tracking Architecture of GNSS Receivers, University of Minnesota PhD Dissertation, April 2012, available at http://www.aem.umn.edu/info/spotlight/bhattacharyya_thesis_final.pdf, accessed 25 September 2015.
12. P. Henkel, K. Giger, and C. Günther, "Multifrequency, Multisatellite Vector Phase-Locked Loop for Robust Carrier Tracking," *IEEE Journal of Selected Topics in Signal Processing*, Vol. 3, No. 4, 2009, pp. 674–681.
13. A. S. Abbott and W. E. Lillo, "Global Positioning Systems and Inertial Measuring Unit, Ultra tight Coupling Method," US patent 6,516,021, February 4, 2003.
14. D. Gustafson and J. Dowdle, "Deeply Integrated Code Tracking: Comparative Performance Analysis," Proceedings of the 16th International Technical Meeting of the Satellite Division of The Institute of Navigation (ION GPS/GNSS 2003), Portland, OR, September 2003, pp. 2553–2561.
15. D. H. Hwang, D. W. Lim, S. L. Cho, and S. L. Lee, "Unified Approach to Ultra-Tightly-Coupled GPS/INS Integrated Navigation System," IEEE AES Magazine, March 2011, pp. 30–38.
16. D. E. Gustafson, J. R. Dowdle, and J. M. Elwell, "Deeply-Integrated Adaptive GPS-Based Navigator with Extended-Range Code Tracking," US patent 6,331,835, December 18, 2001.

A

THEORETICAL FOUNDATIONS

Satnav engineering, especially aspects related to satnav signals and satnav receiver processing, draws extensively upon techniques and results in signal theory, signal processing, and mathematics that may be unfamiliar to some readers. This appendix provides a quick reference to assist readers who seek to refresh skills or understand more background. Notation, terminology, and conventions used in the book are defined as well. In most cases, the intent is not to provide rigorous mathematics with proofs, details of existence, and other formalities. Instead, it provides, for the most part, conceptual mathematics intended to introduce or refresh knowledge of key concepts.

Section A.1 defines useful functions and their properties, while Section A.2 reviews Fourier transforms, which are key for developing the frequency-domain results employed widely in satnav engineering. Section A.3 introduces concepts in signal theory and linear systems, while Section A.4 summarizes important aspects of stochastic processes. Section A.5 provides some specific results for keyed waveforms, including the somewhat obscure topic of cyclostationarity. Section A.6 introduces and compares measures of bandwidth, while Section A.7 summarizes matrices and matrix algebra. Section A.8 describes Taylor Series and linearization in one dimension and in multiple dimensions, while Section A.9 provides an overview of some coordinate systems used in satnav.

Engineering Satellite-Based Navigation and Timing: Global Navigation Satellite Systems, Signals, and Receivers, First Edition. John W. Betz.
© 2016 The Institute of Electrical and Electronics Engineers, Inc. Published 2016 by John Wiley & Sons, Inc.

A.1 SOME USEFUL FUNCTIONS AND THEIR PROPERTIES

The imaginary unit is denoted $i \triangleq \sqrt{-1}$ with $i^2 = -1$. A complex number z can be written in terms of its real part x and imaginary part y as $z = x + iy$, with x and y real valued. The complex conjugate of z is defined as $z^* = x - iy$.

The sinc function is defined as $\text{sinc}(x) \triangleq \sin(x)/x$, with $\text{sinc}(0) \triangleq 1$. Some authors use a sinc function with π embedded, but this book does not. Useful properties of a sinc include its symmetry $\text{sinc}(x) = \text{sinc}(-x)$ and its area and squared area

$$\int_{-\infty}^{\infty} \text{sinc}(x)dx = \int_{-\infty}^{\infty} \text{sinc}^2(x)dx = \pi$$

The rect function is defined to be a unit height rectangle with variable width:

$$\text{rect}_a(x) \triangleq \begin{cases} 1, & |x| = a/2 \\ 0, & \text{elsewhere} \end{cases}$$

Interestingly, it has similar properties to the sinc in terms of symmetry $\text{rect}_a(x) = \text{rect}_a(-x)$, its value at the origin $\text{rect}_a(0) = 1$, and its area and squared area $\int_{-\infty}^{\infty} \text{rect}_a(x)dx = \int_{-\infty}^{\infty} \text{rect}_a^2(x)dx = a$.

The Dirac delta function is not formally a function but can be treated as such in an engineering context. Its heuristic definition is

$$\delta(x) \triangleq \begin{cases} +\infty, & x = 0 \\ 0, \text{elsewhere} \end{cases} \qquad \int_{-\infty}^{\infty} \delta(x)dx = 1$$

It can also be defined as a weak limit of a unit-area function that approaches infinite height and zero width, such as the two shown here:

$$\delta(x) \triangleq \lim_{a \to 0} \left[\frac{1}{a} \text{rect}_a(x)\right], \qquad \delta(x) \triangleq \lim_{a \to \infty} [a \, \text{sinc}(ax)]$$

The Dirac delta function's properties include symmetry $\delta(x) = \delta(-x)$, scaling $\delta(ax) = \delta(x)/|a|$, and sifting $\int_{-\infty}^{\infty} g(x)\delta(x-a)dx = g(a)$.

The Kronecker delta is a discrete-time version of the Dirac delta function defined as

$$\delta_{mn} \triangleq \begin{cases} +1, & m = n \\ 0, \text{elsewhere} \end{cases}$$

A.2 FOURIER TRANSFORMS

A function $x(t)$ has Fourier transform and inverse Fourier transform

$$X(f) \triangleq F\{x(t)\} = \int_{-\infty}^{\infty} x(t) e^{-i2\pi ft} dt$$

$$x(t) \triangleq F^{-1}\{X(f)\} = \int_{-\infty}^{\infty} X(f) e^{i2\pi ft} df$$

In general, the Fourier transform is a complex-valued quantity, even when the time series is real valued. There are many references on Fourier transforms; one is Reference 1. In general, the Fourier transform of $x(t)$ exists if $x(t)$ is an integrable function with $\int_{-\infty}^{\infty} |x(t)| dt < \infty$.

Table A.1 provides a useful collection of Fourier transform pairs that can be derived from the Fourier transform definition.

Finally, Parseval's Theorem [1] states the equivalence of power in the time and frequency domains: $\int_{-\infty}^{\infty} |x(t)|^2 dt = \int_{-\infty}^{\infty} |X(f)|^2 df$.

A.3 SIGNAL THEORY AND LINEAR SYSTEMS THEORY

In this book, signals are usually represented using complex envelope notation, with the RF signal described by

$$s(t) = \Re\{x(t) \exp[i2\pi f_0 t]\}$$

Here, $s(t)$ is a real-valued signal, $x(t)$ is the complex envelope of the signal, and f_0 is the carrier frequency. Complex envelope notation is most useful when $s(t)$ is a bandpass signal—the carrier frequency is much greater than any of the frequencies in the complex envelope. The complex envelope of a waveform is sometimes termed the baseband waveform.

It is often useful to describe the complex envelope in terms of its in-phase component $x_I(t)$ and quadrature component $x_Q(t)$, $x(t) = x_I(t) + i x_Q(t)$. The analytic signal is $\tilde{s}(t) = x(t) \exp[i2\pi f_0 t] = [x_I(t) + i x_Q(t)] \exp[i2\pi f_0 t]$ and the amplitude envelope is $r(t) = |x(t)| = |\tilde{s}(t)| = [x_I^2(t) + x_Q^2(t)]^{1/2}$.

A linear time-invariant (LTI) system can be characterized by its impulse response $h(t)$; if the input to the system is a Dirac delta function, then the output is $h(t)$. In general, if the input is $x(t)$, the output is the convolution of the input and the impulse response $x(t) * h(t) \triangleq \int_{-\infty}^{\infty} x(\tau) h(t-\tau) d\tau$. The Fourier transform of the impulse response is called the transfer function $H(f) = F\{h(t)\}$ and the Fourier transform of the linear system output is $F\{x(t) * h(t)\} = X(f) H(f)$.

TABLE A.1. Fourier Transform Pairs

Name	Time Domain	Frequency Domain
Forward transform	$x(t)$	$X(f) = F\{x(t)\} = \int_{-\infty}^{\infty} x(t)e^{-i2\pi ft} dt$
Inverse transform	$x(t) = F^{-1}\{X(f)\} = \int_{-\infty}^{\infty} X(f)e^{i2\pi ft} df$	$X(f)$
Complex conjugation	$x^*(t)$	$X(-f)$
Time reversal	$x(-t)$	$X^*(f)$
Amplitude scale	$ax(t)$	$aX(f)$
Time delay/phase shift	$x(t - \tau)$	$X(f)\exp[i2\pi f\tau]$
Time compand	$x(t/a)$	$aX(af)$
Multiplication in time/convolution in frequency	$z(t) = x(t)y(t)$	$Z(f) = X(f) * Y(f) = \int_{-\infty}^{\infty} X(\phi)Y(f - \phi)d\phi$
Sinusoid	$\exp[i2\pi f_0 t]$	$\delta(f - f_0)$
Modulation of a sinusoid	$x(t)\exp[i2\pi f_0 t]$	$X(f - f_0)$
Convolution in time/multiplication in frequency	$z(t) = x(t) * y(t) = \int_{-\infty}^{\infty} x(\tau)y(t - \tau)d\tau$	$Z(f) = X(f)Y(f)$
Correlation in time/conjugate Multiplication in Frequency	$z(t) = \int_{-\infty}^{\infty} x(\tau)y^*(t + \tau)d\tau$	$Z(f) = X(f)Y^*(f)$
Dirac delta function	$\delta(t)$	1
Impulse train	$\sum_{n=-\infty}^{\infty} \delta(t - nT)$	$\frac{1}{T}\sum_{k=-\infty}^{\infty} \delta(f - \frac{k}{T})$
Rect/Sinc	$x(t) = \text{rect}_T(t)$	$X(f) = T\,\text{sinc}(\pi fT)$
Sinc/Rect	$x(t) = B\,\text{sinc}(\pi tB)$	$X(f) = \text{rect}_B(f)$

If an LTI system's transfer function $H(f)$ can be written as $H(f) = H_{LP}(f - f_0)$, where the highest frequencies of $H_{LP}(f)$ are much less than f_0, then $H_{LP}(f)$ is called the low-pass equivalent transfer function and has low-pass equivalent impulse response $h_{LP}(t) = F^{-1}\{H_{LP}(f)\}$. The LTI system's impulse response is $h(t) = h_{LP}(t)\exp[i2\pi f_0 t]$. If the complex envelope of the input is $x(t)$, the complex envelope of the LTI system's output is $x(t) * h_{LP}(t)$.

The transfer function can be written as $H(f) = |H(f)|\exp[i\phi(f)]$, where $\phi(f)$ is called the phase response of the LTI system, in general a frequency-dependent quantity. From the time delay/phase shift Fourier pair in Table A.1, the delay of a sinusoid at frequency f is $-\phi(f)/[2\pi f]$, so the phase delay is given by

$$\tau_p = -\frac{\phi(f)}{2\pi f}$$

When a narrowband signal is input to the LTI system, a measure of the delay imposed on the complex envelope is the group delay, defined as

$$\tau_g = -\frac{1}{2\pi}\frac{d\phi(f)}{df}$$

A.4 STOCHASTIC PROCESSES

A comprehensive presentation of probability and random processes is found in Reference 2 and many other texts; only a brief review is provided here. For ergodic stochastic processes (where time averages provide the same result as ensemble averages) $x(t)$ and $y(t)$, the means are $m_x(t) \triangleq E\{x(t)\}$ and $m_y(t) \triangleq E\{y(t)\}$, where $E\{\cdot\}$ is the expectation operator (the average over the ensemble). Their crosscorrelation is $R_{xy}(t,\tau) \triangleq E\{x(t)y^*(t+\tau)\}$ and the autocorrelations are $R_x(t,\tau) \triangleq E\{x(t)x^*(t+\tau)\}$ and $R_y(t,\tau) \triangleq E\{y(t)y^*(t+\tau)\}$. Two stochastic processes are uncorrelated processes if their autocorrelation is the product of the means: $R_{xy}(t,\tau) = m_x(t)m_y^*(t+\tau)$. Two stochastic processes are orthogonal processes if their autocorrelation is zero everywhere: $R_{xy}(t,\tau) = 0, \quad \forall t, \tau$.

Two stochastic processes are statistically independent if their joint probability density function can be expressed as the product of the density functions for each stochastic process. If $x(t)$ and $y(t)$ are statistically independent processes, then the correlation function of $z(t) = x(t)y(t)$ is $R_z(t,\tau) = E\{z(t)z^*(t+\tau)\} = E\{x(t)y(t)x^*(t+\tau)y^*(t+\tau)\} = E\{x(t)x^*(t+\tau)\}E\{y(t)y^*(t+\tau)\}$ so $R_z(t,\tau) = R_x(t,\tau)R_y(t,\tau)$.

If $x(t)$ and $y(t)$ are uncorrelated processes, then the correlation function of $z(t) = x(t) + y(t)$ is $R_z(t,\tau) = E\{z(t)z^*(t+\tau)\} = E\{[x(t)+y(t)][x^*(t+\tau)+y^*(t+\tau)]\} = E\{x(t)x^*(t+\tau)\} + E\{y(t)y^*(t+\tau)\}$ so $R_z(t,\tau) = R_x(t,\tau) + R_y(t,\tau)$.

For wide-sense stationary processes, the means do not vary with time, and the correlation functions depend only upon lag τ, not time t: $E\{x(t)y^*(t+\tau)\} = R_{xy}(\tau)$, $E\{x(t)x^*(t+\tau)\} = R_x(\tau)$. For a wide-sense stationary process $x(t)$, the Fourier transform of the autocorrelation $R_x(\tau)$ is the PSD $\Phi_x(f)$.

If $x(t)$ and $y(t)$ are statistically independent and wide-sense stationary processes, then the correlation function of $z(t) = x(t)y(t)$ is $R_z(\tau) = R_x(\tau)R_y(\tau)$ and its power spectral density is the convolution $\Phi_z(f) = \Phi_x(f) * \Phi_y(f)$.

A complex-valued process $z(t) = x(t) + iy(t)$, with $x(t)$ and $y(t)$ each real valued, is circularly symmetric when it has the property that $e^{i\theta}z(t)$ has the same distribution for any real θ [3]. Circular symmetry holds if and only if $x(t)$ and $y(t)$ are zero mean, statistically independent, and identically distributed.

Complex-valued white noise has zero mean, constant PSD N_0, and correlation $N_0 \delta(\tau)$. It is Gaussian and circularly symmetric. Its real part and imaginary part each have PSD $N_0/2$ and correlation $N_0/2\delta(\tau)$.

When a stationary process $x(t)$ having PSD $\Phi_x(f)$ is the input to an LTI system with transfer function $H(f)$, the output has PSD $\Phi_x(f)|H(f)|^2$. If complex-valued white noise is input to the LTI system, the output has PSD $N_0|H(f)|^2$.

In general, a nonstationary process $x(t)$ has autocorrelation $E\{x(t)x(t+\tau)\} = R_x(t,\tau)$, that depends upon t. In this general case, nothing more specific can be said about the second-order statistics.

A zero-mean (second-order) cyclostationary stochastic process is a special case of nonstationarity: its autocorrelation has a specific form—instead of being constant over time as is the autocorrelation of a wide-sense stationary process, the autocorrelation of a second-order cyclostationary process is periodic in time. The reciprocal of the fundamental period is called the fundamental cycle frequency. A real cyclostationary process $x(t)$ has autocorrelation $E\{x(t)x(t+\tau)\} = R_x(t,\tau)$ that is periodic in t, and thus can be expressed as a Fourier series [1]: $R_x(t,\tau) = \sum_{m=-\infty}^{\infty} \chi_m(\tau)e^{i2\pi mt\phi}$, where ϕ is the fundamental cycle frequency. The Fourier coefficients are found from

$$\chi_m(\tau) = \phi \int_{-(2\phi)^{-1}}^{(2\phi)^{-1}} R_x(t,\tau)e^{-i2\pi m\phi t}dt$$

The frequency-domain characteristics of the cyclostationary process $x(t)$ are described by its complex spectrum, defined as the two-dimensional Fourier transform of the autocorrelation:

$$\Phi_x(f,\alpha) = \int_{-\infty}^{\infty}\int_{-\infty}^{\infty} R_x(t,\tau)e^{-i2\pi f\tau}e^{-i2\pi\alpha t}d\tau dt$$

$$= \int_{-\infty}^{\infty}\int_{-\infty}^{\infty} \sum_{m=-\infty}^{\infty} \chi_m(\tau)e^{i2\pi mt\phi}e^{-i2\pi f\tau}e^{-i2\pi\alpha t}d\tau dt$$

$$= \sum_{m=-\infty}^{\infty} \left[\int_{-\infty}^{\infty} \chi_m(\tau)e^{-i2\pi f\tau}d\tau\right]\left[\int_{-\infty}^{\infty} e^{i2\pi mt\phi}e^{-i2\pi\alpha t}dt\right]$$

$$= \sum_{m=-\infty}^{\infty} X_m(f)\delta(\alpha - m\phi)$$

where $X_m(f) \triangleq \int_{-\infty}^{\infty} \chi_m(\tau)e^{-i2\pi f\tau}d\tau$

This two-dimensional Fourier transform consists of "sheets" that exist at integer multiples of the fundamental cycle frequency, each continuous in frequency f. Clearly the sheet at zero cycle frequency is the PSD.

The stationarized statistics of a cyclostationary process are given by

$$R_x(\tau) = \chi_0(\tau) = \phi \int_{-(2\phi)^{-1}}^{(2\phi)^{-1}} R_x(t,\tau)\,dt$$

$$\Phi_x(f) = X_0(f) = \int_{-\infty}^{\infty} \chi_0(\tau) e^{-i2\pi f \tau}\,d\tau$$

$\chi_0(\tau)$ is called the stationarized correlation function and is obtained by averaging the cyclostationary correlation function over a period. This is the correlation function typically used in satnav treatments of keyed waveforms, and its Fourier transform $\Phi_x(f) = X_0(f)$ is the conventional PSD widely used in satnav signal analysis.

A.5 SOME RESULTS FOR KEYED WAVEFORMS

A keyed waveform has complex envelope $x(t) = \sum_{k=-\infty}^{\infty} a_k g(t - kT_c)$ for a sequence of complex values $\{a_k\}$, symbol $g(t)$, and symbol rate $1/T_c$. If the corresponding analytic signal at center frequency f_0 is the input to an LTI system with low-pass equivalent transfer function $h_{\rm LP}(t)$ at the same center frequency, the output is

$$y(t) = \left[\sum_{k=-\infty}^{\infty} a_k g(t-kT_c) e^{i2\pi f_0 t}\right] * \left[h_{\rm LP}(t) e^{i2\pi f_0 t}\right]$$

$$= \int_{-\infty}^{\infty} \left[\sum_{k=-\infty}^{\infty} a_k g(\tau - kT_c) e^{i2\pi f_0 \tau}\right] \left[h_{\rm LP}(t-\tau) e^{i2\pi f_0 (t-\tau)}\right] d\tau$$

$$= e^{i2\pi f_0 t} \sum_{k=-\infty}^{\infty} a_k \int_{-\infty}^{\infty} g(\tau - kT_c) h_{\rm LP}(t-\tau)\,d\tau$$

$$= e^{i2\pi f_0 t} \sum_{k=-\infty}^{\infty} a_k g(t - kT_c) * h_{\rm LP}(t)$$

The complex envelope of the output is a different keyed signal $\sum_{k=-\infty}^{\infty} a_k g_F(t - kT_c)$, whose symbol is the original symbol filtered by the low-pass equivalent transfer function $g_F(t) \triangleq g(t) * h_{\rm LP}(t)$.

Keyed waveforms are cyclostationary. For many, but not all, applications in satnav, however, they can be treated as wide-sense stationary. The following discussion shows that they are actually cyclostationary, and indicates conditions under which they can be treated as wide-sense stationary. For simplicity, the development is restricted to real-valued processes; extensions to complex-valued processes are straightforward.

Consider a unit-power baseband biphase keyed waveform $x(t) = \sum_{k=-\infty}^{\infty} c_k g(t - kT_c)$, with the $\{c_k\}$ an ideal long spreading code. Since the spreading sequence bits are equally likely $+1$ and -1, the keyed waveform is zero mean.

Its correlation function is

$$R_x(t, \tau) = E\{x(t)x(t+\tau)\} = \sum_{k=-\infty}^{\infty} g(t - kT_c)g(t + \tau - kT_c)$$

which is clearly periodic in t with fundamental period T_c.

The Fourier coefficients of the correlation function are

$$\chi_m(\tau) = \frac{1}{T_c} \int_{-T_c/2}^{T_c/2} \sum_{k=-\infty}^{\infty} g(t - kT_c)g(t + \tau - kT_c) e^{-i2\pi mt/T_c} dt$$

$$= \frac{1}{T_c} \sum_{k=-\infty}^{\infty} \int_{-T_c/2 - kT_c}^{T_c/2 - kT_c} g(u)g(u + \tau) e^{-i2\pi mu/T_c} du$$

$$= \frac{1}{T_c} \int_{-\infty}^{\infty} g(t)g(t+\tau) e^{-i2\pi mt/T_c} dt$$

and the stationarized correlation function is

$$R_x(\tau) = \chi_0(\tau) = \frac{1}{T_c} \int_{-\infty}^{\infty} g(t)g(t+\tau) dt$$

Observe that the power is $R_x(0) = \frac{1}{T_c} \int_{-\infty}^{\infty} |g(t)|^2 dt$

Now

$$X_m(f) \triangleq \int_{-\infty}^{\infty} \chi_m(\tau) e^{-i2\pi f\tau} d\tau$$

$$= \frac{1}{T_c} \int_{-\infty}^{\infty} \left[\int_{-\infty}^{\infty} g(t)g(t+\tau) e^{-i2\pi mt/T_c} dt \right] e^{-i2\pi f\tau} d\tau$$

$$= \frac{1}{T_c} \int_{-\infty}^{\infty} g(t) \left[\int_{-\infty}^{\infty} g(t+\tau) e^{-i2\pi f\tau} d\tau \right] e^{-i2\pi mt/T_c} dt$$

SOME RESULTS FOR KEYED WAVEFORMS

$$= \frac{1}{T_c} G(f) \int_{-\infty}^{\infty} g(t) e^{i2\pi ft} e^{-i2\pi mt/T_c} dt$$

$$= \frac{1}{T_c} G(f) G^*(f - m/T_c)$$

and the PSD is

$$\Phi_x(f) = X_0(f) = \frac{1}{T_c} |G(f)|^2$$

When the spreading symbol is the unit-power rectangle function corresponding to BPSK-R $g(t) = \text{rect}_{T_c}(t - T_c/2)$, the correlation for chip rate $\phi = 1/T_c = 1/1.023$ μs is shown in Figure A.1. In this figure, the plot continues infinitely in the t direction, corresponding to the periodicity in time.

The magnitude of different Fourier coefficients is plotted in Figure A.2. The coefficient with $m = 0$ is the stationarized correlation.

The stationarized correlation is shown in Figure A.3 on the same time and delay axes as Figure A.1. This is the standard triangular function usually associated with a BPSK-R spreading modulation, but it is really a time average of the autocorrelation function of the cyclostationary process.

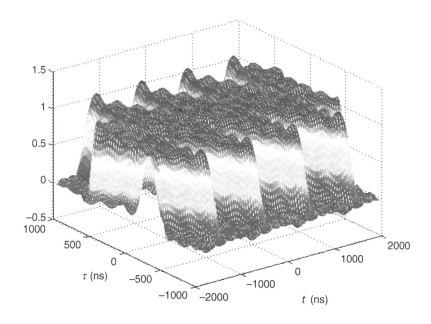

Figure A.1. Cyclostationary Autocorrelation of BPSK-R(1) Waveform

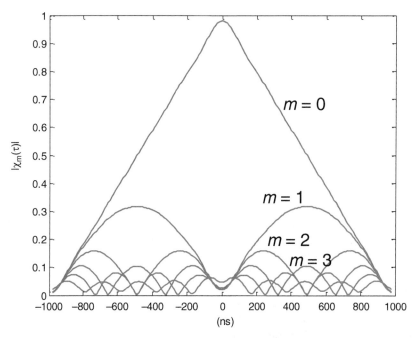

Figure A.2. Magnitude of Fourier Coefficients for Autocorrelation of BPSK-R(1) Waveform

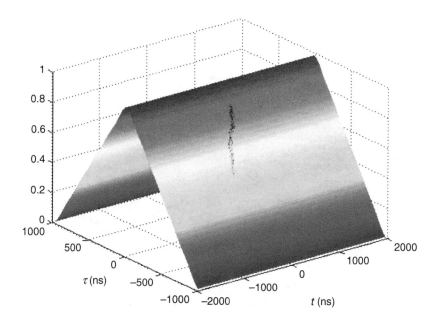

Figure A.3. Stationarized Autocorrelation of BPSK-R(1) Waveform Plotted for Comparison to Figure A.1

Although keyed waveforms are not stationary, in some key aspects of satnav they can be treated as stationary, and their stationarized statistics can be used. In particular, satnav receiver processing is often based on correlations computed over milliseconds, whereas satnav signal chip rates (and consequently the associated cycle frequencies) are of the order of MHz. This correlation processing averages over hundreds or thousands of cycle frequencies, so the stationarized statistics apply. However, cyclostationarity needs to be accounted for in sampling keyed waveforms [4] and in taking the product of keyed waveforms (see Review Question QT3.16).

A.6 BANDWIDTH MEASURES

It is common to characterize a PSD or transfer function by its bandwidth. Here we consider lowpass equivalent transfer functions or PSDs of a waveform's complex envelope. While the term "bandwidth" is very commonly used, there are many different meanings of "bandwidth," some of which are described below. Observe that these bandwidth measures are independent of amplitude scaling in the PSD or transfer function.

A common measure of bandwidth is the null-to-null bandwidth, defined as the frequency span between the closest frequencies, above and below the maximum of the PSD, where the PSD is zero. Another common measure is the 3 dB bandwidth, the frequency span between points on the PSD that are closest to the maximum but half the power in the maximum. The 10 dB bandwidth is the frequency span between points on the PSD that are closest to the maximum but one-tenth the power in the maximum. The 90% power bandwidth is the frequency span between points equidistant from the center frequency that contains 90% of the power in the PSD. All of these are defined as two-sided bandwidths, and are divided by 2 to represent one-sided bandwidths. All of these have been stated in terms of PSD but apply equally to a transfer function $H(f)$ by replacing the PSD with $|H(f)|^2$ in the above definitions.

Another two-sided bandwidth measure, the equivalent rectangular bandwidth also known as the equivalent noise bandwidth, is defined as follows. For a PSD $\Phi(f)$, the one-sided equivalent rectangular bandwidth is

$$B_{\text{eqrect}} = \frac{1}{2} \frac{\int_{-\infty}^{\infty} \Phi(f) df}{\max_{f} \Phi(f)}$$

While normalizing by retaining the maximum PSD is a common normalization, other normalizations can be used including retaining the same PSD value at band center or some other selected frequency.

Since common use of this bandwidth measure is as a one-sided bandwidth, this convention is used here as well. The equivalent rectangular bandwidth is thus the width of a rectangle that has the same height as the maximum in the PSD and contains the same power as the PSD. For an LTI system, the equivalent rectangular bandwidth is defined as above with $\Phi(f)$ replaced by $|H(f)|^2$. The equivalent rectangular bandwidth

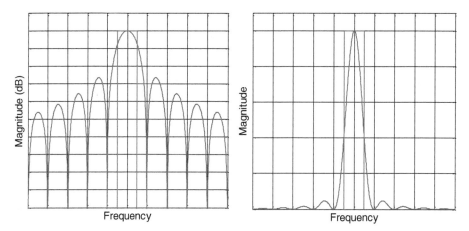

Figure A.4. Equivalent Rectangular Bandwidth of Sinc-Squared PSD

of an LTI system then describes the width of an ideal "brick wall" filter that has the same maximum gain and outputs the same output power as an arbitrary LTI system, when the input is white noise. Figure A.4 portrays the equivalent rectangular bandwidth of a sinc-squared PSD over an infinite frequency range.

Finally, the (two-sided) RMS bandwidth, sometimes known as the Gabor bandwidth after the electrical engineer and physicist Dennis Gabor, is

$$\bar{\beta}_{\text{rms}} = \left[\frac{\left[\int_{-\beta_r/2}^{\beta_r/2} f^2 \Phi_x(f)\, df \right]}{\int_{-\beta_r/2}^{\beta_r/2} \Phi_x(f)\, df} \right]^{1/2} = \left[\int_{-\beta_r/2}^{\beta_r/2} f^2 \bar{\Phi}_x(f)\, df \right]^{1/2}, \quad \int_{-\beta_r/2}^{\beta_r/2} \bar{\Phi}_x(f)\, df = 1$$

The frequency-squared weighting of the PSD in the integrand emphasizes higher frequency parts of the PSD even if the power is much lower at those frequencies. PSDs associated with BPSK-R spreading modulations and BOC spreading modulation fall off with frequency-squared, so their RMS bandwidths approach infinity as the rectangular bandwidth β_r approaches infinity. Normalizing the PSD in the expression for the RMS bandwidth removes the power lost due to bandlimiting, so this definition of RMS bandwidth quantifies only the effect of PSD shape; power loss due to bandlimiting is accounted for separately.

Figure A.5 shows examples of these bandwidth measures for a bandlimited sinc-squared PSD given by

$$\Phi(f) = \begin{cases} \text{sinc}^2\left(\pi f/f_c\right), & |f| < 2.046 \times 10^6, \quad f_c = 1.023 \times 10^6 \\ 0, & \text{elsewhere} \end{cases}$$

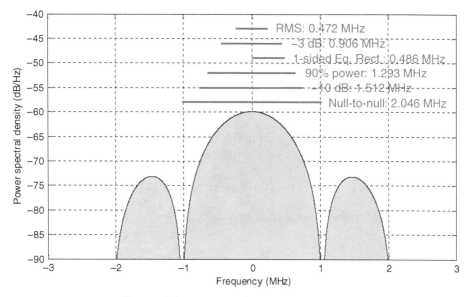

Figure A.5. Examples of Bandwidth Measures

A.7 MATRICES AND MATRIX ALGEBRA

A matrix (see Reference 5 for many more matrix properties) is a two-dimensional m by n array of numbers

$$\mathbf{A} = \begin{bmatrix} a_{1,1} & a_{1,2} & \cdots & a_{1,n} \\ a_{2,1} & a_{2,2} & \cdots & a_{2,n} \\ \vdots & \vdots & \ddots & \vdots \\ a_{m,1} & a_{m,2} & \cdots & a_{m,n} \end{bmatrix}$$

The matrix transpose is denoted $\mathbf{A}^T$ and given by the n by m array

$$\mathbf{A}^T = \begin{bmatrix} a_{1,1} & a_{2,1} & \cdots & a_{m,1} \\ a_{1,2} & a_{2,2} & \cdots & a_{m,2} \\ \vdots & \vdots & \ddots & \vdots \\ a_{1,n} & a_{2,n} & \cdots & a_{m,n} \end{bmatrix}$$

The matrix Hermitian is the complex conjugate transpose

$$\mathbf{A}^H = \begin{bmatrix} a_{1,1}^* & a_{2,1}^* & \cdots & a_{m,1}^* \\ a_{1,2}^* & a_{2,2}^* & \cdots & a_{m,2}^* \\ \vdots & \vdots & \ddots & \vdots \\ a_{1,n}^* & a_{2,n}^* & \cdots & a_{m,n}^* \end{bmatrix}$$

Matrix multiplication is defined between an $m \times n$ matrix $\mathbf{A}$ and an $n \times p$ matrix $\mathbf{B}$, producing the $m \times p$ matrix $\mathbf{C}$

$$\mathbf{C} = \mathbf{AB} = \begin{bmatrix} a_{1,1} & a_{1,2} & \cdots & a_{1,n} \\ a_{2,1} & a_{2,2} & \cdots & a_{2,n} \\ \vdots & \vdots & \ddots & \vdots \\ a_{m,1} & a_{m,2} & \cdots & a_{m,n} \end{bmatrix} \begin{bmatrix} b_{1,1} & b_{1,2} & \cdots & b_{1,p} \\ b_{2,1} & b_{2,2} & \cdots & b_{2,p} \\ \vdots & \vdots & \ddots & \vdots \\ b_{n,1} & b_{n,2} & \cdots & b_{n,p} \end{bmatrix}$$

$$= \begin{bmatrix} \sum_{k=1}^{n} a_{1,k} b_{k,1} & \sum_{k=1}^{n} a_{1,k} b_{k,2} & \cdots & \sum_{k=1}^{n} a_{1,k} b_{k,p} \\ \sum_{k=1}^{n} a_{2,k} b_{k,1} & \sum_{k=1}^{n} a_{2,k} b_{k,2} & \cdots & \sum_{k=1}^{n} a_{2,k} b_{k,p} \\ \vdots & \vdots & \ddots & \vdots \\ \sum_{k=1}^{n} a_{m,k} b_{k,1} & \sum_{k=1}^{n} a_{m,k} b_{k,2} & \cdots & \sum_{k=1}^{n} a_{m,k} b_{k,p} \end{bmatrix}$$

The identity matrix is a square matrix with 1's on the main diagonal and 0's elsewhere

$$\mathbf{I} = \begin{bmatrix} 1 & 0 & \cdots & 0 \\ 0 & 1 & \cdots & 0 \\ \vdots & \vdots & \ddots & 0 \\ 0 & \cdots & 0 & 1 \end{bmatrix}$$

such that for any matrix $\mathbf{A}$ with the same dimensions, $\mathbf{AI} = \mathbf{IA} = \mathbf{A}$.

A square matrix whose rows and columns are linearly independent is called a square full-rank matrix.

If $\mathbf{A}$ is a square full-rank matrix, it has an inverse $\mathbf{A}^{-1}$ with the property $\mathbf{AA}^{-1} = \mathbf{A}^{-1}\mathbf{A} = \mathbf{I}$.

When the following matrices exist and have appropriate dimensions, the following matrix identities apply.

$$(\mathbf{AB})^T = \mathbf{B}^T \mathbf{A}^T$$
$$(\mathbf{AB})^H = \mathbf{B}^H \mathbf{A}^H$$
$$(\mathbf{A}^T)^{-1} = (\mathbf{A}^{-1})^T$$
$$(\mathbf{A}^H)^{-1} = (\mathbf{A}^{-1})^H$$
$$(\mathbf{AB})^{-1} = \mathbf{B}^{-1} \mathbf{A}^{-1}$$

The Sherman–Morrison formula provides a useful way to update the inverse of a matrix. If $\mathbf{A}$ is an invertible square matrix and $\mathbf{u}$ and $\mathbf{v}$ are column vectors, then if $\mathbf{v}^T \mathbf{A}^{-1} \mathbf{u} \neq -1$,

$$(\mathbf{A} + \mathbf{uv}^T)^{-1} = \mathbf{A}^{-1} - \frac{\mathbf{A}^{-1} \mathbf{uv}^T \mathbf{A}^{-1}}{1 + \mathbf{v}^T \mathbf{A}^{-1} \mathbf{u}}$$

This formula provides a computationally efficient way to find the inverse of a matrix augmented by a new vector outer product, if the inverse of the original matrix is already known. It also can be useful analytically.

Matrices provide convenient ways to solve linear systems of equations having the form $\mathbf{Ax} = \mathbf{b}$, where $\mathbf{A}$ is $m \times n$, $\mathbf{x}$ is $n \times 1$, and $\mathbf{b}$ is $m \times 1$. If $\mathbf{A}$ has an inverse, the solution is $\mathbf{x} = \mathbf{A}^{-1}\mathbf{b}$.

If $m > n$, there are more equations than unknowns, and unless they are redundant, there is no solution to this overconstrained system of equations. However, approximate solutions can be found. One common approximate solution is one that minimizes the mean-squared error $(\mathbf{Ax} - \mathbf{b})^H(\mathbf{Ax} - \mathbf{b})$. The ordinary least-squares (OLS) solution is $\mathbf{x}_{LS} = (\mathbf{A}^H\mathbf{A})^{-1}\mathbf{A}^H\mathbf{b} = \mathbf{A}^{\#}\mathbf{b}$, where $\mathbf{A}^{\#} \triangleq (\mathbf{A}^H\mathbf{A})^{-1}\mathbf{A}^H$ is called the left pseudoinverse [6] of $\mathbf{A}$ for the overdetermined case.

If $\mathbf{R}$ is an invertible $m \times m$ matrix, the previous result can be generalized to minimize the weighted mean-squared error $(\mathbf{Ax} - \mathbf{b})^H \mathbf{R}^{-1}(\mathbf{Ax} - \mathbf{b})$. The weighted least-squares (WLS) solution is [6] $\mathbf{x}_{WLS} = (\mathbf{A}^H \mathbf{R}^{-1} \mathbf{A})^{-1} \mathbf{A}^H \mathbf{R}^{-1} \mathbf{b}$, which reverts to the least-squares solution when $\mathbf{R} = \mathbf{I}$.

A.8 TAYLOR SERIES AND LINEARIZATION

Let $f(x)$ be an infinitely differentiable function in the neighborhood of $x = a$, with derivatives $f'(x), f''(x), f'''(x), \ldots$. The Taylor Series in this neighborhood is then

$$f(x) = f(a) + (x-a)\frac{f'(x)|_{x=a}}{1!} + (x-a)^2\frac{f''(x)|_{x=a}}{2!} + (x-a)^3\frac{f'''(x)|_{x=a}}{3!} + \cdots$$

and a linear approximation to $f(x)$ near $x = a$ is $f(x) \approx f(a) + (x-a) f'(x)|_{x=a}$.

The multidimensional Taylor Series generalizes this result. Let $f(\mathbf{x})$ be a scalar real-valued function of multiple variables expressed as the row vector $\mathbf{x} = [x_1 \ x_2 \ \cdots \ x_n]$. If all first and second derivatives and partial derivatives exist in a neighborhood of $\mathbf{x} = \mathbf{a}$, then the second-order Taylor polynomial is the multidimensional Taylor Series in this neighborhood:

$$f(\mathbf{x}) \approx f(\mathbf{a}) + (\mathbf{x}-\mathbf{a})\ \nabla f(\mathbf{x})|_{\mathbf{x}=\mathbf{a}} + \frac{1}{2}(\mathbf{x}-\mathbf{a})\ Hf(\mathbf{x})|_{\mathbf{x}=\mathbf{a}}\ (\mathbf{x}-\mathbf{a})^T$$

where the gradient is

$$\nabla f(\mathbf{x}) = \begin{bmatrix} \frac{\partial f(\mathbf{x})}{\partial x_1} \\ \frac{\partial f(\mathbf{x})}{\partial x_2} \\ \vdots \\ \frac{\partial f(\mathbf{x})}{\partial x_n} \end{bmatrix}$$

and the Hessian matrix is

$$Hf(\mathbf{x}) = \begin{bmatrix} \frac{\partial^2 f(\mathbf{x})}{\partial x_1^2} & \frac{\partial^2 f(\mathbf{x})}{\partial x_1\, \partial x_2} & \cdots & \frac{\partial^2 f(\mathbf{x})}{\partial x_1\, \partial x_n} \\ \frac{\partial^2 f(\mathbf{x})}{\partial x_2\, \partial x_1} & \frac{\partial^2 f(\mathbf{x})}{\partial x_2^2} & \cdots & \frac{\partial^2 f(\mathbf{x})}{\partial x_2\, \partial x_n} \\ \vdots & \vdots & \ddots & \vdots \\ \frac{\partial^2 f(\mathbf{x})}{\partial x_n\, \partial x_1} & \frac{\partial^2 f(\mathbf{x})}{\partial x_n\, \partial x_2} & \cdots & \frac{\partial^2 f(\mathbf{x})}{\partial x_n^2} \end{bmatrix}$$

A.9 COORDINATE SYSTEM OVERVIEW

This section provides some concepts of coordinate systems and introduces some coordinate systems commonly used in satnav. Many different coordinate systems are used in navigation and geodesy. The science of geodesy addresses coordinate systems in extensive detail [7]; the quick summary here is merely an introduction to basic terminology and concepts.

Three-dimensional Cartesian coordinate systems are commonly used, with axes at right angles to each other and a point's location specified by an ordered pair of three signed numbers indicating the projection of the point onto each of the three axes, often denoted (x, y, z). In a right-handed coordinate system, if the x-y plane is horizontal and the x-axis is pointing toward an observer and the y-axis is pointing toward the observer's right, the z-axis is up. This convention follows the "right-hand rule" that if one uses one's curled right hand to point from the x-axis toward the y-axis with palm facing left, the thumb points in the direction of the z-axis. A left-handed coordinate system follows the opposite convention.

An inertial coordinate system or frame of reference is one where a body with no forces acting upon it does not move. For objects on or near the Earth, the stars are so far away that they appear to be immobile. Thus, the stars define an approximate inertial coordinate system for terrestrial use.

An Earth-centered inertial (ECI) coordinate system has its origin at the Earth's center of mass; its z-axis passes through the North Pole, in particular, the mean position of the Earth's rotational axis between 1900 and 1905. Its x-axis is parallel to the equator and passes through the Prime Meridian at $t = 0$, and its y-axis is orthogonal to both of these axes, in the sense that a right-hand coordinate system is formed. Satellite motion is commonly described in an ECI coordinate system.

An Earth-centered, Earth-fixed (ECEF) coordinate system also has its origin at the Earth's center of mass. The most common ECEF coordinate system is the conventional terrestrial system, whose x-axis is parallel to the equator and passes through the Prime Meridian. The z-axis passes through the North Pole. The y-axis is orthogonal to both of these axes, in the sense that a right-hand coordinate system is formed. Once every 24 hours the ECEF coordinate system is aligned with the ECI coordinate system, when the Prime Meridian aligns with the location of the Prime Meridian used to define the ECI coordinate system at $t = 0$.

It is often useful to describe coordinates relative to one's own position. One such coordinate system is the NEU coordinate system, which places the x-axis toward north (strictly speaking, the astronomical north, or the direction of the conventional terrestrial pole), the z-axis "up," or opposite to the direction of gravitational pull and away from the Earth's center of mass, and the y-axis perpendicular to both of the other axes, or east. This is a left-handed coordinate system.

The NED coordinate system is a right-handed coordinate system that differs from NEU only in that the z-axis is "down." Navigation of flying bodies often is based on the NED coordinate system.

In contrast to Cartesian coordinate systems, ellipsoidal coordinate systems define the location of a point in terms of latitude, longitude, and height from the origin on an ellipsoid separately specified. The geodetic latitude of a point is the angle from the equatorial plane to the intersection of a line normal to the reference ellipsoid that passes through the point. The geodetic longitude of a point is the angle between a reference plane and a plane passing through the point, when both planes are perpendicular to the equatorial plane. The geodetic height at a point is the distance from the reference ellipsoid to the point in a direction normal to the ellipsoid.

REFERENCES

1. E. Oran Brigham, *The Fast Fourier Transform and Its Applications*, Prentice-Hall, 1988.
2. A. Papoulis and S. Pillai, *Probability, Random Variables, and Stochastic Processes*, 4th edition, McGraw-Hill, 2002.
3. "Circularly-symmetric random vectors," http://www.rle.mit.edu/rgallager/documents/CircSymGauss.pdf, accessed 5 December 2012.
4. W. A. Gardner, "Common Pitfalls in the Application of Stationary Process Theory to Time-Sampled and Modulated Signals," *IEEE Transactions on Communications*, Vol. COM-35, No. 5, 1987, pp. 529–534.
5. K. B. Petersen and M. S. Pedersen, The Matrix Cookbook, Version: November 14, 2008, available at: http://orion.uwaterloo.ca/~hwolkowi/matrixcookbook.pdf, accessed 9 February 2013.
6. G. H. Golub and C. F. Van Loan, *Matrix Computations*, 3rd edition, The Johns Hopkins University Press, 1996.
7. R. M. Rogers, *Applied Mathematics in Integrated Navigation Systems*, 2nd edition, AIAA Press, 2003.

INDEX

1.5 bit quantization, 335
10 dB bandwidth, 619
2.5 bit quantization, 335
3 dB bandwidth, 619
90% power bandwidth, 126, 619

Absolute differential system, 570
Accuracy, 2, 3, 6, 12, 24, 26, 44, 67, 72, 84, 106, 112, 113, 119, 141, 143, 150, 158, 161, 163, 165, 166, 168, 192, 201, 202, 210, 215, 217, 220, 261, 271, 287, 291, 305, 307, 309, 365, 369, 450, 476, 480, 484, 486, 487, 494–497, 499, 509, 511, 516, 521, 524–527, 529, 531, 537, 538, 545, 555, 563, 566, 570, 573, 576, 588–592, 602–604, 607
Acquisition, 11, 81, 122, 125, 129, 136, 189, 272, 295, 364–402, 429, 430, 440, 542, 545, 547, 548, 607
Active antenna, 294, 300, 301, 323
Active mixer, 316
Active resistor-capacitor (RC) filter, 309
Advanced multipath mitigation approaches, 566
Advanced Receiver Autonomous Integrity Monitoring (ARAIM), 530
Aeronautical Radio Navigation Service (ARNS), 67, 205
Aggregate gain of interference (G_{agg}), 134
Airports Authority of India (AAI), 203
Algebraic representation, 43, 87
Aliasing, 308, 324–327, 334, 340, 345, 358–361, 541, 542
Allan deviation, 319–321, 572
Allan variance, 319
Alternative Binary Offset Carrier (AltBOC), 54, 56–58, 242, 243, 245, 248, 262, 263, 437

Altitude hold, 525, 526, 593
Ambiguity resolution observation model, 584–585
Amplitude envelope, 39, 611
Analog tracking loop, 407
Analog-to-digital conversion (ADC), 13, 294, 299, 306, 313, 318, 324–327, 329, 333–362, 364, 373, 437, 541–543, 545
Analog-to-digital converter (ADC), 13, 294, 295, 298, 299, 306, 313, 318, 324–327, 329, 333–362, 364, 373, 437, 541–543, 545, 549
Analog-to-digital converter (ADC) histogramming, 545, 549
Analytic signal, 38, 63, 90, 309, 438, 611, 615
Angle off nadir, 31, 32
Antenna directivity, 301, 302
Antenna directivity pattern, 301, 302
Antenna efficiency, 105, 107, 301, 302
Antenna electrical size, 304
Antenna electronics (AE), 304
Antenna element, 10, 66, 300, 301, 304, 305
Antenna gain, 66, 102, 105–107, 109–111, 134, 300, 302, 312, 322, 395, 549
Antenna gain pattern, 105, 109–111, 133, 134, 300–304, 561
Antenna total efficiency, 301
Antialiasing filter, 299, 308, 324, 325, 345, 540
Argument of perigee, 22, 272
Ascending node, 21, 22, 272
Assisted start, 368, 369
Asynchronous code division multiple access (CDMA), 72–74, 144, 170, 177, 185, 187, 189, 194, 206, 209, 213, 216, 217, 220, 221, 222, 232, 238, 239, 241, 243, 259, 273, 275, 276, 278, 279, 283, 286, 288, 401

Engineering Satellite-Based Navigation and Timing: Global Navigation Satellite Systems, Signals, and Receivers, First Edition. John W. Betz.
© 2016 The Institute of Electrical and Electronics Engineers, Inc. Published 2016 by John Wiley & Sons, Inc.

628 INDEX

Automatic gain control (AGC), 11, 294, 299, 325, 334–338, 340, 343, 348, 362, 373, 542, 548
Auxiliary circle, 21–23
Availability, 12, 27, 28, 65, 141, 161, 201, 202, 228, 267, 268, 271, 272, 291, 305, 328, 394, 509, 538, 567, 574, 586, 588
Availability enhancement signal, 161, 202, 267, 268, 271, 272, 281
Axial ratio, 69–71, 108, 119
Azimuth cut, 302

Backlobes, 302, 303, 548
Balun, 304
Bandlimiting, 52, 123, 129, 136, 295, 334, 336, 341, 343, 344, 346, 348, 349, 351, 354, 356, 362, 494, 500, 620
Bandlimiting, sampling, and quantization (BSQ), 295, 334, 338, 340, 341, 343–346, 348–358, 362, 408
Bandlimiting, sampling, and quantization (BSQ) loss, 295, 340, 343–358, 373, 376, 388
Bandpass sampling, 326
Bandpass signal, 38, 39, 60, 611
Baseband, 9, 11, 87, 89, 122, 123, 218, 239, 241, 277, 294, 297, 299, 316, 324, 325, 333, 335, 336, 364, 366, 380, 408, 438, 616
Baseband waveform, 364, 611
Baseline, 151, 152, 208, 209, 214, 229, 572–577, 583, 584, 586
Baseline length, 572, 581
BeiDou Experimental Satellite Navigation System, 253
BeiDou Ground Base Enhancement System, 256
BeiDou system (BDS), 160, 161, 252–264
Between receiver single differencing (BRSD), 579, 580
Between satellite single differencing (BSSD), 579, 580, 581, 583, 584
Bi-binary code, 218, 219
Binary coded symbols (BCS), 54, 55, 58
Binary offset carrier (BOC), 47–58, 131, 166, 167, 176–178, 181, 187, 197, 222, 223, 228, 232, 233, 263, 274, 275, 288, 289, 349–351, 354–357, 359–361, 370, 380–383, 387, 390, 393, 397, 402, 438–441, 471, 475–479, 485, 486, 488, 493–498, 500, 559, 562–564, 620
Binary offset carrier (BOC) ratio, 50, 131, 237, 494
Binary phase-shift keying with rectangular symbols (BPSK-R), 45–50, 55–57, 60, 64, 84, 124, 128, 131, 166, 170, 185, 189, 193, 197, 206, 209, 210, 221–223, 240–242, 245, 255, 258, 262, 263, 271, 273, 276, 278, 281, 345–351, 353–356, 359–361, 370, 374, 375, 379–382, 389, 390, 438, 479, 383, 485–489, 491, 492, 497, 499, 500, 557, 558, 561–565, 617, 618, 620
Biphase keyed signal, 39–41
Bit error probability, 458–460
Blanking the replica, 436, 440
Block coding, 78
Block forward error control (FEC), 77–80, 82, 287
Block interleaver, 80
Block processing, 588, 592, 593
Blocker, 313
Boltzmann constant, 311, 321, 339
Boresight angle, 300
Bulk acoustic wave (BAW) filter, 309
Burst error, 79, 80, 182

Carrier frequency, 10, 13, 38, 47, 49, 50, 52, 60, 64–68, 72, 81, 84, 91, 103, 122, 124, 129, 165–171, 176–179, 185–189, 193, 194, 202, 204–206, 208–210, 215–217, 219–223, 229–232, 237–239, 241, 243, 245, 255, 258, 259, 262, 270, 271, 273, 275, 276, 278, 279, 284, 286, 288, 292, 294, 295, 298, 315, 317, 318, 320, 321, 323, 326, 327, 359, 366, 367, 370, 371, 373–375, 390, 407, 409, 428–430, 435, 442–444, 480, 504, 506, 514, 523, 541, 576, 578, 579, 581–583, 611
Carrier phase measurement equation, 528
Carrier tracking, 13, 73, 81, 129, 136, 294, 319, 365, 407, 409, 428–463, 467, 468, 480, 486, 511, 523, 529, 537, 542, 547, 548, 566, 567, 579, 586, 588, 604, 607
Carrier tracking loop, 292, 295, 296, 427, 428, 429, 442, 470, 523, 595, 601, 603
Carrier wiping, 123
Carrier-aided code tracking, 296, 480–481, 485, 499, 511, 581, 603

INDEX **629**

Carrier-based differential system, 573, 576–586
Carrier-smoothed code (CSC), 511, 512
Cavity filter, 308
Celestial reference frame, 157, 158
China Geodetic Coordinate System 2000, 257
Chip antenna, 306
Chip rate, 42, 44–47, 50, 51, 56, 61, 88, 90, 91, 124, 128, 131, 163, 166, 186, 220, 222, 223, 255, 354, 361, 480, 482, 617, 619
Choke ring antenna, 561
Chu-Harrington limit, 305, 306
Circular Error Probable (CEP), 149
Circularly symmetric process, 123, 613, 614
Clock and ephemeris data (CED), 182–184, 367, 369, 460, 462, 515, 516, 518, 525–527, 529, 531
Clock curve fit error, 141
Clock error, 141, 143, 151, 508, 509, 513, 579, 580
Clock estimation error, 141, 142
Clock hold, 525, 593
Clock prediction error, 141, 142
Clock stability error, 142
Coarse time, 75, 76, 364, 369, 526, 527, 591–593
Coarse time navigation, 75, 76, 526, 527, 593
Coarse/Acquisition (C/A) code signal, 165
Code combining, 293, 462
Code division multiple access (CDMA), 72–74, 144, 213, 216, 217, 200, 222, 283, 401
Code measurement equation, 528
Code shift keying (CSK), 81, 275
Code tracking, 13, 130, 144, 166, 241, 248, 292, 294–296, 319, 366, 407, 409, 432, 434, 437, 462, 467–501, 505, 507, 511, 512, 531, 537, 542, 547–549, 556, 557, 559, 561–563, 567, 571, 573, 575, 576, 578, 581, 584, 586, 593, 595, 601–603
Code tracking jitter, 472, 478, 511, 581
Code tracking loop lock criterion, 485
Code tracking spectral separation coefficient (SSC), 127, 128, 130–133, 135, 136, 344, 345, 491, 492, 542
Code-based differential system, 152, 573, 574–576
Code-carrier divergence, 65, 512
Codeless processing, 185

Code tracking loop, 292, 295, 296, 366, 426, 467–470, 476, 480–482, 484–489, 493, 494, 499, 510, 511, 515, 557, 563, 593, 601, 602
Coherent code tracking discriminator function, 474
Coherent early-late processing (CELP), 475–477, 482–492, 565
Coherent integration, 549
Cold start, 368, 369, 396, 607
Comm/nav integration, 526, 537
Commensurate sampling, 338, 375
Commercial Service (CS), 4, 229, 231
Communications/navigation integration, 270, 369
Compass, 252, 365, 370
Compass Navigation Satellite System (CNSS), 252
Compatibility, 159, 160, 166, 169, 267
Complementary error function, 456
Complex conjugate, 610, 621
Complex envelope, 38, 39, 42, 43, 46, 52, 57, 60, 61, 63–65, 81, 315, 316, 318, 325, 335, 336, 365, 371, 373, 388, 407, 409, 467, 611–613, 615, 619
Composite binary offset carrier (CBOC), 54, 58, 59, 232–235, 248, 351, 352, 358, 382, 440, 441, 442, 479, 563
Composite clock, 168
Composite signal, 59, 84–86, 191, 241, 247, 248, 429, 435, 459
Constant envelope signal, 39, 84, 222
Constant false alarm rate (CFAR), 396
Constant-envelope alternative binary offset carrier (AltBOC), 54, 56–58, 242, 243, 245, 248, 262, 263, 437
Constraint length, 78, 189, 194, 206, 209, 232, 239, 243, 273, 278, 279, 455, 458, 459
Continuity, 12, 554
Continuous-update tracking loop, 407
Controlled reception pattern antenna (CRPA), 304
Conventional Terrestrial System, 624
Convolution, 49, 611, 612, 613
Convolutional coding, 78, 188, 189, 194, 206, 209, 221, 232, 239, 243, 273, 278, 279, 286, 455, 458–460
Convolutional forward error control (FEC), 77, 78–80, 82, 287

Copolarization gain, 302
Correlated interference, 540
Correlation properties, 74, 181, 236
Correlator output signal-to-noise ratio (SNR), 18, 122–136, 334, 339–344, 362, 374, 377, 388, 392, 393, 395, 436, 440, 444–447, 449, 450, 452, 482–484, 494, 560, 563, 565
Correlator tap, 432, 462, 469, 472, 478, 495
Cospas-Sarsat, 167, 213, 229
Costas loop, 292, 295, 366, 428, 447–450, 451, 453, 455, 462, 463, 523
Costas loop lock criterion, 426, 446, 449, 450, 453, 485
Cramér-Rao lower bound, 483, 566
Crest factor, 339
Cross-ambiguity function (CAF), 370–375, 377–381, 383–388, 390–398, 401, 402, 588, 590, 592, 593
Cross-ambiguity function (CAF) output signal-to-noise ratio (SNR), 374, 377, 388, 392, 393, 394
Cross-channel aiding, 596, 603, 604, 606
Crosspolarization gain, 302–304
Crystal oscillator (XO), 320
Cycle slip, 511, 512, 529, 575, 579, 602, 604, 606
Cyclic redundancy check (CRC), 77–80, 182, 183, 189, 192, 207, 236, 247, 273, 275, 278, 279, 286, 287, 292, 293, 460, 462
Cyclostationary stochastic process, 614

Data chip rate, 88, 90
Data component, 81–83, 87, 88, 91, 132, 176–178, 188, 189, 193, 194, 208, 222, 232, 241, 248, 276, 292, 391, 393, 398, 402, 429–432, 435, 439, 440, 445, 454–456, 459, 460, 463, 468, 469, 592
Data component waveform, 87–90, 190, 233, 234
Data message bit sequence, 87
Data message symbol rate, 87, 88, 189
Data message symbol sequence, 87
Data overlay bit duration, 88
Data overlay code bit rate, 88
Data overlay code bit sequence, 87, 88
Data overlay code length, 88
Data overlay code waveform, 87, 88
Data overlay rect sequence, 88
Data spreading code bit rate, 88
Data spreading code bit sequence, 87, 233
Data spreading code length, 87
Data spreading sequence, 88, 89
Data spreading symbol, 87, 91, 190, 191, 233, 431
Data spreading symbol sequence, 87
Data spreading waveform, 87, 88, 232, 233, 432, 455
Data spreading waveform (pre-overlay), 87–88
Data symbol, 41–44, 83, 87, 88, 181, 184, 233, 241, 242, 295, 454, 455, 460, 462, 566
Data symbol combining, 184, 295, 455, 460, 462
Data symbol duration, 42, 87, 88
Data symbol rate, 42, 43, 233
Data symbol rect sequence, 87
Data symbol waveform, 41, 87, 88, 233, 241
Data wiping, 388, 397, 591, 592
Decimation, 221, 308, 326
Deep integration, 596, 605
Delay-locked loop (DLL), 295, 366, 467
Delta range, 522–524, 531
Desensitize, 313
Device interference, 312, 540
Dielectric resonator filter, 308
Differential demodulation, 457–458
Differential Global Navigation Satellite System (GNSS), 5, 14, 15, 67, 159, 184, 193, 202, 204, 272, 297, 333, 406, 504, 589
Differential satnav, 13, 139, 150–153, 158, 505, 527, 552, 570–586
Digital attenuator, 337, 338
Digital gain control, 338, 543
Digital intermediate frequency (IF) architecture, 325
Digital tracking loop, 295, 407
Digitizing correlator, 295, 334, 337, 340–362, 373
Digitizing correlator guidelines, 343
Dilution of precision (DOP), 27, 141, 146–149
Dirac delta function, 128, 610–612
Direct conversion, 316, 326, 327, 329
Direct sampling architecture, 324, 326, 327
Direct sequence spread spectrum (DSSS), 17, 18, 37, 38, 40–44, 55, 56, 72, 73, 92, 123, 128, 255
Direct-conversion architecture, 327

Direction of periapsis, 22
Direction vector, 518, 522
Discrete Fourier transform (DFT), 383, 384, 386
Discrete-update tracking loop, 407, 408–415
Discriminator gain, 409, 411
Dispersive medium, 61, 63
Distance root mean-square (DRMS), 149, 150
Distress Alerting Satellite System (DASS), 167
Distributed gain filtering, 307
Dilution of precision (DOP) matrix, 148, 153
Doppler shift, 17, 26, 33–35, 37, 59, 60, 66, 74, 92, 132, 181, 364, 388, 438, 523, 551, 590
Double delta processing, 563, 565
Double difference equations, 577, 581
Double difference integer ambiguity, 580
Double differencing, 576–578, 580, 584
Downconversion, 294, 297, 299, 307, 313, 316, 324, 326, 328, 371, 430, 432, 433, 435, 436, 443
Downconversion mixer, 316
Drooping crossed bow tie antenna, 305
Dual use, 163, 164, 212, 252
Dynamic range, 304, 309, 311, 313, 315, 318, 327, 328, 337–339, 362, 437, 543, 547

Early correlator tap, 469
Early measurement, 469
Early-late spacing, 469, 475–479, 482–492, 494, 499, 557–559, 561–563, 565, 567, 581
Earth Orientation Parameters (EOP), 157, 184, 192, 193, 279, 280, 287, 288
Earth-centered Earth-fixed (ECEF) coordinate system, 7, 8, 21, 153, 184, 192, 215, 574, 624
Earth-centered inertial (ECI) coordinate system, 21, 624
Earth-centered, Earth-fixed (ECEF), 26, 29, 30, 33, 516, 522, 570
East-North-Up (ENU), 7
Eccentric anomaly, 21, 22, 23, 508
Eccentricity, 20, 29, 153, 268, 272, 508, 509
Effective C/N_0, 18, 122, 129–136, 295, 321, 343, 362, 392, 395–397, 402, 413, 439, 445, 446, 449, 450, 456–460, 462, 463, 492, 524, 540, 542, 545–547, 549, 557, 592

Effective isotropic radiated power (EIRP), 104, 106, 107, 312
Effective noise temperature increase, 321
Effective spectral separation coefficient (SSC), 344, 345
Effective temperature, 321–323
Elevation angle, 27, 28, 31–34, 62, 65, 106, 108–112, 133, 134, 169, 206, 215, 231, 258, 266, 271, 286, 288, 300–303, 305, 509, 510, 549, 561
Elevation mask, 27
Ellipsoidal coordinate systems, 625
Ephemeris curve fit error, 142
Ephemeris error, 141, 142, 151, 572, 602
Ephemeris estimation error, 142
Ephemeris extension, 526, 589
Ephemeris prediction error, 142
Epoch, 3, 7–9, 22–24, 76, 175, 186, 189, 215, 230, 257, 284, 334, 336, 575, 592
Equatorial prograde orbit, 21
Equatorial retrograde orbit, 21
Equivalent noise bandwidth, 619
Equivalent rectangular bandwidth, 308, 339, 412, 416, 619, 620
Ergodic stochastic process, 613
Even correlation, 74
Excess delay, 62, 63, 291, 553–559, 561, 563, 567
Expectation operator, 613
Extended Kalman filter (EKF), 599, 603–606
External interference, 540
Extra state theorem, 521

F/NAV data message, 247
Far echoes, 554
Far field, 103
Faraday rotation, 70, 108
Fast Fourier transform (FFT), 378, 379, 383, 384
Fault Detection and Exclusion (FDE), 27, 529, 531
Federal Aviation Administration (FAA), 202
Fine time, 368, 369, 591–593
Finite impulse response (FIR), 308
Fixed reception pattern antenna (FRPA), 304
Fixed solution, 585
Float solution, 584
Forward error correction (FEC), 77–80, 82, 287

Fourier transform, 38, 40, 43, 49, 52, 55, 60, 73, 358, 378, 383, 611–615
Four-quadrant arctangent, 443, 444, 451, 452
Four-thirds Earth radius model, 104
Fractional bandwidth, 437
Fraunhofer distance, 103, 304
Frequency division multiple access (FDMA), 72, 73, 144, 216, 217, 219, 222
Frequency plan, 169, 205, 215, 227, 230, 258, 270, 284, 323–328
Frequency-locked loop (FLL), 366, 428, 442–446, 457, 463, 524
Frequency-locked loop (FLL) lock criterion, 426, 446, 449, 450, 453, 485

Gabor bandwidth, 500, 620
Galileo, 2, 4, 12, 28, 29, 33, 81, 160, 161, 176, 203, 205, 226–249, 254, 262, 267, 278, 280, 380, 513, 527
Galileo System Time (GST), 230
Galileo Terrestrial Reference Framework (GTRF), 230
Gallium Arsenide (GAs), 315
Gate time, 319
Generalized Marcum Q-function, 392, 393, 395
Geodetic datum, 157, 158, 160, 168, 214, 270, 284
Geometric dilution of precision (DOP), 148
Geometry matrix, 147, 518, 521, 522
Geostationary orbit, 24, 256, 266, 268, 282, 283
Geosynchronous orbit, 24, 161, 256, 283, 516
GLObal NAvigation Satellite System (GLONASS), 2, 4, 12, 29, 30, 33, 66, 71, 72, 77, 80, 81, 84, 108, 109, 133, 157, 159, 160, 203, 205, 212–224, 226, 252, 278, 291, 294, 296, 349, 351, 379, 380, 401, 442, 455, 456
Global Positioning System (GPS), 2–5, 12, 14, 15, 23, 27–30, 37, 65, 66, 71–73, 75, 77, 80, 81, 84, 107–112, 119, 123, 132, 133, 140, 142, 153, 159–161, 163–197, 202–210, 212, 213, 215–220, 222, 223, 226–228, 230, 242, 248, 249, 252, 254, 256, 258, 266–274, 280, 284, 286–288, 291, 294, 296, 305, 379, 380, 383, 387, 401, 456, 508, 509, 513, 515, 520, 527, 530, 531, 561, 572–574, 591, 596

Gold code, 74, 132, 171, 172, 173, 175, 181, 189, 259, 272, 287
GPS And Geo-Augmented Navigation (GAGAN), 203, 283
Grid point, 261, 288, 585
Ground track, 24, 29, 134, 214, 256, 257, 266, 268, 269, 283, 284, 561
Group delay, 63–65, 76, 141, 142, 144, 151, 182, 193, 261, 272, 287, 307, 309, 313, 318, 613
Group delay correction (GDC), 142, 506
Group delay error, 141–145, 151, 152
Group delay stability, 142
Guidance 2, 5, 113, 203, 530
Gyrostabilized inertial measurement unit (IMU), 600

Hamming distance, 458
Hard decision, 454–456
Hatch filter, 511
Helix antenna, 305, 306
Hemispherical gain pattern, 106, 303, 304
Heterodyne architecture, 324–326
Highly elliptical orbit (HEO), 24
High-side injection, 316
High-side mixing, 316–318
Histogramming, 398, 400, 401, 545, 549
Homodyne architecture, 326, 327
Horizontal dilution of precision (HDOP), 148, 149
Hot start, 366, 368, 369, 607
Hydrostatic delay, 144

I/NAV data message, 229, 236
IBOC/QBOC processing of BOC signals, 381, 382
Ideal long spreading code, 40, 41, 43, 46, 55, 56, 58, 59, 75, 89, 90, 128, 132, 133, 170, 171, 177, 185, 187, 193, 232, 237, 240, 241, 244, 248–250, 259, 260, 431, 434, 436, 616
Ideal repeating spreading code, 75, 171
Identity matrix, 622
Image filter, 318
Image frequency, 317, 318
Imaginary part, 123–125, 335, 336, 374, 414, 610, 614
imaginary unit, 28, 610
Immediate data, 219

INDEX **633**

Impulse response, 308, 553, 563, 564, 566, 611, 612
In-band interference, 18, 73, 131, 135, 217, 220, 315, 318, 320, 540, 542, 548
Inclination, 21, 24, 28–30, 214, 227, 229, 256, 268, 272, 283
Inclined geosynchronous orbit (IGSO), 24, 28, 29, 161, 253, 254, 256, 257–260, 283
Incommensurate sampling, 338
Indian Regional Navigation Satellite System (IRNSS), 12, 157, 162, 282–289
Indian Regional Satellite System (IRNSS) Restricted Service (RS) Signals, 286, 288–289
Indian Regional Satellite System (IRNSS) Standard Positioning Service (SPS) Signals, 286–288, 401
Indian Space Research Organization (ISRO), 203, 282, 283, 284
Indoor MEssaging System (IMES), 271, 272
Inertial aiding, 494, 602
Inertial coordinate system, 624
Inertial Measurement Unit (IMU), 599–603, 605, 606
Inertial Navigation System (INS), 600
Initial frequency uncertainty (IFU), 367, 369, 370, 379, 383, 387, 388, 393, 396, 402, 549, 590–592
Initial synchronization, 10, 11, 13, 75, 79, 82, 294, 295, 319, 364–367, 369–402, 432, 434, 437, 439, 442, 445, 463, 470, 493, 526, 542, 549, 588, 590, 604, 606
Initial time uncertainty (ITU), 67, 227, 367, 369, 378, 379, 383, 387, 393, 398, 402, 549, 590–592
Innovations, 1, 3, 4, 598
In-phase component, 38, 194, 279, 611
Inphase/quadraphase mixing, 325, 326
Insertion loss, 294, 307–309, 317, 318, 322–324
Integer solution, 584, 585
Integer solution search space, 585
Integer-valued ambiguity, 584
Integrity, 2, 12, 77, 201, 202, 204, 210, 229, 230, 236, 248, 260, 261, 272, 283, 296, 505, 530, 537, 538, 570, 586, 588
Integrity information, 77, 202, 228, 570, 586

Integrity monitoring, 283, 529–531
Integrity Monitoring Service (IMS), 229, 231, 248
Interface Control Document (ICD), 26, 56, 78, 82, 212, 216, 218, 220, 231, 232, 234, 235, 247, 254, 257, 262, 263, 283, 286, 364, 369, 398, 409, 431–434, 506, 508, 513, 516, 518, 520
Interface Specification (IS), 13, 24, 110, 163, 169, 170, 207, 271
Interleaving, 77, 79, 80, 87, 176, 182, 261, 460
Intermediate frequency (IF), 11, 294, 297, 299
Intermodulation components, 56, 57, 222
International Civil Aviation Organization (ICAO), 202, 203, 204, 271, 273, 280
International Telecommunications Union (ITU), 67
International Terrestrial Reference Frame (ITRF), 158, 230
Interoperability, 24, 71, 85, 159, 167, 205, 207, 210, 228, 232, 267, 272
Interplexing, 86, 436
Inter-signal correction (ISC), 506
Inter-signal group delay, 142
Intersystem interference, 540
Intersystem multiple access interference (MAI), 38, 41, 44, 67, 72–74, 82, 107, 122, 123, 128, 133–136
Intervisibility, 572, 573, 576
Intrasystem interference, 540
Intrasystem multiple access interference (MAI), 133
Ionosphere, 8, 13, 38, 59, 61–65, 70, 76, 77, 143, 145, 146, 152, 202, 287, 437, 507, 512, 513, 571, 575, 602
Ionosphere-free pseudorange estimate, 511, 513–516, 534
Ionospheric error, 67, 72, 76, 143, 145, 148, 150, 213, 291–293, 508, 512–515, 522, 531, 580, 583
Ionospheric grid point, 208, 261, 274, 287
Isotropic antenna, 105, 300, 303

Jamming, 130, 228, 392, 540, 549
Japan satellite navigation Geodetic System (JGS), 270

Kalman filter (KF), 538, 593, 596–599, 606
Kalman gain, 598, 599
Kasami code, 221
Kasami sequence, 221, 277
Kepler's Laws, 12, 20–25, 35
Keplerian parameters, 20–25, 27, 28, 35, 261
Keyed signal, 39–41, 44, 92, 343, 436, 615
Kronecker delta function, 75

Late correlator tap, 469, 495
Late measurement, 469, 472
Launch on replace, 167
L-band experimental (LEX) signal, 275
Leap second, 158, 168, 215, 230, 257, 270, 284
Least-squares AMBiguity Decorrelation Adjustment (LAMBDA), 585
Left hand polarization rotation, 68
Left pseudoinverse, 519, 623
Left-hand circular polarization (LHCP), 69, 71, 303, 561
Left-handed coordinate system, 624, 625
Legendre sequence, 178–180
Lever arm, 601, 602
Limiter, 298, 310
Linear analog-to-digital conversion (ADC) guidelines, 338
Linear analog-to-digital converter (ADC), 334, 337, 338–340, 348, 362
Linear region of the S-curve, 409, 426, 446, 450, 463, 484–486
Linear time-invariant (LTI) system, 611–615, 619, 620
Lissajous figure, 68
Local oscillator (LO), 299, 315, 316, 411, 527, 540
Local-area differential system, 573, 574
Location server, 589–591
Logical representation, 43, 82, 178, 186
Long term clock and ephemeris data (CED), 182–184, 367, 369, 460, 462, 515, 516, 518, 515, 525–527, 529, 531
Long-baseline differential system, 572
Long-term ephemeris, 589, 591, 592
Long-term orbit, 589
Loop gain parameters, 410, 477, 478
Loop signal to noise ratio (SNR), 18, 79, 122, 427
Loose coupling, 538, 601, 606

Loss of lock criterion, 295, 450
Low density parity check (LDPC) code, 79, 176, 182, 455, 460–462
Low Earth orbit (LEO), 25, 574
Low noise amplifier (LNA), 298, 299, 307–309, 315, 318, 322
Low ultra-high frequency (UHF), 66, 576
Lower L band, 67, 116, 216, 230, 248, 541, 581
Lowpass equivalent impulse response, 612
Lowpass equivalent transfer function, 612, 615, 619
Low-side injection, 316
Low-side mixing, 316–317

Majority voting, 85, 86
Matched spectrum interference, 130, 131, 491, 541
Matrix, 80, 147, 148, 153, 182, 387, 408, 518, 523, 524, 578, 598, 621–623
Matrix Hermitian, 621
Matrix multiplication, 622
Matrix transpose, 621
Maximal-length-sequence (m-sequence), 74, 172, 173, 181, 185, 186, 191, 193, 195, 218, 221, 259, 277
Mean anomaly, 23
Mean delay of a multipath channel, 554
Meander sequence, 216, 217, 219, 401
Measurement matrix, 518, 522, 597, 599
Measurement model, 597
Measurement noise, 578, 579, 597, 604
Medium Earth orbit (MEO), 3, 25, 28, 32–34, 132, 153, 161, 227, 253, 254, 256–260, 516
Memory code, 74, 178, 232, 234, 236
Michibiki, 268, 270, 274, 276
Microstrip antenna, 306
Mixed time-domain/frequency-domain initial synchronization algorithm, 384–387
Mixer, 299, 304, 309, 315–319, 322–326, 329
Mobile station (MS), 589
Mobile station (MS)-based positioning, 589–591
Modified replica multipath mitigation, 563
Monolithic microwave integrated circuit (MMIC), 315
MTSAT-based Satellite Augmentation System, 203
Multidimensional Taylor Series, 154, 623

INDEX 635

Multifunctional Transport Satellite (MTSAT), 203, 271
Multilateration, 6
Multipath, 4, 8, 13, 27, 44, 73, 125, 141, 144, 166, 201, 291, 303, 304, 407, 456, 476, 507, 510, 527, 528, 530, 537–540, 551–568, 574, 580, 593, 606
Multipath error, 144–146, 151, 152, 551–552, 557, 559, 561–565, 567, 578–580, 585, 602, 604
Multipath-to-direct path ratio (MDR), 553–562, 565
Multiple access interference (MAI), 18, 38, 41, 44, 67, 72, 83, 82, 107, 122, 123, 128, 133–136
Multiplexed BOC (MBOC), 58, 167, 228
Multiplexing efficiency, 85, 86

Narrowband Doppler approximation, 60
Narrowband interference, 73, 75, 82, 128, 130, 291, 292, 293, 359, 373, 492, 541, 545
Narrowband interference excision, 548
Narrowlane pseudorange measurement, 582, 583
Nationwide differential GPS (NDGPS), 573, 574
Navigation, 1–4, 5, 8, 67, 75, 203, 206, 224, 221, 229, 247, 260, 271, 283, 387, 369, 397, 505, 517, 525, 526, 537, 538, 567, 624, 625
Navigation processing, 11, 294, 542
Navigation Warfare (Navwar), 165
Near echoes, 554, 562, 567
Network-based positioning, 589–591
Networked differential system, 572, 573
Neyman-Pearson criterion, 392
Noise and interference errors, 144
Noise factor, 321–323
Noise figure, 298, 313, 315, 321–323, 328, 329, 620
Noise input matrix, 597
Noncoherent early-late processing (NELP), 478, 482, 483, 485–487, 489–491, 557–559, 561, 562, 565
Noncoherent integration, 383, 388–392, 402
Non-hydrostatic delay, 144
Non-return to zero (NRZ), 44, 45
Nonstandard code, 169, 170

Normalized discriminator for code tracking, 477, 499
North-East-Down (NED) coordinate system, 625
North-East-Up (NEU) coordinate system, 625
Notification Advisories to QZSS Users (NAQU), 270
Null-to-null bandwidth, 22, 43, 128, 171, 237, 357, 437, 492, 548, 619
Number of correlators, 371
Numerically controlled oscillator (NCO), 365, 495

Odd correlation, 74
One dB compression point, 310, 311, 313–315
Open Service (OS), 160, 214, 216, 228, 231, 254
Open signal, 4, 74, 75, 82, 146, 160–162, 176, 248, 254, 258
Orbital altitude, 24, 25, 28, 29, 33, 153
Orbital plane, 21, 22, 27, 28, 32, 34, 166, 167, 214, 216, 229, 257, 268
Ordinary least-squares (OLS) solution, 519, 525, 529, 623
Orthogonal processes, 613
Osculating orbital elements, 26
Out-of-band interference, 315, 317, 318, 320, 323, 325–329, 540
Output signal-to-noise ratio (SNR), 18, 122–136, 334, 340–343, 362, 374, 388, 392–395, 436, 440, 445, 446, 449, 450, 482, 494, 560, 563, 565
Oven-controlled crystal oscillator (OCXO), 321
Overlaid spreading code, 82, 132, 234, 241, 242, 397
Overlay bit duration, 88, 90
Overlay code, 41, 57, 82, 83, 88–90, 181, 183, 193, 218, 233, 234, 242, 245, 263, 275, 365, 375, 388, 397–402, 409, 428, 432, 435, 442, 460, 469–471, 498, 526, 527, 576

Page, 184, 236, 247, 272, 517
Parameter rate tracking, 427
Parametri Zemli (PZ), 214
Parseval's Theorem, 127, 611
Passband, 123, 307, 308, 327, 334, 338, 482
Passive antenna, 294, 300, 301

Passive inductor-capacitor (LC) filter, 309
Passive mixer, 316
Patch antenna, 305
Performance enhancement signals, 267, 268, 271
Periodic terms, 26, 35, 142
Phase center offset, 305
Phase center variation, 305
Phase delay, 63, 578, 584, 613
Phase jitter, 512
Phase noise, 9, 66, 292, 319, 320, 388, 407, 424–427, 444, 445, 447, 448, 452
Phase quadrature combining, 84
Phase response, 307, 612
Phase-locked loop (PLL), 295, 320, 366, 435, 447, 450–454, 463, 523
Phase-locked loop (PLL) lock criterion, 453
Pierce point, 65, 202, 288, 512, 574, 575
Pilot component, 3, 65, 81–83, 87, 89–91, 128, 132, 176, 178, 179, 181, 188–190, 193, 208, 210, 222, 232–234, 241, 242, 244, 263, 275, 391, 397, 398, 402, 429, 430, 432–439, 447, 450, 455, 456, 462, 463, 468, 469, 592, 607
Pilot component waveform, 89–91, 190, 234
Pilot overlay code bit sequence, 89
Pilot overlay code waveform, 83, 89, 90, 234, 241
Pilot spreading code bit sequence, 89, 90
Pilot spreading sequence, 90
Pilot spreading symbol, 89–91, 190, 191, 234, 432, 434, 435
Pilot spreading symbol sequence, 90
Pilot spreading waveform (pre-overlay), 90
Planar crossed bow tie antenna, 305
Polar orbit, 21
Polarization, 17, 65, 68–71, 92, 107, 108, 110, 111, 119, 206, 215, 299–304, 540, 548, 561
Polarization loss, 70, 71, 108, 110, 111, 231, 258, 271, 286
Polarization mismatch, 70, 301
Position dilution of precision (PDOP), 148
Postamble, 77
Postcorrelation C/N_0, 546
Power delay profile (PDP), 553, 554
Power early-late processing (PELP), 476–479
Power flux density (PFD), 104–108, 222

Power spectral density (PSD), 38, 40–47, 49, 51–59, 73, 75, 82, 89, 90, 122–124, 126–129, 132, 136, 170–172, 177, 185, 187, 188, 190, 193, 208, 217, 220, 232, 234, 235, 237, 238, 240, 241, 242, 244, 259, 260, 311, 313, 321, 339, 345, 358, 359, 374, 425, 481, 482, 492, 541, 542, 545, 548, 613, 614, 617, 619, 620
Preamble, 176, 192, 207, 260, 261, 275, 400
Precise point positioning (PPP), 152, 278, 505, 527–529, 531, 573, 576, 586, 607
Precise Positioning Service (PPS), 163, 167
Precision (Encrypted) (P(Y)) code signal, 165
Precorrelation bandwidth, 123, 124, 126–129, 131, 176, 312, 339, 342, 343, 345–361, 382, 439, 475–479, 484–489, 491, 500, 501, 540, 541, 545–547, 561–563, 566, 567
Precorrelation C/N_0, 546, 547
Precorrelation filter, 126, 295, 308, 312, 344, 440, 441, 482, 548, 557, 561, 562, 565
Precorrelation filter bandwidth, *see* precorrelation bandwidth
Prefilter 298, 307, 308, 322, 323, 327
Primary code 234
Principal value of arctangent, 443
Probe waveform, 546
Process, 597
Process matrix, 597, 599
Process model, 597, 599, 606
Process noise, 597, 599
Processing gain, 122, 127–129, 136, 397
Prompt correlator output, 432, 433, 454, 477
Prompt correlator tap, 432, 495
Prompt tap, 456, 462, 471, 472, 474, 478, 493
Pseudodelay, 7, 507
Pseudoinverse, 519, 623
Pseudo-noise (PN), 73
Pseudorandom noise (PRN), 73
Pseudo-random number (PRN), 169
Pseudorange, 7–9, 11, 63, 116, 118, 140, 152, 163, 202, 204, 367, 467, 493, 505–515, 517–519, 522–525, 528, 531, 566, 574, 578, 580–586, 592, 593, 597, 603, 604
Pseudorange correction, 508, 574, 575
Pseudorange rate, 505, 522–524, 593, 603
Public Regulated Service (PRS), 4, 227, 228, 231, 237–239, 241, 242, 248, 380
Pull-in region, 369, 445

INDEX **637**

Pulse blanking, 548
Punctual tap, 462

Quadraphase keyed signal, 39–41
Quadrature component, 38, 55, 56, 258, 611
Quadrature phase-shift keying with rectangular symbols (QPSK-R), 46
Quality factor (Q), 307
Quantization, 13, 111, 295, 334, 335–349, 354, 361, 362, 454
Quantization set point, 335, 337, 343, 344, 346–349, 362
Quantization step size, 335, 338–340, 343, 346, 347, 350–358
Quasi-coherent integration, 397
Quasi-Zenith Satellite (QZS), 268, 272
Quasi-Zenith Satellite System (QZSS), 2, 266–281
Quasi-Zenith Satellite System (QZSS) time, 270

Radio determination satellite system (RDSS), 160, 161, 253–257, 262
Radio frequency (RF), 1, 3, 79, 84, 159, 166, 298, 539
Radio frequency (RF) conditioning, 298
Radio horizon, 103
Radionavigation Satellite Service (RNSS), 5, 67, 169, 541
Ranging code, 73, 259
Real part, 123–125, 335, 374, 471, 472, 474, 610, 614
Real-time kinematic (RTK), 538, 573, 573, 607
Real-valued ambiguity, 584
Receive antenna, 11, 17, 27, 31, 32, 61, 66, 69–71, 104–111, 113, 119, 133, 134, 140, 144, 169, 206, 215, 291, 298–300, 302–305, 312, 318, 322, 395, 505–507, 520, 527, 528, 543, 547–549, 576, 597, 601, 603
Receiver Autonomous Integrity Monitoring (RAIM), 529–531, 566
Receiver channel, 7, 11, 294, 365, 407, 428, 429, 438, 442, 446, 454, 455, 460, 467, 468, 506
Reciprocal mixing, 320, 328, 329, 542
Reciprocity principle, 299

Rect function, 42, 610
Reference oscillator, 11, 66, 319, 320, 324, 369, 370, 590
Reference receive antenna, 107–111, 119, 133, 134
Regional-area differential system, 573–575
Relative carrier phase, 380, 553, 556, 557
Relative differential system, 570
Relative satnav, 573
Restart, 368, 369, 370, 401, 445
RF amplifier, 298, 309, 326, 327
Right ascension of the ascending node (RAAN), 21
Right hand polarization rotation, 68
Right-hand circular polarization (RHCP), 68, 69, 71, 169, 206, 215, 231, 258, 271, 286, 301, 303, 548
Right-handed coordinate system, 624, 625
Root mean-squared (RMS) bandwidth, 176, 232, 295, 437, 476, 481, 483, 485, 490, 499–501, 566, 581, 620
Root mean-squared (RMS) delay of a multipath channel, 554, 555
Root-mean square (RMS) bandwidth, 176, 227, 232, 295, 437, 476, 481, 483, 485, 490, 491, 499–501, 563, 566, 567, 581, 620
Root-mean square (RMS) horizontal position error, 149
Root-mean square (RMS) positioning error, 149
Root-mean square (RMS) three dimensional (3-D) error, 149
Root-mean square (RMS) time error, 149
Root-mean square (RMS) vertical positioning error, 149
Rover receiver, 570

Safety of Life (SoL) Service, 229, 231, 236
Sagnac effect, 153, 516, 517
Sampling rate factor, 345–348, 354
Satellite group delay error, 141, 142
Satellite-Based Augmentation System (SBAS), 2, 160, 201–210, 231, 253, 266, 271, 274, 278, 280, 283, 574
Satnav augmentations, 537, 570, 593
Saturation point, 310
Scalloping loss, 374–377, 379, 380, 381, 382, 388, 391, 396

Scintillation, 65
S-curve, 409, 410, 426, 443–453, 463, 474–479, 481, 482, 484–486, 488, 492–494, 498, 499, 563, 576
Search And Rescue (SAR) Service, 228, 229
Search and Rescue GPS (SAR/GPS), 167
Secondary code, 82, 234
Secular terms, 26
Selective Availability (SA), 141, 165, 574
Selectivity, 307, 309, 318, 324–327
Self-SSC, 131, 132
Semicodeless processing, 165, 185, 205
Semimajor axis, 20, 21, 22, 69
Semiminor axis, 20, 21, 108
Semi-synchronous orbit, 25
Serial search, 370, 383, 384, 396
Service region, 26–28, 35, 572
Shadowing, 4, 13, 201, 291, 527, 530, 537, 552, 553, 556, 566, 567
Sherman-Morrison formula, 622
Short cycled, 186, 195, 247, 277
Short-baseline differential system, 152, 572, 574
Short-time correlation, 385–387
Sideband processing of BOC signals, 380, 382, 383, 439, 440, 494
Sidereal day, 24, 25, 29, 167, 268, 283
Signal in space ranging error (SISRE), 140–143, 145, 150–153, 192, 571, 572
Signal-to-interference ratio (SIR), 128
Signal-to-noise ratio (SNR), 339, 340, 342, 344, 394, 427, 444, 447, 452, 494
Signature sequence, 73
Silicon-Germanium (SiGe) heterojunction bipolar transistor, 315
Sinc function, 42, 610
Small-signal approximation, 309–310
Small-signal gain, 310, 311, 313–316, 318
Soft decision, 454, 455, 460, 462
Solution separation method, 530
Space-based positioning, navigation, and timing (PNT), 2
Spatial correlation of errors, 571, 573
Spectral separation coefficient (SSC), 18, 127–133, 490
Spectrum-specific factor, 490–492
Specular reflection, 555
Spherical Error Probable (SEP), 149

Spreading code, 8, 9, 39–41, 43, 44, 46, 55, 56, 58, 59, 65, 73–76, 79, 81–83, 88, 90, 123, 128, 132, 133, 152, 166, 169–175, 177, 178, 180, 185–196, 207, 208, 217, 218, 220–223, 232, 234, 236, 237, 240–242, 244, 245, 247, 248, 254, 259, 267, 277, 287, 291, 292, 364, 366, 369, 370, 372, 374, 379, 392, 397, 398, 400, 431–436, 592, 616
Spreading code bit sequence, 42, 44, 75, 87, 89, 90, 233
Spreading code length, 74, 75, 79, 82, 83, 85, 87, 90, 259, 379, 397
Spreading code type, 74, 191
Spreading modulation, 8, 13, 38–59, 65, 72–74, 84, 92, 128, 130, 131, 143, 166, 167, 176, 181, 193, 205, 208, 227, 228, 237, 241, 242, 262, 271, 295, 296, 340, 343, 345, 358–361, 364, 370, 372, 374, 377, 380, 429, 432, 438, 439, 462, 464, 467, 474, 482, 484, 485, 487, 489, 491–494, 496–501, 557–559, 560–567, 581, 586, 617, 620
Spreading symbol, 42–47, 51, 52, 55, 58–61, 73, 85, 87, 90, 91, 176, 177, 181, 187, 188, 234, 239, 242, 277, 375, 382, 400, 409, 431, 432, 434–436, 440, 482, 564, 617
Spreading waveform, 42, 176, 398–400, 431, 456, 467, 471
Square full-rank matrix, 622
Squaring loss, 248, 445, 449, 450, 453, 482–485, 491
Staircase approximation for cross-ambiguity function computation, 385, 386
Standard point positioning (SPP) equations, 519, 523
Standard Positioning Service (SPS), 163, 167, 282
Standard Positioning Service Performance Standard (SPS PS), 164
State, 191, 367, 597–599, 601, 604, 606
Static one-multipath model, 556, 557, 559, 561–563, 565, 567
Stationarized correlation function, 615, 616
Stationarized statistics, 615, 619
Stationkeeping maneuver, 25
Statistically independent processes, 613
Stopband, 124, 307, 308, 318, 327, 334, 338

INDEX 639

Strapdown inertial measurement unit (IMU), 600
Submeter class Augmentation with Integrity Function (SAIF), 272
Superheterodyne architecture, 324–326
Surface acoustic wave (SAW) filter, 308–309
Swept tone, 541
Synchronization code, 82
Synchronization sequence, 1 93
Synchronous code division multiple access (CDMA), 74
System for Differential Correction and Monitoring (SDCM), 203–205
Systematic code, 275

Taylor series, 154, 310, 315, 319, 517, 609, 623–624
Temperature-compensated crystal oscillator (TCXO), 320, 321
Terrestrial reference frame, 157, 158, 230
Third-order intercept point, 314, 315
Tiered code, 234
Tight coupling, 538, 596, 601–603, 605, 606
Time dilution of precision (TDOP), 148
Time division multiple access (TDMA), 73
Time to first fix (TTFF), 292, 365, 401, 588
Time to initial fix (TTIF), 292, 293, 593
Time to subsequent fix (TTSF), 370
Time-companding, 61
Time-frequency cell, 370–374, 377–379, 383, 384, 388, 390, 391, 396, 592
Time-frequency tile, 371, 383, 386, 387, 391, 394, 396, 549, 590
Time-multiplexed BOC (TMBOC), 54, 58, 176, 181, 274, 351, 382, 439–441, 479, 485–489, 562, 563
Time-to-alert, 12
Timing receiver, 525
Total electron count (TEC), 61
Total electron count units (TECU), 61–63
Tracking loop bandwidth, 480, 485–489, 581, 606
Tracking loop closure, 425–426, 603
Tracking loop damping, 295, 410, 411, 416, 418
Tracking loop lock criterion, 485
Tracking loop order, 410
Tracking loop pull-in, 369
Tracking loop transfer function, 411

Transfer function, 63, 64, 123, 306, 308, 334, 411, 412, 414–416, 418, 424, 557, 611, 612, 614, 619
Transition band, 307–309, 324, 325, 327, 334, 338
Triangulation, 6
Trilateration, 6, 32, 161, 253, 254, 256, 262, 283
Tropospheric error, 143–145, 150, 151, 508, 510, 513, 522, 528, 575, 580
True anomaly, 22, 23, 28
True root mean-squared (RMS) delay of a multipath channel, 554, 555
Turbo code, 79, 460
Twice the distance root mean square (2DRMS) error, 149
Two-body orbital model, 20
Two-quadrant arctangent, 443, 447, 448
Two-ray model, 112–116, 547

Ultra-high frequency (UHF), 66, 573
Ultratight coupling, 538, 596, 605
Uncorrelated interference, 540, 551, 557, 567
Uncorrelated processes, 613
United States Naval Observatory (USNO), 168
Universal Coordinated Time (UTC), 3, 158
Upconversion mixer, 316
Upper L band, 67, 84, 206, 215, 216, 222, 227, 228, 230, 236, 541, 581
User equipment error (UEE), 140, 141, 143–146, 151–153, 571
User equivalent range error (UERE), 140, 141, 145–147, 150, 152, 153, 571

Validity interval, 26, 35, 141, 526, 591, 592
Variable gain amplifier (VGA), 335, 337
Vector tracking, 596, 603–606
Vector-locked loop, 538, 603
Verification processing, 393–395
Vernal equinox, 21, 22
Vertical dilution of precision (VDOP), 148, 149
Very high frequency (VHF), 66, 573

Walker constellation, 28–30, 227
Warm start, 368, 369, 397, 607

Warm up time, 321
Weakly nonlinear, 310
Weighted least-squares (WLS) solution, 519, 521, 523, 524, 529, 578, 623
Weil index, 178–180
Weil sequence, 178–180
Weil-based sequence, 74
Welch bound, 75, 181
White noise, 123, 124, 126–130, 136, 343–346, 349–358, 361, 372, 412, 468, 481–492, 499, 597, 614, 620
Wide Area Augmentation System (WAAS), 160, 202–204
Wide-area differential system, 573, 575
Wideband interference, 541

Wideband processing of BOC signals, 380, 382, 383, 439, 440, 494
Widelane carrier phase measurement, 582, 583
Wide-sense stationary process, 613, 614
Wiping off the carrier, 123
Wiping the data spreading waveform, 432
Wiping the overlaid pilot spreading code, 435
Wiping the pilot spreading code, 433

X1 epoch, 172, 175, 186, 189

Z count, 175
Zero-intermediate frequency (IF) architecture, 324, 325

CPSIA information can be obtained
at www.ICGtesting.com
Printed in the USA
BVHW010825250220
573209BV00004B/9